Hospitality Sales and Marketing

An Evolutionary Journey with Howard Feiertag

Hospitality Sales and Marketing

An Evolutionary Journey with Howard Feiertag

By Howard Feiertag

Apple Academic Press Inc.
3333 Mistwell Crescent
Oakville, ON L6L 0A2
Canada

Apple Academic Press Inc.
1265 Goldenrod Circle NE
Palm Bay, Florida 32905
USA

© 2020 by Apple Academic Press, Inc.
Exclusive worldwide distribution by CRC Press, a member of Taylor & Francis Group
No claim to original U.S. Government works
International Standard Book Number-13: 978-1-77188-789-2 (Hardcover)
International Standard Book Number-13: 978-0-42905-720-5 (eBook)

All rights reserved. No part of this work may be reprinted or reproduced or utilized in any form or by any electric, mechanical or other means, now known or hereafter invented, including photocopying and recording, or in any information storage or retrieval system, without permission in writing from the publisher or its distributor, except in the case of brief excerpts or quotations for use in reviews or critical articles.

This book contains information obtained from authentic and highly regarded sources. Reprinted material is quoted with permission and sources are indicated. Copyright for individual articles remains with the authors as indicated. A wide variety of references are listed. Reasonable efforts have been made to publish reliable data and information, but the authors, editors, and the publisher cannot assume responsibility for the validity of all materials or the consequences of their use. The authors, editors, and the publisher have attempted to trace the copyright holders of all material reproduced in this publication and apologize to copyright holders if permission to publish in this form has not been obtained. If any copyright material has not been acknowledged, please write and let us know so we may rectify in any future reprint.

Trademark Notice: Registered trademark of products or corporate names are used only for explanation and identification without intent to infringe.

Library and Archives Canada Cataloguing in Publication

Title: Hospitality sales and marketing : an evolutionary journey with howard feiertag / Howard Feiertag.

Other titles: Hotel management.

Names: Feiertag, Howard, author.

Description: Articles originally published in Hotel management. | Includes bibliographical references and index.

Identifiers: Canadiana (print) 20190072547 | Canadiana (ebook) 20190072563 | ISBN 9781771887892 (hardcover) | ISBN 9780429057205 (eBook)

Subjects: LCSH: Hospitality industry—Marketing.

Classification: LCC TX911.3.M3 F45 2019 | DDC 647.94068/8—dc23

Library of Congress Cataloging-in-Publication Data

Names: Feiertag, Howard, author.

Title: Hospitality sales and marketing : an evolutionary journey with howard feiertag / Howard Feiertag.

Description: Oakville, ON, Canada ; Palm Bay, Florida, USA : Apple Academic Press, 2019. | Includes bibliographical references and index.

Identifiers: LCCN 2019007137 (print) | LCCN 2019016138 (ebook) | ISBN 9780429057205 (ebook) | ISBN 9781771887892 (hardcover : alk. paper)

Subjects: LCSH: Hospitality industry--Marketing.

Classification: LCC TX911.3.M3 (ebook) | LCC TX911.3.M3 F454 2019 (print) | DDC 338.4/791068--dc23

LC record available at https://lccn.loc.gov/2019007137

Apple Academic Press also publishes its books in a variety of electronic formats. Some content that appears in print may not be available in electronic format. For information about Apple Academic Press products, visit our website at **www.appleacademicpress.com** and the CRC Press website at **www.crcpress.com**

Contents

List of Columns .... *vii*

About the Author .... *xxvii*

Howard's Story .... *xxix*

List of Abbreviations .... *xxxiii*

Foreword .... *xxxv*

Preface .... *xxxvii*

1. General Managers' Involvement with Sales 1
2. Developing Leads and Prospects for Hospitality Sales 65
3. The Working of a Hotel Property Sales Operation 129
4. Techniques for Increasing Sales 159
5. Working with Group Markets 193
6. The Travel Agent and the Leisure Travel Market 249
7. Sales Planning and Sales Calls 275
8. Hospitality Sales Training 309
9. The Negotiation Process in Sales 365
10. Attributes of Successful Salespersons 395
11. Helpful Tips for Hotel Sales Staff 417

12. Action Plans for Marketing and Sales 499
13. Working on Contracts for Groups 551
14. Everyone at a Property Is Involved in Sales 577
15. A Professional Approach to Hospitality Sales Along with Networking 611
16. Understanding the Features of Your Product, Especially F&B 631
17. HR in Connection with Sales Staff Employment 655

List of Columns

Title	Publication Date	Page
CHAPTER 1: General Managers' Involvement with Sales		
For Owners and General Managers Only	January 1980	1
High Visibility: A Great Management Sales Tool	March 1983	2
You Do Not Need a Sales Department . . .	April 1983	3
Let the Salespeople Sell	July 1983	4
New Year's Resolutions for Sales-Minded General Managers	December 1983	5
What Successful Sales Directors (and General Managers) Do	February 1984	7
How to Be the Best Leader of Your Sales Force	August 1984	10
The Problem with Hotels	December 1984	11
Management Responsibility to the Sales Effort	August 1985	13
General Managers Need to Get into the Sales Act	February 1986	14
Sales Department Responsibility to the General Manager	June 1986	15
Opportunities for Sales Lead Development for the Smaller Hotel/Motel	March 1987	17
Salespeople Need to Be Evaluated	February 1998	18
Managing the Sales Effort in a Distressed Market	May 1988	19
Sales-Minded General Managers Need New Year's Resolutions Too	January 1989	20
Our Most Neglected Sales Tool . . . Our Employees	March 1989	21
Managing the Sales Effort	April 1989	23
Turnover in Sales . . . A Real Problem	August 1989	24
1990 Will Not Be a Great Year for Some . . .	January 1990	26
Just How Productive Is Your Property Sales Program?	August 1992	27
You Never Miss the Water Until the Well Goes Dry	April 1993	28

Title	Publication Date	Page
General Managers: It May Be Time for a Sales Audit	April 1993	30
Are We Paying Enough Attention to Our Salespeople?	August 1993	31
When You Don't Have a Sales Manager . . . You're It!	November 1994	32
Empower Your Salespeople to Be More Productive	March 1996	34
How about Those Weekly Sales Meetings? Do They Work?	February 1997	35
Measuring Performance and Productivity of Sales Personnel	May 1997	36
Are Salespeople Really Producing "Good" Sales?	February 1998	37
Bill 'Em Fast and Collect 'Em Fast . . . Maybe!	May 1998	38
How to Sell Better without Lowering Rates? A General Manager Quiz	March 2000	40
Hotels Should Evaluate Commissions Paid to Third-Party Meeting Planners	January 2001	41
Another Way to Increase Room Revenue	April 2001	42
Improving Sales Productivity without a Sales Department	October 2001	43
What to Do about Salespeople at the Local Level in 2003	December 2002	44
Improve Your Market Share in 2003	January 2003	45
How General Managers and Directors of Sales Can Better Manage the Sales Effort	April 2003	46
The New Twist on Sales Consulting . . . Sales Coaching	September 2003	48
Customer Service: Are We Ever Gonna Get It Right?	October 2003	49
Motivating Sales Staff Is a Key to Improving Sales Productivity	June 2005	50
Managing the Sales Effort to Improved Productivity	July 2006	51
Developing Sales without a Sales Department	September 2006	52
Look Forward to the Need for Aggressive Selling in 2008	April 2007	53
Property Sales Operations and the General Manager	May 2007	54
The Hiring and Employment of the Right Sales Staff	November 2007	56
Smaller Properties Have a Distinct Need for Sales Plans	November 2008	57
So What's Going on with Your Sales Staff?	April 2009	58
Who Is the Chief-in-Charge of Sales?	October 2010	59
The General Manager's Responsibility to the Sales Effort	November 2011	60
Hotel Sales Folks Need to Be Great—Not Just Good	April 2014	61
How Owners, Operators, and General Managers Can Improve Group Revenue	November 2014	62
From Where Will All the Profit in Hotel Properties Come from This Year?	March 2015	63

Title	Publication Date	Page
An Easy Way for General Managers to Improve Profitable Sales	November 2015	64

CHAPTER 2: Developing Leads and Prospects for Hospitality Sales

Title	Publication Date	Page
The Prospect Survey Sheet and How to Use It	August 1980	65
Handling That Important Lead: The Customer Inquiry	January 1982	68
Don't Miss the Business That's Right under Your Nose	May 1984	69
Sales Opportunities That Are Missed	November 1985	70
Profitable Sales Tips	December 1985	72
Handling the Business That's Being Called-In	April 1986	73
Finding Business Leads Can Be Easier Than You Think	June 30, 1986	74
The More People You Meet, The More Contacts Made	February 2, 1987	76
Going After Local Business Will Boost Your Group Sales	February 27, 1987	77
Closer Look at Newspaper Can Deliver Valuable Leads	November 23, 1987	78
Cold Calling Does Work	January 1988	79
Let's Sell More Receptions	February 1988	81
How to Get and Use Referrals	October 1988	82
Improve Sales with Proper Telephone Practices	November 1988	83
Keep Panning for Gold	May 27, 1991	85
Mishandling of Inquiries Loses Sales	March 22, 1992	86
The Check Is in the (Phone) Mail	April 6, 1992	87
The Handling of Inquiries . . . We Do It All Wrong!	August 16, 1993	88
Mishandle the Inquiries and Lose a Lot of Business	September 6, 1993	89
Good Prospecting Intro Letters Can Help You Improve Your Sales	February 1994	91
Prospecting for New Business Is a Never-Ending Job	June 1994	92
The Inquiry Is Still the Hottest Lead You'll Ever Get . . .	July 6, 1994	93
Prospecting Is an Ongoing Job for Successful Sales Development	July 5, 1995	95
Some Tricks of the Trade to Increase Profitable Sales	March 6, 1996	96
Every Sale Begins with a Prospect	May 20, 1996	97
Gimme the Fax Please, Just the Fax	September 16, 1996	99
Prospecting Is Still a Key to Improved Profitable Sales	October 7, 1996	100
How Are You Responding to Sales Inquiries?	November 9, 1996	101
Doubletree National Sales Office: A Well-Oiled Machine	June 2, 1997	102
Turn Inquiries into Selling Opportunities	July 20, 1998	104

Title	Publication Date	Page
In Selling, Do What the Politicians Do	February 1, 1999	105
Can We Shorten the Sales Cycle to Bring in More Business?	May 3, 1999	106
You Can Be More Productive If You Know What Prospects Want from You	January 2000	107
Not Handling the Telephone Inquiry Properly Will Hurt Your Business	March 6, 2000	108
Want to Book More Business? Respond to Inquiries Promptly!	November 2000	109
The Hottest Sales Lead We Will Ever Get . . .	November 2001	111
Improve Sales by Standing Out Against the Competition	December 13, 2005	112
Stop Losing Those Telephone Inquiries!	January 2006	113
Improve Your Prospecting for Better Business with Referrals	December 2006	114
Improve Productivity When Prospecting for New Business	February 18, 2008	115
Methods and Resources for Developing New Sales Leads	July 21, 2008	116
Let's Get Back to the Real Selling Mode to Build Business	May 18, 2009	117
Proper Prospecting Prevails in Producing Profit	September 2010	118
Read Your Local Newspaper and Book a Lot of Business	February 2011	119
Sales Is Where It's At! So Let's Get to It Right Now!	February 2012	120
How the Easy Method of Data Mining Brings in Profitable Sales	October 2012	121
Get Free Student Help in Prospecting for New Profitable Sales	February 2013	122
How to Find Prospects and Book Groups in Hotels—The Easy Way	January 2014	123
The Telephone Inquiry and How We Mess It Up	May 2014	124
Finding Leads to Develop into Qualified Prospects Is the Business of Sales	April 2015	125
Sources for Leads to Help Qualify Prospects for the Group Market	May 2015	126

CHAPTER 3: The Working of a Hotel Property Sales Operation

Small Operators May Need Direct Sales Effort	January 1980	129
Pump $25,000 into Operation to Get Sales Effort Rolling	July 1980	130
First Calls: Taking the Plunge into "Cold" Waters	September 1980	131
A Place for Everything: Setting Up a Filing System	October 1980	133
Stacking the Cards in Your Favor: The 'Trace Card' File	November 1980	134

Title	Publication Date	Page
The Job Is Not Finished Until the Paperwork Is Done	January 1981	135
The Function Book: An All-Purpose Record-Keeping Tool	February 1981	137
Examine All Profit Centers When Booking Groups	April 1981	138
Weekly Meetings—a Positive Way to Inform Direct Staff	June 1981	139
Sales Reporting System Measures Staff Productivity	August 1981	141
Sales Professionals Need Recognition	October 1987	141
So You're the New Director of Sales?	November 1992	142
The Hospitality Sales Office of the Future	October 1993	144
The Small Group and Meetings Market Is for All Size Properties	July 1994	145
Low-Cost Sales Automation System for Small Hotels ... It's a Breeze!	May 1995	147
When Should You Hire a Salesperson or Add to Your Sales Staff?	October 1997	148
What's the Job of a Salesperson in Our Industry?	November 1999	149
Some Things Have Not Changed in 20 Years When It Comes to Sales	February 2000	150
Sales Training for Limited Feature Properties Pretty Limited	April 2000	151
It Takes More Than Just Sales at Limited Service Properties	June 2000	152
Technology Will Help Sales Effort, But Won't Replace Salespeople	August 2000	153
Let's Help Salespeople Become More Productive	October 2001	154
Small Sized Properties May Need Direct Sales Effort	August 2008	155
Recognizing the Value of the "Small Meeting" Market	July 2009	156
CHAPTER 4: Techniques for Increasing Sales		
Some New Year's Resolutions	January 1983	159
The Student Blitz: A New Technique for Building Sales	June 1983	159
Sales Blitzes: A Look at Some Actual Results ... and How You Can Arrange a Sales Blitz of Your Own	October 1983	160
Closing the Sale Ain't Easy ...	October 1990	163
Sales Resolutions for 1991	November 1990	164
How Not to Sell	November 1990	165
How about Promoting the GM to Director of Sales?	February 1993	166
Increasing Profitable Sales in the Hospitality Market Place	July 1993	167
Time to Temper Telephone Tactics	October 1993	168

Title	Publication Date	Page
Sell Value Not Price and be the Winner	November 1993	169
Measuring the Sales Performance	October 1994	170
Improve Sales by Knowing How to Overcome Sales Resistance	September 1995	172
How're We Gonna Sell Those Higher Rates Next Year?	December 1996	173
It May Still Be a Seller's Market for Hotels, But . . .	January 1997	174
Your Most Important Sales Tool—Knowing Your Customer	July 1997	175
How to Get More Room Reservations from Your Central Reservation Service (CRS)	September 1999	176
Build It and They Will Come; Or Will They?	July 1999	177
Salespeople Should Be Spending More Time Selling	May 2001	178
For Hotels in Crisis, Sales Training Is Critical	January 2002	180
Outsource Your Sales Department? Why Not?	February 2004	181
Relationship Selling: The Key to Hospitality Sales	August 2004	182
Let's Make Hay While the Sun Shines (2004)	December 2004	183
Let's Start Writing Some Better Sales Letters	April 2005	184
One Person Out of Four Can Sell	August 2005	185
To Improve Hotel Group Sales, Talk Less and Consult More	September 2005	186
Improve Profitability by Adding Ancillary Charges for Group Meetings	August 2007	187
Build Your Sales via Consultative Selling	June 2009	188
Knowledge Is the Key to Making the Sale	October 2009	189
Know the Attributes of Your Property to Help Make a Sale	April 2015	190
CHAPTER 5: Working with Group Markets		
Aggressively Selling Specific Groups Aids Growth in the 80s	March 1981	193
Property Operators and the Small-Meeting Market	July 1981	194
Concentrating on the Small-Meeting Market	April 1982	195
Getting Started in the Small-Meeting Market	February 1983	196
Boosting Group Sales Volume Is Every Staffer's Business	April 1984	197
The Meetings Business Keeps Growing	November 1984	198
The Meetings Market Means a Big Job for Small Properties	December 1984	199
Sales Tips on Selling Meetings and Conferences	February 1986	202
Working with the Meeting Planners	April 1986	203
Brainstorming Can Boost Your Group-Meeting Bookings	December 1986	204
Small Meetings Are for Everyone	December 1989	205

Title	Publication Date	Page
What Do Meeting Planners Expect about Quality from a Hotel?	November 1992	206
The Whole Hotel Should Service Meetings and Other Group Business	September 1993	208
How to Sell to and Work with the Novice Meeting Planner	March 1995	209
Government Business Could Be Big Business for You	December 1995	210
There Will Be More Corporate Meeting Business to Book This Year	February 1997	211
We Need More Education and Training to Better Serve the Disabled	April 1997	213
What Does the Corporate Traveler Expect from Hotels?	April 1997	214
Build Group Sales by Training Your Local Event Planners	September 1997	215
Do We Really Understand Meeting Planners' Concerns?	November 1997	216
The "Third-Party" Meeting Planner . . . Is It a Problem?	July 1998	218
How Hotels Should Value Group Business	November 1998	219
Meetings Are on the Rise—Take a Look at the Outlook for 2000	December 1999	220
Meeting Business Is Expected to Increase This Year	March 2001	221
In These Times We Need All the Business We Can Get!	June 2001	222
Military Reunions—The Only Niche Market That Is Growing	October 2002	224
Business Is Out There . . . All We Gotta Do Is Go Get It!	February 2003	225
Group Business Is Still Out There—Go Get It!	May 2003	226
Hotels of All Sizes Need to Get Some of That SMERF Stuff	May 2003	227
Start Now to Look for 2005 Off-Season Business	June 2004	228
Is the SMERF Market for You? Think about It!	July 2004	230
We May Be Getting Back to FAM Trips to Help Build Group Business	May 2005	231
Managing Your Room Block for Groups	November 2005	232
Are You Getting Your Share of the Military Reunion Market?	August 2006	233
Another "Niche" Market for You to Explore	September 2006	234
Groups and More Groups Is What Salespeople Need to Book	January 2007	235
This Is Indeed a Sellers' Market in the Corporate Travel Business, But It Won't Last Forever	February 2007	236
Corporate Meetings Getting to be Run by Corporate Travel Managers	March 2007	237

Title	Publication Date	Page
Relationship Building Matters Most in Booking Group Business	March 2007	238
Military Reunions Could Be Big Business for You	February 2008	239
Group Business Is Getting to Look Better for Next Year	November 2009	241
Understanding Needs Is Critical in Getting a Group Booking	February 2010	242
Many Sales Opportunities Are Being Missed in Group Bookings	July 2010	243
Getting Ready to Improve a Bottom Line with Increased Group Sales	January 2011	244
Military Reunions: An Untapped Market for Business When You Need It	April 2011	245
For Productive Sales, Rate Should Be the Thing in 2015	January 2015	246
Engagement Is Critical in the Business of Booking Groups	September 2015	248
CHAPTER 6: The Travel Agent and the Leisure Travel Market		
START Working Now for Next Year's Bus Tour Business	May 1983	249
The Travel Agent as a Sales Arm	September 1983	250
Travel Agents Are Still Important	January 1994	252
Tips on How to Work with Travel Agents	March 1984	253
Is the International Market for You?	September 1986	253
Travel Agents Still a Key to Incremental Business	September 1987	254
Are You Getting Your Share of the Travel Agent Market?	March 1988	256
We Need to Do More Business with Travel Agents	January 1989	258
Let's Get Back in Business with Travel Agents	June 1990	258
Corporate Travel is Still Big Business	November 1990	259
Here Are Some Travel Agency Complaints about Hotels	February 1991	261
Travel Agents Still Upset with Hotels	April 1991	262
The Hotel Code of Commission Fair Play	December 1991	263
Are Phony Travel Agents Getting Your Freebies?	February 1992	264
New Travel Agent Directory Lists Chains That Pay Commissions	September 1992	265
Improve Sales Effort with New Technology to Attract Travel Agents	April 1996	267
Phony Travel Agents Get Their Due	August 1997	268
Isn't It Time to Start Courting Travel Agents Again?	October 1997	269
Travel Agents Are the Best Bet for Selling to the Affluent Market	December 1997	271

Title	Publication Date	Page
The Travel Agency Business Is Still Alive and Well	June 2002	272
CHAPTER 7: Sales Planning and Sales Calls		
Start the New Year with a Fresh Sales Program	December 1981	275
Incentives Will Help Boost Your Weekend Occupancy	October 1982	276
A Successful Salesperson Always Makes the Right "Call"	July 1986	277
Time Again for *Sales Clinic's* New Year's Resolutions	January 1987	278
More Sales Calls Outside Means More Business Inside	February 1987	280
Backyard Bookings Help in Boosting Group Business	February 1987	281
Cold-Call Sales Approach Will Result in Hot Prospects	January 1988	282
Selling Is Not a Spectator Sport	April 1990	283
Managing Sales Calls	March 1991	285
Let's Get the Facts Straight Right Up Front—Then Sell	November 1991	286
A Remedy for Hard Times? How about Looking at the Basics Again?	February 1992	288
Think about Ways to Maximize Sales	March 1992	289
There's No Substitute for Making More Sales Calls	July 1996	291
Have a Different Pair of Eyes Take a Look at Your Sales Effort	October 1998	293
Preparing for the Year 2000	October 1998	294
Let's Not Jump the Gun on Lowering Those Room Rates	September 2001	296
Why Don't We Get Business from Our Vendors?	March 2002	297
How Do Salespeople Spend Their Time?	April 2002	298
Why Are Salespeople Still Making Sales Calls?	May 2002	299
Plan Now for Your 2003 Sales Activity	October 2002	300
Trends in Sales: Cold Calls Are Out, Consulting Is In!	March 2003	301
Watch for Higher Room Rates in the Next Couple of Years!	April 2003	302
Salespeople Need to Spend More Time Selling	September 2003	304
Selling Opportunities Are All Around US—Make the Most of Them	January 2004	305
The Easiest Way to Improve Hospitality Sales	February 2005	306
Cold Calls by Suppliers Are One of Meeting Planner's Biggest Complaint	July 2007	307
CHAPTER 8: Hospitality Sales Training		
A Blueprint for Bigger Payoffs from Your Sales Efforts	February 1984	309
Mastering Sales Methods—A 'Do-It-Yourself' Proposition	January 1986	310

Title	Publication Date	Page
When the Going Gets Tough, the Tough Get Positive	September 1988	311
Idle Chatter May Result in Silence When Closing a Sale	April 1989	313
Asking Key Questions Can Help Boost Your Group Sales	September 1989	314
Good Salespeople Have Opportunities to Become Great	September 1989	315
Training, Training, Training—It's Still Important	February 1990	323
You're Wanted on Stage	April 1990	325
You'll Never Know How Much Business You've Lost Because . . .	June 1990	326
Time Management for Hospitality Salespeople	October 1991	327
Are Our Sale People Being Trained?	May 1992	328
Do We Have to Live with Discounting?	October 1992	330
Discounting: The Downward Spiral!	December 1992	331
You Really Should Get into Consultative Selling	March 1993	332
Be Sure to Get as Much Detail as Possible When Booking Meetings	May 1993	333
Things Are Looking Up Because of Improved Sales Efforts	April 1994	335
Training Improves Profitability of a Property	August 1994	335
Continuing Education in Sales Is Ongoing and Vital to Success	March 1995	336
Training Is Still Vital to Increasing Profitable Sales	June 1995	337
Closing the Sale Is Critical for Individual Reservation Inquiries	September 1995	338
Take a Look at Your Group Sales Policies and Maybe Make a Change	October 1995	339
Knowledge Is Power When It Comes to Selling and Closing Sales	November 1995	340
Salespeople: It's Time to Get on the Information Highway	January 1995	341
If You Have a Phone on Your Desk, You Need Training	January 1996	343
Let's Start Writing Better Sales Letters	May 1996	344
When Someone Doesn't Return Your Call, Whose Fault Is It?	June 1996	345
Communication Is the Key to Staying Out of Trouble with Customers	November 1996	346
What Do We Mean by Selling "Value"?	November 1997	348
Price/Value Relationship Is Almost Always a Consideration in Buying	November 1998	349

Title	Publication Date	Page
What's Wrong with Our Salespeople These Days?	January 1999	350
How Is Your Telephone IQ?	February 1999	352
"High Touch" Still Beats "High Tech"	March 1999	353
Poor Voicemail Technique Can Hurt Your Sales	September 2000	354
Distance Learning—The Latest in Technology for Self-Education	October 2000	355
Apex Is Working for You, Making Your Challenges Easier	September 2004	356
How Boring Was the Last Meeting You Attended?	November 2004	357
Better Improve Your Communication Skills for Selling in 2005	November 2004	358
Presentation Skill Training Is Critical in Selling to Groups	April 2004	359
Hotel Sales Managers Fail to Respond to Customer Inquiries	November 2007	360
Salespeople Just Need to Start Doing a Better Job	September 2008	361
Hotel Salespeople Talk Too Much	May 2012	362
The One-Hour Property Sales Meeting	June 2012	363
CHAPTER 9: The Negotiation Process in Sales		
Don't Let Sales Talk about "Freebies" Until They Have To	January 1984	365
Giveaways Are Not Essential When Negotiating Group Sales	February 1986	366
In Negotiating Group Business, Sell Value Before Price	July 1988	367
Travel Managers Fret Over Yield Management's Impact	July 1989	368
Successful Lodging Salespeople Are Good Negotiators	March 1990	370
Improve Profitable Sales with Good Negotiating Practices	April 1991	371
What's Negotiable with Lodging Facilities . . .	June 1991	372
Understand the Negotiating Process Before Making the Deal	June 1992	374
When Negotiating with Groups, Hold Off on Rate Quote	May 1993	376
Sales and Negotiation Tips for Working with Groups	December 1993	377
In Selling, When It Comes to Negotiations, Knowledge Is Power	August 1997	378
Tips for Negotiating Successfully with Business Prospects	October 1999	379
Overcoming Objections in Hotel Sales Is a Tough Job	October 2002	381
Be Better Prepared for Negotiations with Groups	May 2004	382
Negotiations Are Still the Hottest Issue When Booking Groups	June 2004	383
Knowledge Is Power When It Comes to Negotiating	May 2005	384
Understanding a Prospect's Position in Negotiations	April 2006	385

Title	Publication Date	Page
Be Prepared for Negotiations When Selling to Groups	May 2008	386
Overcoming Objections Is the Critical Phase of the Selling Process	August 2012	389
Now Is the Time to Be Great Negotiators in Booking Group Business	May 2013	390
Overcoming Objections Is a Key to Closing Sales on Group Business	June 2014	391
Negotiate to Your Best Advantage When Working Group Prospects	June 2015	392
More of What You Need to Know When Negotiating with the Group Market	July 2015	393
Hold Off on Negotiating Rates with Group Market Prospects	October 2015	394
CHAPTER 10: Attributes of Successful Salespersons		
Aggressiveness Can Cure Even the Worst Sales Ills	July 1984	395
Characteristics of Successful Salespeople	October 1990	396
In Tough Times Good Salespeople Prevail	March 1991	397
Consultive Salespeople Bring in More Business	June 1991	398
Be Seen and Be Heard to Sell Better	January 1993	399
You Can Capture Accounts by Capturing Data	March 1993	400
Relationship Selling: A Good Sale Benefits the Buyer and Seller	June 1994	402
Listening Is Still the Key to Better Selling and Closing of Sales	October 1995	403
What Is It about Salespeople That Make Them Great?	February 1996	404
Selling May Not Be So Easy—If You're in It You Gotta Love It	October 1996	405
The Basics Still Work When It Comes to Hospitality Sales	June 1997	407
Sales Certification Proves Your Professionalism	July 1999	408
The Successful Salesperson of the Future	July 2001	409
Barriers to Successful Sales Communications	April 2003	410
Build Sales by Hard Work and Being a Good Listener	July 2003	411
Effective Consultative Selling in Hospitality Sales	October 2004	412
Salespeople Should Be Heard and Seen!	January 2010	414
What Makes a Salesperson Sell Well?	February 2012	414
Building Your Confidence Will Build Your Hotel Sales Productivity	September 2013	415

Title	Publication Date	Page
CHAPTER 11: Helpful Tips for Hotel Sales Staff		
Understanding Body Language Can Assist Sales Efforts	June 1985	417
Employing Formula Should Give Boost to Sales Activity	July 1985	418
Courteous Telephone Manners Will Help Make the Sale	November 1985	419
Listening Carefully Is Often Unheard of in Sales Offices	April 1986	421
Paying Attention to Meeting Details Can Pay Off in Sales	April 1986	422
Salespeople Should Learn about Hotel's F&B Facilities	May 1986	423
Breakfast May Be the Best Time for Booking Group Business	April 1987	425
Today's Salesperson Needs to "See the People" More	September 1987	426
Titles of Certification Are Signs of True Professionalism	October 1987	428
Salespeople: Use These Tips to Help Improve Your Skills	March 1988	429
Use an 'Action Calendar' to Bring Hotel Sales Up-to-Date	May 1988	430
'Probe' Method Maximizes Sales, Minimizes Objections	July 1988	431
Positive Sales Attitude Leads to Positive Performance	September 1988	433
Interviews with Satisfied Customers Can Help Spur Sales	October 1988	434
Start the Year Right by Reviewing Tips for Better Sales	December 1988	435
Sales Opportunities Depend on You, Not on the Market	November 1989	437
Become a Star on Videotape to Refine Sales Technique	November 1989	438
Stop Selling . . .	May 1990	439
Don't Let the Time Wasters Get in the Way of Selling	May 1990	440
Move Faster and Succeed Faster	July 1990	442
Cancel Your Next Sales Meeting . . . Unless . . .	July 1991	443
Visual Enthusiasm Helps in the Selling Process	July 1992	444
How Important Is It to Know Your Competition?	February 1993	445
Let's Make Hay While the Sun Shines (1995)	February 1995	447
Some Tips Provided by Meeting Planners for all Types of Hotels	May 1995	448
Salespeople: Discounting Doesn't Build Customer Loyalty!	February 1996	449
If You Wanna Be Good in Hotel Sales, Then Be a Good Listener	March 1997	450
Relationships Selling: The Key to Getting and Keeping Customers	March 1997	452
Seek Out Business for When It Is Most Profitable to You	September 1997	453
Tips for Improving Profitable Sales in 1993	January 1998	454
How Much Business Did You Lose Today?	March 1998	455

Title	Publication Date	Page
Let's Talk Phone Mail . . . A Vital Communications Link	April 1998	456
Let's Not Unload Unprofitable Business on Our Staff	April 1998	457
Writing the Way You Talk May Improve Your Sales Efforts	June 1998	459
How Near-By College Students Can Help Your Sales Efforts	June 1998	460
Make Better Use of the Basic Technologies and Improve Your Sales	September 1998	461
The On-Site Visit—Turn It into a Successful Sale	September 1998	462
Websites That May Be Helpful to You in Sales and Marketing	May 1999	463
How about Joining an Online Reverse Auction to Fill Rooms & Space?	February 2000	464
We'll Be Doing More and More Business Online, So Let's Get to It	June 2000	465
Loads of Sales Tips Available in Newsletters	July 2000	466
Let's Not Forget All the ". . . ings" When It Comes to Selling	June 2001	467
Sell . . . ing Has So Much to Do with a Lot of ". . . ings"	August 2001	468
Hi-Touch Is Still Better Than Hi-Tech in Sales and Service	September 2001	470
Sales Strategies for Survival and Recovery	December 2001	471
Salespeople Gotta Make Contacts to Make Sales	February 2002	472
Sales Tips Are Available All Over the Place	April 2002	473
We Need to Start Writing Better Sales Letters	October 2002	475
Those Sales Emails, How Are We Handling Them?	March 2003	476
Selling Value Is the Key to Improving Profitable Sales	February 2003	477
What Meeting Planners Say about Hotel Sales Communications	September 2003	478
Tips for Sales Folks on Email	March 2004	479
Voicemail Procedures Reflect Sales Professionalism	March 2004	480
Let's Stop the Whining and Get Out and Sell	April 2004	480
Meeting Planners Complain Loudly about Hotel Salespeople	May 2004	482
Improve Sales with Better Voicemail Techniques	July 2005	483
50 Little Tips That Make a Big Difference	September 2005	484
APEX Time-Saving Reports Helps Your Sales Effort	June 2006	485
How about Using Your Business Card as a Sales Tool?	November 2006	486
Most Hospitality Sales Letters Still Need a Lot of Work	February 2007	487
Handling the Issue of Salespeople Not Responding to Call and Emails in a Timely Manner	July 2007	488

Title	Publication Date	Page
How Many Electronic RFPs Have You Been Receiving Weekly?	September 2007	489
Don't Cut Rates Even at the Expense of Occupancy	October 2007	490
Boosting Sales Activity in Troubled Times	August 2009	491
Cutting Rates May Lower Perception of Value, Just Sell Better	June 2010	492
We Are in a "Seller's" Market; It Is Time to Start Thinking about Sales for 2016	August 2014	493
Measure Sales Staff Group Business ROI by Productivity, Not Sales Activity	September 2014	494
Are Request for Proposals (RFP) Driving You Crazy?	October 2014	495
Help! Where to Find Sales Help When You Need It	February 1981	496
CHAPTER 12: Action Plans for Marketing and Sales		
Letter Writing Hints for the Small Lodging Operator	May 1982	499
Using Your Business Calling Card as a Sales Tool	November 1982	500
Success at Trade Shows Starts with Marketing Strategy	March 1985	501
Trade Directories Offer Valuable Advertising Exposure	October 1985	502
Giveaways Aren't Needed for Effective Hotel Marketing	September 1986	503
Small Properties Have Special Needs for Marketing Plan	October 1986	504
The Best-Laid Hotel Marketing Plans Never Go Astray	May 1988	506
Marketing Plan Must Be More Than a Collector of Dust	July 1989	507
The Right Approach to Direct Mail	January 1990	508
Look for ROI on Trade Show Exhibiting	July 1991	510
What's Marketing All About?	January 1992	511
What Works and What Doesn't Work in Ads for Meetings	June 1992	512
Common Sense: The Key to Improved Profits and Service	September 1992	513
Do We Need to Spend More Time Selling?	October 1992	514
Environmentally Conscious Hotels Can Create New Sales Markets	February 1994	515
Public Relations Should Play a Big Role in Your Marketing Effort	April 1995	517
Let's Make Sales Letters Work Better for You	June 1995	518
How to Make Your Direct Mail Sales Pieces Work Better for You?	August 1995	519

Title	Publication Date	Page
The 10 Top Marketing Tips for the 10 Basic Principles of Marketing	November 1995	521
Are You Communicating by E-Mail or Fax?	September 1996	523
Build More Sales Through Public Speaking	February 1998	524
How Well Is Direct Mail Working for You?	March 1999	526
Exhibiting at Trade Shows Is a Lot More Than Just Showing Up	June 1999	527
An Expert's Advice on Marketing via the Internet	May 2000	528
Marketing Tips for the Professional Hospitality Sales Executive	July 2000	529
Marketing vs. Sales or Maybe a Hybrid	February 2002	531
The Power of the Press Can Help Build Your Business	March 2002	533
Sales Newsletters Can Provide Some Good Tips	June 2002	534
We Need to Start Looking at New Ways to Handle Our Sales and Marketing Activities	July 2003	535
Your Business Card Is an Effective Sales Tool	October 2003	536
Good PR Can Help Increase Sales at Low or Little Cost	September 2004	537
Your 2005 Sales Action Plan	January 2005	538
Create a Sales Plan for Your 2006 Marketing Plan	October 2005	539
Working a Trade Show? Here's How to Make It More Productive	July 2006	540
Exhibiting at Trade Shows to Build Prospects and Business	June 2007	541
The Sales Plan as Part of the Marketing Plan	January 2008	543
To What Extent Do All These Social Networking Sites Bring in New Business?	February 2009	544
Revenue Management Is the Key to Profitable Sales	August 2010	545
Exhibiting at Trade Shows a Good Way to Find Leads for New Business	January 2011	546
An Easy Strategic Plan to Maximize ROI on Sales Programs and Activity	January 2006	549
CHAPTER 13: Working on Contracts for Groups		
Agreement Letter Important When Booking Hotel Space	September 1985	551
Hotel Construction Projects Causing Buyers to Beware	November 1986	552
We Gotta Get Tougher on Groups That Cancel	March 1990	553

Title	Publication Date	Page
Remember the Basics of Written Contracts When Booking Groups	March 1998	555
Cancellation Clauses in Contracts Work for Both Parties	August 1998	556
Take a Good Look at the Meetings & Convention Contract	December 1998	557
You Need to Have Contracts for Meetings and Events at Your Property	April 1999	558
Watch Out for Contracts with "Third Party" Planners	September 1999	559
When Contracting with Groups, Make Sure You Have an Attrition Clause	November 1999	560
Let's Do a Better Job on Drawing Up Contracts on Groups	November 2000	561
Renegotiating Existing Corporate and Group Contracts	May 2002	562
Be Sure to Protect Yourself on Attrition Clauses in Your Group Contracts	July 2002	564
Understanding the Value of an Attrition and Cancellation Clauses in Group Contracts	September 2002	565
There's a Move On to Change Your Contract Clauses for Meetings	June 2003	566
Attrition Clauses in Group Contracts Need to Be Clearly Addressed	June 2003	567
We're into a Sellers' Market Again, So Watch Those Contracts	November 2004	568
What's Important about Cancellation Clauses in Group Contracts?	May 2006	569
Contracts on Group Business Getting More Complicated	November 2006	570
Group Meeting Contracts Keep Getting More Complicated	June 2007	571
Understanding the Attrition Clause in Group Contracts	October 2007	572
Caution Is Needed When Signing Group Contracts	March 2008	573
New Construction and Renovation Projects May Cause Group Buyers to Beware	February 2016	574
CHAPTER 14: Everyone At a Property Is Involved in Sales		
Turning Every Employee into a Salesman/Marketer	October 1981	577
Sales Income Often Depends on Incoming Phone Calls	December 1985	578
Don't Blame Sales Staff If Customers Aren't Returning	August 1986	579
In the Lodging Industry, Staff Enthusiasm Spawns Success	November 1986	580
Accounting Department Can Help Sell and Rebook Meetings	June 1987	581

Title	Publication Date	Page
Let's Make More Use of Our 'Front Office Salespeople'	April 1988	583
Properly Handled Phone Calls Can Ring Up Bigger Profits	November 1988	584
Biggest Key to Better Sales Is Improved Guest Service	November 1988	585
Make Workers Feel Important and Watch Sales Zoom	March 1989	587
Selling and Servicing or Just Selling?	September 1990	588
All Employees Really Do Help Sell	June 1991	589
Internal Communications Is a Key to Successful Sales	October 1991	590
Can We Improve Our on Property Reservations Operations?	January 1993	592
Poor Customer Service Will Ruin Your Sales Effort	June 1999	593
Good Communications Lead to Improved Profitable Sales	July 2001	594
Front Office Staff Attitude Training Is Very Critical in These Tough Times	November 2001	595
Proper Telephone Handling Will Improve Your Sales	September 2002	596
Customer Service Week Should Run for the Whole Year	November 2003	598
You Never Get a Second Chance to Make a First Impression	October 2004	599
The Wearing of the Name Badge	July 2005	600
Quality Customer Service Supports the Sales Effort	June 2006	601
Is Customer Service Part of the Selling Process?	October 2010	602
Sales and Operations: A Perfect Team for Building Profitable Sales	February 2013	603
Who Else at a Property May Be Helpful to the Sales Effort?	April 2013	604
How Much Revenue Do You Lose Daily, Unknowingly?	June 2013	605
Service Recovery Has an Important Role in Sales Productivity	August 2013	606
Performance Enhancement Helps Improve Sales Productivity	July 2014	607
Great Customer Service Ensures Future Group Bookings	August 2015	607
Staff Enthusiasm Starts the Success in Building Sales	February 2016	608
Turning Every Employee into a Salesperson via Exceptional Guest Satisfaction	March 2016	609
CHAPTER 15: A Professional Approach to Hospitality Sales Along with Networking		
Do Business by Solving Other People's Problems	August 1983	611
Let's Not Get Too Technical	December 1983	612
Networking—It Works!	May 1989	613
It Is Time to Worry about Competition	July 1990	615
Let's Make Site Selection Easier for Prospects	August 1990	616

Title	Publication Date	Page
How Meeting Planners Select Hotel Properties	November 1991	617
How about That Sale That Was Lost?	April 1992	618
Proper Networking Will Add to Your Personal Development	October 1999	619
Networking: Old Idea, New Drive	December 2003	621
The Selling Process, Let's Make it Better	April 2004	622
Make Your Hospitality Sales Meetings More Effective	March 2006	623
Just Keep Selling in 2009 to Maintain Profitability	January 2009	624
Improve Your Networking Skills to Build Business	March 2009	625
Coopetition Preparing for What Is Next in Hospitality Sales	November 2010	626
Networking Is Not Selling	March 2012	627
Revenue Management: The Guide to Improving Profitable Sales	November 2012	628

CHAPTER 16: Understanding the Features of Your Product, Especially F&B

Title	Publication Date	Page
Facilities, Not Room Rates, Will Net Meetings Business	November 1984	631
Salespeople: The More You Know, The More You'll Sell	March 1986	632
Salespeople Should Learn about Latest Trends in Food	March 1997	633
More Creativity Needed in Selling Hotel Bar Functions	April 1987	634
Group F&B Functions Have Potential to Increase Profits	May 1987	636
Planners Need to Know the Benefits of Hotel Features	July 1987	637
Suggesting Receptions Can Help Improve Groups Sales	February 1988	638
Basic Knowledge of F&B Is Big Help in Boosting Sales	May 1989	639
More Knowledge of Banquet-Bar Service Can Hike Sales	June 1989	641
More Creativity, Aggressiveness Needed in F&B Sales	October 1989	642
Catering an Important Part of the Sales Effort	August 1991	643
Are You Sure You Want a "Podium"?	May 1992	645
Cocktails Are Still in at Meetings—But So Is Drinking Light	September 1994	646
Let's Start Charging for Audio Visual Equipment and Supplies	July 1995	647
Helping with the Group Menu Planning	March 2005	648
Increase Catering Sales with Better Prospecting Functions	October 2005	649
Product Knowledge Is a Key in Improving Group Sales	May 2006	650
Improve Catering Sales with More Prospecting	October 2006	651
Product Analysis: A Key Activity of Sales Personnel	March 2011	652
Product Knowledge = Knowing All the Features of a Hotel	November 2013	653

Title	Publication Date	Page
CHAPTER 17: HR in Connection with Sales Staff Employment		
Outline Job Description and Compensation Before Recruiting	February 1980	655
Recruit Salespeople from Existing Staff, Friends, and Suppliers	March 1980	656
Hiring: Look for Confidence and Aggressiveness	May 1980	658
Train New Salesperson with Immediate Field Experience	June 1980	659
Poor Hiring Practices Aggravate Employee Turnover Rate	August 1989	660
Performance Appraisals in the Sales Office	February 1990	662
The Selection Process	April 1991	663
Recruiting, Selection, and Job Description	May 1991	665
Sales Rep Companies Can Work Well for You	September 1991	667
We Could Be Doing a Better Job Working with Our Younger Employees	November 1993	669
Tips on Hiring the Right Salespeople	October 1994	670
Get Sales Incentives to Work Better for the Property	May 1998	671
How Do We Motivate Our Staff to Do the Things We Want to Be Done?	December 1999	673
Resolving Challenges with Meeting and Event Planners	April 2000	674
Good Help Is Hard to Find, Especially in Our Business	May 2000	675
We Need to Honor Our Sales Folks Who Bring in the Business	February 2005	676
Bits and Pieces of Studies and Reports That Affect Us	April 2005	677
The Three R's of Motivation: Recognition, Rewards, Retention	June 2005	678
Keep Your Salespeople from Saying "Goodbye"	March 2006	679
Will the Independent Salesperson Replace Property Sales Staff?	September 2007	681
You May Not Need a Sales Department	March 2010	682
We Are Customers to Each Other in This Business of Hotel Sales	February 2013	683

About the Author

Howard Feiertag
Department of Hospitality and Tourism Management,
Virginia Polytechnic Institute and State University,
Blacksburg, VA, USA

Howard Feiertag, a well-known hospitality industry veteran, is a member of the faculty at Virginia Polytechnic Institute and State University, Blacksburg, VA, in the Department of Hospitality and Tourism Management. The department was recently renamed the Howard Feiertag Department of Hospitality and Tourism Management in his honor. Before joining Virginia Tech, where he has been teaching for many years, he gained an extensive background of almost 30 years in hospitality, including work in convention bureau management, hotel operations, food & beverage, sales & marketing, catering, hotel operations, meetings and convention management, and as well as tour and travel. Mr. Feiertag has been the group and meetings editor for Travel Trade Publications as well as a regular monthly columnist for *Corporate & Incentive Travel* magazine. He has also been a regular monthly contributing columnist for *Hotel Management Magazine* for 35 years, 1980 to 2015. He is also the co-author of the book *Lessons from the Field: A Common Sense Approach to Effective Hotel Sales.*

Howard Feiertag is a Certified Meeting Professional (CMP), a Certified Hospitality Marketing Executive (CHME), and a Certified Hotel Administrator (CHA). He has received numerous awards during his career. As a member since 1962, he was appointed to the Hospitality Sales and Marketing Association International's (HSMAI) Hall of Fame and has been awarded a lifetime membership. As a charter member of Meeting Professionals International (MPI) since 1972, he has received the President's Award as well as the organization's lifetime membership. He received the Students' Choice Award at Virginia Tech from the Student Alumni Associates in 1997. In 1998, he was voted Educator of the Year by the Professional Convention Management Association. In 1999, Feiertag was inducted into the Event Industry Council's Hall of Leaders as a lifetime achievement award in recognition for his leadership and contributions as a role model for the meetings and convention industry. In 2001, he was appointed a Fellow of the American Hotel & Motel Association's Educational Institute. Mr. Feiertag is also the recipient of the 2001 Arthur Landstreet Award from the American Hotel & Lodging Association for significantly advancing the quality of hospitality education and training. In 2004, he received the Virginia Tech Award for Excellence in Outreach. Mr. Feiertag received HSMAI's Lifetime Achievement Award in 2013, and during the same year, he received the Virginia Society of Association Executives Executive Career Service Award. In 2017, he received the Virginia Tech, Pamplin College, HTM Department Lifetime Achievement Award, and in July 2018, the University renamed the HTM Department to the Howard Feiertag Department of Hospitality and Tourism Management.

Mr. Feiertag started his career in convention bureau management, then moved to the hotel business, working in sales and catering departments, and has worked his way through to executive positions. Prior to joining Virginia Tech in 1989, he was Senior Vice President of Operations with Servico, Inc., a Florida-based hotel owner, developer, and manager of 60 hotels. Prior to that, he was Senior Vice President of Operations at American Motor Inns, Inc., a Virginia-based owner, developer, and operators of 50 hotels and restaurants.

Howard's Story

This is a story of a man's almost 60 years in the hospitality business, who had never planned, or even thought of, being in this business. In fact, getting involved in hotels and conventions was a complete accident. Throughout his career he never had a job he did not like—and never applied for any employment. In all his years he had always been recruited by companies for employment.

It all started in July 1928, when Howard was born in Brooklyn, New York, of immigrant parents who already had nine children. This was in an area pretty much described as "the melting pot," with residents, all poor, from many countries, languages, race, religions, where people lived in tenement houses, not homes. He had a very happy childhood, being taken care of by older brothers and sisters. In this family the work ethic was very important; everyone had a job to contribute to the family's living conditions.

Howard's first job was at the age 12 when he worked daily 7–8 a.m. selling newspapers for two cents to people traveling to work and stopping at his newsstand, which was in front of a candy store. After a period of time he was moved into the store, which had an open window where people on their way to work would stop for a paper, cup of coffee, and a cigarette (being sold for a penny a piece). The next step for him was when he moved, further into the store, to the lunch counter, where he helped serve breakfasts to customers.

Moving onto high school he continued to work but this time at a larger luncheonette, serving early morning breakfast to guests. Sometimes in the after-school hours he would work there making ice cream sundaes, sodas, etc. Toward the end of his high school days, he was working evening hours helping to prepare dinner meals for customers.

Right after high school graduation, Howard volunteered for military service. This was in 1945, just before the end of World War 2. He had three brothers already serving in the army, and he thought it was his responsibility to join as well. During his three-year period of service, he had military police training and served in a criminal investigation department. He then went on to become an investigator in the security and intelligence division. In 1948 he was discharged but remained in the military reserves.

At this point he was ready to find a job and go to work, but he had no vision or plan of what he was going to do to make a living. His commanding officer had told him that he needed to go to college. This was far from his mind. All during his high school period, the discussion of going to college never came up in his family or at school. Many students upon graduating went to a trade school to learn about being a plumber or an electrician, etc. Going to college was just never part of any discussion in school. Everyone thought you had to be rich to go to college. Howard's commanding officer insisted and took over the job of searching for an appropriate university where Howard could follow in what he had learned and did while in service. The C.O. did the paper work, wrote letters, etc., and was able to get Howard into The School of Public Service, with a major in police administration, at Michigan State College.

Here was an opportunity for the government to pay all expenses under the G.I. Bill, and it even provided an allotment to Howard's family and a small wage for him. Howard also worked as a dormitory counselor while at Michigan State. He was there from 1949 to 1952, graduated with honors, and through his military reserve status was awarded a commission in the army reserves as a second lieutenant.

Here we go again, like "now, what am I going to do for a job?" Right after graduation he was sent to a prison in Jackson, Michigan, the largest maximum security prison in the U.S. Without his knowing it, one of his professors had already made arrangement for his employment. He was now a prison counselor.

This lasted a pretty short time; the Korean War had started, and military reserves were called to active duty. Howard was ordered to report for duty with the assignment as Commanding Officer, 56th Military Police Prisoner of War Camp in Korea. It seemed that the military had it all figured out, that with Howard's experience as an enlisted man in the military police, a degree in police administration, and working in a prison, it all added up to this new military assignment.

After two years, the war was over, and Howard was ready to return to civilian life. Again, it was "now what do I do for a job?" After discharge he joined a friend who lived in Atlanta and thought he would get a job there. His friend's father knew about Howard's background and experience and offered him a job working for his insurance investigation firm. Within six months Howard became assistant manager of that office, and was then promoted to manager of the Savannah, Georgia, office. After a couple years he was promoted to run the company's regional office in Charlotte, North Carolina. During his period in Charlotte, Howard became pretty active in the community, belonging to organizations, taking on leadership roles in the local chamber of commerce, and becoming president of the junior chamber of commerce. After about six years with the company his next move would have been to manage the New York City office. When the local newspaper ran the story of Howard resignation as president of junior chamber of commerce and his planned move to NYC, he began to receive a number of phone calls and luncheon dates with propositions of employment to keep him in Charlotte. One call was from the chamber of commerce with an invitation to lunch to offer him the job as the first convention bureau manager for the city. The chamber had been working to start a convention bureau and had been searching for someone to run it. Howard had absolutely no idea of what this was all about. "How does one run a convention bureau?"

This was his entry to the hospitality business. Needless to say, he was most successful, even without knowing where to begin. About two years later, he received a call from the manager of the Orlando, Florida, Chamber of Commerce. They wanted to talk to him about taking over the running of a convention center they wanted to start. Here we go again, no application or resume. So, that was the next move. However, that lasted only a short time because he received a phone call from a New York real estate company that had purchased two hotels in Charlotte. He went up to meet those folks, and they wanted him to be Director of Sales for the two hotels, with a five-year contract and nice income. So that was the next. You need to understand that in all those three moves Howard never really applied for those jobs nor did he offer, nor was he asked for a resume.

So now it was back to Charlotte and being very successful in sales for the two hotels, when an incident developed at one of the hotels in which Howard was accused of "stealing" business from a competing hotel. This is a most interesting story that you need Howard to explain to you. The hotel management company that registered the complaint came to Charlotte to face Howard with their complaint. It ended up with that management company asking Howard to join them as Director of Sales of the 22 hotels they owned and operated. The company was American Motor Inns. Again, there was no application for employment or resume.

It was around 1967, and the move was made to Roanoke, Virginia, home office for AMI. During the next seven years a good deal of different things had happened. In 1970, Howard and his partner (at that time they were together for 10 years) went into the cattle farming business. They bought a farm outside of Roanoke, bought Black Angus steers, and started a "feeder" operation, raising the steer calves from around 400 pounds to 800, then selling them on the market. It was quite a successful operation.

At AMI Howard had started organizing sales teams at all of the properties. He set up a standard system of sales operations for the company and hired and trained all the sales teams. He was promoted to Vice President of Sales and was given the marketing responsibilities for the company. In 1972, Howard, along with several meeting planners from the around the country, organized a Meeting Planners Association, which was intended to be a small club of planners. However, this was the beginning of Meeting Professionals International, which now has a membership of 60,000 meeting and event professionals. Meanwhile during this period, AMI doubled its size in the number of hotels owned and operated. Also at that time Howard got involved in writing a monthly column for *Travel Trade* magazine, a publication for travel agents. The theme of his monthly articles had to do with travel agents working with meeting planners to place business meetings.

All was going very well for Howard, as he always loved his work. Then came along another hotel management company just getting started in Memphis, Tennessee. Howard was recruited again, no application or resume, just hired as Senior Vice President of Operations for Servico, a new company put together by three independent owners of hotels. So, the farm got sold, and he move made to Memphis. This did not last but a year or so when Howard got a call from the president of his former company, AMI. "Ya gotta come back, everything is so screwed up." So, no questions asked, Howard went back to Roanoke, bought another farm raising Black Angus steers, and added a flock of sheep. Now he was Senior Vice President of Operations. There are many good stories Howard can tell about now operating 50 hotels, including a new building in St. Thomas, which he personally named Frenchman's Reef, a well-known hotel.

During this period, back with American Motor Inns, he developed a subsidiary company for AMI, helping motor coach tour companies locate appropriate housing at stops for their tours. He continued writing monthly columns for *Travel Trade* and also started writing meeting planner columns for *Corporate and Incentive Management* magazine. In 1980 he started the "Sales Clinic" column for *Hotel Management* magazine, which ran monthly for 35 years. At this time he was writing monthly columns for three publications.

At one point he took over the 50 restaurants at the various hotel locations the company had operated because they were losing money. He turned them around in one year.

In 1983 he was contacted by Servico, the hotel management company for which he briefly worked in Memphis. They had moved their company headquarters to West Palm Beach, Florida, and were operating 60 hotels. They wanted him to come back. Of course, this was Howard's next move. He sold the farm, moved to Florida, and took up operations as Senior Vice President of Operations for the company. This next chapter in his business life lasted six years until he was recruited by Mike Olsen, HTM Department Head at Virginia Tech, to join the teaching staff in Blacksburg. So in 1989 his teaching career began. Interestingly, again there was no application or resume. Of course, he and his partner got back into cattle and sheep farming business.

On July 1, 2018 a really big honor was bestowed upon him. The name of the department at Pamplin College of Business had been renamed the Howard Feiertag Department of Hospitality and Tourism Management in recognition of his contributions to the hospitality industry as well as Virginia Tech. Howard is still teaching there.

In 2001 he coauthored the book *Lessons from the Field: A Common Sense Approach to Effective Hotel Sales*.

List of Abbreviations

AH&MA	American Hotel & Motel Association
APEX	Accepted Practices Exchange
ASAE	American Society of Association Executives
CAE	Certified Association Executive
CHA	Certified Hotel Administrator
CHME	Certified Hospitality Marketing Executive
CHSE	Certified Hotel Sales Executive
CHSP	Certified Hospitality Sales Professional
CIC	Convention Industry Council
CLC	Convention Liaison Council
CMP	Certified Meeting Professionals
CTC	Certified Travel Counselor
EI-AHMA	Educational Institute of the American Hotel and Motel Association
HSA	Hospitality Services of America
HSMA	Hotel Sales Management Association
HSMAI	Hospitality Sales and Marketing Association International
HTI	Hotel & Travel Index
ICTA	Institute of Certified Travel Agents
MPI	Meeting Planners International
MPI	Meeting Professionals International
NARE	National Association of Retreat Executives
NTA	National Tour Association
PCMA	Professional Conference Management Association
PCMA	Professional Conference Managers Association
RFN	Reunion Friendly Network

RFP	Request for Proposal
ROI	Return on Investment
SGTP	Society of Government Travel Professionals
TRN	The Reunion Network
UETA	Uniform Electronic Transaction Act
USTTA	United States Travel and Tourism Administration

Foreword

When Howard Feiertag called me in 1979 to lament what he perceived as a lack of focus on sales topics in *Hotel & Motel Management,* he modeled the best behavior of a hotel sales professional. He was well informed, cordial, respectful, and direct. He identified a need to be filled, and in his own gracious way, asked for the order. How could I have done anything other than giving him an opportunity to pen an article for the magazine?

Howard's first column, published in 1980, was so well received that I hired him to be a regular contributor. Little did I know that he would retain his column, despite multiple management and ownership changes, 35 years later. His longevity with the magazine is a testament to the wisdom of his words. Imagine how many sales professionals have gained knowledge and received career guidance through Howard's sage insight and advice.

This compilation of Howard's work assures its legacy and illuminates the timeless tenets of hospitality.

—**Mary Gendron**
Former Editor of Hotel Motel Management
(now called Hotel Management)

Preface

It is incredible how things happen in life that comes up never having been expected. Never in my earlier lifetime would I have ever thought that I would have been, or even could have been, a columnist for the hospitality industry's largest circulating publication, called *Hotel Management*. Well, it happened in 1980 and kept going for 35 years.

This most unexpected event started with my one day complaining to that magazine editor about my disappointment with the publication never having anything published covering the business of sales in the hotel industry. My concern was related to Mary Gendron, then editor of *Hotel Management* magazine, sometime during the last quarter of 1979. The response I received during the phone call from the editor was quite a surprise when she suggested that I write something about the subject that they could use in one of their issues. Little did I realize that I was even capable of writing anything for a magazine. At the time I was Senior Vice President of Operations for American Motor Inns, a management company operating 50 hotels in the U.S., Puerto Rico, and the U.S. Virgin Islands. Of course, I had worked my way through to that title after joining the organization in 1967 as Director of Sales with the job of hiring and training sales representatives for all of our properties. Never did I ever have the thought that I would be writing a column on sales, but I agreed to take a shot at it.

Frankly, I never did expect to get paid for it, but it was a big surprise when I started getting a check from the publication in the amount of $100 for each column that was published. Little did I realize that I would end up writing a column called the "Sales Clinic" for each issue, 21 times a year. (Yes, they ran that many issues each year for about 15 years, then reduced their publication to 15 issues a year for a few years before dropping that number to 12 issues a year.) The Sales Clinic column has now been running for 35 years, without any issue missing my column, with the pay going up little by little over the years. All the columns ran at about 500-words, with some even running up to and over 750 words.

Looking back at some of the earlier issues, I had a good laugh about some of the material. I think it was in my first column, in January 1980, where I had written that a requirement for hiring someone for sales at a hotel was that the prospect had to have a car and know how to type. In those earlier days we did not even have an electric typewriter, and without a secretary, sales folks had to type their own letters and contracts. Of course, a car was needed to move around town, as well as out of town, looking for business. As you get into reading some of these columns from 1980 through the 90s, you will get a big kick out of some of the material. However, you will also realize that so much that was written during those past years still holds true for today's management of the sales effort.

Without the benefit of the advancement in technology, in those days the person-to-person contact for sales development was communicated via the telephone (non-mobile devices) or by U.S. Mail, as well as attendance at meetings or personal contact via sales calls. Does anyone remember mimeograph machines? That was the only way to reproduce multiple copies of written material for distribution; copy machines had not yet been developed. Also, to make copies of letters we had to use carbon paper between two sheets of paper to make a copy of a letter or contract. I still have a collection of hard copies for every column written for *Hotel Management* magazine, about 700 pages. Even after the computer came into being, I continued to keep hard copies through to today.

So, even after being recruited to join another hotel management company, operating 60 hotels, I continued writing my Sales Clinic column. Then in 1989, I was hired to join Virginia Tech as an instructor in the Department of Hotel and Restaurant Management, which ultimately was called Hospitality and Tourism Management, where I am today, teaching courses in Hospitality Sales as well as Meetings and Convention Management.

> **For a wide variety of forms and tip sheets that are applied to sales, please visit my blog at www.howardsez.blogspot.com.**

CHAPTER 1

GENERAL MANAGERS' INVOLVEMENT WITH SALES

FOR OWNERS AND GENERAL MANAGERS ONLY

January 1980

With a new calendar year approaching, the universal concerns in the hotel/motel business would generally be centered around sales. What will next year bring? Will the rooms still be around to eat up market 'glut' of hotel shares? How well prepared are the sales teams to bring in the new accounts and help increase occupancies and rates?

Perhaps in anticipation of fretting over the above, the owners and the General Managers should concern themselves with what their salespeople are thinking and what are their thoughts? Are they hampered in productivity as a result of corporate or an owner's or manager's lack of interest in what the salespeople think?

As a result of just a rough sampling, here are some thoughts on why salespeople leave one company for another, and some of their general concerns:

▶ They are looking for more promotion within. Some sales managers have been around for a while, and they think they are doing a good job and getting good reviews. Then the sales director leaves, they want the promotion badly, and some think they are going to get the job and even take over the responsibilities for a period of time. Then BANG, the General Manager brings in a brand new body from outside the company. No explanation even being offered.

▶ Many feel they are not recognized for what they do. They get little direction and supervision and feel unappreciated.

▶ Salespeople, in many instances, are not consulted about decisions that are made which affect the sales effort. They are left out of knowing 'what is going on' and don't feel part of the team.

▶ Training is almost non-existent. This is especially true with new salespeople, and even experienced salespeople feel the need to be further trained. Many feel management won't spend the money to send them to sales training seminars.

▶ Very few, if any, feel there are many career opportunities if they stay in sales. They don't feel it's the right path to a future General Manager's job. There's

no management development by staying in sales at one property.
▶ More and more salespeople are starting to look for monetary, short-term incentives for the business they bring to a property.

The list can go on, but these are the highlights, so?—What needs to be done to correct some of the situations?

In fact, even if it does or does not exist at the property, owners and managers need to become more involved with the sales staff. These are the things that are really on their minds.

To improve sales productivity, someone needs to pay attention to the salespeople! Time needs to be invested by talking to them—find out what it is that can be done to help them do a better job. Involve them in the decision-making process as it relates to sales. Talk with them frequently formally and informally, let them know how they're doing. Suggest how they can improve productivity. Discuss their work habits, their sales calls, their bookings, and their attitude. There's a great deal to talk about.

Help them, and it will help you. You'll have a better sales year coming up.

HIGH VISIBILITY: A GREAT MANAGEMENT SALES TOOL

March 1983

Observers of management practices in lodging operations should be able to report that those properties which reflect a high degree of management visibility are the ones that also reflect sales success.

Regardless of property size, whether a 2.5-unit or 2,500-unit, high visibility of management personnel is directly related to sales as well as profit success. We can all relate stories of visiting properties where you hardly see a manager 'out front.' That should become a red flag danger signal. If the owner/operator/manager or whoever the top person may be, if that person cannot be available to see what's going on with the guests and how they are treated, then someone else, an assistant or other responsible members of the management team should be visible up front. Management these days of troubled times with high costs of doing business and drop off in sales just can't afford to be out there where the business is being conducted. There's plenty of time during the stay to be stuck in the office doing bookwork. The time for high visibility on the property is really all the time but to pinpoint specific hours, let's just list the critical times.

▶ During check-out hours. Be right there at the desk, personally checking people out and thanking guests for staying at your property—giving them your calling card and asking them to come back as well as sending their friends and associates to stay with you.
▶ Help pour coffee in the restaurant. A great opportunity to let guests know you are there 'looking after them'—a little chit-chat from the manager in the morning goes a long way in building goodwill.
▶ During check-in time. Greeting the guests as they come into the lobby. Walking them up to the front desk to

check-in. Pleasant words of welcome can start off a guest's visit with the right tone and make them feel glad they are staying at your property.
- In the lounge at Happy Hour time, not drinking with the guests but just 'checking around.' They'll know you are there and will appreciate your attention into seeing that you are taking care of your business.
- A check through the kitchen to make sure all is going well and the food is being served properly. Then cruise through the dining room welcoming the dinner guests.
- Anytime during the day when there are meetings in the house, the manager should be visible to the meeting attendees. Being 'around' when they go into a meeting room, come out of a meeting room, or taking a coffee break.

Now, here's what happens when you do all of this:

- Guests will recognize you as being a good business person.
- Guests know you are looking after their interest.
- You will observe problems that can be taken care of immediately before they grow out of proportion.
- You'll come up with ways of improving service to your guests.
- Your employees will know you are 'always' around.
- Employee productivity will improve.
- Business will be better.
- Profits will be better . . . and most of all, you'll feel good!

YOU DO NOT NEED A SALES DEPARTMENT . . .

April 1983

If it doesn't improve results. It is not the effort that counts but the results of the effort. There are just too many salespeople and sales departments around in hotels and motels that are just not getting the job accomplished. The productivity of salespeople in many cases is not what it could be. A number of factors create this situation:

- Lack of direction and supervision
- No sales plan or procedures
- Poor time management
- Other functions delegated to sales staff
- Staff communication poor
- Untrained sales staff
- Too comfortable offices
- Non-responsive management
- Low sales budgets
- No job description

The solution to the problem is for the lodging owner or operator to accept the responsibility of being the chief in charge of sales. Objectives need to be established for the sales department, objectives that are results oriented. For example, goals for improved room occupancy, room revenue as goals for food/beverage functions. Goals that are measurable need to be established. Pie in the sky doesn't work; all goals should be realistic.

Remember the objective must be an improvement. The measurement of the quality of a sales team is in its ability to improve sales regardless of the obstacles that be in the way. Usual arguments from salespeople for not improving sales come from a long list of *reasons why we can't get business*:

- Many complaints about food service
- Highways have changed traffic patterns
- Product is dated
- Hotel or motel needs refurbishing
- Budget not large enough to do the job
- Too much changeover in managers
- Staff not courteous
- Complaints on housekeeping
- Need more help in the office
- Other jobs delegated to sales
- Newer competition in the area
- Lower prices at other properties
- Meeting room lighting is poor
- Meeting rooms, not sound proof
- Poor parking, etc., and the list goes on and on.

The items on the list sound plausible as excuses for not getting business, and many of the excuses or reasons are correctable by management. However, if the sales department is convinced that these are some of the reasons business can't be developed, then there is no choice but to save some money by getting rid of the sales department and get business catch-as-catch-can.

More logically the salespeople need to be convinced that regardless of the property's shortcomings there is a business to be had.

Proper management direction, better communication, a job description, a sales plan, and procedure, hard work, improved time management practices, some sales training, hard work, interested department heads, staff motivation, outside sales calls, telephone calls and lead development program, reasonable budgets, analysis of business needs, hard work sales reports reviews, discussions with staff on how to improve results, enthusiastic people on payroll, hard work, entertaining on property, civic affairs involvement, leading by example, hard work, good guest services, etc., are just some ideas on how to help maximize results from a sales department.

LET THE SALESPEOPLE SELL

July 1983

It's happening all around us in the hotel business. Wherever we go, it can be spotted almost immediately upon entering a hotel or motel. Salespeople are busying themselves doing the non-sales productive activity. Just think of all the things that salespeople have had to do (whether voluntarily or not): work the front desk, be part-time hostess or cashier, bus tables, wait tables, answer switchboard, temp bar, clean meeting rooms, schlep tables, run room service, pour coffee at banquets, hire banquet waiters/waitresses, make out banquet bills, go to the bank, go to the post office, run errands, make table decorations, change the function board, talk to salesman/purveyors, interview applicants for various jobs, write advertisements, produce promotion pieces, run to the printer, stuff envelopes, type envelopes, and the list can probably go on and on.

None of these functions really is the job for salespeople. They all need to be done, but certainly not by sales. Good management on the property will want to see sales personnel out selling at every chance possible. In-house staff should be used for those little extra jobs that have been delegated to sales. Of course, it is not entirely management's fault this occurs, in many cases, salespeople make themselves very available and look for reasons to stay in under the pretense of 'helping' out when necessary.

Management's responsibility is to make sure that salespeople spend their time selling. The continual development of new business for the hotel/motel is what will keep a property healthy. Salespeople must be continually searching for and developing new prospects for business hotels/motels can't survive on repeat business alone—no matter how good a product or service there is never enough repeat business to keep an operation healthy without new business. If we can all think about the sales department's responsibility being that of bringing in the new business, we'll be more successful in our operations.

In some smaller properties the sales manager is somewhat of 'jack-of-all-trades.' However, management in those types of properties should try to develop a program whereby other staff people assist the sales manager with various 'in-house' details so that more time may be spent selling.

General Managers should start taking over responsibilities for advertising and sales promotion. Too much time is spent at the property level, in all sizes of hotels/motels by the salespeople in these endeavors. Unfortunately, we have misused the title of marketing director or marketing manager and have saddled some very excellent salespeople with a title that now dictates they have responsibility for all marketing activity at a property. At very large properties, this is understandable as long as there is someone left to go out and sell.

Goals are an essential incentive for selling, and it also establishes an opportunity for measuring performance. Salespeople should be given targets to reach in increased performance periodically, as it relates to sales goals should be reviewed. This is one good way of getting the salespeople to spend more time selling and less time doing other people's work.

NEW YEAR'S RESOLUTIONS FOR SALES-MINDED GENERAL MANAGERS

December 1983

Want to improve sales in 1984? Want to add to the personal development of your sales team in 1984? Then General Managers and owners take heed—and make the following New Year's Resolutions:

Resolved—That in 1984 I will:

▶ Take time to visit with the Sales Department Personnel on a daily basis formally or informally to find out what's going on and to offer assistance.
▶ Do away with comp meals for Sales Managers and Sales Directors unless they will be eating with prospects. I will encourage on-premise entertaining whether for breakfast, lunch, dinner or cocktails—the best-selling job is accomplished on the property.
▶ Make sure there is a sales plan for the year for the month and for the week

- and follow-up to make sure it is being followed.
- Review weekly activity sheet for each salesperson with the sales director and provide direction for improvement of activity.
- Establish goals for each salesperson and provide an incentive program for exceeding goals short-term goals work better.
- Be sure to see that salespeople get out to make sales calls on a daily basis. The 'old fashioned' face-to-face sales call still works best.
- Take time periodically to make sales calls with the sales personnel to get a feel for the activity as well as provide direction to improve the quality of sales calls.
- Be time-management oriented and pass on to the salespeople tips on improving their time management practices.
- Review the reader file daily to see how sales proposals and letters are being written and to offer constructive criticism on their improvement.
- Not let accounting policies and procedures interfere with the development and booking of business by the sales department too many restrictions hinder sales.
- Allow sales personnel freedom in negotiations when working with meeting planners, within certain guidelines, to maximize profitable sales.
- Give all sales personnel the opportunity to improve their skills by joining, attending meetings and participating in organizations such as Hotel Sales Marketing Association International and Meeting Planners International.
- Encourage sales personnel to improve their knowledge of hotel operations by exposing them to the activity of other departments on their own when it does not interfere with, or substitute for their own sales activity.
- Improve sales department performance by using other personnel to perform duties that may take away from direct sales activity (direct mail programs, letter typing, social director activity, envelope stuffing, addressing envelopes, secretary's club, handling reservations, writing brochures, working on advertisements, handling in-house promotions, etc.).
- Support sales personnel in their efforts to improve profitable sales through the development of the group market by making sure groups are serviced properly and that the services are delivered as agreed upon by the sales department.
- Have a formal meeting with each salesperson quarterly to discuss individual performance and make a written evaluation.
- I will be honest in the evaluation, so sales personnel know how they are doing and where improvement is needed.
- Review sales salaries to be sure that appropriate remuneration is made commensurate with performance and the competitive situation in the marketplace.
- Communicate my feelings about the sales situation on a regular basis with salespeople, so they know how I feel about something before they hear it from a third party.
- Treat sales personnel as a 'partner in business' allowing them to feel comfortable

in sharing their thoughts and ideas with me at any time.
▶ Consider myself as the chief executive of sales and put some effort into sales activity myself by calling on friends and people with whom I do business.
▶ Make five telephone contact calls daily to existing room business accounts thanking them for business and asking for more.
▶ Observe sales techniques at the front desk and offer suggestions for improvement.
▶ Take a look at inquiries called into the property to see how they were handled and how—as well as when—they were followed up. The hottest lead for business we can get is the inquiry.
▶ See if all the departments on the property are doing their share of selling.
▶ Be sure to have a weekly sales meeting with all department heads.

With the above as a starter see how many more resolutions you can make for 1984. Even though you may own or operate a property without a sales department, a review of these proposed resolutions will reflect that most can relate to you anyway.

WHAT SUCCESSFUL SALES DIRECTORS (AND GENERAL MANAGERS) DO

February 1984

In all sizes of lodging operations, the success of the sales effort rests in the hands of the Sales Director and General Manager of a property. How well they perform in soliciting, booking, and servicing reflects the overall success of a property. Although the following listing is by no means complete, it covers basically the 12 areas of knowledge, performance or involvement that will make people who are involved in the sales effort more successful.

1. **Good Market Knowledge:** Who are the customers, where are they coming from, how did they get to select your hotel, why are they here, how long do they stay, how much do they spend, etc. Good market knowledge helps you direct your marketing effort—getting the information is not enough—using it is what's important.

2. **Plan a Good Market Mix:** The mix of business reflects how much of what type of business you are doing. What percentage of your total room sales comes from meetings, wholesaler, retail travel agent, corporate, government, transient, etc. These categories of types of business usually have different rate structures. Naturally, by planning and going after a certain mix of business, the sales generated can be more profitable.

3. **Knowledge of the Competitive Edge:** In competing for business with other hotels in your area, knowledge of the other hotels can help you size up your property to see where you can compete to your advantage factors are location, price, size, product, service, amenities, etc. The idea, of course, is to sell your positives.

4. **Multiple Price Policy:** There's nothing wrong with selective discounting. Hotels have been doing it for years. Off-season

rates are usually lower as are weekend rates then there are corporate rates, group rates, wholesaler rates, and a wide variety of discounted rates for special users such as senior citizens, government employees, military, etc. Are they all really necessary? And are they generating business for you? Review all your rates periodically and determine what the multiple price policy should be.

5. **Good Business Contacts:** Many salespeople, and General Managers, have good business contacts, but they just don't bring their business to the hotel. Get the business contacts to work for you. Be sure they are frequent users of your property—use them as referrals ask them for business and ask them to provide you with leads of others who can give your business. Good business contacts are only good if they can help you.

6. **Willingness to Try New Things:** Most successful entrepreneurs would not be where they are today if they did not take a chance and try new things. Come up with some new ideas to promote business and then try it. Develop new techniques in selling to book rooms—be creative in making proposals for groups. If only half the things you try to work, you are still ahead of the game.

7. **Attention to Costs:** Spending more than you take in any business is dangerous. Cost-effectiveness in selling for a hotel is very important. As total sales expenditures start to creep up, we must continue to expect a greater return from the sales effort. Budgeting for sales is a must and monitoring the budget against results is also a must.

8. **Good Management Procedures:** Basic techniques of management are fundamental when it comes to a sales operation. We need to be concerned with doing a better job in: communicating, developing, training, motivating, planning, organizing, directing, and controlling. It's not enough these days to hire a salesperson and say 'go out and sell.'

9. **High-Quality People:** Successful managers look for and employ high-quality people. Of all the factors that represent 'quality,' the most important is sincerity and credibility. We're not looking for snobbery or fine family background, or high educational levels, or show-offs. What we want is down to earth, friendly, well dressed, well-mannered people, who will represent the hotel well.

10. **Realistic Growth Plans:** Where are we today and where do we want to be next year, the year after, etc. What are the possibilities for growth for the sales director and/or General Manager? There should be a plan. Also, growth plans for the amount of business that should be booked in future years. Each year a salesperson is on a property more business should be booked than the previous year.

11. **Technical Efficiency:** Without the technical know-how of a job, very few people can be successful. The sales director or General Manager has to have the knowledge of getting where to business and how to get the business involved. The sales director or General Manager must get involved in some sort of a continuing education program to become more

proficient in the business. Two organizations that are geared to helping salespeople become more technically proficient are the Hotel Sales Marketing Association and Meeting Planners International.

12. **Aggressive Selling:** Energetic, enthusiastic, hotel salespeople generally are the ones that are aggressive—and they book the business. You can still be friendly, credible, sincere and at the same time aggressive. Going after the business repeatedly asking for the booking, following up regularly, and making sincere and believable proposals gaining customer confidence making many sales call working hard and smart is what aggressiveness is it works!

If you have a property that's doing so well you are number one in your market don't act like you're number one, that is to say, don't brag about it but be humble. If you act like you're number three, you'll be still fighting to keep the number one slot.

You can't stop working hard, pushing, and hustling. You can't rest on your laurels and believe that since you're number one you're going to stay there—not so! It just doesn't work that way.

Number one's stay on top by continually improving the product. Better maintenance repair and replacement, conscious of continually improving guest services, constant awareness of quality, and just as important: continuously pushing the advertising and sales effort.

Reducing advertising and sales costs when everything is going well and you're number one won't keep you in that slot too long. This is time to increase the effort (you can probably afford it better, also) in all marketing efforts (Direct Sales, Advertising, Public Relations, Research, Promotions, Merchandising). When you're number one, the morale is usually at a really high level and this is the time that you got more for your marketing dollars. With the proper 'push' direction and motivation from top management, the sales effort will be more productive than at any other time. People generally do better when there is a healthy mental attitude about their job performance and this positive feeling almost always occurs during a period of success.

Management of people is the key here. The owner, General Manager all have to act positively and pass the feeling out through the ranks. Everyone has got to feel good about number one, but everyone needs to act like they are number three and still pushing for the top slot.

Are the switchboard operators still acting like they really want to serve the people calling in? Are the front office salespeople still checking guests in and out making the guests feel happy they came? Are the greetings guests get in the dining rooms and lounges preparing for an experience? Is the housekeeping department oh top of making every room clean? Are salespeople still out knocking on doors making new contacts following up leads and booking business? Is the marketing action plan still being followed? Is the General Manager taking the time to talk with the employees?

When you think about it, there's never any time that the sales and operation effort can 'let-up.' No matter how great you are now, there's always tomorrow and next year. Let the guests know you are great you are.

But tell yourself you're still number three and act like you are number three.

HOW TO BE THE BEST LEADER OF YOUR SALES FORCE

August 1984

No matter what size hotel or motel property the responsibility for sales still lies with the General Manager. Of course, the sales function could be delegated, but the head leader of the sales force still has to be the General Manager. To be the best leader: there are certain general business practices that need to be followed to be successful in sales—one of these business practices is the good managerial procedures which include:

- **Planning**: Is there a business plan for your property—and is there a marketing plan as well as a sales activity plan? These are all basic in the marketing of a property. The plan is a map to lead the way to a successful operation.
- **Organizing**: Management must be adept at putting together a team and placing people as well as 'things to do' in the right places. Providing a structure for operations with standards and productivity measurement helps add to the success of an operation.
- **Directing**: Good, sound-leadership requires the manager to provide proper direction to the hotel/motel staff—it's not enough to say: I'm paying out to do your job to do it! Daily direction for successful sales operations is the most important. This applies to all other departments as well.
- **Controlling**: "Where did you go today"; "who did you see"; "what did you do"; "let's talk about who said what." These are all proper questions to ask of salespeople in providing a control over a sales operation. Unless the manager knows what's going on daily, there will be no control.
- **Motivating**: A most important tool in getting the sales job done the little time it takes to talk with the sales staff provide direction review the work and give the sole slap on the back pays off exceptionally well in motivating salespeople to do their jobs well.
- **Communicating**: Two-way discussions between managers and salespeople where all parties are honest with their comments to create the best communication. Expanded communication through all departments within a property concerning the sales effort will immeasurably improve the success of a sales program.
- **Training**: An ongoing training situation is also essential for success in improving sales productivity. Salespeople generally want to learn more and are very much open to continued training take advantage of outside training opportunities through Hotel Sales Marketing Association or Meeting Planners International for programs that will help train the sales staff. Cross-training in other departments helps the staff to better understand the workings of a hotel/motel and is another opportunity for improved productivity development.

Sales personnel must have an opportunity to grow within an organization or they will be lost to a competitor an opportunity should always be available for self-development with the help of management.

Try to answer the sales manager's question: 'Where will I be five years from now?' Provide the opportunity for growth and development within your own organization. There need to be realistic growth plans for the property as far as sales are concerned. And, there need to be realistic growth plans for the sales staff keeping in mind what the objectives and goals are of the salespeople and helping them achieve the goals/objectives will create success in sales for a hotel/motel.

THE PROBLEM WITH HOTELS

December 1984

More and more smaller size properties are getting involved with meeting planners in the booking of small meetings. Managers of meetings for companies and associations are finding excellent opportunities exist for them to deal with roadside, suburban, downtown and airport properties that have 150 rooms or less. This is a big boom for this size lodging operation for excellent short-term bookings of meetings. Most meetings being held these days number less than 50 people. This makes the smaller property a logical location, particularly for the size gathering. Dates and locations are usually selected within 90 days of the meeting date, and the site is generally selected by a local representative of a company or association.

However, problems do exist that are a great concern to the meeting planner working with the smaller operation (most of the problems also exist with the larger lodging operation as well). We have selected some of the most common problems to outline along with brief suggestions on either how to correct the problem or at least make it livable for the meeting planner.

- ▶ **Problem**: Salesperson turnover in a breakdown of communication. Planners have to sometimes start from "scratch" with a new salesperson.
- ◊ **Solution**: Since turnover in our industry is almost inevitable—lodging operators should make sure that the details of arrangements with planners are put in writing and proper files and records be kept. The property needs to stay in touch with the planner on a regular basis.
- ▶ **Problem**: Verbal commitments: too many times planners are being acknowledgment in person or on the telephone what the hotel/motel sometimes do not know—planner does not get what has been promised.
- ◊ **Solution**: Put everything in writing for an acknowledgment from the meeting planner.
- ▶ **Problem**: Meeting room set-ups: salespeople sometimes are not knowledgeable on various types of set-ups and room capacities.
- ◊ **Solution**: The new salesperson at a small property must be taught appropriate meeting terminology and learn how to set up rooms in different ways. Room

capacities must be accurately figured with dimensions made available on a rooms layout chart.

▶ **Problem**: Food planning: many salespeople are not knowledgeable about food products, pricing, or menu planning and work only from pre-printed menus.

◊ **Solution**: The manager should have a food/beverage manager or the chef into discussions with the meeting manager when planning food functions.

▶ **Problem**: Location of function boards: some lodging operations don't have any and some are located where attendees can't find, in many cases the boards reflect "yesterdays" functions.

◊ **Solution**: If the property is meeting business there needs to be a functioning board and its location needs to be a lobby or entrance area the public can readily see. Someone must be assigned the task of changing the board daily (preferably late at night so it is ready early morning viewing). A printed function listing should be prepared and provided at locations of guest tact (front desk, switchboard, dining room hostess station, lounge, etc.).

▶ **Problem**: Rates quoted differently than what is actually paid (refers rates as well as banquet prices).

◊ **Solution**: When quoting rates and confirming in writing, be predetermined what taxes need to be added or gratuities and charges for set-ups; there should be no surprises at departure time.

▶ **Problem**: Bar brands, prices, and measures: usually price quoted by the bottle, by the person, per hour but often no mention is made of the size of drink or brands (meeting planners are also at fault for not asking).

◊ **Solution**: Printed price lists for bar service should indicate prices per brand and size of drink service, try to sell up to "name" brand. Some people don't drink, but everyone reads.

▶ **Problem**: Poor quality of public address system; many planners do check this out before booking the meeting they're upset when the system doesn't work properly and never rebook.

◊ **Solution**: Salespeople and property management aware of public address capabilities. Determine needs microphone, etc., and then be honest about the local audio/visual company that can supply the planner with what is needed. Salespeople need to be instructed on the uses and capabilities of the various audio/visual equipment.

▶ **Problem**: Planners have to deal with too many different people at hotel/motel for rooms, banquets, set-ups.

◊ **Solution**: If the salesperson at a property will not be able to handle all the details in connection with a meeting (usually larger properties) then the planner must be advised front—who will be working for the group on all the arrangements. It would be logical for one person assigned to be the contact for handling all the arrangements.

▶ **Problem**: Accounting/billing problems exist, for many when a bill finally arrives the first complaint is that it was too long for them to get a bill. Other problems include separate statements for functions, counts on functions to which they do not agree, charges they

do not understand and credits not given on deposits.
◊ **Solution**: If credit had been established billing than statements should be filed immediately following the meeting. Each function during the meeting a banquet billing should be signed by the planner reflecting actual counts or guarantees price breakdown. All appropriate bills for food/beverage functions, meeting room rental, and other charges that have been approved need to accompany the statements.

MANAGEMENT RESPONSIBILITY TO THE SALES EFFORT

August 1985

The success or failure of a sales department at a hotel/motel property depends on management; the General Manager of a lodging operator is, in fact, the "real" sales director and is responsible for the sales effort. There are other areas where the General Manager is responsible; there is a responsibility to the owner, customer, employees, community, self, and to the sales effort.

▶ **Management Responsibility to the Owner**: The General Manager must operate to guarantee the owner a profit and must operate within the guidelines and rules established by the owner (or company).
▶ **Management Responsibility to the Customer**: The customer visits a lodging operation by choice; the pleasures and satisfaction the customer enjoys are a direct result of a General Manager's attention to responsibilities.
▶ **Management Responsibility to Employees**: The General Manager must take the time to talk with employees. Open communication must be provided. It keeps productivity at the highest level. The employee looks to management to keep his family alive and well.
▶ **Management Responsibility to the Community**: General Managers must let it be known that the people who run the property are part of the environment in which they operate, and local involvement in community activity is essential.
▶ **Management Responsibility to Self**: General Managers have the responsibility of looking after themselves and their families. Physical fitness, good health, family involvement, coping with stress and looking after personal and business development is all part of this as well as looking after one's own income and worth.
▶ **Management Responsibility to the Sales Effort**: General Managers are responsible for the entire sales function and must guarantee support of the sales effort and deliver the product as sold. Management must provide attention, direction, supervisions, and motivation.

◊ **Attention**: General Managers must be aware of what is going on in the sales department and pay attention to each salesperson and the progress each is making. Attention must be given to daily sales calls and sales activity on the part of each person and problems encountered daily by the staff. Attention must be given to the major accounts in the house; even making a few daily phone calls to the top accounts would help the sales effort; of course, paying attention to guests' comfort is a priority. Attention must also be given to increase or decrease room counts. If regular corporate accounts are providing more or less in daily rooms sold attention most certainly must be given here.

◊ **Directions**: General Managers provide sales staff direction by telling people what to do, how to do and why to do. The direction is also provided by going out on sales calls with the salespeople and critiquing the work. When General Managers sit in on client meetings they are in a position to provide direction—the same goes for participating in client entertainment.

◊ **Supervisions**: Good supervision has 'to be accomplished on a daily basis by General Managers. Reviewing daily activities is a key to productivity, this need not be a lengthy daily job; brief review of who did what, when, where, why, etc., is all that is needed. Taking a look at follow-up by salespeople is also important—this is accomplished by a daily review of the reader file. A periodic random pulling of files to see what is being done is a good supervisory method.

◊ **Motivation**: Salespeople need encouragement, particularly after a long day in the field without appreciable results. The General Managers need to encourage when needed, when good is done a compliment is in order—verbal and/or written it goes a long way. Reward when appropriate in some manner, whether it be monetary or by providing special perks. Reprimands are necessary, as well, when appropriate. It should be in private and handled in a non-demeaning manner. Properly rendered reprimands can be a motivational factor.

If the General Manager of a property, in fact, will take over complete responsibility for the sales effort and work with the sales staff by providing proper leadership—success will follow.

GENERAL MANAGERS NEED TO GET INTO THE SALES ACT

February 1986

OK GMs! It's your turn! You have to get into the sales action. Please take the following simple test, write down your answers then compare them with answers/comments at the close of the article.

1. List by the name of the company (and the contact)—10 of your best room night accounts. When did you last speak to them?
2. How many outside sales calls did you make last week?
3. How often do you go out with your salespeople on calls?

4. What did you do last Tuesday between 11:30 a.m. and 12:30 p.m?
5. How often do you review sales calls with the salespeople?
6. When did you have your last sales department meeting? At what time of day?
7. How many hours were spent by salespeople last week on non-productive sales activity?

Here are the sales clinic responses to the questions:

1. Any General Manager should be able to name the property's best accounts; and call or visit or invite for a luncheon the contact at the company. All top accounts need to be contacted somehow at least once a month. The GM does this even though the salespeople may be calling on them anyway.
2. A goal of at least ten good sales calls should be made weekly by the General Manager.
3. Time needs to be taken to work with each salesperson by going on joint sales calls. It creates an opportunity to offer direction and supervision as well as helps with performance appraisal.
4. General Managers should get the habit of checking-out group functions. Check the kitchen, the room set-up, etc.—just generally take a look at all the functions going on in the hotel. Try to check on all three parts of the day when group feeding takes place. Meeting planners love to see General Managers checking things out.
5. Hopefully, salespeople get to review their daily sales calls with the General Manager at the end of each day. It really doesn't take long to do and it sure helps in providing direction to the sales force. Each day something valuable will sometimes be learned salespeople pick up comments that will help in improving a hotel's operation. Also, it's a great opportunity to get to know the sales staff better.
6. Sales department meetings should be held weekly. Here is where the General Manager gets to learn what's happening? Who booked what? Are goals being met? Clear away obstacles to improving performance. Meetings should never be held during productive sales time. This means let the salespeople do their selling 9 a.m. to 5 p.m. and conduct the meetings before 9 a.m. or after 5 p.m.
7. Some General Managers delegate jobs to salespeople that have absolutely nothing to do with productive sales activity. The list is endless. Let's keep salespeople SELLING and have the General Managers worry about getting the other jobs done.

SALES DEPARTMENT RESPONSIBILITY TO THE GENERAL MANAGER

June 1986

The responsibility of salespeople to management doesn't end with just going out on sales calls. It's the actual generation of sales or, more aptly put, the generation of profitable sales that count.

Salespeople need to concern themselves with the mix of business, working on accounts where there is more potential, increasing the penetration in existing accounts, looking for new accounts in the same business category and looking for new opportunities for the sale of rooms, food, beverage.

Being a professional in hotel/motel sales comes from a continuous process of learning and being aware of what's happening in the marketplace. Learning is never ending—learn what the competition is doing learn what different group accounts need and/or want; learn how to present the hotel/motel features and relate them to benefits; learn about creative food & beverage functions and how they may be used as a selling tool—above all, salespeople need to learn how to be enthusiastic about what is being sold.

The business mix: salespeople need to meet with the General Manager to examine and discuss the current mix of business by analyzing how much business is being generated from each market segment—and at what rate it will become obvious if a change in the mix is desired. Most hotel/motel operations have a system for daily recording of this information. Typically, the market segments will include: transient, meeting groups, corporate, government, group tours packages, and any other category where there is a special rate. The objective is to work on developing more business from the categories (or market segments) which bring in the best rate or could be more profitable to the operation. If the meetings group rate is better than the group tour rate—then it is obvious that more effort should be put into selling more group meetings.

Work on accounts where there is more potential: some types of business for a property generate more profit than others. Whereby some groups, for example, might generate good room sales they may not provide any help to the food & beverage outlets of the property. Student tour groups at some geographically popular destinations may fill a hotel/motel with quad occupancy. However, most of these types of groups are notorious for not eating any meals on property. They seek out other low-cost food operations for breakfast, lunch, dinner. Certainly, they don't generate any lounge or bar business. The objective is to generate sales of market segments that not only rent rooms, but will provide sales to all the property outlets.

Increase the penetration into existing accounts: it is amazing the number of salespeople that make sales calls on corporate accounts looking for the individual corporate traveler only—or, calling on prospects to join a secretary club. The potential for additional business from an existing corporate (room) account is great. Learning as much as possible about an existing room account will provide the leads necessary to secure additional business but the questions must be asked: "Does the company have meetings?," "Who handles the meetings?," "Which other departments in the company have people visiting who will need overnight accommodations?," "Who takes care of the company's social activities and Christmas parties?," "What clubs or trade groups do the company's executive belong to? (Can they influence the decision to bring a meeting to the property?)." Generally, an existing account can and will provide more business to a property if the right contact is made and the business is solicited.

Look for new accounts in the same business category: there is a tendency for like businesses to do like things. If the hotel/motel does business with the U.S. Army recruiting people for meetings and meals, then it is most likely that the Navy, Air Force, and Marines are also having similar type meetings someplace. If the hotel/motel has a weekend conference of an association of piccolo players then it is most likely that similar associations of drummers, piano players, violinists, accordionists have weekend meetings. If there is a weekend gathering for a 10th reunion of the class of 1976, then there is great potential for similar reunions for the class of 1966, 1956, 1946, etc., as well as other high schools in the area. If there is a party for a new car showing put on by one dealership—then there may be a similar type of party that could be solicited from other automobile dealerships.

Salespeople need to keep on the lookout for ways of developing new business for rooms, food, beverage. Opportunities abound for all types of business, knowing what's going on in the marketplace, checking on the competition, exploring with the General Manager new sources of profitable sales is the key to success.

OPPORTUNITIES FOR SALES LEAD DEVELOPMENT FOR THE SMALLER HOTEL/MOTEL

March 1987

A wide variety of opportunities exist for the development of sales leads. Leads are available from many sources for all types of business for all types of hotels and motels. Even the very small motel operations those without sales departments can take advantage of lead development for room business (commercial and leisure) as well as for the small meetings that may be available for the smaller hotel.

The obvious leads, of course, come from the existing client base. People that have already stayed at the property, files on groups that have held meeting and banquet functions previously and local business contacts. The operator of a property without a salesperson needs to take the time to seek out new accounts for business. In addition to local contacts, local newspaper leads, local purveyors, etc., there are companies that provide listings that could lead for individual and group room business. These companies vary in services they provide mostly there are computerized printouts available, as well as address labels for direct mailing promotions. The detail of information varies with the company providing this information.

A relatively new source for hotel/motel operators is a company that has been in the business for 35 years, providing directories and mail lists for other industries. The Salesman's Guide, Inc., 1140 Broadway New York, NY 10001 (800/223-1797) (212/684-2985) now has a nationwide directory of corporate meeting planners which lists 10,500 corporations that hold off premises meetings (many of these are perfect for the small hotel/motel since there are a good number of small companies listed).

The Salesman's Guide has a brand-new directory just off the press that lists 10,000 meeting planners for over 6,500 associations that hold small and large meetings in all states. They also have a national directory of incentive travel buyers. This listing includes over 7,000 potential buyers of incentive trips.

The company has available for sales, a breakdown of meeting planners by the state which provides names of people responsible for making meeting arrangements, number of meeting held, size of meetings, number of days held, months held and basic geographic destination.

Similar lead services are available from many other companies and as they become known to this column, they will be reported to the readership.

Many publications that have travel managers and meeting planners on their circulation list also provide lead services to their advertisers. There are also companies that maintain a database of leads in all forms that are available at a subscription service to hotels and motels. This information will also be reported in this column as information becomes available.

SALESPEOPLE NEED TO BE EVALUATED

February 1998

Hotel/motel owners, operators, managers should establish a program or system for evaluating the effectiveness of a sales department.

In many companies or properties, sales personnel are evaluated by a performance appraisal completed by a supervisor once a year. The report then, to some extent, helps determine if the individual should be considered for a salary increase. One goes hand-in-hand with the other, and sales personnel look upon the evaluation as the time to get a raise or not using the appraisal as a tool for salary adjustment may not be the most effective use. An appraisal of performance, or evaluation, should best be used to determine in which areas of performance the individual does well and which areas need improvement. Then, through consultation between the sales supervisor or General Manager and the salesperson, the direction is provided to improve performance.

To improve sales performance, the evaluation system needs to be an ongoing project. Instead of an annual review, which is generally tied to a salary review, sales personnel need to be appraised at least quarterly. This provides more frequent feedback so that steps may be taken to improve the sales operation. It also provides a more frequent opportunity for management to see if improvement is made from period to period.

The key to better appraisals is to tie down the responsibility of the salespeople being evaluated definitely. There are many salespeople working at hotels and motels that do not have a written job description. Salespeople need to know specifically what area of responsibility they have assigned. Then, they need to be evaluated on how well they perform in that area.

In performing the evaluation, there are certain areas that need to be considered as the job-related functions. These may include, but need not be limited to:

◊ Room nights booked vs. actual room nights;
◊ Dollar volume of sales closed vs. actual dollars volume;

- ◊ Relationship of bookings to goals established; and
- ◊ Catering volume compared to goals (if this part of the job description).

Of course, the big concern just about always involves "how much business is put on the books."

There are other areas of performance that need to be looked at and evaluated. For salespeople to be successful, they need to be concerned with a variety of components necessary for selling. These are selling skills which need to be developed in all salespeople.

The evaluation process should also include which are salesperson's strong points and in which are further training or development is necessary. Managers should consider "rating" the salesperson in the following areas:

- ◊ Appearance (dress and demeanor);
- ◊ Time management;
- ◊ Organizing work;
- ◊ Enthusiasm;
- ◊ Pace of work;
- ◊ Persistence;
- ◊ Closing techniques;
- ◊ Prospecting lead development; and
- ◊ Following up.

There may be other areas to be added to this list, but the idea is to be able to measure a person's strong points as well as weaknesses. Falling short in any of these areas listed does not, by any means, reflect a poor sales ability. Primary consideration should always be given to "the bottom line" it's the bookings and actual sales that really count.

However, improving sales skills will always improve sales. Sales skills should not be taken for granted. They need to be evaluated frequently and discussed with sales personnel. With proper direction, perhaps some retraining, these skills can be improved.

Salary increases should be related to performance and it not necessarily based on an evaluation at the end of a year. Small salary increases during the course of a year may well pay off from a motivational standpoint.

MANAGING THE SALES EFFORT IN A DISTRESSED MARKET

May 1988

It's just too easy for lodging owners and operators to accept a variety of reasons (or excuses) for lack of sales, "the supply is more than the demand;" "too much competition," "we're in a distressed economy;" "the market is soft;" "we're overbuilt;" etc.

In getting to the "meat" of this problem of declining sales—owners and managers need to think of the existing situation as an "underutilized" condition. Meaning, of course, that business is out there somewhere—it needs to be brought in to create better utilization of the lodging property.

Easy to say . . . but hard to do. Granted! However, with owners and operators taking a more aggressive sales supervisory involvement, there will be a better utilization of a property's facilities. It just takes a strong commitment to get involved, on a daily basis, with the sales team at a hotel or motel.

First, there has to be a plan. Most properties are doing marketing plans. Whether this is being done or not—at least there needs to be an action calendar. This is a guide that is prepared for each month of the coming year. The calendar describes a type of action that will be taken during that monthly period (even by a specific date) and who has been assigned to get it done. For example, if a local sales blitz is to take place in September, the planning may be started in July. In the action calendar an entry might read; "September Sales Blitz Planning Meeting" then, enter initials or name of the person who will be responsible for getting the meeting organized.

All sales trips, meetings, special projects, trade show participation, need to be entered on the action calendar. The overall program of the year needs to be reviewed by the General Manager and perhaps even changed or enhanced, as may be necessary. Of course, once approved, the plan not only needs to be followed, but should also be updated as necessary. From time to time, situations may change requiring an adjustment in the action calendar. The General Manager must accept the responsibility to review the calendar on a regular basis and make sure appropriate action is taken.

The plan may even include how many sales calls and/or presentations, or proposals are to be made in any one month period, by each of the sales staff.

From a daily involvement standpoint, the General Manager or owner/operator needs to visit the sales office sometime during the day. The visit should be to find out "what's going on." A brief communication session involving who's doing what, a review of the reader file report on what's been booked, a review of an hour daily sales calls, etc. Certainly in a day to do this is definitely a good investment of time.

Periodically, the General Manager should join salespeople on their outside calls to get a feel if some direction may be provided. Getting feedback from accounts during the sales calls could be very enlightening. A General Manager can learn firsthand where the property needs to improve its operation.

Calling five local room accounts a day could end up being very productive—a "thank you," "how're we doing" goes a long way with an account in reassuring the hotel's interest in that business.

Lodging owners and operators need to understand that the sales effort at a property doesn't start and end with the sales department. It starts and ends with the "chief in charge" of the hotel or motel the operator.

SALES-MINDED GENERAL MANAGERS NEED NEW YEAR'S RESOLUTIONS TOO

January 1989

Hotel/motel managers and owners need to be more sales minded and involved in marketing more than ever before. With competition as it is with operation costs what they are we've got to bring the dollars in at the top if we're to bring it down to the bottom line.

More marketing involvement is necessary from top management and owners. Help improve profitable sales this year. Help with the development of your sales team. Here are some New Year's Resolutions to consider:

◊ Be sure there is a Marketing Plan that the sales team helped to create.
◊ Have Action Calendars for all departments and make sure they are being followed each month and updated when necessary.
◊ Review Weekly Sales Activity Reports for each salesperson and ask questions.
◊ Provide direction to the sales team on a regular basis.
◊ With the sales team establish individual and team goals for bookings, actual sales, proposals, sales calls, etc.
◊ Take time to talk with salespeople daily informally or on a formal basis. Offer assistance, encourage on-property entertaining of prospects.
◊ See that salespeople meet people on a daily basis.
◊ Periodically make sales calls with the salespeople. Provide direction to improve sales call quality.
◊ Give salespeople opportunities for self-improvement through membership in Hotel Sales and Marketing Association International and Meeting Planners International.
◊ Look out for accounting policies and procedures that interfere with the development and booking of business by the sales department.
◊ Allow freedom in negotiations by salespeople with proper guidelines.
◊ Deliver the product as sold. Back up the sales department with great service and proper attention to their booked business. Employee appraisals and salary reviews need not go hand-in-hand. Keep them separate. Do appraisals quarterly. Salary increases should be given on merit and not necessarily in connection with a review of performance.
◊ Keep salespeople informed on "what's going on" in the company and property. They shouldn't find out what's going on from the street.
◊ Treat salespeople as partners so they'll feel comfortable in sharing thoughts and ideas with you. See how inquiries are being handled at the switchboard and how the transfer of calls are being made.
◊ Check on inquiry follow-up.
◊ Are all departments doing their share of selling and providing leads to the sales department?
◊ Take time to call major accounts to be sure they understand that top management is concerned. Accept the role of "Chief-in-Charge of Sales" for the property and help develop prospects through friends, neighbors and business contacts.

OUR MOST NEGLECTED SALES TOOL... OUR EMPLOYEES

March 1989

Informed employees work, think and relate to customers better than those working in a vacuum. The most often overlooked audience in a hotel is within our own building. Sell to them and we may just manage to be able to sell to the rest of the country.

It's not surprising that there are many cases where hourly employees on the staff of a hotel and restaurant don't really know what the people in the sales department do. There are many jokes about salesladies going out to the hairdressers in the afternoon or going shopping or salesmen going to the movies. And how many times have you heard an hourly employee giggle when a salesperson says, "I'm going out to make sales calls."

The message that we're spreading outside the hotel to bring business into the property needs to be spread within our own organization at every level in the hotel. We seem to ignore the most important element of our marketing strategy the employees at the hotel.

You probably don't realize that the staff of the hotel really doesn't know what's going on in the sales office and they don't know what our objectives are or how we go about doing our business or how important it is to bring in new guests. And they can't possibly know unless we take the time to tell them.

We can't take it for granted that they understand the sales and marketing strategy at the hotel.

Does anyone in sales or management ever really make an effort to explain to desk clerks or waitresses or bellmen or van drivers or housekeepers what takes place when a prospective group buyer comes to the hotel for a site inspection?

Who takes the time to explain to the switchboard why inquiries are important or what is supposed to happen when someone calls in and says, "I want to book a meeting."

Whoever explains to the front desk clerk that when a guest checks-in and the reservation is stamped "travel agent," what has actually happened in order to get the guest there.

Does anyone ever explain to the front office staff why contract business is charged much lower than rack rate and why that particular type of business is important to the hotel's profitability? Is the banquet staff ever clued in as to the importance of a piece of meeting or banquet business in the hotel; how it originated; how we got it; how much time and money it took to get that piece of business into the hotel and how important it is for that piece of business to want to return.

Is the importance of the preparation of meals as it relates to making customers want to come back ever discussed with the kitchen staff; how important is the hash brown potatoes served with the eggs in the morning are in getting a guest to return to eat again in our dining room.

Has anyone ever discussed with busboys the importance of refilling water glasses or cleaning tables promptly as it relates to making people want to come back and eat in the dining room?

The very people who are in a position to help the hotel get repeat business are the ones who are generally neglected from a sales standpoint.

Many employees in a hotel probably see their function as operating independently from anything else in the company. They have specific tasks to perform and they work for one department and that's it. Employees need to hear the message that your hotel wants its customers to hear. Employees have to get beyond their own individual tasks and understand how their own individual jobs reflect whether the customers come back or not and whether they get paid or not.

When the departments have their own individual departmental meeting each week every once in a while the General Manager or assistant manager or sales director or sales manager could show up at the meeting. They should talk about sales and explain to all employees how important individual tasks and jobs are and how they relate to the overall sales strategy of the hotel. They can explain what it means to the hotel from a business standpoint when they do their jobs properly.

Employees have to think of themselves as part of the overall sales strategy of the hotel, not just service, or busboys, or kitchen help, or bellmen, etc.

The next time you take a guest in for a meal in your dining room the waitress, busboy, dining room manager and the cook will all realize that this guest is where their salary originates.

MANAGING THE SALES EFFORT

April 1989

Sales managers, sales directors and General Managers of lodging properties throughout the country seem to use a wide variety of methods in managing their salespeople.

In all probability, there are many properties in which a General Manager has no method or system to manage the sales effort. In many cases, it is just a question in these types of facilities for hiring a salesperson and telling him, "just get me some business."

We're not discussing a marketing plan or an overall action plan for the hotel or motel to create additional sales. We'll concern ourselves here with just managing the sales effort on the part of a salesperson.

Management must take the time to work with salespeople on a daily basis to make sure the job gets done properly. A salesperson's work must be reviewed daily—as well as weekly—in order to manage properly.

On a daily basis, a salesperson should produce a very simple report reflecting who was contacted that day, with the name of the organization and a one-line comment on results of the call, such as "good prospect" or "not a prospect." Also, there should be a space on the form to indicate how the contact was made—telephone, outside call or inside visit.

The report should not be typed, but completed as the salesperson makes the calls. At the end of each day, management reviews the daily activity with each salesperson. There are several things that should be included in the review:

▶ How much time was spent outside on the calls and how many contacts were made? Appointment calls will generally take more time than prospecting or cold calls. If cold calls are planned in a proper geographic manner, a salesperson could make three to four such calls an hour. On appointment calls, a period of 30 minutes to an hour for each could be acceptable.
▶ How many leads or prospects were developed? A ratio of three leads to 10 cold calls would be good.
▶ Was good information developed? Did the salesperson get who, what, when, where, why and how?
▶ What was the feedback from contacts on your hotel product and services compared with the competition? On

appointment calls with local accounts, this information will generally surface without any or much probing.

The entire scenario of reviewing the daily work of a salesperson need not take more than 30 minutes. By doing this daily, the person may be better managed and directed to improve performance. Productivity will unquestionably improve.

Weekly reports and reviews are also important. The manager needs to know a number of things on a weekly basis:

- What activities were performed and how much time was spent on things other than sales? These may include working the front desk, the dining room, catering or helping out anywhere in the property.
- Where does the hotel stand as to group-room bookings compared to the plan or compared to last year? Is the hotel on track to meet its goals? Compare room-nights booked and projected revenue.
- Compare the average rate for group-room bookings against the plan and against last year.
- Review the number of site inspections that took place during the week and the results of those visits. Who was entertained?
- How many trace files were worked and what were the results of telephone follow-ups?
- Review proposals that were sent during the week.
- Review reader-file letters and memorandums.

By getting involved in the daily and weekly sales activities, management not only has a better handle on what's going on, but can more effectively manage the sales effort.

A review process is an important tool for better management of the sales effort.

TURNOVER IN SALES... A REAL PROBLEM

August 1989

Turnover in sales personnel is a genuine problem. Looking at it realistically, the salespeople, themselves, are not the ones to be blamed. It is the 'system' that lacks something.

Many studies have been conducted over the years by human resource groups on the subject of turnover. The results all about tell us the same thing: the reasons why people change jobs.

I'm sure there are some specific studies about why hotel salespeople change jobs. In all probability, the list of reasons may read something like this (although not in any particular rank or order):

- No opportunity for advancement
- Poor or little direction from superiors
- Not trained on the job
- Not knowing what's expected
- No incentive to produce
- Personality clashes
- No recognition
- Too much paperwork
- Required to do things other than sales
- Burn-out
- Too much control
- No review system

▶ Not enough pay
▶ Too many hours

Certainly, there are other reasons other than the above, but it would be interesting to see what a professional study would uncover.

Perhaps we need to explore, first of all, the recruiting, selection, and hiring process. Many times General Managers or sales directors will look to hire a salesperson without really knowing what is needed. It would seem logical that if the 'right' person was hired to fill a specific slot, this would be 'a' step in reducing turnover. Too often people hired for sales are just not right for the job. The proper selection process would help eliminate some of the problems in that area of concern.

Before a decision is made to employ someone in sales, there should be a written job description that needs to be agreed upon by all supervisory people involved, That may include director of sales, General Manager, regional supervisor or corporate officers. Then, sales goals for that particular job needs to be established, followed by criteria for selection.

A plan should be written so that there is a guide for how to get people to apply for the job. Probably, the easiest and most professional route to go would be through an executive search company. Of course, there is a fee involved for this service; however, there are many advantages mostly the saving of many hours in screening applicants. According to Phil Al, Senior Principal with Innkeeper's Management Corporation, a New York City-based executive search firm: "Too often hotels will place an advertisement in their local paper and hope it will attract the right candidate. This is fine if you are looking for trainees, (hotels should first look in-house for that level). If the desire is to find an experienced candidate who can open new markets, then there is a need for a good search firm. They would canvass potential applicants, some of whom could beat competitors and present candidates who fit the specific profile designed for the position. Turnover in sales causes client confusion and a stalled sales action plan which results in a revenue debilitating recapture period. As a result, a good search firm can be very cost effective."

Turnover could be further reduced with proper employment or hiring practices. Whatever the deal is it needs to be written and clearly defined so there is no misunderstanding of what the job requires, what goals need to be achieved, what salary, benefits and incentives are available. Promises should never be made on what may not be fulfilled.

To further eliminate turnover there needs to be continued excellent communication between the salesperson, the supervisor, and the General Manager. It's been observed that the hotel properties that do not seem to have a sales turnover is where the General Manager is personally involved, visits with salespeople and shows interest and concern for them. General Managers need to be sure that there is an open-door policy so that contact is available at any time.

Training on the job is also vital in reducing turnover. Good salespeople are always eager to learn more and improve themselves. There should be ample opportunity to learn more about food/beverage, front office, convention services, etc. They generally will do a better job in selling when they know more about the workings of a hotel.

From time to time, salespeople who can handle it will need to be given more responsibility—this helps improve their self-esteem. It also helps in the burn out the situation, even a change in the sales function would work. Cross training in large sales departments works well for large hotels. The travel industry salesperson can get exposed to corporate sales, etc.

Probably one of the biggest concerns with newer salespeople is the "How Am I Doin'" syndrome. These new people need feedback constantly good or bad. They really want to do better and they need to be told where they stand. Performance reviews are tied to this. What we do wrong in our system is to provide annual reviews related to salary increases. Why don't we do periodic reviews, perhaps quarterly, and never tied to salary? They should be strictly performance reviews to help the person improve. Salary reviews may be annually but tied to a bunch of other things including, perhaps, all the reviews during a year.

We'll always be faced with hotels stealing salespeople from each other. With new properties opening up all over there is a need to recruit experienced salespeople naturally, they have to come from another hotel. Just don't let it be yours.

1990 WILL NOT BE A GREAT YEAR FOR SOME . . .

January 1990

More and more owners and operators are getting interested in marketing. They are asking a lot of questions and are concerned about results. This year we will see the "hierarchy" getting tough on productivity in the sales area as well as with results. The dust will get blown off those 100-pages marketing plans that were sitting on shelves throughout 1989. A quick review of last year's plan will cause many Q&A sessions with General Managers and sales directors, whereas the ROI will be the big question. Some General Managers and salespeople will be heroes, of course. However, many will not have good, acceptable answers. Some of the responses will have to do with overbuilding (of course), poor product, low A&P budget, poor service, turnover, etc. How many of such answers will stand up to the scrutiny of the owners is hard to tell. Owners will want to know: how much time is our General Manager devoting to supervising sales? How much new business did each salesperson book? What are we doing about low occupancy periods? Why is the average group rate not higher? And the list of questions will go on and on.

We'll see search companies looking for top marketing people to replace those that are "not making it." There will be people coming out of marketing ranks from other businesses and industries to take over top marketing jobs. They will be implementing new and innovative techniques in hotel sales and marketing.

General Managers will have to start getting more involved in property marketing efforts, for self-preservation, if not for anything else. We will see General Managers taking over the preparation of the marketing plan. We'll see them handling the advertising and public relations end of the business. Salespeople will be salespeople again, not marketing people and there will be a rush to reprint calling cards and letterheads (it's about time!)

More time will have to be devoted to prospecting and selling by all salespeople and new sales revenues will be measured. Salespeople's performance will have to be measured. Activity alone will not suffice in the measurement of good performance. It will be the new sales generated that will determine how well a person performs. This will be measured against the cost of sales to see if the return on investment is adequate.

The '90s will be the time for salespeople to "put up or shut up." It's going to be a performance that counts not the number of sales calls, proposals sent, conferences attended, letters written, people entertained, advertising placed or whatever. It'll be the sales closed and the amount of dollars generated.

Owners and operators will insist on salespeople getting out more and doing more face-to-face selling. There will probably be more latitude given salespeople to make "deals" when business is needed. No question about it! The competition will be tough, which will require some more sophistication in our sales approach and methods of doing business. Buyers of lodging services have become more knowledgeable and particular when it comes to dealing with properties. Association and company meeting planners are learning more and more about how to negotiate with hotels and so are travel agents and wholesalers. Are our hotel/motel salespeople keeping up? Are we tuning up our negotiation skills? Chain organizations and management companies will be devoting more dollars and time to training salespeople, outside trainers will be brought in to establish seminars and workshop programs to teach operators and salespeople the techniques in bringing additional sales to a property.

The next decade will be exciting for the industry. Sales and service will be the hot topics throughout the country at all of the conventions and meetings in which we are involved.

JUST HOW PRODUCTIVE IS YOUR PROPERTY SALES PROGRAM?

August 1992

As an owner, operator, General Manager or sales director can you step back and look at your sales program and tell if it is working the way you want it to work?

From looking at your marketing plan (if you have one) all the way through the various processes of getting the business, servicing it, and taking the profits to the bank, you'll see room for improvement. No matter how well we plan and execute, there is always room for doing more and doing it better. There's a lot of money being left on the table because we're not doing everything we could or should be doing to improve business and bottom line profit.

There are so many things to look at like:

▶ How are we prospecting for new business? Are new leads being developed for markets we may not have served previously? Who else on the property beside the sales staff can deliver leads to us? How do we follow-up on the leads? What's the system?

- What is the process when someone makes an inquiry? Who does what? What are our conversion rates on inquiries, not only for group sales, but for individual room sales as well as catering sales?
- What kind of a return on investment do we have on our sales budget? What should it be? Should we hire more salespeople to get more business?
- How do we manage our accounts? Should salespeople be assigned geographically or by market segment? What do salespeople do? What proportion of time is spent on various duties? What kind of help do salespeople need? To what extent are they empowered to make decisions? Are files up to date?
- What sales strategies are being employed to close more sales? What are the techniques used? Are sales calls being made properly? Are we on target with our trace system?
- To what extent do catering staff solicit new business? How can we get at least a 15% to upsell on catering from groups brought in by the sales department? What is being done to improve banquet room rentals? How about a la carte restaurant and bar business?
- How does our interdepartmental communications affect our sales effort? How do our department heads help the sales department?
- What kind of job are we doing with our advertising, public relations, market research, merchandising and direct sales?
- Do we really know our market mix? How much business comes from each market segment and how do they rank from a revenue standpoint? Can we change the mix to improve the revenue?
- What's our market share and how do we relate to competition on closing deals?

These are some of the things we need to take a look at, on a regular basis. Of course, there are other items you can add to the list, but the idea is to improve your sales operation through all the departments ... reservations, front desk, sales, food/beverage, recreation, etc.

It's a question of being aware of what's going on around you and constantly looking for ways to improve the operation. Sometimes we are so close to things on a daily basis that we "can't see the forest for the trees." We just have to take a step back to get a good look.

Ray Swan, President of the Educational Institute of the American Hotel and Motel Association has some thoughts about this. In a recent conversation, he indicated: "sometimes you need a second pair of eyes, from someone not connected with the operation, to take a good look for you." With this in mind, he is considering having the Educational Institute provide a Sales Audit service for owners, operators and General Managers.

YOU NEVER MISS THE WATER UNTIL THE WELL GOES DRY

April 1993

How true this is when it comes to salespeople at a property. We find some General Managers constantly complaining about their sales personnel: always out when you need 'em, don't know where they are, takes too long for the lunch hour, doesn't get in the office early enough, goes home too early, doesn't make enough sales calls, can't get the right information for booking, should not have booked that particular group, how come lost that particular group, spends too much time out of office, spends too much money entertaining, gives away the house . . . Well, it goes on and on.

When you get a group of salespeople together and ask the question: What is the first thing your General Manager is apt to ask you, when obviously you have had a very bad day in the field, come back in disheveled, in a bad mood, tired and worn out, with a piled high list of phone messages to return? The answer is almost always, in unison: "What did you book today?"

But, when the sale person is out sick, or on vacation, or decides to leave for another job, watch out. The place is in turmoil. How are we ever gonna get along without so and so? It seems that salespeople just can't win. God bless 'em. Can't get along with 'em or without 'em. So wouldn't it be better to try and make things work right? In other words, instead of complaining about what salespeople do or don't do, why not establish a program that ensures that both the General Manager and sales personnel together are on the right track? Every job has some elements that cause stress and everyone is affected somehow by interpersonal relationships. Here are some suggestions for avoiding some stressful situations when it comes to a property a sales operation and salespeople's relationships with General Managers and others:

▶ A job description is a must, without one all would know who is supposed to do what . . . then don't deviate. The salesperson and General Manager must agree on the job description (watch out for that line and other duties ad necessary).
▶ If someone is not suited for the job, then break clean. There is no use for anyone to be doing something they can't do or don't like to do.
▶ Salespeople need to report to one person only. Even if it appears that there are multiple bases it should be very clear to who is responsible for the supervision and direction of a salesperson.
▶ Everyone should ask for and respect other people's opinions and input.
▶ Feedback and constructive criticism should be furnished and expected.
▶ Always give credit where credit is due. Don't take credit for someone else's work.

One of today's big problems in the hospitality industry is employee turnover. From a cost standpoint, it is very expensive. From a customer relationship standpoint, it is also very expensive. So far we have not been able to determine the real cost of turnover in a property salesperson. From a dollar standpoint and even from a client standpoint. We are all of the superstar salespeople who have been able to move the business from one property to another just by changing jobs. Meeting planners talk a lot about this business of salespeople turnover. They do not like, they prefer to deal with the same person all the time, particularly if the good relationship is there.

The best opportunity of all is to keep what you've got and need to determine winy salespeople leave a property and got to work elsewhere. Many surveys have been done in this area and I think all the true answers are still not known. We do know however, that is not always money that makes people leave. Although this has been given as a reason in many cases, we tend to find out there are many other factors that created a situation that made it more comfortable to leave for more money. Sometimes "opportunity for advancement" was a reason (couldn't we have provided that?). Another reason was "personality clash" (couldn't we have fixed that if we would have known?).

Let's strive to keep the property environment happy, so everyone enjoys the work and relationships. We almost always miss 'em when we lose 'em.

GENERAL MANAGERS: IT MAY BE TIME FOR A SALES AUDIT

April 1993

Is your sales system paying off for you? How did you fare in 1992 and what are your plans for the rest of this year? Are you satisfied with your return on investment when it comes to sales performance? Do you think you need to look into what's going on in sales to make sure your system is maximizing profitable sales? How do you know if there are better ways to improve productivity resulting in more business and better bottom line profit from your sales program?

Maybe it is time to take a closer look at what's going on in sales and examine ways of doing things better. Conducting a property level sales audit could pay off in high dividends for you and get you more comfortable about being ready to approach the remainder of 1993 and 1994 with highly productive sales systems and procedures. This is something you can do yourself or with your own director of sales; but you should consider that doing it in-house would not be as objective or as meaningful as an outside source. It is always better to bring in a fresh pair of eyes of someone who is familiar with your type of operation and has the experience necessary to dig deeply into your current system and procedures and then come up with recommendations for improvement.

The Educational Institute of the American Hotel and Motel Association has just come up with a new EI Sales Audit program that does just that for you. It may be a bitter pill to swallow when you discover the many ways sales activities could be improved at your property because some are just so obvious, but it takes an "outsider" to see them. You will probably come up with many "why didn't I think of that?"—thoughts when you look at the outcome of the audit.

The EI Sales Audit itself can take many directions depending on the focus required, type of property, and even what you think you want to be examined. Depending on the size of your property, a good audit should take from 2 to 4 days with some advance work being accomplished before the actual visit to the property.

Here's what could be file systems and records are examined sales methods are observed; account files are reviewed; sales staffers are interviewed; goals and accomplishments are analyzed; sales activity reports are examined; the marketing plan and strategies are reviewed; the marketing mix evaluated; and a host of other related activities are performed including the testing of hot inquiries are handled. Other areas that may be examined could very well be budgeting, job descriptions, sales proposals, and contracts, sales travel and expenditures, the public relations effort, advertising, sales promotions, etc. The audit actually looks for a business that slipped through the cracks. The whole idea is to make sure that you are maximizing potential sales through your reservations department, front-office sales department, catering department, food service outlet, as well as all other profit centers, etc.

Upon completion of the visit, a report is filed that presents an objective picture of what was found to be good; what may be improved upon; recommendations on what to do to improve sales productivity and how to do it.

Even though we usually think sales activities are great there always seems to be room for improvement. You still have time to fix the rest of 1993 and plan better for 1994. You want to get the occupancy you need at a rate that is profitable for you. Is there room for improvement in catering sales? How can you get some average check increases in your food and bar outlets? Can you further upgrade the room sales through your reservations department and front desk? Are you maximizing potential sales in all areas? Take another look and see if you are really satisfied with that is being done or could be done in your overall sales operation.

ARE WE PAYING ENOUGH ATTENTION TO OUR SALESPEOPLE?

August 1993

Salespeople, generally, are a rare breed. Although it may not be obvious, we are really very sensitive, have a tough time taking criticism, need recognition of accomplishment as well as reassurance on a regular basis. Salespeople have a need of being wanted and consulted on most matters concerning the operator of a property. Those General Managers and corporate operations people who understand this know that by staying in close touch in sales, offering suggestions, commenting on sales figures, having a satisfactory working environment which improves morale, gets salespeople to more enthusiastically reduce sales turnover.

One of the most powerful management tools in the area of motivation is the use of recognition in persons' accomplishments. We just don't do enough of it. We can do it verbally at a meeting; we can do a complimentary letter with copies to appropriate corporate offices; we can provide a gift which doesn't need to be expensive; or how about an extra day off? Sometimes we make a mistake by providing too quick a promotion, such as taking a great salesperson and putting him or her in charge of other salespeople. We must remember that in managing people, there is a big difference between being a productive salesperson and being able to supervise other people.

Sometimes we recruit from other facilities great salespeople and hire them for a sales supervisory position. In many ways, it could work, but be aware it could create havoc with your sales organization. It takes a special talent to be great in sales production and have the ability to manage other people at the same time. There is always the risk of losing that sales productivity when we promote into supervision that person who has the skills to do both jobs.

Perhaps we need to consider offering a special job title that just may be a suited supervisory role. Along with the new title (something "executive sales rep" or "senior sales rep"), they can get special treatment such as: engraved business cards, personal stationery, some other little extras that other salespeople do not have. Although these people may not be in a position to supervise other salespeople they could be helpful in many ways to other junior salespeople on the staff by providing selling tips. The idea is to keep great salespeople selling. You want to be sure they do not lose contact with their accounts and the planners, association executives, motor coach tour operators, corporate travel planners, travel agents, etc.

Some other things to consider is to provide these people with more freedom of movement, more opportunity for them to be creative, perhaps even a chance of reducing paperwork (most good salespeople just hate paperwork). Allow them the opportunity to sit in on high-level operational meetings and let them offer comments; use them as sounding boards for ideas that come from your executive committee.

We must remember, of course, that with the title change we still must stay in close touch and provide the reassurance and recognition that salespeople need so badly. Here is an opportunity to provide some sort of incentive compensation that goes along with the title. This will also be an encouragement for newer and junior salespeople so that there is something to shoot for.

Above all let salespeople know that they are needed and wanted and are appreciated for what they do.

WHEN YOU DON'T HAVE A SALES MANAGER... YOU'RE IT!

November 1994

It appears, from recent reports (HMM July 25, 1994 Standing Strong Special Report) that the economy and budget groups are lagging in revenue per available room when compared with other hotel segments. Now that the growth of new hotels is down considerably and demand has increased we sure are in a better position to make some hay while the sun shines. We don't know how long this good time will last before we see another period of overbuilding. So what are we doing about improving profitable sales in the limited service, economy and budget segment?

Most of us involved in this segment of the lodging industry do not have the luxury of a person assigned specifically to sales to bring in new business, mostly because we feel we cannot "afford" to put, someone on the payroll. What's the alternative? Look in the mirror and decide who should be responsible for the activity that will bring in more customers through the door. Of course, it has to be you, the General Manager or Operator of the property. Certainly, we need to provide great service so that we get the repeat guest, that's a given, but what we also really need to do is get the new customer. Here are some tips on how to do it the least, costly way and the easiest way:

▶ The first thing you do is make sure your telephone is being answered promptly and by someone who has an enthusiastic voice. I've tested hundreds of properties on how long it takes for the phone before it is answered and the level of enthusiasm by the person answering. The results have been shameful. There is no telling how much business is being lost at properties because inquirers do not have their call answered promptly or pleasantly. This is something you can fix.
▶ Again, with the phone, we are not trying to close sales. When one makes inquiries regarding rates and even offer to buy signals, it is very, very seldom that, someone will as to make the reservation, ask for my name, phone number, etc. It seems that we train our front office people to answer questions over the phone, but not to say: "May I hold that room for you?" Central reservation offices at franchise chain operations do very well at this, but, when a local property is being called is where we fall down on the job. Many times when I asked if there are other places that charge less. Believe it or not, but true I am offered not only names of competing properties that charge less, but, are also given phone numbers? Can you believe that? Believe it, it's true! They've got the tapes to prove it.

▶ People do business with people they know, respect, and like! Are you well known in the community? Do people like you? Are you respected? Get involved in the community, join a civic club; get a couple of your staff involved with organizations. Be highly visible, work on committees, get people who know and respect you because of your community involvement. Locals will then want to send their visitors to your property.
▶ Have you read the newspaper lately? Of course, you have, but, did you read it with an eye towards developing new acts and new business? A daily paper, any size, town or city, will have more leads than you can handle. Look for companies moving to town, companies promoting executives and new construction, wedding announcements (there's a Comfort Inn in my area that does a booming business just about every weekend with overnight guests attending a local wedding), sports teams coming to town, entertainment groups putting on a show, look at the obituary column many times people come in from out of town to attend funerals, just about any section of the paper from sports, to local news, business, social, finance, etc., will have leads for overnight room business you just gotta look for them. You probably won't find any leads on the comics page, but read them anyways you never know.

The point is, as General Manager or operator of a property, if you do not have a salesperson getting new business, then you just have to be the one that makes sure it comes in somehow. I wouldn't be surprised that if we handled the telephone inquiries properly, we could show a nice little increase in occupancy.

EMPOWER YOUR SALESPEOPLE TO BE MORE PRODUCTIVE

March 1996

A situation that always seems to come up when salespeople get together, especially during our training sessions, is the topic of how far should a property manager go in empowering the sales staff. Of course, most sales professionals would prefer to be "free wheeling" to make all the decisions necessary to close a profitable piece of business. It just doesn't happen that way at most properties. When it comes to negotiations, generally, salespeople can go only so far without getting an "OK" from a General Manager.

It seems to me that we are burdened with "policies." In discussions with sales personnel, we find that there are a wide variety of these "rules" which must be followed during the selling and booking procedures. For example, there are policies about complimentary rooms for groups. Comp policies seem to range somewhere between one comp room for 50 room nights sold and on down to 10 room nights sold. Why do we need to have a policy or even offer comp rooms at all? Many properties actually print up a sheet reflecting their comp policy, along with group rates and other things that may be "given away." In this period of a seller's market let's do away with any policy with regard to giving anything away and let's empower the salespeople to make the most profitable deal they can be based on how badly the business is needed, how profitable the business may be, and how badly the prospect needs the facilities.

Recently I made a telephone inquiry for a group booking at a resort hotel and received an American plan quote—a single room rate with three meals included. I had indicated that it sounded like a very fair price for that particular property, for the time I wanted to bring my group there. I asked about green fees for the golf course and was advised that there was no charge for that. Even though I offered no objection, and expressed interest, I was told that they would also throw in a free cocktail party the evening of arrival. This was the policy of the property, I was told. Great! Are we getting to the point that swelling benefits to create sales has become a question of just offering a lot of extras just to make the deal?

When it comes to room rates—and we all know that this is probably the most negotiable item in group bookings, why do we usually start off with a group rate—something below the rack rate, before we even know how profitable the piece of business may be? Most professional meeting planners and group tour arrangers use that rate as the starting point for their negotiations. It seems that we could do much better if we did away with any policy about group rates and let our salespeople start off with quoting rack rate, and . . . if necessary, negotiate to a lower rate based on the quality of the business. And, I do not mean just dropping the rate arbitrarily. We drop the rate only with concessions . . . getting something extra . . . a trade-off.

Then we have this business of giving away the meeting space. More and more properties are able to get space rental even though groups book large blocks of rooms. Meeting planners around the country are getting to realize that free or reduced rate meeting space is something that needs to be negotiated. There just needs to be a charge for everything and it should be up to the salesperson making the deal to determine what is in the best interest of the property.

It just comes down to management at properties do not have the confidence in the sales staff to let them make the best deal they can, without having to follow policies or to "check" with the GM for an OK. There may be good reasons for this. There are so many hotels, motels, conference centers, and cruise lines that have sales personnel who have not been professionally trained. Many have been promoted into sales from secretarial jobs or front office duties or whatever, but without any formal training in sales. Not everyone is suitable for sales; many are not very good at closing a sale; some are intimidated when it comes to negotiations. Many can accept the responsibility of being empowered to make decisions on behalf of the property. It is up to the property management, franchisers, management companies, and franchise organizations to start thinking about doing something to help get the salespeople trained and then empowered to close more profitable sales.

HOW ABOUT THOSE WEEKLY SALES MEETINGS? DO THEY WORK?

February 1997

When we are dealing with a profit center of a hotel operation (and of course, our sales departments should be profit centers not cost centers), we want to make sure that our profit centers are productive. The whole idea behind a meeting is to make the people involved in the meeting more productive. Therefore, "getting together to communicate" (isn't that kind of the definition of a meeting?) needs to be very effective. Sales meetings need to have a measurable impact on business; there needs to be a purpose and objective. Why are we having this meeting and what is it we want to accomplish? Of course, there are many reasons for having a meeting, but they need to be good communication sessions with a good reason for having it. Many hotel salespeople have mentioned to me (at our Hospitality Sales Workshops) that their weekly sales meetings are unproductive and a big waste of their time.

To make your sales meetings more effective, measurable, and enjoyable take a look at some of these suggestions:

▶ If we want our salespeople to be more productive, then let's not schedule meetings during their most productive selling time (9:00 am to 4:00 pm). Make it more convenient for the folks doing the selling than for the GM or F&B Director.
▶ It has always been my thinking that a General Manager of a property is "Chief-in-Charge of Sales;" therefore, this is the person that needs to manage the meeting.
▶ Be sure there is always a reason for having the meeting. Why should we have one this week and what is it we want to accomplish? And, even put it in writing, so there is no misunderstanding.

- Develop an agenda and send it out in advance to each person who is to attend. There should be no mystery about what will be discussed. This gives everyone an opportunity to prepare for the meeting.
- It would be good to include a little something about product knowledge on the agenda. Even invite a different department head each time to tell the salespeople something about the department operation that may help in the sales effort.
- Try to include a little time for sales training. Let the salespeople take turns each time you have a meeting to share a tip on sales techniques.
- Set a time limit for the meeting and keep it as short as possible, but give everyone a chance to contribute something. Communication is a two-way street.
- Opening up the meeting with a little humor could set the tone.
- Conduct the meetings in a manner that is not intimidating, and participants should not be worried about being "called on the carpet" for any reason. Salespeople should look forward to attending.
- Wrapping up the meeting with something motivational is always good ... a little "rah, rah" is always helpful for salespeople.

If you are determined to have a weekly sales meeting, then it is not a bad idea to have them scheduled on the same day and time each week so participants may arrange their schedules. Just be sure the meetings are meaningful, productive, and salespeople look forward to attending. I've got a sample, generic agenda for a one-hour hotel sales department meeting, you may want to see. It may give you some ideas. Just fax me at 540-231-3746 or email to howardf@vt.edu for a copy.

MEASURING PERFORMANCE AND PRODUCTIVITY OF SALES PERSONNEL

May 1997

How good are we at measuring the performance and productivity of the folks who are responsible for bringing in new business? We are probably pretty good at measuring productivity, determining how much business was booked by individual salespeople, that's pretty easy. The tough part is being able to measure performance, that is, what is it that the salespeople are doing with their time so that they may be more productive. With the availability of technology today it doesn't have to be much of a problem. Those properties that are fortunate enough to be able to use automated sales systems such as Newmarket Software's Delphi or similar products, it is pretty easy to measure not only production but performance. That is, provided the right data is entered appropriately, and management makes use of the various reports that the system can kick out.

But, the problem is that for the most part, our industry is still very much behind in automation and we are still dealing with paper files and systems. So, if you are not automated, what system is in place at your property to measure the performance and productivity of the sales staff? What information is available to Directors of Sales and General Managers to provide information so that there is some measurement of the performance of sales personnel? And, how do we do it?

The logical approach is to get periodic reports from the sales staff on what they are doing with their time along with revenue being generated in terms of definite future bookings. Here we are discussing group business, whether business or leisure, rather than individual business. This sounds like something very basic that is already going on all over the industry. However, in discussions with General Managers, sales personnel and conducting property sales audits it has become a "hit and miss" situation. My guess is that well over 75% of properties with sales and catering sales personnel are not automated and of those, less than half are using any kind of written, documented periodic reports of any kind on the activities of salespeople.

It would seem logical that directors of sales and General Managers would want to have sort of a weekly report of performance on sales staff. Not only to measure productivity, but to use it as a guide for further training, coaching, and development of the individuals. Some of the major brands do provide direction in this area, but there are so many properties that do not take advantage of it. Further, there are many properties being run independently, without the benefit of a major brand or management company that should implement such a measuring device.

Here are some of the things a report could reflect that would help in determining how effective a salesperson may be:

▶ How is daily time spent (how many hours spent in the office, out of the office, or doing something else ... or maybe doing something else that doesn't even apply to sales.
▶ How many calls were made each day outside, by phone or with persons visiting in-house?
▶ A report on definite future bookings reflecting projected income from rooms, food, beverage, meeting room rental.

A simple report form, reflecting this information for each day of the week, with daily and weekly totals, should be enough to be able to evaluate the performance.

ARE SALESPEOPLE REALLY PRODUCING "GOOD" SALES?

February 1998

Some time ago I was doing a sales audit for a very large, four-star, convention type property. I noticed that the salespeople were on an incentive program based on room nights booked. The sale department included eight salespeople. All had great advanced bookings, and many groups were booked a number years out. And, all were doing exceptionally well on the incentive plan. The trouble was that the property was paying out dollars based on bookings, not actual paid business. We took a look at the actual rooms booked against the original blocks (from which the incentives were paid). When we showed the figures to the General Manager, he was shocked. None of the groups we checked reflected more than 70% pick-up, and some were under 50%. Needless to say, the incentive program was changed immediately. Of course, the answer is rather basic: if you have an incentive for sales personnel it should be based on actual paid business. And, it should be on dollar revenue pick-up, not room nights. It's the dollars you take to the bank, not room nights.

Is every sale made a good sale? We hope so, but too often, in the eagerness of sales personnel to put the business on the books, we find that from time to time there may be some bookings that are just not profitable, or perhaps, there could have been more profit shown had we taken the time to analyze the situation.

The answer to doing a better job in closing sales is to analyze the hotel's position as well as the position of the prospect. This also has much to do with how a salesperson will conduct the negotiations. How far can you go in giving something away . . . be it a reduced rate, comp rooms, upgrades, free this or that? All too often we find that we are offering reduced group rates to just about any group without analyzing the situation. We offer the basic comps—one for fifty paid rooms (what genius ever thought up that idea?). Why give comps at all without first analyzing?

In understanding the hotel's position, there are two very important things to consider. One . . . how badly do you need the business? Does the prospect need the space during a very low occupancy time? If you need the business very badly, you may want to consider "how far you can go" in negotiating the deal. Then, again, you may not need the business so badly, but the business could be important to you. It could be a very visible group which could generate additional business in the future, or bring favorable publicity to the hotel. Or it may bring in very heavy food and beverage business that would help the profitability. Or provide heavy deposits up front; or give you multiple year contracts.

In understanding the prospect's position, these are the things to find out and consider: How badly is your property needed? For example, if the decision was made to meet in your city, but space is not available at other hotels, then of course, you are very badly needed. Or, if you are the only "act" in town with the space that the group needs . . . OK for you. Further, you also need to check-out the group's history . . . where they have met before, what they paid, what they have for a budget, what the room pickup was at previous meetings, what food and beverage revenue was generated, what problems they had, etc. Get all the information you can which will help you analyze the situation.

This will help you figure out how badly the prospect needs you. Of course, you need to weigh all this information to determine how far you can go in negotiating the deal. The idea is to start high (not offering any discounts, comps, freebies) and know how far you can go to make the business profitable. You want every sale to be a good sale.

BILL 'EM FAST AND COLLECT 'EM FAST . . . MAYBE!

May 1998

It just sounds so logical, that if our accounting departments at hotel properties will send out billings faster then our collections on receivables will also be faster. Of course, there are always some companies that will take a month or more to pay their bills. I guess we will never get away from that; however, there are many that want to pay their bills as fast as they can. This is especially true when hotels are dealing with meeting planners of companies that run multiple conferences and events year-round and want to settle-up so they may finalize their budgets on each meeting. After all, in these cases the money is readily available to pay the bill, there is no need to wait for any collection on their part. Recently I had a discussion with a friend who runs 60–70 one-day workshops a year, all over the country. These are small meetings with about 25–50 in attendance. They have two refreshment breaks and a luncheon. In making the arrangements with the hotels being used it was requested that the billings for the event be sent out the next day for payment. Sometimes he gets it, but most of the time he doesn't. He wants the billing fast, and he has the money to pay and needs to reconcile other accounts for that event. One recent experience reflected much difficulty in getting the bill from the hotel . . . it was for two refreshment breaks, and a luncheon for 27 people. Two days after the event a call was made to the hotel requesting the bill, or at least a fax of the copy. Eight days later, still nothing. What's the problem?

Over the years meeting planners, of small and large organizations seemed to have had the same problem . . . and it hasn't yet been fixed. Why can't an accounting department in a hotel get out a billing for catered events the next day? Be it 27 or 2,700, it should not be a problem. Even on conventions with master account billing covering rooms, food/beverage, gratuities, meeting room rental, phone calls, etc., there should not be a problem getting the billings out within a day or even two days. These days with such wonderful technology available it should certainly be workable. But it is not happening in many instances. It seems to me a hotel would want to collect its receivables as fast as possible. Maybe someone needs to look into what's happening in the accounting departments of hotels.

Yes, there is also a good deal of questions being raised by meeting planners once they get the bill: questioning meal counts, room charges, taxes, gratuities, phone charges, etc. But one of the things I did when I was planning meetings was to meet with the property controller each morning to go over all the previous days' charges on the master account. Any differences were settled right then and there. By the time I checked out there was no question about anything on the bill . . . and it got paid right away.

Here's a suggestion for property managers, salespeople, and controllers: IN working with groups discuss that idea with the person planning the event . . . set up a daily meeting, early am, to review the previous day's charges to the master account. Any differences of opinion need to be settled at that time, one way or another. When the group checks out, there would be a final meeting between the planner and the controller to go over last minute billings. Now there is no question about the bill and there should be no reason why it could not be paid right away. Try this and see how wonderfully clean your accounts receivables will look.

I expect to get mail from some disgruntled accounting folks about this issue, with all kinds of reasons why the billings cannot be available the day after an event. Let's hear about it!

HOW TO SELL BETTER WITHOUT LOWERING RATES? A GENERAL MANAGER QUIZ

March 2000

All too often these days we see and hear a good deal about properties lowering rates to ward off competition. Or, to follow the lead when a competitor lowers its rates. It is happening quite regularly in areas where additional hotels are being built in a particular market. Of course, occupancy is taking a little bit of a dip in some of these areas due to the increase of rooms being made available. However, demand is still strong and probably will be for sometime, and rates are staying up, so profitability is still good. Sometimes managers tend to get nervous too soon and start discounting unnecessarily. The whole idea in selling is to stay ahead of the game, and to be in front of the competition by knowing as much about what is going on at your own property's reservation office and front desk operations. Here's a GM Quiz to help you better understand what's going on with room sales at your property: (thanks to Barry Anderson, VP Marketing, Peabody Hotel, Orlando, Florida for putting this together).

1. Do I visit the Reservation Department on a daily basis to check rate and inventory status?
2. Do I know what rates Reservations are selling for my hotel and why?
3. Do I participate in daily informal yield management discussions with Reservations?
4. Do I participate in weekly formal yield management discussions involving Reservations, Front Office, and the Director of Sales?
5. Do I participate in relevant training with Reservations agents in stressing overall selling skills, overcoming objections, and product knowledge?
6. Do I maintain involvement with Front Desk and Reservations in assuming complete knowledge of guest room/suite configurations in selling the hotel's strengths and overcoming product weaknesses?
7. Do I take an active role in observing Front Desk personnel with upselling techniques?
8. Do I keep my Reservation Department well informed on their progress each month in achieving budgeted ADR and rooms occupied for each market segment?
9. Do we have a Front Desk and Reservations incentive plan in place?
10. Do I review my local competitor's anticipated weekly occupancies to assist in yield management decisions?
11. Is my competitive rate survey updated quarterly?

Here are some General Managers Actions/Solutions to the overall quiz: Daily visits will allow you to get a pulse on what's happening, allows you to share your philosophies and offer direction, motivation, and guidance. To listen-in will help you to coach and council reservation agents on selling techniques, overcoming objections, and upselling, as well as product knowledge. Monitor denial and lost reservations to get instant feedback on why, and to offer solutions to maximize better conversions: training is the key. Your staff is better informed when having daily supply and demand conversations on hotels and local area. It works to have formal weekly rate strategy meetings to discuss 30/60/90 days yield management of inventory. At the same time review competition for anticipated weekly occupancies to assist in yield management decisions.

So what we are saying here is not to jump, and make rash decisions about discounting the rate until you are sure it is something you need to do, after you've done your homework. Get the staff to understand what it is they are selling—the value of what they are selling, to whom they are selling what, and make the sale. One good way to ensure improvement in conversions is to keep a daily record for each Reservation agent on the number of bookings and denials and take a look at the ratio. It automatically improves.

HOTELS SHOULD EVALUATE COMMISSIONS PAID TO THIRD-PARTY MEETING PLANNERS

January 2001

Since the Ritz Carlton Hotel group recently decided to reduce its commission payment to some independent meeting planners, there has a lot of discussion on Internet listservs, discussion boards, as well in various meetings trade publications. There has been a heavy increase in the use of independent planners by companies and associations over the past number of years to help in the meeting planning process. Some of these independent companies provide full-service meeting planning activities, while some only do site selection on behalf of their client, the meeting sponsoring organization. Many of these third-party meeting planners require a commission payment of 10% of room revenue generated to the hotel that is selected. Ritz Hotel management has reported that those independent companies providing only site selection services will now be paid 3% commission instead of 10%.

The question that needs to be raised now is should any third-party planner be entitled to any commission at all. These independent planners are engaged by sponsoring organizations (companies and associations) to represent them in selecting properties which best meets the needs of the organizations' meetings or conventions. Some independents only source of income is the commission paid by hotels; others get a fee plus the commission; and, of course, some ask for no commission. Are there some situations whereby a property, which may be the best choice for a sponsor's meeting, but because the hotel refuses to pay a commission, the hotel is not selected? Probably so, I would guess! In fact, some independents have already gone on record, in trade magazine interviews, indicating that they will not again use Ritz Carlton Hotels.

Does it not seem logical that if organizations hire independent planners to represent them in the process of selecting hotels and negotiating on their behalf, that it be the organization who should pay them? After all, they are being engaged to do the best job possible in representing the folks who hired them. At the same time, the independents are saving the sponsoring organization a good deal of time, effort and expense. Of course, it is time-consuming, and hard work for the independents, but why should a seller have to pay a buyer to do business?

There is another consideration. If an independent planner brings a piece of business to a property, and if it is a meeting which the property may not have been able to get on its own, is it not worth the 10% in commission? Many times a hotel can recover the commission payment in added fees, somewhere along the line in charges for rooms, space rental, food/beverage, etc. The whole thing boils down to how badly a property needs the business. Will it replace something else that would bring in as much revenue without any commissions to be paid?

Has the Ritz Carlton group started something that will spread? Will other hotel brands or independents also reduce commission percentages, or even decide not to pay any commissions to independents? It all remains to be seen.

Hotel sales staffs need to be cautioned about doing business with independent meeting planners who refuse to indicate the organizations they represent. It is always a good idea to ask independents for letters from the sponsoring organizations reflecting that they are being represented by the independent. Further, if any commissions or fees are paid to independents that those conditions are disclosed to all parties involved just be sure it is reflected in the contract.

ANOTHER WAY TO INCREASE ROOM REVENUE

April 2001

We are seeing it happen all over the place. Owners and operators getting concerned about RevPAR are looking for new ways to increase revenue through various extra charges; and, there is nothing wrong with that, in fact, a very good idea. We've seen resorts come up with an additional daily "resort fee" covering a variety of things such as newspaper, free local phone calls, free coffee, exercise room, etc. And, it works! Then in the banquet department, we see extra charges for resetting meeting rooms for conferences; charges for setting up audiovisual equipment; extra fee for a carver working the buffet table. And these work too. Some properties, which provide Internet, access for laptops (most access via a local phone call) provide this for free or for a small local call charge. However, we now see some properties providing the access line free, but for only 20 or 30 minutes, after that, there are charges on a per minute basis. And it works.

Guests and meeting planners are getting used to paying these extra fees. It is just a learning experience that people get used to doing. General Managers and sales staff should not be fearful of creating opportunities for these types of extra charges. After all, it is getting more and more costly to operate hotels, and as long as we can provide something extra for the guest, there should be no problem in charging extra.

Here's the latest, and I think it is a grand idea which really provides something extra for the guest, but there is absolutely no extra cost for a property. It is a new concept called Garden Suites which creates a new profit center and other benefits for hotels (and restaurants) without any capital investment or any cost at all for maintenance. Here is where a typical hotel room is turned into a lush, tropical garden-like setting through the installation of a vast array of colorful live plants. The in-room presentation of 30–40 plants includes floor planters, hanging baskets, wall sconces, dish gardens, and table planters. What this now becomes is a new amenity for the guest to enjoy a very impressive Garden Suite hotel room. Guests who upgrade to a Garden Suite pay a surcharge, thereby creating a new profit center for the property.

Most interesting about this new concept is that the company setting this up assumes all capital and operating expenses with establishing the Garden Suites, The company gets its income from sharing the up-grade surcharge. A real partnership. Another interesting part of this concept is that it is made possible by the company's patented technology involving the growing of live plants in "Millennium Soil." The plant is in a Planter Pouch, which contains Millennium Soil, a clean, environmentally safe, space-age soil substitute that absorbs and holds water like a sponge. There is no dirt required, no mess, and the plants don't require watering for several months.

So, this looks like another opportunity, but not only for creating a new profit center, but improving operating ratios (RevPAR), improving guest satisfaction, having another marketing tool, and a new, unique guest amenity.

IMPROVING SALES PRODUCTIVITY WITHOUT A SALES DEPARTMENT

October 2001

There is no question about it; many small or limited service properties around the country are operating without a person specifically assigned to the sales function. So, of course, in these cases, it is the General Manager who needs to figure out who will be doing the sales work to bring in new business. Maybe taking a look in the mirror will indicate to a GM who that person may be. With proper organization, there should be no reason why General Managers can't put in an hour or two daily on the sales effort. Especially in these times, someone needs to be looking for a new business just about all the time. If there is no one else on the payroll to do this, then it is up to the GM. Making three to five contacts a day should be an easy task for a GM. The idea is to make it easy by starting off contacting folks with whom the property is already doing business, right in your own backyard. Just take a look at the accounts payable list to get an idea of how many local businesses from which the property is making purchases. All of those may not be prospects for new business, but we never know until we ask. There are probably a good number of out of town visitors to those companies, we just find out about that by making contact, and asking the right questions. I have a form that lists all the necessary questions to ask. Just email me for a copy (howardf@vt.edu).

After you've gone through all those accounts payable leads start out with taking a good look at your own business neighborhood. Make calls on all the firms in the area, letting everyone know you are seeking business from people coming in from out of town. At the same time, you are creating visibility and awareness of your property. Internally, you can talk to your own staff for business leads. All your folks have neighbors, relatives, and friends who may be in a position to refer leads to you. Then, of course, you have your local newspaper that provides a wide variety of leads for business, on a daily basis. People getting married generally require overnight accommodations for out of town guests. With the wide range of sports activities going on today, there are local teams hosting other teams for tournaments, and they need housing. Check on grade schools, high schools, and colleges for listings and contacts. Entertainment groups coming to town for special events need housing, so check the paper to see what's going on where overnight rooms may be required.

Overflow housing from nearby convention or meeting type hotels is always a good opportunity for extra room nights. Establishing a good relationship with all the other property General Managers and sales folks will surely help in this area. Joining your local chapter of the Hospitality Sales and Marketing Association International does help.

Then, there is the opportunity a good deal of referral business from guests already using your property. Making contact, one on one, with guests does help, but you can't meet every guest personally. Get your front office sales associates to help. From time to time, when the opportunity presents itself, with regular guests, we should be asking for referrals.

We get into this subject in a lot more detail on my new set of audiotapes. Sixteen hot topics on how to improve sales productivity are covered. For more information visit www.hsa.com and click on my name.

WHAT TO DO ABOUT SALESPEOPLE AT THE LOCAL LEVEL IN 2003

December 2002

Hire more, pay them more, provide incentives, manage them properly, provide training and measure their performance. NHUSSS is an acronym that has been going around this world for many years, and it is still true: NOTHING HAPPENS UNTIL SOMEONE SELLS SOMETHING.

It seems that we hear a good deal about sales folks, at the local property level turning over too much. Not only do we hear it from hotel managers and management company executives, but more importantly from people with whom we do business. It upsets meeting planners greatly if the salesperson who booked their business is still not on the payroll at the time of the meeting. Things are getting better according to HSMAI (Hospitality Sale & Marketing Association International). According to a recent survey turnover in the hospitality industry sales ranks is 24.7%. However, it looks like there is a 35–40% turnover during the first year of employment. So what's the problem? From my experience alone, I feel the biggest problem of all lies in how we recruit, hire, orient, train, manage, motivate, and measure performance. Sometimes, it appears as if we hire the first warm body that shows up answering a newspaper advertisement. Other areas where we fall down on the job in retaining salespeople are when it comes to:

- job description (so many salespeople do not have a written one),
- salaries (for the responsibility, so many are really underpaid),
- incentives (if there is one it is so complicated that salespeople don't even try to keep track),
- communication (so many breakdowns between sales department and other departments).

Managing the sales effort is the responsibility of the property General Manager.

I think we would all agree on this. Probably not enough time is relegated to this function, with too much delegation to other folks. In 2003 sales needs to be a top priority in our industry. Surely, there will be some sort of economic recovery that will help business, but we should not have to wait for prospects to come to us. We need strong sales teams seeking new business. Some properties do not invest sufficient budget funds for additional salespeople. There is no rule on how many salespeople should be working for a property. The whole idea is that there needs to be a return on investment in everything we do in sales. Each salesperson needs to be responsible for generating enough revenue to satisfy his or her cost of employment and expenses. Incentive compensation does a good deal to motivate people, but they should be easy to compute. A percentage of new sales over goal is easy enough to manage and understand. Managers need to listen more to salespeople, and communication lines within all departments at a property need to be open.

One of the things that came out of the HSMAI survey was very interesting: "A large amount of direct selling is being attempted by corporate sales and marketing staff. If too much of this role is assumed by the corporate office it, can dilute the level of accountability for the sales process by the unit-level GM and sales team" So often over the years, this has been exactly what I have been preaching. I've gone both ways on this issue in my lifetime of hospitality sales. We just cannot turn away from putting the responsibility on the local General Manager.

IMPROVE YOUR MARKET SHARE IN 2003

January 2003

OK, so we really don't know when our economic recovery will come about. The companies that monitor our hospitality industry keep changing their minds about when we will see the recovery. For the past year, the economic situation has kept changing so much that it has been really hard for anyone to estimate when we'll get to see improvements in the demand so that hotels will be back in the drivers' seat. So, right now we are being told to look for recovery towards the end of the third quarter. This estimate will probably change by the time this Sales Clinic column will get published.

Anyway, I believe that if many owners, managers, and salespeople will spend more time bringing in new business and less time griping about the economy, overbuilt situations, drop in business travel, and competitors cutting their rates, we would see those properties increasing their market share. Owners and managers just need to spend more time in being concerned about their sales folks, and get interested in what salespeople do. Talk to them, help them figure out what to do to get more profitable sales into the property, without cutting the rate. What are we doing to help motivate salespeople to get them to spend more time in pro-active sales and less time on mostly unproductive cold calls? What system is in place to measure productivity? What type of incentive program is in effect? I am a firm believer in incentives; they must be simple to understand and implement, and need to be achievable. Incentives also can go a long way in retention.

A participant in a recent workshop I conducted for a major brand, who is a property owner, approached me about how to retain salespeople. It seems that a long-term sales director at her property just left for another job at a local competitor's hotel. This employee was bringing in over a million dollars in room sales a year and left for an increase in salary plus commission. It was too late for this owner to now say "I'll pay you more and give you an incentive if you will stay."

However, had that employee been paid and incentivized properly, to begin with, she probably would not have left. The owner is not only faced with trying to replace this valuable salesperson, but there is also the risk of losing, to this competitor, a good portion of the business that had been built by her. The owner agreed that had she been providing a simple 1% of the revenue of new business generated by this salesperson, over a period of time, she probably would not have lost that employee.

It is unfortunate, but true that so many properties have missed a myriad of opportunities to build market share during this rather brief economic downturn. Salespeople are critical to our business, but they need to be managed, directed, supervised, and motivated. Here is where managers and owners need to get involved. The business is there, and we just need to do less complaining and more selling to beat out the competition. Let's get our salespeople to spend more time on proactive sales; that is, calling on qualified prospects, rather than going out "knocking on doors, making cold calls. In these times, increasing market share is critical.

Let's go get it.

HOW GENERAL MANAGERS AND DIRECTORS OF SALES CAN BETTER MANAGE THE SALES EFFORT

April 2003

It happens all too often, in our business; we hire salespeople and then tell them to go out and book business. If we get someone with prior hospitality sales experience it may work out well, or maybe not. It just seems logical that someone, a director of sales at a property, if there is one, or more importantly, the General Manager, needs to accept the responsibility of managing the sales effort. By this we mean, not only providing a job description, orienting and training the person employed in sales, but also provide Attention, Direction, Supervision, and Motivation. Here are some tips that may be helpful to those who are responsible for the activity of sales personnel:

▶ **Attention:** Here is where we need to know what's going on in the sales department. How about a quick five-minute stand-up meeting in the sales office, every day, and just asking that question? Further, we need to be concerned about salespeople. If someone is not feeling well, or having a bad day, perhaps it would be better to have them take the day off. Next day will probably be better for them, and they wouldn't be in the proper state of mind to sell anything anyway. We also need to pay attention to sales folks' daily sales activity; how are they spending their days? All this is paying attention to what's going on.

▶ **Direction:** Providing direction to any employee is critical in getting them to understand and do their job. The whole idea is to make sure we tell people what it is they are supposed to do, how they should do it, and why it has to be done. Sounds simple, but important; this is easily accomplished by accompanying salespeople on prospecting calls and sales calls on qualified accounts. Also, by sitting-in on meetings that salespeople have with prospects, on-site, or joining in on client entertainment at the property, is another way of observing performance and providing direction afterward.

▶ **Supervision:** In supervising we are talking about reviewing the activity of a salesperson to check what is being done, to make sure the person's performance is in line with corporate, or property sales procedures. Here you are reviewing, with a salesperson, their weekly sales activity report. You also want to take a look at the follow-up on accounts, to see if the trace system is working accordingly. Pull some files now and then to see how accounts are being handled.

▶ **Motivation:** Just because you think you have a salesperson who is "demotivated" don't feel that motivation isn't needed. Just about everyone is motivated by different things. We motivate to get people to behave in a certain manner. Sometimes people need encouragement when things are not going properly; a pat a back could do the trick or even kind word. Managers need to understand that a compliment, when good is done, is important. Everyone appreciates a compliment. Once, a General Manager told me why someone compliments for doing what someone gets paid to do? That was not the right attitude on the part of management. Of course, reprimands might be necessary, but how it is done could also be motivating. We should also consider rewards, when appropriate. It does not have to monetary, although that is good too. A letter to the salesperson's home, with a copy in the personnel file would work. Or a gift certificate for dinner with a spouse in the property restaurant or at a friendly neighbor's restaurant, a comp weekend at another property operated by the same management company or brand; any of these work well as a reward.

Improving productivity, and improving sales is why it is important to make a strong effort in managing the sales effort.

THE NEW TWIST ON SALES CONSULTING . . . SALES COACHING

September 2003

For years, in our industry, we have employed consultants to help properties with development, management, marketing, and a host of other functions in the interest of improving operations and profitability. These consultant firms or individuals worked on specific situations and challenges, adding their own expertise in their particular field. From a sales standpoint, we have had consultants work with sales folks to provide training programs, create sales plans and provide assistance in learning the techniques of selling to various market segments. And, all this has worked well for our lodging industry. But we have not heard or learned about this new business of sales coaching.

There is a lot to be said about salespeople having coaches. Actually, it is a form of consulting; however, the coach is someone who stays with the salespeople to help implement new skills, and goals to make sure they really happen. And, the coach is also the mentor, someone who takes the time to provide the benefit of their experience and wisdom as well as encourages and motivates the sales team to be productive.

According to Neil Salerno, a professional sales coach and practitioner, who has coached and mentored salespeople at all types and sizes of properties, coaching does the following:

▶ Helps salespeople set better personal and business goals, and reach those goals.
▶ Assists salespeople to become more focused to do more than they would have done on their own, and to produce results more quickly.
▶ Provides the tools, support, and structure to accomplish more.
▶ Creates synergy between the coach and salesperson to create momentum.
▶ Helps with new skills and techniques which translate into more successes.

It is interesting, that what we see today, in so many properties, where there is someone who is directly responsible for the supervision of salespeople, that the business of providing supervision and direction is very limited. Usually, that person is so busy with his or her own work, that time is not available to provide the extra help that is needed by sales personnel. We also find that a good many salespeople at properties are being supervised and directed by General Managers, who themselves have had no or very limited sales experience. This is where there is a need to have someone provide the coaching and mentoring to help the salespeople succeed in their functions. Some properties have even adopted the philosophy that the General Manager is also the Director of Sales and the salespeople report directly to that person. In these cases, an outside coach would certainly be appropriate.

Using coaches is a good way to follow-up on training provided to sales personnel. It works well to have a coachwork with the sales staff to ensure the techniques learned during the program are implemented. In fact, it is a pretty good idea to have a coach come to the property to provide on-site training, then follow-up.

CUSTOMER SERVICE: ARE WE EVER GONNA GET IT RIGHT?

October 2003

No question about it, in our industry, customer service is really getting better all the time. But we are still getting customers to be unhappy, and the concerns and complaints are getting louder; particularly from the standpoint of those who plan meetings and events.

A few months ago, meeting planner Bonnie Wallsh moderated a panel discussion for the Charlotte, N.C. Chapter of HSMAI (Hospitality Sales & Marketing Association International). Her panel members were all meeting and event planners. In a nutshell, here are some of the panelists' concerns and issues regarding customer service in hotels. "Finding a hotel/citywide team that I can trust to deliver services;" "Poor communication, inability to carry through and gaps within hotels;" "Making initial contact and not getting a response, only 1 of 6 properties responded to initial contacts;" "Boiler-plate contracts;" "Not taking the time to do your homework, and research to know and understand the client;" "Reducing size of meeting space;" "Passing the buck;" "Breakfast for a group not set up and meeting planner had to help serve the meal;" "Hotels should have contingency plans for emergencies." At the conclusion of the meeting, panelists stressed the importance of pre-con and post-con meetings. Better communications were stressed. Wallsh, a Certified Meeting Professional (CMP) is an industry veteran specializing in the meeting and event planning, and is a well-known speaker for hotel and meeting planning groups. She can be contacted at: bwacmp@carolinaaxxcom.

Starwood Hotels and Resorts have the right idea with their new meeting program designed to improve the consistency of its meetings at their properties. They have come up with new standards and tools, with new communications ways of improving customer service. It all starts off with getting their General Managers more involved in clients' meeting experience by having them accept more direct responsibility. This is great, and I have been preaching this for years. Starwood properties will offer pre-con and post-con meetings to all groups and will employ a "Star-Meeting Concierge" at every hotel who will work directly with a planner while on-site. They are standardizing forms to make it easier for communications between hotels and planners. Further, there will be an upgrading of standard VIP treatment for groups, and all planners will be provided with Nextel walkie-talkie equipment for instant communications with hotel staff. And, there's a lot more to it than what we described here. Congratulations, Starwood, and Dave Scypinski, Senior Vice President of Industry Relations, Ya Done Good!

Then, of course, the CUSTOMER IS ALWAYS RIGHT mentality could be very frustrating. Nancy Friedman, President of Telephone Doctor, Customer Service Training says: "It should be the customer always THINKS they're right." She is getting ready to write a new book "Customers from Hell" and wants to hear stories from "the other side of the coin," meaning what has been your customer horror stories and how were they handled? Do you have any nightmare customers?

MOTIVATING SALES STAFF IS A KEY TO IMPROVING SALES PRODUCTIVITY

June 2005

In our last issue, we discussed the three R's (Recognition, Rewards, and Retention). Of course, we have to agree that perhaps there is more to motivating sales staff than that. Based on how salespeople (as well as all employees) are treated and respected determines to what extent they are motivated to do their jobs.

Many years ago, while with American Motor Inns (a hotel management company with 50 properties), our personnel director (they are now called human resource folks) decided to survey our salespeople to determine what they needed and wanted from the company. At the same time, we surveyed our General Managers and regional vice-presidents to find out what they thought the salespeople wanted or needed. Almost 20 years later, when I was with Servico (a hotel management company with 60 properties) did just about the same thing to find out what we had to do to motivate our 120 sales department staff at the properties.

Times do not seem to change what people think they need from supervisors, and what also doesn't change is what supervisors think their salespeople need or want. Twenty years later, we were still looking at the same results of 20 years prior, with different companies, different types of properties, different geographic locations; it all came out about the same.

General Managers and Supervisory people thought that the first few things on the "wanted" list would be better pay, job security, and a chance for advancement. However, what sale people listed for the first three items were: (i) acknowledgment or appreciation of the job they were doing (a "thank you" now and then); (ii) being kept informed on what's going on (being included on meetings about the company); (iii) management support (providing resources to get the job done).

Good pay, job security, and better working conditions followed these. From the supervisor's standpoint, it was pretty much the same, but it looked like they thought the top three issues were good pay, promotion opportunity, and interesting work.

If we did a similar survey today would things change? Probably not, we still think that sales employees leave companies mostly for pay. Sure, most would leave if the increases in salary and benefits were high enough. However, there are so many other considerations that would keep employees well motivated to stay on the job. Proper salary and benefits do have a great deal to do with it as well as providing a worth-while sales incentive compensation plan. Hopefully, this will get them to forget about the money and concentrate on the job of improving sales productivity. Then we have to treat them right, with great respect, and show appreciation for their work; acknowledge accomplishments. We need to be very clear on what is expected of them and keep them informed about what's going on. Involve them in plans for the property and company, and listen to them.

Motivation has all to do with how bosses (i.e., managers, supervisors, company officials, etc.) treat employees and provide them with opportunities to do a better job. A comfortable, friendly workplace, with supervisory support, goes a long way in helping sales staffers to be more productive in building greater sales.

MANAGING THE SALES EFFORT TO IMPROVED PRODUCTIVITY

July 2006

For years, we have all talked about how we manage our sales force at the properties. It is time we took the advice of Peter Drucker, which has to do with managing people. In his "Management Challenges for the 21st century," he states: ". . . the knowledge of the individual knowledge worker becomes an asset and, in more and more cases, the main asset of an institution. What does this mean for personnel policy? What is needed to attract and to hold the highest-producing knowledge workers? What is needed to increase their productivity and convert their increased productivity into performance capacity for the organization?"

What we have to think about at the property level is not only how we manage the people bringing in profitable sales, but also how we help them, and keep them, through appropriate leadership. This should a major function of a property General Manager, or at a corporate level, the senior sales official or operator, and sometimes by an owner as well.

In providing management and leadership for a sales staff, it is all about understanding the needs of the staffers in relations to their job function. Management needs to provide the direction, supervision, and motivation, and as well as to provide the proper work environment, and assist with goal setting, and performance reviews with suggestions for improvement. Coaching is another area where management can provide leadership, and we really do not see enough of this going on at the local property level, but coaching is a very important function to help the sales staff in improving production. Here is how it works.

Marketing Coach: "So what is the basic coaching philosophy anyway? Simply put, that we humans are great, that we're all discovering what we really want and that we get can what we want faster and easier by having a coach who's been there and who can help us.

How is coaching different from consulting? Coaching is a form of consulting. But the coach stays with the client to help implement the new skills, changes, and goals to make sure they really happen.

- Coaches help people set better personal and business goals and then reach those goals.
- Coaches ask their clients to do more than they would have done on their own.
- Coaches focus their clients better on producing results more quickly.
- Coaches provide the tools, support, and structure to accomplish more. Coaching works for several reasons:
  - Synergy between the coach and the client creates momentum.
  - Better goals are set ones that naturally pull the client toward the goal rather than goals that require the client to push themselves to the goal.
  - The client develops new skills, and these skills translate into more success.
  - Some clients hire a coach to help them accomplish specific goals or projects. Usually, however, the client keeps working with the coach after that because there are even more interesting things to accomplish.

More and more people are beginning to become focused on this interesting twist on consulting called coaching. Coaching is even more effective when combined with training in some form.

My advice, for everyone in our industry, is to change your paradigm and check into coaching. It could be well worth the investment.

Contact me (at howardf@vt.edu) for a copy of Salerno's recent article on coaching or contact him (at nsalerno@hotelmarketingcoach.com) directly for links to his articles on the subject or further information on coaching.

DEVELOPING SALES WITHOUT A SALES DEPARTMENT

September 2006

One of my favorite expressions has always been: "The sales department is not the whole hotel, but the whole hotel is the sales department;" meaning, of course, that every employee is a salesperson, and everyone is responsible for getting business, and providing the service that gets guests to return. However, most properties have salespeople for the purpose of developing new business, and most sales staffs concern themselves with locating and bringing in groups. This works very well for the larger, limited service properties and full-service properties that have meeting space, and can handle catered events. Many smaller, limited service hotels, without much meeting space, do not have full-time staffed sales departments, but most of these properties can handle some types of group business that do not require meeting space or catered events. It makes sense for someone at these properties to seek out and try to book group that can use a block of rooms. In all probability, it would be the General Manager or some staff member who can spend some of their time going after this kind of group business.

Group business in on the increase in this time of improved economy, and the smaller, limited service facilities need to get into prospecting for small meetings and group travel. Even without a dedicated meeting space or a food-service outlet, there is still an opportunity to get this type of business. Most limited service properties have some space dedicated to the usual, complimentary continental breakfast service. This is a space that could be used for meetings anytime, the rest of the day, following the breakfast time period. For food service, most local nearby restaurants would be pleased to provide appropriate lunches, delivered to the property for a group. Or, most groups would welcome the idea to take a break and find lunch off the property. The whole idea is to seek out a business that would be able to provide group room nights. Here are some tips on where to look:

The local newspaper is just loaded with leads for small meeting business as well as incoming travel business.

▶ Look for meetings already being held at competitors, and try to get them for next time.
▶ Where will out-of-towners stay when coming in to attend weddings?
▶ The same for funerals.
▶ Seek out sports activities to be held, where will the visiting teams.

Your list of vendors, from whom you are purchasing products.

▶ They have major brand salespeople calling on them, where are they staying?
▶ Don't they have small meetings from time to time?
▶ They have families, friends, relatives. What about their reunions, weddings, and other events where people come to town?

And the telephone directories.

▶ Look at the listing of associations and clubs. Make phone calls and ask about their meetings and events that bring people into town.
▶ Catering halls that do a variety of events which may need sleeping rooms that they do not have.

All we are mentioning here are just a sample of the opportunities that are available to small properties to seek out group room business. Of course, we go on and get into family reunions, school reunions, military reunions, tour groups, etc. It just takes about an hour or two a day to seek out a variety of business opportunities for any type of hotel property.

LOOK FORWARD TO THE NEED FOR AGGRESSIVE SELLING IN 2008

April 2007

Lodging operators, General Managers, and Sales Directors better start thinking of their marketing plan, and specifically, their sales action plans for 2008 and 2009. Business is still great now, and we all have great predictions for next year, but lots of things can happen to make the situation change any time between now and then. Rates have gone up considerably in the past couple of years; how much longer could we keep raising the rates. History has already told us what happens in our industry right after we have a very successful year with increased occupancies and rates. The developers get busy with investors, major brands and hotel management companies and the pipeline grows. Could we be hit with another period like the mid-70s, and mid-80s, with a severe overbuilt industry and massive rate undercutting all over the country? Sure, it is possible! There is probably no question about it, but we also need to look forward to continual increases in the cost of operation. Technology keeps changing, for the better, of course, but those costs will surely rise. Cost of marketing will continue to be expensive as will labor costs, along with employee benefits.

The best protection is to place greater emphasis on improved employee productivity, greater cost control, better quality of service, and most importantly, more personal sales activity. Property managers should start looking to the reassessment of marketing plans to accommodate the change in needs and directions for the coming years. Sales staff functions may even need to be reassessed due to drastic changes that will be forthcoming with regard to the online booking of groups, and changes that are taking place with third-party independent planner companies. For sure there will be new market segments emerging which may require different selling opportunities for the sales personnel. Programs may have to be developed that will reach special market segments which may have never been previously tapped. As an example: Senior Citizen travel more now than ever, and have more time and money available, and they are living longer, and more are willing to spend their money on vacation activity. Also, the trend for companies will be toward meetings in smaller cities at smaller lodging locations. Companies, as well as associations, will have more meetings of 50 people or less than they have of larger groups, thus reducing travel costs. The smaller lodging operations will have more of a chance to go after this type of business. We already see tremendous growth in the family travel market, and this will even get greater. Even with higher gas prices today, families are taking to the road for vacation trips.

Even as rates may continue to increase this would not necessarily reflect that all properties will meet with success in the next couple of years, An aggressive, profitable sales program is necessary regardless of the size of the property; large or small. Responsibility for the sales effort lies in the hands of the owner or operator.

PROPERTY SALES OPERATIONS AND THE GENERAL MANAGER

May 2007

No matter what size the property, and regardless of how the sales function may be delegated, overall responsibility for sales still begins, and ends with the General Manager. Because General Managers have such a responsibility, they should be greatly concerned with a number of crucial areas involving sales activity:

▶ **Planning:** A business plan, a marketing plan, and a sales activity plan are all basic tools in the marketing of a lodging property, regardless of the size or the market it serves. A good plan is a map that leads the way to a successful sales operation. The time and attention devoted to planning can pay off many times over, just as neglect in this area can greatly hurt a property's sales effort.

▶ **Organizing:** Management must be adept at putting together a team, and placing people, as well as the "things to do" in the proper places, providing a structure for sales operations, setting standards, and then measuring productivity, all contribute to the success of a sales operation.

▶ **Directing:** Good, sound leadership requires the General Manager to provide proper direction to the sales staff. It is not enough to say: "I'm paying you to do the job, go do it." Daily direction for successful sales operations is the most important. This applies to all other departments as well.

▶ **Controlling:** Being directly involved is critical. "Where did you go today?" "Who did you see?" "What did you do?" "Let's talk about who said what," are all proper questions to ask of a salesperson in providing control over a sales operation. Of course, with multiple sales personnel, this is something that would be delegated to a Director of Sales, who would then brief the General Manager as necessary.

▶ **Motivating:** The little time it takes to talk with the sales staff, provide direction, review work, and the old "slap on the back" pays off exceptionally well in motivating sales personnel. Further, incentive compensation, of some sort, to reward a salesperson's improved productivity also works in motivating the staff.

▶ **Communicating:** Two-way discussions between General Managers and salespeople, in which all parties are honest with their comments, makes for the best communications. Expanded communication throughout all departments within a property, concerning their involvement with sales effort, will immeasurably improve the success of a property's sales program.

▶ **Training:** An ongoing training environment is also essential by success in improving sales productivity. Salespeople generally want to continue learning, and are very open to attending training programs. Sometimes these opportunities are curtailed by management for a number of reasons, but General Managers should budget for, and encourage sales personnel to participate in training programs provided by organizations such as HSMAI and its conferences as well as its online workshops. There are many other training opportunities available through this organization and others. Cross training with other departments helps the sales staff to better understand the workings of property and helps improve relationships, as well as productivity.

Clearly, the General Manager should never forget about how important it is to have realistic growth plans for the sales staff; and always to be aware of the objectives and goals of sales staffers to help them vigorously strive for them.

THE HIRING AND EMPLOYMENT OF THE RIGHT SALES STAFF

November 2007

Our business has been great the past few years, with increased occupancy, a much better average daily rate, and increased profits. Naturally, in such a favorable climate for business, there is no doubt that we will see a pretty big increase in new properties being built in most markets. Already we have seen figures for the number of new rooms opening this year, along with news reports of the major brands expanding with new hotels, and new companies getting into the hotel business. The pipeline of new constructions planned for 2008, 2009 and 2010 is moving right along with pretty large numbers. Therefore, all this will result, without question, the hiring and employment of sales staffs to work to bring business to all these new operations.

So many costly mistakes have been made in the hiring of sales staffers over the years, and it appears that we have not learned much about how to handle the process to management's advantage. The big question to get answered is: "Does this candidate have the TALENT to be a successful salesperson?" In their book, *How to Hire and Develop Your Next Top Performer*, the results of hundreds of thousands of salespeople assessments with actual sales performance measurements. His results: "55% of the people earning their living in sales should be doing something else" and "another 20% to 25% have what it takes to sell, but they should be selling something else" (www.mysalestest.com).

So, this tells us something about what we should be seeking in those persons we want to hire in our business of selling for a hotel property. What are the talents we need in people we want to hire for sales? Of course, there are the basic resumes we look over, and we make some calls to gather information on a prospect for employment. Perhaps the prospect may have hotel sales experience that may look good, but are we really looking for experience or talent? The interview process could be the key to determining if there is a FIT for the prospect in your organizations. There are certain characteristics that may be a key to whether the prospect would have the talent to be successful in hospitality sales. For example, see if you can determine if the candidate for the job reflects enthusiasm, professionalism, a sense of humor, is motivated, has communicative skills, is interesting, has transferable skills, is willing to learn, has the ability to be a good listener, and appears to be able to build relationships.

Certainly, there is a lot more to the hiring and employment of sales staffers. Before a decision is made it would be important for a good deal of information to be made very clear to the prospect We should be very specific about the details of the job function so that there is no misunderstanding of what is expected; there should be a written job description provided. Certainly, an explanation of salary, benefits, and incentive program, if any, along with any sales goals that are required; how their performance will be measured and how salary increases are made. Often the management fails to cover some of these issues.

SMALLER PROPERTIES HAVE A DISTINCT NEED FOR SALES PLANS

November 2008

There is no question about the fact that more hotels are being built each year. While the rate of construction varies by market and location, competition is still growing almost everywhere. Not many hotels are being converted into other uses these days, nor are many closing their doors. Add all this together, and one finds that the pressure to find and keep customers continues to grow.

Owners and managers of smaller properties must look at time and effort as a realistic cost of reaching success in their hotels. Because many smaller hotels do not have a full-time (or even part-time) salesperson, the owner or manager must undertake that responsibility. Leaving it to the "brand" alone to bring in profitable sales is not sufficient. Of course, brands do a great deal in the area of sales and reservations for their franchisees, but there must be some local, hands-on sales effort, in addition, to bring success to a property. The smaller hotel management must be able to identify meaningful market research, determine avenues for effective public relations and promotions, and still find the time for some direct sales effort. To accomplish this effectively, there needs to be a plan. A basic plan can identify the kinds of business segments (markets) that currently exist at a property, such as: corporate overnight guests, youth sports groups, motor coach tours, leisure vacationers, truckers, etc. The plan also needs to identify which of these different market segments would be the most logical and profitable business to go after for future business. Once the markets that are accessible are identified, the strategy can be developed to promote those market segments that are meaningful from a volume (occupancy) and profit perspective.

Some hotels have been hiring local college students or parents seeking part-time employment while their children are in school. These options have proven to be very cost-effective, and there are success stories in virtually every brand and region of North America.

Another area of involvement for owners and managers to get more business is to become active in local community activities, and to make a certain number of sales calls by phone and in person on a daily basis. Involvement in local activities gives one the reputation on being the local hotelier or go-to person. It will require some volunteer work on committees or public service projects, but it reflects one's interest in the community and is a valuable public relations tool.

If a manager/owner will commit enough time to make 5 to 10 calls per day, that would equal up to between 250 to 500 personal contacts per year, this personal networking, while cost-effective, should likely yield substantial business. One of the problems in connection with this is that management/owners at many small properties become so involved in operations that they end up working the desk or housekeeping or other areas of operation so frequently that they seldom have time for community involvement or sales. This may be some saving of labor costs; however, it is at the risk of losing the long-term profitable effect of marketing and selling.

Effective operators of smaller properties will dedicate at least 25% of their time (week-day, prime contact hours) to networking and selling. Hoteliers who do not either handle this critical responsibility themselves or ensure that it is handled by someone competent, run the risk, of their investment failing. It is an unnecessary failure.

SO WHAT'S GOING ON WITH YOUR SALES STAFF?

April 2009

Bill Eckstrom, President of the EcSell Institute, wrote a letter to his subscribers with the headline "The Role of a Sales Manager." In the letter, he quoted Curt Coffman, co-author of "First Break All the Rules: What the World's Greatest Managers Do Differently." And he said: "The ability to build organic growth and the future viability of any organization will be in the hands of their sales managers." That's a real truth that so many of our property owners and managers probably do not believe. After all, it has often been said: "Nothing Happens Until Someone Sells Something." And, don't we all really believe that it cannot fall to the bottom line until it comes in on the top line." Sales are where it's at folks! Ya gotta believes it.

If ever the time was to believe this, it is right now. Even with everyone talking doom and gloom, with lower occupancies, lower revenues, we really do have a bright side, there is a recovery with sales. Let's get busy and talk sales. What's going on with your sales staff? Who is managing the sales effort? How about the General Manager of a property doing the job in hiring the right people in sales, then providing the proper orientation and training? Following that, salespeople need to have direction, supervision, and motivation. What are we doing about that? What is being done to provide the resources to get all this done, particularly the training?

Sales staffers need to be able to participate in professional development opportunities via their membership in organizations like HSMAI (Hospitality Sales and Marketing Association International) and MPI (Meeting Professionals International) and be able to attend their meetings and conventions. This is where they get their professional development to be able to do a better job in developing their opportunities to improve their productivity. Of course, there are other organizations to which they could belong that would provide similar opportunities. Resources must be made available to assist the sales effort in this manner.

Just recently I received an email message from a salesperson who was told by her General Manager that business was very bad and she needs to go out and make some cold calls. Ugh! What a way to treat a salesperson; now is that really providing direction? Further, where is there motivation any more on the part of a company to help a salesperson achieve production goals? Other than a monetary incentive, which is a good idea, how about providing the opportunity for a salesperson to attend a training workshop to learn skill improvement and techniques that would work to bring in more business? Or provide the opportunity to attend conferences and conventions where prospective buyers may be met. We need to do things to help sales staffers get excited about their jobs, and motivated to bring in more business so we can all make it through these somewhat difficult times. We just have to remember that sales are where it's at.

WHO IS THE CHIEF-IN-CHARGE OF SALES?

October 2010

OK, the reports are in that business is picking up, occupancy will be back up a little bit by the end of this year, and rates may be very slowly picking up, but RevPAR is down quite a bit. Of course, we knew this would happen when hotels started to compete by cutting rates, all over the place the past two years, one of our biggest problems was the fact that product was up considerably with new rooms on the market. Other than the economic situation, we had an oversupply of rooms; however, now the report is that demand is picking up (maybe by 3%) and the supply of rooms will not be more than #2960. We are indeed in a better position to start making hay while the sun is shining.

We do not know how long these "good times" will last before we see another supply of overbuilding. So, it only makes sense that we should be doing something about improving profitable sales, especially in the limited-service, economy and budget segment; types of properties that may not have the benefit of a full-time sales staff, Most properties in these segments do not have the luxury of a person assigned specifically to bringing in new business because of budgeting for staff. What's the alternative? Look in the mirror; the answer, of course, is you. The General Manager or owner/operator of the property.

Here are some basic tips that do work. The first thing we need to do is to make sure your telephone is being answered promptly, and by someone who has an enthusiastic voice. Yes, right! We do have our reservations coming through our Website, and of course through those central reservations systems operated by our major brands and independent systems. However, telephones are still an important method used by people to make reservations. In all probability, no one really knows how business is lost in a day by the phones not being answered promptly! This can easily be fixed by management paying more attention to it.

Another telephone related tip: "when people call about rates and even offer buying signals, how about the desk clerk, or front office manager, whoever it may be, start asking key questions like name, phone number, company, etc., and make the important statement: go ahead and hold that room for you." Central reservations offices do this very well, and it is at the property level where we fall down on the jobs.

People do business with those they know; respect and like. How wed-known are you in the community. Do people like you? Are you respected? You need to be highly visible and respected, being active in organizations and project is a key to doing more business.

The point is that if you are a General Manager or owner/operator of a property, it is likely that you have to be a person of business.

THE GENERAL MANAGER'S RESPONSIBILITY TO THE SALES EFFORT

November 2011

It really doesn't matter how large or small a property may be, in any case, regardless of type or size of a hotel, the one person, accepting responsibility for profitable sales is the General Manager. All too often, we see entirely too much delegated to the director of sales or the director of marketing. No question about it, there should be the burden of the job being accomplished by those heads, and their staffs. Of all the operating departments in a property, sales is the one that generates the business to bring in the revenue. It would seem that a General Manager would be vitally interested in knowing what is going on in that area. One big question with which a General Manager should be interested is: "How productive is our sales team, and how could we make them more productive?" One big thing that should be done is to provide the research upon which sales staffers can create strategies to seek more business from various market segments.

We have written earlier about the business of market segment analysis based on information gathered, on the property, of the various market segments, and the amount of business generated by each. Then this is followed by an analysis to determine which segment needs to have its business increased to create more profitable sales. From that, strategies are then determined on how to do this.

What also needs to be examined is what are the competitors doing in regard to those same market segments? As an example, if it is the corporate group market that appears to be the most profitable, and a property has been getting 10% of its total business by that segment, then perhaps this is the segment that needs to be increased by another 10%, to make it 20% of total business in the coming year. However, how much of this market segment is also being received by competing properties? This information is difficult to get, unless another source can do this for you. Having been in the business for over 10 years, TravelClick (www.travelclick.com) has specialized in generating a wide variety of data to member hotel properties, such as providing comparative figures of revenue generating sources among a competitive set of hotels. This way a hotel that is getting a certain percentage of total business from a particular market segment, can get TravelClick's comparative figures on the same segment of their competitors. This helps provide information to a sales department in helping to determine how much more business for a market segment that may be "out there." Further, a General Manager would want the sales department also to find out, "why" the competitor is getting more business, and "how" are they doing it?

According to Tim Hart, (former) Executive Vice President of Information Intelligence of TravelClick, General Managers can find a wide variety of additional research information available from TravelClick, including a quarterly report that focuses on demand performance of booking channels, segments of hotels, and loyalty programs.

HOTEL SALES FOLKS NEED TO BE GREAT—NOT JUST GOOD

April 2014

There are many things that go into making a hotels salesperson great. Sometimes we will hear somebody say: "That person is a born salesperson." We probably do not believe that is entirely true. Energy, looks, motivation, personality; all of this helps (and perhaps genes have a little to do with it). Whether or not someone has the right genes, there are many opportunities for someone to make themselves great in the business of hotel sales.

Here are some areas that are very important in improving a person's ability to become great in our business of selling for hotels and resorts. A little practice in each of these areas will significantly increase a person's ability to make a sale:

▶ *Listen:* Listening more and talking less makes one appear intelligent. Just ask some good, key questions, and keep quiet until it's your turn to talk again. And, don't oversell. Silence sometimes can be a powerful tool. Learn to let "them" do the talking.
▶ *Features/Benefits:* It's OK to discuss features of your property—provided they can be discussed as a benefit or value to a prospect. There is no point in bragging about parking for 500 cars if everyone in a group will be flying in. Knowing a prospect's needs, and applying for benefits, and advantages as a result of your features is what works.
▶ *Looks and Answers:* You can create a good or bad image by your looks, your voice, and your ability to communicate effectively. People make instant deductions from what they see. How you appear upon contact with a prospect should be a prime consideration. An energetic and enthusiastic tone of voice is an important attribute of good salespeople. And, knowing what to say, and how to say it is another skill that they acquire.
▶ *Motivation:* Great salespeople are usually great motivators of others. In the Business of selling it is important of taking on the job of motivating the prospect to do the buying. Learn how to encourage the prospective buyer to make "the buy."
▶ *See and Be Seen:* Salespeople need to get out and be visible in the business Community, bc active, and join groups, give talks to groups, be quoted in the newspapers, and be interviewed on broadcasts, and be sure to be seen at important business events. People like to do business with people who are well-known.
▶ *Be a Good Communicator:* Learn to be articulate, and use proper grammar. Avoid using slang expressions. Get rid of the bad habits of repeatedly saying: "ummmmm," 'Cya know," "like," etc. Good speaking habits add strength to your sales position.
▶ *Know Your Stuff:* You must have the knowledge, not only of your own property,

but of your competition, and the needs of the prospect.
▶ *Have Confidence:* Sellers' confidence translates into buyers' confidence. Without confidence in yourself or your hotel, you cannot be successful in sales. For a copy of characteristics of successful salespeople, email: howardf@vt.edu.

HOW OWNERS, OPERATORS, AND GENERAL MANAGERS CAN IMPROVE GROUP REVENUE

November 2014

It sure does look like this coming year will be one of the better years for Demand, Occupancy, and Rates. How well are your sales staffs producing in the group arena? What is it that could be done to have sales associates be more involved in Sales Productivity? At many properties, we do see those in sales getting involved in other marketing activities, such as: Advertising, Promotion, Public Relations, Social Media, etc. If we could just get them to spend an additional 20–25% in pro-active sales the better sales figures would soon be there.

So here is how YOU do this! Develop an Incentive Program for salespeople to get rewarded for getting more group business on the books. Of course, many incentive compensation programs are already out there, but some do not really do the job. Some are based on budget; "meet your budget and you get so much compensation." Many are very complicated and are based on issues that should not concern a salesperson, who is responsible for booking group business.

A simple rewarding program is to offer a 1% incentive for each salesperson based on total revenue generated by group business. This is not a group program where all those in sales share on all the business booked. Further, it is not paid on bookings, but only after the group has met, and all the bills have been paid. It should reflect group revenue on the master account, including room rates, group food and beverage, space rentals, recreation and other activities generating incremental revenue; everything excluding group gratuity, taxes, and commissions or fees paid to outside entities. Revenue generated by individuals in the group who charge purchases to the room folio should also be included.

Now here is what all this would do for your property:

▶ **Retention:** By making incentive compensation payments only after a group has departed the property, and paid all the bills, even if groups are booked 1, 2, 3 years, or more ahead, the well-performing sales associates will want to stay around until they collect, rather than hunting for other sales jobs or being recruited by another company.
▶ **Salary Increases:** This is pretty much of an uncomfortable position for a sales associate as well as a General Manager. It looks like everyone expects a raise every year when it comes to an annual performance review. For some reason, a raise may not be due because of performance or budgeting situations. With this 1% incentive program sales associates are really creating their own raises as

they go along all year long. A decision about raises after an annual review is not necessary.

Increase Group ADR and Sales Motivation: Sales associates will know they have control over their own income. The more revenue generated per group puts more in their pockets. Payments being paid on a monthly basis reflect that they can receive their incentive compensation at about the same time the groups make payment on their bills. This alone is terrific motivation.

FROM WHERE WILL ALL THE PROFIT IN HOTEL PROPERTIES COME FROM THIS YEAR?

March 2015

Of course, generally, we took to the chief-in-charge of operations of a hotel or resort property, the General Manager, for profitability. All agreed! But where does it all start? It always starts at the top, meaning of course, with revenue generated by sales. It has to come in at the top to fall to the bottom line. Once all the costs are paid, the bottom line (or profit) is what is left. The more we take in, and the less we pay out is what we call profit.

So, we need we need to look at what we take in at the top, and make sure it covers the costs, plus. We all know that the costs of property operations continue to increase. There is really, absolutely nothing that costs less these days (maybe except gas for your car and other uses). Just think about it, debt service, leases, rentals, taxes, utilities, maintenance, supplies, advertising, franchise fees, management fees, contractual services, repairs, benefits, and the list keeps going on. This year we may even be hit with minimum wage increases, and in all probability, it may very well result in many other wages being increased. So, with all this in mind, it tells us that we just better be in the business of doing a better job in hotel sales.

According to Travel Click, we can expect about a 4.5 ADR increase over last year, and of course, this is good. We just need to be sure that our sales teams are on board with the understanding that in the job of booking groups we need to work on getting better rates. Sales staffers need to get better in negotiating deals to make sure that the properties represented do maintain the increase in ADR that is expected. It appears that ADR from group bookings have always laid behind transient travel. That, of course, brings down the total ADR a couple of points for a property. Sales departments should be geared up to secure better group rates this year.

Here is where the General Managers need to include some training opportunities during the usual weekly sales meeting. The sales performance of the staff needs to be monitored to see what works, and what doesn't. We need to look at what procedures are being followed in getting prospects to make the BUY decision. Are there objections from prospects? How are they being handled? Many real objections are not rate related, but perhaps many objections come forward as rates, but are they real? Sales staffers need to understand the business of overcoming objections. Quite often we feel that because of a group booking we need to offer a rate lower than published rates. Why do we do this? We need to reduce the effort of offering discounts right at the beginning of an inquiry. There really should not be a reason for this. We may want to offer other valued benefits other than rate discounts. Send for a free copy of our "Overcoming Objections" tip sheet.

AN EASY WAY FOR GENERAL MANAGERS TO IMPROVE PROFITABLE SALES

November 2015

OK, we do know that the General Managers at hotel properties are kept very busy every day, and just about all day; however, maybe we need to think of what those folks were doing years ago that helped get repeat business, more business, as well as learn firsthand what is really going on at a property.

What we are talking about is let's see more of the hotel management and see what is going on "up front," in front of the guests. It used to be that we would regularly see top management pretty much visible to the guests in the house. Sometimes at the front of the lobby as guests enter, with a very nice smile and welcome greeting. Or even spend just a little time at the front desk to observe the check-in process, and how the staff relates to the guests. How about the guests seeing the General Manager, once a while walks on through the dining room during meals hours to greet guests, and of course observe the operation. These days we do not see much of this; why not? Guests are truly very impressed when they get to see top management being around in a property; it makes them feel good to get a welcomed "Hello" from the top dog.

Being visible does a darn good job on showing employees the value of smiling, greeting guests, trying to be helpful, etc., and it is catching. If the GM does it, so do the employees, and that is what creates the repeat business as well as referrals from existing guests to their contacts. If an owner, operator, General Manager or whoever the top person may be, cannot be available to see what's going on, and to greet guests, then someone else, an assistant or other members of the management team should be visible "up front."

High management visibility really should be required during certain hours on any day when guests are arriving, departing or using the services of a hotel property. It is important for guests to know that management cares. People do talk about this when they chat with friends and associates regarding their recent stay at a hotel. This helps makes people return on a future visit to the area. A business card is a very low-cost marketing tool. Management folks need to get used to passing them out to guests whenever appropriate; this also impresses the guests, and they do remember that gesture. Also, it is not a bad idea to ask the guests to return, and to send their friends and associates to the property.

Another important area to show visibility is group sales, when the sales department is working with a prospect, or on a site visit. Making an appearance for a welcome to show interest could be very valuable to the group sales effort. Meeting Planners often talk about never having seen a General Manager when visiting a property on a site inspection. This could be very valuable in helping the prospect of making a BUY decision.

CHAPTER 2

DEVELOPING LEADS AND PROSPECTS FOR HOSPITALITY SALES

THE PROSPECT SURVEY SHEET AND HOW TO USE IT

August 1980

A simple form for taking notes on the key questions asked is important when a salesperson goes prospecting.

In order to properly use the survey sheet (*see* the box next page), here is a step-by-step approach to its application. Feel free to use the form as it is or change it to suit your needs.

Company or Association: Write the correct company or association name here. Do not use abbreviations. Instead of "IBM" use "International Business Machines." This is important when setting up a file and follow-up system.

In using association names, it is important to identify the file as a local chapter, state, regional or national. For example, using the "Chamber of Commerce" is not enough identification. Since the call is probably being made on the local Chamber of Commerce, it should be identified as such—"Orlando Area Chamber of Commerce." Eventually, the files may include "Florida Chamber of Commerce" and "United States Chamber of Commerce."

Address: City address should always be complete, including zip code. If a post office box is used for mailing purposes by the prospect, it should be included, but a street address must always be shown.

Contact and Title: This is the full name of the person being interviewed. Try to get complete given name rather than initials and indicate Mr., Mrs., Miss or Ms., if preferred by the prospect.

The spelling of names is extremely important. Never hesitate to ask specifically for the spelling of a name. Get a business card from the person, and this will answer the spelling question.

Titles are very important and should not be assumed. The "secretary" being interviewed may really be the president, owner or executive Vice President. Ultimately, letters might be written to these contacts; therefore getting correct spellings and titles now may prevent any embarrassing moments later.

Phone Number: Include area code—it's a good habit to be complete.

Other Contacts and Titles: The person being interviewed should be able to provide

PROSPECT SURVEY SHEET

COMPANY OR ASSOCIATION _____

ADDRESS _____

CONTACT & TITLE _____ PHONE NUMBER _____

OTHER CONTACTS & TILES _____
_____

EXPLAIN TYPE OF BUSINESS _____
_____

MONTHLY ROOM NIGHTS USED _____ WHERE? _____ RATE _____

RESERVATIONS MADE BY _____

MEETINGS OFF-PREMISES:

WHEN? _____ HOW MANY ATTEND? _____

SPACE REQUIRED: _____

FOOD & BEVERAGE SERVICE: _____

WHEN IS DECISION MADE FOR MEETING LOCATION: _____

BY WHOM? _____

PREVIOUS MEETING LOCATIONS: _____

SOCIAL ACTIVITIES: (Retirement Parties, Christmas Parties, etc.) _____
_____
_____

BUSINESS ENTERTAINMENT: (Prospect of breakfast, lunch, dinner, lounge) _____
_____
_____

HOW OFTEN? _____ WHERE? _____

ASSOCIATION MEMBERSHIPS AND CONTACTS: _____
_____

REFERRALS: (Other Prospects) _____
_____

DATE: _____ MADE BY: _____

names and titles of other people in the organization who could be helpful in assisting in the development of business for the property.

Always ask for other contacts. A simple, "Who else in your company could be helpful to me in getting some business?" will probably get an excellent response. Many companies and associations have various departments that can generate business for a lodging property. Included may be the personnel department, sales department, advertising department, etc.

Explain Type of Business: This is part of "knowing your customer." A successful salesperson gets to know something about the prospect-a feeling about what the company or association does helps establish a rapport and leads to mutual understanding and respect.

The experienced, successful salesperson gets to know a lot about the prospect. There will be an opportunity to offer suggestions for conferences or social activities that could help meet the objectives of the company or association.

Explain briefly type of activity, such as "manufactures door stops," "new car dealership," "association of people who own flower shops," etc.

Monthly Room Nights Used: An assumption could be made that everyone has someone coming to town sometime and needs to have a room reservation. Many companies and association do not have their own people coming to visit them and may indicate that they have no one for whom to make room reservations.

However, they may have people from other companies visiting them to sell products or services. If this line of questioning is pursued, then a positive answer might be developed for this question. The answer can certainly be "none" and should be so indicated if that is the case.

If it is developed that room reservations are being made for out-of-town guests, get a reasonable estimate of the number of rooms used monthly. Then ask for the name of the lodging location being used and try to get the rate being charged.

There is nothing wrong with asking for the rate—so don't hesitate on this question and don't "beat around the bush"—come straight out and ask. In many cases, the contact will not know or may even refuse to answer. Don't push it.

Follow this with getting the name of the person making the reservations; the contact being interviewed may very well not be the one to make reservations.

Meeting Off Premises: Many companies and associations conduct meetings at their own office locations, but what needs to be developed is information on meetings held outside their own offices. Try to get specific times meetings are held, such as "monthly," "twice a year in summer and fall," "every January and August," etc.

Ask for an estimate on the number of people who attend and what requirements they have for meeting space (do they need exhibit space, how large a room do they need, how many different rooms they use at one time, etc.).

Always find out what food and beverage services they need: breakfast, lunch, dinner, reception, coffee breaks, etc.

It is important to know when and by whom decisions are made for meeting locations so the salesperson can adequately follow-up on this prospect. Previous meeting locations are good to know since the successful salesperson establishes good relationships with other lodging operators to verify details of previous meetings held at their properties.

Social Activities: Some companies, whether they have meetings or not, will have social activities. These activities may be handled through an employee club or various departments in a company. Always ask about Christmas parties, retirement parties, promotion parties, special events, showers, birthdays, etc. Companies are an excellent source of social business.

Business Entertaining: Business people usually entertain someone at one time or another. Try to find out something about this practice so that it can be further developed.

Association Memberships and Contacts: Just about everyone belongs to something. Companies and company executives belong to trade and professional associations—even associations belong to and are affiliated with other associations. Always ask for the names of these groups and the name of a contact, phone, and address for future follow-up.

Referrals: Successful salespeople get in the habit of asking for referrals. It never hurts to ask the contact for names of other people (not even in their own company) who may be in a position to give business to a lodging operation.

Sometimes the salesperson will get leads this way on someone getting married who needs a place for a reception. Or the name of a neighbor who happens to be president of an association. Or the name of a wife who is active in a local garden club. It never hurts to ask!

HANDLING THAT IMPORTANT LEAD: THE CUSTOMER INQUIRY

January 1982

Coulda—woulda—shoulda—didn't ... That's the old story when it comes to following up a sales lead.

Leads are available from a variety of sources, including referral services, computer printout services, mailing lists, Chamber of Commerce lists, newspaper articles, a variety of directories, old files, friends, neighbors, and chance meetings.

However, the hottest, most important lead is the inquiry. Unfortunately, this is also the most abused, least attended to and worst handled by most sales departments and hotel/motel/resort operators.

Lodging operators spend big bucks continuously on advertising, sales departments, promotions, and representatives just to bring in a qualified lead. The inquiry, whether it be by letter, phone or walk-in, means those expenditures have been effective; they've gotten the business prospect to make the inquiry. But too often the inquiry gets lost, misplaced, ignored, put off, and just isn't handled promptly or properly.

THE LETTER INQUIRY

The scenario often goes like this: Someone writes to the property for information pertaining to a prospective visit to the city, or expresses an interest in holding a meeting or social function. This inquiry, in many cases, sits around or gets passed around to different people for a response. The response is often just a brochure mailed back to the prospect. Or, if a letter is sent, too many times it's poorly written or a form letter.

The inquiry must be handled *immediately*. Make it a property policy that requests for information should be sent out the same day they're received. How urgent the inquiry is will determine how it should be answered. The person who writes to find out about one room several months in the future or wants general information about the property may be responded to by letter. However, the letter should be a personal, individualized response, answering specifics referred to in the inquiry. Keep the letter brief and to the point, and include a brochure or descriptive literature about the property and the area in which you are located.

If the letter inquiry pertains to a group movement where there is a potential for a sale of a block of rooms, *use the phone*. Most people are very favorably impressed with a long-distance phone call in response to a letter, so it helps in the selling process. The cost of the call is nominal compared to the potential revenue from a block of rooms.

Every effort should be made to close the deal while you have the prospect on the phone. You must be prepared to discuss the availability of rooms, meals, prices and a variety of specifics that may be needed by the group.

All telephone calls must be followed up with a letter confirming everything discussed.

THE TELEPHONE INQUIRY

This must be responded to by phone. The prospect went to the trouble and expense to call because the information was needed right away. Same day response by phone is essential.

Again, every effort should be made to confirm the piece of business, and the conversation should be followed up with a letter.

WALK-IN INQUIRY

Never let them get away! When prospects visit a property to make inquiries, you can be sure they're serious. Treat these people royally. Never make them wait in the lobby and don't send them into the dining room for a cup of coffee while they wait for you. Someone—sales manager, General Manager, food and beverage manager, whoever—must greet these people with a warm welcome and spend the time necessary to visit, look over the property, sometimes sample the food, review their needs and—hopefully, book the business. Following the visit, a letter needs to be written covering the details of what was discussed.

In every case of an inquiry, a follow-up letter must be written, even if the business was not booked. Once a prospect, always a prospect—you may be able to get them the next time.

DON'T MISS THE BUSINESS THAT'S RIGHT UNDER YOUR NOSE

May 1984

There's no telling how much business slips through our hands. In fact, we're getting pretty sloppy in looking after our business.

To start with, think of all the phone calls—many from good prospects—that don't end up amounting to anything.

Think of those that don't get answered promptly—or, worse yet, get answered with an "XYZ Hotel. Hold, please!"

Or those that are answered promptly and cordially and then transferred to an office where they ring off the hook until the caller, in desperation, hangs up.

Or those that result in messages that don't get passed along to the proper person promptly—or at all.

Or those that result in messages that make it to the proper person right away, but which aren't returned promptly—or at all.

Then, think of all the written inquiries that meet the same fate!

Really, do you like good business passing you by?

Stopping the Leaks

Other missed pieces of business stem from not knowing who is registered in the house. If a survey were ever taken on the subject, we would probably discover that, for every 10 people registered in a hotel at any given time, one would be a prospect for a new piece of group business.

There's sad irony in the fact that salespeople often go to great lengths to develop prospects—taking trips out of town, hiring lead service organizations, buying mail lists, conducting blitz programs, advertising in trade journals, etc.—but frequently miss hundreds of live prospects already on their properties every week!

Who at the property is responsible for looking over the arrivals list to see if any names, titles, and companies appearing there might be good leads for the sales department?

And who at the property reads the paper daily to pick out leads for the sales department to work? In most cities, a careful reading of a daily newspaper will result in a minimum of ten excellent sales leads.

Along with this line, here's an excellent game for General Managers to play with department heads: Give each a daily paper and a highlight pen. Then give them 30 minutes to carefully read the paper and highlight all the leads for business. The game is fun, very interesting and most worthwhile, with each department head competing to come up with the most leads.

Another source of lead development is your own staff. Sometime, when each department has its own staff meeting, a salesperson should visit and talk about sales leads. Let's face it: just about everyone belongs to some kind kind of group list or has relatives and friends list who belong to a club—from a bowling team to a church group.

Ask your staffers to suggest leads to your sales office. In our business, everyone needs to sell.

But the encouraging thing is the leads are all around us.

SALES OPPORTUNITIES THAT ARE MISSED

November 1985

It's happening at the switchboards and in the sales offices—telephones that are not answered promptly. People who call to do business should not expect to be kept waiting while the phone "rings off the hook." Friendly and courteous people handling phone calls at a lodging operation help make the sale. No one really wants to do business with a grouch. From time to time, salespeople should place calls to their own hotel or motel just to see how long it takes for the switchboard to be answered and then, how long it takes for someone in the sales office to answer once the call is forwarded. Do the people who answer the phone sound as if they are people with whom to do business?

How long does it take to return telephone calls? Salespeople need to get in the habit of responding to every telephone message immediately. Salespeople who have the reputation of promptly returning calls seem to do more sales than those who delay responding to messages. Consider every phone message a prospect for business. The name or company on the message may not look like a prospect, and it may even appear to be someone trying to sell something rather than buy. But, one never knows. A telephone inquiry is probably the hottest lead any salesperson can get. It shouldn't be ignored and a response shouldn't be delayed.

The same holds true for letters of inquiry. The best response to an inquiry letter is an immediate phone call—especially a long distance call. Most people who write to hotels/motels inquiring about meeting space or room blocks are probably writing to several locations. A quick response by phone is generally appreciated, gets questions answered and results in bookings.

Probably one of the most important prospects for group business is someone who is already a guest in the hotel or motel. Salespeople should establish an alert system with the front office manager. This should work out so that the sales department is advised of the "P" factor (prospective prospects) people who are expected arrivals each day.

When reservations are made, certain information is generally obtained which could be a good "tip-off" to a "P" factor person. A meeting of the General Manager, Sales Manager, and Front Office Manager could result in a list of factors which could help determine if an expected arrival could be a prospect for future business. Certainly, corporate executives, association executives, company salespeople, travel agents, tour company representatives, airline representatives would be a good starter for the list. Salespeople should take the time to explore this possibility with General Managers.

Of course, once the guest is identified as a "P" factor person, then contact has to be made to explore further whether or not that person is, in fact, a good prospect for future business. This may be accomplished with a nice note waiting for the guest upon arrival, indicating a warm welcome and a request to call the sales manager (or General Manager) for a very brief visit. A visit in person, or even by house phone, for only a few minutes could help identify whether that guest is a prospect.

Sometimes, a phone call to the guest's room, immediately after arrival will work just as well. What needs to be determined is:

1. the guest someone who decides or influences decisions regarding hotel accommodations for other people?
2. the guest involved in helping to determine locations for meetings?

3. the guest involved in a leadership role in professional, trade or social organizations?

A "yes" to any of these three questions reflects that this is truly a person that the sales manager or General Manager should personally meet. An invitation for cocktails or dinner or even breakfast the next morning would be in order. More details may be developed at that meeting, as well as a tour of the property facilities could be conducted. This is the time to get as much information as possible which will help in making a sale. Develop names and addresses of other people associated with this guest who may be in a position of placing business in the hotel.

Sometimes we spend too much time and money looking for business miles away—when the prospects are right there—where they can easily be developed.

PROFITABLE SALES TIPS

December 1985

Meeting planners are continuously amazed by the lack of professionalism displayed by Hotel/Motel salespeople in the performance of their jobs. From the handling of an inquiry to the solicitation effort through to the booking stage and follow-up, salespeople need to be better performers.

The Telephone: From the start, we tend to put people in a bad frame of mine—with regard to incoming calls. It all starts with the switchboard. Well-operated hotels/motels sometimes are reflected in the promptness in the switchboard being answered as well as the attitude of the hotel being displayed through the manner in which the call is taken. As a starter, people calling in may be "turned on" or "turned off" by how the switchboard is answered. Then, when the caller is connected to the sales department, the same opportunity exists to either impress or displease the caller. How many times does the sales department phone ring before it is answered—once and not more than twice, hopefully? Then, what is the attitude displayed by the secretary or salesperson answering the phone?

If the caller has to answer too many questions before he or she gets to the person wanted—the call (and possibly a lead or a piece of business may be lost forever). Too often a caller has to answer a quiz such as: "Who may I say is calling," "What is the name of your company" or even, "What is the nature of your call"—These could all be real turn-offs. If the call is for a particular person in the sales office—the calls should be taken immediately—without questions. Salespeople should take every call without screening. Sometimes it is much easier getting through to a president of a large corporation than it is to get to a sales manager.

Sales personnel should also examine how many calls they don't get right away—and how they are handled by secretaries. If salespeople are out on the street making sales calls—that is understandable by the caller. If the salesperson being called is busy with a prospect—showing the hotel or helping them plan a conference—that, too, is understandable. However, when a prospect or client calls and the salesperson is "in a meeting" with other hotel people—they should be interrupted to take the call. In fact, a good policy for salespeople to follow would be to have all in-house staff meetings before 9:00 AM or after 5:00 PM. All other time in between needs to be available for selling.

Another area regarding the telephone is how long does it take to return phone calls. Some calls naturally have priority over others. This is usually recognizable by the name of the caller or the company name or messages left. Many calls are made and messages left by what may appear to be "bothersome" such as people trying to sell something in which there is no interest. However, one never knows. The one call never responded to—which appeared to be a magazine sales representative trying to sell magazine advertising—could very well have been that person trying to set up a sales meeting for the people from the magazine. It's good to prioritize telephone messages, but at the same time, professional salespeople always manage to respond to phone messages right away—and always within 24 hours.

The salespeople at a property need to look at themselves and the manner in which they answer the phone—enthusiasm shows up very easily as does the lack of it. Then, they need to look at the sales secretaries and other people in the department and the switchboard. Let's all be enthusiastic and sell.

HANDLING THE BUSINESS THAT'S BEING CALLED-IN

April 1986

All of us in hotel/motel sales should wonder how much business we are losing because of telephone inquiries are not being handled properly. A good percentage of meetings and catering business comes through the telephone. Of course, it may have been initiated through a personal sales call or letter or advertisement; however, the prospect will generally call by phone to check on availability of space and dates.

Many times, whoever answers the phone in the sales office is not prepared to take inquiries or answer questions—and in many cases not qualified enough to ask specific pertinent questions; or perhaps not even trained to listen properly. See if this scenario is familiar.

Prospect: "Hi, my name is Howard Feiertag with Servico, and I'd like to know if you can handle a meeting for me on June 25th for fifty people."

Sales Office Person: "I'm sorry, I didn't get your name!"

Prospect: "Howard Feiertag—FIRETAGE!"

Sales Office Person: "OK, Thank you. When did you say you wanted a meeting?"

Prospect: "June 25th."

Sales Office Person: "Thanks. How many people was that for?"

Anyway, you get the idea! It can go on and on this way, and usually does.

The answer, of course, is for everyone in the sales office to be trained to listen carefully, ask the right questions, be knowledgeable and exude confidence. The prospect calling in must get the feeling that the person at the other end of the phone is a professional.

By using a banquet function sheet or other types of worksheet which covers all the questions that need to be answered—the sales office person will not only get all the questions answered, but will also sound very professional. Avoid using writing tablets and pads of blank paper—eliminate these all together and use only worksheets. Develop a proper worksheet form with questions on it that covers all situations—this way almost anyone can get the information on an inquiry.

The worksheet should include: prospect's name, title, company, address, phone number, name of the group for which the inquiry is made, dates of meeting or functions, number of sleeping rooms that may be needed, number of nights of stay, total number of people in attendance, schedule of activities each day, group functions needed (breakfast, lunch, dinner, coffee breaks, receptions), social activities, types of set-ups needed for each meeting; any exhibits, entertainment, demonstrations, type of people in attendance (all men, women, estimated age group, spouses, any children, special dietary considerations), budget available, who will make decision on location of the meeting, when can they come to visit the property.

Of course, this reflects an elaborate worksheet. It can be simplified, and it can be expanded. For just a local luncheon meeting all these questions need not be asked. The idea, of course, is to listen carefully to determine the prospects' needs.

If answers to the prospect's inquiry are not readily available, then the prospect is advised that someone will return the call very soon. The worksheet is then passed on to someone who can handle the sale. Before the call is returned all the information is gathered so that the prospect can get the information needed without having to answer the questions again.

Listening is hard work—and all available information about the prospect and the group should be recorded. It is important not to interrupt when the prospect is talking—very often the things they say and how they say them can provide good tips in helping to close a sale. Never feel uncomfortable about asking questions—the more information available, the better suited is the salesperson to close a deal.

Listening is one of the most important elements in closing a sale and yet one of the least seldom practiced skills in our industry. Listening, unlike talking, never lost a sale.

FINDING BUSINESS LEADS CAN BE EASIER THAN YOU THINK

June 30, 1986

Success in sales is often a result of developing good business leads, and some salespeople travel far and wide to track them down.

But the truth is, there are business leads right on your hotel's premises may be more than you'd find by going out of town.

Here are four excellent way to develop business leads without even leaving your office:

- **Current Guests:** People who stay at your property are prime prospects for other types of business, such as meetings, conferences or social activities.

   A quick check of the reservation cards will reveal which guests may be logical prospects for more business. Look for

corporate officers, association representatives, sales managers, personnel managers, etc. They may be in a position to decide where to hold their next meeting.
- **Employees:** Everybody belongs to some kind of organization—a church, garden club, bowling team, PI'A, educational group . . . the list goes on and on.

    Salespeople should take advantage of this by meeting with maids, bellmen, restaurant and bar employees—anyone who works at the property. Ask them for their help in providing leads for meetings and social events.

You even start an incentive program. Offer employees dinner for two at the restaurant of their choice, a weekend at the hotel, or a cash award based on the revenue generated as a result of the lead.

Morale Booster

There's a hidden benefit to having employees provide leads: improved staff morale. Almost immediately, you'll see employees become more involved in working together. Just having salespeople speaking to other staff members will do the trick. The message here is "Everybody sells."

- **Existing Files:** Probably the best source of leads is already at everyone's fingertips-right in the file drawer. Unfortunately, files get buried so deep after a conference or meeting that no one takes the trouble to book the group again.

    Going through a few files each day is a good way to generate business from groups or associations that have booked your property in the past. If they had a positive experience at your hotel once, there's a strong possibility that they'd be glad to return. All you have to do is ask them!
- **Newspapers:** Everyone reads the newspaper at some point during the day. But does anyone read them with the purpose of developing business leads? There's a difference between simply reading a newspaper and reading it with a specific purpose in mind.

    Try an experiment. Ask three or four people to review a local newspaper for anything that might lead to room business, meetings, parties, dining-room and bar business, etc. See who can find the most leads. You'll be astounded by the results. The "readers" will come up with more sales leads than you can contact in a day.

The idea, of course, is to train people to read the newspaper with the thought of looking for business. Once they do this exercise, they'll look for leads every time they read a newspaper, and new business will be popping up automatically.

Get Everyone Involved

Who does the reading for the leads? Everyone can do it. Or it can be assigned to different people on different days. Leads can be cut out and passed on to various department heads for follow-up. The sales manager should get the leads for group meeting business; the front-office manager should get the leads for local contacts on room business; the catering manager gets the leads for social functions; the food and beverage manager gets the leads for restaurant and lounge business.

And the General Manager's job is to make sure e system works.

THE MORE PEOPLE YOU MEET, THE MORE CONTACTS MADE

February 2, 1987

It just stands to reason that the more people you met—the more contacts made—that more prospects can be developed. Prospects lead to presentations, which lead to property visits (or vice-versa) and ultimately, more closings.... Which, of course, means more business for the hotel/motel. It's not that difficult to figure out, and the law of averages comes into play. The more sales call made—the more business booked.

It is so difficult to understand why so many industry salespeople won't (not DON'T, but WON'T) go out, making contacts, doing prospecting and looking for deals to make for their properties.

The old excuses are still around: "Too much paperwork;" "Other Jobs" imposed upon salespeople; "No one to take phone calls;" "No one to REALLY make sure the group coming in will be well taken care of;" etc. Those salespeople that don't get out can certainly add more excuses to the list.

Of course, there are those properties where people just walk in or call in to book business—and, someone has to be around to close the deal. If this is the best time usage of the property salespeople, then the General Manager needs to review with the salespeople, their job functions. It would seem logical that part of the job would be to develop a new group business, as well as book long-term, future business. Perhaps there are other people on the property that can answer the phone and take care of walk-ins that want to book business.

If each salesperson can make ten new contacts a day—each working day—this could add up to a lot of new business for the property (even if only 10% actually book some business).

There's still a place for telemarketing, direct mail, directory advertising, media advertising, sales promotion and whatever other methods there are to avoid outside calls. But, there's still nothing like the basic "face-to-face" visit that will make friends for the property, close sales, create referrals and develop long-lasting relationships for the hotel/motel.

These outside calls don't necessarily all have to be presentation calls. For the most part, they probably should be prospecting calls. Just calling on as many businesses as possible and qualifying them (finding out if they are prospects for business for the property).

Using one page, a brief questionnaire, the salesperson asks key questions (writing down the answers) to determine if that organization could use the property for any kind of meeting, parties, room business, etc. The entire interview need not take more than five minutes. The selling starts (and, the closing, as well) as soon as it's determined that a qualified prospect has been developed. This is when it really starts to get interesting—since more detailed information needs to be secured before the salesperson may be able to make a logical presentation.

It gets most exciting when a "cold call" develops such a prospect, and the selling techniques start to unfold and a prospect actually does get "sold" on the spot and a piece of business is closed.

It does happen—just ask around some experienced hotel/motel salespeople. There are some good "war stories" to be told. It's a very exciting and motivational experience to have this happen.

However, it'll only happen if salespeople go out and make it happen. Ten new contacts a day is not too much to ask of a sales person—maybe it should be 15 or 20.

GOING AFTER LOCAL BUSINESS WILL BOOST YOUR GROUP SALES

February 27, 1987

Is it really necessary to take those "sales" trips all over the country and to foreign countries as well? These trips certainly are not a waste of time—particularly participating in trade shows—where sales time can be very productive.

However, for most hotel/motel properties—80% of the room business (corporate travelers, as well as groups)—probably could be generated right in the "back yard." Unless the property is located in a very rural area—the salespeople (or manager, if there are no salespeople at a hotel/motel) should be able to generate most of the business through local contacts. Constant calling on local businesses—asking the right questions, qualifying, following up, inviting prospects to see the property—will turn cold calls into leads ... then, into prospects ... and then into closed sales.

Each local company than generates overnight, individual travelers is a prospect for meetings. Even local companies that don't provide the property with overnight stays are prospects for some type of group movement to the area.

Most business people belong to some type of organization—probably at least three different types—such as trade or professional; civic; fraternal; religious; or social. Getting the local business people to help bring their organizations to meet at the property is the approach that needs to be taken.

Just calling on the hotel/motel present purveyors should keep a sales department busy with the prospects that could be uncovered. Make a list of all the companies (and individuals) that do business with the property (such as bank, post office, lawyer, advertising agency, dry cleaner, plumber, electrician, air conditioning service company, contract cleaners, landscapers, pool maintenance, printer, all food products, beer—wine—liquor distributor)—of course, the list could go on and on.

Then, there are the institutions, such as: hospitals, clinics, nursing homes, high schools, colleges (all teachers in all departments usually belong to at least one educational association), libraries, and thus this list can also be enhanced.

There are also the civic clubs, and this list, with contact names, can generally be obtained through the local Chamber of

Commerce. The list will probably be pretty long, but each local club usually belongs to a state organization which is then followed with regional and/or national groups.

Any music story could probably provide a list of clubs that involve particular instruments (groups of pianists, accordionists, drummers, trumpet players and so on). These clubs usually have statewide or regional gatherings with sub-gatherings by age groups.

Local military reserve units also have meetings, especially active military recruiting departments. Veterans groups meet statewide at least annually, and these lists are usually available through the local reserve unit.

Just think of all the potential meeting business that can be generated locally—and all of the companies and groups locally that have not been called on yet. Then, think twice, before buying that airline ticket to . . . wherever.

CLOSER LOOK AT NEWSPAPER CAN DELIVER VALUABLE LEADS

November 23, 1987

As unbelievable as it may sound, hundreds of great leads are available weekly to hotel/motel operators and salespeople. And the cost for these leads is nominal.

The source is your local newspapers. Just for fun (and for a good opportunity to develop some business), conduct a contest. Ask each front-office salesperson (desk clerks) to read the same editions, on the same day, of a local paper. Ask each person to highlight what would be considered good sales lead for rooms or catering.

The person that identifies the greatest number of leads would get a prize of some kind.

The results will amaze you. Salespeople could be doing this as well; however, they may overlook some very basic information that a front-office salesperson would identify as a good lead.

GETTING CREATIVE

When lodging operators read a newspaper with the objective of searching for leads, the creative juices really start to flow.

In the business section, food section, society pages, sports section, classified ads, even in the obituaries, leads can be found throughout the paper. With careful reading and good concentration, leads almost seem to "pop-out" at you.

The newspaper is certainly the easiest and least costly method of getting new business for rooms, meetings, and banquets.

Here's what you should look for:

- new personnel assignments;
- executive promotions;
- company or club meetings;
- announcements of meetings that have already taken place at competitors' facilities or future reference);
- business firms relocating;
- sports teams coming to town;
- entertainment groups scheduled to perform;
- funeral announcements (indicating that people may be coming to attend the services);

- announcements of engagements and weddings;
- announcements of civic-club meetings; and
- auto dealerships announcing new-car showings.

With some creative thinking, the sales staff could add more items to the list.

Of course, getting the leads is not enough: Immediate action needs to be taken to follow-up with visits, phone calls, letters whatever it takes to qualify the leads right away.

This should be done quickly, since tomorrow's newspapers bring more leads, and they stack up quickly if not handled right away.

EXPECT 10 PERCENT

No one should expect all the leads to develop into a piece of business. In all probability, less than half may even end up qualifying-but even that amount will create a lot of work for the sales team. It's reasonable to expect that at least 10 percent will actually end up as bookings for the property.

Another opportunity for leads rests at the property level. Visiting the various hotel departments when they have their weekly departmental meetings can develop good, qualified leads.

Take housekeep as an example: Spend five minutes at their meeting and explain the importance of sales to the hotel. Then ask them to seek out leads among their neighbors, friends, and relatives.

This will bring results, particularly if there's a little incentive thrown in. Why not offer five percent of the revenue generated by any group suggested by any member of the hotel staff? All an employee needs to do is provide the lead to the sales department.

These days, just about everyone—including your hotel's staff—belongs to some organization: PTAs, church groups, garden clubs, and civic groups. Get leads there, too. Follow-up immediately, qualify the leads and do a good job of selling. And then, book the business!

COLD CALLING DOES WORK

January 1988

Much has been discussed about cold calls over the years, and it appears that more and more "marketing" types "pooh pooh" the practice. There are many arguments against cold call selling—some may be valid; however, it still beats most other methods of developing new prospects for business. The face-to-face method of gathering facts, establishing relationships and putting the business on the books may still be "old fashioned" to some people—but it works.

Some salespeople are fearful of cold calling because there may be some rejection—good salespeople know how to accept rejection. These are the types that would rather work within their own comfort zone—calling only on existing accounts and people they know. This is important also; however, the real function of the salesperson is to develop NEW business. "Ya gotta make calls if ya wanna get results"—so the old saying goes.

Perhaps if we didn't think of cold calls as being sales calls, it could be swallowed a little easier. Actually, the cold call is really a "prospecting" call or a "research call." The objective is to gather information to qualify a prospect. It must be determined during the call, if that particular company, association or group has any potential business for the salesperson. The process is very simple—the salesperson just asks a series of questions WITHOUT trying to sell anything. After all the necessary questions have been answered and recorded (create and use a fact sheet form)—the information received immediately indicates whether a prospect has been developed. Then, the selling may begin.

To maximize a salesperson's time in cold calling—the idea is to make as many calls as possible within a small area, so that time is not wasted in travel. The amount of time needed to get in and get out with information should not exceed fifteen minutes. It is most unlikely that a sale would be made while cold calling: however, it has happened and many hotel salespeople can relate some interesting stories about how it happened.

Some salespeople will say that it is a waste of time, or not professional, or not polite, or to make calls without an appointment doesn't look good. This is not a valid argument at all. It would be most unusual if any hotel ever lost any business because a salesperson made a cold call.

It would be most unusual if anyone ever became angry with a salesperson because of a cold call. The worst that can happen is that someone might refuse to see the sales person—or, they may suggest an appointment should be made.

Actually, it really doesn't matter if the information gathering process is developed through questioning the "boss" or the secretary. In fact, anyone at the place of business who can furnish the information is the person who could be interviewed.

One of the best ways to train new salespeople is to have them make cold calls every day. It gives them the practice in getting out and seeing people—and when salespeople see people—more selling can be done. It gives the new salesperson practice in penetrating tough situations. It helps the new salesperson develop good attitudes about selling. Salespeople being trained is to start and develop their own techniques in interviewing. They learn how to influence people (convincing a secretary to answer questions is good practice). Meeting new people and new situations are always good practice for new salespeople.

To be successful, a proper mental attitude will many times outweigh technical skills or ability. When it comes to selling, the first step in developing the proper attitude is to eliminate the fear or negativism toward cold calls. A positive position toward this method of doing business will work. It is probably the best use of a salesperson's time for developing new leads. If cold calling is not being practiced at a particular hotel or motel—it is suggested that a testing period be conducted. Just one week of calling on local area business firms—all day—by one or more salespeople—should result in 30 calls per day, per person. And . . . a minimum of 1/3 of those calls (10) will be new business prospects.

Try it . . . and let us know.

LET'S SELL MORE RECEPTIONS

February 1988

At most hotel/motel locations, sales personnel are selling rooms—and rightfully so. However, when it comes to selling group meetings, the sales department should also concern itself with improving the hotel's group bar sales. This should go hand-in-hand with the selling of the group room block.

Sometimes the group has not intended to have a reception or wasn't even thinking of having one. Here's where the sales department can assist and show them how a reception can enhance the entire meeting and help meet its objectives. In this way, additional revenue can be developed for the property.

Salespeople should always suggest the reception for a group prior to the opening of the meeting. It can be used as an "icebreaker" to get the people together. Then when the meeting starts the following day, attendees will already know each other and feel comfortable in each other's company.

From a profit standpoint, as well as a service standpoint, it's usually in the best interest of the hotel to sell a cocktail reception by per person, per hour system. This is where the meeting planner pays a flat fee per person who attends the party for a specific amount of time, usually one hour. The hotel then provides a completely open bar where the meeting attendees can obtain any type of drink (except perhaps cream drinks). The bar will serve a wide variety of cocktails along with beer, wine, and soft drinks.

From a bar cost standpoint—the hotel benefits because usually with this type of set-up, the bar cost is lower as compared with selling liquor by the drink or by the bottle. When selling liquor in this format, it's important that the cocktail reception start on time, never earlier than the designated hour and close on time, never later than the designated hour.

One very good selling point for this form of buying a reception is that the meeting planner knows exactly what the cost will be upon conclusion of the party. By multiplying the number of people in attendance by the price, he or she will know in advance what the cost will be. With other forms of setting up receptions, the meeting planner does not know until the end of the party what the cost will be, since there's no way of telling how many drinks the people will drink or how many bottles will be used during the hour reception.

The sales department should always consult with the catering department or the food & beverage director with regard to having the reception set up in this manner. Also, it's always in everyone's best interest to serve brand names rather than unknown names—even if it does cost a little bit more.

The food service during the cocktail reception could also be sold in the same manner—per person, per hour—and offer a continuous serving of food. An ideal way to handle this would be to provide butler or tray service.

This is when you have waiters or waitresses walking around the room continuously, with small trays of hot and cold "pick-up" food. The servers walk throughout the reception room, giving everyone an opportunity to select items from the serving tray. One server for about 20 guests would be appropriate when serving food butler style. This also gives the kitchen an opportunity to present a variety of foods which can be served in small quantities on individual trays. It gives the kitchen a chance to be creative in the items that are presented.

In pricing out the open bar, one-hour service, the average drink consumption would probably be 2.5 drinks per person, with an evenly mixed group of men and women. A satisfactory and acceptable highball will contain 1.25 oz. of liquor.

For pricing out the food to be served at the reception, it could be figured that if you have an evenly mixed group of male/female, each person will consume 8 to 10 pieces in an hour.

From the labor standpoint, to adequately service a group of 100 guests, two bartenders should be sufficient, considering that almost the entire group will arrive at about the same time and stay through the entire one-hour reception.

Salespeople need to keep in mind that the property must help the meeting planner meet the objectives of the meeting. Sometimes, the suggestion of a reception, a cocktail hour might very well not be in the plans of the group—and in fact, in some cases may even "turn them off." Therefore, salespeople need to be careful about how the proposal is made and they should be certain to know the background of the organization, the group, and the meeting planner before such suggestions are made.

Next to sleeping rooms, the most profitable part of the hotel business is in the sale of hard liquor. By suggesting the cocktail receptions, it will help maximize profitable sales for the hotel.

HOW TO GET AND USE REFERRALS

October 1988

The best sales pitch in any business probably is a favorable referral from a satisfied user. Somehow, I feel that there are only a small percentage of hotel/motel salespeople using this wonderful and least costly sales tool

Many properties keep a "complimentary letter" file and do use this as a referral tool. That's good. Generally, the letters come unsolicited—copies go to the boss, home office, parents and even the staff gets to read them. Then, they go into the file and hardly ever are retrieved to show to prospects.

Would it hurt to ask every group account to send a favorable letter that you may use as a reference for other prospects? Of course, not. There's nothing wrong with soliciting complimentary letters. When received they go into a binder that can be carried around on sales calls to show prospects when a doubt is raised about the hotel's ability to perform.

Now, how about taped messages? The taped interview with satisfied customers could be a great tool. This is one way to prove to someone that you're proud of the hotel/motel's record of performance.

Here's how it works:

1. Locate some accounts (three four, five or six) that have been happily doing business with you over a period of time. These are the people that you are in touch with on a regular basis and with whom you have developed a good, long-lasting relationship. They should represent

different segments of business such as: corporate account; meetings account; motor coach tour; travel agent; contract; etc. If you've done your job in a professional manner—if they are happy working with you—if you've made them feel important—then there should be no problem getting them to agree to a taped interview with you.
2. Set up the interview location. At the hotel/motel would be better than in their office. But, whatever the customer prefers. Explain the purpose of the interview and how it would be used. Ask them to think of some of the reasons they selected your property; how does your services meet their needs; what are some of the things that they like at the property (friendliness, food, location, rooms, etc.).
3. Now it's time to record. Start the tape and proceed; "My name is (your name), and I'm here today with (client name), who is (title) of (company). (Charlie), how come you selected our hotel for your meeting and please tell us how satisfied you are with your decision."

The idea is to make sure your interviewee makes some positive comments about the property or service. During the course of the interview, it's perfectly OK to bring in important remarks in the event the interviewee does not make mention of them. For example; "I remember, (Charlie), after the last meeting you mentioned how especially pleased you were with our banquet set-up staff. Tell me what happened that pleased you so much?" Of course, your remarks have to positively relate to an actual situation.

4. Following the close of the interview, ask the client for a reference letter relating to the comments made in the interview. Following this, a "thank you" note and a small gift would be appropriate for the client.
5. The tapes that are collected are used as a back-up for sales presentations and negotiation sessions. If a situation arises with a prospect that questions the ability of the banquet staff being able to handle unusual set-ups, then it's appropriate to play (Charlie's) interview.

The variety of interviews with different types of accounts with different situations will come in very handy to help with overcoming some other objections. There's nothing you can say that would be as powerful as the taped interview and/or letter of recommendation.

Even before a presentation is made to a prospect, a tape can be played to re-enforce the salesperson's position and the quality of the service or product. Or, a tape may be played at the close of a presentation to reinforce the credibility of the salesperson.

This is just another low-cost method of sales support that may be used in developing professional sales skills.

IMPROVE SALES WITH PROPER TELEPHONE PRACTICES

November 1988

There's no telling how much business is lost by Hotels/motels because of poor telephone practices. Probably millions of dollars annually move into competitors hands because we "missed the boat" by not handling a call properly.

The inquiry is the hottest lead any lodging property can get. Most inquiries are made by telephone, and unknowingly, we mess it up. The first shot we get at messing it up is when we don't answer the switchboard right away. Four . . . five . . . six rings could feel like an eternity to the person making the call. If there is no prompt answering, the call probably goes to a competitor.

The second shot we get at losing an inquiry is when the telephone is answered by someone who is intelligible because of spitting out what sounds like a "100-word greeting." Or, by someone who sounds like you've interrupted their morning coffee, or by a very unenthusiastic voice.

The third shot we get is when the caller starts to explain why they are calling and our operator does not know what to do with the call (this often happens when we use trainees, or a brand-new entry-level employee to answer the incoming calls.)

The fourth shot is when the call is transferred to the sales department or catering, and it takes almost forever for someone to answer. The switchboard operator never comes back on the line, and the inquirer finally hangs up.

Let's say someone does answer promptly in the sales or catering office, probably a secretary, who could very well explain, "There's no one here right now." Sometimes messages are taken, sometimes the caller is asked to call back "later." Sometimes the call may be transferred again to another department (front office or reservations) and, guess what? . . . A disconnect!

Anyway—the scenario can go on and on. Does all this really happen? Of course, it does! The bad part about it is that NO ONE KNOWS WHAT'S GOING ON. How often will a switchboard operator or secretary tell someone that a call was lost—even if they did know?

How can we clean up our act and not lose inquiry calls? We start with *not* letting inexperienced people answer the telephone. Before a person takes on that job, they should be knowledgeable about the property; know who works in which department; understand the working relationships between departments; be screened for good, intelligible, enthusiastic telephone voice—and understand the importance of answering the incoming calls promptly (never more than three rings).

The person responsible for taking incoming calls should only be responsible for directing the call to the proper department. This person should not be put in the position of answering questions—this ties them up on one call and delays the answering of other calls. If there is any department in the hotel/motel that will not have anyone on hand to answer the telephone, the switchboard operator needs to be informed as to who will be taking those calls. Or, there could be a call forwarding system installed to take care of this. In any event, if the department being rung does not answer in three rings, then the call needs to go somewhere to someone who could answer and, at the least, ask questions.

Perhaps we also need to place mirrors next to the switchboard. In fact, perhaps next to all administrative extensions in the hotel/

motel. The mirror needs to be placed in such a position that people talking on the telephone can see themselves and their expressions. Putting on a happy, smiling fact comes across just that way, by voice, to the calling party. I don't believe anyone who has tried this will ever say it doesn't work.

Let's answer our telephones promptly, with a smiling face. Be enthusiastic, speak clearly, transfer calls properly, make sure people are available to answer transferred calls, and take messages accurately. It's just that easy.

NOT LOSING ANY CALLS FOR <u>ANY</u> REASON = MORE INQUIRIES RECEIVED = MORE SALES.

KEEP PANNING FOR GOLD

May 27, 1991

If we were around during the days when prospectors were out seeking fortunes from gold nuggets, we would have learned a good lesson in sales. Prospectors knew that in seeking gold in creek beds that the nuggets would not jump right out of the creek into their hands. In order to find the gold, they had to pan for it, endlessly. They had to scoop up earth from the creek bed, in a metal pan, let the water wash away the sand and then pick out the gold from the gravel that settled to the bottom of the pan, if there was any to be seen.

Prospectors knew that each time they scooped up a panful of earth, it didn't mean there would be any gold found. But, they did know that if they kept on panning, they had better chances of finding those gold nuggets.

Aren't we in the same situation when we are looking for additional hotel sales? In fact, aren't we prospectors looking for the gold (the booking)? How many phone calls, letters, and knocking on doors does it take to close some sales? Quite a few, probably, but it has got to be done. It would be wonderful if those sales would just jump right out of the creek for us . . . it just doesn't happen that way. We, in hotel sales, are in fact, prospectors. And going out looking for business is prospecting.

Of course, there is a certain amount of business coming into a property without any direct sales effort. And, very often salespeople get tied down taking care of this business. But, any successful, professional salesperson knows without a continuous, concentrated effort made on prospecting, sales without a doubt, will diminish. Finding prospects to whom to sell is basic to a sales function. This business of prospecting, as in the old gold prospectors' days, needs to be an endless task. It needs to be done daily, on a scheduled basis.

There are unlimited lead opportunities for salespeople that need to be prospected. With the right probing (asking questions) the gold nuggets will be separated from the gravel. Salespeople need to take time and list all the lead opportunities that exist internally and externally at a property. There is a never-ending list that could be compiled. Regardless of the types of properties . . . from a one-star to five-star . . . from budget to luxury . . . whether downtown, airport, suburban, resort, offshore, conference center . . . and whether large or small, leads will always continue to surface . . . if you are looking for them.

Here are some examples of sources of leads that will generate names of people and companies or associations that need to be prospected as potential buyers of hotel services: Existing accounts (for more business or business from other departments of that account); guest registration cards; property staff members (they do belong to organizations and they have relatives, friends that also have connections); sales files which have not been looked at for a long time; people who are in attendance at meetings in your property (one out of three probably belongs to an organization that could have a meeting *or* function); suppliers of hotel products and services; daily newspapers (unbelievable amount of leads—but you gotta look for them like you are panning for gold); various meetings trade magazines (look for names and companies of meeting planners mentioned in stories and announcements); cold calls; sales blitzes; inquiries coming in by mail, phone and walk-ins and responses to advertisements and direct mail. There are also a wide variety of directories and mail lists available.

By now you get the idea. You cannot afford to wait for the gold nuggets to jump right out of the creek and into your hand.

Let's start doing some endlessly panning and watch the nuggets collect.

MISHANDLING OF INQUIRIES LOSES SALES

March 22, 1992

The problem with lost sales due to the mishandling of inquiries is that we never know how much business we have lost. There is no way of tracking and no way of following up if we never get the calls *or* the information from the inquirer.

Here's a typical example of how we lose business. Recently, in preparation for a training session I was going to conduct, I called eight hotels to inquire about holding a reunion. The positive side of the experiment was that in seven of the calls the phone was answered within two rings, with pleasant greetings announcing the name of the hotels. That was great! Six of the calls were routed to the catering department. This was what I expected since I had asked to "speak to someone about a party for a family reunion." On two of the calls I was asked for a name and FAX number so the hotels could send me information right away (only one hotel did follow-up with the FAX, the other did not). One contact did ask about the use of sleeping rooms. Only one of all the contacts sounded interested and enthusiastic about the reunion I was planning. All asked about the dates and proceeded to check availability . . . I had to hang up on one because I had to wait too long for someone to come back to the phone.

So let's look at what was learned from this experiment:

- Telephones need to be answered promptly, at all times.
- Four rings or more can lose a prospect.
- Routing the call to the correct department is a must then someone in that department must answer the phone promptly.
- All persons answering phones must do so enthusiastically and with "smiles."

- Name of caller, company, address and phone number should always be taken immediately upon contact.
- Always get to understand the purpose of the inquiry, what is needed and when it is needed.
- Do not keep prospects on hold while checking availability. You can always get back to them later on this.
- It is amazing how often prospects will switch dates if space is not available when they want it. Of course, they have to feel you are the right person to deal with and the property is the right site.
- Keep discussing the event, in a positive way and provide consultation. Suggest how the event may be accomplished at your property.
- Do less talking and more listening by asking open-ended questions.
- Steer away from offering rates and prices until the entire picture is developed.
- Be alert for tips on what the caller likes or dislikes.
- Discuss benefits rather than features when relating to prospect's needs.
- Always try to close the deal or at least offer to hold space.
- If a close cannot be made then find out when a decision will be made and by whom. Find out when it would be appropriate to follow-up . . . and then be sure you do.
- Get the inquirer to visit the property for an inspection. That would be your time to make a sales presentation. That is, if the caller is local or nearby or if the caller is planning a conference and would expect to make a site inspection.
- Be sure to thank the caller for the inquiry.
- Provide appropriate written communication immediately by FAX or overnight mail. If a FAX is sent also send it by regular mail.
- Create a file and trace card.

How does your operation stack up when telephone inquiries are made? Are you losing any good, hot prospects?

THE CHECK IS IN THE (PHONE) MAIL

April 6, 1992

Some love it . . . others hate it! Whoever dreamed up the idea of phone mail really came up with a winner. The system certainly does the job in solving many communications problems and could be a time saver for some, but it also appears to be a "pain in the _____" for many. Frankly, I love it. From my standpoint, it does the job it is supposed to do. When I'm away from the office—and since I have no secretary—it takes my messages so that I can respond the same day (most of the time) to any callers since the messages are retrievable from any telephone in the world. And, when I return calls, if the other party is out-and-they also have phone mail, I can respond to the message left for me. A wonderful time saver with no wasted time with "chit-chat." For the same reason, I love phone mail installations for guest rooms in hotels.

So, what is the problem? Well, it seems there are some phone mail installations that try your patience. These are the ones where the system provides a message that seemingly goes on forever giving you options 1, 2, 3, etc. Some don't even give you a miscellaneous option or the opportunity to speak to a real live voice. You can really feel like a prisoner with absolutely no control whatever in being able to talk with someone—anyone!

A real disaster, for any lodging property, in my opinion is to have phone mail hooked up to a sales or catering or reservations department. Believe it or not this does exist. Here's the scenario: A prospect (meeting planner, travel agent, etc.) have a real immediate need for a resolution to a problem and needs to be in touch with someone to hold rooms or space for a particular date. The property telephone operator receives the call and transfers it to the appropriate department. After four rings the phone mail system comes on and advises the caller either: (1) "No one is here right now, etc., Please leave a message, etc." or (2) "We're on the phone right now, etc." If there is no urgency the caller might leave a message. However, we will never know how many good inquiries are lost by prospects hanging up and calling a competing property.

It seems to me that if we are in the business of selling, surely there has to be someone available at a property to take calls personally—particularly in the departments that are responsible for booking business like reservations, catering and sales. Of course, there are times when everyone in a department is on the phone or out of the office. Wouldn't the telephone operator taking the call, initially, be aware of this and forward the call to some management person on duty? What about using a telephone call forwarding" apparatus?

I think courtesy and good business practices dictate that when you are in the business of selling a product, a caller should be able to speak to a real live person even it has to be the GM.

THE HANDLING OF INQUIRIES . . . WE DO IT ALL WRONG!

August 16, 1993

Generally, we do a very poor Job in our industry of handling inquiries that come in by phone. There is no practical way to measure conversions (inquiries to bookings). I challenge anyone to be able to tell us how much business is lost because of poor handling of telephone inquiries. It all starts with the phone ringing at the switchboard, or the front desk, depending on where and who answers the telephone for the property. No one can really tell us how much business may be lost to competition when we do not answer the phone promptly. These days many of us are too impatient to hold on to a phone through more than four rings.

So, let's say the phone does get answered at the switchboard or front desk within two rings. That's great! Now what happens? The caller states the situation, such as: "I'd like to talk to someone about a family reunion." Where does the call get forwarded? To sales? To catering? Or to reservations? Unless appropriate questions are asked we could be shoving the inquirer from department to department, again, perhaps losing the call or even losing the confidence of a prospect. The next question is . . . even if the call does get directed to the right department, how many rings will it take for someone to answer the forwarded call? What if the line is busy or there is no one available to take the call. Do we tell the person to wait, or call back, or do we ask for a name and number? Again, do we have another chance of losing a prospect?

OK, let's say someone does take the call in the sales or catering departments. Here is where we have a big problem. Based on my experience of having made hundreds of these types of inquiries over the years, for training purposes, I have found that generally we do a poor job. If I was really trying to book something, like a family reunion, I would not have booked into over 90% of the properties that I called. Here are some of the reasons:

- Put on "hold" too long while waiting for a sales rep, when a secretary answers the phone and tells me I have to wait to talk to someone else.
- Waiting too long on "hold" while being transferred to another department since "we only handle the function rooms and you have to talk to someone else about sleeping rooms."
- Salespeople and/or secretaries are most unenthusiastic about my call. Sometimes I feel like apologizing for having interrupted them.
- One of the first questions I get asked is about "dates." When a date is given I'm on hold for a very long time while someone checks the book.
- If I ask about rates, again there is a long wait until someone gets a hold of the rate sheet.
- Very seldom does anyone invite me by to see the property (they do not even know I'm calling long distance since they haven't yet asked for my name, address or phone number).
- There is hardly ever a suggestion to "hold the space." Usually after all my questions are answered there is a long pause. I guess some people are fearful of asking for the business. When I get the pause I say "thank you, I'll let you know." The response is then "OK thank you for calling, let us know if you are interested." No request for name, address, phone number, etc., so someone could follow-up.

So, if all this does take place on a daily basis at most properties (and, by the way, the five-star hotels are not any better at this than all the others) how do we know how much business has been lost?

Work on a program at your property to eliminate all the possibilities of losing telephone inquiries. Find out how you are now doing it and come up with some solutions. All you need is a common sense approach to doing business. Call your own property every once in a while to check out your procedures. Once you have to wait for four rings before someone answers you'll know what to do.

MISHANDLE THE INQUIRIES AND LOSE A LOT OF BUSINESS

September 6, 1993

It's hard telling how much business is lost daily because we do not take care of inquiries properly. It all starts with how the inquiry is received. For the most part, inquiries are made by phone. Of course, there are some that come in by mail and others come walking in through the doors.

Inquiries have to be handled fast. When someone comes through the door and asks about rooms or *a* meeting or a catered function, we just have to be sure that someone gets to speak with them right away. If the salespeople are out then just make sure there is someone there to make contact. There is nothing wrong with even the General Manager taking care of this. In the case of queries coming in by mail, these need to handled the same day. A phone call response is most appropriate no matter where the mail originated.

The biggest problem we have is with the telephone—when someone calls into the property and asks about room rates *or* space availability *or* planned meal functions. Whatever your system of taking phone calls . . . at the front desk *or* with a switchboard operator, no matter; however, it is done we need to be sure that the phone is answered within two rings. Once the phone is cheerfully and enthusiastically answered the call needs to be directed to the right department. Depending on your operation you need to put it through to reservations, catering *or* sales. What happens when no one is available to answer the transferred call? Does it ring off the hook? How many prospects have hung up the phone waiting through too many rings?

The idea is to develop a system at your property which makes sure that all inquiries are taken care of properly. This means there has to be someone available at all times to take the calls. It is reasonable to understand that the sales staff cannot be sitting around waiting for the phone to ring. However, certainly there is someone on property that could get specific information such as who is calling, full name, title, company name, address, phone number, and then get into the specifics of their needs such as dates, number of people, number of rooms needed, etc. The inquirer needs to be advised that someone will call back within a specified period of time and then make sure someone really does call back at that time. If the caller needs to know something right then and there, we just have to be sure that we can find someone to take the call . . . how about the General Manager?

Even if the call does get to a salesperson or a catering manager one of the problems we have is putting the caller on "hold" while we "check the books" for space availability. We need to find out specific needs, show how the property can meet those needs, be enthusiastic about taking care of the caller and then advising that rather than put them on hold while space availability is being verified, "we'll call you right back." Then, of course, be sure to call right back.

To help reduce the possibility of losing business from callers be sure that everyone near a telephone is able to answer questions correctly and is capable of asking specific questions that are necessary so that someone call follow-up with the caller later. Provide scripted information sheets to everyone that has a phone. This way just about anyone can take a call and know what questions to ask. Train telephone operators (whether front desk people or switchboard staff) to answer calls promptly, properly, and enthusiastically. We do not want the caller to feel that they have interrupted someone. Here is a chance to promote goodwill and set the mood for your salesperson or whoever will respond to the inquiry. We need to be sure that we let the prospects know that their inquiry is appreciated.

The next place where we drop the ball is with the follow-up. After the details are developed and the conversation is concluded we need to be sure that we have some written communication sent to the prospect. Whether it is a proposal, or confirmation let's get it to them right away; by fax or overnight delivery or by regular mail.

I don't have the statistics but I'm willing to bet that over 50% of people who make inquiries at our properties either don't get through to the right person, hang up because they are "on hold" too long, get disconnected, aren't received enthusiastically, don't get called back or never receive the information in the mail that was promised.

GOOD PROSPECTING INTRO LETTERS CAN HELP YOU IMPROVE YOUR SALES

February 1994

Aside from a good job, we do in constructing and writing solicitation letters and proposals, how often do we take the time to write prospecting introduction letters? Don't forget, we're not talking selling, just prospecting . . . that is, gathering information to determine if someone or *a* group is a prospect for business at your Inn, Motel, Hotel, Resort, Conference Center, Attraction, Cruise Ship, Restaurant or Destination. Yes, we can prospect with letters. But here we are not discussing the use of pre-prepared, computer-driven, formatted letters. People know when you are using that type of letter and they figure you've just sent out 1,000 of them so you are really not interested in them, personally. And, we're certainly not referring to the standard letter that accompanies a form for someone to complete to determine if they are a prospect ugh!

Here's what I mean. In a recent issue of Corporate Meeting & Incentives magazine, I found 19 suspects for group business. That is, there were articles that contained information on 19 different companies reflecting they had meetings, the location of the last meeting or incentive trip and the name of the person in charge.

In every meeting, type publication you are bound to find these types of articles where meeting planners are discussing what they do, how they do, where their groups have been, etc. If you are getting these types of magazines, and you should be getting all of them, since you are probably a good advertising prospect for any of them, you 11 have an unending listing of suspects (leads). When I was working in hotel sales I would jump all over this kind of stuff.

Here's how you approach it from the personal letter aspect. You can clip the article and include it with a brief letter that goes something like this: "Dear Charlie (or Mr. Smith) . . . enjoyed your comments in the January issue of _____ magazine. You were really on target by you what you said about _____. Would like to get to know you and your meeting wants and needs. Will call you the week off and hope you can spare a couple of minutes to chat with me." Then, of course, be sure to follow-up with the promised phone call. You see, this is no longer a "cold" call that you'll be making, it is almost an appointment call and Charlie will be expecting it. It will only take a few minutes when you make the call to determine, if, in fact, Charlie is a prospect for your property.

How about your attendance at trade shows or conferences where meeting planners are attending? How do you follow-up with those "suspects" you didn't get to meet? The same type of letter will work. "DARN IT, Charlie (or Mr. Smith), missed meeting you at the _____ conference (or trade show, meeting, etc.).

Sure would like to get to know you and your meeting wants and needs. Will call you week of _____."

Or how about the person you ran into, met briefly, but didn't get to talk enough and to find out if the person was a good prospect for your property? "Sure was delighted, Charlie, to meet you at the conference. Sorry we didn't have more time to chat. Will call you the week of _____ to get to know you better and find out about your meeting wants and needs."

Too many of the prospecting letters I have seen are burdened with product information and selling copy. The only time you really want to sell is after you have found out that the contact is a real prospect and that you've done your homework and have gathered all the information necessary before you get into the selling mode.

PROSPECTING FOR NEW BUSINESS IS A NEVER-ENDING JOB

June 1994

Wouldn't it be wonderful if every guest we've ever had in our properties and every piece of group business returned time and time again? It's the ideal situation and it would probably do away with all the sales jobs that we now find necessary in hotels, motels, resorts, conference centers, cruise ships, etc. It's just a fact that all the business we've ever booked just doesn't come back time and time again. So, it is important that we follow David Green's "three r's: recruit, retain, rebook." That is: recruit new business, retain the business we've got, and of course, rebook the business we've lost. Sounds simple? Sure! Providing the right service should retain the business we've got and going back after business we've lost is kind of obvious. It's the going after new business, recruiting, is what prospecting is all about.

One of these days we'll all be into database marketing—that's David Green's thinking on the Sales Office of the Future; when salespeople will not have to do any prospecting, but will have all the information necessary so that all their time is spent in the selling process rather than prospecting. Great! And it is even happening right now with some properties and chain operations. But, for those who are not or cannot get into database marketing how do we go about recruiting new business?

The whole idea about prospecting is to find out who could use, or needs your particular property for overnight stays or meetings or group incentive travel programs. It certainly is a lot easier to identify meetings, convention and incentive groups of all sizes than to hunt down individuals. From a sales point of view it may also make more sense to spend our time going after this group market, for business when we need it, than to try to identify individual travelers. For the most part, one of the best ways to go after the individual business traveler is through a corporate travel manager at corporations. But for now, let's talk about business groups.

What we want to do in prospecting is to locate companies and associations that have meetings that would fit our profile. There's no point looking for a group of 500 when we can only accommodate 50 in our meeting rooms. So first, we need to identify what it is that we are after. Even the very small motels, with limited meeting space, should be going after this meeting market. These days there are more people attending conferences in groups of 35 or less than in all the other meetings put together in any single day. The small meetings market is big business.

You can identify the market that is best for you in several ways: Look through your own files. We have to agree with David Green who states that 80% of your business comes from 20% of your customers. You can also go out in your neighborhood and knock on doors. That is, calling on local businesses to find out if they have meetings or know of people who have meetings. There are also a wide variety of directories that may be purchased that lists associations, meeting planners, etc.

We know of three directories that could be helpful. There is the Directory of Corporate Meeting Planners, which lists more than 12,500 companies in the U.S. and over 400 Canadian companies. The listing provides information about when meetings are held, how many attend, the area of preference for meetings and the name, title, address of the person planning the meetings. The listing is made geographically, by location of the company. So if you have a property in Virginia you'll probably want to start contacting the meeting planners in and around Virginia first.

The Directory of Association Meeting Planners and Conference Convention Managers is a listing of 8,000 national associations that hold off-site conventions, seminars, and small meetings (under 50 people).

There is also *The Directory of Premium Incentive & Travel Buyers* which will identify for your companies that use group incentive travel programs. It includes travel destinations, group size, and average cost of programs that have previously run. This would work well for resort properties, offshore properties, and cruise ships. All three directories are published by *The Salesman's Guide*. The guide will also provide mailing labels by geographical area.

THE INQUIRY IS STILL THE HOTTEST LEAD YOU'LL EVER GET . . .

July 6, 1994

. . . and we still seem to mess it up. Time and time again we've made phone calls to properties and had meeting planner friends write to hotels about booking a meeting and the situation just doesn't seem to get any better. You would think that in this day and time, with big bucks being spent for sales and marketing efforts and group business just about being needed by all types of properties, we would be doing a better job of taking care of inquiries.

Here's what going on (and it ain't much different than what was going on thirty years ago—even with all the new technology we've got): The Walk-In . . . it still happens, particularly when a local catering event is being sought. How long does the party have to wait in the lobby until someone from sales or catering, or anybody, decides to talk with them? A lot of wedding type walk-ins happen on Saturdays. If there is someone available from sales or catering—a great shot to get the business. But, what if no one is around? When prospects walk to another property?

Mail inquiries still come in from time to time, when some meeting planner or social group planner is not in any hurry to make a decision, but does want information. Most responses get back to the inquirer by mail, without a cover letter. If a cover letter is included it is always just too general and obviously has come off as a pre-programmed computer piece. Sometimes it is just a brochure and maybe some floor plans are included as may be some banquet menus. Very few properties have someone pick up the phone to call the inquirer. Of course, the best shot you have to make the call as soon as the inquiry is received. This way you have the opportunity to gather more information, get some dates and maybe even get a tentative booking, before you even send anything to the people inquiring.

When I am checking on reservation offices to see how they handle individual requests, there seems to be no problem when I go through a central reservation office of a chain operation. These people who handle the 800# phones do a great job in gathering information, quoting rates, and trying to make the booking. One of the saddest times I have is when I call properties to try to get information about booking a room.

Generally, there is no enthusiasm, no energy, no interest in my call. Sometimes the phone doesn't get answered right away, sometimes (and I'm calling long distance) I'm asked to "hold.' Most of the time there is *never* an effort to find out my needs or a try to hold a room for me. I'll get a rate quoted and then there is a long silence . . . I guess someone just waits for the caller to make the next move. Many times, particularly at resorts, the person answering the phone will give me a long list of amenities that are available at the property, without even knowing my interest. Just about everyone is nice and I always get thanked for calling. Every once in a while I get thrilled when I find a real professional actually finding out my needs, being enthusiastic and wants my business and tries to make the reservation for me.

When it comes to trying to book a group . . . it kind of gets worse, sometimes. If the call gets transferred it may take some time before it gets answered—then maybe, there may not be anyone around to answer my questions about availability for rooms or meeting space. Most do ask me to leave a number for someone else to call back, which is good, but there are times when no one ever calls back. We get a lot of different people responding to the call: salespeople, catering folks, assistants, receptionist, secretaries, all who are generally nice, but lack sales skills as well as property knowledge to handle the call properly. Most offer to send or even fax information, which is good.

It may not be a bad idea for General Managers of properties to set up their own training program to cover the area of how to handle an inquiry. Maybe every couple of weeks a one-hour session could be held with all the front office, reservations, sales, and catering staff discussing this and helping each other do a better job. It's not difficult to establish some standards.

The most uncomfortable feeling about all this is that we never know how much business is lost because of not handling the inquiry properly.

PROSPECTING IS AN ONGOING JOB FOR SUCCESSFUL SALES DEVELOPMENT

July 5, 1995

Unless your property is running full all the time you should come up with a program to make sure that someone at the location is doing some prospecting every day. Under normal circumstances, it may very well be the people assigned to the sales department who are responsible for bringing in new business. However, at limited service properties where there may not be any staff assigned to the sales function, then it should be someone ... the manager, front office staff, night auditor (yes, why not?) or just about anyone who can spend a couple of hours a day looking for prospects. Why not some part-time students from a local college?

Prospects are people or organizations that may be in a position to provide you with rooms or meetings or restaurant business, but the problem is we do not know that they are prospects until we make contact and ask specific questions. So we have to find these contacts first and here's how we do it.

Your own files: Somewhere on the property you've got files on people or groups that have had business with you previously, but they may not have been contacted again. There may even be some names of those that made inquiries, but never booked anything. Search out these files and information and start calling.

Your own staff: You'll probably be surprised when you start asking questions of your staff about who they know, the organizations to which they belong, their spouse connections, etc. Just your housekeeping staff will probably belong to more groups than you would ever expect ... religious groups, school organizations, garden clubs ... a variety of special interest activities. And don't forget ... everyone has friends, neighbors and relatives.

Your regulars: They stay with you on a regular basis and are in a great position to refer individuals and companies to you who may be prospects. You've just got to ask.

Current guests: Just "eyeball" the registration cards each day and see who's there. Information on the card may be able to tip you off if your current guests are prospects for future business or have connections with companies that might be prospects for other business opportunities.

Your vendors: If you take a look at your own accounts payable list you'll probably be surprised at length, and shouldn't some of these be prospects for business? After all, you are doing business with them why wouldn't they not only do business with you, but refer their contacts to you?

Meeting participants: There's a pretty good chance that one out of three people attending a meeting or an event at your property may be a prospect. Whether it is a local civic club event; or a company meeting or social activity . . . find out who is in attendance to make contact.

The cold call: Don't be afraid of it . . . it's still probably a pretty good way to find prospects. Even if you can only devote an hour a day . . . you can do this while on the way to the post office, or bank or insurance office or even when shopping.

The newspaper: Probably the easiest way to find contacts that may be prospects and certainly the local paper is the best source. In any daily paper you will find many opportunities for prospecting . . . in just about every section of the paper from sports, to business, to social, the classified, even in advertising you will find tips on whom to follow-up. You should learn to read a paper with an "eye" towards looking for prospects. With a little practice, these will pop out at you.

Anyone of these opportunities for prospecting may be accomplished by just about anyone on the staff . . . in fact, be let the salespeople do the selling (if you have anyone assigned to sales) and get you and your staff to do the prospecting. Once you've found a likely prospect all you need to do is the typical six questions of: WHO, WHAT, WHEN, WHERE, WHY, and HOW to find out if, in fact, these are people or organization who can do business with your property.

SOME TRICKS OF THE TRADE TO INCREASE PROFITABLE SALES

March 6, 1996

What is it that some salespeople have done to make them stand out from the rest when it coming to building profitable sales? Mainly it is probably knowing where to go to get the business, how to develop prospects and then, of course, asking for the business. The idea is to drive business from new accounts as well as repeat clients.

Here are some down-to-earth, and not-so-secret, but essential tricks of the trade that successful salespeople must practice in order to succeed:

- Don't count on all your business to come from your advertising or direct mail or property promotions. Of course, everything helps build awareness, but ya gotta do the prospecting the determine who could use your facilities.
- Prospecting is an everyday endeavor. Individuals and groups who could use your property are located right in your own backyard, for the most part. Of course, this would not hold true for destination resorts or off-shore properties, but sales personnel for those types of properties need to prospect also.
- Leads are all over the place . . . start with your old files; get leads from employees from other departments at the property;

- get referrals from existing accounts; your daily newspaper will be loaded with leads each day; your property accounts payable list; trade publications may very well unearth some leads for you; then of course, you've got neighbors, friends, and relatives.
- Many leads will develop into prospects if you make the inquiries and ask the right questions. You will want to find out if your property can meet any needs of the individual or group.
- Typically, ask the "Who, What, When, Where, Why, and How." The answers will tell you if the leads become prospects. For every 10 new contacts (leads) you should develop three prospects for business. Try it see how it works for you. The more contacts made the more prospects developed, the more business booked.
- Understand your markets not every hotel, motel, conference center, is suitable for every market segment.
- Be sure that your marketing plan includes a sales plan for each market segment along with goals and strategies for going after each market.
- Strive to learn how to sell better and you should provide training so that your sales team sells better.
- Measure the number of inquiries for business against how many you were able to close. Keep a daily record, a ratio of closings to inquiries . . . See where you stand and set a goal for improvement.
- If you are not working with a sales representative firm seek one out and see how they can be helpful to you. There are many all over the country. Some work on a monthly fee basis, plus commission, some work on commission only.
- Who does your advertising? If your sales personnel are involved in writing advertisements and placing them, or writing brochures and promotional material it's probably a mistake. Seek out a professional advertising/marketing organization and turn it over to them. It will cost a fee, but will probably be more productive for you. Let salespeople sell.
- Networking is critical to salespeople. Participate in industry events, depending on the markets you may be seeking. Be active locally with the Chamber of Commerce and/or Convention Bureau. Become well known in the community. People will want to do business with you if you are well know and respected.
- Don't feel you have to close very piece of business. Sometimes you will need to walk away from a deal if it may not be profitable or in the best interests of your property.
- Understand this business of negotiating and prepare yourself to negotiate effectively when you are trying to close a piece of business.

EVERY SALE BEGINS WITH A PROSPECT

May 20, 1996

If there is no prospect, there is no sale. It doesn't matter whether you are making cold calls, appointment calls, having site inspections or answering inquiries, until someone is sold on putting business into your property, that someone is only a sales prospect. Not until they are sold do they then become a customer. So, in order to make a sale we have to have prospects.

Prospects are developed through leads which are developed through many sources. We get leads by reading the local newspaper; they are referred to us by our own staff; they may be referred to us by existing customers; we find leads in our old files that have not been traced in a long, long time; trade publications provide leads when articles reflect names of companies having meetings and the name of the contact; leads are developed through our own advertising efforts with ads in publications or direct mail programs; taking a look at your list of vendors will provide leads; front office staff can develop leads for you when they scan the reservation cards of expected arrivals for the day; some properties use a business card collection box at the front desk or in the dining room for a give-a-way program; many times persons attending local meetings at your property are leads for business; cold calls either by phone or in-person create leads; local sales blitz programs can get you a large volume of leads. The opportunity for developing leads is unending, but remember these are only leads and they need to be prospected. That is, every lead needs to be contacted and asked specific questions to determine, if in fact, the person or group may be in a position to use your property. Every lead will not turn into a prospect for business, but unless you ask the questions you will never know. Across the board in a variety of situations about three of every en contacts will produce a prospect. And one of the three will probably book something with you within six months.

One of the better techniques of lead development that will provide a better percentage of "hot" prospects is referrals. This is where you ask your current customer for referrals—people they know who may be in a position to book business with your property. Be careful to use open-ended questions—that is, questions that require more than just a "yes" or "no" answer. Stay away from questions such as: "Do you know anyone that could use our hotel?" Even if the answer is "yes," you still haven't gotten any names . . . so how about asking: "Who do you know that . . . ," etc. It leaves it wide-open for your customer to think and choose. And, asking the customer to recommend your property is not getting you any leads, so there is no follow-up for you. Further, it even isn't a good idea to ask your customer to come up with a list of names of people who may be in a position to use your property. This is putting the workload on your customer rather than on you and that's not a good way to do business.

All you really want is one good lead from an existing customer. Start with: "Who do you know that . . ." and continue with focusing your question around a benefit that you may provide for the lead. Here's an example. Let's say you have a parking area that is exceptionally large; the largest of any hotel property in the area. Your question to a customer for a lead referral might then be: "Who do you know that has meetings which require a lot of parking for their participants?" Get the idea? You are offering a solution to someone that may have a specific problem. Tell your customer how you can solve a problem for someone that may be referred to you.

Referral leads should get you a higher ratio of prospects than any other type of lead development.

GIMME THE FAX PLEASE, JUST THE FAX

September 16, 1996

In the July 22nd issue of HMM, in our Sales Clinic column, we offered readers a copy of a Prospect Survey Sheet form which could be used in prospecting or taking information from an inquiry. The response by fax was great ever 200 requests. However, I was amazed by the lack of quality of the information in many of the fax transmittals. Technology is great! We often wonder how we got along prior to the fax machine. A great way to move a message, next to the email. But, our problem is that too many people do not understand how to make effective use of this technology. The transmission itself, we do not even have to understand how it works. You stick a letter or pages of information into the machine properly—punch in the correct (we hope) telephone number — push a "send" button—and it gets to a receiver machine almost instantaneously. So what's the problem? People are the problem (isn't it always?). It is what we put into the machine.

On the faxes, I received the most were legible and understandable. A good many, however, were too difficult to read and were frustrating when we wanted to respond. Here' what we found: On fax transmittal cover sheets:

- Too much "cutsie" stuff that didn't make sense and left little room for notes.
- Too much property promotion information that had nothing to do with the message.
- A few had the name of the hotel, but no address, location or phone number—unable to contact the sender when the fax number didn't work.
- Hotel name, address, phone number and fax number too hard to read. The form itself was probably several generations old by the original form having been copied over and over again.
- Many had hand-written notes that were undecipherable, especially the signature. With no typed or printed name we didn't know to whom to address the response.

If you are using a fax machine to communicate with prospects, to solicit or confirm business, it would probably be in your best interest to take a look at what you are using for a fax cover sheet. Here are some thoughts:

- When sending a typed letter—using your own letterhead—you do not need a cover sheet.
- Use a cover sheet when the information being transmitted is not a letter. Maybe even make your letterhead into a cover sheet.
- When it is necessary to put information on a cover sheet, have it typed or printed in block letters, so that it is very readable.
- You do not have to use a signature or any cover sheet print or type your name and title.
- Use a cover sheet designed specifically for fax purposes and reproduced so

that the property name, address, location, phone number, and fax number are prominent and easy to read.

We always want to look good whether we are soliciting, booking or confirming business or even just communicating with friends. We surely want our faxes to look good too—after all, the faxes represent us.

By the way ... we also need to look great when we leave phone mail messages: This has been discussed previously in this column, but it is worth repeating. When leaving messages—start off with your name and phone number, speak slowly and clearly, and even spell your name. Leave a message (please be brief) and say your name and phone number again. It is most frustrating for me when I get phone mail (or voice mail) messages and I cannot decipher the name or even the phone number.

PROSPECTING IS STILL A KEY TO IMPROVED PROFITABLE SALES

October 7, 1996

The industry is enjoying its best year ever; we're still in a sellers' market with high occupancies and improved revenues just about all over. Will it remain that way? Probably not! The developers, lenders, and operators are out there looking for new locations; major franchise groups adding more brands to compete in just about all markets, and little by little we see competition sneaking in all around us—and not only in limited service categories. Although business is great, salespeople report that it is getting a little tougher in some areas when it comes to competing in the group market. We're not giving as much away as we used to—we're sticking with rates, but maybe throwing in something extra here and there. Recent reports from the companies that track the hotel business indicate that the rest of '96 will be great and even 1997 looks good, but not much is being said about 1998. If the economy holds the way it is now we may still be in a seller market, but who knows? How prepared are you for a repeat of the mid to late 1980s? Will we be in an over-built situation in your area? Will there be an economic situation, where companies reduce travel, where the loss of income and discretionary income will curtail pleasure travel, will associations see a decline in convention attendance and where you will find yourself under great pressure for performance?

No matter how well you are doing now, you must protect yourself for the future. That means—not counting on the business you are doing today, but looking for new business for tomorrow. That brings us to the subject of the importance of prospecting for new business, on a regular basis, every day, even when business is great. We have always preached in this column that the job of salespeople is to bring in new, profitable business—not only for now, but for later as well. We always have to be on the lookout fur individuals and groups who have not been with us before. We cannot rely on all of our business always coming back to us, even if we provide the best service in the world. There are other reasons for losing repeat business.

Prospecting every day needs to become a habit—time must be taken on a daily basis to locate leads that may be qualified to determine if, in fact, they are prospects for future business. It sounds like "back to basics," but experienced, professional and successful salespeople understand this. Leads are available all around us, but that doesn't mean each lead becomes a prospect for business. Once a lead is found it must be developed by questioning to find out the "who, what, where, when, how, and why." The answers to these critical types of questions determine if the lead turns into a prospect. Then, of course, there needs to be some selling done to make the sale. Like my good friend and super sales trainer, Tom McCarthy, always says "first you have a suspect (a lead) which could become a prospect, then becomes a customer"—So let's get out and find those suspects.

Ten good leads should develop into three good prospects, and one of those prospects (if you do your job right) should become a customer within the year—and probably even much sooner. So get the mindset to spend some time each day looking for leads. Just about all properties that can handle group business will have old files that have not been traced in years. Pull some of the old ones and follow-up. When was the last time you asked your fellow staff members (housekeeping, front office, kitchen, dining room, maintenance, recreation, parking, bell service, etc.) about who they may know that could be in a position to bring business to the property? They all have friends, neighbors, belong to organizations and have a spouse who works for a company that may have meetings. Does the front office or reservation manager provide you with names of some of the day's arrivals who may appear to be "suspects"? Then, of course, there is the newspaper, with loads of leads, but you already know about that. And, don't forget there is always your existing business who should be providing you with referrals of others who may be prospects for you.

HOW ARE YOU RESPONDING TO SALES INQUIRIES?

November 9, 1996

Two days prior to my start of writing this column I received a package in the mail from a hotel which was in response to an inquiry I had made. Postage on the envelope was $4.25 and the cost of the collateral enclosed would probably run about $5.00. It was really great looking stuff—brochures, meeting plan kit, promotional pieces of all kinds explaining packages available and a beautifully done banquet menu kit. But, no letter or note along with it.

Here's the scenario: about a month ago, conducting one of my little tests of how hotels handle inquiries, I called a property to inquire about having a meeting there. The call went through OK to the sales office. The person answering the call asked a few questions about my needs, but was not able to quote rates . . . "a sales manager had to do that." OK. Meeting space was available for the dates that I wanted, but she could not check on the availability of sleeping rooms . . . "sales manager had to do that." Sales manager was not available so that person took my name and telephone number and address and stated that someone will call me, but in the meantime, she would send me some information about the hotel. Fine.

So here it is about a month later . . . the information arrives, but still no phone call from anyone at the hotel. What happened to the inquiry? Is this unusual in our business of hospitality sales? I should say not. It goes on every day.

Why did it take so long to get something to me in the mail? And, was it necessary to send me everything they had printed about the facility . . . I really got it all, except the kitchen sink. Was there anything in the package of information that really helped me meet my needs? Well, some of it was helpful. However, I still do not know the availability of sleeping rooms for my group and I still have no information on rates for anything. Wouldn't it have me feel pretty good to have gotten a letter from someone (like the next day), thanking me for the inquiry and advising that someone will call? How about getting someone to call me the next day? Also, why wasn't someone around at the time of the call to be able to talk to me about sleeping rooms and rates? I can understand that a sales manager may be out on calls or busy with other prospects or eating or even going to the bathroom. But, where is the General Manager? I even asked the person ". . . isn't anyone around that I could talk to about rates." Wasn't there someone in charge of front office or reservations that would be available to discuss sleeping rooms availability and rates? No . . . the easiest thing to do is get me off the phone and tell me I'll "hafta" talk to someone else. Remember, prospects for business don't "have to" do anything . . . they can always go somewhere else.

After not receiving a call back on my inquiry and not getting anything in the mail, I completely forgot about it. Of course, if I was really trying to book a meeting, I surely would have called another property.

So, what's the point? There are several. We lose a good deal of opportunities for business by not having telephone inquiries handled in the proper manner (we've covered this before in previous columns). Even if salespeople are out why wouldn't there be someone else around that is in charge to talk to someone who may be interested in booking a piece of business? Further, why can't information be put in the mail the same day or at least the next day after an inquiry? And, what's the big deal about adding a letter or note along with the collateral being sent out?

Probably, we all need to rethink our procedures in taking care of people who are interested in doing business with us.

DOUBLETREE NATIONAL SALES OFFICE: A WELL-OILED MACHINE

June 2, 1997

What a wonderful experience it was, getting the chance to visit Doubletree's national sales office in Philadelphia. Last year, four of our Virginia Tech Hospitality and Tourism graduates were hired to work there to help develop sales for the brand. A surprisingly large operation with about 30 staff members, all rather recently out of college; all highly motivated, full of energy and most enthusiastic about Doubletree's products.

What was most interesting to see was how these young sales enthusiasts follow the basic principles of prospecting and selling, which we have preached in this column for years, and put millions of dollars worth of new business on the books for the Doubletree brand properties. The key word around the office seems to be "prospecting" ... go look for more new business for our hotels. And, they all do it by phone. All are very computer literate and operate their Global Delphi programs like something with which they were born.

Jack Ferguson, Sr. Vice President of Distribution for Doubletree, started this New Business Development Office about three years ago to search out new prospects for their brand partners. And it's working like a well-oiled machine. This office is developing about $10 million in new business each year through a strong company marketing effort where outbound calls are made all day long. Sometimes a booking may be made directly with a specific property and other times the information is just forwarded to the appropriate location for follow-up. This is not a reservation center; the function here is strictly the development of both transient and group business from key market segments. The salespeople here spend most their time actually looking for new prospects through an extensive database. Their number one goal is to qualify accounts and pass them on to a property or to national directors of accounts.

Responsibility for developing new accounts is divided among the staff by geographical territory and by market segment. Every segment of the business is touched by someone. One of our last year's graduates, Scott Robison, has assigned to him the largest independent meeting planning organization in the country. Just out of college for 9 months and he has already contributed a bunch of dollars in new business. On the day of my visit, he had received six new requests for proposals from that account as well as four phone calls from them for information on other business. Scott attributes his success in dealing with this account to his ability to build relationships with the client's staff. He knows they need fast attention and he gives it to them.

In another area of the office, there are two (also recent graduates) working on a special promotion. Doubletree started a campaign identified to get new prospects by running an advertisement in trade publications, offering the famous Doubletree cookies if meeting planners would fax in a completed form indicating their meeting requirements. This is a way of creating a database of new prospects. Once a form is received the cookies are sent and one of the team members makes a call to further prospect the account. Although the program only just started in late February, through April the two fellows handling this program have booked over a $200,000 in new business for Doubletree properties.

The air of excitement and energy around the Philadelphia office was such a wonderful feeling that I felt it was worth sharing in this column. If something like this could be catching, like a cold, we sure would love to pass it around the country to all the local sales offices operating in hotels, motels, conference centers, and resorts.

TURN INQUIRIES INTO SELLING OPPORTUNITIES

July 20, 1998

We have written about this before, several times, but it sure is worthwhile doing it again since we have not improved much over the years in handling inquiries properly. It just comes down to training the people who answer the phone to know what to say when people call to inquire about room rates, meetings, social activities, etc. There is no telling how much business is lost because at so many properties there are no instructions or trained personnel to respond to inquiries.

For the most part, just based on our experience in making telephone inquiries around the country, as a test, we find that when we call to inquire about rates for room reservations, most properties fail in trying to make the sale. Generally, we find no selling being done, rates are quoted, but no questions are asked, no value explained. And, in so many cases when we indicate that the "rate is too high" and ask if there are any other properties nearby that charge less, we not only get the names of the competition, but phone numbers as well.

Training is the answer to improving sales on telephone inquiries.

When it comes to inquiries for groups and meetings, we also fail by not asking the right questions or not getting to the right person on the staff who can handle the call, or getting put "on hold" for too long a period. At most hotels, when we ask to talk with someone about a meeting, the call gets transferred to the catering department. But instead of someone first asking the critical questions such as: name, company, address and phone number—we usually get asked the date of the meeting and number of people, as well as "configuration" desired for seating. Then we are put "on hold" for what always seems to be a very long period of time while the function book is being searched to see if space is available. You would think by now that all properties would be up-to-date with an automated system such as Delphi or Breeze so availability of space may be searched in seconds. Many times we are told that space is not available and we get a "thank you for calling." No further questions are asked—not even to see if we can switch dates. Since we didn't get to leave a name, number, etc . . . .there is no way the property staff can follow-up at a later date to see if they can handle the business in the future. It just doesn't make sense not to ask those critical questions so that a file may be created and traced at a later date for business.

And, even if space is available for the date requested and additional information is obtained by the catering staff, when we ask about sleeping rooms, we get a response such as: "you'll hafta talk to reservations about that," and we get transferred to someone else. Sometimes we are told that we need to speak with someone in sales about the sleeping rooms. Why wouldn't we have been put through to sales at the start of the inquiry? And, even if we do get someone is sales, it is so disappointing to find out that so many do not know the right questions to ask, or how to talk about the property features as benefits or how to get the prospect committed to letting the hotel hold the space. And, then of course, there is always the situation of getting voice mail if we are put through to the sales office. No need to go into that again . . . we've done it so many times in previous columns.

IN SELLING, DO WHAT THE POLITICIANS DO

February 1, 1999

Recently, there was a segment on "60 minutes," a CBS television show, which got into how politicians do a "selling" job on the public. Political parties engage the services of polling organizations to get information on how the public feels about certain issues. Pollsters also find out how people feel about certain words, expressions and even body language. Once it is determined which words make people "feel good" the speechwriters tend to use those words or expressions as much as possible in the material to be used by the politicians. Likewise, they avoid words and terms that have a negative effect on people.

Listening and watching how this works for politicians had a very positive effect on me. It definitely reinforced what we are teaching hospitality salespeople. We've been "preaching" for a very long time that it is not enough for our salespeople just to go out and make presentations on our products and try to convince people to buy our services because we think we have wonderful properties with all kinds of features that people want. The job of selling is first to find prospects . . . that is, individuals and groups that have a need, and could use our properties because we can meet those needs. But, of course, that is not enough. The next step is to put those individuals, and groups into a buying position. As politicians do, we want to find out what people what to hear and feed that back to them . . . Making a purchasing decision is a very emotional thing . . . There has to be a good feeling, in order for someone to decide to buy something. Think of the last time you bought something . . . anything! Whether it be a house or car or even a piece of jewelry, or suit, or tie, or dress, or a gift for someone, or even something to eat in a restaurant. A decision was made by you with some feeling, and there was some emotion involved. The same is true when someone makes a decision to use a room at your property or book your hotel for a group. Just the same as when you hear a politician speak on issues . . . you either like it or not, based on what they have to say, and how they say it.

So, it has to be the same with your presentation in selling. Don't just sell products and features! You need to find out what "turns them on." Part of the selling process is to gather information. "Knowledge is Power" has been my favorite expression for many years. It is not enough to know the basic needs like how many people, dates, space needed, budget requirements, etc. You need to get into "feelings." Therefore, the process of selling needs to involve an information gathering process called "probing." That is, digging deep into feelings with questions of the prospect that make them think, evaluate, and even speculate. You need to ask open-ended questions which get prospects to talk about things which will give you hint of what they feel good about or even with what they feel uncomfortable. Questions like: "Tell me what you really want to accomplish with this meeting." Or, "What was it about your last meeting that you liked very well." Or, even, "What happened at your last event that didn't go very well" (of course, only ask this question if the previous event was held at a competing property).

The whole idea of asking these questions is to be able to relate to the situation with something positive about your property, as far as benefits are concerned, that would make the prospect feel good about using your hotel. If you can relate to something that would help the group meet its objective (in response to the first question mentioned above), it surely could make the prospect feel good about what you have to offer. Probing is asking the open-end questions to get the prospect to talk about how they feel about something. The response is then followed with you relating a benefit of using your property which would make a prospect feel favorable about what it is you are presenting.

CAN WE SHORTEN THE SALES CYCLE TO BRING IN MORE BUSINESS?

May 3, 1999

During a visit to Chicago recently, I had dinner with David Green. He is Chairman of David Green Companies. Somehow or other we got around to talking about sales (what else?), and David told me a story about General Electric Company. It looks as though GE got involved in a program where they embraced shortening the sales cycle and quality control through telemarketing and profiling their better customers. Sounds pretty good. Apparently, they set up telemarketing crews to locate prospects via their existing customer base and to get as much information as possible through profiling. Sales folks then knew enough about what's what to go to the prospects to make presentations and confirm some business. The whole idea is to shorten the sales cycle by providing salespeople qualified prospects rather than having them cold calling or looking for leads, who then still have to be qualified before a sale may be made.

Can we apply the same thinking to what we are doing in group sales at our properties? David reminded me of the old 80/20 rule which we have all talked about over the past many years. Eighty percent of our business comes from twenty percent of our customers. Wouldn't it be more productive for us to go through our old files (customer base) and get someone to do the profiling for the sales team; then provide the salespeople with only very qualified prospects; creating an opportunity for them to make real sales calls which could end up with some business.

At our HMM sponsored Sales Workshops, I usually ask for a show of hands of salespeople who spend more than 50% of their time in actual sales work. Very rarely do I get more than one or two people to indicate they spend that much of their time selling. It seems that some do work other than sales (that's another story which we will handle at another time), but most are looking for leads, knocking on doors making cold calls and out prospecting for business. Yes, we can consider that sales work too, but is it productive sales work? If we can take that portion of their "sales" work away from them by providing very qualified prospects upon whom to make presentations, they should be able to close more business. If an excellent salesperson can spend an additional 20–30% of his/her time calling only on qualified prospects it seems that there would be no question of sales being improved.

It seems that a good many General Managers "push" their salespeople out the door and tell them to make "calls." What does that mean? We find salespeople going almost door to door in the community dropping off brochures and telling people: "if you have any meetings coming up or anything we can help you with, please give me a call." Even if salespeople spent more time at each call and did a little prospecting, probably only three out of ten calls would develop into a prospect . . . and that's not bad at all. But still, nothing has been sold. Take that same amount of time calling only on qualified prospects—those that need what you have—and more sales will be made. So, how do you do it?

Well, David Green can do it for you. His Database Marketing Company does just that.

They can do it with their telemarketing crew to profile your better customers, and come up with plenty of detailed information which can give you enough power to have the sale folks at your property spend their time calling on only qualified prospects. The whole idea is to have salespeople spend more time selling than doing other things like going out to "knock on doors." Call Bill Lynch, Vice President of Sales, The David Green Organization, 312-440-1727 to get more information on how this works.

YOU CAN BE MORE PRODUCTIVE IF YOU KNOW WHAT PROSPECTS WANT FROM YOU

January 2000

Over a number of years, in regular contact with meeting planners, travel agents, leisure as well as corporate travelers, I have had the chance to learn what these prospects like or don't like about their contacts with hotel sales personnel. The following is a random selection of comments from notes I have taken, from time to time, during visits and meetings with prospective buyers:

Many salespeople have told me that they need to be on-site during an event they booked to "make sure everything goes well" or because "the planner feels better if I'm there." According to most professional meeting planners, this is not necessary as long as someone (conference service manager, catering manager, etc.) is assigned to oversee the event, and he/she has the information. Planners indicate they would prefer being in contact with that person well in advance of the meeting or event date.

Personal "thank you" notes, hand-written by sales folks are more appreciated than the computer-driven, form letter.

Want to get more local prospects to visit for site inspections? Invite them to breakfast instead of lunch. They also want to meet the General Manager while they are there. Planners indicate they can tell a lot about how a hotel is run by meeting the GM (even for just a short period of time to say "hello").

What's more important than price? Service, of course. Most planners will want to negotiate price, not because their budgets can't afford it, but just to make them feel they have made a "good deal." However, service is really more important. In fact, you'll be seeing more service performance clauses added into your contracts by planners.

Sales folks just need to do a better job in answering phone calls; same for catering offices. This has been a hot item for a long time. Now, even hotter, people doing business with hotels prefer doing it via email. Sales departments without email will be missing the boat.

Unannounced sales call on corporate travel managers, meeting planners, and travel agents are taboo. They don't like it, don't have the time, and feel it is discourteous. They prefer having salespeople make appointments.

Many salespeople just don't seem to have the "big picture" of the hotel. They should know a good deal more about the operation of the property . . . facts and figures . . . size of sleeping rooms, number of connecting rooms, number of parking spaces, etc.

Enthusiasm sells! Some planners indicate that sales presentations lack emotional involvement. They like to see enthusiasm and feeling put into presentations . . . make it exciting! Salespeople talk too much! Some "show off" too much, and some are even obnoxious. Even though a property can meet the needs of a group, it's the salesperson who can kill the deal.

It is in the hotel's best interest to have a review of the meeting or event with the planner once it is all over. Most salespeople do not ask for this. Many planners think it is a good idea for the hotel.

This is also a good time to go over the master account to make sure all the charges are correct. And, they want their bills within a week after departure.

Good salespeople, when moving to another property, are followed by planners.

When visiting a hotel for a site inspection, it is always a good gesture to have a "welcome note" greeting from the General Manager.

Pre-conference meetings are a "must" for professional meeting planners. The key staff members of the hotel need to be in attendance, and there should be a review of all activities planned.

I hope some of these random comments will be helpful in improving sales productivity in this new year.

NOT HANDLING THE TELEPHONE INQUIRY PROPERLY WILL HURT YOUR BUSINESS

March 6, 2000

Do you know how much business you lost today because of how your phone was answered (or not answered)? There is no way of telling. Inquirers could have hung up if the phone was not answered within a couple of rings. Or, if the voice on your end was intelligible, it may not have encouraged anyone to do business with you. Or, if the call did get answered and the inquirer was put "on hold" without getting a chance to say anything, they could have hung up. Then, perhaps, when a call gets transferred to the sales or catering department, it feels as if it takes forever for someone to answer it. Or, it gets answered with voice mail. Anyway, this can go on and on, but the message is that it is a very good bet that a bunch of business was lost today at your property, but you don't know about it.

We know all this happens all too often because at many of my sales workshops we do some tele-testing. We make calls to various competitors of the properties represented in our audience. We try to book a room for overnight, or a wedding reception, or a meeting, etc. The participants get to hear the conversation via a speakerphone. And, you have to believe it, more than 90% of the inquiries we make, even if we get to talk to a real live person, are handled in a most unprofessional and unfriendly manner. So often, when we finally get connected to someone in reservations, or sales or catering, they do such a poor job in asking the right questions. It seems to me that if a caller is a qualified prospect someone should be asking for a name and telephone number, at least. It looks like nobody gets any telephone training.

I thought it was about time to call my friend the Telephone "Doctor" ... who is in reality, Nancy Friedman. Nancy has been all over the country to talk to groups about customer service training, particularly in the area of how to work with a telephone. A few years back I sat in on one of your presentations, and it was terrific. We all learned quite a bit. Nancy agrees with me, in our business of hospitality, we do a poor job in using the phone. She says that whenever she does a hotel program and asks the question "how many here have some sort of new employee orientation program on customer service and telephone skills." She is shocked how few raised their hands. Nancy says that guests are frustrated in every area of a hotel, not only at the front desk, where the telephone is usually answered, because of lack of telephone training. Not only do new employees need telephone training, but especially employees in sales and catering departments ... that's where so much business is lost because of poor telephone call handling techniques. One big problem which Nancy recognizes is when a "sales assistant" answers a phone call and tells the Inquirer: "she's not in, can I take a message." I agree with Nancy when she says this is weak and ineffective. Every inquiry is a sales opportunity. We just miss the boat too often. There is a lot more to be said in this area of sales ... and the person to do the talking is the Telephone "Doctor" ... and of course, she is Nancy Friedman, a writer, speaker, an author on topics of customer service and telephone skills. You probably could use her help ... she can be reached at 314-291-1012 or by fax at 314-291-3710 ... and by email (teldr@aol.com). Or take a look at the Telephone "Doctor" website: www.telephonedoctor.com I know they have some good stuff they can send you for free.

WANT TO BOOK MORE BUSINESS? RESPOND TO INQUIRIES PROMPTLY!

November 2000

It seems to be happening more and more lately; company meeting planners, association executives, and event organizers are complaining about hotels not responding to their inquiries promptly. We hear about it at association meetings and training sessions; read it on listservs and discussion boards. Things are just moving faster and faster as time goes by. The biggest complaint from prospects is that their telephone inquiries are not getting a fast enough response. You would think that if we wanted to be really serious about doing business, we would take whatever steps are necessary to get back to inquirers as soon as possible.

There is another problem which we seem to have, according to many group planners. Their complaint is that they are not getting responses from hotels on their Request for Proposals (RFPs). Some group planners don't get a response if the hotel, for any reason, may not want the business. Or space is not available on the dates requested. Or a property just cannot meet the various needs of the group. In these cases would it not be prudent for a property salesperson to make contact with the inquirer anyway to try to help in some manner, or at least gather additional information for future meetings or group business? After all, if these are prospects now, aren't they prospects at some later date as well?

Debra Smith, CMP, President of CONCENTRA, Conference Management Services in Boulder, Colorado just sent me a recap on RFPs which were sent electronically, on September 29th, to 61 facilities. On the same, day she received one response by fax which declined the business. On October 2nd she received 27 responses; 25 by fax and 2 by email; 6 declined the business, 7 acknowledged, and 14 responded with date and rate availability. On October 3rd there were 6 responses, all by fax, with date and rate availability. So of the 61 requests for proposals, Deborah received 34 responses within 5 days. She hopes the rest come in soon. It is interesting to note that only 2 responses were by email, the rest were by fax. Does this tell us something? Why will salespeople not respond electronically?

Here are some other reasons why we don't seem to respond quickly to inquiries: Some properties are very segmented when it comes to market coverage. That is, salespeople are assigned to either geographic markets or product market segments. When an inquiry comes into the facility, once identified to be from a certain market, the call, or fax, or electronic message is given or forwarded to the salesperson assigned to that market. And, if the salesperson is not available, there sits the inquiry until his/her return. At some hotels, there are sales executive sessions held regularly to determine which pieces of business being presented by the salespeople would be best for the hotel. This means that many inquirers cannot get a definite answer from a salesperson until that meeting is held. Again, here, we have delays in responses. There are, also, salespeople who are not empowered to make decisions regarding room allocations, space or rates, until there is an approval by the Director of Sales or General Manager. Here, again, we have delays.

It is difficult for people who are trying to locate space for groups to understand why they cannot get responses to inquiries right away. They want fast answers. We just have to figure out a way to do so.

On the brighter side, Lisa Marshall, Manager of Event Planning for Sonoco told me: "We just started using electronic RFPs." I've completed two and responses began coming in only hours after the RFP was sent, on both occasions. I hope this continues to be the case as this is an efficient way to get the job done.

THE HOTTEST SALES LEAD WE WILL EVER GET . . .

November 2001

In our last couple of Sales Clinic columns, we covered opportunities for building sales leads into prospects, with or without a sales department. However, what we did not cover was the hottest lead we'll ever get, and how we screw it up. It is the most valuable lead coming into us, and yet we treat it so poorly. It is the telephone inquiry.

Here we have people calling the properties, trying to find out something, because there is a need; whether it is to book rooms, or get information on availability, or get a place for a meeting or a party. And, in so many cases that inquirer gets the worst treatment ever. You should be agreeing with me because it is so obvious. We just do a lousy job in this area. There is no telling how much business is lost daily at properties because of mishandling of telephone inquiries. The bad experiences to which callers are exposed starts off with the phone not being answered promptly. Sometimes it has to ring four times or more (or endlessly until the party hangs up and calls a competing hotel) before it gets answered. Then, what's the attitude of the person answering the phone, and does the person speak clearly enough for the caller even to understand what was said?

It would seem logical that everyone who answers telephone calls should be tested to determine how each handles incoming calls. It may sound trivial, but it is important that inquirers get the right message, immediately, when calling a property. The prompt pick up of a ringing phone, the tone of voice, the attitude and enthusiasm projected of the person answering the call, all help create a picture in the mind of the caller. And, the picture could be favorable or not for the property.

If we are OK to this point, then what happens to the call? Is it transferred to the appropriate department to handle the call? Then, how long does it take that department to answer the call, or will the call get voice mail? Well, the story goes on and on. We probably won't ever get it fixed 100% because this is not a new story; it's been going on for years. However, right now, since it appears we are getting into some tough times, we better get it fixed. Of course, the answer is in training, ongoing training, not just one training session. And, it is not that difficult to analyze what's going on in this area at your own property, and to take some corrective measures. If you want some help, you can contact the Telephone Doctor to get her free newsletter, which is loaded with tips, as well as tap into her website which is: www.teldoc.com

When telephone inquiries get transferred to a property's sales office, then there is a whole new series of screw-ups. Of course, not all sales offices do a poor job of handling those calls, only about nine out of ten. Those of you who have attended my two and half day hospitality sales workshops know that to be true. It seems that sales office personnel just don't seem to ask the right questions when someone is making an inquiry for groups. Knowledge is power in selling, so it is logical to ask a series of questions first, prior to answering any questions about space, dates, and rates. Probably many sales offices do not have a form upon which there are standard questions, which need to be answered, that would help in handling an inquiry. I'll be happy to send you my Phone Lead Inquiry form. If interested just let me know via email: howardf@vt.edu.

There is a lot more to be said about this business of handling telephone inquiries, and you can hear about it on my new set of audiotapes covering 16 different topics. Check it out via www.hsa.com and click on my name.

IMPROVE SALES BY STANDING OUT AGAINST THE COMPETITION

December 13, 2005

In the business of hospitality sales, we should always be looking for a competitive advantage. Of course, we look at things like the physical attributes of our property, location, the wide variety of features offered, our rates, and we do a competitive analysis for our marketing plan. We then come up with a plan that we hope would be somewhat of an advantage over competition. All too often it becomes a case of rate and the "deals" we willing to make for groups. However, what we neglect to do in most cases is to get our staff to do a better job of making a profitable sale. The easiest, quickest and least expensive is the training of the folks on the front line. Here we are talking about our reservations staff, front office staff, and sales staff. Most of the time, it is only the question of getting our staff, in all departments, to understand how to handle the inquiries that we get every day. This is where we miss the boat whether it is with people who call in, send requests for proposals or walk-in to "look over" the property.

The telephone inquiry is the hottest lead for business that we can ever get, and yet there is no telling how much business we lose every day because we do not handle it properly. If we were able to measure the dollar value of business lost each day, just because the phone is not answered right away, it would be unbelievable.

We really do not know how many people are calling to inquire about a room reservation, or meeting, or social event, but hang up before anyone answers the call because it almost kept ringing off the hook. But, that's not all that helps us lose business. It could also be the person answering the phone reflects a poor attitude with the manner in which the call is handled, by sounding tired, unenthusiastic, busy or uninterested. Of course, one of biggest problems that exist at too many hotels is the voice mail system used; where the phone is answered right away, but then we seem to get into a system where we have to punch one, two, three, etc. It looks like we are getting into the business of "self-service." If a caller does get to someone who can represent the property, whether it be a front desk salesperson or someone on reservations department, we usually are pretty good at providing information like availability of rooms, rates and amenities, but we need to do something better about making a sale. This is where training comes in to play. It could be done locally by the General Manager or someone on the property staff who may be capable of doing it, but it must be continuous, not just one time. Having a weekly one-hour session will work and you can measure improvement.

Managers should set up a system of reflecting the ratio of bookings against inquiries, on a daily basis, for each front desk salesperson as well as reservations staff. The ratio would show what percentage of inquiries were actually confirmed. This could very well be an excellent motivator for the staff to increase bookings.

To demonstrate how ineffective some properties are with regard to handling the telephone inquiry, we make live test calls during our sales workshops to competitors of our participants. We have proven, every time, that over 90% of properties called fail this test.

STOP LOSING THOSE TELEPHONE INQUIRIES!

January 2006

Yes, we are still losing millions of dollars a year of business that is going into competitors' hands because of the poor job we are doing with handling telephone inquiries. Even with all the wonderful technology available to us (most of it being misused) we are not better off in taking care of people who call in to book business at our hotel properties. The day prior to writing this column I demonstrated this to a group of owners and operators of hotels, and their sales department staff, at a workshop, by making phone calls to competitors in an attempt to book group events. We made eight calls to a variety of types of properties, selected by those in attendance at the workshop. All person takings the test calls failed even to make an attempt to book my business; only one sales office staff person asked for my name, but not my phone number or address. It was interesting to note that the reasons that all these calls failed to result in business prospects for any of the properties has not changed over the years. Just the same old reasons, but something new has been added as a result of technology; and that is voice mail. We are so hung up on this that it is costing us business. We also have been doing these test calls, for many years, at most of my sales workshops around the country, and it always comes out the same way. Just about 99% of the calls to properties fail the test, because of the manner in which the calls are handled, or we should say, "mishandled."

This is going on all over the country, on a daily basis, and the unfortunate part of this is that management is not aware of it because there is no opportunity to measure the results of these inquiries. Particularly the ones that are made by people who hang up the phone because it rings too long before it gets answered. Or, perhaps, a machine answers the phone with a long list of options to be punched. Or, the attitude of the person answering the phone, or a lot of other reasons. And, probably not very believable, but surely does occur pretty frequently; a response by someone who answers the phone in the sales office with: "sorry, there is no one here right now, could you call back later?" Sometimes, with that kind of response, we ask to speak to someone, maybe the General

Manager, and we are told: "No, he can't help, you hafta talk with the sales manager." WOW, what a response!

So how do we clean up our act and not lose inquiry calls? It all has to do with training. We should start out with a procedure with regard to answering the phone at a property. Who answers the phone and how do they answer it? Will the call be transferred to the correct department for handling? If there is voice mail, what is the message and how can information from the inquirer be captured? How can we be assured that a phone will not ring more than three times before it gets answered? Of course, there is a lot more to this, but the message has been delivered to our readership.

IMPROVE YOUR PROSPECTING FOR BETTER BUSINESS WITH REFERRALS

December 2006

Of course, there are many ways to prospect for new business (and we have covered this in our column on previous occasions); however, the best prospecting results are through the securing of referrals. Here, we do not mean just getting someone's name, and maybe a phone number, but a real, honest to goodness referral, from a happy person who has done business with you. Sometimes, in hospitality sales, upon completion of a very successful conference, convention or banquet, we ask the person responsible for the event to refer someone to us who we may contact for similar business. Now, that would be a step in the right direction, but still not good enough. We need to take it a step further if we really want to be successful in prospecting with qualified prospects.

The key to doing the right "asking" for a referral is in the ability to get the satisfied customer to make contact for us. This could be done by having your customer make a phone call to the person being referred, on your behalf, for an introduction to you. You could ask for an email message to be sent, or a letter, or probably, even better, have your customer invite the referred to lunch with the both of you. In this manner, it is just not a question of making a "cold call" on an unknown suspect for business. The referral, of course, would need to be on a contact; that is, in a position to use your property for a conference, convention, banquet or contract on sleeping rooms. Immediately upon completion of a successful event, the planner of the event normally is very excited, happy and feels very motivated. This is the time to be asking for referrals, but it has to be done in such a manner that the customer has to feel that this would be doing a favor for the person being referred. Salespeople need to learn to do this, but most salespeople shy away from asking, thinking that they are imposing upon their customer.

In a recent survey of 2,500 salespeople in the United States and Canada, some interesting facts about referral prospecting were developed by Houston, Texas-based McCord and Associates, who have conducted this research every year for the past four years. In past surveys, they found that about 15% of those surveyed relied on referrals to generate about 25% of their business. The latest results showed a downward trend, with a drop to 10%. This probably relates very well to what's happening in our hospitality industry. Occupancy, sales, average rate, and profits have been up, reflecting excellent business which appears to continue through 2007. It just seems that when business is good, there is a tendency to drop in our sales effort and reduce our prospecting activities. It is more important, during good times, to maintain a heavy drive in prospecting for future group meetings and catering. History reflects that these good times will be followed by growth in new rooms and properties, with a decline in demand. That means we will be back to fighting for business . . . Let's put more business on the books now, in preparation for the future. We need to be prepared.

IMPROVE PRODUCTIVITY WHEN PROSPECTING FOR NEW BUSINESS

February 18, 2008

It is a shame that so many hospitality salespeople spend too much time seeking out new business by running around "knocking on doors." It appears that they are being told by management and owners "get out and sell." So, rather blindly, many of those involved in seeking new business either get out of the property and start going into office buildings, blindly calling on companies, asking if they have any business for their property. Of course, many do not go out, but get hold of listings of companies and organization, and blindly, make calls asking for business. Just minutes prior to my writing this column, I received a phone call from a salesperson in a hotel located in North Carolina (I am located in Virginia), actually asking me for my meeting business. I teach at a university and do not control any meetings going anywhere. The salesman, when asked, advised me that he was given a list to make some calls. This happens a couple of times a week, with calls coming from hotels all over the U.S. asking for my business.

Now, isn't there a way that these sales can be more productive in seeking qualified prospects? Well, of course, there is. Without even leaving the office, any person, and they do not even have to be salespeople, can find qualified leads right from the property's own sales files. Certainly, there are files, either paper files in file cabinets or on a computer via a software program that sets up a sales and catering file system, something like Delphi by Newmarket software company. If someone would just take a look at those files, there will be hundreds of files on groups that have met or could have met at the property years ago. Generally, there will be a very high percentage of files that have not been followed up over long time periods, seeking other business from those companies or organizations reflected in the files. These are files that have not been traced, or just been overlooked by the sales staff. The whole idea would go through each old file, and make follow-up calls. After all, these files represent groups that have been there or could have been there before, so they have been prospects, and may very well still be prospects for new business.

Then, we have salespeople who participate in attending conventions and conferences where they are exhibiting for their properties. Most of the time there is the opportunity for show attendees to drop off their business card because of a prize drawing or an offer of some kind. When computerized name badges are being used for conference attendees, it is just a matter of scanning the badge to get information on the trade show booth visitor. Are all these people who get scanned, or leave their business card in a position to do business with that particular property which is exhibiting? Of course, note, but over the time and time again, salespeople use the information to make phone calls soliciting business from folks who are not real prospects. So, when exhibiting, use some form of a Prospect Survey Sheet to ask pertinent questions of booth visitors to determine if, in fact, which are prospects for business. The idea is not to burden your time, files, and resources with information on people who are not in a position to do business with you.

When checking back on old file, calls have to be made to the persons, whose names appear in the file, using a Prospect Survey Sheet, and ask questions to determine what business may still be available from that account. For a free copy of this form just email: howardf@vt.edu.

METHODS AND RESOURCES FOR DEVELOPING NEW SALES LEADS

July 21, 2008

What prompted me to write this very basic sales column was two email inquiries I received just last week from two, new, hotel salespeople, who each asked for help with a question, like: "What do I do to get business for my hotel." One was at a small hotel in Bulgaria and the other worked at a small hotel in Germany. I am sure the same question is being raised at many hotels in the U.S. by brand new salespeople without any prior experience in hospitality sales. There is very little training going on, around the country, except for seminars, and workshops being held in conjunction with the various annual meetings of the major hospitality associations. However, we do not see very many new salespeople attending these national conventions; budgeting is probably the big consideration. Of course, from time to time we are seeing Web seminars on hotel sales being offered, and at very reasonable pricing, so these are a big help for the new, inexperienced salespeople.

It seems that both sales representatives from Europe were more interested in getting information on advertising and direct mail, thinking that was the best way to get business. My response was to forget about that because of the cost involved, and stick to some basic, no-cost methods of seeking leads for business. Of course, the best bet for getting new business is to prospect old files. This was covered in detail in my February 8 column. A great resource for seeking new business would be from the companies with which your own property is doing business. Get a listing from the property accounts payable department. All hotels do a lot of business with local firms providing supplies and products to the properties. Some of those companies would probably have people visiting your city location, and most would have some sort of catering opportunity when they have meetings; retirement events or holiday parties. Further, some wonderful lead opportunities are missed by so many hotel salespeople, and it is just unbelievable. In any one day, you could find 20–50 possible leads in any daily newspaper. Look for announcements of meetings being held at competitors' properties, then try to get them for next time. Seek out articles in the paper covering special events coming up; make contact with the operator of those events to get listings of expected attendees, and try to get them to stay at your property. The same applies to all sporting events; participants in the events, families, as well as sponsors, need places to stay. Check out engagement announcements to get leads for catering business, not only for the wedding, but rehearsal dinners, bachelor parties, and overnight stays for out of town visitors. Look through the business section to see if there are any companies moving to town, new building under construction, and announcements of any promotions, or honors given to officers of companies. All these are good lead opportunities for room business as well as catered events.

Most of what has been written above are very basic to hotel sales; however, sometimes we forget the basics and get too complicated in trying to figure out how to find resources for developing new sales leads.

LET'S GET BACK TO THE REAL SELLING MODE TO BUILD BUSINESS

May 18, 2009

To some extent, because of the economy, unemployment situation, less people traveling, companies cutting back on meetings and travel expenses, owners and operators of our hotels are most concerned about a drop in occupancy and average daily rate. However, there is another issue, and that is the fact that about 200,000 new rooms are in the market place, resulting in an overbuild in many areas of the country. Were it not for the new rooms; perhaps, our situation may not be as bad as it now appears. There are still folks out there traveling and attending meetings.

What our management people have to think about is doing better with their market share. Occupancy around the country may very well drop about 3–4% or even more in some areas, but that tells us that there is still business out there for the taking.

What we need to do is get back to the real selling mode to maintain, and build market share. Salespeople need to spend more time prospecting for new business to replace the business that might have been lost. For a number of years, up to recent times, business has been good, and to some extent, it has just been the opportunity of doing business by just answering the phone, and salespeople getting busy by doing things other than sales. It is just an opinion, based on what been said about what salespeople have been do with their time, but only about 50% of their time has been on pro-active sales. Many salespeople claim they are being hampered by some time-consuming activities beyond their control that prevent them from spending more time selling pro-actively. Right or wrong, it pays to take a look at some of these activities to determine where more time might be converted to selling pro-actively.

It seems once we got into the electronics mode of maintaining records and reports that it would eliminate paperwork. In many instances, it may have created more paperwork. One of the big issues that concern salespeople is the number of reports they have to complete, and submit for management and corporate offices. Now, new computer-driven, automated sales account systems, such as Delphi, and others, reports needed by management should be available by accessing the system. There is no reason why management cannot gain access to a salesperson's activity; number of bookings, revenue generated; tracing lost business; call reports; letters written; proposals and contracts sent. It is all there in the computer. Why should salespeople have to spend so much time putting out these types of reports?

Another time-consuming issue is about salespeople involved in the servicing end of the business they have sold. Why do they have to be around to make sure everything "goes well" with the business they have booked, when there is other staff taking care of the details?

Let's start working with the sales staff to help them spend more time in going after new business to replace any business that has been lost. That means doing a good job in prospecting. Leads are all over the place, starting right at the property via old files, reservations departments, front office operations, accounting department, and other property associates. Just by checking the local paper every day will turn up large amounts of leads. Keep Salespeople Selling!

PROPER PROSPECTING PREVAILS IN PRODUCING PROFIT

September 2010

The first step in understanding this business of "prospecting" is to be sure what the term refers in dealing with our hospitality industry. There is so much confusion among hotel properties with regard to how managers and salespeople interpret the term, "prospect." So often we hear comments, such as: "I just got a good lead for business." Well, was it a lead or a prospect? Let's explore some very easy and clear definitions with regard to the distinction between terms, in our industry. Because, when we talk about "Prospecting for business," do we really mean looking for leads or actually finding prospects?

The distinction is easy; leads are suspects, and prospects are those who are in a position to do business with us, if they chose to do so. Therefore, once we find a lead, we have to be able to develop the lead into a prospect. Of course, we do this by asking appropriate questions to see if this is someone who could do business with us. Leads are all over the place; these are just people who might be "connected" to a business that could use a particular hotel, but we do not know that until we ask pertinent questions. We can find leads in newspapers, telephone books, business listings, from different hotel staff and offices, chamber of commerce membership lists; on the Internet via various sites, etc. Or even going out and going door to door in office buildings, knocking on doors; we call all this "cold calling," which is really pretty much nonproductive, a big waste of time.

Many salespeople busy themselves in seeking leads via social networks, or getting a listing of meeting planners and making phone calls; again, this is cold calling. Of course, some of these ways work, and maybe it will take a dozen phone calls, or visits in business offices, to find a lead that might become a prospect.

The idea for salespeople would be to figure out where is the best and most productive way to find leads that could be developed into prospects. The best place to do this would be to go back into the old files of the property. In sales, we all know the value of follow-up, and tracing our files, to make sure that a follow-up is conducted periodically. We should admit that we do not trace all files, all the time, and we miss a good deal of good business because we do not do that regularly. Therefore, if we take the time to go back to old, we will find "leads" from the old business that should be contacted. This is the best source. After all, the business has been there before, they know the property, and may very well be developed into a prospect again. Salespeople should be spending more time doing this than getting on the phone to make "cold" calls or trying to "connect" via social networks. For a Prospect Survey Sheet form to use to help gather the Information from prospects just send a request to howardf@vt.edu.

READ YOUR LOCAL NEWSPAPER AND BOOK A LOT OF BUSINESS

February 2011

It is really amazing to know that how many groups, as well as an individual business, could be developed just by carefully reading a local newspaper. If the same amount of time were put in on reading and analyzing the daily paper as we put in on the computer doing social networking, we would be booking more business. In most cities, of any size, in any one day, most anyone would be able to identify at least 25 good, healthy leads that would encourage a salesperson to follow-up with phone calls. It is just a little bit of information that could identify a lead, then, of course, a follow-up by making the phone call would certainly lead to the development of additional information. There is no question about it; to some extent, a number of the calls would reflect not enough to call those leads "valuable." However, some would develop into qualified prospects. It is just a matter of making the calls and asking appropriate questions to determine if that particular lead could develop into a prospect.

The questions need to be open-ended, that is, to be asked in such a manner that we get the person on the other end of the phone to do some talking. We do not want to ask any questions that call for a "yes" or "no" response. The questions to be asked depend on the type of lead found in the newspaper. If there is an item in the paper mentioning a company meeting being held elsewhere, then the questions would want to gather responses concerned with company meetings. If the newspaper item ha to do with social events, then the questions would want to reflect responses regarding the specific events. For an article about a company moving into town, then the questions would be about the organization and its housing of staff members moving into the city. Questions would certainly vary depending upon the particular activity mentioned in the paper.

Without question, every section of the local newspaper would have any number of happenings which could reflect some sort of business for a hotel property; for example: in the Sports Section ... How about looking for information about sporting events to be taken place on some future date in your city? There probably would be some teams coming in to participate in the event. The teams and supporters will certainly need a place to stay; teams will need special attention for meetings and meals. Look for a name, a contact in the newspaper report and make contact. It may very well lead to another name or two before reaching the right person.

Companies moving into the city would certainly need rooms for staffers overnight. Further chances are, there will be some company meetings to be held and some entertainment for meals.

Then there are engagement announcements; which of course, should lead to wedding receptions. And, unfortunately, there are obituaries. However, people from out of town come to funerals, and there are meals involved when families get together. Best contact on these would be funeral directors.

Any, the list really goes go on and on. Just do a test one day; see how many leads could be developed in just one-half hour reviewing a daily paper. Then, take another hour just to make phone calls seeking prospects from the newspaper leads developed. The whole process to do the test would take less than two hours.

If you would like a form to use upon which to write down the responses to your questions, just let me know by email, and I will send you a form.

SALES IS WHERE IT'S AT! SO LET'S GET TO IT RIGHT NOW!

February 2012

All the studies and report we have seen lately reflect a very good year is coming up for increases in occupancy, demand, average daily rate, and total revenue. All of this should lead to improved profitability.

Here is the problem. It seems that almost throughout the entire hotel industry, during times when business is good, we get busy taking care of the business, the productive direct sales effort falls short. What happens at so many properties is that the sales teams get busy doing things to "help out" in operation and service, leaving less time for productive sales activity. It just seems like the right thing to do when an operation gets busy; sales folks just want to be helpful in providing improved service to the guests.

Sales efforts need to be conducted regularly during the course of the year. General Managers need to be concerned that everything is covered properly so that sales staffers are not called upon to help out. This is a time for working on bringing in more future business. An increase in guests creates so many more opportunities for a salesperson to develop new leads that may be developed into new prospects which would create more business for the future. Spending more time prospecting is the key to increased productivity which directly relates to more future business.

The idea is to gather information on those who may look like "leads" for new business. Not all leads will convert to qualified prospects, but there is a good chance that three out of ten will. These three will need to be followed-up with phone calls to gather more specific, additional information to determine to what extent that contact is a real prospect. Then, the selling process may begin. There is a good chance that one out of the three will convert to a prospect.

First, to find the lead, information needs to be gathered from the advance reservation of the guests. This is available from the property reservation department. Here is where you get names, titles, companies represented, the name of the contact that made the reservation and phone numbers. This is a good start in figuring out, from the information, which may become a lead; it is just a "suspect." It could not yet be determined if this could be a prospect until more information is gathered which is what a good follow-up is all about. Phone calls need to be made to the company or organization of the guest to gather additional specific information to determine if this is a qualified prospect. Once that is determined, then the selling process can begin. It does not pay to try to do any selling to anyone who may not be a prospect. It helps to use a Prospect Survey Sheet when asking the questions of Who, What, When, Where, Why, and How. All of that kind of information will reflect the needs of a prospect, and if a property can meet those needs. That's what selling is all about.

A sample prospect survey sheet is available by request via email to howardf@vt.edu.

HOW THE EASY METHOD OF DATA MINING BRINGS IN PROFITABLE SALES

October 2012

We all agree that finding and developing connections that will be able to bring us the business for our properties, is really what this business of profitable sales is all about. Where do we start is always the issue that needs to be resolved. In reality, it all is concerned with marketing. But where do we begin? In marketing, we consider all the facets involved like: advertising, promotion, direct sales, merchandising, public relations, and research. Each of these segments (and maybe a couple more) will, of course, be broken down into a variety of methods used to ultimately bring in more business. It would seem logical that the appropriate response to where do we begin would be the area of research. This brings us to data mining, which means digging for information that has already been collected and available in a property's existing files.

The premise is that a hotel's best opportunity for leads for new business is parked within the files of their own internal customers. Salespeople are always seeking leads for the business that may be developed into prospects, and are then followed up with personal calls and visits to make the sale. Data mining is just a matter of digging into a property's sales files to gather information of a previous customer. Past history provides a good deal of lead information when followed up with personal contact; it could make that former customer a qualified prospect for the new profitable business.

The big question with regard to data mining, with hundreds, even thousands of old files, who could do the research into those files? It is really not the salespeople; their job is to follow-up on the prospects developed, to make contact, and ultimately, make the sale. The sources available to do the data mining could very well be students studying Hospitality at a local college, part-time help, or other property staff; of course, all would need to be trained to do the job. There is also the opportunity of out-sourcing to companies, with trained staff, to do the job of developing leads which are contacted, and then converted, if possible, to qualified prospects. Experience in this area has reflected a high percentage of return on investment.

David Green, a long-time hospitality sales veteran, and President of Integrated Hospitality Sales Solutions agrees that: "salespeople should sell, nurture relations and book business. Prospecting for new business, rejuvenating former customers into repeat customers, should be outsourced. If done properly, you could increase productivity and revenue by 25%." David's company has been doing this. Following are quotes from letters sent in by a couple of properties he represents: ". . . The results realized from this relationship have exceeded my expectations. We consider them to be a valuable addition to our team and our initiatives." And another: "The sales managers have commented on the quality of leads, and we continue to book business from their prospecting efforts, and more importantly, will see future business as well."

Dave . . . Every semester, when I teach Hospitality Sales, I offer our students to hotels to do some data mining for a couple of days. I do not ask the hotel managers to send me any kind of report to determine how effective the students do this work. However, some time ago, one manager of an Atlanta, GA property sent me a report on the activity of four students, in two days, and the follow-up, by salespeople, with results four months after the work was done . . . the ROI was amazing . . . The report was broken down by a number of files searched (800); number where contact was made, additional number information developed, etc. Then, by the name of each salesperson, the number of prospects followed up with personal contact, and breakdown of new business booked (rooms, F&B, etc.). The amount of definite business booked was over $200,000. And, of course, additional possibilities further out.

That's how outsourcing actually works. I can offer a copy of this report requested.

GET FREE STUDENT HELP IN PROSPECTING FOR NEW PROFITABLE SALES

February 2013

It is amazing the number of educational Institutions that include some type of hospitality programs for students. In just about any city, or nearby, in the United States there may be some type of educational facility. Of course, most Universities would have some hospitality, or hotel management programs, perhaps located in their Business School. However, there are a variety of colleges and community colleges where these types of courses are being taught.

Property managers and hotel sales staffers could make inquiries at these institutions to find out if such programs exist in their curricula. If so, contacts need to be made with the appropriate department head or faculty members who teach such a course. The idea, of course, is to discuss the opportunity of your property being able to provide some hands-on support for the students to learn what goes on in the business of hotel sales. What you would provide is the student experience of making prospect contacts, by phone, to inquire about possible future business for the hotel. A Prospect Survey Sheet could be designed to serve as a questionnaire for the students to use. The property salesperson or General Manager would need to provide some brief training on how to make the telephone inquiry. Specific questions need to be asked to gather information which would reflect some future business possibilities. For samples of forms that could be used, as a guide, send a request to howardf@vt.edu, then create your own form to suit your property. Names of contacts to be called would be found in old sales files supplied by the property. The best prospects for future business are the contacts from groups that have used the facility previously.

CAUTION: Everyone involved in this project, the sales staff, General Manager, and especially the students, need to understand that this exercise is experience in developing prospects, in preparation for the actual sales process. This is not a SELLING exercise. The selling is accomplished by the sales staff at a property, using the Prospect Survey Sheet to follow-up with a prospect in order to make a sale. Before a sale is attempted, a salesperson needs to have the appropriate information which reflects the needs of a prospect.

Students should be able to devote 2–3 days for this learning experience. However, it is remarkable the number of contacts made that develop into real, live prospects. This exercise has been going on, very successfully, at Virginia Tech's Department of Hospitality and Tourism Management for over the past 25 years. Student evaluations of the exercise have always been very positive, and the program has been in great demand by hotel operators near and far. Hotels provide housing, where necessary, meals and auto mileage reimbursement. Everyone in this partnership is a winner.

HOW TO FIND PROSPECTS AND BOOK GROUPS IN HOTELS—THE EASY WAY

January 2014

It is amazing how these "old time" methods of seeking new prospects for business always seems to work. Here is a good example: A sales director for a mid-priced property located in a State Capital city called one day to seek advice in being able to get new business for his property. He mentioned all the reasons for not being able to find more groups; economy, too much competition, the government cut back on travel expense (they were heavy in government business), reduction in group meetings, etc.

The advice was pretty simple: go back to your old files and review them, especially the ones that, for whatever reason, were never traced. This sales manager was given the advice to test this method. He was to spend just one week, two hours each day, going through and reviewing old files. On each file, he was to make telephone contact with the person listed in each who was the contact person at the time of the event. He was to ask appropriate questions to determine what group events were being planned, by whom, when, etc., At the conclusion of that week's activities as was advised, the results were reported for inclusion in this column as follows:

- Accounts prospected: 42 (did not quite get to 50) + 3 wedding planners.
- Past accounts from 2010 to 2013 with good potential.
- Past corporate and government meetings mostly.
- Past groups rebooked—3 (2 government, 1 corporate).
- 81 room nights at $8,389.00, ADR $103.57 (competitive set for area ADR is $101).
- New business booked—3 (2 SMERF, 1 corporate).
- 215 room nights, $23,285.00 room revenue; group ADR $108.30.
- Catering connected to group meetings booked.
- $3,300.00 F&B revenue; and $1700 room rental revenue.
- Totals:
  - 296 room nights
  - $33,674.00 in room revenue
  - $5,000 in catering revenue

The point is that time needs to be set aside, perhaps even on a daily basis for someone to get back into the old files to do some prospecting. This is not a difficult task. In fact, salespeople need to get someone else to do the prospecting so that sales staffer could better spend their time in actual selling to the prospects that are being developed. It would not be a bad idea to allow some front office personnel to review some of the old files, and make the phone calls to ask pertinent questions to determine if a prospect could be developed. Using a Prospect Survey Sheet is very helpful in making sure that appropriate questions are being asked. The completed survey sheet is then sent over to the sales department for a follow-up call by a sales representative; hopefully, this way some new business may be developed from some "old" business. For a free Prospect Survey Sheet form contact: howardf@vt.edu.

THE TELEPHONE INQUIRY AND HOW WE MESS IT UP

May 2014

No question about it, the "hottest" lead we will ever get at a hotel or resort is the telephone inquiry. Whether it be for a single reservation or a group prospect, there is no telling how much business is lost because we handle it so poorly. Just think about the number of things that can, and do, go wrong in the manner in which this lead is handled. To start with, how long does it take for someone to answer the call? How many "rings" does it have to take to get someone to answer the call? Do we know how many people "hang up" if they have to wait for more than three rings for someone to answer the phone? How many "hang-ups" do you get in one day?

Then we have to be concerned with the manner in which the call is actually answered. Has the person who will take the call been trained on just "how" to answer the phone? Just think about how many ways there are that can negatively impact the person who is making the call. Of course, not all calls are inquiries for business, but you never know which call is, in fact, an inquiry; so all need to be answered appropriately. In so many operations we are getting the automated voice mail. Ouch! We may already be in trouble, again. An appropriate, real voice, would be much preferred.

So, what does this voice sound like? Does it sound like someone has been Interrupted by the call? Is the person answering phone sound like they are in a hurry? And, how about those very long marketing messages . . . "thank you for calling the XYZ hotel, blah, blah, blah, blah . . . ," and on and on we go telling someone all about our wonderful facility. Too much marketing stuff up front on the call doesn't make a caller feel good. How about a nice, warm welcome, reflecting a smiling face?

Do we show Interest in the caller? Do we have the patience or do we sound rushed? How about forwarding the call to someone who could be of service? Do we really know to whom this call needs to be referred? Are we in a position to answer questions? What questions do we need to ask the caller to be able to handle the inquiry? If the call needs to be referred to the sales or catering department, or reservations, do we know that there is someone at that point who can take the referred call? If not, do we have a form available, from which questions may be asked so that correct information is received to be forwarded to someone who may be able to follow-up with a returned call? Every property should have some type of telephone inquiry information sheet available at the location where telephone inquiries are received. If you want a sample Phone Lead Questionnaire form to use, just send an email to: howardf@vt.edu.

FINDING LEADS TO DEVELOP INTO QUALIFIED PROSPECTS IS THE BUSINESS OF SALES

April 2015

The first step in building new business for hotels, resorts, conference centers, etc., is to find leads that could be developed into qualified prospects. After all, a "lead" is only a suspect, that is to say someone or some group may very well be in a position to do business with a property, but we do not know that for sure. Not until we make contact with the lead, and ask a series of questions can we determine if, in fact, that lead may very well be in a position to do business at a property. If so, then the lead becomes a Prospect. Therefore, the idea is to find "suspects" who may be logical sources of businesses, organizations, or individuals who may end up being prospects.

One may be surprised to learn that there are loads of leads sitting right there under our noses. All that is needed is to make phone calls and ask appropriate questions to determine if those leads we have found could be turned into prospects. So what is the secret to finding these leads? The absolute best source for the most productive leads is right there in the sales office. Just take a look at all the old files, whether they are computerized or kept in folders. There will be many files well over 2-, 3-, or 5-years-old that have never been traced or looked at in those many years. All it takes is for someone to open the file, get information on the previous contact, make the phone call, and start asking appropriate questions. Here, we are not discussing trying to make sales. All that is needed is to get all information possible to determine if, in fact, this contact could be a prospect. Remember that a qualified prospect is someone or a group that is in a position to do business with the property they chose to do so. It would be advisable to use some sort of form upon which to write the responses to specific questions to be asked. The selling doesn't really start until another contact is made with the appropriate person who may make the decision of using the property.

How about the idea of getting leads involving employees from all departments of the property? Many employees belong to a variety of organizations like civic groups, and social clubs; some have a spouse whose employer has company meetings. Sales folks can arrange with department heads to attend one of their staff meetings and deliver a short presentation to explain about developing leads for new business. Have handouts for the employees reflecting what types of business for which they may provide leads to the sales department.

SOURCES FOR LEADS TO HELP QUALIFY PROSPECTS FOR THE GROUP MARKET

May 2015

In the last month's issue of *The Sales Clinic,* there was a discussion of the importance of locating leads (suspects) in order to develop those leads into qualified prospects. First, we find a lead, through various sources, then make the phone call to actually determine whether that lead could be a qualified prospect, that is, someone (or group) who could do business with you, should they chose to do so. Using a Prospect Survey Sheet listing questions to ask and the responses will help make the determination.

Our first suggestions as using old files to make the phone calls, following that we suggested that the property's own staff could very well be a good position to offer leads via their families and connections. Another great opportunity for leads would come from your daily newspaper where one can find logical sources for businesses, organizations, or individuals who may end up being prospects. Carefully read every section of the paper, and seek leads such as: companies moving into your area; people with companies being transferred into your area; organizations managing sporting events that would attract sports groups or spectators. Newly engaged couples reflected in the social columns of the paper would certainly be good leads for wedding events. Local associations and organizations may reflect names of their officers or list their activities which may lead to local and regional meetings. Look for meetings and events being held at other properties in the area would be excellent leads for developing prospects for next year's meeting. Just checking out a daily newspaper could generate as much as 100 leads. Just select what looks like the most promising lead to call to develop as prospects.

Referral prospecting could be very promising. Here is where you make contact with those prospects already doing business with you. Satisfied clients in many cases would not mind offering you leads of those they know who could become prospects, if followed up. Further, just because a property may be doing business with a company or organization, for one particular meeting or event, consideration should be given to seeking out other business coming from the same company or organization. Make the inquiry.

In the event of calling a lead which does not become a prospect, it should be determined if that contact would be able to furnish other leads for business. Ask the question.

The best lead, of all leads, would come from the telephone inquiry. This is where we lose some very good leads, but never know about it. It is the manner in which the phone is answered, and the appropriate questions that need to be asked, as well as where the call is transferred. Properties probably lose most of their phone lead by allowing the phone to ring too long, as well as by using voice mail in answering calls rather than a real, live, friendly voice.

For a free copy of a Phone-Lead Questionnaire send a request to: howardf@vt.edu.

CHAPTER 3

THE WORKING OF A HOTEL PROPERTY SALES OPERATION

SMALL OPERATORS MAY NEED DIRECT SALES EFFORT

January 1980

Current economic conditions and the fuel availability situation are producing a decline in occupancy at roadside and suburban lodging locations. If these two conditions don't improve in 1980 to restore the normal reasonable flow of hotel/motel traffic, what does the smaller lodging operator do to maintain these sales? Consider hiring a full-time salesperson. Regardless of the size of the property, a full-time salesperson doing an honest job, with proper direction, will return at least 10 times the investment made in the person's salary. There can't be any substitute for a direct sales effort in an operation that needs the additional business. A direct sales effort means having a person on the staff who goes out to contact sources of new business that will profitable to the property.

That person gets a new business by making sales calls to prospect companies and organizations that are in a position to put guests in rooms and restaurants. These calls are made in person, by phone, and by mail. The direct sales effort is a very meaningful way to bring in additional business and could very well be the most productive way for a property to become more profitable.

WHO DOES THE JOB?

Now, who does this job for the property? Remember, the operator is really responsible. A decision must be made whether this effort will be undertaken by the operator, personally, or by a special salesperson. In a small property where the operator may feel that he can't afford to hire additional staff, then, of course, the operator must find the time to do this job, even if it is just a couple hours a day.

However, be aware that there are problems with this kind of setup:

- ▶ The operator still has all the other responsibilities of running the property—so a couple hours a day may not be available.
- ▶ Daily sales calls probably will not be made consistently.
- ▶ Leads may not be followed up immediately.
- ▶ Correspondence may fall behind.
- ▶ A good file and record system may not exist.

The lodging operator personally handling the direct sales effort is the best approach to setting up a sales operation—provided all the other daily functions of operating a property are turned over to someone else. If the operator of a property cannot handle this personally, then he should add a salesperson to his staff who will handle sales only. Don't try to mix jobs—the salesperson should not end up being a part-time desk clerk, part-time secretary, or part-time anything. A salesperson should stick to sales. Responsibility for the sales effort ultimately lies with the lodging operator—whether the operator is called the manager, General Manager, innkeeper, managing director or owner, that person should truly be the "general sales manager." The lodging operator cannot shirk this responsibility—even if the property has a full-time sales staff—the top guy still has to make sure it works, that it is effective and brings in additional revenue.

In this new series of columns, we will explore sales problems and opportunities to come up with answers and recommended courses of action to improve business through a sales effort.

PUMP $25,000 INTO OPERATION TO GET SALES EFFORT ROLLING

July 1980

Mr. Feiertag,

Our firm owns and operates four metropolitan-area motels of from 80 to 116 units in the East. Two of our properties were acquired in a somewhat rundown condition, and we have or are renovating them. We have something of an image problem, but none of my managers have expressed any interest or aptitude in making cold sales calls. In one property in Indiana, we approached a desk clerk and asked her if she might be interested in sales. She was enthusiastic, so we sent her to a sales orientation seminar. She started making calls and loved the work. One month later she was offered a job at more than twice what she was making. We couldn't match the salary.

My question to you is, what do you think of the following alternatives: (1) hire a sales development or marketing person to handle the four existing motels and any which we may acquire in the future; (2) seek local sales or marketing organizations in each of the cities to make a local sales effort. I am anxious and willing to devote as much time as possible to our marketing and sales effort. I don't enjoy sales and am not comfortable with it. But I feel I could work well with someone who enjoys this type of contact. Are there any other plans which you feel we might consider? We know this is an important and crucial area and we are barely scratching the surface.

Morton Kimball
Halifax Square Associates
Washington, D.C.

Dear Mort,

The situation you mentioned in your letter is certainly not unusual. There are many small motels around the country that are operated by groups such as yours and the same question always comes up: "What can we do about sales?" My suggestion would be to seek out an experienced *motel* salesperson who has a couple of years experience in developing sales for a small motel. The function of this person would be to work on your motel properties, one at a time, to develop a sales program. Then, get the local management interested and involved in the sales program, or someone else at the properties who can make sales calls on a part-time basis.

I'm guessing that this could very well take about three months to do this job properly and start bringing the business in. Once the program is set up in a property, and you and the new salesperson are satisfied with the progress of the program, relocate your salesperson to another one of your properties and repeat the same program. Within the year, all your properties should have some sort of sales program. The person you hire to do this work now becomes the director of sales over your group of four motels. If you bring more motels into your group, the same steps should be followed for them.

The director of sales, of course, continues to supervise and work with each of the properties, as well as going out and making personal sales calls on behalf of all the properties. He or she could also work on developing business with bus companies, convention groups, etc. This will cost you about $15,000 to start with, for the salary standpoint, and possibly another $10,000 in expenses. So for about $25,000 total investment, you could certainly get a worthwhile program going on each of your properties. You had indicted also in your letter that none of your managers are anxious to make sales calls. Take the one manager who feels most strongly that he is not cut out for sales work and sit down with him for one hour. Have him call some people who have done business with him in the past and ask them to come again.

I am sure will find that all of a sudden a miracle occurs and he finds that he is a very capable salesperson. Believe me, Mort, it does work. We've tried it over and over again and these non-believers become believers very rapidly. It is really not that difficult. Regarding the alternative, you offer of hiring a sales development or marketing person to handle the four existing hotels. I believe I answered this above. However, I do caution that you do stay away from a "marketing person." As for the second alternative, I really don't feel that there is much strength for you in trying to hire a local sales or marketing organization to do your selling for you. I honestly feel this is not the place for an outside company to do your work.

FIRST CALLS: TAKING THE PLUNGE INTO "COLD" WATERS

September 1980

The first two weeks on the job for the new salesperson are very critical. This is the time when the lodging operator and the new employee can determine if the job will work out. A good many people indicate an interest, in hotel sales because they "like people," but it turns out that making outside sales calls isn't what they really wanted. During the early stages of employment, it is important for the lodging operator to have daily debriefing sessions with the new salesperson.

The daily calls should be reviewed—who, what, when, where, why, etc.—and discussed for each sales call made that day. The session should be held the same day the calls are made, immediately upon return to the property. It is too easy to get into the trap of the salesperson "turning in" the completed call report sheets for the manager to "review" and discuss at some later date. Reports seem to get piled up, never discussed, communication breaks down, leads are not followed up, and the property ends up with no sales program.

Management must find out how this person feels about making outside calls. If management feels that the salesperson is not comfortable making calls, can't seem to develop enough information, doesn't make a sufficient number of calls, or just generally reflects an uncaring attitude about the job, this could mean the wrong person was selected. To do well, people have to be happy in performing their daily tasks. It is not an easy job making sales calls. There are many frustrating experiences involved in calling on company offices in any city and some people cannot cope with this type of work. This is why it is very important to have daily sessions to review the salesperson's activity and to get a feeling if the person is really "cut out" for sales.

Make Calls Continuously. The sales trainee needs to concentrate on continually making calls on businesses in the community to develop information. This is to qualify leads so that appropriate records can be maintained to ensure that a follow-up is made to sell and book business. Use of the lead sheet (described in the August 1980 issue) is important to accomplish this mission. It also puts the sales trainee in a position of "making a survey" rather than having to immediately sell a product with only limited knowledge and experience.

These types of visits to businesses should be made "cold" (without appointments). The theory is to make as many "cold calls" as possible with a geographical area to maximize time and productivity. The important aspect is to gather information from any source at the place of business. Ultimately, it will have to be known if that company or person contacted is a source of business for the hotel/motel/property.

A key to maximizing use of time is to preplan for the cold calls. A city directory or street guide is very useful. Daily activity may be planned by one street at a time in the business sector of a city. The objective is for the sales trainee not to miss any address. It is sometimes natural for a salesperson to stand outside a place of business and try to anticipate or guess if "this place would be a good source of business," each address should be considered as a prospect for business, and the contact there should be identified on the lead form.

No Time Wasted. Naturally, there will be some contacts made that will reflect absolutely no business for the hotel or motel. This should not be a problem since it will be discovered within five minutes so very little time is wasted. However, the person being interviewed should be asked if he or she knows of anyone personally that might be in a position to give the hotel some business. The answers might be surprising. Many leads can be developed this way for social functions, such as weddings, school reunions, club dances, relatives visiting, etc.

Maximizing use of time is important to make volume calls; therefore, the sales trainee should be in and out within fifteen minutes, barring unforeseen circumstances. Considering that calls should be made between 9:00 a.m. and 11:30 a.m. and then from 1:00 p.m. to 4:00 p.m., this would provide 5 ½ hours per day of solid calls—reflecting a minimum of twenty completed lead sheets. In a heavily concentrated calling area such as a downtown office building, the average time per call will be reduced since the walking distance between businesses is shortened.

During the review of call sheets, those that reflect absolutely no opportunity for any type of business may be discarded. The remainder of call sheets are used as a basis for starting files and maintaining a follow-up system.

A PLACE FOR EVERYTHING: SETTING UP A FILING SYSTEM

October 1980

Hiring, training and supervising a salesperson must be followed by putting in a procedure and system for files and records. The procedure outlined herein is a guide to a very simple, basic, flexible system of maintaining sales files and records for a small hotel or motel property. A metal file cabinet is needed to store files. These are manufactured with one to five drawers. The cabinet should accommodate letter-size (not legal-size) files. A locking device on the cabinet is not necessary. Manila file folders are needed to house individual files. These should be purchased letter size, with either a right or left cut. (This makes for eye ease when searching for a particular file.) File dividers from "A" to "Z" are used to separate the files alphabetically for easy searching "A," "B," "C," etc.

FILING GROUP BUSINESS

There should be a file for every piece of group business: only potential, already booked, in the process of being booked, or has the opportunity of being solicited for some type of future business. The simplest way to locate a file is to keep the files strictly alphabetical by the name of the company, association, corporation, or organization—all together in the same file cabinet. Each group has its own file folder.

Abbreviations must not be used if a strict alphabetical system is to be followed. Instead of "FBI," the file tab should read: "Federal Bureau of Investigation." Instead of IBM, the folder should read: "International Business Machines," etc. There are some companies or organizations that have only initials. In these cases, the files are placed alphabetically is a strict sequence. In filing folders on organizations and associations that have local, state, or national scope, it is important to strictly identify the organization's file accordingly. Just to identify a file as "Chamber of Commerce" is not satisfactory. The file tab should read "Tampa Chamber of Commerce" or "Florida Chamber of Commerce" or "United States Chamber of Commerce," depending upon which Chamber of Commerce organization your file information reflects.

There are times when different departments of the same company will be sales prospects for a property. The specific department of a company should be identified on the file folder tab, such as "American Telephone & Telegraph Long Lines Division" or "General Electric Corporation Industrial Division."

File Contact's Name

There is a varied collection of files reflecting types of business that ordinarily would not repeat and, for the most part, would reflect only banquet business. These miscellaneous files are in the name of the individual rather than in the name of a group and include weddings, bar mitzvahs, retirement parties, engagement parties, anniversary parties, etc. The files for these types of activities are also kept alphabetically by the name of the individual for whom the party is planned. These miscellaneous files are kept in the file cabinet right behind each divider file for each letter of the alphabet. The file folders should read: "Miscellaneous A," "Miscellaneous B," "Miscellaneous C," etc.

Miscellaneous files should always be filed by the last names, and again in strict alphabetical sequence. On wedding files, use both last names of the wedding couple with the bride's maiden name first. If Sally Jones is marrying John Smith, the file should read "Jones-Smith Wedding." The same procedure is followed for engagements—"Jones-Smith Engagement Party." Once the function has been completed, bills paid and file closed, it should be put into the "Inactive File."

One large folder houses all inactive files and is located at the end of the alphabetical files, behind the "Z" section. This set of files includes completed miscellaneous files and other files where no further follow-up is necessary. These files may include companies out of business, organizations discontinued or other reasons for no possible future business.

Every group and the miscellaneous file is a prospect for future business; therefore, each folder in the file is always "active" and must be worked continuously until such time that it is designated as "inactive" and is relocated behind the "Z" section with other "inactive" files.

STACKING THE CARDS IN YOUR FAVOR: THE 'TRACE CARD' FILE

November 1980

The main and most important files are the ones kept on prospects, existing business, and completed business. The files are continuously worked and followed so that business can be repeatedly booked by the sales department. Other files and records to ensure the smooth running of a sales department are necessary. These include the following:

Internal Files—These include any type of information for a file that is not considered material for sales files. Internal files are kept to assist the sales department internally, and contain internal memos concerning policies of management, information on the competition, mailing list resources, copies of sales activity reports, information on local points of interest and activities, etc. These files are kept separately from the sales files, preferably in another cabinet or drawer.

Trace Card File—A reminder system to follow-up on prospects is very important in sales activity in any type of business. In hotel/motel sales, there is a need for constant follow-up on individuals and groups so that future bookings may be finalized. A trace card system is very easy to maintain and is an excellent method for reminding salespeople every day as to which files need to be "pulled" for follow-up.

The system calls for the purchasing of a supply of 3" x 5" index cards (with ruled lines); a metal file box to hold 3" x 5" divider file cards, a set of 3" x 5" divider file cards for each month (January through December) and a set of 3" x 5" divider cards for each day of the month (1–31). One 3" x 5" card is identified for each sales file in the cabinet. The name of the group appears on the index card at the top left and is typed or written exactly as it appears on the file folder tab.

No other information appears on the trace card except for a date when the file needs to be "pulled" or traced again. (Trace cards are also made out for miscellaneous files.) The person working a file must decide on what date it must be "pulled" for follow-up before returning it. The date is entered on the far left of the card on the first line under the group name. The same date is then put on the front of the file folder. On subsequent pulling of files and trace cards, the old date is crossed out and the new date entered just below the old date. Once the trace date has been decided and entered on the trace card and on the front of the file folder, the file is immediately returned to the cabinet. The trace card is placed into the trace card box behind the divider card for the month when the file is to be "pulled" again.

The divider cards for days of a month (1–31) are used only for the current calendar month. In November, all trace cards for the month will be behind the divider card for November. All the remaining trace cards are kept for future months behind each respective divider card for the months the files are to be traced. At the end of each month, the daily divider cards are placed in sequence (1–31) behind the divider card for the next month, and all the trace cards for that month are then placed behind the daily divider cards as indicated on the trace card.

Whenever a sales file is pulled, it should never be returned to the cabinet until another trace date is recorded on the trace and file folder. This ensures that a subsequent call is made for follow-up. The trace card file needs to be checked daily to see what files need to be worked. Sometimes the workload is very heavy and files cannot be worked. In this case, all files for that day should be reviewed anyway. If it is determined that a follow-up may be put off for another time, the new trace date must be established and entered on the file folder and on the trace card.

THE JOB IS NOT FINISHED UNTIL THE PAPERWORK IS DONE

January 1981

Unfortunately, there will always be "paperwork" in order to do the job right. There are many hotel/motel salespeople who would prefer to only sell and not be "bothered" with paperwork; however, that's just not possible. Besides the various reports and records previously covered in this series, there is still the basic communication necessary between the lodging property and the customer. Communication in the written form is the only sure way that everyone involved can know what is going on. Not only is it a courtesy for the salesperson to write to a prospect or customer confirming arrangements, it is also an important business practice. The well-used philosophy, "If I didn't put it in writing, I didn't say it," should be the salesperson's code.

After a visit to a prospect, whether it be on the property, of property or by telephone, it is very important to write a letter confirming what was said. If dates or rates were committed or any service was offered, put it in writing to make sure that there was no misunderstanding.

Property managers and salespeople should develop a property booking sheet which may be used as a worksheet for the salesperson to record all the details of a booking. Once the sheet is completed with all the requirements of the booking, it should be reproduced and mailed to the customer with a cover letter for approval. Getting the form signed and returned to the property confirms the piece of business indicated on the booking sheet.

Sheet Details Requirements

The booking sheet may take any form but should include detailed requirements and prices committed for sleeping rooms, meetings, and banquet services. There are a wide variety of these kinds of forms now in use throughout the industry, and they have many different names. They are called booking sheets, booking reports, worksheets, convention or conference sheets, function sheets, etc. Although every lodging and restaurant property uses a somewhat different form, basically the information needed to complete the form is the same.

Not only is the form used to acknowledge the details of the booking to the customer, but it may also be used at the time information is taken on an inquiry, tentative booking, or a definite booking. It is most important that the salesperson or property manager try to get as much information as possible from the person making the inquiry or booking. Even if it is only an inquiry, information should be taken on: Date, Name, Address, Phone Number of Person Inquiring, and anything else pertaining to the group's needs.

From this information, the salesperson may start a file so that appropriate follow-ups are made to book the piece of business. This information form does not need to be typed, but completed in pen or pencil. The main thing is to *get the information*. Any information given to the prospect (the person making the inquiry) needs to be indicated on this "worksheet" so that there is a record in the file. This may include any rates quoted for rooms or food/beverage functions, availability of space, commitments of service, etc.

This same "worksheet" remains in the file so that all subsequent changes, additions, or deletions may be recorded. In this manner, all information pertinent to a booking is on one sheet, which makes easy reference. When the booking is confirmed and all information (prices, rooms held, function details) is recorded, then it may be neatly typed so that it may be forwarded with a letter to the customer for verification and signature. The original, with a copy, is sent to the customer with a request that the original be signed and returned, and a copy is kept in the file.

THE FUNCTION BOOK: AN ALL-PURPOSE RECORD-KEEPING TOOL

February 1981

Establishing a file and record procedure for a sales department is never complete without controlling the profitable use of public space. A function book is used for this purpose. At properties with meeting and banquet facilities, a very careful record must be kept so that an easy referral may be made to determine what public space has been committed for functions. The function book is a simple record keeping system. It is a book with a page allocated for each day of the year. On each page, sections indicate different times of the day or just, "a.m.," "p.m.," or "breakfast," "lunch," or "dinner." All meeting and banquet rooms are identified on the page. Larger properties with many meeting rooms can use a double page.

The meeting/banquet rooms are identified by name, number or letter and these are listed down the left side of each page. Across the top of the page are sections for periods of the day. It is here that every function requiring meeting or banquet space is recorded. Function books vary throughout the industry. They come in all sizes and styles, and some printing companies specialize in making these books to suit a particular property. In fact, many properties put together their own books. Whatever form it takes, this book is a must. It is the only sure way to block space for a function without resulting in a double booking.

Use the following guidelines regardless of kind of book used:

▶ Make all entries in pencil. Many changes are usually made between the time a room is blocked and the time it is actually used. When a change is made (or a function is canceled); the information may be easily erased and the new information recorded.

▶ The name of the group should be entered as it appears in the file so that there is no difficulty in locating the sales file for reference.

▶ It is important to enter the time the function is to start and end. This information is helpful if there is an opportunity to book another piece of business immediately following the conclusion of a function. The person making the entry should record his initials and the date the entry was made.

▶ Tentative bookings are marked with a "T" and definite with a "D." This allows the salesperson the opportunity of checking for a prospect before denying any space that may have been previously blocked for a tentative booking. The tentative booking may never develop into a definite. Therefore, it would not be practical to turn down booking opportunities before checking the tentative.

▶ Sometimes a prospect is not sure of dates needed for a function and the sales manager may need to block space tentatively on more than one date for the same function until a decision is made by the prospect. In this case, it is most important that all entries for each date held include a pencil note to this effect. When one date is finally selected, the note will remind the sales manager to erase all other entries.

The function book may also be used by property management for forecasting future business and scheduling labor. As you can see, the book is a very valuable but simple tool. The major supplier of function books is Hodges and Irvine, 1900 Sinclair St., Clair, Michigan 48079, (313) 329-4787. All books furnished are imprinted with a date on each page, national holidays, and function-room names or designations. A wide variety of book styles are available as well as books for tennis court reservations, tee-off times for gold and specialized books for hospitals and universities. Lodging properties purchasing function books from Hodges and Irvine receive a proof copy before it is printed. Usual production time is six to eight weeks. The company does accept collect telephone calls.

EXAMINE ALL PROFIT CENTERS WHEN BOOKING GROUPS

April 1981

When considering group bookings, take a close look at the total amount of business that can be brought in by a group. All the profit centers of an operation need to be considered. Some groups generate more sales and bottom-line dollars than others. Because of this, the sales department must know the background and history of all business prospects. Before booking a group, the experienced salesperson must know more than that a group will generate "x" number of rooms per night at "$" rate. Some groups are heavy restaurant or lounge users, while others are just "coffee drinkers."

The profit centers of a hotel/motel include sleeping rooms, meeting room rentals, restaurants, lounge, vending machines, room service, game room, valet, newsstands, gift shops, leased shops, and telephone. A large group of long-distance telephone callers could be very lucrative since long-distance telephone charges are commissionable to a hotel/motel. Knowing history also helps the sales department in their negotiations when securing the booking. Part of the negotiation process could mean giving extra complimentary rooms or "giving in" on a special consideration the client wants in order to confirm the business. If the group has a history of generating a large amount of lounge business, it may be very worthwhile to provide something "extra" to close the deal.

The object, of course, is to maximize profitable sales. There are groups that are booked into a hotel/motel but have meals and functions at a civic center or other hotel. This situation could create an empty dining room and lounge, when it otherwise would be doing some business from the regular transient guests. This is a very important consideration when booking motorcoach business. If possible, book this type of group with the understanding that they will arrive prior to dinner (and not stop on the highway for dinner prior to reaching the property). Also, it should be understood that the group will depart the following breakfast. In the case of very early departures, it might mean opening the restaurant earlier than scheduled.

Game rooms are an excellent source of added revenue. Working out the meeting and reception schedules with those in charge could help generate this kind of business. Many times attendees at a business meeting can be seen in game rooms or the lounge if there is a break. Sales departments should always try to work in at least a one-hour break between the time a meeting ends and a reception begins. This time usually results in additional revenue for game rooms, lounges, gift shops, newsstands and in long-distance telephone calls.

Extending the group's length of stay is an important sales technique at a property. Regardless of how many nights a group is booked, always attempt to get at least part of the group to arrive one day early or stay one day later.

Again, this is accomplished by knowing the group's history and working very closely with those in charge of scheduling the activities. If a meeting will be starting in the morning or around noon, some of the attendees will probably arrive on the day of the meeting. The objective is to encourage the group to start its activities the night before the meeting begins with a pre-meeting social activity. This would create additional room sales and some catering business. The sales department may have to do some hard work on this. Help get a sponsor for the party if the group doesn't have the budget. It can be done and is well worth the effort.

Meetings that conclude after lunch usually mean that attendees will check out to return home. The sales department should come up with ideas which extend the program so that guests will find it too late to start home in the evening. Some creative thinking will come up with the answers. All too often hotel/motel properties "give away" meeting rooms. Renting meeting space is an important and lucrative way to maximize sales. Meeting and banquet space is crucial for a property to bring in groups who need this space and will rent rooms. The negotiating process determines how much of the space should be given rent free because of the revenue from sleeping rooms. An arbitrary giveaway of all meeting space is not necessary. A reasonable number of rooms is expected by a group, but based on the type of setups, changeovers, etc., some rental fee should be charged.

Also, with costs of operation skyrocketing daily, it becomes more important than ever to analyze the amount of dollars generated for a meal catered in a meeting room. In some instances, it may be wise to establish a rental charge for the room, crediting the amount of food and beverage charges in that room towards the rental.

WEEKLY MEETINGS—A POSITIVE WAY TO INFORM DIRECT STAFF

June 1981

A weekly sales meeting is an absolute must for any size property. The objective of this meeting is strictly communicative so that all department heads will know what is happening. The purpose is to keep staff personnel informed of the business expected in the property the following week and to direct the staff members on the fulfillment of their respective duties.

The following outline for the weekly "sales" meeting (probably better referred to as the weekly "communication" session) will be helpful to any size property management:

Who attends—All key staff department heads should be in attendance at the meeting conducted by the property manager. These include sales manager (or director of sales); food and beverage director (or restaurant manager); housekeeper; front-office manager; chef; various food and beverage room managers such as dining-room manager, bar manager, or banquet manager.

When the meeting is held—Having it held on a regular weekly basis, at the same time, on the same day, will make it a habit. When the meeting becomes a regularly scheduled event, all attendees will not have to be reminded and will block out the time so there will be perfect attendance. It should start on time and end on time. A suggested schedule would be for Thursday afternoon from 2 to 4 p.m. This would not interfere with most dining hours, check-ins or check-outs and, generally, overlaps shifts. A Thursday meeting gives the staff enough time to plan and schedule for the following week's activities.

Where to have the meeting—NOT in the manager's office. A meeting room environment is needed without distractions. An office with a desk should not be used. The atmosphere should be relaxed, and light refreshments should be served (coffee, tea, soft drinks). The department heads attending need to feel comfortable and not put under any pressure—this is a communication session and should not be used as a "chewing-out" session.

What should be covered—After a general welcome by the manager with various congratulations for such events as birthdays or anniversaries (if appropriate), an agenda should follow. Participants may review:

▶ Profit and loss statement or where property stands on sales and profits
▶ Recent general memos from management or corporate headquarters
▶ Overall payroll
▶ Any reports, audits, or inspections recently conducted
▶ Guest comments or letters

Every opportunity should be given to attendees to comment during this session and provide feedback. Each department head should have an opportunity to verbally report on his or her activities and needs. The sales manager, for example, should review all bookings coming in for the following week and hand out function sheets, noting the particular needs of each group. The front-office manager should discuss room flow for the following week so that all departments are aware of staffing needs, especially in housekeeping and restaurant areas.

Following the report of department heads, the manager can allot time for a brainstorming session covering ideas for better employee productivity, problem-solving, and marketing suggestions. Effective internal communication will lead to satisfied and productive employees which, in turn, leads to customer satisfaction. And that leads to increased sales which leads to improved profits.

SALES REPORTING SYSTEM MEASURES STAFF PRODUCTIVITY

August 1981

The following comments are excerpts from a letter I recently wrote to Gunter Muller, resort General Manager of the Wintergreen (VA) resort. Muller had inquired about setting up a form for reporting on activities in the conference sales office. The real measure of productivity in the conference sales office is the bottom line. The profitable sales that are generated as the result of sales activity are what we are really after. There is a wide range of activity in a successful sales office which leads to profitable sales. The operator's objective, of course, is to measure this activity to see how productive the sales staff may or may not be, and how much time is spent on each activity.

I would suggest that you start by listing these types of activities. These include outside sales calls, in-office visits (prospects visiting your property), letters written, telephone solicitations, direct-mail programs, working on conferences (i.e., time spent on the property helping out during a conference), and "other than sales."

MAINTAIN DAILY LOG

Each salesperson would maintain a daily log of the amount of time spent on each activity along with numbers where appropriate, such as a number of outside sales calls, telephone calls, in-office visits, etc. Property management then gets a weekly sales activity report from each person on the sales staff. The form should also reflect information on business booked each week plus year-to-date with a comparison of last year, year-to-date. Only definite, confirmed and deposited business should be recorded. Tentative is not included, but there is no reason you could not include this item as a separate entity so you can see how tentative measure up against definite.

The accompanying chart model can be revised to suit your own needs. You can add in a section of numbers of rooms along with dollar revenue estimated, if it could be meaningful to you.

BOTTOM LINE GOALS

The whole objective, of course, is the bottom line. How well is the sales staff meeting its sales projection—monthly, quarterly or annually? The activity report also gives you insight on activities so that you may provide continual direction. Also, if you implement such a reporting system and the salespeople maintain a daily log, you will have automatically installed a time management tool that will guarantee improved productivity and performance.

SALES PROFESSIONALS NEED RECOGNITION

October 1987

Hospitality industry people are realizing the importance of becoming more professional in their specific fields. It takes time, hard work, experience, interest, love of work, peer respect, association involvement, sharing with others, and developing relationships, among other things, to be a professional. To be recognized as a professional is truly an honor in any industry. When it comes to tourism and hospitality, there are already several areas of recognition through a certification program sponsored by specific organizations. Travel agents who qualify can be certified as a CTC (Certified Travel Counselor) by ICTA (Institute of Certified Travel Agents). In our hotel industry, professionals may qualify to become CHA (Certified Hotel Administrator) by the American Hotel and Motel Association. The CAE (Certified Association Executive) is the professional designation awarded by America Society of Association Executives (ASAE) for its members. Meeting Planners may qualify for the CMP (Certified Meeting Professional) by the Convention Liaison Council (CLC) and, of course, hotel/motel salespeople have the opportunity to qualify for the coveted CHSE (Certified Hotel Sales Executive) recognition award by Hotel Sales and Marketing Association International.

A certification program by any organization is designed to provide professionalism recognition for persons who qualify within a particular industry. Usually, in order to qualify, certain points need to be collected, based on experience and contributions made within the specific job category or industry. An application of qualification, examination and generally a published article may be part of the requirements.

More and more, it is becoming important for hotel/motel salespeople to achieve the deserved recognition of a CHSE. The buyers of hotel services who, themselves, are professional in their own areas, recognize what the CHSE designation means. Buyers really want to do business with industry salespeople who are recognized as professional—and more likely than not, properties with CHSE staff members have a better opportunity of doing more business. CHSE is the designation of professionalism in the sales and marketing area of the lodging business.

It serves to identify to both industry peers and customers that the individual entitled to this special designation has achieved a high level of competencies in sales and marketing experience, education and service. To qualify as a candidate, applicants for CHSE must be actively employed in the lodging industry. They must complete a pre-examination form and then pass an examination component. The pre-examination component includes industry and industry-related experience, association service, and education requirements; with a required total of 250 points attainable from any combination of factors related to experience, association service or education. The examination component contains two parts: a written examination and an original research paper prepared specifically for CHSE consideration.

Although the certification program is sponsored by the Hotel Sales and Marketing Association International, it is not a requirement to be a member of the group. Qualifying for the CHSE designation is an important step in education and career development for any hotel/motel sales professional. It is a goal worthwhile attaining.

SO YOU'RE THE NEW DIRECTOR OF SALES?

November 1992

There's nothing new about the problem we are having in producing profitable sales these days. As a new DOS, you are added with a lot of problems—coming into a new property—or even getting promoted from within. Who does what? Do we change market segment assignments? Does the file system need changing? Are the salespeople happy? What about turnover? Should we change the market mix? Are sales goals too high? Is the hotel providing good service to our group accounts? Is there too much paperwork? Do we have enough secretarial help? What can we do about public relations? Can we make our advertising more effective? These and many other questions come to mind it is difficult to determine where to start.

Here's where you start! Make yourself a hero. Get right to the point of building increased profitable sales . . . right away. It's wonderful being in charge, but don't forget you are a salesperson too. Use your contacts, your experience, your ability, not only to manage people, but to also create additional sales—yourself. Of course, you do not want to forget your people who work with you in the department. So you want to get them to be more productive as well. However, you do this mostly by setting an example. You should be the best salesperson in the department. Your first priority has to be to get more business on the books. Of course, by using the multiplier effect (getting other people to sell better) you can generate more sales. So while you are doing your own thing in the area of closing deals you must also pay attention to the needs of the other sales personnel. Don't worry too much about the internal workings of the organization—let the General Manager do that. Try to get everything out of the way of making sales. Don't try to change the filing system, just make it work the way it is. Don't worry about changing forms and reports—use what you've got. Let the General Manager take care of public relations and advertising. Selling and getting your people to sell is your job.

Go out with your salespeople, observe what and how they are doing. Meet daily with each person to review daily activity. Provide direction and supervision. Be sure to read all their correspondence, make improvements. Observe them on-site inspections. Sit in on negotiation sessions. How do they relate with customers? Counsel with them and train them, one on one. Be sure they are maximizing their selling time. Implement some sound time management principles.

Now, once you have proven you can bring business into the property and you have worked with your sales team to make them better salespeople and the sales reports show it—you can go on to other things. Set up a time management system for yourself. Allocate a certain amount of time weekly to your various functions. You should never stop selling, so take about 50% of your time to do just that. Don't forget, you are still the best salesperson on the staff. You have got to continue to supervise and train the staff so allocate about 30% of your time to that. Now there is time to review the marketing plan and make adjustments as necessary; see how to improve the flow of paper, if necessary; try to reduce paperwork to make it more workable; examine market segment assignments and see if there is any need for changes now that you have had the time learn more about the sales staff; discuss budgeting considerations with General Manager to review cost of sales to overall revenue; review opportunities for increased profitable sales through changes in market mix and there are other areas for your involvement from a sales standpoint.

The thing to remember to be successful, however, is to stay in sales. Once you get involved in other things that may even be delegated to you by the General Manager, you will lose control of the sales situation. The job will take 100% of your time. Just don't let other things get in the way. Improved profitable sales are the most important thing right now to involve your time.

THE HOSPITALITY SALES OFFICE OF THE FUTURE

October 1993

No Virginia, we are not doing away with sales personnel. Sales offices will not be run by robots and computers will not replace people. What we are talking about here is creating a more sophisticated sales situation where more time will be spent in the actual closing of sales rather than prospecting for new customers. It's anyone's guess how much of a salesperson's time, today, is being spent on productive sales work that is, making real sales calls. We've heard guesses anywhere from 30% to 50% of the time. What if we can create a situation whereby we can increase conversions and thereby make salespeople twice as effective and twice as productive? It can be done, and in fact, we will start seeing this more and more as we look towards the sales office of the future.

The concept is really simple ... it has to do with information and an integrated system driven by information supposed by a tele-research environment and designed to impact a sales organization. Information is the key to success in any sales operation whether it be in the hospitality arena or elsewhere. If your sales personnel would work only on actually qualified prospects, that is, only those accounts that we know could, or want to, or have to meet at your property, wouldn't you see your group sales soar? So where is this information? A good amount of it is already in your hands, available, but not being utilized. In the sales office of the future we will see our properties leveraging their past customers' database; going back into past customer files and maintaining an interactive dialogue. Simultaneously, as you understand, and your knowledge increases of your customer base you can then begin the process of also building a universe of new customers who look like and behave like your present customers. Salespeople do agree that there is a great value in maintaining an interactive relationship with past customers which leads to repeat sales, builds customer loyalty and the acquisition costs of capturing a new customer is seven times less. We spend a ton of money and a good deal of time going after new business.

What we need to do, and we will see it happen in the sales office of the future, is develop a profile of the perfect customer which will drive and build a database for our salespeople that will make sure they are talking to the right audience. But it would not only be database marketing all by itself that will make us more productive. According to David Green, Chairman of the David

Green Organization, of which Database Marketing Research is a division, we will develop the concept of integrated marketing in our sales offices. The components of integrated marketing for the sales office of the future will include direct sales (including telesales), direct mail, research, advertising, public relations, special events as well as the database. All the components will be interrelated so that the prime object is the gathering and management of information. Salespeople would be used to close definite business and all non-selling activities would be eliminated from the sales process. Green's theory is that currently, at most properties, 80% of our business comes from 20% of our files. Our salespeople should be working on that 20%. There are our best accounts and he calls them our "A" accounts. Put the rest of the files into a telemarketing environment, label the next possible best accounts into the "B" category. Then the "not so hot" accounts could be put into a "C" category. Now our objective, through research gathering via telemarketing and telesales, is to move "C" accounts into the "B" category and move the "B"s into the "A" category. Salespeople keep busy working the "A" accounts. This creates the constant lead generation which will have to take up for accounts that are lost for various reasons. Because of more information available, as well as building relationships over a period of time the losing are done more easily and rapidly. Green has been working on this concept with his clients over several years, developing the database for them and helping them move into the sales office of the future. When you talk with David Green about this concept, you hear of terms such as "actionable database," "dialogue marketing," "relationship selling," "data management." He says we should be learning from the consumer product giants who have been into database marketing for several years, but they too are still learning. We need to move into the 21st century when it comes to hospitality sales. Our sales office of the future will be an organization that spends all its time actively selling. Salespeople will spend time only with potential customers, closing sales rather than finding new business.

THE SMALL GROUP AND MEETINGS MARKET IS FOR ALL SIZE PROPERTIES

July 1994

The size of your property, whether it be a 25 room (or even smaller) limited service hotel or a full service 1,000 room resort, you can use and can get business from small groups. No one should ignore this very important, and very large, market that is generally available to everyone regardless of where you may be located. Airport, downtown, roadside, suburban, resort, conference center, convention hotel, tourist lodge, time-share resort, whatever and wherever you are ... there are some small meetings and small leisure group business for you. Meetings now present a potential for virtually any property.

When we think of the "association market," generally, we're visualizing the very large city-wide convention or the sizeable regional type of conference. We really need to start looking at the smaller meetings being taken place all over the country for the board of directors and various committees within these groups of associations. Most of the time we do not think of going to associations for meetings if we have a small property, but start thinking in that direction and let your state associations know that you can accommodate those very special small meetings. Then, from the business meeting standpoint, just contacting local companies in your area could very well generate small meeting business. We often don't call on companies thinking that because we are a small property, they wouldn't consider us. You may be surprised. You don't need to have a sales department in order to make sales calls. Anyone can make a call, in person, or even by phone, inquiring about the meeting needs of a company. Just take the time to make three phone calls a day and even throw in one or two in person call while you are going out to the bank or post office. How about calling on your vendors with whom you are doing business?

From the leisure market standpoint, there are more and more travel incentive programs being conducted by companies, rewarding their best employees with trips, made up of small groups of people going, not only to major tourist destinations, but to all parts of the country. Many of these are handled by travel agencies. If you are a small boutique type of hotel located in major cities or at resort locations calls on some of the companies that may already be sending you some room business ... after all they may not be thinking of your property for the small group incentive trip.

Also from the group leisure standpoint, there are the family reunions, military reunions, school reunions. Leads for these are easily generated through local contacts, veterans groups and the different schools in your areas. Small groups of people also travel to cities to attend funerals, which is a market that not too many small properties think about. Call on the local funeral directors who probably would appreciate knowing about your being able to cater to families attending a funeral.

There are hundreds of specialty tour groups, those because of the nature of some of the groups, are seeking out small, unusual types of properties in various locations. These tour groups are put together by tour wholesalers from all parts of the country. Recently I had the opportunity to observe a very small, luxury, boutique type of hotel in group action, the Inn At The Opera in San Francisco. With only 46 rooms and no meeting space and very limited group dining facilities, they have located meeting facilities within walking distance that could be made available to them for small meetings. There is a private area to the rear of their award-winning ACT IV restaurant that is just right for a group of about 25 persons for a reception and dinner. At the time of my visit, I witnessed such a gathering at the property having a wonderful time in that area. It is perfect for local companies that are bringing people into San Francisco for a small meeting ... a place to meet, with a great address (right in the cultural area of the city), luxurious rooms and service, a very attentive staff and superior dining facilities (they also provide complimentary breakfast bar for overnight guests). Here's a small hotel in a major city that is just perfect for the small leisure or business groups. Of course, the larger hotels also need small meetings and group market to fill in the gaps left by the larger groups meeting at those properties.

LOW-COST SALES AUTOMATION SYSTEM FOR SMALL HOTELS ... IT'S A BREEZE!

May 1995

For a long time now, since the sales automation systems have been in place, it has been apparent that only the larger properties with multiple sales personnel were able to afford the cost. For smaller hotels, limited service hotels, hotels with smaller budgets. It has never truly been affordable, or cost justifiable to install sales automation technology. However, with the ever-increasing need for additional new sales, repeat sales, and measuring performance, it is just about time now to take advantage of new technology in sales at a reasonable cost.

Group sales have always been a challenge for the small property without a sales staff, when it is the General Manager who has to wear that special hat along with all the other hats. Most managers are so busy that there has almost never been time to make sales calls or even follow-up on the inquiries that have been made by phone or letter or the follow-up comes about too late to book the business. It is amazing how many small or limited service properties fail to have any kind of sales file or trace system. Most do not have. "to do" list for solicit business. In examining sales activity at these types of properties, we find no evidence of any organized system of maintaining detailed information on contacts who have previously booked business at the property. Isn't it usually a satisfied former customer who is your best new customer? We seem to fail to follow-up in a systemized fashion for additional business.

If you had available to you a low-cost automated system that is very user-friendly, that is ... easy to learn, understand and work, it would put you in a better position of becoming aware of whom to contact, when to contact them, for what kind of business. You would have an easy, paper-less account file set up, that at the touch of a fingertip will tell you all you need to know about former business contacts, prospective contacts, when you really need or could use certain types of business. You would also have a built-in automated day timer to remind you daily of what sales type chores you have scheduled for yourself each day ... whether it be making personal calls, telephone follow-up calls, keeping appointments, and most importantly—reminders show up automatically advising you when to make a follow-up call on a prospect ... it would really be a breeze.

Actually, the name of this brand-new system, just on the market this month is called a "breeze." The product has been under study for a long time by Newmarket Software, the Delphi people. The colorful, inexpensive sales automation system is built around three components ... The Schedule Manager, The Activity Manager, and the Business Manager. Only a limited time is necessary for training (three days in the test properties). Newmarket came out with this new system to accommodate the small or limited service properties, with the idea to streamline sales and booking processes so more time could be spent on selling.

Reporting information is another area that is made easy with Breeze, such as performance reports on how well the property sold last month; revenue reports, market segments reports; booking details reports; activity reports; trace reports; etc. Also, Breeze includes a history file that keeps track of all past bookings so you can track how much business a particular group brought into your hotel or if they lived up to their promise of booking a certain amount of rooms each month. Having this power enables you to control to whom you are offering negotiated rates.

There is a good chance you can get a return on your investment within six months by booking additional business that under normal circumstances would have fallen through the tracks without the benefit of something like this. I was impressed with what the Breeze can do when I saw a demonstration at the recent AHMA Annual Meeting in Atlanta . . . it was a breeze.

WHEN SHOULD YOU HIRE A SALESPERSON OR ADD TO YOUR SALES STAFF?

October 1997

Quite often questions are raised concerning whether a small property should hire a salesperson or how many salespeople should there be at a large hotel. There are many smaller, limited service type properties where managers and owners have often indicated that the property is "too small" to have a salesperson. Maybe so, but is the size of the property really an indication of whether or not hiring a salesperson may be productive and profitable? It really boils down to ROI, return on investment. This would apply to any size facility regardless of size, location, or type of business. It all has to do with owners' or managers' level of expectation with regard to how much more business should be generated for each dollar invested in a direct sales effort.

Many years ago we experimented with this idea while I was with American Motor Inns. In the late 60s we had 26 Holiday Inns and grew to 50 by the mid-70s. Many of these properties were under 150 rooms, with some even less than 100. At the smaller size hotels, we came up with the idea that if we could generate $10.00 in new sales for every $1.00 invested it would certainly be worth the effort. To play safe, we started out at properties where we did not have any sales personnel and hired part-time people. We used college students to work three days a week, maybe three or four hours a day; housewives with children who only wanted to work part-time; and retired folks. All worked for minimum wage and a small percentage of new sales. The system worked exceptionally well at all locations. As the business grew, we found that we needed support for these part-timers and had to hire part-time secretarial help to handle the telephone, booking reports, correspondence, etc. These were also part-timers. At all but the smallest properties, this grew into a full-time job for a salesperson, with the need for part-time secretarial assistance. The whole idea behind answering the question of whether or not a property can "afford" a salesperson was to prove, by experimentation, can a sales effort produce a return on investment of 10 times the amount spent. For us, it worked and by the mid-70s 90% of our properties had full-time sales representation. And, a good thing it was for us. By the mid-70s we suffered an oil crisis in this country, had a recession, and it looked like people just stopped traveling. Business and leisure travel was down and most hotels around the country were suffering for business. At our locations, we had a tough time as well as anyone else; however, having salespeople on board who were responsible for developing new business, we got through a couple of years without missing a mortgage payment or having any serious financial problems. We were ready!

Now, the question is . . . are you ready if, and when, this "sellers" market turns around as a result of an over-built situation, and maybe even a turn-around in the economy resulting in a cut back in business as well as leisure travel? And, even if we continue in this wonderful, profitable, scenario we are now enjoying, would a salesperson added to your staff help increase profitable sales? What about the question of how many salespeople should you have at a property? Well, it really doesn't matter—as long as each is pulling his or her own weight and providing the property with a decent return on investment. Owners and managers need to come up with their own formula of what is a good ROI and start measuring sales personnel productivity. Is it five times salary you want in new business? Ten times? Or what?

WHAT'S THE JOB OF A SALESPERSON IN OUR INDUSTRY?

November 1999

Over the years I have seen many job descriptions for salespeople in our business. Most are very detailed, outlining the various functions of a salesperson. Not too often have I seen the word "sales" or "selling" in the job description. And, have rarely seen something that says: "bring in new business" or "increase sales by xx%." Get a group of salespeople together and ask them what they do. You'll get the details of the job: making sales calls, writing letters, following up with clients, going to trade shows, etc. Then ask the question concerning their responsibility, and you'll probably get the same answers. Shouldn't salespeople be responsible for bringing in NEW business to a property? Of course, we want a rebooked business as well. Hopefully, that will come with great service by the property staff. However, not all business repeats, even with wonderful service. Somehow, we need to replace what does not come back with NEW stuff. And, that should really be the responsibility of the sales staff. Why can't we get the idea across that salespeople need to spend all of their time in proactive sales? Many claim they want to do that, but are involved in other duties, or have other responsibilities at a property. Hardly ever do we get a show of hands when I pose the question at my sales workshops: "how many of you spend 50% of your time or more on actual selling to prospective accounts?" When we take a count on the amount of time spent it seems that most of the participants fall into the 35–50% slot.

It seems that so many salespeople (or at least they carry the title) spend a good portion of their time doing things other than sales. They complain about paperwork. It always just seems to be the number one complaint (sometimes valid and sometimes not). The next complaint has to do with servicing a group. Although many properties now have conference service managers, salespeople still think they have to stay involved with the group when they are in-house. Not very valid in my book! Making "cold calls" is still high on the list of things that salespeople do. Nothing wrong with that there is a time and place for cold calling. However, there are many opportunities for salespeople to get qualified prospects upon whom to make real sales calls. Isn't it more productive for a salesperson to call upon qualified prospects, to make a sale, than to busy themselves with cold calls when maybe they only uncover one or two prospects for every ten cold calls made? The idea is to let someone else do the cold calls to qualify individuals and groups who may have business for the property. Then the information gathered on qualified prospects are turned over to the sales staff to make the sale.

Take a look at the property staff. When is there some downtime at the front desk or reservations? Can those staff personnel be trained to do the cold calling? Of course, they can. In fact, there may be several other property staff personnel who would be able to do this with just some little training on how to do a "survey." After all, cold calling is just survey taking to gather information to determine who may be a prospect for business.

Another opportunity would be to use local college students. Perhaps some could be hired part-time, at minimum wage (if you can get them) to do this work. Compare those costs to the cost of a full-time salaried sales staffer. Come up with a Prospect Survey Sheet for these II survey takers 11 to use. It should have all the critical questions which need to be asked to determine who may be a prospect for business at the property.

SOME THINGS HAVE NOT CHANGED IN 20 YEARS WHEN IT COMES TO SALES

February 2000

Recently, due to increasing competition as a result of overbuilding in some areas of the country, I have had a good many inquiries from small, limited service properties, on how to get a better share of room sales. Twenty years ago the problem was similar; however, it was a result of economic conditions and fuel availability. To some extent, except for technology, things haven't changed much when it comes to the question of how to increase sales. In some recent columns, I have touted that managers of properties need to get involved in the sales effort. With this in mind, I would like to let you read my very first column I wrote for Hotel & Motel Management back in January 1980. Yup, been doing this column for 20 years, and without missing an issue. Here it is . . .

SALES TRAINING FOR LIMITED FEATURE PROPERTIES PRETTY LIMITED

April 2000

The over-building situation, in some markets, in the limited feature segment, is getting a little scary. With competition from new products opening all around us, many owners and operators of the smaller, under 100 room properties, are getting somewhat nervous. We have heard of a good number of managers cutting rates because of new competition. (Probably not a good idea, yet). Many have contacted me with questions about what to do to get more business when they need it, and how to combat competition.

We have looked, and asked around, to see what type, if any training is being conducted for owners, operators, managers of the smaller, independent (and even brand) limited feature properties. Hospitality Services of America does an outstanding job of putting on one-day programs, all around the country, for front office, guest service, and revenue management staff. However, there did not seem to be any other sales training programs available, except for some of the brands offerings.

With this in mind, we are considering setting up a series of sales training programs specifically for the under 100 rooms segment of the industry. There are a wide variety of topics which can be covered to help managers and owners in improved profitable sales. For example, there is no telling how much business you may have lost due to mishandling of telephone inquiries. As a matter of fact, this goes for the larger properties as well. Hospitality Services of America (HSA) does an excellent job in this type of training. However, we expect to cover it as well, with some good tips on how to determine what you are losing, and how to correct it.

We will also want to get into the business of group sales. Just because you have a small, limited feature property does not mean you should not go after groups. There are a wide variety of groups that rent rooms and do not need meeting space, and they are pretty easy to get, if someone will just make the attempt to get after it. We'll show you how. There is also the matter of how to go after local connections to build new room sales . . . even without a salesperson on the payroll. We will demonstrate how you can operate a sales department, without having one. There is also an opportunity for you to develop new business from markets you probably did not know existed.

How about getting some low-cost, or even no-cost sales help? It's out there for you. You can also develop new business through some local public relations and publicity, of course, without cost. Packaging with local attractions is something else you can get into. What's the best way to handle walk-ins so that they do not walk out? Are you getting any overflow from other hotels in your area? Even though you have a small hotel, can the local convention bureau be an asset to you? Of course, it can?

What about doing some low-cost direct mail to solicit some room business? Who can do some tele-prospecting for you without any additional labor cost? Then, of course, there's the question of how do you get more room business from existing guests through their referrals? And, how about building favorable relations with your own staff to help the sales effort? You also need to look at your best market segments and figure out how to get more of that kind of business.

We have raised some questions here for you to get answered to help you build your own business, and not to worry about the new competition. Now I want to ask your opinion. Is this a good idea to pursue putting on a series of sales training programs for owners, operators, and managers of the small (under 100 rooms) limited feature properties? Please let me know what you think of the idea, and perhaps even offer some other topics that need to be covered.

IT TAKES MORE THAN JUST SALES AT LIMITED SERVICE PROPERTIES

June 2000

Our April column ("Sales training for limited-service properties pretty limited") reflected on the growth of this under 100 rooms segment of our industry, and the need for sales training. Since then I have had the opportunity to participate in the Annual Convention of AmericInn International. Here is a fast-growing brand of properties which is really into sales training. It was impressive to review the work being done by this brand in providing not only sales training opportunities to its franchisees, but making available self-training tools to the owners, operators, and managers of these properties. Then, there is a large marketing staff headed by Jon Kennedy, Bill Hobbs and Mark Paulson who arrange for sales assistance and training at the property level by visits to their franchisees.

Most operators would agree that we all know the value of a good sales effort at any type of property. However, when it comes to the limited service product, in many cases, it is the General Manager who has to get it done. In discussing this with General Managers around the country, sales is always viewed as a top priority. When it comes to actually taking the time to get it done, it falls to the bottom of the list. It seems that at all properties of under 100 rooms the General Manager is the chief cook and bottle washer. Probably a day doesn't go by, at all types of brands, in all parts of the country, in all types of locations that folks running these properties always seem to be "putting out fires," and taking care of customer service. So, the question is: "How do we get to build more business without having to go out and make sales calls, if we don't have the time?"

In discussing this particular issue with attendees at the AmericInn International Annual Convention, I was impressed with the work being done to build business through public relations and customer service, without having to leave the properties to make sales calls. With the guidance of the brand's marketing staff, and materials provided, these properties have been busy increasing occupancy and revenues at this upscale limited service brand by taking better care of their guests as well as getting involved in community projects.

Any property can build business through a good public relations plan. The idea is make sure that a property gets known in a community through its involvement in local projects. Most of this gets done with very little or no cost at all. Just doing things to get the name in the paper is what counts. Offering the lobby as a collection point for a variety of good causes gets visibility. Getting to know the city editor or business editor of a daily or weekly newspaper makes it easier to plan human interest stories about staff members, or guests. Inviting local schools to bring students to the property for a tour and explanation of how hotels work also gets great visibility. These are some of the stories we heard from AmericInn owners and managers. Great stuff to build visibility.

When it comes to guest services, these folks were not slackers either. They all seem to know that when it comes to taking care of guest issues . . . they get resolved on-site. A review of the 1999 guest comment cards reflected a very favorable ratio of returns. When it came to discussing customer service, most managers reflected on the little things that are being done which make very favorable impressions on guests. Things like answering the phone within three rings; having an "upbeat" attitude; smiling; never saying things to guests such as: "no" or "yahafta."

I had a great feeling visiting with this group. Managers are busy working the property, but getting the staff trained in PR and Guest Services helps make things work for them. It was very interesting to note that when I mentioned the name of Mark Paulson (their filed marketing guy), from the rostrum, a big cheer went up. They sure know and like him. He's getting the job done for AmericInn.

TECHNOLOGY WILL HELP SALES EFFORT, BUT WON'T REPLACE SALESPEOPLE

August 2000

In a June issue of this publication, there was a special section on Technology in the hotel business. One of the articles was contributed by a representative of a hotel consulting firm, who was extolling the values of technology in our business. No question about it, I think we all agree, technology is and will continue to be a great value when it comes to marketing our properties and helping bring in the sales. One thing that bothered me was the writer's prediction that ultimately, because of technology, an owner can save a good amount of money by eliminating sales offices at properties. The same was said many years ago when we started to see central reservations offices set up by brands; and when the Holidex was introduced by Holiday Inns. Brands also started to create national sales offices to sell to the group markets. Word was out about that time that properties would ultimately get rid of their sales offices. Well, of course, it did not happen; and today we have more people working in hospitality sales than ever before.

It is hard to believe that technology will replace people when it comes to sales in our business. Yes, more and more individual reservations are being made via the Internet. Some recent figures I saw indicated that in 1998 1.8% of room reservations were made through the Internet (no personal contact with anyone). The projection for the year 2003 was that it would jump to about 8%. However, this was individual reservations only. It just seems plausible that we would have to maintain people on the staff to sell individual reservations since a very high percentage of business comes to hotels through local contacts. The local contacts for any property is critical. I am continually impressed with the amount of business going to hotels because of the local hotel reservations staff or front desk staff who take these calls. People, generally, like to deal with people. It is the relationship building, attitude, attention, friendly staff people who really are the ones who bring in a good percentage of the rooms only business.

When it comes to groups, it is about the same. People buy from people whether it be social groups or business groups. I recently wrote about the reverse auction process whereby meeting planners can put up with their meeting business for bids from hotels via a few new dot-com organizations. Yes, the request for proposal and the bidding process is all done via the Internet. But I do not believe there is even one meeting planner who would ultimately enter into an agreement with a property without making a site inspection or at least talking with the sales department on the phone. The only way I can see the Internet handle all phases of the booking process is if the buyer is very familiar with the hotel, has been there previously, and knows the folks in the sales department. So, I agree it could work in the case of someone booking multiple meetings at the same hotel, but only if there has been earlier personal contact with a real live person.

Frankly, I am not concerned that any owner will be doing away with sales departments. When it comes to selling, high-touch works.

E-commerce is here to stay and it could be a very important contribution to a property sales effort, once we understand how to use the Internet to help in the sales effort. Right now, for the most part, we are amateurs, and we need to learn that the Internet could be a valuable tool in communication for prospecting, soliciting, public relations, as well as for maximizing visibility and awareness. We'll be offering a lot of tips on these topics, in this column, as I learn more, myself, about this business of E-commerce.

LET'S HELP SALESPEOPLE BECOME MORE PRODUCTIVE

October 2001

There are so many hospitality sales folks out there with the title of Director of Sales and Marketing. I would guess that most are really salespeople, and the marketing is left up to someone else. There is a big difference in the function of sales and the function of marketing. Yet, we observe salespeople being given the job of creating advertisements, purchasing advertising space, designing brochures, handling public relations, doing research, as well as coming up with promotions. Let's make the sales staff responsible for bringing in new business and pay them well for doing so, with incentives for increasing revenue. Leave the marketing effort up to those much more qualified, and properties will get more bang for their buck. Simply, let's just consider marketing as the function of delivering a product to a market, involving all the activities mention above, as well as the direct sales effort. We should then think of sales as being the delivery of the product for a price. It's just that simple. Salespeople should sell and leave all the other stud to the General Manager, or a real marketing person, or an agency.

In operating lodging properties, no matter what the economic situation is in this country, in the best of times, and in the worst of times, we will always need guests to fill up our rooms. Even during great economic conditions, we don't have properties getting 100% occupancy every night. So, it stands to reason we will always be looking for more business. And, when occupancy does get great, what should we do? We know we have to increase the rate, and go after business which would be willing to pay that higher rate.

When would happen if every one of your guests, individual and group, always returned to stay at your property? If that did happen, but it doesn't, you would be full every night. You see, not everyone always returns to your property, even if they do return to your area. That's the way it works. There is always somewhere else to stay, no matter how good your service may be. Just take a look at your own repeat factor, that is, how many of your guests return to your property for any reason? You will hardly ever get more than 25%, if even that high. Check it out yourself. So, what we are saying here, our job is to make sure we keep getting new business to replace the business that does not come back. That's where the function of sales comes in to play. We need to learn and implement the best ways to go about this.

There are several issues concerning the activities and production of sales personnel. The big one is how much time is being devoted to a sales effort at your property? Whether you, the General Manager might be doing it by yourself, without the help of a sales department, or with a sales force. It doesn't matter; we should always plan on how much time will be devoted to this. In our sales workshops, we have been doing around the country we have found out that most salespeople spend under 35% of their time on pro-active sales. Surprised? Well, that is what they say. More and more people assigned to sales are doing things other than sales. Most get involved in servicing functions, which really should not be part of their job. We find salespeople "helping out" at the front desk, dining room, banquet department, and just about all over the place. When it comes to groups in the house, we find the salespeople staying in to look after their groups just to make sure everything goes well. This is an important issue. If a salesperson is good in sales and he/she spends 25–35% of the time doing it, think how more productive it would be to spend an additional 25% of the time in generating more business. Let's start thinking about eliminating some of the stuff being done under the guise of sales that is not productive in bringing in new business. We will follow-up on this in my next column with some more thoughts on how properties can be more productive in the sales effort.

SMALL SIZED PROPERTIES MAY NEED DIRECT SALES EFFORT

August 2008

Current economic conditions and the fuel price situation may be producing a decline in hotel use at some roadside and suburban lodging locations. The industry is coming off one of the best situations there has ever been for increased occupancy as well as rates for the smaller properties. If the next couple of years do not continue to provide the business, what do the smaller lodging operators do to maintain sales?

Of course, the various Internet providers of business have been doing well for these types of properties, but isn't the business on a decline from those sources? With more properties being built, and in the pipeline, there are more and more rooms coming online. This, of course, means more competition on the way; those properties will also be seeking business from Internet providers. So, maybe it is time to seek out hiring a salesperson. Regardless of the size of a property, a salesperson, doing an honest job, with proper direction, should be able to return at least ten times the investment made in the person's salary.

There is really no substitute for a direct sales effort. All types of marketing may bring the attention of a property to the attention of the public, but it is the salesperson who delivers the product for a price. A direct sales effort means having a person on the staff that goes out to contact sources of new business that will be profitable to the property. That person gets a new business by making sales calls to prospect companies, and organizations that are in a position to put guests in rooms, and restaurants. These calls are made in person as well as by phone, and of course, via the Internet by email. The direct sales effort is a very meaningful way to bring in additional business, and could very well be the most productive way for a property to become more profitable.

Who does the job? In a small property, where an operator may feel that hiring an additional person on the staff to do the job is not affordable, then, of course, it could very well be the property manager who does it, even if it is just a couple of hours a day. The manager or lodging operator, personally handling the direct sales effort may be the best approach for setting up sales operation, provided all the other daily functions of operating are turned over to someone else. Another thing with which to be concerned would be not to hire someone to do sales on a full-time, or part-time basis, and end up having that person being a part-time desk clerk, or part-time secretary, or part-time anything. Make sure the salesperson sticks to a direct sales effort. Responsibility for the sales effort ultimately lies with the lodging operator; whether that person is called the General Manager, managing director, or even the owner; that person is truly the general sales manager.

RECOGNIZING THE VALUE OF THE "SMALL MEETING" MARKET

July 2009

Regardless of the size of a hotel property today, it can substantially boost its overnight rooms business, and its overall profitability by focusing on the small-meeting market. Surveys taken at almost any time in recent history reflect that the most common type of business (or social) meetings involve less than 50 people. In these times of budget cutting, local business meetings off-site, frequently take the place of more exotic locations favored in times of plenty.

While many associations have their annual meetings in large convention hotels, most committee, and board meetings can be housed in smaller properties that may not be in large cities. While these smaller meetings do not fill entire hotels, a block of 10–25 rooms for several nights can be a pleasant bonus, filling valleys in higher demand periods.

Most hotels have at least some space that can be used for meetings. It could be multi-use space, such as that used for breakfast or social hours. It can be a lounge that is used only in the evening. It could be a suite parlor or even a traditional room that has a built-in hide-away bed. If there is absolutely no space anywhere in a property, there is still the opportunity to partner with another business, adjacent to the property. Many of today's restaurant operations, whether they are chains or locally owned facilities, have separate sections of their stores to meet and feed 10–50 people.

Then, of course, is the issue of finding possible, potential sources for this type of meeting business. The list is extensive, and many are in a property's existing base of contacts. They include a hotel's list of suppliers (right out of the accounts payable listing); the trade organization to which some of the hotel staff belong; local service organizations such as Rotary Club or Lion, etc. Let's not forget all those social networks to which many of the property staff enjoy connecting. A property's staff is another base for referrals. Think about the groups to which a hotel's staff belongs, such as the Parent Teachers Association, garden clubs, condo groups, sporting leagues, churches, and a variety of social groups. These groups have a variety of meetings of their own. However, it is the people involved in all these groups, and the companies with which they are employed that could be the best possible prospects for the small meeting market being sought. It takes some time and effort, but the idea is to work the network.

The whole idea is to get the entire property staff involved in this effort. Have meetings in each department to explain how easy this could be for them to search out their own networks for possible business, and pass on the leads to the property sales department or to the General Manager. The Sales Department is not the Whole Hotel . . . but the Whole Hotel is the Sales Department.

CHAPTER 4

TECHNIQUES FOR INCREASING SALES

SOME NEW YEAR'S RESOLUTIONS

January 1983

Be it resolved that in this new year of 1983, I (name) the person responsible for the profitable and successful operations of the name (i.e., the name of hotel or motel) will faithfully and enthusiastically perform the following functions to ensure the continuous development and growth of my business. Be on hand at the front desk early enough in mornings to personally thank my guest as they check out and be on hand at the front desk in the early evening to personally welcome my guest as they check-in.

Make at least five telephone calls a day, Monday through Friday to new or old accounts to say "thank you" and ask for more business. Join Hotel Sales Managers Association International (HSMAI)—a professional organization which proves to be a very low-cost asset for improving my sales skills. Spend some time in a restaurant in greeting and visiting the guest during the mealtime hours.

Motivate my staff and set an example by being pleasant and smiling. Be sure that inquiries and leads of business are handled immediately. Read my local daily newspaper for possible leads for all the services my property has to sell Freely distributing my calling cards and ensure that property sales staff does the same Talk "sales" all the time to all of the employees expressing its importance to the future of all the property and the staff. Ask all the employees to get involved in getting business for the property. Personally visit the people in charge of the group and the business that comes to the property-up on arrival and departure Signed: Now that you have resolved how many more resolutions can you add to this list?

THE STUDENT BLITZ: A NEW TECHNIQUE FOR BUILDING SALES

June 1983

You all know what a sales blitz is: a group of people fanning out after sales leads, making as many calls as possible in a limited area within a short span of time. But have you heard of a new low-cost type of sales blitz known as "the student blitz."

YOUTHFUL ENERGY

Energetic college blitzers can cover a lot of ground generating sales leads by face-to-face contact. The student fills out fact sheets for hotel's full-time sales staff to follow-up on later, besides scattering upon promotional flyers of the hotel's food and beverage offerings.

Recently, one such blitz in Chicago on behalf of a large national chain stirred up 178 firms sales leads from 986 calls made by 10 students. Immediate results totaled $156,000, and the leads are being still worked!

In a similar Chicago blitz, another group of 10 students yielded 107 leads and over 200,000 in business resulting from a batch of 1,012 calls. In both cases, the company involved provided the students with five double rooms for three nights, as well as meals, incentive awards and mileage reimbursement. One plus connected with using students blitzers is that a student can in many instances get in many doors that an experienced salesperson can't. A smiling nervous student can disarm secretaries and executives by appealing a parental instinct or by convincing subjects that the cause of education is a stake. Usually, a group of 10 people can make a 1,000 calls within three days. It is important that time is not wasted in indiscriminate running around. Calls must be planned routes must be marked out, assignments made, information sheets developed, the collateral material selected briefings given and participants rehearsed.

Out of 1,000 calls 200–300 result in some type of sales lead. Although students are doing the legwork, it is important to have your full-time sales staff follow-up to make sure that no lead goes for naught. If well-planned and implemented, a student blitz can keep your employees in the hotel, while the outsiders drum up immediate sales lead for guest room group and social business.

SALES BLITZES: A LOOK AT SOME ACTUAL RESULTS... AND HOW YOU CAN ARRANGE A SALES BLITZ OF YOUR OWN

October 1983

A recent "sales clinic about student sales blitz programs created a good amount of interest and reaction from our readers. Our intention was not to give the idea that student blitz programs are new. In fact, using college students for sales blitz programs goes pretty far back to early days of hotel sales management Assn. (HSMA). The new twist to student blitz programs is to engage the services of an outside source, for a fee, to handle the entire operation. This takes the administrative burden of doing a blitz away from the hotel sales staff, thereby allowing them to continue their usual daily sales activities without taking time away from selling.

Purdue University's Hotel School, to do a student blitz at new Sheraton Crown hotel and conference center located at Houston Intercontinental airport. Dr. Meyer recruited

10 of his students for the project, and blitz was conducted subsequent to the appearance of the sales clinic article. The Hotel staff only involvement included throwing a pre-blitz orientation dinner, providing collateral material, arranging for incentive awards and providing the service of the sales manager each evening of the blitz period. Each day upon the return of the blitzers the sales manager reviewed each blitz call report to determine the number of calls and leads that were turned in. Perry Sahikh, General Manager of the Houston Sheraton Crown, expressed complete delight and satisfaction with the program. Nine hundred and eighty-eight calls were made which generated 400 leads. The business potential available to the party amounted to $600,000. Two definite pieces of business were booked by the student blitzes while the calls were being made. Total cost to the hotel for the entire program: Under $6,000. In addition to the development of the leads, the amount of goodwill and promotion benefit for the hotel is virtually immeasurable. From Dr. Meyer's and students viewpoints, the experience gained by making the calls and exposure to the hotel operations could only benefit the students in their future studies and career paths.

And how you can arrange the sales blitz of your own. A sales blitz is an intensive survey of a given geographical area to determine its market potential. The idea is to lay the groundwork for a sharp increase in business by gathering information a sales staff needs to his job well. The key to a successful blitz is to completely canvas an area in as little time as possible, with a few people as possible. To make that happen proper organization is essential.

THE BLITZERS

Anyone on a hotel' payroll can participate. Since the objective is not to sell but gather information, the blitz team need not be limited to members of the sales department. The idea is to extend the department's reach not to overload it with a new responsibility. But there is one thing to remember: the same people have to be used throughout the length of the blitz. Switching horses in the midstream will only complicate the matters. The staff members chosen to participate have to be under the firm and direct control of one person. The director of sale is a good candidate, but the need not be the case. The prime consideration is finding a blitz director. Who can keep things under control? The blitz director needs to start planning the operation 30 days in advance of blitz date. Logical days to select are Tuesday, Wednesday, and Thursday, when the contacts are more likely to be in town. Three days of blitzing should be maximum. For smaller properties with a limited amount of blitzer time available, a one or two day blitz can be satisfactory. In fact, a one-day multi blitz would be a good place to start experimenting with the process.

30 CALLS AN HOUR

In six and a half hours each person should be able to conduct 30 calls. Depending upon the area assigned the number of calls will vary. In a downtown office building much more 30 calls should be made, however in outlying industrial areas far fewer calls could result. After the day calls are completed a blitz recovery period should be scheduled. It is during this time that the forms are turned in reviewed by the blitz director or a member of the hotel sales department. It will be appropriate to serve cocktail and snacks to the blitzers to help them unwind and discuss day's work. Grades are also assigned at the period when the forms are reviewed. A score chart is kept where each blitzer is credited with a number of calls and number of leads developed. Once tabulated a daily winner of the contest should be announced, and a prize should be awarded. After the last blitz day, a dinner may be planned with husbands and wives and guest of the blitzers included in the festivities.

The real key to a successful blitz program is proper motivation. Management and owners need to be behind the program as much if not more than participants in the program. The blitzers need to be excited and psychologically up throughout the program. Breakfasts with music, signs and banners awards and other mood elevators will help bring in great results.

Since the objective of the blitz is to develop the sales leads, a good deal of time and money would be wasted if a poor follow-up job is done. After each day's calls have been tabulated the blitz forms need to be stored by the grades given each blitzers. The highest grades are obviously the forms with the best leads and usually the ones which require immediate attention. The sales director should determine which leads need to be followed up the very next morning and arrange the remaining forms in priority order for follow-up calls based on information received. Even while blitzers are out the next day, the sales department personnel should start on follow-up of the previous day's work. It is important to check files first before the follow-up to see if more information on prospect might already be available.

Although a successful blitz requires a lot of time and effort on part of the management team, it always produces excellent results. Besides developing leads, it helps promote the property to the business community. In addition, it becomes a motivational tool for staff members that have participated -a team effort which creates a healthy cooperative attitude that lasts a long time.

In organizing the calls to be made the most helpful system is to use a city directory. The directory lists all the addresses in a city by street designation. It is important to concentrate calls in one small area at a time, so blitzers are not spread tooth in.

BLITZ PREPARATION

Index cards should be used to record the addresses of all the cells that have got to be made. These cards are then assigned to the participants based on geographical territory. For a three day blitz each person may be assigned 100 calls (a good day's work would be 30 calls per person) logically in a business area of any city the calls would be located next door to each other or in the case of office buildings, all in same building blitzers going floor to floor.

An adequate supply of collateral material and forms are necessary. Padded blitz survey forms should be plentiful) you will need at least one per call) In addition, property brochures, fact sheets and promotional material (about restaurant and lounge) should be given out on each call. Low-cost advertising gifts may be distributed, but they are certainly not necessary.

Blitz Training

An indoctrination dinner should be held the evening before the blitz day. Since this will be the first time all the blitzers are assembled, the blitz director should use the opportunity to explain the objective and purpose of the blitz. The blitz forms should be reviewed and explained and a demonstration given to show how the survey call is made, starting with introduction asking of questions and closing. This should be followed with blitzers being paired off to try out the question on each other. The blitz director should then explain how the completed forms will be reviewed and graded.

The Blitz

The blitzers should use the fact sheet to answer questions and not try to sell anything during the interviews. If it appears that the interview may appear in a hot lead, the blitzer should react by saying: "Someone from our sales department will be in touch with you" it may be important enough to immediately call the hotel sales department and put someone in touch with the prospect right then and there. No time can be wasted during or between calls if the blitz is to be effective. The first call should be made at 9 am, with the pace continuing until 12:00 noon, and running then from 1:00 to 4:30.

CLOSING THE SALE AIN'T EASY . . .

October 1990

But who said it was supposed to be easy? We are not discussing here the "no brainer" booking where someone comes to the property or calls to book a group or meeting. Probably most of our business comes to us this way, anyway. The type of sale we are discussing is where solicitation action is taken by a salesperson on a new prospector, an account that is using another property. Here is where we go through the normal steps of prospecting, selling, negotiating and closing the deal for a profitable sale.

According to an old survey by Sales and marketing management magazine, most sale people give up trying to close a sale after a first try. This may not hold true for lodging salespeople since I am sure that after we spend time prospecting, selling and sometimes entertaining, we just don't give up that easily, or do we? After all, our files are set up with an automatic trace system which requires that we continue to follow-up—no matter how many times we get a "no." To reduce the number of follow-ups, we could get a quicker "yes" let's look at some of the reasons people are not ready to close:

▶ We may not have the right product (#rooms, meeting, space, location, etc.), which may mean we never gathered enough knowledge about the prospect in the first place to determine if this was a real prospect or not.
▶ We may not understand the real needs of the group. This requires a good deal of questioning upfront like: What is the purpose of your meeting? What do you want to accomplish? The type of product you have may not be suitable based upon the response to that question.

- There are objections which have not been overcome satisfactorily, such as: "Your price is too high" or you are too far from the airport, or we need to be near shopping, etc. Objections have to be taken once at a time and overcome by a variety of techniques—the major one being to sell "Value."
- Sometimes we do not get the real reason for a "no" which could very well be a result of a poor presentation by the salesperson. The prospect just was not convinced, and no one is going to tell you did a lousy job on presenting the product.
- Do you think there may be a question of perceived value? Or dependability? These are areas that must be explored with the prospect to see if there are such underlying reasons.

Many group planners will always question price; it just seems to be second nature to want to buy something at less than the going rate. The idea here is not to give the deal away at a price lower than you need, just to get the business (we do too much of this throughout the hospitality industry). This is where successful salespeople sell "Value." During the negotiation process, it is generally always best to save this item, of price for last—while you keep selling your facility and service and get the deal "sold" before you get into the rate part of negotiations. You may even have to make concessions here and there on other items but not the rate you need. Keep building perceived value, get the prospect motivated to buy—appeal to needs and wants.

Being a good listener is always a key to successful sales. Ask the right questions, do a good job of gathering information in advance of selling and you will have a pretty good idea of budget requirements and where the group has stayed previously. These two items alone will provide you with an idea of what is actually readily available from the group for the rates.

Closing the sale is always not an easy thing to do. It takes courage and conviction along with good selling techniques to be able to close at the right price. Keep Practicing.

SALES RESOLUTIONS FOR 1991

November 1990

Next year will be tough; there is no question about it.

Even though there has been a cut back in lodging construction with less rooms coming into the marketplace than in previous years, there is still some under-utilization of rooms. Owners and operators will be still looking for a return on investments, and the only way to bring the dollars to the bottom line is to make sure the business comes in at the top. And this is what sales all about. It is time now to consider what will it be that needs to be done, from a sales standpoint to improve profitable sales starting next month. Here are some thoughts for sales personnel to consider to start the New Year off in the right direction . . . Can you resolve to:

- Prospect every day, Whether you do it in person by phone or by mail . . . unless we keep bringing new business,

we cannot succeed. Establish a program for making new contacts each day ... set yourself a daily, weekly and monthly goal for contacting prospects for new business. The more calls made, the more prospects for new business closed ... it does work.
▶ Answer inquiries immediately. This is the hottest lead you will ever get, whether they come in by phone or mail get on it right away.
▶ Sell from the top down-When negotiating with groups we lose so much by jumping to conclusion that they will not pay the price, in most cases, they will, if they perceive in a price/value in what you are selling. Let us not start off with offering a group rate, comp rooms, free this and that. Start at the rack rates and based on how badly you need the business, at the time, you can negotiate up to a point as long as the business is profitable.
▶ Review the trace system: So much business is lost because we neglect to follow-up on our files and contact particular groups when appropriate. We seem to get bogged down from time to time with overload on files being pulled for tracing. Then we just go ahead and retrace without making the contact (and sometimes not even looking at the file).
▶ Make sure travel agents get paid their commissions. A big problem facing industry next year will be with travel agents taking lodging properties to quote for not paying commissions. If you advertise to pay travel agents' commissions, then do so on negotiated rates. With the industry needing all the support it can get till no time to alienate the travel agency community. Agree to pay agents 10% commission on all business, all the time (Not only when we need).
▶ Improve internal communication: Make sure everyone knows what the sales department is doing. Bring staff up-to-date at the weekly department meeting. Discuss how each department staff member can help in the sales effort. Get them to remember that EVERYONE SELLS.
▶ Include training Dollars in your budget: We seem to neglect to add in from 2–4% of salaries for training. The way things are going these days we must keep on top of ways to improve ourselves. Take time to invest in education and training. Attend seminars and workshop by the variety of organizations in our industry. Become active in Hotel sales and marketing association international and meeting planners international. Both offer a great seminar from time to time. Get information from the educational institute on their programs, literature and training videos. Be sure to participate in key customer contact conferences such as HSMAI's affordable meetings.

HOW NOT TO SELL

November 1990

We can always learn something about selling. And we probably learn more by observing other salespeople "screwing up" the works. There's little salesman I met recently who represent a well-known travel product. He was new to the area and wanted to meet some key people who were in a position to influence others in using his product. As a courtesy, we offered a bunch of appointments with the people I know that would be a good prospect for him. I was able to establish a meeting for 30 minutes each, and one was for 60 minutes. I went with him to the first appointment, as a courtesy, but left him on his own with a map and appointment schedule. The salesman knew these people were just led but had to be qualified to determine if, in fact, they were good prospects. All were using a major competitor-all were busy but agreed to the appointment as a courtesy to me.

Here is the scenario for the first visit: The first five minutes were spent in "what I would call a "Garbage chit chat" I guess some people think that it is important to do something or say things that create rapport before they get into the purpose of visit. Frankly, there was not enough time for this—and had I not interrupted, it probably would have gone longer. Twenty minutes were spent on a monologue how great the product was, improvements that were being made, etc. No attempts were made to find out the needs or what or what problems would the use of the product could solve. If the prospect himself did not interrupt the sales pitch to volunteer a couple of needs, I am sure no information would have been developed. No notes were taken even after prospect started outlining specific uses and important facts figures found out late that he did not have any) and he did not ask the prospect for a card. Before leaving, however, the prospect (Being a nice guy) gave him his card. I sure he did better with the remaining five appointments.

Now, what can we learn from this?

1. When time is limited and you are meeting with a busy person get right to the point.
2. Ask key questions to determine needs and qualify to determine if, in fact, this is a prospect.
3. On this kind of initial contact do not try to sell the product. There is plenty of time to do so once there is an understanding of the specific needs.
4. Take notes.
5. Offer a business card and ask for one.
6. Tell the prospect when you get back to him with a proposal(if appropriate). In this case, there was a question of rates which never get answered.

I'm very curious now to take a look at the follow-up to see how that was accomplished. Anyone taking any bets? By the way, please do not pass on a copy of this column to anyone in the travel industry. I know the person I'm telling on does not get HMMM, and I sure would be embarrassed if he was to get to read this.

HOW ABOUT PROMOTING THE GM TO DIRECTOR OF SALES?

February 1993

In times like these, when pressure for performance is getting very heavy to bring in more profitable sales, why not have GM in charge of that all-important department? You say there is not enough time to run a property and do sales as well? OK, the get someone as an assistant to help take care of the operations-perhaps some sharp front office manager or housekeeper, or someone from food and beverage.

Think about it! If the General Manager supposedly paid, most knowledgeable and experienced person on the staff, and the head chief in charge of everything) would run the sales department there is no question why it would not be successful. The GM would spend 75% of time directly in charge of making sure that profitable sales keep coming in beside managing the sales effort by providing attention, direction, supervision, and motivation, the new DOS (GM in reality) can also be productive by making sales calls and handling major accounts.

The job description would include the marketing plan done, employment of sales staff, training them and supervising them. At the same time, the GM would also be responsible for solicitation, selling and closing on new sales as well as repeat business. The GM would also take care of the promotional and public relations effort of the property and the advertising. The remaining 25% of the time would be spent overseeing the remainder of the property operations.

How well a sales department functions and how well it brings in new, profitable sales directly affects all other operations of a hotel. Everyone employed at a property needs to understand the value of a sales department and what better way to show it than with the General Manager promoted to the position of directing that department. With this example being set the entire hotel staff may get to learn everyone sells . . . and by providing the service that brings back the guest every time they visit the city, the staff member has helped more in bringing the business.

Granted there are some General Managers who will absolutely refuse to get promoted or involved, personally, with sales.

There are a number of reasons with this: Never had the experience; do not like having people to approach to ask for business (that's a biggie) not persuasive enough. Won't put the amount of time necessary to get the job done. And of course, a host of other reasons. There is still an opportunity, however, to take over as chief executive in charge of sales without making sales calls. Just making sure the job gets done properly to bring in more profitable sales would be good enough. Just being a good business person and running the department like a business will work.

Today, we see more and more salespeople getting promoted to the General Manager slot and becoming very successful. There must be something to it. Just take a look to some of the big names of people in executive slots today who came from a hierarchy: Joe McInerney, Charlotte St. Ray Schultz, and many other managers today who have come from the sales ranks and who know exactly what to do to improve profitable sales at their properties. It would be interesting to find out just who these people might be and perhaps create a success factor analysis, so that we can pass the word out that: Yes, it does work . . . General Manager can be in charge of sales.

INCREASING PROFITABLE SALES IN THE HOSPITALITY MARKET PLACE

July 1993

Even with all the discounting of room rates going on all over the industry business is looking for. It seems that good reports are coming in from various parts of the US concerning more business being booked by salespeople. It is just a question about whether all business being put on the book is really profitable to the properties. We have done a pretty good job training our group business prospects in how to ask for and get lower room rates. The extensive advertising that has taken place the last five years with all kinds of properties (even the 5-star operators) reflecting discounted rate programs has created the awareness among prospective buyers that no one has to pay regular room rates. It is just time now that things are getting better to get back to work on increasing profitable sale in the hospitality marketplace. Appropriately enough we can help!

For a long time now many of our leaders have requested that in connection with this column, Hotel and Motel put on a series of sales clinic workshops. Well, the time has come and that is exactly what we are going to do. Our first will be held September 7–10, 1993 on the campus of Virginia Tech Blacksburg, Virginia. HMM will be sponsoring a workshop in cooperation with Virginia Tech's department of the hotel. Restaurant and institutional management and division of continuing education. We have headlined the program as increasing profitable sales in Hospitality Marketplace. This will be a two and half day highly interactive and fast-paced workshops covering topics that will help salespeople improve their selling and closing techniques. We will get into the subject of sales and operation relationships; how to negotiate to your best advantage; how database marketing gets you right to your buyers; we will take a look at a prototype sales office of future as well as other topics. Because we are interested in delivering the right workshop product to you, when and where you want it, please take a minute to complete the survey displayed with this column and fax it back to us. This will help greatly in preparing other future HMM sales clinic workshop.

TIME TO TEMPER TELEPHONE TACTICS

October 1993

Here is something we absolutely cannot do without ... our telephone; a wonderful business tool if we use it correctly many of us have the opportunity to use the phone, mail or voice mail. This is a terrific piece of equipment which may be used to leave messages and retrieve the messages when we are not available to take calls. BUT WE DO MISUSE IT!

It becomes most annoying to callers particularly those who want to inquire or even book business at properties when they cannot get through to anyone except a recorded voice asking them to leave a message. Let's be honest about this, don't we sometimes program out phones to be answered by phone mail even when we are sitting right there in the office? We get busy doing a wide variety of things and do not want to be disturbed by a phone call. The renowned sales trainer, Jim Cathcart, has always said: A customer is never an interruption to our work, but the cause of it." Do we ever lose business because someone cannot get through a live person? Of course, we do. Can we afford to let this happen?

What's interesting about some recorded messages is the one that goes like this: "I'm sorry I'm either on phone or out of the office, but if you would like to speak touch star or zero and you will get my secretary . . ." So you punch star and zero and what do you get? You guessed it! The secretary is also on the phone mail. You can't win, so you call another hotel. How about those of you who travel a great deal? What's wrong about leaving a message where you can be reached in case someone really needs to talk with you right away. Call me when I'm out of town and see how this works. There is nothing wrong with leaving phone messages for incoming calls telling callers exactly what's what, where you are when you will be back a name and number of someone else the caller may want to contact. At least this way tot becomes more personal and shows some degree of interest.

Now from the other side, when calling someone who happens to be on phone mail or voice mail, how good are we at leaving messages? Let's take salespeople who are prospecting for business. My big problem is with someone who speaks so rapidly it is too difficult to understand who they are, where they are from and the telephone number they leave. Sometimes I had to reply messages three times to try to get the correct name of the caller. Recently I have received a phone mail message from someone whose name I could not understand, but I did get the name of the hotel and phone number. I returned the call asking for "someone that sounds like . . ." There was no one there by that name, but the person did offer to check around with their salespeople to see if they could find the calling person. The next day someone from a property called and said that they could not locate anyone that called, but they were probably calling to see if I had any meeting business to book at the hotel. How's that for a gem? And, that's just a sample. It goes on and on and on, and I am sure it has happened to you with prospective buyers calling and leaving similar messages. You can avoid this by having a real live person answering the phone.

So those tactics do you need to employ when calling someone who is not in, but does have phone mail? Speak slowly, repeat your name and spell the last name, at least. Advice the name of your property and explain the purpose of the call, but keep it brief. Long-winded explanations could be upsetting when someone has ten messages on phone mail to retrieve. Leave your phone number, with area code and repeat it. Advice when would be the best time to return the call. If you are selling something it is good business to either leave an 800 number or suggest they call you back collect." Why should a prospect pay for a long distance call to speak to you if you are trying to sell them something?

Remember, don't rush, speak softly and distinctly, leave good messages . . . it's gotta pay off with more business.

SELL VALUE NOT PRICE AND BE THE WINNER

November 1993

Over the past few years, we have heard so much about discounting room rates because the "other guys are doing it." In too many cases our own salespeople saying that price is what people consider the key in their decision making. Of course, this is true in some cases, but not all. Many prospective buyers may want to give us the impression that price is the only critical thing they are concerned, but my feeling is that this is just a way to get a lower price . . . even if they are really satisfied with the price being quoted. It is a kind of a natural thing to do . . . to try to buy cheaper. Just recently I purchased a new car with a trade-in. I was thrilled with the amount I got for trade-in, and the cash I had to pay was much lower than I anticipated. However, I figured it would not hurt to negotiate a little further which resulted in reducing the price another $1000. Could I have gotten more? Perhaps, but then again the salesman could have held out for his price and gone a little further in showing me the advantages in buying at the price he quoted and I would have gladly gone ahead and paid the amount we originally discussed.

What's the point? Do not give in too readily to people that push for discounted room rates. Just do a better job of selling; Find out the prospect's real needs; how can you meet those needs what is their position with regard to the booking with you; where do you stand against the competition from a product and service viewpoint; how badly do they need you; what do you have that the competition does not have; how badly do you need the business (enough not to be able to make money on the sale) find out what other objections the prospect may have with regard to using your property. Think about why people want to stay with you in the first place. What initiated their call or inquiry or why was there some interest shown in your property? There must have been some thought of benefitting by staying with you. Could it be location; service; reputation; quality; facilities etc.? You have got to play up whatever the reason might be.

I think to overcome this business we have to change our mindset. We have to be proud of what we sell. We have to believe in our product and believe that there is a great price we are offering. Then we must do a better job of selling the value.

MEASURING THE SALES PERFORMANCE

October 1994

We all agree in the value of the sales, or rather profitable sales. To remain in business, we need to have people staying in our rooms, using our meeting facilities and eating in our restaurants and ballrooms. So sales in very valuable to us whether it just happens or we have people working with us that makes it happen. There is always room for improvement in the area of the sales-we could use more occupancy a better rate and more repeat business and more group business and more leisure business whatever. Sometimes there are problems that need fixing, like if we have a sales team can we do something to make their time more productive; or even if they are productive is their productivity very effective; that is, we can spend a lot of time making more sales calls, but do they work?

In looking at productivity perhaps we need to identify what it is that we want our salespeople to do. Is there a job description that specifically indicates the function for the sales representative, sales manager, catering manager, sales director and for the people with all those other titles which really means they are responsible for bringing in the business? If we spell out what proportion of their time needs to be spent in direct sales, sales planning, sales administration it would give us a better gauge when measuring what it is the salespeople do with their actual time ... how are work hours typically spent? If we do not specify we cannot measure. This means we have to come up with some standards and methods to monitor how work hours are spent.

If we find that too many hours are being spent on things that are administrative, then it is time to rethink what's being done that can be eliminated or delegated to someone else so that the salesperson spend less time in that activity. What we need to do is to identify those who are not spending as many times as they should (according to standards set) on building business and work on eliminating factors that affect that performance. In other words, if the job description indicates that 65% of salesperson time should be spent on direct sales of building and we find that only 30% of person's time is spent in this area, we gotta do something about this. A good weekly sales activity report would reflect hours spent weekly on various functions. The idea is to get rid of the activity that does not contribute directly to building customer relationship or actual sales.

When it comes to effectiveness, we are talking not about how someone's time is being spent, how many calls are made or the number of proposals written. Here we get into actual sales performance. In order to measure this standard have to be set in the job description. It is not enough to indicate how many room nights are booked—this does not reflect the real dollar value. What we need to do is to measure actual revenue created by sales. Although measuring someone's performance based on advanced booking is good and necessary, it is the actual revenue collected that tells the tale. When salespeople do not achieve the agreed upon goal, then we must examine the reasons. You have to start examining the strength and weaknesses of the sales department to determine where help is needed. In all probability, it will be in the area of training.

Some typical problems that affect the closing of sales lie in the understanding of customer needs, developing prospects, product knowledge, presentation skills, consultative skills and skills in the negotiations. When sales activity is not reflected in the increased sale, then there is a problem with effectiveness. Training and education is the only medicine to cure those ills. This may be accomplished in-house with training conducted by management or with staff personnel; through educational materials which are made available through sources such as the Educational Institute of the American Hotel and Motel Association or at attendance at seminars and workshops provided by HSMAI and other institutions like at the Virginia Tech, improving profitable sales in hospitality marketplace."

IMPROVE SALES BY KNOWING HOW TO OVERCOME SALES RESISTANCE

September 1995

Somewhere along the line whether you are trying to sell a single room, a block of rooms, a meeting or a banquet there will be an inevitable objection to the sale. It could be the location of your property, the type of facility, parking, quality . . . and most of the time . . . probably the price. Whatever reason anyone may have to resist buying. It could be a lack of knowledge (Perhaps we have not told them about the benefits or they do not have enough facts) maybe they want something extra Some sort of price advantage or extras) there is also a possibility that they are making comparisons with competitive properties (do we know all we can about our nearby competitors?).

Of course, if we can avoid objections, then we really will have it made and it would be a lot easier to close a sale. One way to decrease resistance to a sale is to try to anticipate the objections. Understanding the prospect's needs, gathering facts about the prospects (history, background, budget, likes/ dislikes, etc.) will make it a lot easier to include information in your sales presentation that would overcome the objection even before it becomes one.

Start out by making a list of typical objection which you have already encountered and come up with ideas on how to overcome them. For the most part, you will find that you will have to emphasize the benefits of whatever features you have that would minimize the objection. You should never be surprised by any reason for someone resisting the sale. And, you should always be prepared to be able to handle it. You can handle it very well if you try to find out what prospect really means. You want to find out the real reason and you want to get to analyze the objection-along with the prospect by asking questions. It is not always best to answer an objection directly. This would put you in a defensive position, when you want to be able to take the offensive. Questions will make the prospect think a little bit, they will have to analyze situations by themselves. Always go for the open-ended questions which require more than just a Yes or No. Let them explain. You should also try to follow their explanations with another question. Let them do the talking and you do the listening. Sometimes there would be no real explanation for the objection and the prospect may even realize this. If the prospect is not clear about something or does not understand the value or benefits then this is the time to offer more details to show how the benefits create an advantage for the prospective buyer. In all cases, you should show some understanding of the situation—although you do not have to agree with the objection. ("I understand what you are saying Mr. Smith and I'm glad you mentioned that because . . . etc.). Acknowledging an objection shows respect for the prospect and appeals to a person's ego. Sometimes you may even want to admit if an objection is obvious as long as you can concentrate on overcoming it by creating more of a value ("yes, we are ten miles from the airport, Mr. Smith: however, by using our free, courtesy vehicles we can get you to our property within 15 minutes, etc.").

Every professional salesperson understands and even anticipate an objection. We don't enjoy them and for the most part, hate to deal with them, but it is all part of the selling job. We just need to line up strategies for handling these opportunities when they come up. In fact, without resistance to buying there would be no need for salespeople. Resistance to buying is the most important hurdle every salesperson has to get over. You just have to know and understand what is keeping people from buying so you can bring up the right benefits to minimize resistance.

HOW'RE WE GONNA SELL THOSE HIGHER RATES NEXT YEAR?

December 1996

Yes, there is every indication that room rates will again go up next year. With the demand still high and supply of rooms not yet, but almost, approaching the percentage increase in demand in most areas around the country, the talk is that room rates will be higher. In some areas where competition has increased with new rooms and new brands coming on board and where the occupancy has not gone up, rates will probably remain flat. Of course, competition is creeping all over the place, little by little, but it looks like another year of a seller's market for us in 1997.

Some salespeople are getting concerned about how much higher can we go on rates and what is it we have to do to be able to do a good job in getting the rates that our owners and operators want. Higher budgets have been set for an increase in revenue and profits and now it's up to the people "on the line" to make sure the budgets are met. Now it will be up to the property managers, personnel working in reservations, front office and sales to figure out "How're we gonna do it."

There is a basic Buy premise in a sale that People Buy From people. And, People Buy From People They Like and Trust. This means, that those of us who are in contact with likely prospects for sales in rooms, food/beverage (yes dining room serves too), space, entertainment, recreation, and any other profit centers, just gotta brush-up up on our relationship selling.

What we're talking about here is the business of getting and keeping customers. It has to do with making a "good" sale which is a value to the customer and a benefit to your property. Both parties come out happy and pleased with the deal. Just a couple of week ago I went to a local restaurant, very early in the morning, on the way to the airport to get some breakfast. There was one customer in the place and one server on the floor. I requested a seat in one area of the dining room at a window, but was told that that section was closed. When I inquired about the reason I was advised that they only had one server on the floor, and that I would "hafta" be seated where they wanted me to sit. Needless to say, I left the restaurant and ate elsewhere (and they didn't seem to care that I was leaving). The point here is that the restaurant didn't benefit from a possible sale and certainly I didn't benefit. There was no relationship selling. Just think of some of your own experiences. You probably never bought anything from someone you didn't like. Whether it was a car salesman, real estate agent, clothing store salesperson, etc., we just don't buy from people we do not like or from those that do not want to serve us the way we want to be served. None of us like to be exploited or manipulated, so we do not buy. When you think of the millions of dollars spent continually by hotel brands, management companies and properties in our business, just to get customers to try our products and services and then so often just one person at any location can mess up the prospective purchase by not practicing the fundamentals of relationship selling. According to Jim Cathcart, author of *Relationship Selling* (published by Perigee Books, The Putnam Publishing Group), "poor human relations can ruin the best marketing effort you could ever put together." And, you will agree, this all basic stuff we're talking about. The attitude of a salesperson can make or break a sale. Just what is being done at your local property, or within your management group at your brand headquarters to help get people to understand that it is this business of the right attitude that keeps your business going doing it the right way?

IT MAY STILL BE A SELLER'S MARKET FOR HOTELS, BUT...

January 1997

In a recent issue of *Corporate Meetings & Incentives* magazine (a former sister publication of HMM), the editor, Barbara Scofidio, quoted a segment from one of my articles (I had a monthly column with them for over five years). It was in a 1991 issue that I wrote, in addressing meeting planners: "In your negotiations, keep in mind that the hotel has to make a profit to stay in business. Be reasonable, and leave something on the table for you negotiating counterpart. It's possible that at another time, the shoe will be on the other foot-and when you're the one desperate for a deal, you'll want the hotel's management to remember when you were generous with them."

Now, bear in mind, this was an article for meeting planners when they were in a buyers' market. The hotel business was way down, and hotels were losing money, now, the shoe is really on the other foot and we are in a sellers' market. How long will it last like this with increased occupancy and average daily rates? Who knows if anyone does, they are not saying. It is my guess and, only a guess, that we better watch out in the second half of 1998 and 1999. With the way development is going with new properties all around us, it may even be sooner. I'm just saying that the business we may be taking advantage of right now is the business we may be legging for down the road. Meeting planners, tour operators, corporate travel managers, travel agents and all those other buyers of our products may very well have good memories.

Speaking of memories, for many years I've thought that someone should create a database of the high-end hotel user market. How does one get hold of a list of travel agents, incentive buyers, meeting planners that deal with individuals and groups who want extra special types of accommodation and services and are willing to pay the higher price? After all the four-star and five-star hotels, as well as cruise lines, certainly would be interested in marketing to that segment. Well, it has finally happened. There's a new trade publication entitled "Premier Hotels & Resorts" that focus exclusively on the luxury hotel and resort market. The publication has the opportunity to feature 300 hotels and resorts in their 1997 editions. In addition, they have programs to allow you to access their expansive database of "carriage-travel" travel agents and corporate planners, plus interactive features that highlight your special programs, brochures and seasonal packages throughout the year. They also have a website to expand your reach. Premier Hotels and Resorts is part of the Travel and Leisure Group of Advanstar Communications (Parent Company of HMM). Our very well-known hotel friend, Melinda Bush, is heading up this new operation as Executive Vice President.

YOUR MOST IMPORTANT SALES TOOL—KNOWING YOUR CUSTOMER

July 1997

We all know that before you get a customer you need to get a prospect and even before questioning the leads become prospects or non-prospects. Once you get a prospect the questioning continues so that you may develop specific needs and wants and you get to know what to sell and how to sell. And hopefully, a sale is enough information at your fingertips to be able to relate to the caller with specific information to reflect that you know something about the organization or group? Technology is already in place to be able to do that for you. Here's a scenario created by David Phillips of the Virginia Association of Realtors. (Used here with his permission) that was published in a recent issue of Cardinal News, a newsletter of the Virginia chapter of Meeting Professionals International:

Hotel: "Sales Department, this is Rick. How may I help you?"

Dave Phillips: "Hi Rick, this is Dave Philips. We met at the last VAMPI meeting and I wanted to see if you could check some dates for me."

H: "Oh Dave, how are you doing? Let's see, you're with the Virginia Association of REALTORS, right?"

DP: "Yes, you sure have a good memory" Actually, Rick didn't have a clue who I was, but he had pulled my name up from his computer using contact management software.

H: "Are you looking for dates for you 1999 Annual Convention? That's in September, isn't it?" Little did I know that Rick had accessed my association's file on his computer.

DP: "That's right. But we can go to October if necessary.

H: "Has the agenda changed much since you were here in 1993?"

DP: "Not much, just a few minor things."

H: "OK, let's look at some dates then I have September 20th through the 24th open, or the 13th through the 17th.

DP: "Great! We prefer the 13th through the 17th. What kind of rate can you offer?"

H: "Well, I'd have to see the exact agenda before I can quote a firm rate, but based on the $85 rate you paid in 1993 I'd say we are looking at around $100. If you email the new agenda, I'll get a proposal put together for you."

Well, that's the story and you know what? This same scenario is played out throughout the country every day between customers doing business with hotels, resorts, conferences that are up-to-date with technology that will help them do a better job of selling. Of course, there is the question of cost for new hardware and software programs, and training to get off and to run. We have to think of these costs of long-term investments to make the sales effort work better. Sales and catering personnel of properties already using systems such as Delphi, Breeze and others already know the value of having information readily available at their fingertips. Accessing account file information along with the availability of rooms and meeting space without making a telephone inquirer wait is critical to being professional and doing business. Being online via email is another plus for hotel salespeople wanting to do business with meeting professionals like David Phillips and according to him: "knowing your customer is the most powerful tool in this age of technology. I'm not talking about what my birthday is or that I like to play golf, rather, I'm talking about knowing the details of my group and being able to use that information to help make booking a meeting more efficient."

HOW TO GET MORE ROOM RESERVATIONS FROM YOUR CENTRAL RESERVATION SERVICE (CRS)

September 1999

If your property is flying a flag of a major brand, then it is pretty certain that some of the hotels with which you are competing are also flying the same flag. And, those competitors may not even be in your own backyard, but could be 100, or more, miles away. There are hundreds of reservations agents working for you at central reservation system offices around the country who probably have never seen, or talked with, anyone from your property. These folks have the responsibility of booking rooms through the system when people call in to inquire about space and rates or to make room reservations. Of course, up-to-date information is readily available about your property to these reservation agents via the computer or hard copy directories. However, there is just so much that written material can provide for selling purposes. Sometimes it is difficult to express real value to a prospective guest without seeing or being told more about a property.

If you had the opportunity of visiting your brand's central reservation system offices to visit with all the sales agents, you surely could convey to them a good deal more information about the value of the property's features so that they may do a better job of selling your hotel. Many years ago when we had 50 Holiday Inns (at the time we were their largest franchise operator), we had a good deal of competing Holiday

Inns all around us. Twice a year we would visit the central reservation system offices, sponsor a coffee break, and make a "pitch" for our properties. It worked like a charm. Immediately, actually, the same day, we saw an increase in reservations coming from the office where we were visiting.

Reservations agents are salespeople, and salespeople sell better when they know more about what they are selling. It seems like a very logical thing to do for General Managers, salespeople or front office managers to visit their brand's CRS offices to make a "pitch." Here is how it works: First, get an OK from the CRS office manager. You do not want to take any of the agents away from their positions to listen to you. But, you can make a presentation during their breaks and lunch period. You can provide coffee, juice, donuts for a morning break. For lunch, you arrange for sandwiches, pizza, salads, drinks, etc. All this may be done in the office cafeteria, or break rooms, with prior approval. Sponsoring some food service is a great way to win friends, and get people to listen to your presentation, without taking time away from their job function. Imprinted give-a-way items could also make a big hit. Leaving something behind helps the agents remember your property long after the presentation is over.

The more the agents know about your property, and area, the better. You can do this through a narrated slideshow or use a video presentation. Just be sure you leave enough time to answer questions of the agents before they go back to work. Also be sure to leave behind a very detailed fact sheet which shows specific features of the property which may reflect value to the prospect.

Of course, the big question is being able to take the time to make the trip. Also, to some extent, there is some uneasiness about doing something like this if you have never done it before. There is an opportunity for you to let someone else do this for you. At one of my sales workshops, I met Joe Manly, who does just that. He has a great deal of experience in making presentations at central reservation system offices so he started his own company. It doesn't matter which brand's flag you are flying, he will represent you to the reservation agents of your brand.

BUILD IT AND THEY WILL COME; OR WILL THEY?

July 1999

In these days of great U.S. economy, and a very healthy hospitality and tourism industry, we see investors, lending institutions, franchisors, and developers still looking for opportunities to add to the 3.7 million rooms we already have in the U.S. And, there is nothing wrong with that, provided the demand keeps even with or exceeds the growth. So far, it looks pretty good. Demand is still high, and predictions are that companies will be having more meetings, incentive trips, and business travel than a year ago. There will also be more association type meetings and conventions. And, it looks like the discretionary income is very much available to most families to take vacation trips. So, we are probably looking at continued success in our industry through the year 2000, as long as nothing happens to hurt the U.S. economy.

Since the industry has done so well, there probably has been an influx of newcomers getting into the business looking for a share of the profits to be made. However, before profits show up there has to be headed in beds, and something on the bottom line. Sometimes the vision of new owners/operators of properties is that by getting a good location, a brand name, and even great bricks & sticks, people will just flock to the door and beg for rooms to rent. It just ain't so! Or, at least, it generally doesn't work that way. What we see happening quite a bit around the country (or we hear a lot about it at our sales workshops) is that new properties coming into an area are looking to steal business from competitors by reducing rates. Cutting rates to get rooms business does not make sense. The whole idea behind doing business is to offer some benefit or value for the price, and having some sort of sales effort to bring in new business to the area. There are just too many new products out there with managers (or owners) who think that if they build it, they will come. Something needs to be done to get people to use a property, and that something is a sales effort. Trying to save labor dollars by not having a sales staff is one big mistake. The idea is to get a good return on investment by hiring people who can bring business to the property. You can come up with a formula which should reflect how much new business is expected from a salesperson for a certain amount of salary. A ratio of one to ten may work or even if it is one to five. That is, a property with high-priced rooms, it may mean a person should bring in ten times salary. For a property with lower price levels, it may mean five times salary. Or even something in between.

A good way to bring more business to an area is for hotel properties to work together through a convention and visitors bureau. As an example, the Jacksonville & the Beaches Convention and Visitors Bureau conducts regular "University" sales workshops for its members. At a recent workshop which I Presented, we discussed ways to work together to bring new business to the area from outside the state, as well as from other parts of Florida. The idea was to get the message across. There can be more room sales made from outside the area than trying to compete with each other for local business. The whole idea is to use the Convention and Visitors Bureau as the catalyst to get the whole area to work together. And, according to Kitty Ratcliffe, President of the bureau, it's working. Getting a brand franchise along with its central reservation system is just not enough to be successful. It is a question of what kind of sales effort will be conducted at the property level.

SALESPEOPLE SHOULD BE SPENDING MORE TIME SELLING

May 2001

It is amazing that so many salespeople tell me that they spend less than 50% of their time in pro-active selling; many spend even less than 25% of their time. What's going on here? We've had such a good run since 1992 with a great economy, more leisure travel, more meetings and conventions, wonderful increase in demand for rooms, that maybe we have forgotten how to sell. The phones have been ringing "off the hook," and all we had to do was just answer the phone and book the business. Now, with the current economy, the over-built situation in some market areas, and in some cases, a cutback in the small meeting market, it is time to get back into the real selling mode. That means salespeople need to spend more time searching for (prospecting) for new business to replace the business that may have been lost.

Many salespeople claim that they are being hampered by some time-consuming activities that are beyond their control. And, these things prevent them from spending more time on pro-active sales. Right or wrong, it pays to take a look at some of these activities to determine where more time may be converted to pro-active sales:

Paperwork: We all thought that once we got into working with computers, this new technology would eliminate the paperwork. In many situations, for whatever reason, it has created more paper flow. One of the big issues concerning salespeople is the number of reports that have to be completed for management and corporate offices. It seems to me that with new computer-driven automated sales accounts systems, like Delphi, Breeze, and others, all reports that may be needed by management can be accessed via the system. There is no reason why managers cannot gain access to a salesperson's activity to find out about a number of bookings, amount of revenue generated by bookings, tracing, lost business reports, call reports, letters written, proposals and contracts sent, etc. It is all there just by knowing how to access the information. Why all that paperwork?

On-site servicing: Why do salespeople have to be involved in servicing the business they have sold? It just does not seem logical for them to have to be around, at the property to "make sure" everything goes well for an account during a meeting. In many cases, it is the sales folks who do not trust the staff to provide the proper service, so they want to be there to assure that things go well. In other cases, it is management who feel it is proper for them to be in attendance to help out; of course, who booked the event was there. Salespeople are hired to sell, not service. We have conference service managers and catering managers who should be very capable of handling all the details of a meeting or event. With proper internal communications, there should be no need for sales folks to be involved.

Meeting: Why, oh why, do meeting have to be so long. It seems that there are weekly sales meetings at just about every property. OK, yes, meetings could be important, but only if they are productive. So many are not. Why do they have to be held during productive selling hours? Why do they have to last three or four hours? Why do items, not pertaining to sales need to be discussed? Let's think about arranging the meeting time not to conflict with selling time. How about meetings at 8:00 a.m. or 4:00 p.m., maybe even on Saturday (ouch!). Meetings really do not have to last more than one hour. All meetings should have an agenda, and all those who are invited should have a copy of the agenda well in advance. I have a one-hour sales meeting agenda that should work for most properties.

All we need to do is figure out how we can get more time available for salespeople to spend more time in selling so that they can bring in more new business to replace the business that is being lost. More time for prospecting, more time for selling to qualified prospects. That will work to your best advantage.

FOR HOTELS IN CRISIS, SALES TRAINING IS CRITICAL

January 2002

Many people in our industry will agree, for the most part, sales folks at the property level have very little or no training for their jobs. Usually, the route into sales, internally, comes via the front desk or some administrative or secretarial function. Some are already working in the sales or catering departments, in administrative jobs, and have learned how to handle bookings, answer the phone, and even make prospecting calls. Externally, we hire, for selling jobs, folks who may have had been a salesperson in another industry, or even had already been in a sales function at another hotel. We also bring in recent college graduates and put them in sales after some sort of brief orientation or sales familiarization program. Whatever training that has been provided in the area sales, however, has been "on the job training." Perhaps, in many cases, one person without any formal sales training, showing a new person "the ropes." Now, there is nothing wrong with this. We have a great many very qualified salespeople in our industry, who have learned, on their own, through their experiences, and they do a superior job.

As long as business is good, as it has been for about the past eight years, many sales folks have been getting by very well by just answering inquiries, and taking orders. Soliciting business during a "sellers' market" is certainly not as tough as trying to build business during a "buyers' market." And, this is where it appears to be today. As a result of the economy, overbuilding, strong competition, and to some extent, because of the September 11th situation, we have found ourselves in a real crisis situation. Occupancy is down, profit is down, less people are traveling, and we see properties getting involved in a rate-cutting scenario. Now is the time we look to sales departments to fight for additional business, improve the occupancy and bring the properties back to profitability.

Maybe now is the time to start looking at how well our salespeople have been trained to sell in these hard times. When business is good, it is easier to sell than well business to not so good. Now training becomes critical when it comes to improving sales.

"Faced with the greatest RevPAR decline since the sixties, Sales Training is the survival edge in troubled times," according to Date Turner, President of Turn Key Hotel Advisors. In an article published by Hotel On-Line, Turner goes on to say: "An ill-trained Sales Managers, or more likely, one who has had little to no formal training at all, cannot compete for market share in a downside economy. Many Sales Managers in our industry today have only been taught how to sell in the good times. Sales Managers today are ill-trained in matters of prospecting, cold-calling, telemarketing, negotiating, persuasion tactics and most importantly, closing the sale. It is in these areas that a sales training program should be focused and if there were ever a time when these skills were needed, it is now."

Sales training opportunities are available all around us. Turn Key Hotel Advisors is a Dallas-based hotel consulting group, specializing in hotel management, operations and sales training (www.turnkeyhoteladvisors.com). Also, there are a wide range of training programs as well as training materials through the Hospitality Sales and Marketing Association International (www.hsmai.org). The Educational Institute of the American Hotel & Lodging Association has training materials and programs, as well (www.ei-ahla.com). Then there are folks like Tom McCarthy, who conducts sales training program and just introduced a new Web-based training model "Handling The Inquiry Call" (ttmccarthy@hotelpros.org). Other sales trainers include Ed Iannarella (eid@wideworl.net). Bill Flor and his newsletter entitled: "Sold out" (www.no-vacancy.com).

OUTSOURCE YOUR SALES DEPARTMENT? WHY NOT?

February 2004

To outsource, seems the way to go these days in just about all industries just in the business of meeting management we see corporations and associations getting involved with companies that will take care of their housing, physical meeting arrangements, exhibits, ground transportation, registration for events, and even site selection as well as negotiations. There is certainly a distinct advantage to outsourcing part of a business function to companies that can specialize in particular aspects of on business.

One of the things that come up most often during our HMM sponsored sales workshops is the fact that most directors of sales find that there is so much to get done in the business of selling at a property, that it all just does not get done. In the function of booking business for a property, so many salespeople have to get involved with a good deal of marketing as well as sales details. It is almost unbelievable the wide variety of things with which a sale person needs to get involved. Just prospecting for new business via telemarketing, alone, could be a full-time job. Then, of course, there is the maintenance of accounts, building relationships with existing clients, keeping up to date n files, conducting the actual sales calls on prospects (and that is also a full-time function). Lately, because of the Internet, We find ourselves more and more involved in spending time responding to requests for proposal, coming in at a rapid speed to most properties with sizable meeting space. Now we also have salespeople that have to deal with the third party meeting planner, as well as those Internet travel agencies. Of course, there is also the conducting of site inspections, advertising, public relations, direct mail, promotions, arranging a sales blitz or familiarization trip and sales trips out of town. The list just goes on and on. There just isn't enough time to get everything done properly in order to improve sales productivity. It all needs to be done.

We're getting to the point in our industry that in order to get more sales and marketing effort accomplished we should start to outsource. There are wee-experienced hotel industry sales and marketing people that are setting up shop just to do this kind of work for you. Hospitality veterans Ken Benjamin and Tony Diarmondo have seen the need for some hotel organizations and properties to outsource some of their sales activities. Their company "Hospitality Taskforce" (www.hospitalitytaskforce.com) provides temporary sales help. They also have a program to train General Managers of small properties to better understand the sales process and help direct the sales effort. Limited service type properties also need to figure out how to get their sales to work accomplished when they do not have a sales department to make things happen. So why wouldn't we find ourselves seeking an opportunity to outsource some of these duties; it sure would make sense?

RELATIONSHIP SELLING: THE KEY TO HOSPITALITY SALES

August 2004

How good can you be in hospitality sales if you got all the breaks, did your very best, and applied yourself? Each of us has to answer that question about our own ability in building business for the properties we represent. Whether we are the owner, General Manager, sales manager, front office salesperson, or involved in any other aspect of guest contact, one area of great concern should be where we are in the business of Relationship Selling. A relationship selling professional is the person who believes that selling by persuasion alone is not the key to the long-term business building. We surely are not in the business of taking people into buying just one time. The idea is to build the relationship with guests so that they buy again, and again, become loyal to the property or brand, and provide referrals.

So, how do we go about doing this? The key to relationship selling is to build competencies in a variety of skill areas. According to Jim Cathcart, author of The Eight Competencies of Relationship Selling, there are eight skill areas, as follows, that is necessary to develop and implement in professional selling:

▶ **Preparation:** What skills or habits could you acquire in raising your sales readiness to its highest level?

▶ **Targeting:** What resources, strategies, system, and habits could you cultivate to assure that you are always focused on the best prospects for your business?

▶ **Connecting:** Which skills and traits could you cultivate that would cause others to want to do business with you?

▶ **Assessing:** What qualities and expertise could you develop in order to be aware of not only what someone needs, but also what they want?

▶ **Solving:** Which attributes could you develop that would increase your ability to convince others to take the actions that were best for them?

▶ **Committing:** What behavior and processes could you encourage in you that would cause others to trust you implement that would make people remain satisfied with their decision to do business with you?

▶ **Managing:** What systems, habits, resources, and schedules could you put in place to assure that you were always doing well and consistently getting better as well?

When we take a look at these eight areas of competencies, they look very obvious to us all and we know that we have been practicing a variety of skills reflected up these areas. To what extent, however, have we been successful in building long-term business for our properties by effectively using these skills? How good we are in sales really needs to be measured by the lifetime value of an account. It is not only a one-time piece of business, but also the relationship we have built with a client to get repeat business as well as referrals. People, with whom I have done business over my years in hospitality sales, are still in touch with me, today, even though most of them have been retired for quite some time. Not only have they been long-time uses of the properties I have represented over the years, but have become lifetime friends. This is what relationship selling is all about.

LET'S MAKE HAY WHILE THE SUN SHINES (2004)

December 2004

With all these reports coming out lately about our economic recovery, more business and leisure travelers, more business meetings to be held next year, increase in demand, increase in occupancy, and not much of an increase in new room construction, it sure appears like a no-brainer for profitability in our industry. It also appears that the industry has done pretty well in maintaining expenses. So, now we need to spend more time thinking about how to go about improving that bottom line. And, the answer is in improving sales productivity. No question about it, you cannot bring anything to the bottom without bringing it in at the top line. What this means is that now is the time to do a better job in sales revenues generation.

Who is doing anything at the property levels to make sure that we maximize revenue in all areas of hotel operations? These are some owners, operators, General Managers and even salespeople who are very optimistic about the business coming in because of the favorable reports of business getting better. It ain't so! Sure, business will get to the properties via the Internet or the third party independent meeting planner, but all at a lower rate, probably than what you can get with better sales production. What is being done at the front desk in the area of up-selling room rates? How about the dining rooms sales; how do we get our services to improve revenue production by up-selling appetizers, desserts, etc. Who is doing anything in our catering departments to get those staff members to up-sell banquet items or even being pro-active in going after local social events? Then, of course, we have to look into our sales offices to see what is going on with revenue improvement in that area.

All too often property management and operators look at the salespeople and want detailed reports on how they are spending their time. Too much paperwork is required on what kind of activities in which sales folks are involved. No survey has done on this but I would be willing to bet that most property salespeople spend less than 35% of their time in pro-active sales. By this we mean, calling on, developing relationships with only qualified prospects. If a salesperson is good at what he or she is doing with regard to booking profitable business, then does it not make sense, that if they spend more time on pro-active selling, they would increase revenue production? With regard to improving revenue production, it is unbelievable that there are many properties and management companies that are still using the old-fashioned way of providing incentives to salespeople, of providing goals and measuring on room nights booked. We cannot take room nights or occupancy figures to the bank. Why wouldn't we measure sales productivity by the amount of revenue generated? As a salesperson, being measured by room nights booked, it would a motivation to book business at almost any rate rather than the highest rate possible. No wonder a good many salespeople are giving-in one negotiations and offering lower rates with the excuse of "building relationships." Let's get with it folks, if we want to Make Hay While the Sun Shines, let's do a better job of bringing in more at the top line.

LET'S START WRITING SOME BETTER SALES LETTERS

April 2005

We really do write some pretty poorly written sales types letters in our industry. We all get loads of them from someone who wants to sell us something. A day doesn't go by that we don't get a few solicitation letters from hotels asking for our business. What we see in these letters some very poorly trained hotel salespeople who are using computer-driven letters that are too long, not creative, do not address needs, are not warm and friendly, do not appeal and nothing is said that makes us feel good. And, to top it all off we are not even prospects for any hotel business.

Why are we writing to anyone who may not be a prospect for business for the property? Where do we get our mail lists, and what is the purpose of writing these letters anyway? It seems that we should have a reason for writing someone a letter. If we want to solicit someone's business, then a letter would be appropriate, but only after we have determined that the address is a real prospect. By this we mean it should be someone who is in a position to do business with you, should they choose to do so. Let's not waste our time and money on letters to people who we do not know if they are in a position to bring business to us. When we write to a prospect we need to address their specific needs and how they can be accommodated. Why try to tell that prospect everything about the hotel? Just concentrate on the specific needs. Who cares if you have a ballroom to seat 500 people if that prospect has meeting needs for 75 people?

Who are we trying to impress with some of those big words we use in letters, but never use in oral communication? When was the last time you spoke the phrase: "Pursuant to our recent." Or "as per our agreement." And why do we include phrases in our letters such as: "enclosed herewith" or "please don't hesitate to call" We all need to take a look at the letters we have planted in our computer. The chances are they were written some time ago by some previous sales director who is no longer employed at the property. Look at all those letters and try to make them more believable, look less computer-driven, more creative and meaningful.

All letters should be short and limited to one page. Use the first paragraph to get someone's attention. And, you do not do that with an opening such as: "let me introduce myself. I am new." Write something that will grab the reader and make them read further, you have to be creative. Make sure the content of the letter reflects information with regard to what the prospect needs; and get to the point fast. Your last paragraph should be a call to action; that is, they are asked to do something or some indication that you will do something. And, it doesn't hurt to add something with feeling, which makes the reader feel good.

ONE PERSON OUT OF FOUR CAN SELL

August 2005

Some years ago, a person conducting a sales seminar made that remark about "only one person out of four can sell." Surely, anyone can sell, and of course, some may sell better than others, but what is the big deal about being able to make a sale? In our hospitality business, we have front office personnel as well as reservations persons selling rooms. In our restaurants, our servers sell food and bartenders sell drinks. Just about everyone is selling something or other in our business. However, when it comes to putting together a sales department team, a property owner or manager is looking for proactive sales from the staff. This means, of course, going after and bringing in new business. Whether it be soliciting and booking major corporate rooms contracts or tour operators or meetings of all kinds, or sports teams, etc.; putting a new business on the books is what counts. So this is where we have the differences, and maybe the point of only one person out four can sell has more to do with the staff we hire in our sales departments. There is, of course, a big difference in selling when it comes to someone "taking orders" or someone who has to locate prospects, and then attempt to sell them on using a property.

Someone who comes to a property, like a person calling in to make a reservation, or walks in to rent a room, or comes into the restaurant to order a drink or meal makes it kind of a lot easier to make a sale. Most of our staffs do that pretty well. When it comes to selling, to bring in group business or trying to get a corporate travel manager to enter into a year-long agreement to send their travelers to a property it gets somewhat tougher. The same applies to tour groups, government contracts, sports teams, etc. Property salespeople going after these markets spend a good deal of time in relationship selling; that is, getting to know the prospects, and how to approach them, understanding their specific needs, as well as building their confidence and trust in the salesperson and the property being represented. Many salespeople will have some good stories to about after having called on prospects, time and time again, sometimes, over the years, they finally did get the business.

For a property seeking new sales generation, it all starts with the proper hiring of salespeople. The idea is to find prospective sales staff employees that possess certain characteristics which have proven to be a mark of successful salespeople. These may include, but certainly would not be limited to: persistence, persuasiveness, patience, ego drive, empathy, enthusiasm, energetic, outgoing, has initiative, customer oriented, positive attitude, passion and a desire to help others.

In the business of selling, we need to understand the reasons people buy. People make buying decisions based on certain things which could include any or all of the following: they like the salesperson, see the value, feel of a "fit" with the property and the salesperson, feel the salesperson is committed to them. There is no question about it, if the property fits the needs, then it is the salesperson who can make the deal.

TO IMPROVE HOTEL GROUP SALES, TALK LESS AND CONSULT MORE

September 2005

Hotel sales staffers, generally, do talk more than necessary, and because of that, could kill a sale rather than confirm one. It appears that many salespeople tend to talk more than necessary, and try to convey to a prospect how much they know about a property, and attempt to tell just about everything there is about their product; this is not all necessary. The idea in productive sales is to talk less, but consult by helping a prospect solve an issue or concern.

In hotel group sales, knowledge is the key. We need to have knowledge of our product, knowledge of the competition, and knowledge of the needs of a prospect. We do not seem to have a problem with our own product knowledge as long as we understand all the features of the property and how each could be a value to a prospective buyer. We also would not have a problem with knowledge of competition as long as we understand how our product features compare to a competitor's features from a value standpoint. The area where we probably have some difficulty is having detailed knowledge of the needs of a prospect. Just knowing that someone makes an inquiry for a meeting of 100 people during a certain period is just not good enough. We need to dig deeply and ask a lot of questions to determine more specifically, the needs of the group, the organization and the person making the inquiry.

The whole idea about being a consultant is to be able to ask critical questions to get a prospect to think, analyze, and even speculate in responding to those questions. This consultative selling process requires less talking on the part of a salesperson, and getting answers to important questions that reflect specific needs or "hot-buttons" which will allow the salesperson to then respond by presenting a product's feature which will reflect a real value to the prospect. All questions need to be open-ended so that a prospect can have the opportunity to do a lot of talking rather than answering with a "yes" or "no." Questions need to start with something like: "Tell me about." or "Please explain.," or "What was."

One of my favorite activities at a convention is to walk around the exhibit floor and eaves-drop at a booth. It is amazing the number of salespeople who tend to tell passers-by about all the features of a property, without even trying to find out if the person is a prospect. It seems to me the best opportunities salespeople have when participating as an exhibitor would be to try to find prospects by asking critical questions to determine if someone is in a position to do business with them. There is no point trying to "sell" someone without first determining if that person is a prospect. Next time you attend a convention walk around the trade show floor and listen-in on how your competition "works the booth."

IMPROVE PROFITABILITY BY ADDING ANCILLARY CHARGES FOR GROUP MEETINGS

August 2007

Profit is not a dirty word. We need to start thinking about Owners, Managers, and Sales staffs' responsibility for protecting a property's assets. Operation costs have increased considerably this year, and it is expected to continue increasing in the coming years. We can no longer think that high room rates alone will be the way to profitability, but profits should also be considered in adding ancillary charges for meeting and convention groups.

Naturally, there has been a concern that because of the great business we have had over the past few years, there will be an increased in new hotel products in a coming couple of years, reflecting an over-supply of rooms and creating increased competition. Yes, more rooms are being built, and we will probably see a small drop in overall occupancy; however, it does appear that the growth will be in secondary and tertiary cities with probably limited service products with limited meeting space. Further, reports have reflected there is expected very limited growth of new properties in the 25 major markets. Room rates in those markets are expected to at least, stay the same as now, and with the continued increase in demand, will probably go higher. In addition, a recent study report from Meeting Professionals International (www.mpiweb.org) indicated a continued growth in corporate as well as association meetings as well as conventions. Further, the report reflected a considerable increase in the meeting budgets of these organizations.

So, how do go after these extra charges for meeting groups? For many years, we, in the industry have been providing free meeting space for groups with sizable room blocks. More recently we find that many properties have been charging meeting room rentals based on a sliding scale of rooms rented out of the block. Now is the time, as long as there is continued demand, for sales staffers to be stronger in negotiations, and to charge full prices for the use of meetings rooms. Additional fees for meeting room rentals should be standard if we want to increase profits. There should also be an opportunity for special charges for resetting meeting rooms. Then there is a so-called policy of complimentary sleeping rooms when a certain amount of rooms in a block are used. Why do we have to continue this "policy"? How about rejecting some group business that may not be as profitable as another during a demanding season, when you know there is another group that may be sought for space? Another critical area is that of attrition, and how well do we handle the amount to collect when a block of rooms is not filled? In the attrition clause of a contract how well do we protect ourselves when it comes to food and beverage counts, and meeting room rental, when space is not used? Also, we need to consider how much in additional income is expected by each guest, daily, in a hotel during his or her stay. Should we not consider this amount of revenue not being collected as a result of a room block not being filled? This too should be included in the attrition clause.

Of course, all this needs to be worked out during the negotiation process and there are a number of things based on what the buyer can offer to gain consideration. However, we must remember in the process of negotiation we should not be "giving" anything away, without getting something in return. Relationship building also comes into the process and this is understandable. However, improved profitability is the key to staying in business.

BUILD YOUR SALES VIA CONSULTATIVE SELLING

June 2009

As a salesperson, if you have the experience, knowledge, and confidence, regarding your ability to be good at what you do, then you need to get into the business of consultative selling. This is not a new concept. Salespeople have been providing advice to prospects every time in making a sale, but perhaps most have not considered themselves as consultants.

A consultant is someone who provides professional or technical advice. Every hospitality salesperson is really a consultant, but perhaps we do not go about our business as such. Perhaps we are too intent on selling something that we forget to provide advice and counsel. One of the biggest complaints meeting planners have had over the years is that some salespeople are so aggressive about making a sale that they forget to find out if the meeting planner is really a prospect. There is no point in trying to hit someone over the head with knowledge of a product unless that person is a prospect for the property. One characteristic of a good consultant is being able to listen intelligently.

Unless you know and understand the needs of the individual or group, you cannot figure out how to help them. A consultant asks specific questions to determine needs or wants, and then listens intelligently to look for clues and "hot buttons." Next, he relates to the response by providing specific information that meets those needs or wants. When you think about it, providing professional or technical information or advice really is just sales.

Before you try to sell your lodging or meeting facility, you need to find out as much as you can about the individual or group: and it is not question of only "googling" or searching the Internet to get the information you want, You need to get more involved with the prospect, personally, if you want to provide consultative services. For example, in the case of meeting planners looking for an appropriate location for a meeting, think about asking a question that makes the person analyze, speculate and evaluate. Also, some questions that stimulate emotion will help draw out feelings from the prospect. These feelings will provide you with clues to help show the prospect of the benefits of using your property.

Ask more than yes or no questions, such as: "Tell me what is most important to your group in the selection of a property," or "Please tell me what you liked best about your last meeting." Questions like these keep the prospect talking, gets you to listen, and keeps you from doing too much talking (a great big fault of most salespeople). The important thing about being a consultant is solving people's problems; helping them make decisions and getting them to feel comfortable about having you around. Building long term relationships are very important in consultative selling. Maintaining good contact with key prospects over a long period of time, and letting them know they can call on you anytime for advice (even if they are not booking your property) makes you a better consultant. Build your network of contacts, and be reliable, trustworthy, honest, informative, helpful and loyal. Getting into the consultative-selling process will do more for you in confirming sales than anything else.

KNOWLEDGE IS THE KEY TO MAKING THE SALE

October 2009

For many years the approved way of selling had been "hitting them over the head with the product." That is, keep telling the prospect all about what it is you want to sell them, explaining the "wonderful" features, and using glowing terms about the product. Even today we find many salespeople, mostly new staff folks, without the experience or training of sales strategies, still going after probable prospects (and some non-prospects), telling them all about their wonderful hotel property. They explain how big it is, where it is, the amenities, "terrific" service, food, etc. Doing it this way hardly ever got the booking. Potential buyers of hotel services these days do not need to be "pushed" into using a facility just because the salesperson tells them how great the place might be for them. What they really want to know is "why." "Why should I use your property? What will it do for me or my company? What benefit is there in meeting my needs?" Let's not misinterpret the word "benefit" to mean "rate reduction." We are doing too much of that in our industry these days. Just giving discounted rates does not answer the question of benefit, all by itself.

Of course, there is always the prospect that will call a property, and proceed to make a group booking without even getting involved with any salesperson approaching them. These are the people who already know the property, have used it before, and do not need a sales pitch. What we are really discussing here, however, is the prospecting, the going out after the business. Salespeople who take the time to gather information (and we should be doing that quite well via the internet) before making any sales pitch, have a better chance of confirming sales. Knowledge is power; knowing as much as you can about the prospect; what they need or want, previous experience, how buying decision is made, what properties they have used previously, who makes decisions, what kind of budge they have for rooms and functions, the purpose and objectives of the meeting or group function, and the list goes on. Salespeople should be able to construct their own checklist of questions that need to be answered before a sale is attempted.

This whole idea of gathering knowledge is to be able to approach the sale from the standpoint of providing a product that is needed and will benefit the prospect. The idea is not to try to make a presentation before learning as much as possible about the prospective buyer's hidden agenda. It just does not make sense to come in with a presentation that doesn't meet needs or even budget. The point is not to try to sell someone beer if they have a champagne budget. Gently asking open-ended questions to get the prospect talking about their organization or group is what is needed to be done. Lead-in questions should start with "how," "tell me," "what." As a prospect responds to one of these questions, then the opportunity may arise to relate to something about your property that may be a feature that would provide a benefit for the group. The more knowledge we have about prospects the better chance we have of presenting a proposal that best meets their needs, and the better chance we have of confirming a deal.

KNOW THE ATTRIBUTES OF YOUR PROPERTY TO HELP MAKE A SALE

April 2015

Yes, of course, hotel sales staffers know their product, to some extent. However, do they really know how the features really relate to prospects' needs to help them make the "buy" decision? The key to closing a sale has more to do with how well we can describe how certain features of a product meet certain needs. It is never enough to tell a prospect, for example, that your property has an exercise room, or 10 connecting rooms, or "large" rooms, or a swimming pool, etc., What we need to do is describe a particular feature that meets a particular need of a prospect or the attendees of a group event.

All too often salespeople like to tell just too much about the property in order to make a sale. Talking too much really does not help. What is needed is first to find out about needs. What specific needs are necessary for a particular group's activities, its meeting planner, and the attendees? This means there has to be a lot of questions which need to be answered. Only after a salesperson understands the needs is there an opportunity to explain in detail what and how certain features meets those needs. So, the idea is to make sure the members of the sales team have made a complete product Analysis of the property being represented. It is in the best interest of any hotel, resort, conference center, etc., that this is completed by all sales staffers who have not yet done this exercise. The idea would be to have them tour the property, not only to identify all the features in every area (rooms division, food service, bell service, parking area, etc.), but to understand how each feature may meet the needs of a prospect. There is no purpose in trying to explain the value of all the features, but only in those that meet needs (or wants). For example, all prospects may not at all be interested that you have 10 connecting rooms; however, if the group has an attendee who may bring their children to the event, then, in this case, the teams having a good number of rooms with two double rooms could be important for doubling up.

Expressing features as values are what is important in the business of selling. Features of a property should be explained as it fills a need, and the value has to be detailed. It would not be enough to indicate that you offer a complimentary breakfast. It needs to be explained how it saves time; attendees would not to leave the property to find a place to eat; how it saves them the cost of a breakfast, how it keeps everyone together, etc. Every salesperson should ask to do a product analysis and to create a factsheet on the property.

Chapter 5

Working with Group Markets

AGGRESSIVELY SELLING SPECIFIC GROUPS AIDS GROWTH IN THE 80S

March 1981

Should a property increase its rate to compensate for a downtrend in occupancy and help improve its profitability? Can a profit improvement be seen with a discounted rate? This is the dilemma which is being faced today as a result of a tremendous increase in the cost of operation plus a decline in hotel occupancy. Operators will be faced with cost increases continually throughout 1981. The recent minimum wage increase, along with its ripple effects, will create additional payroll costs of about 15% to 18%. With this will be dramatic rises in employee benefits (which include added FICA costs to the employer, vacation pay, sick benefits, insurance costs, and a variety of other employee benefits provided by various food and lodging companies). Other increased operating costs will be mainly in the area of utilities (expect increases in energy costs in the area of about 15%) and maintenance (at least a 10% rise in the area).

For properties that operate restaurants, the same increased costs of operation apply along with additional problems that arise in the area of food purchases.

Pass Along Costs

Industry-wide there will be great pressure to pass along all added costs to the consumer with higher menu prices and room rates. This is inevitable and has to happen for an operation to continue to be profitable. However, most times when there are increases in consumer costs, it is reflected in a drop-off in sales. This applies generally to all types of businesses. The key is to be able to pass on cost increases without losing business. The successful operator in 1981 will have to place greater emphasis on improved employee productivity, greater cost controls, better product quality and most importantly, more sales activity. Properties with sales staffs will have to immediately reassess their market plan to determine if it is up-to-date. Sales staff functions may have to go after specific markets more aggressively to develop increased occupancy and sales. Programs may have to be developed that will reach special market segments which may have never been previously tapped. Some of the changes in the market trends are:

Senior Citizens—They travel more now than ever, and have more time available and in many cases, more money available for travel. Great opportunities exist for the lodging operator to get a better share of this market by soliciting motor-coach tour business. This group also takes advantage of reduced rate airline fares to destination locations. And packages sold through travel agents are attractive to senior citizens.

Small Meetings—The trend is toward meetings in smaller cities at smaller lodging locations. Companies and associations have more meetings of 50 people or less than they have larger groups. Rather than have participants travel great distances to major hotels, they spend less on travel by holding smaller meetings at close locations. Small properties can now go after this market aggressively.

Family Travel—Increased airfares have created a return to automobile travel even with higher gasoline prices. Families will be taking to the road again. With the purchase of smaller cars, the family market is getting used to the high price of fuel and will be back on the road this summer. However, the trips will be shorter. This mobility will create great opportunities for the small operator.

Increased rates alone will not necessarily mean that lodging operators will meet with success. An aggressive, profitable sales program is necessary no matter what kind of sales staff the property has. Responsibility for the sales effort lies in the hands of the owner or operator. This responsibility cannot be passed down to someone else.

The ultimate answer to bottom line profitability lies with improved, profitable sales. Raising rates may decrease occupancy and still produce profits. Selective discounting could reduce average rates but it may improve occupancy and bottom line profit. Whichever way a property goes, new business is a requirement, and can be accomplished through a plan and aggressive selling.

PROPERTY OPERATORS AND THE SMALL-MEETING MARKET

July 1981

Many operators of small motels, hotels, resorts, and inns overlook the meaningful, important and potentially lucrative market of small meeting groups. Some operators feel that their property is not suited for meetings. Or facilities are not attractive enough. Or they will have to hire a salesperson to develop this market. Or they have no food service. These are all the wrong reasons.

More "small" meetings of less than 25 people are held in the U.S. today than any other type of meeting. These small meetings include regional or district company meetings, all types of management or board meetings, association committee sessions and training seminars. It is this type of group that the small lodging operator should try to book. The easiest source of this business for a small property usually is just around the corner. There are a lot of companies, for example, looking for a nearby meeting site, something "handy" to their own corporate offices.

Your Merchant Friends

In addition, the same people that the property buys from should be solicited. Consider: the bank that handles your account; the local insurance company contact that has your coverage; the people to whom you pay taxes (i.e., government employees); the meat wholesaler that supplies your restaurant—as well as the dairy, produce company, bakery, wholesale grocer, beer and wine distributor, for that matter. In fact, all the people that serve your property are excellent prospects for small meetings.

These sources make easy and excellent starts for soliciting this type of business. You know with whom you are dealing; you usually know these people personally, so it should not be difficult to ask for the business. In turn, they are in a position to want to do business with you. In case they don't have their own meetings, these people usually belong to a district or region of their own parent company that has meetings and your local contact should want to help you get that meeting business. And, they probably also belong to some type of local association that meets periodically.

No Advertising Costs

Interestingly enough, this market is easy for you to solicit yourself, making it unnecessary to hire a salesperson. It is also the type of market to go after on a personal basis, without incurring any advertising costs. Another nice thing about the small meeting market is that the booking lead time is generally short, usually a month or less. That's when you need the business—now—not next year. Space requirements for the small-meeting user are generally easy to meet. If the property does not have rooms set aside exclusively for meetings, there are alternatives: part of a dining room where tables can be put together; the cocktail lounge during the day; a parlor room; or even a guest bedroom, with some moving of furniture or perhaps set up with extra tables. The opportunities to generate additional sales through the small-meeting market are endless. The business is there. Just as for it.

CONCENTRATING ON THE SMALL-MEETING MARKET

April 1982

No matter how small a lodging property or its sales staff may be, it can greatly boost its occupancy rate by focusing on the small-meeting market. Surveys taken time and time again have always indicated that the most common type of business meeting involves fewer than 50 people. And get-together of that size are apparently recession-proof; as the cost of travel continues to rise, many corporations are relying on a series of small regional meetings rather than the full-scale national convention of the past.

Small meetings obviously don't draw the volume of conventions, but they very easily can fill anywhere from 30–50 of your rooms for at least one night. And that return comes from a minimal investment. You won't need a full banquet staff or an army of bartenders to provide an adequate—and profitable—amount of food and drink. Nor is sophisticated audio-visual equipment needed to accommodate a small group.

Even a lodging operator without a sales force should have no trouble cashing in on this market. A good place to begin is with suppliers he already knows, such as the local beer distributor, dairy operator, produce wholesaler, banker, insurance agent and all the others with whom the hotel/motel regularly does business. These are people who already know the property and should feel close enough to management that they know they will get good service.

Even if that particular supplier does not need any meeting space himself, each one probably belongs to a trade and professional association or a civic organization, a social group or a religious group. They could very well be in a position to recommend the property for future meetings of those organizations. Employees of a property are also good sources of meeting business. Most employees will belong to some sort of organization, such as garden clubs, PTA's, or religious groups that meet regularly. The site might as well be your hotel.

Once the General Manager or lodging operator has made all of these contacts, then he can start calling on everyone else in the neighborhood. It is really just a matter of walking down the street and knocking on the right doors. Once the lodging operator has gotten involved in the local meeting business and has developed some confidence and know-how in selling, he can try to reach out-of-town meeting planners through different sales and marketing techniques.

GETTING STARTED IN THE SMALL-MEETING MARKET

February 1983

No matter what size property you operate, there is always a meetings market for it. The corner of a dining room; a sleeping room or parlor with the furniture moved around a little bit; part of a lobby screened off; a lounge does not use during the day; even the manager's office, if need be—somewhere in your property is enough space for a small meeting. The objective of an inn, of course, is to sell all sleeping rooms, and even a meeting of ten to fifteen people could very well end up getting you five to ten rooms plus food and beverage sales. For a small property, that sure does help.

And that's exactly what happened at the Imperial 400 Motor Inn in Minneapolis, Minnesota, according to John Canney, sales representative for the Imperial chain. The property moved bedroom furniture out of a sleeping room to provide space for a meeting of fifteen people which resulted in the booking of ten sleeping rooms. Now, this kind of business doesn't just walk in by itself. Someone from your property has to go look for it, find the prospect, and then sell him. Fortunately, though, it's not that difficult to do.

Think of yourself as a meeting consultant helping your prospect to meet his or her needs. Most small properties that don't have salespeople or advertising budgets are not thought of as "meeting sites" by local businesses and they are not aware that even small properties can accommodate meetings. In fact, many lodging owners and operators themselves do not look at their own properties as meeting sites. It is certainly worthwhile to look over your property and assess your assets as a meeting site, then list all the positives or pluses there are to use as selling points. For example, if you only have one room or space available to accommodate only fifteen or twenty-five people, then that could be a big plus to a company or organization that could feel you will be giving that group your full attention by being the only meeting in the house. If you can provide food and beverage service in that room, it's another plus. If parking can be given free, stress that as another plus. Just make yourself a list of all the favorable assets, and that's your sales pitch. No expensive sales plan, no brochure, no advertising budget, no sales salary—just you and your list telling everyone in town by telephone and in person that you can handle small meetings.

Where do you go from there? The place to begin is with the people and companies with whom you are already doing business. The local banker, insurance company, dry cleaner and laundry, various wholesalers from whom you purchase—make your list of everyone you already know who could conceivably book some meeting business. Put down the name of the company, name of the contact that you know and the telephone number. When you have the list, start telephoning and tell them very honestly that you can handle small meetings and you need their help in getting started. Also seek out local contacts that may be in a position to help you get training meetings, which more and more companies are using these days.

BOOSTING GROUP SALES VOLUME IS EVERY STAFFER'S BUSINESS

April 1984

Most guests really don't expect VIP treatment; however, they do expect—and are entitled to—a smile, a friendly gesture, or some simple acknowledgment of their presence. Doing a good job in sales is really a challenge. If the people in a particular group aren't pleased with the stay, regardless of the reason, it's very unlikely the members of the group will "bad mouth" the hotel for years to come. It's a serious loss of future business for the property. Lodging operators need to have a strong sense of responsibility for honoring commitments made by the sales department as well as ensuring that all business is handled in such a manner as to be completely satisfying to the guest. If the client is satisfied with the manner in which the hotel serviced the group, then the meeting planner should be urged to let the salesperson rebook the group—right then!

If the client is satisfied with the way the hotel serviced the group, then the meeting planner should be urged to recommend the hotel and the salesperson to other meeting planners; in fact, the salesperson should write down the referral leads—right then!

CONSULT WITH OPERATORS

Operations people are a great help to the property's sales team by letting the team know about problems as they arise in handling a group. A salesperson is much better prepared to rebook a group if he or she knows what went on during a meeting—both good and bad. Good communications between operations and sales is a must. This is true regarding any size of the meeting—whether a convention for 1,000 people or a simple luncheon for merely 10 people.

Periodically, the property sales team should meet with operations people who handle or get involved with attendees at a meeting or group movement. This means that from time to time meetings need to be held with the housekeeping staff; the front office staff; the dining room and lounge staff; the kitchen staff; and the banquet department staff, besides with engineers, maintenance people and bell persons.

These meetings need not be lengthy, but they need to discuss the importance of groups to the property's success. With brief remarks and a question and/or comment period the sales spokesperson should relate:

- How the sales department operates;
- How groups are booked;
- How much time and effort it takes to make a sale;
- How future business counts on service;
- How future business can be lost because of something that appears to be insignificant;
- And how each staff member can help the sales department sell.

Clearly, every job and every person working for your property is vital to your success. Therefore, a property's sales team needs to sell internally as well as externally.

THE MEETINGS BUSINESS KEEPS GROWING

November 1984

By the end of 1984, the meetings business will have concluded a banner year, according to Mel Hosansky, Editor of Meetings and Conventions Magazine. Over $30 Billion will have been spent on a variety of types of meetings, conferences, trade shows, training seminars, conventions, etc. This dollar amount represents a considerable increase over 1974 when the meetings business was measuring at about $9 Billion.

Hotel and motel owners, managers and sales personnel should take note that a good portion of those dollars goes for sleeping accommodations, meeting space, food functions, receptions, and equipment rental. All of which reflects in a property's bottom line. Next, to transportation expenditures, lodging operators are the biggest recipients of those meetings dollar.

With hotel construction still going strong, and an estimated increase of over 70,000 new rooms on board in 1984; there is great competition in the meetings market. Still strong contenders for this business are the downtown and airport locations. Suburban and roadside type properties are up and coming, also beginning to get a good share. Resorts have always been strong and will continue to do so. In fact, there's not any type of location that is out of the running for meetings business.

Surprisingly enough, it's this writer's opinion that well over 50% of the meetings business gets to a property because of a buyer's inquiry rather than as a result of sales solicitation. Location, location, location is still the cry of meeting planners in determining where the conferences and conventions will take place. After location, the next emphasis is on facilities to determine which site would be selected. It's not only the availability of rooms and meeting space, but product quality and guest satisfaction are prime considerations.

Some Hotel/Motel sales personnel still perceive room rates as being a major consideration of meeting planners in selecting a site. As planners become more professional and sophisticated, they find that rate is no longer a major factor in the selection process. More importantly, they are concerned with meeting the needs of the group. Accessibility, facility, guest services, friendliness of the staff, cleanliness of rooms and food service rate high on their checklist. Meeting attendees will remember good (or bad) food service longer than another else about a meeting.

Room rates certainly enter into the picture, but companies and associations are not as rate sensitive as they used to be. It's still on the checklist, but not as high up. Some rate negotiation will almost always take place, but salespeople should try to sell the product, staff, and value. Trying to close on rate alone is not working anymore. Maximizing sales means get the best possible rate for a property at all times. The average daily rate for group meeting business should be the next highest rate after transient business.

Salespeople should make a very strong effort in not discussing rates until the very last in closing a sale. Get the meeting planner convinced that the property is the best possible choice for a wide variety of reasons. And then when it is determined that the property meets the needs of the group—it's time to close the deal at the best possible rate. Soliciting and selling to the meetings market is important to all types and sizes of lodging facilities. Establishing a plan to go after this business should be a must for 1985. The meetings business will continue to grow in the future—and at a very rapid rate. The largest growth will be in the small meetings market which consists of conferences for under 50 persons. Cities, convention bureaus, hotels/motels, conference centers, civic centers, and even free-standing restaurants are scrambling for this business. The competition is great, but a good job of selling the product and services is what will determine the successful sales operations.

THE MEETINGS MARKET MEANS A BIG JOB FOR SMALL PROPERTIES

December 1984

Smaller sized properties—150 rooms or less—may be experiencing boom times in the meetings business as meeting managers for companies and associations are finding these roadside, suburban and airport locations ideal for the 10–30 person gathering. But "small in size" doesn't necessarily mean "small problems." Here are some of the things that can go wrong and what you can do about them:

▶ **Problem**: Verbal commitments: Too many times planners are being told what

the hotel/motel will do, but the planner does not get what has been promised.
- ◊ **Solution**: Put everything in writing and get an acknowledgment from the meeting planner.
- ▶ **Problem**: Meeting room setups: Salespeople sometimes are not up-to-date on the facilities offered at their properties.
- ◊ **Solution**: The new salesperson at a small property must be taught appropriate meeting terminology and learn how to set up rooms in different ways. Room capacities must be accurately figured with dimensions made available on a rooms layout chart.
- ▶ **Problem**: Food planning: Many salespeople are not totally familiar with food products, pricing or menu planning and work only from pre-printed menus.
- ◊ **Solution**: Include the food and beverage (F&B) manager or the chef in discussions with the meeting planner when planning food functions.
- ▶ **Problem**: Rates quoted differently than what is actually paid: (This refers to room rates as well as banquet prices.)
- ◊ **Solution**: When quoting rates and confirming in writing, taxes, gratuities or extra charges should be determined so there are no departure time surprises.
- ▶ **Problem**: Too man people to deal with: The planner has to speak with someone about rooms, another about banquets, yet another about setups.
- ◊ **Solution**: If the salesperson at a property will not be able to handle all the details then the planner must be advised up-front who will be working for the group on all arrangements. At a small property, it would be logical that one person handles all the arrangements.
- ▶ **Problem**: Accounting/billing: It always takes too long for a planner to get the bill, and then problems exist for many planners when a bill does arrive. Other problems include separate statements for rooms and food functions, counts on functions which don't agree with the planner's count, charges planners do not understand and credits not given on deposits.
- ◊ **Solution**: If credit has been established for billing then statements should be mailed immediately following the meeting with each item clearly identified. If only rooms are being billed to the meeting planner, charges on each folio for rooms only should be transferred to the master account with copies of the folios as a backup.

Attendees need to pay incidentals, if not originally indicated, to be picked up by the planner. After each function during the meeting, a planner should sign a banquet billing reflecting actual counts or guarantees and price breakdowns. All appropriate bills for rooms, food/beverage functions, meeting room rental and other charges that have been approved need to accompany the statement.

SOME CAREFUL DIGGING WILL HELP UNCOVER LOTS OF PROFIT

The salesperson's responsibility to management doesn't end with going out on sales calls. It's the generating of sales—or more aptly, of profitable sales—that counts.

Being a professional in hotel/motel sales comes from a continuous process of making yourself aware of what's happening in the marketplace. Learn what the competition is doing. Learn what different group accounts need or want. Learn how to present your hotel's features. Learn about creative food and beverage functions and how to use them as selling tools.

Specifically, salespeople should learn all they can about their property's business mix; they need to explore market segments that have more profit potential; they need to tap the potential of existing accounts; and they need to solicit accounts similar to the ones they're already servicing.

Let's take a look at each area:

▶ **Business mix:** Salespeople need to meet with the General Manager to examine the current mix of business. By analyzing how much business is being generated from each market segment, and at what rate, it will become obvious if a change in the mix is desired. Most properties have a system for recording this information daily.

Typically, the market segments will include transient, meeting groups, corporate, government, group-tour rate, then it is obvious that more effort should be put into selling group meetings.

▶ **High-profit segments:** Some types of business generate more profit for a property than other types. Student tour groups, for example, generate good room sales, but little in the way of food & beverage revenues.

The objective here is to generate sales in market segments that not only rent rooms, but provide revenue to all property outlets.

▶ **Tapping existing accounts:** There are many salespeople who make sales calls on their corporate accounts only to look for the individual business traveler. The potential for additional business from an existing account is great, but it requires the salesperson to learn as much as possible about the account and to ask questions like: Does the company have meetings? Who handles the meetings? Which other departments in the company have people visiting who may need overnight accommodations? Who takes care of the company's social activities and holiday parties? What clubs or trade groups do the company's executives belong to? Can they influence a decision to bring a meeting to the property?

Generally, an existing account can and will provide more business to a property if the right contact is made and business is solicited.

▶ **Similar accounts:** There is a tendency for similar accounts to do similar things. If, for example, a property does business with Army recruiting people for meetings and meals, it is likely that the Navy, Air Force, and Marines are doing pretty much the same thing—somewhere else. If a property hosts a weekend conference of piccolo players, then it is probable that there are associations of drummers, piano players or violinists who have similar meetings. If there is a party for a new-car showing put on by one auto dealership, then there may be a similar party that could be solicited from other dealerships in the area.

Salespeople need to keep on the lookout for ways of developing new business for rooms and food & beverage. Opportunities abound for all types of business. Knowing what's going on in the marketplace, checking on the competition and exploring new profit sources are important keys to success in sales.

And remember: be enthusiastic about what you're selling!

SALES TIPS ON SELLING MEETINGS AND CONFERENCES

February 1986

All too often, salespeople tend to start "giving away" too much unnecessarily. In the area of group sales (meetings, conferences, incentives, etc.). There are many meeting planners for companies and associations who are more concerned with being at a particular site location with an excellent product that provides quality service. The question of getting better rates, complimentary rooms, free things, etc., many times doesn't even come up. Salespeople need to learn to speak less and listen more—and learn when and how to negotiate in the best interests of the property.

Room Rates: Generally, one of the first things brought up during negotiations—or when an interest is indicated by a planner in using a particular hotel or motel. Hotel salespeople should learn to put off discussing rates until the very last possible moment. The salesperson's objective is to satisfy the needs of the meeting planner and the group first. Selling the features and benefits of the property and convincing the prospect that, "This is your best choice for the meeting," needs to precede the negotiating process. Also, it is not always necessary to sell at discounted rates. Many times, the regular rack rate could very well be accepted because the meeting planner's needs are perfectly met by the product and the budget. Salespeople need to learn prospects' history: where they met previously; how much they paid for rooms; what was room block pick up, etc. To what extent rate should be negotiated depends upon a wide variety of items—most of all . . . How profitable will the group be to the property? . . . How badly is the business needed? . . . What other business will the property get other than rooms? (Food, beverage, meeting room rental.).

Complimentary Rooms: This is another area where salespeople start giving away before it is necessary. It appears that one "comp" room for fifty sold seems to be the expected standard—but why? This should not be a standard practice—more often than not—planners making arrangements for meetings do not ask for complimentary rooms, but are being offered them anyway by salespeople, unnecessarily. Providing complimentary rooms should be a part of the negotiating process and not be offered, but should be used as a "trade-off" for something else. It is not necessary to be consistent—the idea is not to give unless necessary to make the deal.

Letters of Agreement: Everything that is agreed upon during the making of a deal with a meeting planner should be put in writing and agreed to by both parties. Salespeople should be certain that the person with whom they are negotiating is authorized to make the agreement. Generally, this will be established well enough in advance when information is developed during the prospecting and further development of an account. Dates, food, beverage functions, meeting space, cut off dates, guarantees, who pays for what (if billings are separate), when payments are to be made, cancellation clause, penalties on cancellation should also be spelled out in detail.

Salespeople will do much better for a property in closing a deal on a meeting by knowing all that is possible about the group. Previous history, purpose, and objective of the meeting are readily available by asking. Listen a lot, talk little and do not commit any extras without a "trade-off."

WORKING WITH THE MEETING PLANNERS

April 1986

Professional meeting planners want to do business with professional hotel/motel salespeople who understand the meeting planner's role and their needs. Sales personnel need to impress upon their General Managers, as well as all department heads—the importance of taking good care of meeting details so that the property may develop a reputation for handling meetings and conferences. Just in the area of guest registration and housing, there are situations that need to be discussed with the meeting planner and then properly handled by the hotel/motel staff. These items could apply regardless of the group size.

▶ **Rooming List:** Ask the planner to provide a rooming list two to three weeks in advance of the group's arrival. VIPs on the list need to be identified. Every attempt should be made to have all rooms ready for check-in. However, depending upon the scheduled arrival time (if early), rooms may not be ready due to heavy occupancy the night before, without enough early checkouts. Professional meeting planners understand this and this should be explained to them in advance. Alternative arrangements need to be made for early arrivals for checking luggage and working with meeting planner for scheduling some type of activity. In any event, sales personnel should work with the hotel/motel to try to get enough rooms ready so at least the VIPs will have their rooms ready. Provide the meeting planner with two copies of the room list, one by room number and one alphabetically.

▶ **Check-In; Check-Out Areas:** Depending on the size of the group, it would be in the best interest of the group and the property to provide a special area—away from the front desk for meeting attendees to check-in (and, perhaps checkout). If extra personnel is required, most meeting planners would understand this and may be willing to pay for this extra service. This should be discussed well in advance of the meeting date. It is a good convenience for the group as well as other hotel/motel guests.

▶ **Bell Staff:** It is important to be aware of all arrival times of group participants. Adequate bell staff must be made available to move baggage to rooms. Also, the property staff should make sure that luggage waiting to be delivered should be in a relatively out-of-sight area from the group guests. Well in advance, details should be worked out with the meeting planner on luggage tags with names printed clearly with the name of the group. If possible, they could put large numbers on the tags which could correspond to the rooming list—this saves times in trying to read names. An automatic (blanket) gratuity system for luggage should also be discussed at the time the system is explained to the meeting planner.

▶ **In-Room Amenities:** A good opportunity for creating additional revenue for the property would be to generate a printed listing of in-room amenities that the meeting planner may purchase for his group. This could include fruit, cheese tray, wine, nuts, dried fruits, liquor bars for VIPs, turn-down service, candy, flowers, etc. Prices for each item should be on the list. This will turn out to be a very good selling tool for the sales department.

▶ **Accounting:** Suggest to the meeting planner that each day, the accounting department will provide, for inspection, all available master account charges as well as a few individual folios. This is to be sure posting is being handled according to agreed-upon instructions. The time this is done should be mutually agreed upon. By doing this, it helps both parties. When the meeting is over, there should be no discrepancies in the billing and it will encourage quick settlement of the account. Most meeting planners need to pay the account as quickly as possible to close their books and get busy planning other meetings. They do not want to delay payment due to some inaccuracies in the billing. Resolving differences on a daily basis gets the bills paid promptly.

▶ **Post Meeting Summary:** The sales department should be provided with some sort of internal worksheet. This should show actual room pick-up by day and by category—compared to the room block. Gross food and beverage revenue and meeting room usage. This provides a good record in the sales file for future bookings of the same group. The sales department may also send a copy of this to the meeting planner with an appropriate "Thank You" letter. The planner will appreciate the summary and will know that the group has been dealing with a very professional sales staff.

BRAINSTORMING CAN BOOST YOUR GROUP-MEETING BOOKINGS

December 1986

Meeting planners are turning to hotels and motels for help in making meetings more interesting. Generally, planners have no problems with the technical details of meetings. However, there seems to be an increasing need for creativity in the planning of F&B functions. People are trying of the "rubber chicken" circuit—even prime rib is becoming old hat. Many professional meeting planners are finding themselves having to come up with innovative ideas for planned meal menus. While the creativity such people display is admirable, they shouldn't have to be teaching the food business to food people. Salespeople, F&B managers, and catering managers are learning from meeting planners—but the shoe should be on the other foot.

Constant Change

We are in a constant mode of change, and we cannot continue to do the same thing we did last year or even five years ago. In many cases, it isn't a question of price; rather, it's a question of meeting the needs of the group buyer. People attending a meeting remember the food functions long after they've forgotten the meeting's program content. They also remember the site of the meeting—the hotel or motel—largely because of its food functions. Salespeople should learn to consult with the appropriate staff members to prepare for a presentation or site inspection. Getting together with the chef, catering manager or F&B director to hold a brainstorming session can yield valuable results, and can pay off later in repeat business.

Important Knowledge

Knowing the previous experience of a group regarding its functions, knowing more about what they "go for" or knowing the demographics of the group is certainly important in preparing for the brainstorming session. (It sure wouldn't hurt to know the budget, either.) In addition to different menu items, there should be some discussion about the capabilities of the staff to deliver certain products. For example, theme parties still sell well. To prepare for one, the "brainstormers" should discuss what's available locally for props; what types of room setups are available, and what kind of equipment is available for those setups.

Conducting a brainstorming session helps the sales situation in a number of ways:

▶ It gets the staff in a selling mood;
▶ It creates excitement;
▶ It gives people a chance to be creative;
▶ It can become a motivational tool for the entire staff.

In short, the brainstorming team becomes enthusiastic—and enthusiasm is contagious.

But most importantly, with the information and knowledge provided by this approach, the salesperson does a much better job of selling during presentations or site inspections.

SMALL MEETINGS ARE FOR EVERYONE

December 1989

There's no question about it ... small meetings are on the rise! The small meeting (50 people and under) has had a steady increase since the beginning of this decade. More companies are having more meetings and even large associations are having more smaller meetings. Earlier this year the American Society of Association Executives completed a meetings trend survey which reflected lesser numbers of large meetings. Last year there was an apparent decline in the number of meetings and the number of people attending those meetings? What is happening is that probably, associations are finding that they could have smaller meetings around the country and get better attendance with the same budget. They probably have also found that the smaller meeting provides better personal attention to the participants.

However, aside from the associations, it appears that companies and organizations realize the need of smaller group educational and training sessions. Sophisticated technology has just about dictated the need for more training. According to the American Society for Training and Development, the companies are spending $30 billion a year on training done in a formal setting (workshops, seminars, meetings).

What this means to us is that if we're not into going after this lucrative market—no matter how small a hotel or motel we have—we're "missing the boat." The business of training meetings is BIG business. Even our very large, mega hotels, which for years have almost exclusively gone after large conventions, are now going after these 25–50 size training meetings. Right now, if salespeople of any hotel, of any size, would check on the number of small meetings being held at the property—the result would be that about 50–75% of the meetings are 50 people or under.

The people who plan these meetings and select site are an elusive group. For the most part, they don't plan meetings full time—so they are hare to identify. They are presidents of companies, secretaries, salespeople, personnel managers, etc. A lot of them are right in your own backyard. There's an excellent chance that most of the regular corporate accounts that provide you with room business can also book meetings with you.

Salespeople need to have a plan to go after this special market. At companies, we need to look for meetings of the Board of Directors, strategic planning sessions, financial planning sessions, sales meetings, training sessions, etc. At associations we need to go after committee meetings, Board meetings, strategic planning sessions (contacting the association executive will probably result in a long list of types of small meetings held.

The SMERF market is the other place to look for small meetings (Social, Military, Educational, Religious, and Fraternal). This particular market generally is looking for affordable meetings. They use hotel/motels offseason or during a period or days when a property needs the business.

WHAT DO MEETING PLANNERS EXPECT ABOUT QUALITY FROM A HOTEL?

November 1992

If you haven't seen the latest report on quality that meeting planners expect from a hotel you better try to get a copy. It sure could be helpful to you in understanding what meeting planners want. It has always been tough to quantify what quality means. Certainly, to a great extent, it has a lot to do with perception, but also has to do with living up to certain expectations of the receiver of hotel services. Recently Meeting Planners International, along with the American Society of Association of Executives and Meeting News magazine put together a series of questions that would sniff out what meeting planners expected from a hotel with regard to product and services. They used Beta Research Company, a marketing-research firm, to conduct the survey by telephone of 125 Certified Meeting Professionals. Fifty questions were used covering a variety of aspects of meetings management in connection with hotel services. Here are some of the highlights, of a very extensive report, which will give you an idea of what is expected from hotels:

- According to 78% of the planners surveyed hotels salespeople should return calls the same day received.
- With regard to the availability of meeting space—the majority of the planners expected a response the same day.
- Complete proposals were expected within five working days.
- 97.6% indicated a problem should be remedied by the Conference Service Manager, who should be authorized to do so.
- A majority of the planners expect the General Manager to be visible from time to time during a conference. More than 1/3-rd expect the GM to be at the pre-conference meeting.
- Having a written disaster-preparedness plan was either significant or of high priority to 92% of those surveyed.
- It appears that planners look for sleeping rooms that are at least 400 square feet, according to three out of five responses.
- 52% indicated that they expected a banquet server to take care of two rounds of ten. The rest indicated 1 or 1.5 per round of 10.
- For bartenders, there should be one server for every 75 guests according to 7 out of 10 surveyed.
- Over 90% of respondents expect that supervisory management and staff receive ongoing customer service and job-specific training.
- Greeting guests by name was significant to a high priority for two out of three planners.
- The majority of those polled expected check-in (as well as check out) at the front desk to take not more than four minutes (included waiting in line).
- Nine out of ten surveyed say that room service should take not more than 30 minutes; with tray, removal to take not more than 30 minutes voted by 7 out of 10.
- 97.6% indicate that hotels should offer non-smoking rooms.
- 88% want 24-hours room service.
- 86.4% expect a gift shop on property.
- 84% want a fitness center.
- 77.6% expect a club level floor.
- Very high priority was placed on hotels having remote control color television (97.6%).
- 92.8% wanted cable TV channels, sports, movies, CNN.
- In-room movies were important to 73.6% surveyed.
- It looks like most meeting planners don't appreciate telephone surcharges. 60% said none was appropriate; 31.2 indicated that 0.25 to 0.50 was OK and 8.8% approved up to $1.00.
- As far as bathroom amenities were concerned everyone seemed to want at least something other than soap and toilet

paper, but they ranked in this order from high priority to low: soap; facial tissues; extra roll of toilet paper; shampoo; hair conditioner; shower cap; sewing kit; hair dryer; skin moisturizer; clothesline; shaving mirror; show sine sponge; toothpaste; cotton swabs.
▶ Being able to send a fax (99.2%) or making a copy (98.4%) was very high on the priority list.
▶ The majority of planners expected to get their hotel bill within one week after the meeting ends. 25% wanted to see it in two days or less. And, more than half expected that billing dispute to be resolved within two days.

THE WHOLE HOTEL SHOULD SERVICE MEETINGS AND OTHER GROUP BUSINESS

September 1993

To a great extent, at many properties, the success of the business has a lot to do with the amount of bookings through group sales. In today's competitive environment just about every type of property would welcome group business at one time or another. We seek group business from meeting planners who work for corporations and associations; motor coach tour operators; incentive houses; travel agents; independent meeting planners; sports teams; social groups; and a host of other entities that will provide us with a block of rooms along with meals and receptions.

Since the whole hotel should be in the business of servicing groups why don't we take the time to orient the entire staff at a property with regard to the needs of the group market and particularly, the value of this type of business to the hotel? We give a lot of lip service to the fact that this is being done, but I don't believe it is happening. The sales department at a property gets very much involved with the group customer and understands their needs, but what about the rest of the staff? If we take a close look at the other department's employee's knowledge of a group's value we probably will find that they have only a vague idea of a group's importance and particular requirements.

The front office staff is generally concerned with rooms so are probably not fully aware of the total needs of the group; catering is concerned with the social functions; meeting room setups may be handled by a banquet department or conference services department while audio-visual requirements may very well be handled by an outside source. From the food and beverage end of the business, each dining room manager and bar manager is concerned with groups only as far as he is involved in any of the functions. Each department is generally concerned with its own day to day operation, but shouldn't they also be as involved in knowing about the groups in the house?

We see more and more hotel sales departments conducting pre-conference meetings. This is where the person in charge of the group, along with the sales director, gets a chance to meet with all the department heads a day or so in advance of the group arrival to go over activities and needs of the group. For some types of groups, there is no contact person arriving at the property, or that person will not be able to arrive in advance. In these cases, the salesperson or catering manager or whoever booked the business could conduct a short meeting with the department employees to review their involvement and to explain the value of having the group in the house. Of course, there are many hotels that have multiple groups on the property at the same time. This should not be an excuse for not keeping the entire staff informed.

The whole idea is that when a staff member comes in contact with a guest who is part of a group the employee should recognize the name of the group, understand its needs, give recognition of the group and appreciate the value of the business. And, of course, provide the necessary outstanding service.

There are many hotels that do a great job at this, but one stands out since I was there recently conducting workshops for Meeting Planners International. I don't know just how they managed to accomplish it with a number of different groups meeting at the same time, but the Hyatt Regency at Gainey Ranch in Scottsdale, Arizona sure did a fine job in orienting all their employees about the different groups in the house. From the time you arrived through to the departure, you felt like you were someone special and every guest in the hotel (whether part of a group or not) got the same treatment. From bellmen to housekeepers, to set up persons, room service attendants, dining room servers, groundskeepers, every employee seemed to be aware of the different groups—who they were, why they were there and what their needs seemed to be. Everyone working at the property appeared to want to do his or her best to please and make the guests happy to be at the Scottsdale Hyatt Regency . . . a smile on every face. A great experience.

HOW TO SELL TO AND WORK WITH THE NOVICE MEETING PLANNER

March 1995

In all probability, you and your staff, at any type of property, are selling to and working with non-professional, part-time, novice meeting planners. It has been reflected in various surveys that most the people who plan meetings do not do it on a full-time basis, but have other responsibilities within an organization, and only about 10% of meetings are planned by professional planners. More than not, we are dealing with secretaries, presidents, sales Vice Presidents, accountants, personnel managers and a host of people with other job titles and functions, who plan meetings as a side-line. After all, someone has to do it. These people generally are out of their element, are not comfortable with the idea of planning a meeting, have little idea of how to communicate with you, do not understand the hospitality business and need your help to a great degree.

Whether you are the General Manager, owner/operator, sales manager, catering manager, front office salesperson or whoever at a property, you want to win this person's present business, future business and recommendations to other people . . . you can, if you would recognize the psychological, emotional and practical needs of these people when they come to you to inquire about a meeting or when you are soliciting their business. Here are some thoughts to consider:

▶ Recognize that these people would probably rather be back doing what it is they usually get paid to do; that they may not be comfortable with the meeting planning assignment. You've got to reassure the people and take the initiative to walk them through the process, being the consultant, getting the information and making suggestions while getting approvals along the way to closing the sale.

▶ Discuss needs . . . of the organization, the boss, other people involved and how to accomplish the needs and what benefits there will be to them. Stay away from terminology used in our business. Do not confuse issues such as talking about square footage in meeting rooms when the interest is only in how many people may be seated. Who cares if your parking lot can handle 500 cars if everyone is arriving by plane?

▶ Do things that will make it easier for the planner. Provide a resume of the meeting prepared in detail. Include information that will help in communicating back to the boss and/or participants.

▶ Stay in constant contact to reassure the people that you are taking care of all details. Communicate in writing and by phone . . . it's almost like hand-holding for a nervous person. The whole idea is to have them gain confidence in you. This will also give you the opportunity to sell more.

▶ Be totally prepared and knowledgeable. You cannot gain confidence if you cannot respond to questions without checking with someone or if you are unable to offer suggestions.

▶ Understand that there probably are people who will be in attendance at the meeting who are special and who need to be impressed. Ask in advance for their identification so that you and the staff may do something for these people that will make the meeting planner look good. Once the VIPs get back to the planner about how great everything was you've made a friend for life.

▶ When the meeting is over do something to help the planner unwind. Perhaps a casual dinner with you to discuss the meeting.

▶ Once a prospect always a prospect . . . there will be other meetings, you may be assured, so stay in touch. Write the thank you letter, make the phone call, trace the file for follow-up. Ask for names of other people who plan meetings, other groups, and organizations to which the planner belongs, friends who may be in a position to give your business. One satisfied customer should always lead to another prospect.

One of the most critical things we seem to forget in trying to make group sales for a property is that we are in the business of solving problems. We need to initially talk with prospects about their needs, their business, what they are trying to accomplish. We need to focus on their problems and how they may be solved rather than pushing the property and its features and how wonderful we are.

GOVERNMENT BUSINESS COULD BE BIG BUSINESS FOR YOU

December 1995

The government travel market may very well offer you a great opportunity to increase room occupancy during periods when you need it. Agencies of the various government entities . . . local, state, federal as well as military, create a travel market of well over $15 billion annually. This includes travel expenses for government employees and contractors on individual business travel as well as those attending meetings. Travelers on government business are restricted to using these hotel facilities that are qualified through offering an approved government per diem rate. The General Services Administration publishes a listing of the authorized rates for hotel nights and meals. Persons on government travel make their own personal choice of where to stay and eat; however, they will only be reimbursed for the amounts approved by the GSA. Hotels that agree to offer rates consistent with the per diem for the area in which the property is located may be listed in the GSA Federal Travel Directory. In most cases, the rate could be considerably lower than the published rack rate of the hotel. Hotels also have to adhere to the provisions of the American With Disabilities Act in order to qualify.

Further, the various Department of Justice's regulations regarding practices of nondiscriminatory requirements have to be followed. A new law that went into effect in September 1992 was that any hotel with over there stories must be equipped with in-room smoke detectors and sprinkler systems before it may be listed in the directory, that they not only returned, but told others what to expect when they stayed at the Radisson in Alexandria. Besides getting the hotel staff to continue to do the basics in providing super service, they had to do a little more than their competition. But they also charged a little more. And even though these travelers on government business may have had to take a little out their own pocket, they kept coming back. To get in this PER DIEM PLUS PROGRAM qualified travelers paid an additional $10.00 over the government per diem rate. And here's the extras they got: Free bull continental breakfast served in a private lounge (Monday thru Friday), free parking, free local phone calls, free shuttle every half hour to National Airport, the Pentagon, three Metro stations as well as to Crystal City. Also, they received free wine, beer, and soda reception, with hors-d'oeuvres every day Monday through Friday. Their indoor-outdoor pool, executive fitness center, and indoor racquetball and tennis courts were also made available to members.

Harris and Wilkens go after this government market by being active in STAG (Society of Travel Agents in Government) as well as in SGMP (Society of Government Meeting Planners). They go after travel agents in feeder cities that have government contracts and make calls on government agencies for their meeting business as well as individual travel. Besides having a strong direct sales effort, the property advertises in various government publications. However, both indicate that their best business comes from word of mouth between government travelers.

THERE WILL BE MORE CORPORATE MEETING BUSINESS TO BOOK THIS YEAR

February 1997

It looks like corporate training budgets took a big jump in 1996 and there is every indication that it will probably continue in 1997 along with other types of corporate meetings. According to a survey conducted by Training magazine, spending on meetings and conferences rose to $3.7 billion compared to $3 billion in 1995. Of course, not all of the training sessions and meetings were held in off-site locations like hotels, conference centers, resorts, etc. However, this has got to tell us something about these market opportunities for all types of lodging properties.

Just take a look at the breakdown of your own market segments and make the comparison between the corporate meeting market and other types of group activity at your property. You will probably find that this business brings in a higher average room rate than most other segments in the group market. Further, most salespeople will tell you that the people that plan these meetings and training sessions how pretty much what it is they need and want with regard to space and function room set-ups; that it takes less time to finalize the booking and much less "selling" has to be done. Also, if you take a look at the lead time, you will probably notice that most of these corporate training and meetings are booked with short lead time; just about all within 90 days of the meeting date and many within just 30 days. The size of this corporate group business usually runs under 50 people with many in the 15–25 size—and all singles.

Here we have an absolutely terrific market for any type or size of the property. Since the business is booked within short-term your yield management program will tell you what room rate you need to get to make it most profitable for you. You should not feel that the meeting space you are providing should be given any without a fee, even if meals are arranged. Corporate meeting planners are getting used to the idea that they will be paying for meeting space. At a recent training session, I conducted for meeting planners at the MPI Institute planners told me that they are having a very tough time just being able to get small meeting space in the short lead time that they have. They do understand that we are in a sellers' market and they are willing to pay meeting room rental.

Properties that have full-time sales staff already know about the value of this market and how to go after this business; however, there are many properties with small meeting space, limited services, and no sales staff. So here are some suggestions: First, you need to find leads . . . that is, contacts that may be in a position to bring this type of business to your property. Once you get these leads you need to make contact with them and ask specific questions to determine if, in fact, they may be in a position to bring you their business. Questions such as how many meetings do you have a year? How many people attend? Where are you meeting now? Who makes the decision? Etc.

Where do you find these leads? Just about all over the place, if you look for them. Start with your own guests. Take a look each day at your expected arrivals. Those that list a company name and a title (owner, president, V.P., sales director, etc.) would be a good tip-off. There is always the business section of the local newspaper. Look for companies having meetings at other locations; people who have been promoted; business firms coming to town; civic leader appointments; etc. Then, how about the vendors with whom you are already doing business? And, of course, your own staff . . . we all have neighbors, friends, and relatives who may become prospects.

WE NEED MORE EDUCATION AND TRAINING TO BETTER SERVE THE DISABLED

April 1997

Yes, we are all in compliance with ADA (Americans with Disabilities Act), or are really? Most lodging and restaurant facilities have done the "basics," perhaps enough for compliance. But when we take the time to read the details of the law we see there is a lot more that needs to be done. The Justice Department has responsibility for enforcing the act, but resources are not there to make inspections. All they can do now is just handle complaints and prosecute where necessary. I'm advised they are backed up on cases.

My eyes were really opened to what this ADA and travelers with disabilities were all about when I attended, as a guest speaker, the First World Congress For Travelers With Disabilities. It was surprising to see that I was the only one on the whole three-day program connected with lodging. That has got to tell us something. Certainly, a major brand President should have been a speaker or a representative of AHMA, NRA, AAOA or someone from a major management company should have been there to represent the industry. It was mostly a travel agent affair with about 350 delegates from the U.S. and 15 other countries. On the program, we have speakers representing airlines, cruise lines, car rental companies, tour operators, wholesalers, educators, government agencies, ASTA (American Society of Travel Agents), ARTA (Association of Retail Travel Agents), and major travel agent consortiums. Our industry just missed the boat. Apparently, the "connections" weren't there for invitations for our major figures to speak or they may have been invited, but turned down the invitations. Curiously enough co-sponsors listed on the Congress letterhead reflected only The Congress was sponsored by SATH Society for the Advancement of Travel for the Handicapped. Murray Vidockler, who organized SATH over 35 years ago has been a proponent of making travel for people with disabilities an opportunity to travel like everyone else.

What did I learn from my visit? There are all kinds of disabilities; not only people in wheelchairs are disabled. In fact, of the over 49 million people who are disabled, only 9 million are in wheelchairs. And, another 9 million have sight impairments. The remainder is ambulatory, but have disabilities that may not be visible. According to a survey conducted by ASTA members 86% of member travel agents receive requests for assistance each month from travelers with disabilities. One of the difficulties travel agents have with hotels is that so many lodging facilities will not guarantee a special room. Agents usually get a reservation or front desk peon to tell them that "we'll put it on request." This may work OK for persons asking for a low floor or high floor room or a room with a view or a king room or even a non-smoking room, but it doesn't work if the inquiry is for a person in a wheelchair or someone who is deaf. There is no question about it, some prospective guests do need these ADA rooms, get confirmations, but actually don't get the room upon arrival—it was rented to a non-disabled guest.

The "war stories" about disappointed and disgruntled people with disabilities went on and on. Most were about lodging facilities that were not in compliance or weren't aware of needs and many about lodging staff personnel who were not aware of their own facility accommodations. Agents complained that many hotels refused to guarantee a specially equipped room; so people with disabilities requiring special rooms take their chances upon arrival.

Compliance with the law is working, in most cases, but only the minimum is being done. There is much more we can do to accommodate this particular market segment. After all, we are talking about 49 million people; there has to be some sort of a niche market there. Why not go a little further than the law requires and really make accessibility available to all people who want to travel. And, work on training your staff to understand this market a little better. If you are really accessible and barrier-free and want to get the word out to this market, tell your story to SATH.

WHAT DOES THE CORPORATE TRAVELER EXPECT FROM HOTELS?

April 1997

Guest surveys at hotels over a period of time pretty much tell about the same thing, when it comes to the corporate traveler—those folks that travel at company expense. We've read about studies on frequent travelers conducted by Doug Shifflet's company and Rolfe Shellenberger's Runzheimer International, as well as the variety of surveys conducted by trade publications, and some of the major chains. So why aren't we paying attention to these people that produce these reports? Oh, I guess there are many hotels that are doing something about looking after the corporate traveler needs and those that are doing something about it will benefit from the additional business it brings in.

It looks like the number one item on top of all the lists is this business of speedier check-in (hasn't that been complained about over a period of many years?). When was the last time you attended a convention at a large hotel? Did you have a waiting problem to check in? If not, you are lucky. People attending a meeting are not concerned with labor problems, labor cost, scheduling, people not showing up for work, etc . . . they just want fast check-in. I understand that Hilton conducted a study some three years ago and quicker check-in turned up as a key priority in the guest surveys. Surely, there is enough management staff around a large hotel that could pitch-in during a busy check-in period, even just a half hour or so would relieve the pressure.

I've noticed a lot of management "floor walkers," looking busy, in and out from behind the front desk, but not lending a hand. They take a quick look and disappear behind a door. Some hotel groups are getting into the automated check-in kiosks. It seems to be working well with Hyatt and some others. I think it's a great idea—some folks don't need personal contact with the front desk. Way back when . . . when I was working with American Motor Inns and then with Servico, at some of our airport hotels we started having people check-in while on the shuttle enroute from the airport to the property. I see more and more of this going on. The next big item on the list of "wants" is more non-smoking rooms. We can remember about fifteen years or so ago most brands started requiring hotels to have at least 10% of the rooms as non-smoking. Now we find that there are as many as 50% of the room allocated for non-smokers. If you are, like me, and get stuck in a room that has been used by a smoker, ugh! Most unbearable! I wouldn't be surprised if we will end up the other way around, with hotels allocating only 10% of rooms for smokers and it would have to be "on request." We see it happening in restaurants. The smoking areas are having less and less seats, and in some states, no smoking is allowed at all in restaurants. Notice what's happening at meetings? Many years ago meeting planners used to reserve the back of the room, or one side of the room for smokers. Now, you hardly see any smoking at all in meeting rooms. And, in some hotels, there aren't any smoking areas available in the lobby area. You gotta go outside to smoke. Of course, if you have an international market, with folks checking in from foreign countries, then you will have to look at their specific needs when it comes to smoking. It is my understanding that gateway-city hotels need lots of smoking rooms.

Interestingly enough lower or reduced rates just seem to come up high enough on the survey lists to be even mentioned. Guess the corporate travelers are getting used to paying the higher rates. Good for us! I'm surprised the mention of better lighting didn't seem to make the lists or maybe even that item didn't have any reference on the surveys. We may be doing better in this area, but lately in many of the hotels where I have stayed lighting could have used some extra wattage. We're still doing HSA and Hotel & Motel Management magazine sponsored Hospitality Sales Workshops. Eight more cities are scheduled.

BUILD GROUP SALES BY TRAINING YOUR LOCAL EVENT PLANNERS

September 1997

For a long time, in this column, we have been preaching that most of your room sales, individual corporate, and group, will be generated through local contacts. This works for all types of properties except for most resort and resort area locations. Even limited service properties, without meeting space, should be interested in developing group room sales. The opportunities are there for a wide variety of types of groups that would need hotel rooms, without meeting or feeding space. Prospects for this kind of business are all over the place, but they need to be hunted down and contacted. Blocks of rooms could be booked for reunions (school, military, family), weddings, funerals, sports teams, entertainment groups, retirement parties, graduations, and there are a few more you can add to the list. There are some key folks involved in recommending or making decisions with regard to where these people will stay when coming to your area. They would be connections through churches, funeral parlors, high schools and colleges, companies, civic centers, veterans' groups.

Getting this type of group business certainly is not restricted to only limited service properties, but even the full-service properties, with meeting space, should also be interested in developing this market segment. We also find that so many people responsible for taking care of arrangements for events (especially those requiring meeting space and food service) are not experienced in working with hotels. Here we have secretaries and department heads working for various companies that have never done a meeting or event planning, but are told to do so. Think about it! How many times have you had inquiries on group bookings and the people making the inquiries don't have the slightest idea of what they need to know? Sometimes it is like "pulling teeth" trying to get information out of them so you could help make the arrangements.

Well, here's an opportunity for you to be a hero, make some good new contacts, expand your network and even book a bunch of new business. Locate these local "connections" and train them! Yes, teach them what they need to know when working with hotels. Make them feel more comfortable in making arrangements when planning events and handling group room blocks. For years, hotels have been forming secretary clubs, for the purpose of getting companies to book their travelers into the properties. Most of the time a club was organized to provide bonus points to secretaries each time a room was reserved, with a collection of bonus points leading to free nights, meals and prizes. From time to time there were social events organized for club members. It would seem most appropriate, these days, to continue in forming these types of clubs, but in addition to company secretaries, include others who are involved in event planning. The focus would not have to be incentives for booking rooms, but education.

The idea is for you to get companies, and organizations to participate in your program so that their people can get trained in not only how to work with hotels, but how to put on better events. You can do your own in-house training, or even bring in professionals in this are to help you do this. A new company has been formed to help hotels do just that. Two professional meeting planners who have been working with hotels all over the country for many years have put together training session outlines for just this purpose.

DO WE REALLY UNDERSTAND MEETING PLANNERS' CONCERNS?

November 1997

Some time ago last year the Professional Conference Management Association (PCMA) conducted a survey among meeting managers to determine what were their greatest concerns with regard to working with hotels. At the same time, the organization surveyed hotel suppliers to find out what they perceived as the meeting managers' top concerns. Here are the top five in each listing as published in Convene, PCMA's monthly magazine.

According to the meetings Managers themselves:

1. Trust/Mutual Respect
2. Legal Considerations
3. Room Rates
4. Lack of Understanding of Member Expectations
5. Incorrect Group History

According to their Supplier Partners:

1. Contract Inflexibility
2. Response Time
3. Trust/Mutual Respect
4. Inaccurate Billing
5. Circumventing the Meeting Manager

Trust and Mutual respect came up in the top five as a great concern on the part of meeting managers and it was also perceived as a great concern of meeting managers by the hotel suppliers. Hopefully, things are getting better between the buyer and supplier; however, according to Ed Nielsen, Executive Vice President, Pennsylvania Academy of Family Physicians (as quoted in Convene) "We gather but we don't talk. We talk but we don't hear. We hear but we don't act."

It really is a question of trust. In my own experiences, in meeting with hotel salespeople during different workshops I get the feeling that perhaps we really do not trust how can we really do a great job in servicing our customers? One of the big things we see is that so many times properties are being asked to block and find out too late that the buyer does not deliver as promised. This is where we are not doing our job in checking history, learning more about the organization with which we are doing business, asking the right questions and sticking to appropriate cut-off dates for room blocks. Building relationships are the key to doing good sales work. Without mutual respect and trust, it just doesn't happen.

Legal considerations (along with contract inflexibility) also showed up at the top of the list. Our contracts are just getting too long and in order for each party to cover themselves, we find so many "protection" clauses. Too often hotels are using their own, regular corporate attorney to review or even write the contracts. This business has become too specialized to use any lawyer. What we need to do is locate any of the several attorneys who specialize in this business of meetings and conventions and seek advice on the writing of contracts.

The question of room rates also was high on the list of meeting managers, but did not appear in the top ten of the hotel suppliers. We can appreciate that people who plan meetings for their companies and associations will be concerned with the pretty high rates that they are being charged as a result of our sellers' market. Hotels are cautious about making room rate commitments too far out and meeting managers keep hearing the statement: "We can't tell you what our rate will be two years from now," but we want them to sign the contract now. Good contracts are always very specific, and rates or a formula for coming up with rates need to be included in every contract.

All this just boils down to property managers and sales personnel need to be better prepared in dealing with prospective buyers . . . and the best way is to better under the customer, their needs, how they operate, what their history has been and working together in a partnership type involvement.

THE "THIRD-PARTY" MEETING PLANNER... IS IT A PROBLEM?

July 1998

A new phenomenon has arisen in our meetings market segment over the past several years ... the third-party meeting planner. If you are into group meetings and have not yet experienced working with one, you sure will soon. It looks like they are almost taking over the meetings booking business. Years ago a hotel sales staffer worked with full time meeting planners or even part-time meeting planners who were paid staff members of a company, association or some other type of organization. Pretty clean cut ... an agreement was reached and the meeting was held. Now we see more and more of these third-party planners getting into the act and as Jerry would say: "Not that there is anything wrong with that." They usually are Independent Meeting Planners, Destination Management Companies, Special Event Companies or Travel Agencies.

Most are very professional, experienced and capable of taking care of all the meeting arrangements for a client. Here's how it works: An independent planner (and there are about 1500 who are members of Meeting Planners International ... a good 10% of the membership) solicits companies or associations to do their meetings so that the staff person may be free to do other company functions. The Destination Management Company usually works with a client to take care of all the meeting arrangements in a particular city where they have knowledge of the area. A special events company generally would handle the special events for an organization, be it entertainment, food functions, or any related activity. A travel agency may get involved in travel and hotel arrangements for the client. All these are in the business of making a profit (and nothing wrong with that) ... so they gotta get paid for what they do. Here's the rub: Who pays for whom for what? Just about all of the third-party planners have been "hitting up" the hotels for their pay (or part of it) via commissions or rebates. And, if hotels want the business bad enough, they pay it.

One of the big issues being covered by the meetings industry publications these days and at meetings conferences is the question of "disclosure." Should the hotel disclose to the user of the services (company or association) the financial arrangements that are made with the third-party planner? To be covered ethically (and maybe even legally) a clause in the contract should cover this. And, in most cases, I would imagine that any rebate or commission paid out would have to be covered in the rate quoted to the user. In this case, would it not be proper and ethical to inform the meeting participants, particularly for an association or organization meeting (where guests pay their own room rate) that the rate includes an amount paid out to ... ? Some hotels are starting to do just that by mentioning it on pre-printed reservation cards.

It just seems so logical that hotels go back to doing what we did years ago when travel agencies booked blocks of rooms for clients. That is, just quote a net, non-commissionable rate. This lets the client set the best available rate for the meeting (which is what they probably wanted to begin with by hiring the third-party planner). The third-party planner should then get whatever fee can be negotiated with the client for handling all the arrangements. I don't know, but I am pretty sure that sometimes the hotel pays and the clients pay too.

It should clean up the whole mess if we just decided to do just that ... quote net, non-commissionable rates. However, if the third-party planner is the decision maker, you may lose out on the business. It always gets back to how badly you want the business or how important it may be to you.

HOW HOTELS SHOULD VALUE GROUP BUSINESS

November 1998

In the business of selling rooms, space, food/beverage, and other profit outlets, salespeople need to be concerned with several things, which should help determine the value, from a profitability standpoint, of the business.

The Market Mix: Here we should be concerned with the rooms' rates being charged for a particular market, and how it compares with other markets available for the same time period. For example: Do we want to fill up the house with all discounted business, if other full-rated business could have been available? I remember very well, many years ago, that one of our hotels in a capital city seemed to always fill up with discounted government rates. At the same time, we were turning away loads of other business that would have paid a higher rate.

Meeting Space Related To Rooms Rented: Many times, we've seen meeting space given away, without charge, because a group was expected to provide a healthy number of rooms. However, after we've agreed to give the space, we find out that the room "pick-up" was not what was expected. Protect yourself from this, in the contract, with some sort of a sliding scale of meeting room rental related to the number of rooms actually rented for the group. In addition, in many cases, we may be jumping the gun by renting meeting space (without rooms) to a group too far in advance. By doing this, we may very well be missing some business for the same space from a group that would also use a block of rooms.

Understanding Lead Time: The general theory in yield management is that when a product is more available (low occupancy projected), a lower price may be offered to start "moving" the product (increasing the occupancy). It is a normal practice in our business to charge the highest published rate when we expect to "fill up the house." In addition, if we project a low period of occupancy, we charge whatever rates we can get to make sure we increase our business. Therefore, it makes sense that when a group wants to book business in a hotel well in advance of the meeting date, and projections for that period is not very good based on how profitable the business may be, we may very well want to discount the rate. As occupancy increases, then the rate increases ... the next group to book for the same period pays somewhat more, etc.

Peak/Valley/Shoulder Seasons: We've all got em! During peak periods, of course, we charge more since we expect high occupancy. During a "valley" period, when we expect occupancy to fall because of seasonal business, the rates are reduced (and sometimes very drastically). However, the shoulder season, which comes between the two other periods, is where most of us can do a better job. Here is where we want to extend, as much as possible, the peak season. If normally, as an example, the peak season ends right after Easter, we don't want to jump right into the valley period rates. The idea is to try to develop some group business that may lap over between peak and valley. This may give us some business in between the two seasons (which now becomes shoulder season) at a rate a little lower than peak, but not quite as low as a valley.

The Days Groups Arrive and Depart The Hotel: The day of the week members of the group arrive at a property and then they depart tells us a lot about how profitable the business might be and if we need to do any discounting. We must always be concerned with the displacement of other more profitable business. The idea is to try to move groups in and out of a property based on when business is needed. To displace more profitable corporate business to take a group at a discounted rate doesn't sound too good—unless, there is an opportunity to bring in business on a day or two that when you can use the occupancy. For example, if a normal corporate type hotel which fills up on Tuesday to Thursday, taking groups on weekends would make a lot of sense. So would a group arrival on Sunday with even a Wednesday departure. There is a little displacement here, but there is the value of picking up rooms on Sunday and Monday nights.

MEETINGS ARE ON THE RISE—TAKE A LOOK AT THE OUTLOOK FOR 2000

December 1999

Meeting Professionals International (MPI) and American Society of Association Executives (ASAE) have published the Meetings Outlook Survey for 2000. Half of the people surveyed were meeting planners for corporations and the other half were planners for associations. Here's a glimpse of what was reported:

International meetings (of the planners surveyed 63% already plan international meetings) are anticipated to increase by 17% over last year. The three most popular countries to hold international meetings were identified as Canada, Mexico, and the United Kingdom. National meetings in the U.S. are anticipated to increase by 15%; while regional meetings are expected to increase by 26%. Average lead-time for meetings planned by associations is one year, but corporate planners' lead-times average 6 to 8 months. Most respondents (64%), from both groups, expect lead-time to remain the same in 2000, while 30% anticipate an increase. Ninety percent of those polled said they use computer controlled, multi-media or LCD panels in the production of their meetings, and 81% utilize the Internet in some fashion.

I've recently signed on to a listserve for meeting planners around the country (and some from other countries). It is most interesting because of the varied topics that are raised for discussion. We see a good amount of comments about what hotels are doing, good and bad. Here are some randomly collected comments from planners, which bring out some good points to be considered by properties:

1. "Referring to the comments about 'rewarding groups for picking up more than the room block.' The point I was trying to make is that when a group does not pick up their block, the hotel is right there to collect. On the other hand, it would be nice to hear appreciative comments from the hotel on how pleased they are that the group went over the projected block. Unfortunately, everything now is the bottom line and relationship building is not always at its best... By the way, how many planners have seen a GM take the time to stop by, introduce themselves and say how much they appreciate your business?"
2. "Actually, I had a two-days meeting held at the Hyatt Regency Houston Airport Hotel and I met and spoke with the GM, DoriFamiliant, about four or five times during and after our meeting. He said over and over how much he appreciated our business and hoped that we would come back again. I also know that this is not typical. I have held meetings at resort properties and needed to speak with the GM and he didn't appear to want to address the problem/issues. There were other problems we had with the hotel and will not use them nor recommend them to anyone."
3. "It doesn't happen every time, at least half of my multi-day meetings. Sometimes the GM will be at the person meeting to say thanks, sometimes he/she makes a point of stopping by our registration table. Maybe my experience is unique?"
4. "Ed Rudzinski at the Marriott Wardman Park is wonderful showing his appreciation to groups. I had a wrap-up meeting for one of our conferences, and he stayed for the entire hour."
5. "I want to echo that in a few cases I have found GMs to be very helpful with presence. Two, in particular, are in the Boston Marriott Burlington, MA and the El Conquistador, Tucson."
6. "It is nice to see some "positive" comments regarding how GMs and some of the hotels. Hopefully, others will see these comments and follow suit. If more hotels would realize how loyal meeting managers are, when we are treated professionally, they will have repeat business and great referrals."

MEETING BUSINESS IS EXPECTED TO INCREASE THIS YEAR

March 2001

It sure looks like the overall number of meetings held by companies and associations will increase by 9% during 2001, according to Meeting Professionals International's annual Meeting Outlook Survey. More regional meetings will be on the books, about 26% more, as indicated by those expecting to increase their meeting business. About 15% of the organizations surveyed indicated to have more national meetings; and 17% more international meetings. Canada seems to be the first choice of associations among international destinations (selected by 50% of the survey participants), followed by Mexico and the Caribbean. Among corporate planners, 54% of those surveyed are leaning toward the Caribbean as the top international venue; followed by Mexico and Canada. The other countries listed in the survey included England, France, Germany, and Italy. The biggest jump in the number of meetings to be held will be by corporations, indicating an increase of almost 15%, while associations will plan 5% more meetings. Of course, although associations plan fewer meetings, their number in attendance is much larger than the corporate type meetings. The 2001 Meeting Outlook Survey is available through MPI's market research area.

What's good about this message, of course, is that there is an increase in demand in the meetings market; particularly in the corporate segment. This means more meetings in the 50 and under attendance category. This should be very appealing to all types and sizes of properties. Even the limited service segment can get into the act of going after the corporate small meeting market. With some locations being overbuilt, it may not be as easy getting some of this market, because of competition. But, that's what sales are all about. Instead of waiting for business to walk in the door, or call in, someone needs to find prospects and do the selling.

With hotel business being so healthy over the past 8 years, or so, maybe we have gotten a bit lazy about going after new accounts. We just don't see the hustling for business as we did about ten years ago. So many sales folks have been used to seeing the business come to them. Has there been a lot of enthusiasm? Someone once said: "A salesperson, minus enthusiasm is just a clerk." When you think about it, staying in and answering the phone is probably a sales clerk's job. Or getting third-party meeting planners coming to you to book a client of theirs into your property; that's pretty easy too. We need salespeople to be more enthusiastic about seeking new accounts; a business which could produce more on the bottom lines. We probably would all agree that it is the enthusiastic salespeople, who believe in their job being more of selling than servicing, and those are the ones who are most successful. Just take a look of many of the hospitality leaders today who got their start in sales (many years ago, of course), and you know who they are. These are all exceptionally enthusiastic people who were, and are, superior salespersons. The business is out there folks; just get your sales teams excited about going after new accounts, and even, new markets.

IN THESE TIMES WE NEED ALL THE BUSINESS WE CAN GET!

June 2001

All too often we hear comments from salespeople, managers, operators, and even owners that the "association market" is not for their properties. The perception seems to be that the association business may be "too large" for their "smaller property." In reality, business from this particular market segment is available to all sizes and types of properties. Many times, a small or mid-size boutique or B&B (bed & breakfast) type facility could be very desirable for certain types of meetings. I can recall many years ago I was called upon to do a training program for a small committee of an insurance association. The meetings were held in Hartford, CT; there were only about 12 or 15 in attendance. When I got picked up at the airport, instead of being taken to one of the many hotels in the area, we ended up at a very nice B&B. There were only 6 of us from out of town, the rest in attendance were local, so we took over the 6 room facility, and had our meeting in the living room of the establishment; breaks and lunch were catered. Not a bad deal for the operators of the B&B.

So, let's take a look at this market segment. There are 27,000 national associations in the United States. Most of these fall into the categories of either being a trade, social, civic, or professional type of association. Just about all have national, regional, state, and local meetings. Of course, generally, the national conventions are very large, are held at cities with convention centers, and provide room business for many hotel properties near the meeting facilities. However, most smaller and mid-size property owners, operators, managers as well as salespeople, may be surprised at the relatively small-size association meetings being held at the smaller, even limited service type properties. Just take a look at the quarterly schedule of medical association meetings listed in every monthly issue of Medical Meetings magazine. There are many of these meetings held all over the country for 50 and under participants.

In addition to annual conventions, all associations have Board of Directors meetings, Legislative sessions, and committee meetings of all types. There is usually a trade show contained, within an annual convention and sometimes in conjunction with regional and state meetings.

What is interesting about this particular market segment is that opportunities for making contact with people connected to associations are right under our noses, but we don't realize it. Think about it! Just about everyone in business belongs to some type of association; whether it be through a local civic club, through a profession or even through a job function (to which do you belong?). I wouldn't be surprised if one out of every three guests at your hotel belongs to some sort of an association. The leads for making contacts with appropriate people who decide where meetings are to be held are all around us, every day. We just do not seem to take advantage by developing these leads into prospects. Members of your own staff probably belong to some group. And, how about your neighbors, friends, and relatives? Then, there are the folks with whom you do business; they surely are members of their own trade association. Let's not stop there! Every day, in your local newspaper, you will probably find an abundance of leads for this type of business for your property. Who just got elected as an officer of a local organization? Who received an association award? Which association is meeting or has met elsewhere?

The list goes on and on. Take a look at the reservation information gathered for your expected arrivals. See anything there which might reflect an opportunity to make contact for an association lead? And, what about just taking a look in your local phone book under clubs or associations. That could be a great place to start. Here you want to find out who makes the decision with regard to their local, state, regional, as well as committee and any other types of meetings.

You see, it's just a matter of taking a little time to make conversation with a contact. The idea is to gather enough information to determine if there is a potential prospect with whom you may be in touch with later, to find out about future business for your property. There is no magic about this way of developing new or additional sales for rooms, space, food, beverage, or other profit centers. It is a question of thinking about it, asking around, and doing it. But, please be careful. Just don't start calling associations and asking: "Do you have meetings?" They always do.

MILITARY REUNIONS—THE ONLY NICHE MARKET THAT IS GROWING

October 2002

Many years ago, when I was the first Convention Bureau manager for the Charlotte Area Chamber of Commerce (years before the CVB became a separate entity), we used to go after this Military Reunion market. And, we packed 'em in the city during the summer months, when we needed to occupancy. It was a great business for us, since we realized that this market segment usually held their reunions during summer months and folks made it their family vacation period. People came to Charlotte from all over the country, most came by car and stayed for about a week.

It is amazing that so few properties, today, go after military reunions. According to The Reunion Network, Inc., this niche market segment represents over six million room nights per year. There are between 14,000—15,000 such reunions held each year, with a median attendance size of 120 people. Of all the reunions 45% have an attendance of 100 people or less; while the rest of the reunions have between 100–200 people participating. The average length of stay is 4.7 days and each attendee spends $185.00 a day.

Probably, the main reason many hotel salespeople don't go after this business is that the meeting planners for this segment are so elusive; they are hard to find. Of all the military reunions being held only about 1,000 or so are planned by full-time meeting planners. All the rest, and that's a whole bunch, are managed by volunteers, and those folks are difficult to locate. Paul Spiewak, CEO of The Reunion Network started the organization because there has been no official source of information about or for the planners of these reunions. The planners represent all of the military service reunions, on a volunteer basis. In 1991, Spiewak recognized that not only did the planners need help in locating appropriate reunion locations, but hotel salespeople also needed help in locating these elusive planners. Spiewak then decided to form a source of information for the volunteer military reunion planner as well as for properties seeking this particular market segment.

TRN now has a database of about 80% of all the military reunions being scheduled, with specific details on contacts and needs of each group. There is a monthly newsletter, keeping planners and suppliers up to date on what's going on in the marketplace. Spiewak also organizes two-days face-to-face conferences with a marketplace so that planners and suppliers can meet, with seven hours allocated to private one on one meetings. Ten such events are already planned for 2003 in different U.S. Cities. Convention Bureaus, individual properties as well as hotel management companies are invited to participate.

What is ideal about this niche market is that it is generally off-season business for many locations. About 16% of the reunions have been held during the August through October period. Just about all of the meeting planners are flexible with dates. Most of the time there is no need for extensive meeting space; mostly, a hospitality room is sufficient. If you have free parking for cars and RV is a good selling point for this market. Go for it!

BUSINESS IS OUT THERE . . . ALL WE GOTTA DO IS GO GET IT!

February 2003

The figures are out there from all over the place . . . for business, association meetings, and leisure travel. Reports are being furnished by associations and a variety of study groups, reflecting that business is not as bad as it may seem. There are still people traveling who need hotel rooms whether it is for business or for vacation trips. Here are some examples: The business of association management continues to grow rapidly. There are 147,000 associations in the United States and there are as many as 1,000 new associations being formed each year, according to a report furnished by the AMC Institute. Altogether, association annual budgets now exceed $21 billion. Considering that, associations generate 35–40% of their budget from meetings and conventions that sure does look like a lot of business being held at hotels, resorts, and conference centers. The other interesting thing about the association meetings market for hotels is that there has been a great interest on their part to move to secondary and tertiary cities for their events. Taking meetings to new venues to save costs is becoming very popular these days.

So, now, how about the leisure traveler? According to a survey completed by USTOA (United State Tour Operators Association) and Travel Holiday magazine, 7 out of 10 travelers (73%) expect to spend more in 2003 vacations than they did in 2002. Now that is good news for hotels. A majority of USTOA members responding to the survey indicated they expected increases of up to 10% in tour and travel packages in 2003. According to the survey, which polled 600 Travel Holiday readers, only 27% of the readers polled anticipate paying less on 2003 vacations. For tour bookings, it looks like travelers are more prone to book vacations 14–30 days ahead. For those hotels working with escorted motorcoach tours, it appears that there was a feeling that those types of tours would increase by 20%.

It seems that even under the most difficult circumstances, people will travel. The numbers move around a little from time to time, based on the economy, the oil availability (as in the mid-70s), or terrorism (as in 9/11). According to the National Tour Association (NTA), last year North American Travelers spent $98.7 billion on trips with the U.S., representing only a 6% decline from the previous year. That tells us something about people wanting to travel. Rates may be down in some area properties due to an overbuilt (or should I say, under-utilized) situation and motorcoach tours are taking advantage of lower rates being offered. However, we should be aware that the tour business is still out there and we need to figure out how to go after it . . . and not by reducing rates. It is a question of just doing a better selling job.

We probably spend too much time talking about the downside of the economy, recovery, and all the other negatives about the industry. After all, people are still traveling by car, plane, bus, train, and ships. There is money out there being spent in all categories of travel, whether it is business or pleasure. Sales are the answer, we just gotta go out after the business. You'll find a good deal of great selling tips in our new book: "Lessons From The Field" authored by myself and John Hogan.

GROUP BUSINESS IS STILL OUT THERE—GO GET IT!

May 2003

It looked like the demand for room nights was growing faster than room supply towards the end of the last quarter in 2002, according to Smith Travel Research. That was indeed a welcoming sign. The Conference Board's survey of 100 CEOs indicated that those top industry leaders were feeling more confident about the state of the economy for 2003. According to a report on the Industry Outlook 2003 by Corporate & Incentive Travel, "the number of corporate meetings and events is holding fairly steady." Therefore, things looked pretty good up to this point, but now that we are at war, there is really no telling what will become of room night demand around the world. Some business might fall off with leisure travel, but there will still be the group market demand, which may surprise us with a continual need for meetings and training conferences.

No question about it, there is a steady volume of meeting business with companies and associations in the sales-driven industries. The importance of sales to any organization or industry is best expressed in the saying, that I have used a number of times in my column and sales workshops: "Nothing Happens Until Someone Sells Something." Just think of the industries that count on sales, and because of that, those companies and organizations must continually work to train and motivate to achieve top performance. These are the businesses that know it is important to keep spending on meetings and incentive programs that contribute to the bottom line. Just think of the number of conferences, training programs, management meetings, incentive trips planned by organizations and companies in the industries of insurance, financial, medical and pharmaceutical, just to mention a few of the most well-known.

There is business out there; it is just a question of doing the sale work. Of course, we can sit back and depend on those connected with those industries to come looking for us, but don't count on it. Some major brands have just recently made some cozy deals with site selection companies to find business for them; on a commission basis, of course, paid by the hotel properties. Even a bunch of Convention Bureaus around the country and looking to site selection companies to bring business to their cities, and of course, who pays the bill on a commission basis? Yup, the hotels pay it when the conventions or meetings are booked. Wonder what their own paid salespeople are doing. Then, of course, we can also sit back and look to the independent planners, and let them go out and get the business for us, and they get a commission too. And, there are more than a dozen Web sites out there wanting meeting planners to use their electronic Request For Proposal so they can search and find meeting sites, and get paid a commission from hotels.

Looks like we will turning our sales staff folks into administrative order takers. With all these out-sources bringing in the business, all we need is someone to process all the business. Or maybe we need to forget about hiring and paying sales folks to get business for our properties. We can just rely on these other sources to do the work for us, and we only have to pay the commission. Oh yeah? Tell me another story.

HOTELS OF ALL SIZES NEED TO GET SOME OF THAT SMERF STUFF

May 2003

In looking at group business for hotels, resorts, and conference centers, we hear so much about the corporate and association market, but very little concerning the most elusive of all market segments, the SMERF market. Even the smaller properties with limited meeting space could very well meet the needs of some of these SMERF groups. What we have here is an acronym for Social, Military, Educational, Religious, and Fraternal groups. More and more property sales departments have been going after this market segment since 2001 when we noticed a lag in the economy, increase in supply over demand, and the general decline in occupancy, which resulted in some cutting of rates.

It seems that we have always looked at this market as being the "low-end" of the business, meaning that their budgets have been less than other markets for hotel rooms, space, and catering. However, that is not entirely correct. For the most part, this business is more social and usually falls during the off-season at most locations, and particularly over weekends. This could be particularly attractive to properties that normally go after the corporate market. Attendees at these events participate on a volunteer basis, being more involved from a social standpoint than business. Participants pay their own expenses; therefore, they usually seek out locations and sites that would be able to offer discounted rates.

What is interesting about this market is that the groups involved come in all sizes, interests, and needs. Many of sub-segments of these types of groups do not even require extensive meeting space, or maybe no meeting space at all, except perhaps of a hospitality room for a social gathering. Although most have national or international meetings that draw very large citywide crowds, there are many that are small enough requiring a very limited number of rooms and meeting space. This could be a very attractive business to the smaller, limited service type of properties.

The SMERF market is pretty elusive; meaning, to a major extent it could be very difficult to know where to go to look for this business. The people who plan meetings within this market segment, for the most part, are volunteers. There are very few full-time meeting planners organizing these meetings, except perhaps for the very large national conventions. So let's take a look at some of these SMERF segments and try to identify where we can find some of this business:

- **Social**: Within this sub-segment, we mean any kind of group event involving family and friends. This would include events requiring overnight accommodations for folks attending weddings, funerals, retirement, major anniversary parties, graduation, etc. This type of business is generated locally by making appropriate contacts with pastors, funeral directors, local companies, and getting leads listed in newspapers.
- **Military**: All branches of the military service have veteran groups that meet, for the most part, during the summer months. Entire families usually attend. This is a very large market for all sizes of groups. The best contacts here are the local veterans who may belong to specific groups. In addition, The Reunion Network is a primary source of information on this market.
- **Educational**: Contact with all schools is necessary. Educators at all levels have a wide variety of meetings going on all throughout the year, particularly when school is out and on weekends.
- **Religious**: This is pretty obvious. Contacts with all religious denominations in the area. Many of these groups are state-wide or district-wide and may require meeting space.
- **Fraternal**: Just take a look at the civic organizations in your own community and you can see how large this market can be. The local Chamber of Commerce would probably have a listing or even look in the telephone book. These civic organizations have local meetings, district, state, and national meetings.

START NOW TO LOOK FOR 2005 OFF-SEASON BUSINESS

June 2004

It seems that we are continually seeking business to improve occupancy for the off-season when it is too late to get the business. This particularly occurs when it comes to getting group bookings. If summer months are off-peak, we start scrambling, looking for groups to book sometime around May or even June. In most cases, it is just too late. We really need to get started with developing prospects for off-peak business at least a year out; and even further out for larger groups.

One particular market segment, with which many hotels are not familiar, is the military reunion market. This is probably one of the most elusive segments of groups business since just about all such groups are organized and run by volunteers. This means the persons involved in planning the events are not full time, or even part-time paid individuals. Generally, they do not belong to organizations with which we are familiar like Meeting Professionals International (MPI) or Professional Conference Managers Association (PCMA) or any of the others. Therefore, the people with whom we need to be in touch, are very hard to identify or to find. Yet, this is a very lucrative type of business for properties that are familiar with military reunions. Salespersons of facilities that get this type of business know where to go, how to find the buyers, and how to go about doing business with them.

Here is a little look at this market:

- Just about all branches of our military service have veteran members who join together annually to meet and visit about "old times."
- Korean Veterans make up the largest group. Cold War Veterans (those serving between Korea & Vietnam) are still increasing and will overtake the Korean groups by the beginning of 2005. There are some still around from WWII participating, and representing about 25% of all military reunions.
- More than half, maybe even about 2/3 of military reunions occur between the end of August and mid-October. June is slow, July is dead, mid-November to mid-March have very few.
- The most recent (2004) count of names and addresses, on file, of people who plan these meetings is 12, 603.
- There are an estimated 15,000 military reunions held each year.
- Average attendance at these reunions is 122.
- Average length of stay is 4.6 days
- 98% of these reunions are held off-season (mostly summer months).
- Only 2–3% of the 55,000 hotels in the U.S. are aware of this market.
- Only 10–15% of the hotels are suitable for this business.
- There are 116 metropolitan areas that do 10 or more military reunions a year.
- Branson, San Diego, Norfolk to more than 300 reunions a year.
- In rough numbers, this represents more than $15 million to the hotels.
- Using a 10x velocity multiplier, it represents more than $150 million boosts to the economy of those communities each year.
- Even a small community like Cedar Rapids, which ranks #116 on a list of popularity for military reunions, it means only $5 million impacts on the community.

This data was developed by The Reunion Network, an organization that monitors and constantly keeps abreast of these figures. TRN also publishes a newsletter and quarterly magazine designed for military reunion planners and suppliers. The organization also conducts several ConFAMs around the country each year where military reunion planners get together for educational programs on meetings management and get to visit an area on a familiarization trip. These are usually sponsored by local convention and visitor bureaus.

IS THE SMERF MARKET FOR YOU? THINK ABOUT IT!

July 2004

Too little has been known about this SMERF meetings market when it comes to how much they spend, revenue generated and destination preferences, and particularly how to go after it. If you could use some weekend business, then this could be the market to seek. The acronym SMERF, just in case some readers are not aware, stands for Social, Military, Education, Religious, and Fraternal (and you could even throw in another "S" at the end for Sports). Earlier this year Prime Media Business Marketing Research Department conducted a survey among its readership of their Association Management publication, who had identified themselves as working for SMERF-type organizations. Prime Media is the publisher of five meetings type trade magazines (www.meetingsnet.com). The results of the study were very interesting and reflect some very good information for hotel sales staffs who may want to take a look at this market to fill holes in room blocks over off-season and weekends. The figures for 2003 were very encouraging and forecasts for 2004/2005 were very positive.

- Attendance for 2003 increased by 30%
- Revenue for 2003 increased by 39%
- Most respondents (70%) indicated typical attendance was under 500
- 43% indicated an annual budget of under $100,000
- 50% reflected that room nights for its largest even was under 500
- 32% plan 1–3 off-site meetings a year
- 27% plan up to 10 off-site meetings a year

When it comes to domestic cities that respondents thought best served their meetings, the top choices were Orlando, Chicago, Atlanta, Dallas, San Diego, and New Orleans. Most popular second-tier cities were St. Louis, Nashville, Phoenix, and Louisville. Regarding projections for this year, more than half indicated their largest event would see increases over 2003, with increases in meeting revenues and expenditures.

We used to think about SMERF groups as a market niche for downtown hotels, campgrounds, and arenas. Now they use convention centers, conference centers, university facilities, suburban hotels, downtown hotels, and resorts. A little less than half of the respondents indicated that their annual event includes an exposition, which means that some exhibit space would be necessary for their meeting.

As expected, one of the biggest challenges facing SMERF planners in the coming years is Attrition. Most respondents reported they expect big trouble in this area since too many members will be booking outside the contracted block of rooms, seeking Internet locations for making reservations at lower room rates. Some other issues raised were not getting enough meeting space without using a large block of rooms; surcharges; affordable audio-visual charges; getting government rates, and acceptance of groups that are not highly marketable (like students).

Of course, the study respondents were all planners, who for the most part were full-time association employees. In this market segment, we are more likely to find most the planners working part-time or even volunteers who are very difficult to recognize and locate. When it comes to the purely social market, military reunions, and fraternal we are seeking some very elusive, hard to locate planners. Seeking local contacts that are affiliated with these types of groups is critical. Local networking is vital to get to identify some SMERF groups and their planners.

WE MAY BE GETTING BACK TO FAM TRIPS TO HELP BUILD GROUP BUSINESS

May 2005

Those of us in sales over the years may well remember the popularity of conducting FAM (Familiarization) trips for meeting planners who would be in a position to bring large groups to our properties. Sometimes the planning and arranging for those prospects to travel to, and stay at our locations were taken care of by our own property sales personnel, and sometimes all the details were undertaken by the local convention and visitors bureau. These trips were very popular promotional activities, which introduced the group buyers not only to the hotel, but also to the area attractions. It was a "sampling" of where their attendees would be housed and entertained during a convention, conference or meeting. We have not seen many of these going on lately except for one particular market segment which relies pretty heavily on FAM trips during all times of a year.

This year 15 such trips have been planned for meeting planners who are in the business of putting on military reunions. They are scheduled for Buena Park, CA; Shreveport, LA; Lancaster, PA; St. Louis, MO; Chicago, IL; Ogden, UT; Dallas/Fort Worth, TX; Kalamazoo, MI; Daytona Beach, FL; Colorado Springs, CO; Chattanooga, TN; Rochester, MN; Cape Cod, MA; Central Florida; Phoenix/Mesa, AZ. All are coordinated via The Reunion Network, which is an organization of Military Reunion Planners, made up of ex-military men and women who have served in all branches of our military. TRN reports that there are 14,000 military reunions being planned each year in the U.S.

What goes on during these FAM trips (they are really called ConFAM trip by this market segment) is two full days of heavy-duty meeting planning training along with a very successful marketplace for suppliers, and attendees are invited to visit the area hotels and attractions to sample what the area offers. The training segments are conducted in a wide variety of topics and techniques necessary to educate the planners on how to put on a successful reunion for their members. The area activities are generally arranged by the local convention and visitor's bureau, the agency that invites TRN to come to their area for the ConFAM.

What makes this type of event for hotel salespeople and area attractions very effective is that every attendee is pre-qualified. TRN puts out a booklet of fact sheets to the marketplace suppliers. Each fact sheet includes where each particular planner had their last reunion, how many people attended, where they are booked the following year and shows their next open date, with the number expected to attend. These are very unique events with 149 having been completed by the end of 2005. Hotel salespeople and convention bureau sponsors have valued the opportunity to meet one-on-one with each planner that attends. Only about 25 qualified, military reunion planners are invited to each ConFAM. Recently I attended the one in Buena Park, CA, just to see what it was like, and I was so very thrilled to sit through the very professional training sessions conducted by Paul Spiewak, TRNs CEO and his Director of Marketing, Charlene Adamson. While on site, two of the planners had indicated to me that they were already planning to book their reunions at that location.

MANAGING YOUR ROOM BLOCK FOR GROUPS

November 2005

Whether your property has meeting or banquet space available or not, at one time or another, you probably will be involved in making arrangements with a group for a block of sleeping rooms. Handling it will be very meaningful to your occupancy, profitability, as well as meeting the needs of the prospect who will be booking the group rooms.

As part of its Accepted Practices Exchange (APEX) initiative, the Convention Industry Council (CIC) has adopted best practices for event housing and registration by approval of its board of directors. This best practices study on is one of seven initiatives conducted by panel groups appointed by CIC to create greater unity and cohesiveness among all of the hospitality industry segments. The panel was charged with developing best practices for collecting, reporting, and retrieving housing and registration information. Already completed, published, and available to the industry are approved reports on terminology, history/post event reporting, and event specifications. The remaining three areas awaiting final approval and reports are requests for proposals, meeting and site profiles and contracts. The study and report on housing and registration reflect accepted practices in the business, templates, and management instructions developed for event registration, housing, and rooming lists. It also includes housing-related terminology, hotel room-block management, and the role of computerized technology in the industry housing and registration processes. The CIC is an umbrella group for 31 leading hospitality-related organizations that represent more than 100,000 individuals and 15,000 firms and lodging properties in the meetings and hospitality industry. The report, including forms, is available at www.conventionalindustry.org.

A new "Guide to Room Block Management," has just been published by Conferon Global Services, a large meeting planning company in Twinsburg, Ohio. It is a 64-pages resource on best practices for blocking rooms, managing registration and housing, and creating incentives for convention attendees, and exhibitors to book within collected blocks. A free downloadable version of the guide, sponsored by Hilton Hotels, is available at www.cgscompanies.com.

A major concern to associations and organizations holding blocks of rooms at properties is booking of rooms outside the block by meeting participants. This is where attendees at an event decide to make reservations at locations other than where the group has committed a block of rooms. Through an attrition clause the group may have with a property, the sponsor of the event is held responsible to deliver a certain number of booked rooms, and may have to provide a dollar amount to the property for the block of rooms not used. This is determined by how the attrition clause in the contract is written. In view of this, meeting professionals involved in blocking group rooms for events, are seeking ways to make sure that their participants make their room reservations with the appropriate contracted properties. There are a variety of incentives being used, and it is also in the best interest of the properties involved to work with the planners in helping to create these incentives.

ARE YOU GETTING YOUR SHARE OF THE MILITARY REUNION MARKET?

August 2006

This military reunion business is really something that is exploding, and it is amazing that so few hotel operations are going after this market. There is such a variety of reunion groups from each of the various services that just continue to grow. Right now there are over 13,500 different groups meeting annually for their reunions. Of course, the number of reunions for WWII veterans has dropped, but there are still 3 million still with us (out of the 16 million who served), they still represent more than 15% of military reunions. Korean vets (about 5 million remaining out of the 9 million who served) represent about 25% of the military reunions. Cold War Vets (between Korea and Vietnam) now represent better than 40% of the military reunions. The Vietnam numbers are growing daily, and this bodes well for the future, because up until two years ago, Vietnam Vets weren't meeting in any great numbers, and of course, in terms of age, they represent the future of military reunions for the next 25–30 years. The Vietnam vet, in general, is also a lot more affluent and a lot more willing to spend, which is really great for the hospitality industry.

Within the next few years, the number of meeting planners for military reunions is expected to reach 19,000, according to TRN (*The Reunion Network News**), which is the source of information for the Volunteer Military Reunion Planner. This niche market has many advantages over other reunions, and even better than some corporate or associations markets. Here are some things to consider about this market:

▶ Very low demand for meeting space
▶ Usually booked off-season
▶ Usually booked off-peak
▶ Dates are very flexible
▶ Total annual room-night count runs about 7 million per year
▶ Total spent on hotel rooms is over $500,000,000
▶ Exceptional extra revenue from food/beverage and recreation

---

* The Reunion Network is now *The Military Reunion Network*, https://militaryreunionnetwork.com.

► Estimate on additional spending runs about $80 per person per day

According to TRN, which continually monitors the military reunion market, only about 3% of the hotels in the U.S. are currently active in this market, leaving the field wide open for aggressive hotel salespeople. There are about 115 communities in the U.S. that host at least 10 military reunions a year, with some hosting as many as 100. It is truly a specialty market. Your potential and actual success in this market depend on two basic factors: (1) What your location can offer military reunion groups. (2) Your effectiveness in attracting and dealing with the volunteer planners who represent and make the arrangements for these groups. Peculiar to this market is being able to locate those volunteers. In the North East States there are 2,633; North Central States–2,873; South Central States–2,199; South East States–3,583; Western States–2,464. If you are new to this niche, or are uncertain on how to approach it, TRN has a new book: *Reunion Business: What Is It and How To Get It.* This was written by Paul Spiewak, founder of TRN, and the outstanding authority on this niche.

ANOTHER "NICHE" MARKET FOR YOU TO EXPLORE

September 2006

Every time we think we have all the possible sales markets covered, we find there is one we missed, or at least never heard of previously. Wasn't the SMERF (Social, Military, Education, Religious, Fraternal) market supposed to cover all those others that were not included in the major markets we have been soliciting? Well, anyway, here is another one to consider the Retreat market. We probably need to take a good look at this and try to determine what it is all about.

Of course, we know that all sorts of companies have retreats; like Board of Directors have retreats to work on their strategic plans. Then there are team-building retreats for volunteer committees working within a wide variety of associations. Just about every department within colleges and universities has faculty retreats. So, when someone says: "Hey, let's have a retreat to study this or that" what they usually mean is let's get outta here and go someplace where we will not be disturbed. Retreat actually means to withdraw, so that means getting away somewhere.

And, we all know that for just about every different segment of business there is an organization. So, would you believe there is now a National Association of Retreat Executives (NARE)? Although the organization Website (www.retreatconnection.com) is up and running is still lacks a good deal of information that would be helpful to hotel sales personnel. It is starting to list facilities, properties, suppliers, trainers, and speakers, and they plan to have a forum for the exchange of ideas of their members. Right now there are over 250 Retreat Leaders listed with the organization and it is growing daily. Retreat Leaders, to some extent, are pretty difficult to locate since most are volunteers for groups and associations.

If this thing gets off the ground properly, it could get to be something big for a new market, which, for the most part, sales staffers just do not seem to go after for business. The challenge with this retreat market is how we identify who to seek out within all the existing markets to find the folks who actually do the booking and the planning. Is there such a thing as a retreat meeting planner? There probably is and it will be this new organization that would be the focal point for the listing. Walt Sibley (walt@retreatsconnection.com) who runs the association advises that an ideal retreat facility is usually in a remote location, rural in nature, wooded, with spacious acreage. Most groups are small, under 100; however, there are some that run 400–500 people. Of course, there need to be room accommodations, meeting as well as dining facilities available. Structured retreats usually last from 3 days to two weeks. There are approximately 600 suppliers to this market already listed with RetreatsConnection, of which 364 are facilities.

Within the past couple of years, we've seen more and more counties in rather rural areas creating Convention and Visitor Bureaus. It seems that this Retreat business would be an ideal market for those areas to go after and develop.

GROUPS AND MORE GROUPS IS WHAT SALESPEOPLE NEED TO BOOK

January 2007

Lots of things keep changing in our hospitality business. Change is what we all have to deal with throughout our business careers. Whether it is pricing, types of properties, locations, technology, management techniques, marketing methods, etc., everything seems to move with lightning speed as far as changes are concerned. However, there is one thing constant in our business of hospitality—hotel sales representatives are committed to spending their time selling to groups.

Although our salespeople get involved in a variety of activities that help in building business for a property, it makes good business sense for sales staffer to spend most, or all of their time in seeking out prospects for group business. The return on investment is substantial with groups in terms of higher occupancy and revenue generated via rooms, meals, and other profitable business outlets. There is every indication that group business should continue to be robust through 2007, and probably 2008. Of course, generally, some think of group business is something with which the larger hotel properties are concerned; however, that is not the case. All properties, regardless of size, or meeting room capacity need to be concerned with seeking group markets. Although the major source of group business may very well be in the association or the corporate market, we must not neglect the opportunity of a variety of social markets that are also responsible for having group activities, large and small numbers of people attending.

We are now seeing a change in occupancy percentages reflecting the difference between leisure and business travel. What used to be a mix of 60% business travel to 40% leisure travel in our hotels is now a split of 50%–50%. This surely reflects the opportunity for many hotels to start seeking group business from the social, or leisure markets. There are more reunions (family, schools, military) taking place now than previously, as are other social groups, and hobby clubs, as well.

The real key in doing well in bringing in group business is the opportunity that sales personnel have in booking these groups as far out as possible. We need to do a better job of putting the business on the books at least two to three years out. This way property management can do some long-term forecasting of business, which helps in determining future rates and occupancies. Although business appears to continue to be great in 2007 and possibly 2008, we can surely be certain that developers are getting busy putting more rooms on the market. Most certainly, with the profits in hotels coming into double digits for 2006, more rooms are out there on the horizon. Sales personnel can help protect the property assets by booking groups for the future. If you are not selling more groups, you are leaving money on the table. Happy Group Selling!!

THIS IS INDEED A SELLERS' MARKET IN THE CORPORATE TRAVEL BUSINESS, BUT IT WON'T LAST FOREVER

February 2007

If you have not been deluged by RFPs (Request for Proposals) from corporate travel managers, they sure will be soon to come. Having been exposed to increases in corporate room rates all over the country by at least 5%–10% (and even higher) company managers responsible for contracting with hotels for rooms are seeking options all over the place. Wherever possible they are, and will be seeking more moderately priced properties. In face of a very healthy hotel industry, exploring as many options as possible is the big function these days of corporate travel managers. Property managers and sales personnel who have possibly never looked to this market for business are now finding themselves being approached to enter into contracts for housing corporate travelers.

The most popular corporate business hotels reject or even disregard the many RFPs that come into its sales office. The rate of demand increase necessitated sales staffer to go after the most profitable business, and there is a reluctance to enter into any long-term agreement that provides reduced rates. Of course, travel buyers who are in a position of offer heavy volume of traffic, and contract for the higher rates will be able to secure facilities in major markets at appropriate hotel properties. Right now, there is probably very little negotiating leverage on the part of corporate travel buyers, but contracts usually will be negotiated at least one year in advance. With the probability of more rooms and properties coming online in 2008 and 2009, and with a decrease in demand percentages compared to growth of new rooms, corporate travel managers may get a breather. That is, hotel sales staff may very well be in a position of willing to seek out the RFPs, and more readily respond to those they receive. This may create an opportunity for the buyers to be in a better bargaining position than they are right now.

Based on our industry history of a cyclical business environment, sales staffers who have been around a while will know that the climate will change within a couple of years. With that in mind, there is, and will be, an opportunity, to work "deals" now, and create better relationships with the buyers, so that when the time of a turn-a-round comes about there would be a better chance of maintaining the accounts. Now is the time to work on creating favorable relations with the corporate travel managers, and one great way of doing this is to ensure prompt responses to their RFPs. Even if properties cannot accept offers of a contract at a low suggested rate, those requests need to be acknowledged. This takes a good deal of time in responding to the many that are being received at properties, especially so many not even coming close to what would be acceptable. However, we never know the one to which we do not respond could turn out to be lifetime account of the year. We need to remember that with corporate travel, it is not just a one-year relationship. Think about being creative in the negotiation process. Maybe there is an opportunity of shooting for a two-year contract rather than just one year. Or, even consider other options than the regular January to December contract period.

CORPORATE MEETINGS GETTING TO BE RUN BY CORPORATE TRAVEL MANAGERS

March 2007

Property sales personnel involved in booking corporate meetings may have to start dealing with the corporate travel manager instead of the usual corporate meeting planner. In the interest of consolidation, some of the larger corporations, who are buyers of blocks of rooms for their corporate business travelers, have started moving corporate meetings buying decisions over to their corporate travel manager. According to a recent member survey of the National Business Travel Association (made up of corporate travel managers), 30% of their respondents said they are already responsible for meetings, and another 11% said they had recently been given control over group meeting expenditures. Another 12% indicated they expect to be in charge of meetings at their companies within two years.

With these changes coming up sales personnel need to re-evaluate their selling strategies with regard to learning how to develop long-term favorable relationships with this new body of buyers. Up until now, these corporate travel managers were involved in working with lodging accommodations on a contract basis for a certain number of guaranteed room nights for their company's business travelers. For the most part, this was a rate conscious group of people who needed to contract for rooms at a wide variety of lodging properties across the country as well as foreign locations. Of course, their objective was always to seek the lowest price they could get through negotiations at the best quality property they could locate in cities where their personnel traveled. The corporate travel market has always been a very competitive business among hotels in the same city.

At most lodging operations, particularly in the larger hotels sales personnel are assigned to specific market segments. For the most part, at most corporations, the corporate travel managers worked separately from the corporate meeting managers, and they probably even worked in different cities. Many hotels sales offices operated the same way, with some sales personnel calling on the corporate travel manager and other calling on the meeting planners.

With consolidation at the two departments into the business travel department, lodging operators may have to start thinking differently about the assignment of sales staffers. Now it appears that we need to work it out so that sales personnel could be working on contracts for individual business travel as well for corporate group meetings. It is getting to the point that sales folks will need to get involved in all facets of selling to different markets whether it be individual or group or business or leisure travel. Those sales department personnel who have had the experience of dealing with both the meeting planner as well as the business travel planner will understand the value in knowing that building relationships with both types of people vary greatly.

The unfortunate part of the decision to merge departments in the corporate world is that a whole new look at the group meeting business will be taking place. It looks like the priorities for the corporate travel planner will now be price probably more than value or fulfilling needs.

RELATIONSHIP BUILDING MATTERS MOST IN BOOKING GROUP BUSINESS

March 2007

With the impact that is taking place with Independent, Third-Party Meeting Planners working directly with property sales department, on behalf of corporate and association meeting buyers, it is a wonder that there is time to build relationships directly with the real buyer of space and rooms. One of the key elements of building the group market business has always been the relationship developed between the facility sales staffers and directly with those who are responsible for signing the group contract. Things really have not changed. According to a December 2006 survey of 240 meeting planners by Meetings and Conventions magazine, it seems that relationship building really doesn't matter. The study reflects that 63% of those surveyed indicated their relationship with a property's salesperson is very important, and 30% said it was somewhat important. Further, 65% said their relationship with the property convention service manager is very important when selecting a site and another 26% said it was somewhat important.

Interestingly enough it appears that the chain brands do a good deal of business via their national sales offices, and mostly due to the relationship building that is being done with the national sales staffers. About 63% of the survey respondents had indicated that they have regular contact with someone at a chain's national office, and 56% of those have met that person face-to-face; another 51% have met that person more than once.

So what happens when a salesperson leaves a property, and takes on a similar job at a competing property or another facility? Would the meeting planner with whom that salesperson has developed a relationship follow him or her to the new property? Well, if the circumstances are right; that is, if the other property can meet the needs of the group, 70% specified they would do so, and another 24% said they would consider the new property for their meeting. Unofficially confirming these figures, our experience at the HMM sponsored Hospitality Sales Workshops over the past several years verifies that information. Just about all the participants in the meeting agreed that is the case in their experiences.

Based on this information it seems that sales personnel would want to concentrate on the relationship building with the real buyer of space and rooms. This would be an improvement for them rather than depend mostly on working with the independent third-party planner, who also gets paid by the property, on a percentage basis, and acts as an agent for the buyer. Sales personnel need to get more involved in organizations, and attend meetings where the meeting planners for companies and associations gather.

Another part of the survey questioned the respondents about how they would like to be contacted by property sales personnel. Some 12% preferred not to be contacted at all, while most, 90% did not want to be contacted by cold calls. Nearly half, 48% indicated they would prefer being contacted via email, while 20% preferred direct mail. In spite of this, we still find salespeople being told by management to "get out and make some calls." What does that mean? To most, it probably means actually going out into the field, visiting business offices, knocking on doors to find companies and organizations that may be in a position of having meetings, and need hotel facilities. That is what "cold calling" is about, and obviously is not very effective in spending one's time.

MILITARY REUNIONS COULD BE BIG BUSINESS FOR YOU

February 2008

Just a few years ago, there was a drop in U.S. Military Reunions held around the country, to about 11,000. Now, latest figures from TRN (The Reunion Network) indicate that these reunion celebrations are just passing 13,600 a year. There was a big drop in reunions of WWII veterans; however, with the increase in veterans groups, it looks like this market is really growing.

I've booked many myself, years ago, and really had a wonderful time working with these volunteer planners. The interesting things about Military Reunions, is that they come to the property when you really need the business. Further, they are not just an "in and out" convention. Most run a full week. The attendees make it a week-long, family fun set of activities, and 90% of the participants attend year-in and year-out, working it around their vacation time, for the most part, is during the summer months.

These are groups of veterans representing all branches of the U.S. Armed Forces, who belonged to a wide variety of military units during the various armed conflicts. Even with the drop in a number of reunions from veterans of WWII, there are still some units that continue to meet, and they say they will continue to meet "even if they have to crawl" to get there; the average age of this group is 87! Vietnam veterans are coming out of the woodwork, and many are joining reunions with veterans who wore the same patch, even at a different time and conflict. For example USS Iwo Jima, originally was one ship, now has veterans from 4 different ships—all called USS Iwo Jima. For WWII, a number of groups are combining and not unusual for a group to represent more than 1 ship; W.E.S.T. group is a combination of 4 ships, each represented by an initial.

These veteran member groups usually get together for a weeklong series of visits with each other, along with their families, and retell stories of their wartime activities, show pictures and documents of their experiences, and share their friendships, which really are life-long friendships. It is a wonderful experience to actually be a part of these types of gatherings, and to just sit-in and listen to some of the experiences. Of course, during the week there is a visit to local attractions, especially, when there are military connection or involvement.

Any properties located reasonably near military monuments, armories, military installations, etc., surely would be a good prospect for a military reunion location in 2008 or 2009. One of the difficulties that sales staffers may have is being able to locate the key personnel in these markets who plan these meetings. Planners involved are all, for the most part, volunteers; and most do it year after year. The best bet for seeking "connections" to find the right person with whom to make contact is The Reunion Network (TRN). This is the only organization of its kind with direct contact with the reunion planners. TRN conducts ConFAMs every year, being sponsored by our Convention and Visitors Bureaus, where these reunion planners come together to learn about an area, its facilities, and attractions.

MORE ON MILITARY REUNIONS AS A GROWING NICHE MARKET

The daily total of military reunion planners continues to grow. The volunteer military reunion planner count of only 600 back in 1991, reached a peak of 10,000 by 1996, and then in July 1997 started to grow again, reaching 10,024 by the end of 1999. This was the forecasted decline in numbers of active WWII veterans. By the end of July 2003, with the addition of Korean and Cold War veterans groups, the total had increased to 11,295. Then there was an enormous surge of growth with Vietnam veterans increasing the number of military reunion planners to 13,200. All this represents a growth of business available to hoteliers of about 20% per year, according to TRN, The Reunion Network. Further, TRN also predicts that the number of these planners is expected to increase to over 19,000 within the next couple of years as a result of veterans returning from the current conflicts. In simple terms, the Military Reunion market is exploding thereby making this niche market a great opportunity for properties within this next 2008–2010 period. However, right now is the time to start seeking this business.

There are many advantages to sales personnel in going after this type of group meeting. In some cases it is even better than the corporate or association markets because there is very low demand for meeting space; these groups are usually booked during an off-season, and off-peak period; they are very flexible on dates, and generally these are nice, friendly folks with whom to deal.

According to TRN, as the average daily (off-season) room rate paid by veterans groups climbs to about $100, and the total annual room-night count used by this sector approaches the 7,000,000 per year level, more hotels, CVBs, and attractions are getting involved each day in seeking this market. The total spent on hotel rooms only, while over the half-billion-dollar mark, is only part of the economic impact. Invariably, there are banquets, breakfasts, and visits as attractions, plus substantial shopping. Estimates on additional spending, over and above rooms and food, approximate $100 per person per day. A substantial portion of this spending could be for alcoholic beverages.

Currently, according to TRN, less than 24% of the Convention Bureaus, and less than 3% of the hotels are active in this market, leaving the field wide open for aggressive sales personnel. Like any other nice market, this requires an investment in time, and marketing. The potential for this business and actual success depends on two basic factors: What your location (destination) can offer military reunion groups and your effectiveness in attracting and dealing with the volunteer planners who represent, and arrange for these groups.

GROUP BUSINESS IS GETTING TO LOOK BETTER FOR NEXT YEAR

November 2009

Now is the time to start getting ready for a better year coming up. Occupancy should pick up by mid–2010, although room rates may not go back up so fast. Many properties surely made some mistakes by jumping the gun, and cutting rates during this year, as soon as there was a drop in demand, and occupancy. This seems to happen every time we get hit with a slump in business. For years we have been advised by accounting firms like PKF and STR that cutting rates does not appreciably help RevPAR. It takes about 6%–7% increase in additional occupancy when the rate is dropped by about 10%. This means, that around the country, because of many properties reducing rates to pick up additional occupancy, there will be a sizable decrease in overall revenue per available room. This then reflects in a decrease in bottom line revenue and profits. So, of course, the idea is to get the rates back up as fast as possible; not an easy chore as we have learned in past years . . . However, there is a great opportunity to do this by starting with the group business market since demand for space, and room blocks are projected in an increase in 2010 through 2012 as reflected in the 2009 Future Watch report of Meeting Professionals International. But, we must not wait to get the job done. There are competitors out there that will beat out the other properties that just sit, and wait for the business to come in.

One thing about sales is that you do not sit back and wait for things to get better. The idea is to help make the market get better. Now is the time to get strategies set; determine which market segments can produce the best business for a property, and how will the sales department go after those markets? A key is to create a Sales Action Plan, and put it to work.

Take a look at all the market segments at a property for this past year, and find the top three or four that produced the best business form a revenue standpoint. Figure what percentage of total business each segment generated for the property (rooms, food/beverage, etc.). Then figure out how the strategies that would be best for increasing business form those market segments. This is the basis for a Sales Action Plan. Come up with a form that will reflect a goal for increasing business in each of the market segments; do this by percentage increase and revenue increase. Follow this with three strategies that should accomplish those goals in a specific period of time. Create the plan by listing what functions needs to be accomplished, by whom on the staff, within what period of time, and the cost for that particular function. Continue these steps with additional functions. Use one form for each strategy, for each goal. Keep adding up the proposed cost for each function that is to be accomplished. This will produce a clear picture of how to accomplish reaching total goals for each market segment targeted. Be on the lookout not to exceed a total cost, not to exceed 10–15% of the revenue goal projected for each market segment.

UNDERSTANDING NEEDS IS CRITICAL IN GETTING A GROUP BOOKING

February 2010

All too often, sales staffers, in just about all industries, and businesses, jump the gun when trying to make the sale. Of course, in most cases, there is great knowledge on the part of the salespeople about the product they are selling, and can probably really relate the value of the product to a prospect. However, where many miss the boat in making the sale is not understanding the real needs of the prospects. There is absolutely no point for any salesperson, in any business, to try to start off to make a sale by telling all about the product. In many cases, a real prospect doesn't need to know about the product, but would be interested in those aspects of the product that would be important to his or her needs.

In our business of hospitality sales, when it comes to group bookings, we find this so true with many sales personnel whether it be in the group rooms end of the business of catering. Right now, most prospects for group business at a lodging property will seek information that is important to them in making a buy decision; such as, size and location of the property, number, and types of rooms, size of meeting facilities, information on food service and other features of the prospective site. Therefore, when inquiries are made via the sales office of a property, via the phone, email, Website, or third-party providers, some knowledge of the facility is already known; however, what is really missing is what value is there for the prospect to use that particular facility, maybe even over another site.

The key to being successful in this business of selling in our hospitality industry is to really understand the particular needs of a prospect before trying to explain what features and amenities are available. The first step is to find out the who, what, when where, and why of many things before even making an attempt to confirm some group business. Once any part, or all of those questions are answered then it should be possible to relate whatever features of the facilities would be a value to the group or prospect. Then those features need to be related to the prospect in such a manner that really does prove the value.

You see, it is not enough just to mention that the property does have a fitness center, once you have knowledge that most members of the group are used to exercising every day. It is not enough to just say that you have one, since the prospect already knows that. What we need to do is express that feature as a value to the members of the group. This, of course, can be accomplished in many ways. The key for sales staffers would be to get together and discuss with each other how to express value to prospects of the different property features. We need to start training ourselves.

MANY SALES OPPORTUNITIES ARE BEING MISSED IN GROUP BOOKINGS

July 2010

Year, after year, in surveys conducted by meetings trade publications, as well as by Meeting Professionals International, hotel salespeople come up short in one big area. The number one issue that many meeting planners always have with hotel staffers responsible for booking groups is that they do not respond in a timely manner or even not at all.

Today, especially these days, with electronic messaging, things seem to be worse than ever before in this area of salespeople responding to inquiries. Although the issue still remains with telephone calls, and email messages, the problem is very heavy in regard to RFPs (Request For Proposal). Understandably, some properties, more than others, receive a very huge batch of these on a daily basis. Most come electronically, either directly from the people actually responsible for planning meetings or from Independent Planners seeking to locate facilities for clients, and probably many come through a variety of software companies that provide a service for their users.

To some extent, this is a burden to some sales offices with a lack of staffing. Further, Sales staffers claim that a very large amount of these electronic RFPs are "just a waste of time" since information is being requested from properties that do not even have the facilities to meet the needs of the group functions being considered, so "why to bother to respond?" Of course, the answer to that question, we all know, is that the inquiry could be coming from some person or business or organization that could be a prospect for some other business, in the future, for the property. Anyway, common courtesy dictates that all inquiries should have a response; it is just a good business practice. Just recently the publication received an email message from a meeting planner complaining about having sent out RFPs to 21 hotel properties in specific cities in Pennsylvania and Texas. After two weeks he received nine responses, and no others have even acknowledged receipt of the inquiries. Of the nine that responded, none have followed up by phone.

In current times when hotel properties are all in need of business, it is almost unbelievable that something like this would even exist. What are salespeople thinking about how to handle RFPs? It would seem that this would be a priority. There is so much group business going down the drain because of lack of response to inquiries. In addition to problems meeting planners have with salespeople not responding to RPFs, there is also the complaint about salespeople not being available by phone; that there are too many voicemail messages also not being returned. The big issue in regard to this is sales training. There probably is not enough of training going on. Sales staffers need to better understand time management, priorities, and that their best prospects are those that come into them via the RFPs. Further, the three best tools for booking business are: Follow-up; Follow-up; Follow-up.

GETTING READY TO IMPROVE A BOTTOM LINE WITH INCREASED GROUP SALES

January 2011

As we move into 2011, it seems there is a reason to celebrate. Recent surveys of corporate meeting planners indicate that budgets are coming back and meetings are happening again. Meanwhile, Smith Travel and other hospitality research firms are predicting significant growth in RevPAR in 2011 as rates will begin to climb and occupancy increase. In many markets, we are also seeing this trend. All this good news should see more improved sales productivity at all types of properties creating better bottom line profitability.

One area of concern that needs attention with regard to group sales is what has been happening during the last couple of years with the business of meeting and banquet room rentals. A good many "deals" have been made with group accounts, in order to get the business. Not only have hotels, resorts, conference center, et al reduced rates, some drastically, but some sales departments, and General Managers have gone further. There had been many locations not enforcing attrition as well as cancellation clauses; cutting prices on banquets; providing a variety of free or reduced cost services; as well as not charging for meeting and banquet rentals.

Perhaps now it is time to get back into business when it comes to soliciting and booking groups. It is the time change our "standard" forms of proposals for groups. Cut out the offer of free banquet space, comp rooms or free anything when offering a proposal, and leave the "give-a-ways" for the negotiating table. The whole idea behind negotiations is to have a "give and take." That is, "I'll do this, if you do that."

In particular, it is really time to get back into the business of charging for meeting and banquet space. This is the area where most of what you take in on rental goes directly to the bottom line. There are hardly any additional costs in this area. After all, labor being used to set up and tear down a room, is likely already on staff, and being paid anyway; same for the cleanup. Utilities and maintenance costs are already covered on a regular basis. Space rental is, or could be a heavy duty profit center.

It is interesting that hotel operators are consistently concerned with room occupancy; nothing wrong with that, and it should be that way. However, there doesn't ever seem to be concerned over space occupancy. It is just hardly ever discussed. There would be some great surprises to owner, operators, and managers if ever they did an occupancy check-up on their meeting and banquet space. It seems that since this, or could be a good profit center, management would want to be checking up on this area. As an example, in its simplest form: if there was meeting space (classroom style) for a capacity of 100. Each day it is booked for a rental fee, it would reflect 100% occupancy. If during a month it was booked only half the time, it would represent 50% occupancy. Even if it was booked for a banquet, it could count toward the occupancy. Of course, with multiple meeting rooms, it is a question of just adding up the maximum potential for seating among all the space to come up to see how many meetings and attendees you will need to represent 100% occupancy at any one time. There really would be some great surprises if anyone cared to check up on their meeting/banquet space occupancy.

MILITARY REUNIONS: AN UNTAPPED MARKET FOR BUSINESS WHEN YOU NEED IT

April 2011

Here is a very valuable market for some properties that could use group business during a downtime. This is somewhat of a hidden source of group business not known, or even recognized by sales departments of hotels. It just doesn't fall into any specific market that we usually go after for business. It is not corporate, government, social, associations, unions; it just doesn't fall into some kind of general category that we seek when looking to go after groups.

However, interestingly enough, some convention bureaus around the country have been familiar with this very big market, and have sought to go after this business for the properties in their area. Convention Bureaus within areas like Branson, MO; Southern, NV; Fort Worth, TX; Washington, DC; Nashville, TN; Bowling Green, KY; Providence, RI; Newport News, VA; Galveston, TX; Jacksonville, FL, have all realized how important the Military Reunion market could be to their members and areas. Each of these Convention Bureau has organized a Con-FAM, which is an all expense paid multi-day event, combining a Reunion Conference and a Familiarization tour of the area. The bureaus have arranged for a room at local properties, meals, admissions at attractions, and local transportation. Instructors for workshops and instructional materials are furnished by the Reunion Friendly Network (RFN). Local attractions and properties have the opportunity to exhibit at these events and network with the attendees who are all military veterans who are planners of meetings for the various military branches of service represented at the event. Those attending only pay for their own transportation to and from the event, and personal purchases. Although the Con FAM lasts only three days, these veterans who attend, with their families usually make a week-long vacation out of this trip, and stay around, visiting that area of the country.

Organizing and putting on such an event is not an easy task. It takes locating those volunteer veteran meeting planners, and the various veteran groups around the country, to create an appropriate listing of those who plan these reunions. Then, of course, communicating with them through various means, as well as with Convention Bureaus to get them to assist in making appropriate arrangements for the events. This is all done by RFN, located in Hollywood, FL. An organization founded by veteran Paul Spiewak in 1991 who saw the need for military reunion planners to have a source for information and training. RFN now has a database of over 8,000 planners of reunions. They have trained more than 6,000 reunion planners at 200 educational conferences and FAMs conducted in cooperation with CVBs in 38 different states. RFN, since 1992 has published a quarterly magazine, which is distributed free to all of the volunteer military reunion planners, as well as RFN member CVBs, hotels, and reunion vendors. The publication is also available online. There is also an online publication called Reunion Planner Memos which is a monthly newsletter for military reunion planners.

FOR PRODUCTIVE SALES, RATE SHOULD BE THE THING IN 2015

January 2015

It sure does look like we are heading to the best year we have had for a long time. All projections from the major accounting firms reflected some great information for us in the hotel business. Reports tell us that travelers' demand should be up around 4% with an increase of room supply of something under 2%. Basic economics would tell us that just that alone would reflect an improvement in rate to be in the area of 4–7%, along with an occupancy increase of just about something over 64%. That would put back to around 2007 when occupancy was about 63%, the highest it has been since the 70s. Of course, we all know that we took a big hit in 2008, and never really got back to the high rate of 2007 until just this past couple of years. However, with everything looking great for this coming year, no doubt hotels, of all sizes and types, should do well.

So, now what is the message? Shall we just sit back, and let the business ride in for us? What is the message for our General Managers, and sales departments? Increase in occupancy is good, but the rate increases and revenue are better. This year we have the best opportunity to show improve profit results from rate improvement. We just have to do better in sales to make those projections come through. Salespeople just need to really get to work for profit improvement by putting more time into proactive sales activities. That means more time to be spent with qualified prospects. General Managers may want to spend more time seeing that sales staffers are spending more time on being productive rather than doing other things.

The group market is what is needed to be beefed up by the sales teams. Recent reports reflect that the ADR for transients is well above than the ADR for groups. Where ADR has been running at around 4–5% growth, it seems that the group ADR is only in the area of about 1–2%. So beefing up the rates quoted on groups needs some attention. Hotel sales staffs need to do more in the area of advance bookings, especially for the next 1–2 years. We have got to pay attention to what will happen within the next couple in regard to competing hotels. The pipeline has been growing; investors, developers, owners and operators will for sure be seeking to develop new properties. This always seems to happen in our industry as soon as we have an excellent year in building occupancy and higher rates. So, we need to be on the lookout for more competition coming in within the next couple of years.

General Managers should be continually talking to sales folks about getting the rate. The rate should be King this coming year; that is what generates profit for a property.

STOP SELLING, AND JUST HELP THE GROUP MARKET PROSPECTS MAKE A BUY DECISION

One of the issues with which we need to be concerned, when it comes to Sales, is to stop selling. What we really mean is to stop Talking and start Asking. The successful salesperson is the one who doesn't start telling all about the property, but starts asking open-ended questions which gets the prospect to do the talking. It is always best to get information than to give it. It is amazing what type of information comes out that could be critical in helping the prospect make the Buy decision.

To get started sales staffer really need to know all about the product. That is, all the specific details of all the features at a property being represented. It is not enough just to know how many rooms, or total square footage of the meeting rooms, or that there is parking for guest, etc. What is needed to know are the details pertaining to all the features. For example, depending on the group market prospect, perhaps a salesperson really needs to know things like how many connecting (no adjoining) rooms; the size of rooms; the number of different types of rooms, and of course, other details involving the room locations, etc. This means that all sales staffers need to conduct their own product analysis; coming up with a detailed description of all the property features that make up the whole product.

The next step would be for the salespeople to understand how each feature may be presented to a group prospect as a value. That is what we are really doing; helping the prospect by showing value to their needs. There is no point talking about something about the property unless it could meet the need of the buyer and is a real value for the group. This whole thing about getting a buy decision is getting the prospect to talk so that something will come out that represents a need for the group or prospect. The once real needs are determined, the features of the property that could represent a real value could be described. Property management needs to start working with sales personnel during their meetings to practice describing various features of the property as value to certain, specific group prospects. Corporate meetings could have different needs than association meetings. Sports groups will have different needs than social groups. If attendees of a group are driving to the property, how would you describe the parking opportunities, showing a value? If families with children are part of a group, how would you describe accommodations, including the connecting rooms?

ENGAGEMENT IS CRITICAL IN THE BUSINESS OF BOOKING GROUPS

September 2015

We all say that we are already doing this in our working sales departments, but how are we doing it? Let's start thinking about our being more of a consultant than a salesperson. The idea is to stop selling, stop talking so much. Stop with the sales pitch and sales presentations. In today's world of information technology, most people responsible for booking meeting venues have already explored options for their needs. We do not need to start telling prospects all about what it is we have to sell them. What we need to do is to engage the prospect by working with them, as a consultant, in helping them make a valued Buy decision. This means knowing all you can find out about their needs as someone in charge of a group; their needs as a company, association, or organization, as well as the possible needs of their attendees, and maybe even their bosses. There is no need to say much about your venue until you know more of what it is they are seeking, actually their needs, and maybe even wants. This is the consultant approach in assisting them in making a buy decision. Engagement is so very critical in developing the information. Just do less talking and more of asking critical questions.

Everyone would agree that having confidence in sales is of the utmost importance. People generally like to deal with those who have confidence. The most successful salespeople are those who have confidence in themselves. So, how does one build confidence? In our business of hospitality sales, the three areas of knowledge that create confidence are: Knowledge of your product; Knowledge of the competitions, and Knowledge of prospects' needs.

With the appropriate engagement of a prospect we gather the needs aspects through the asking of Who, What, When, Where, Why type questions. Of course, sales staffers need to know their product, inside and out! That means to be aware of all the features of the product as well as ow any feature may meet a need of the prospect and their attendees. This means salespeople must create an in-depth product analysis of the venue. The other area of knowledge has to do with the product analysis of the competition, and to understand the comparison of features.

It is really the confidence the salesperson has developed in knowing needs through engagement; the product, as well as the competitor product through conducting the analysis. That creates the consultative approach that helps a prospect in making a valued buy decision.

CHAPTER 6

THE TRAVEL AGENT AND THE LEISURE TRAVEL MARKET

START WORKING NOW FOR NEXT YEAR'S BUS TOUR BUSINESS

May 1983

Now is the time to get your act together for the 1984 motor Coach tour business—after August it may be too late. Escorted tour packages are being put together right now by companies that market tours to the consumer—either directly—or through travel agents. These packages, available to individuals or groups always includes Motor Coach transportation—overnight lodging en route and at the destination. The following are also included to varying degrees: baggage handling; meals; parties sightseeing; admission to attractions; air transportation to a destination; and gifts.

The small hotel/motel operator should be interested right now in getting some overnight room business from their ever growing, multi-million dollar tour business. To go after this market some commitments have to be made:

▶ Rates quoted should be something lower than the normal rack rate.
▶ Rates should be quoted as "net" (not commissionable).
▶ Rates need to be established more than a year in advance (right now–1984 rates need to be committed).
▶ Most tours will not want planned meal functions, but tour guests will eat on their own.
▶ Check-out is usually very early and restaurants are expected to handle early breakfasts.
▶ Tour operators and other motorcoach tour companies will not pay for incidental expenses-separate folios for other than room charges should be made and collected individually from guests before departure.
▶ Reputable tour companies that have not established experience with a property will expect to pay reasonable deposits to hold confirmed blocks of rooms. It's reasonable to expect to collect a deposit thirty days in advance of group arrival.
▶ Rooms reserved are mostly double occupancy-there may be some minimum number of singles, triples quad occupancy.

- Youth type groups in many cases require quad occupancy.
- It's always best to quote rates on a room basis for single, double, triple, quad, and quote applicable taxes.
- When rooms are reserved it generally will be for 25 per bus; however, it may more realistically end up being 15 to 20 per bus.
- Nearby motor coach servicing is desirable, but not necessary.
- Collecting brochures form all available sources that advertise motor coach tours helps determine which companies to solicit. Knowing the tour company's routing helps in targeting your market.
- Mailing Labels of National Tour Association members are available from that organization for a fee.
- Rooming lists should be requested from the tour company at least ten days in advance so appropriate rooms could be blocked. If the rooming list does not arrive call the company to be sure they are still coming to your property.
- Rooms should be locked be handed envelopes for each party that contains the room key and a nice welcome message.
- It is always appropriate to board a bus upon its arrival to present the escort with the keys and to welcome the guests at the same time sell all the services of the property-tell about the lounge, the restaurant, and other features.
- Baggage handling should be quoted on a person basis-a flat rate for coming and going (in and out).
- Outdoor reader boards or sings are always appreciated. A property that has one should use it to welcome tours.
- Be sure to be friendly and provide good service. There's a great opportunity for future business if it is well taken care of—people return to where they have been well treated.

Caution: Because of recent deregulations in the motorcoach industry, there will be many new and inexperienced people trying to organize motor coach tours.

- If there is no experience with a company operating motor coach tours or credit cannot be established is recommended that the company be requested to prepay for rooms. Members of the national tour association are required to have at least $100,000 coverage for professional liability errors and omissions. NTA established a code of ethics that is being enforced to see that members perform on an ethical basis. NTA does represent the established on an ethical basis. NTA does represent the established motor coach tour operators and was organized in 1951.

THE TRAVEL AGENT AS A SALES ARM

September 1983

The Travel Agent is a professional who provides a variety of services in the travel-related fields to individuals, companies, associations and many types of diverse groups. The services range from offering suggestions for a honeymoon trip to planning an entire movement for 1000 or more people for an African safari. And in between, there are individual as well as group arrangements for air travel, car rentals, motor coach movements, attractions admittance, cruise, ships, insurance coverage, Amtrak reservations, sightseeing, meetings, shopping, eating, entertainment and, of course, hotel/motel room reservations.

Income for a travel agency comes strictly from commissions earned as a result of bookings on behalf of a client. The client (individual or group) seeks the services of an agent because of assistance needed in making necessary travel arrangements. The professional travel agent provides the consulting service, with recommendations for the client without any cost to the client. By handling the client's travel arrangements, the agent receives a commission from all the travel components used (airline, car rental, hotel/motel, entertainment, etc.).

At one time, travel agencies were considered a place to go to make arrangements only for pleasure travel. Today's more and more companies and associations are using travel agencies to handle transportation and lodging needs. Over a half million companies in the United States today are allowing agencies to handle their needs. Very recently the U.S. government started an experiment of having travel agencies arrange travel for certain departments. So far, reports reflect that there have been considerable savings in expenses for the government, and practice will spread to other departments and bureaus.

It appears that business travelers are using travel agencies more and more—mainly because of the experience that travel agents have in handling the very complicated business of airline schedules, rate structure, and lodging arrangements. More and more travel agencies are getting into the meeting planning business. Besides arranging for travel to meetings, agents are now getting involved in all the details of conferencing—even to the point of dealing with catering managers on planned food and beverage functions. Group incentive travel is another important area for travel agents, which requires that a good deal of time be dedicated to moving very large groups to resort destinations and making appropriate resort hotel arrangements for groups.

There is no question about it. Lodging operators all over the world can use the help of this very worthwhile intermediary—the travel agent. Doing business through a travel agency is an agency is an important investment for a hotel or motel. Repeat business is almost always certain from a travel agent, provided the clients sent to a property are treated properly and are pleased with the accommodations and, of course, provided the commission to the agent is paid promptly.

Generally 10% of the room rate, for the period of the guest's stay, is what is expected by travel agents for reserving rooms at the property for a client. There are no other costs to a lodging operator for doing business with travel agencies; therefore the cost of sales is only 10%—a pretty good bargain! There are some other costs involved in the operations that choose to promote a property to this travel partner. Promotion to travel agents may be accomplished by:

▶ Direct sales effort: making personal calls or telephone calls to individual agencies.

- Advertising: several trade magazines and directories are targeted to the travel agent audience.
- Co-ops and consortium affiliations: cooperative groups of agents that band together for joint marketing efforts and other mutual interests.
- Packaging the property with other travel components and marketing through wholesalers or tour operators.

Trade show participation: Several are conducted during the year; however, of significant importance is the Business Travel seminar sponsored by Travel Trade Productions and the Hickory Travel System. During the seminar, travel agents learn from a panel of hoteliers and travel agents, "How Importance are Travel agents to Hotels/Motels. "There has been a great deal of misunderstanding and mutual distrust through the years between agents and hoteliers. A platform such as the business Travel Seminar provides an opportunity for sharing of ideas of profitability for lodging operators and travel agencies.

TRAVEL AGENTS ARE STILL IMPORTANT

January 1994

A recent review of various studies made including the 1982 Travel Pulse Survey; 1982 Travel Weekly/Louis Harris Study; 1982 Travel Agent Usage Study by Meeting Planners International and Hotel-Travel Index reflect that the travel agent is a most important source of business Hotels and motels. The compilation of facts of facts and statistics boils down to the following to help lodging operators in their sales plans to develop additional occupancy: 39% of business travelers use travel agents at least once a year- and interestingly enough, 41% of those asking for assistance in hotel selection. About 29% of leisure travelers use a travel agent at least once a year and 75% of those asking for assistance in hotel selection. Altogether 745 of all travelers have ever used a travel agent.

Travel Agencies report account types as follows:

- 91% have corporate accounts
- 58% book group travel
- 49% book conventions
- 59% book meetings
- 35% book incentive trips
- 95% book weekend specials

These facts alone tell us a great deal.

Let's promote weekend specials and give travel agents an opportunity to sell these for us. It also tells us that travel agent are involved in the meeting business so they are good prospects for small as well as the large lodging operator. Important to the travel agent in recommending hotels are: (in no special priority order)

- hotel location
- how the hotel pays commission
- whether or not reservations are honored
- hotel reputation
- whether or not have an 800 #

The attitude of hotel management towards travel agents on the subject of commission we found out that 63% of travel agents are satisfied with transactions with larger hotel chains and groups and that 66% of travel agents are not satisfied with transactions with small chains and independents. The problem still exists with many smaller, independent motel/hotels not paying commissions at all or not paying commissions promptly. Lodging operators must recognize that if they consider travel agent business important to them, they must pay 10% commission on all business referred to them by agencies and that *the commission check should be written and sent the same day the guest checks out.*

Most of the over 20,000 travel agencies are doing billings of over $1,000,000—30% of the agencies are doing between $500,000 and $1,000,000 while only 8% are doing less than $500,000. More than 28% of the agencies are doing over $2,000,000. This tells us there are more agencies today than there were 3 years ago (up about 5,000) and they are all doing more business.

While travel agencies use a variety of source of reference material the most often consulted appears to be Hotel and Travel Index.

TIPS ON HOW TO WORK WITH TRAVEL AGENTS

March 1984

If you advertise in Hotel and travel Index be sure to include all facts helpful to the agent in making a decision rates and taxes; location; miles to airport/downtown; credit cards honored; hotel amenities such as airport courtesy car, information on pool, tennis, in-house restaurants, clubs, etc.; attractions nearby; 800# or if collect calls accepted—the idea is to make it easy for the agent to make a reservation.

▶ Honor all reservations.
▶ Pay commission promptly upon check-out of the guest.
▶ Thank the Travel Agent for the business and *ask for more.*
▶ Monitor Travel Agency business to determine points of origin and frequency of business from different agencies.
▶ Measure the amount of travel agency business during the year and compare with the previous year to see how well your program is working
▶ If you have corporate rates advise travel agencies and pay commission on corporate rates.
▶ Establish budgets for advertising and promotion to travel Agencies—for the individual as well as group business.

IS THE INTERNATIONAL MARKET FOR YOU?

September 1986

Of course, ANY KIND of business coming our way, well take with pleasure. However, going after a particular market, which will knowingly bring in a small return on large investment of time and money—generally does not make sense. Going after the international traveler certainly does make sense for certain hotel/motel operators. There's always a potential for this type of business for a property that is located in a gateway city or popular destination location or resort location.

Getting to the individual international traveler is difficult and costly—but, don't despair—there are other, lower cost opportunities to tap into this market. Travel agents are the key. There are many tour operators located all over the world. They specialize in creating U.S. destination packages for individual travelers from their own countries and market the trips through Travel Agents in that country. The trips run from one to four weeks and longer—and generally include air travel, transfers, hotels/motels, attractions, and some meals.

The tour operators solicit net rates (non-commissionable) from each of the partners in the package and print State council of Tourism, etc. They all have different names but usually come under the Department of Commerce in most states. These are an absolutely great source for developing research information to determine the amount of foreign business already coming to your locations. Most programs may be marketed with the help of the United States Travel and Tourism Administration (USTTA), U.S. Department of Commerce, 14th & Constitution Ave., NW, Washington, DC 20230. The undersecretary of Commerce for Travel and Tourism is the Honorable Donna F. Tuttle (202/377-4752). USTTA has offices in key cities around the world and staff members stay in touch with the foreign tour operators to help them create travel to the U.S.

TRAVEL AGENTS STILL A KEY TO INCREMENTAL BUSINESS

September 1987

With the number of U.S. travel agencies hovering around the 30,000 marks, it's surprising that more effort and budgets are not directed to this market by hoteliers. Granted, except for resorts, of total room sales to any hotel or motel, only 10–15% are generated directly by travel agencies. Agencies use "system" toll-free 800 members more readily than going directly to a hotel/motel.

Opportunities exist for property managers, owners, salespeople to generate more traffic through travel agents are agencies continue to increase their corporate base of business, as well as their leisure business.

According to a recent study by the U.S. Travel Data Center, there are over 33 million people who travel Data Centers; there are over 33 million people who travel on business annually. About 25% of these use a travel agent and 40% of the most frequent travelers (those taking more than 10 trips a year) user agents. Almost 40% of business travelers ask their agents to make a hotel recommendation with some specifying a chain preference, but not a particular hotel. The study, underwritten by Hotel & Travel Index, reflects the important role travel agents play in influencing traveler's decisions.

If agents are, in fact, an influencing factor in the selection of hotel and motels—and evidence reflects that they are—it appears that this is a market that needs more attention from the lodging industry. How share of travel agent referred business?

▶ **First, take good care of the business:** Most hoteliers take it for granted that frequent business travelers select the same hotel on repeat visits. According to the study 40% of US. Business travelers, upon returning to a previously visited city. DO NOT return to the same hotel or motel.

▶ **Second, take good care of the travel agent**: Generally, travel agent doesn't have too high a regard for the way hotel/motel operate. The biggest complaint is the lack of prompt payment of agency commission. According to many agents, hotel/motel personnel blame agents when quests have problems, such as no reservation on file; specific requests no noted or any number of other details not handled properly.

▶ **Third, use directory advertising and use it properly:** Hotel &Travel Index subscribers indicated that in dealing with business clients, agents use directories 82% of the time. Hotel and Travel Index is probably the most frequently used resource by travel agents in recommending lodging facilities for clients. Field representatives for the directory have a specific recommendation for lodging operations on what to include in directory advertising.

▶ **Fourth, include a travel agent program in your annual marketing plan:** Include directory advertising; Travel Trade advertising; direct mail activity; sales calls; public relation efforts and participate in agent type trade shows. With time and budget directed to the travel agent market, an increase in the share of this business can be noticed in a relatively short period of time. Remember that agents are an extension of the hotel sales force. For many years, the incentive Travel business was relegated to the large incentive houses. Not anymore. Now, local travel agents, independents, meeting planners and companies—on their own, are all in business.

Lodging operators need to get in the business also and can do it very easily, economically and can create extra weekend sales for themselves. Small hotel/motel operators, as well as the larger ones, need to get into this incentive travel business. There's a way to do it without going to a great sales and promotions extent. Get other people to do the selling. Here' show to do it: Find a strong, local travel agency that uses outside salespeople. Explain that they can buy rooms at a very reduced net rate from the hotel over certain weekend periods. Naturally, use only weekends when business is needed. Based on the cost of the room to them, plus tax they can build a weekend package (provide a variety of options that can include such as meals, attraction, gifts, etc.). The agency then approaches their corporate accounts to offer local, short-term incentive trips as a reward to the company employees.

Some education and training may have to be offered to the agency staff so they will understand how to sell the travel incentive program. The program may be sold to any number of types of businesses that would benefit from a motivational program for employees. The short-term aspect is good because employees need only to achieve a certain goal to qualify over a period of a week or perhaps a month.

Types of companies for the travel agency to consider approaching are:

- Branch offices of insurance agencies
- Real estate brokers
- Banks
- Automobile dealerships
- Men's and women's apparel shops
- Companies that have route sale peoples.
- Manufacturing plants
- Shopping centers—Promotion department.

With just a little brainstorming with the hotel staff, the list could be almost endless.

In order to help sell short-term weekend incentive trips, the travel agency staff needs to understand that the incentive trips are used to reward employees for:

- Sales Improvement—Safety
- Profit Increases—Attendance
- Reduction in Losses—Referrals
- Work Activity—Complaint Reduction
- Production Performance—Maintenance of Accounts

The advantages for the agency's client would be:

- Weekends—so, no time off from work.
- Less planning time.
- Easy on the budget.
- Drive at own expense since it's local.
- Winners can take the family.
- Motivates for greater productivity.
- Spread the word among employees.

The advantages for the agency include:

- Profit generated from mark-up-they do not have to wait for a commission.
- The agent may create own profit structure
- Minimum administration once sold.
- Open door for the agency to sell bigger and more costly incentives trips.
- Builds company employee desire for travel, developing more prospects for the agency.
- Mostly non-competitive type business for the agency by using a travel agency, the property expands its sales force. In fact, several agencies may be given the opportunity to sell weekend incentive trips to local companies. A bonus may even be offered to agencies based on the volume of weekend business generated.

ARE YOU GETTING YOUR SHARE OF THE TRAVEL AGENT MARKET?

March 1988

The corporate travel market has always been "big" on the list of many lodging operators. This segment of the business has been approached through consumer and trade advertising, directory advertising, direct mail and sales call on corporate travel managers—all with offers of discounted rates, special amenities, upgrades, frequent guest programs, etc. The savvy operators have also been soliciting travel agencies that handle corporate accounts and making deals with travel agency marketing groups.

Although many travel agencies are "into" corporate travel, the biggest part of the business mix at two-thirds of the travel agencies is provided by leisure travel. In fact, more and more commercial travel agencies are expanding into leisure sales. From a hotel/motel standpoint, the leisure, or vacation traveler, is a prime source of incremental revenue.

Now, a new report is available to the lodging operator to help analyze and understand what's happening in the leisure market.

Hotel & Travel Index (Hotel Directory used by travel agents) has sponsored a research study conducted by Plog Research, Inc. of Los Angeles. It's called: "Hotel & Travel Index CTC Front Line Report on Leisure Travel." This report takes a look at the issue of importance to hotel/motel operators, travel agents, and tour operators. Some of the areas explored in the study include: Demographics of the leisure market—what segments show the strongest growth potential; Selecting a hotel—what do travelers look for; Destinations—what's hot and what's not. The study reflected responses to a survey sent to 2,000 Certified Travel Counselors of the Institute of Certified Travel Agents (ICTA). Respondents had an average of 12 years in the business. Seventy-eight% (78%) were owners or managers of retail travel agencies, the others were full-time counselors.

Who's Traveling: According to the study, the strongest growth was in the retired couples and young persons without children. Other growth areas included honeymoons, single, mature women and single young women. Showing the least growths were single, mature men and middle-aged couples with children.

Where to Go: Among those who buy packages, the attractiveness of the destination and price are top priorities. Among other considerations (in order of descending importance): Quality of the hotels included in the package: Reputation of the packager; Length of the trip; Tour attractions included; Preferred airline included. More experienced agents are less concerned with price than quality.

Top Factors in selecting a Hotel: What does the leisure traveler look for in a hotel? Among the top are: upkeep and cleanliness; Desirable location and convenience; Good price;

Recommendation of a travel agent; Courtesy; Good security; Fast check-in and check-out. Among the least important considerations are: Entertainment available; Frequent Guest Program; Good hotel shops; concierge floors.

"The fact that recommendation of a travel agent is ranked at the top makes it all the more important for hotel marketing strategies to take into account the demographics included in this study. If the majority of travelers rely on the advice of agents in booking hotels, it is essential that hotels/motels communicate the right message to the right agent. This study was designed to provide hoteliers with guidelines for doing just that. "According to Stanley Plog, the researcher.

Copies of the study are available free of charge by writing or calling: Ms. Melinda Bush, Publisher, *Hotel & Travel Index,* 500 Plaza Drive, Secaucus, New Jersey 07096. Phone (201) 902–1600.

WE NEED TO DO MORE BUSINESS WITH TRAVEL AGENTS

January 1989

With over 250,000 travel agents working in 30,000 travel agencies, Something is gotta be wrong with hotels and motels not getting more business from them. Most properties don't get more than 12–15% of room sales referred by agents. Resorts and international hotels get much more than that.

What's wrong? Why are 'not we doing more business with travel agents? Why is it there is still somewhat of an adversarial relationship between agents and hoteliers? (However, it's much better than it has been in previous years.) How can we get agents to sell more hotel rooms for us?

Great opportunities exist for improved profitability for both parties if a secure testing partnership would develop. And -we're getting there with the help of our trade associations. For the first time ever, ASTA (American Society of Travel Agents). HSMAI (Hotel sales and marketing association International) and AMHA (American Hotel and Marketing Association on International) and AMHA (American Hotel and Motel Association) are bringing together hoteliers and travel agents ASTA will produce the program called "training fest" which will focus on hotels and marketing in cooperation with HSMAI and AHMA under the sponsorship of Hotel & Travel Index and official Hotel and Resort Guide.

The program is designed so that hotel owners, managers, and salespeople will be able to meet with agency owners, managers and counselors for joint educational sessions, networking and selling opportunities. This will be the most comprehensive look at the hotel industry ever presented by ASTA according to Anne Sullivan, ASTA's educational chairman. ASTA's Hotel chairman, Bernice Rosmarin, CTC, and Anne Sullivan have both been long advocates of improving hotel-agency relationships.

A combination of general session, seminars and workshops will provide hoteliers with plenty of information so that they 'll be able to improve profitable incremental sales through travel agencies. There will also be agent/hotel business exchange sessions and hotel participants will have an opportunity to sell their product during a tabletop marketplace.

Some of the subjects for hotel attendees will include:

▶ What hoteliers need to know to get more agent
▶ Improving travel agent relationship
▶ Meeting the needs of travel agents
▶ Improve repeat sales with agents
▶ How to get agents to sell your property
▶ How to make an agency call
▶ Direct mail tools travel agents notice
▶ How to run effective FAM trips
▶ A good way to pay commission
▶ Tips on advertising to gents.

LET'S GET BACK IN BUSINESS WITH TRAVEL AGENTS

June 1990

Lodging operators' relationships with agents do not seem to be getting better. It may even be getting worse. It appears that there are more and more complaints these days about by-passing, commissions not paid, only partial commissions being paid, calls not being returned, correspondence not answered, and a general lack of interest in the travel agency business.

The complaints of agents are coming in "fast and furious" according to Joel Abels, publisher and editor-in-chief of Travel Trade, a very popular twice-monthly trade magazine, with a very heavy travel agent readership. Mr. Abels publishes the information presented to him by agents who had a wide range of problems with lodging operators. Most of the complaints have to do with collecting commissions. It seems that lodging facilities and hotel companies that solicit travel agents for business and indicate that they pay commissions to end up with some excuses for not paying it. Some of the real "biggies" are being named most of the time. After a property is contacted by travel agents, they end up paying the commission.

It would appear that in this day and time of underutilization of rooms and tough competition, lodging operators would be thankful to get business directed by an agency. Agreed, there are times, when there is heavy occupancy, it would appear that a hotel would prefer to hold rooms for direct business. However, "turning off" a travel agent could very well mean not getting any agency directed business when the property needs it. One of the biggest complaints is this business of "negotiated rates." A travel agent sends a client to a hotel at a rate quoted by a reservation department: when the client arrives, somehow she or he negotiates with a desk clerk for a lower rate when the agency complains about not getting a commission, the response is that the hotel doesn't pay on "negotiated rates." This is really absurd. Why would not the operator, at least, pay a commission on the lower rate? Many agents have just given up on trying to book rooms for clients. They just do the air travel portion of a trip and never even ask if a room is needed.

There are about 34,000 U.S. retail travel agencies and all have great opportunities to generate room business for those lodging operations that want it. Many hotels and motels do very well with travel agencies, are very happy to get this type of business, and are very willing to pay the commission promptly. These are the ones that will continue to get volume from this segment of the business. Marriott recognizes that importance of this market and has advertised to agents that if they do not receive commissions within 48 hours after a client has departed one of their hotels, they will double the commission. Now that is some great commitment.

Of all the ways to get to travel agents, one good method is exhibiting at conference and trade shows where agents gather.

CORPORATE TRAVEL IS STILL BIG BUSINESS

November 1990

With business travel, still the fastest growing segment of the travel market, the lodging sales effort should be targeting more corporate travel managers. According to American Express, business travelers will have spent about $115 billion by the end of the year, with a good portion of it directed by a travel department of a company, or by a travel agent.

Corporate travel departments are becoming more and more involved in directing their company travelers to hotels according to a recent survey by Hotel and Travel Index, a hotel directory primarily used by corporate travel managers, travel agents and meeting planners. Overall, the study revealed the role of a corporate travel manager and how a travel department influences travel decisions. The survey reflected that on a weekly average a typical department in a company books 35 hotels. Obviously here lies a key decision maker in determining which lodging facility gets the business. It appears that 80% of those surveyed are directly involved in recommending hotels and 76% in booking hotels: over half are involved in negotiating hotel rates. When queried as to how hotel information is secured, 9 out of 10 (87%) travel managers indicated they use hotel directories: 53% said chain directories: 18%—brochures: 16%—trade magazine advertising.

It was interesting to note that when it comes to using travel agencies, 67% said they use agents to book hotel rooms. Although 37% of those surveyed did use a computerized reservation system, most of them also referred to hotel directories for additional information before booking on a CRS system. The reason for this seems to be that most central reservation systems are not designed to carry all the information about a hotel that is necessary for making a decision. These systems are designed basically to conduct transactions rather than provide information.

According to the survey, 74% of the travel managers that use hotel directories indicated they use Hotel and Travel index. In using directories, travel managers are more able to get detailed information on location, nearby attractions, proximity to the airport and the direct telephone number, among other details about restaurants, health clubs, etc.

Travel managers are fully aware of how competitive the lodging market is today and they are taking advantage of it. They are learning how to negotiate and make the best deal possible on hotel rates. Lodging sales personnel should be fully prepared to negotiate in the best interest of their properties: to understand the needs of a corporate account and never to feel intimidated in participating in negotiations. Corporate travel is an important market for most lodging operations and a good share of this business must be sought by sales personnel.

Go after this business with caution. Travel managers may be giving promises of the heavy volume of room nights over a period of time, but many do not deliver the goods. It is always a good idea for salespeople to test the waters on such promises. If nights, just do it for a short period of time rather than get into a long-term (annual) contract. There is nothing wrong with providing a reduced rate, provided the volume of business will provide incremental sales. However, once you enter into a contract for a specific rate you are stuck if the account doesn't deliver the volume. If it is done on a short-term basis, as a test, it will provide an indication if the business is really there. I rather like the idea of a corporate rebate rather than a discounted rate. Here's the deal: the account books their people at a regular rate. If they deliver with xx volume, as promised, they get xx% rebated for the total revenue at the end of the year, or designated period.

HERE ARE SOME TRAVEL AGENCY COMPLAINTS ABOUT HOTELS

February 1991

Most agencies have some pretty good reasons for not wanting to book certain hotels and they are passing the word within their own community about the properties they think are not playing "fair." General Managers and Sales personnel who are active in working the travel agent market are pretty well aware of the problems if they have been carefully listening to agents it's not just some hotels not paying commissions outright or some hotels making delayed payments. The problems go further than that, although they are the two most important areas of complaints. Travel agents indicate that if a property doesn't want to pay a commission it certainly is their own business, but agencies do not want to be misled and be given the impression that commissions will be paid and then told after bookings are made that no commission is being paid by the property. Or being told that they only pay commission "off-season." If agencies are aware of properties that have a no commission policy, then they avoid using that location.

Here are some of the biggies:

▶ When agents call a hotel directly or a reservation system to book a room and ask for the "best" rate (they mean the lowest rate) sometimes they really do not get the various rates available and book the best deal they can get. If, for any reason, their client calls the hotel directly, sometimes the client gets a lower rate than the travel agent can get. This makes it look like the client can do a better job than the agent. Further, in many of these cases, the hotel refuses to give commission. A typical response would be: "no commission on negotiated rates."

▶ Many agents indicate that sometimes client gets to the front desk of the hotel to check in and find out that a lower rate is available. The reservation gets changed and the agent loses the commission.

▶ It seems that clients are extending their stays at hotels, but agencies only get commissions on the original booking of whatever number of nights was originally reserved. When a guest reduces a stay by a day or two the hotel reduces its commission to the agency and only pays for the days stayed. It should work both ways.

▶ Sometimes agents book a number of rooms, but use only one name on the reservation. The commission arrives for only that one room.

Here is one we hear a lot: * For a group inquiries, the agency advises the hotel of the name of the group. Salespeople proceed to contact the group directly and offers a better deal if the group books directly with the hotel.

According to any agents, this last item happens more often than anyone would think and it angers the agents just about more than anything else. (Frankly, this a very legitimate complaint and I experience, over the years, that this has been happening at hotels that were part of the company where I used to work).Others areas that hurt the relationship with agencies is the way some clients are treated at the front desk. When the specific room type or location requested is not available it seems that the blame is always about booking honeymooners on a special honeymoon package at a hotel and the couple gets twin beds.

It would seem that as hard as it is to do business these days and with most hotel operations needing help in building occupancy, we would want to get on the ball and work with the agency community. The First hotel group that makes a sincere commitment such as 'WE PAY COMMISSIONS PROMPTLY ON ALL RATES" will be able to see how fast travel agencies can send the business in that direction.

TRAVEL AGENTS STILL UPSET WITH HOTELS

April 1991

It looks like an old story, but travel agents just about all over the country are still upset about the country are still upset about the treatment they are getting from hoteliers. The complaints range from the commission payment -to reduction in commission payment-to no commission at all-to paying commission on "off" season only-to not accepting travel agent reservations when hotel is busy-to hotel personnel not returning phone calls-to promotional material coming in late-to clients getting a lower rate than what the agent quoted-to clients getting "walked" on guaranteed reservations-to desk clerks blaming agents to clients when mistakes are made. In any event, for whatever reason, hotels are getting a bad rap. Many agents have indicated that they just refuse to offer to make hotel reservations for clients.

The first major thrust for an organized effort to help agents to something about this situation was undertaken last year by Joel Abels, publisher of *Travel Trade*, the best-read weekly travel agent magazine. His travel agent readership was so "up in arms" about being by-passed on commissions that he started an agent By-pass column and created as SOP letter for agents to use.

The idea was for agents who have been unsuccessful in collecting commissions to use the SOP letter. They have found that the letter gets results. So far, better than 90% of the letters sent have resulted in agents getting their commissions. If an agent is successful in collecting they can then write to Mr. Ables who then follows up with a call to the General Manager of the property in question to resolve the situation. This action usually develops into most of the problems being resolved. If this does not prove successful then the story is printed in the By-pass column, with names being mentioned. Through this means 60,000 travel agents, who are paid subscribers to *Travel* will be able to see which hotels have been bypassing agents. All complaints are investigated very clearly to be sure there was a legitimate by-pass, which was refused to be resolved before hotel names are mentioned. In many cases, as a result of *Travel Trade's* involvement, lodging facilities and management groups have made policy changes with regard to commission payments. Several groups of hotels that refused to pay travel agents commissions previously have decided to work with agents and are now paying commissions. In these cases, *Travel Trade* has mentioned the names of the properties so that the agent readership of the publications can start sending clients to those locations. In just about all of these "corrected" situations, the operating companies of these hotels have thanks to Mr. Abels for calling this oversight to their attention.

Since travel agencies derive their income only from commissions received from airlines, hotels, car rentals and others, it is logical that they want to do business only with those entities that agree to and will pay commission. Currently, of all travel agency income, only 10% comes from the hotel commissions. There certainly is room for them to increase this percentage of their overall business. All they gotta do is ask their clients to let them make the hotel reservation, book it and hope for the commission to be paid.

THE HOTEL CODE OF COMMISSION FAIR PLAY

December 1991

It had to be coming and sure enough, someone decided to start the ball rolling. Joel Ables, the publisher of Travel Trade Publications, has decided that the time has come for hoteliers and travel agents to get together and establish such a code. According to Ables, in his *Travel Trade News Edition* editorial of October 7, 1991, travel agents are losing $32 million a year in hotel commissions. Abels states: "After all, how can the multi-billion dollar hotel industry hope to significantly increase its share of travel agent-generated business unless or until travel agents feel fairly compensated for their efforts. No one of sound mind, not even a travel agent, can be expected to work for little or no compensation, or be forced to go to great lengths and expense in order to receive commissions that are rightfully due them'.

In 1990 only 12.6% of all U.S. hotel room revenues were generated by reservations made through travel agents according to 1989 *us lodging industry report* provided by Laventhol and Horwath. The remaining reservations were made as follows:

▶ Tour operators—5.6%
▶ Hotel representatives—8.9%
▶ Direct by consumer—45.9%
▶ Hotels chain systems—22.1%
▶ Independent systems—3.8%
▶ Airline systems—1.1%

Mr. Abels feels that if the hotel industry wants to raise its travel agent-generated sales figures it must initiate a fair commission payment policy and recognize the need to provide travel agents with the best available rates to offer to their clients. Travel agents are being placed in the position of offering its clients higher hotel rated that the client would receive by dealing directly with the hotel, in some cases, according to the editorial.

Mr. Abels goes on to say: "While acknowledging the impossibility of developing a policy or program that would make everyone happy, I believe the industry is at the point today at which a 'Code of Hotel Commission Fair Play' could be adopted, initiated and observed. Such a code should contain a list of the rights and the obligations of both parties, travel agents as well as hotels. Implementation of the code would, of course, be purely voluntary."

In order to implement such a Code, for the benefit of consumers, travel agents and hotel alike, Travel Trade will sponsor an open Travel Industry Agent/Hotel Forum, to be conducted in either New York City or Washington, D.C. Travel agent association executives, as well as individual travel agents who wish to attend, will have the opportunity to discuss the proposed Code guidelines. The whole purpose is to provide with a forum at which an equitable code of fair play can be developed. This would give both groups an opportunity to greatly improve their intra-industry relationships, significantly raise their respective sales volumes and substantially increase their individual profits, while at the same time providing the traveling public with better service and greater value for their travel dollars.

ARE PHONY TRAVEL AGENTS GETTING YOUR FREEBIES?

February 1992

For years we have all been providing special privileges to our partners in travel the travel agent, We've given them comp rooms, reduced rate rooms (most at 50%discount), comp upgrades, free meals, free recreation, fruit baskets, wine and a lot of other little goodies. It has only been lodging operations doing this but other suppliers to the travel industry-the cruise ships, airlines and auto rental companies. In the past we have suspected that from time to time an agency would print up some business cards for friends and relatives, This would get the "non-agents" an airline upgrade. A discount on rooms and maybe a free car rental. We were suspicious, but it wasn't a big deal and we certainly didn't want to embarrass a real agent, if we were wrong so we went along with the deal.

Now it appears that agencies themselves are "killing the goose that laid the golden egg." would you believe that there are what appears to be legitimate retail travel agencies out there that are selling agency identification cards to people "off the street" for some amounts up to $500? The gimmick is to tell the consumer that by having an agent Id card they can get many times more than their investment in "freebies from travel suppliers.

In an article that ran in the ***tour and travel news*** (Dec 9, 1991) it was reported that some retires, professional students and business people have two things in common: a desire to travel more cheaply and fraudulent travel agency identification cards. It looks like companies selling these IDs are popping up all over the country. One somewhat legitimate way to manipulate the system is for a travel agency to charge people to "learn" the business-the sign up those that pay the fee as "outside sales agents." There is one such agency in California that claims to have 800 outside sales reps, according to an official of ASTA (American Society of Travel Agents).

ASTA is very much concerned and wants to protect the professional benefits that suppliers afford the travel agents who truly distribute the products and to protect the image of professional travel agents. According to Paul Ruden, ASTA senior vice-president, if the benefits offered to legitimate travel agents spread widely among general travelers it will raise the cost of the incentives markedly and cause suppliers to consider dropping the benefits.

It is going to be tough now for hoteliers now in trying to separate the real agents from the phonies. We have to be careful not to turn off the legitimate card is issued by IATA (International Airlines Transport Association). All agencies, in order to issue airline tickets, must belong to this organization. Agencies send in, annually, an "IATA List" which reflect all their fulltime agents. Outside sales representatives may be included in the "International listing, but may not be included in the "Domestic" listing. The card that is issued to Individual agents represents that the holder is part of the International Airline Travel Agent Network. Now, you should understand that the phony agents may very well get this kind of a card since they are being signed up by legitimate agencies as outside sales reps. It may not be a bad idea when getting a request for special considerations from agents to ask them to bring their IATA Id card along with a copy of the agency's domestic IATA listing, as well as a business card. This may make the phonies a little nervous, but a legitimate agent will understand the request.

You may also want to consider providing special considerations to those agents that are members of ASTA or ARTA (Association of Retail Travel Agents) and publicize special travel agent "deals" through these organizations. If you are offering special discounted rates to agents then you may want to contact those organizations and ask the best way to get the information to their member agents. Of course, ARTA and ASTA both have their own distinctive membership cards.

NEW TRAVEL AGENT DIRECTORY LISTS CHAINS THAT PAY COMMISSIONS

September 1992

Although the travel agent commission situation is getting better, there are still problems out there with properties not making the payments or not paying the agreed amount. Joel Abels, publisher of **Travel Trade** (a travel agent publication with 40,000 plus ABC audited travel agent readers) has created a "Code of Hotel commissions Fair play." Hotel chains and management companies are asked to endorse "The Code" which ensures that the properties represented will guarantee commissions to travel agents. All hotels that do this will be listed in the new directory.

"The Code" stipulates:

1. Hotels should pay travel agent commissions on all bookings made by travel agents unless the hotel advised the travel agent that the rate is not commissionable.
2. Travel agents should receive full and prompt payment of all commissions they have earned.

3. Travel agents should be paid a commission based on the rate the client actually paid for the dates the client actually stayed. This amount could be higher, or lower, than the original booking.
4. Hotels should disclose whether group business that is directed to the hotel by a travel agent will be protected.
5. Travel agents believe that hotel rooms promoted to the general public, and held in the hotel chain's central reservation system, should be available to travel agent-generated clients.
6. In the event a client is a no-show, the hotel should notify the booking travel agent.
7. Hotels should include sufficient information with their commission checks to allow travel agents to reconcile their records.
8. Hotels should respond promptly to all agent inquiries about commission payments.
9. Travel agents should contact hotels within 30 days of receipt of a statement if the commission has not been paid as expected.
10. Travel agents and their clients, as well as direct-booking clients, should receive fair-and-equitable treatment when making reservations with hotel, chains that have already endorsed the code include: Best Western, Days Inns, Embassy Suites, Hampton Inn, Homewood Suites, Howard Johnson, Hyatt Hotels, Intercontinental, IIT Sheraton, LaQuinta, Ramada, Utell international, Westin.

HOTEL CHAINS REALIZE THE VALUE OF TRAVEL AGENT BUSINESS

It seems that more and more hotels, motels, resorts of all types, chain affiliated as well as independents are getting on the "bandwagon" of paying travel agent commissions and paying them more promptly. There are over 4,500 individual hotel properties that indicate in their Hotel & Travel index listing that they pay their commissions to agents automatically. Hotel & Travel index is the largest as well as most preferred directory used by all travel agencies in North America. For the past two year properties listed in the directory have been able to code their listings by using "CASH" in the listing which stands for: "Commission Automatically Settle by Hotels." This helps properties effectively communicate their commission payment policies to agents using the directory. For the most part, this actually reflects those hotels and motels that are affiliated with a franchise chain operation, which makes payments automatically through the chain headquarters for all affiliated properties. Hotel & Travel index plans to include, in future issues, individual hotels which pay commissions automatically. Some chains believe so strongly in the value of travel agent business for their franchisee properties that they actually have a program of enforcement of the payment. Recently, the directory has included a section listing the commission policy and other essential sales information for major hotel chains, which is called: "Hotel programs & Policies At-A-Glance." If there is a "CG" (commission Guaranteed) code listed it indicates that the chain will enforce, if need be, the chain's commission policy with regard to any member property. Currently, there are more than 2,500 hotels listings that incorporate the "CG" code.

The whole idea behind the program is to allow travel agents, when using the directory to quickly see and understand the commission policy of any hotel company. This could very well put any one property at an advantage over another by having the "CG" code listing. According to Melinda Bush, publisher of Hotel & Travel index: "Agents need to know who they can depend on for commission payment, and hotels deserve credit and recognition for their diligent efforts in this area."

Operators of hotels, motels and conference centers should really take a good look at their commission policies. The retail travel agent business is expanding very rapidly into other areas such as group leisure tours, meetings & conferences as well as incentive travel. More than half of all travel agencies indicate that they are involved in arranging sites for meetings and other group travel. The only income for most travel agencies today is from commissions from airline tickets, car rentals, cruise sales as well as hotel stays. When travel agents discover that a property will not pay a commission, they will not use it. Retail travel agencies have been long known as an extension of the sales department of any hotel. The numbers of agents and agencies keep growing which reflects a tremendous sales force available to properties. There are now an estimated 35,000 agencies employing approximately 250,000 individual agents and they are all out meeting the public offering to take care of travel arrangements. Hotel operators should start looking towards the travel agent community as partners in fulfilling traveler's needs.

IMPROVE SALES EFFORT WITH NEW TECHNOLOGY TO ATTRACT TRAVEL AGENTS

April 1996

If it's the travel agency market you are after as an important segment of your room business, for individuals and/or groups, then look to something new in technology that could be very helpful to you. Just out brand new and hot! it is called the "Deals Digest."

ALL too often we get questions from a variety of types of properties about how to go after the travel business. More and more consumers, as well as corporations, want in the market to this segment. But are at a loss as to how to approach it. After all, there are about 35 retail agencies in the United States. Do we make sales calls, send direct mail pieces, send brochures to all of them? Of course not. Too costly, too time-consuming and it becomes just a "buckshot" operation with little results. Through the latest in technology, hotels, motels, conference centers, cruise lines, and anyone else interested in this market can now put their promotions and packages in agency hands at the touch of a button with the help of Hospitality resources worldwide.

This new program just doesn't target the travel agency market with mass mailings. It provides a series of direct marketing and fax publications based on targeted databases aimed at specific specialized markets within the agency business that would suit your needs whether it be the luxury, incentive, group meetings business or any other specialized market. The "Deals Digest" will produce ten monthly broadcasts, plus four quarterly special issues aimed at the group travel, meeting and incentive markets. What this does is allow hotels to transmit details of special promotions to travel agencies within minutes. The information is loaded into a private fax wire. Typically, these broadcasts will feature approximately 10 to 15 offerings, plus a handful of special bulletins addressing meetings and luxury market hotels and resorts.

Here's how it could work for you. First, you get an idea for a promotion to fill space that has just come available as a result of a canceled block of room. Or you are not achieving expected occupancy during a specific time period, or more competition has just opened in your area, or you are running late in getting information out for a special holiday promotion. You want to move fast to make contact with the agency market so they can generate room business or meetings business for you. Through Hospitality Resources worldwide and their Hospitality Resource center, they can include your promotional information in their "Deals digest," getting your idea into an agent's hands and be in a selling mode within a few days. There is also plans to get your information on the internet which means your promotion could very well be seen by tens of thousands of consumers.

Special issues of "Deals Digest" will address the meetings and group segment through "Meetings Digest" and the luxury market through "Luxury Digest." Plans are also underway for corporate and international editions of the product. According to the company, recent independent research shows that fax communications have become the second most valuable source of information for agents, exceeded only by personal familiarization visits to hotels and resorts.

It is important to understand that your promotional information just doesn't go to any group of travel agencies. Agents receiving any of the editions have been carefully screened on the basis of the type of business they produce. For example, "Deals Digest" will be targeted to the 20.000 travel agencies actively producing hotel room business. While the Luxury edition will be aimed at some 6.500, "carriage trade" agencies and the meetings product will be broadcast to 5,500 specialists in the group, meetings and incentive market.

PHONY TRAVEL AGENTS GET THEIR DUE

August 1997

Just about a year ago, in this column, (June 17, 1996 issue "ARE PSEUDO TRAVEL AGENTS GETTING YOUR FREEBIES?"). We addressed the issue of phony travel agents getting comp rooms or reduced rates or upgrades as well as some free meals and all kinds of "deals" from hotels all over the country. It has been a practice in our business of providing freebies to people who can bring us business. We've been doing it for years. Not only for travel agents, but for meeting planners, tour operators, corporate travel managers. Just about anyone who is in a position to book rooms with us. Not a bad idea at all. And why not? We can look at it from the standpoint of "sampling" a product as well as building goodwill. Most professional trade associations like Meeting planners international, American Society of Travel Agents, Association of Retail Travel Agents, American Society of Association Executives and a host of others, however, frown upon their members taking advantage of this courtesy, unless they are in a position to use the facility.

For some time, now, as indicated in the June column, some companies have been advertising to the public, that for a fee (in most cases $495) anyone can be trained to be a travel agent and get the benefits of free and discounted travel. And it worked; of course, most of the people that bought into the program did it only for the benefits and had no idea at all of actually selling any travel. Hundreds, probably thousands, got their "training" kits, but most importantly, received their business card indicating that they were travel agents. All it takes, as many of our hotels, is for someone to show any kind of a card indicating that the holder is an agent and we start bowing down, offering all kinds of freebies.

To distinguish the real travel agents from the phonies, airlines, for a long time, have been accepting only the IATAN (International Airline Travel Agent Network) identification card before any privileges were offered. As a result of our previous column, alerting our HMM readers to the situation, hotels started doing the same thing. Now, because of action on the part of Travel Trade (a major weekly travel industry publication), IATAN, and ASTA, the FTC (Federal Trade Commission) has started to crack down on these card mills with five lawsuits (and more to come) designed to shut them down. FTC has won temporary court orders halting operations and freezing the assets of the companies, while they continue with the cases seeking permanent "shut-downs" According to a report in Travel Trade magazine, in one case against World Class Network Inc., Irvine, CA, consumers were offered a kit that purportedly would allow them to receive travel agent-like discounts and upgrades. FTC charges that purchasers could not receive the discounts and upgrades for personal travel "because many in the travel industry do not recognize the ID card" (Hurrah for our readers) and the kits are inadequate to allow them to open and operate functioning businesses

This is a start. It doesn't mean that we are completely free of companies offering kits and phony ID cards to people who want to "travel like a travel agent." There are probably thousands of people around the country who have bought into these types of deals and are fooling hotel operators into giving them upgrades, discounts, and comps. Again, we are suggesting that lodging operators who are willing to offer courtesies to legitimate travel agents ask to see the approved IATAN agent card. These cards, by the way, have a photo of the bearer and are plastic, not paper constructed.

ISN'T IT TIME TO START COURTING TRAVEL AGENTS AGAIN?

October 1997

Room occupancy and rates have been so good lately that perhaps we have neglected to go after some specific markets that could be very important to us. Of course, we are probably not turning down any travel agent referred business unless we just don't have space available, but what's being done to try to generate room business from this market when we could use this type of business. Travel agents are still the major distributor of travel business around the world. With over 235,000 agents in about 35,000 agencies in the U.S., they sure do have an impact on travel. So why don't we see much activity on the part of hotels to go after this business? Apparently, it is not a question of paying commissions to agencies for referred business. Most properties are happily doing so, especially through the major brands. And, the major brands are still very active in soliciting this business for their franchisees and also making sure that the commissions are paid promptly.

Travel agents are involved as well these days in booking rooms for major corporate accounts as well as the very small companies; over half of all travel agents make arrangements for group travel: both leisure as well as business. More and more travel agencies are getting involved in meetings management for companies as well as associations. And, of course, the bulk of business for agencies is in the leisure travel market.

Travel agent business should be an important segment of your overall room business and there is a place for it in your plan for marketing your property. Do you know what percentage of your overall room revenue is generated from this business? I would bet that most hotels do not know this without doing some research on the amount of commissions paid during any period.

Take a look at the figures (as you should be doing with all your market segments). Also, take a look at the average rate paid by guests that are referred to you by agents. You'll probably find this puts this business pretty high on the list of segments that are giving you the best rates. If this is the case doesn't that tell you that you would go after more of this market? I have not been about to locate any figures reflecting the amount of dollars in room business that has been generated via travel agents. But it has to be in the billions. Sometime back I recall that it represented 16% of total room revenue in the U.S. At that time, while working with service, we measured the room revenue from this source for our 60 hotels and it represented 9% of our total room revenue. We then put on a campaign and within three years we were generating 17% of our total room sales from travel agents. Not bad! So it can be done.

Of major importance in going after this business is to make sure that travel agent commissions are paid promptly. Commission checks should be issued the same day a travel agent referred guest leaves the property; except, of course, if the commissions are paid through a system arranged through via a brand. Properties need to maintain a database of agencies referring business so that mailings can be made to agencies thanking them for the business and asking for more. Hotels should take an active interest in organizations such as the American Society of Travel Agents and Association of Retail Travel Agents. Property participation in trade shows such as Travel Trade Productions and others could be important for building awareness and networking.

The idea here is to come up with an action plan to develop this market to your best advantage. First, determine a goal: how much growth do you want from this market in 1998? You can come up with a percentage increases or even a dollar amount. Then list all your ideas of how to go after the travel agent business, who will do what and even include a budget so that you can measure return on investment.

TRAVEL AGENTS ARE THE BEST BET FOR SELLING TO THE AFFLUENT MARKET

December 1997

We hardly ever see the affluent traveler specified as a market segment in sales and marketing plans. And, we just don't see any strategies laid out in going after this high-end travel segment. Of course, this particular segment of the travel business is not for every lodging property. People who are identified in this category are willing to pay high rates, but demand high-end service. For many years hotels and resorts catering to this market found that it was difficult to identify where this segment existed and how these travelers could be reached. A good many properties maintain its own database of guests and marketed directly to them. Of course, this did generate repeat business, but it was most difficulty locating new guests. We know we can't survive too long counting on existing customers to keep coming back, so we must always seek out new business. The idea is to come up with strategies to be able to secure bookings of the affluent market, and our best bet would be to go after the travel agencies which cater to these travelers.

Certainly, we want to be able to identify the travel habits of these people and then figure out how to be able to get to them for the business. Well, Stanley Plog has made our job easier through his research which has been conducted by his company, Plog Research Inc. This past September the study was completed, entitled "characteristics of the Affluent Travel Market." Some very interesting data was developed which included information on the comparison of the spending habits of the average U.S. travelers with the affluent travelers; travel agencies' level of influence and assistance in selecting hotels and resorts; the top domestic and international destinations sold by agencies; and, defining the affluent travel market.

Without question, a greater portion of affluent travelers relies on the services of a travel professional for their trips and hotel selections than typical travelers. Eight out of ten of these travelers consult a travel agent. Affluent travelers spend many more nights away from home and significantly more money on trips that the average U.S. traveler; on an average, these people spend 37 nights away from home over the course of a year compared to 21 nights for the typical traveler, for business as well as for pleasure. They also spend almost two and a half time more on leisure travel than typical travelers do.

What we are referring to here is the upscale travel market where approximately 9,000 travel agencies specialize in working with these affluent travelers with their travel arrangements. This represents about 25%

of all U.S. travel agencies (almost 34,000). The research reflected that the affluent travelers studied followed the recommendation of travel professionals 99% of the time in the selection of hotels; 82% of the time in destination selections and 80% of the time in package tour selections.

It sure looks like travel agencies are getting pretty busy working more with hotels and resorts since they have been cut back on commissions by the airlines. The recent second cap put on commission payments by month all airlines have made travel agencies think more about the value of booking hotels, selling package tours and making cruise line bookings for their clients.

THE TRAVEL AGENCY BUSINESS IS STILL ALIVE AND WELL

June 2002

Granted, these are difficult times for the entire travel industry, particularly the travel agencies, As travel agents now know, to its regret, the airline industry decided some six years ago that, with the advent of new technology, the carriers no longer felt dependent on travel agents to sell their seats. As a result, they cut agent commissions to the bone. The airlines now evidently believe that, with the introduction and growth of E-ticketing and direct airline Internet booking sites, their ticket sales expense will be cut further, while the agent share of total air ticket sales will continue to be reduced.

There is a very distinct message here for all of us in the hotel, resort, and conference center business. Agents need more outlets paying commissions for the services they provide their corporate and leisure accounts. We are still paying 10% commission to agents booking rooms at our properties, aren't we? Of course, there is nothing new about agents booking hotel rooms; it is still going on. However, some years ago, agents became very disgruntled with hotel facilities that slackened off on paying commissions or not paying commissions promptly. Therefore, many of 32,000 agencies around the country did not even bother booking hotels for that reason, and relied solely on airline, tours and cruise commissions for their operating profit. Right now, because of many of the smaller agencies getting out of the business, there are only 26,000 agencies around, but there are over 250,000 agents working at agencies or out of their home, all are looking for any type of business they can book and collect commissions. In all my travels when working with General Managers and sales folks, during my workshops, I hear almost nothing about anyone going after travel agent business. It seems to me that this could be a very lucrative market segment for just about any type of property. Have we just given up trying to sell to agents? Sure, we are still getting business from them, and for the most part, commissions are being paid promptly, but it looks like, if it comes in OK, if not, that's OK too. Is anyone out there working this market segment?

After all, there are many corporations that are now using travel agencies for their travelers. And, there is still a good percentage of leisure travelers who are using agencies as well.

One way to start targeting this business is to look for cities from which your travel agency referred to business originates. For example, if your property is in Charlotte, NC, and you notice that you are getting some business from agencies in Atlanta, then this could be a good indication that other agencies in Atlanta are also booking hotels in Charlotte, but maybe not yours. Perhaps a direct mail campaign to agencies in Atlanta, or a sales trip to call on agents would help generate more business from that area. Although some brands now pay commissions directly to travel agencies, a property should still send a card or note to the agency thanking them for the business. For those properties paying their own commissions, it is imperative to make sure that the commission checks go out the same day the guests depart; of course, along with a "thank you" note.

LET'S GET BACK TO MAKING THE TRAVEL AGENT YOUR ALLY IN BOOSTING SALES

We do not hear any more talk about it, or see much more of the sales action being involved with getting travel agents to book business with our hotels. The travel agent, once a useful figure to the hotelier, could even be more useful today than ever before, Business is off and we need all the help we can get.

The travel agent today is still very much in business with the handling of a wide variety of arrangements for the business traveler as well as the leisure traveler. There are probably over a half a million companies, along with a large number of federal agencies that have travel agents handle their needs—everything from finding their most reasonable and appropriate air travel and lodging arrangements to booking conferences and meetings. Agents, of course, are into heavily booking cruises, for leisure travelers, but also for companies, and a variety of social groups. Group incentive travel is another important area for today's agent. The complicated arrangements required fit right in with what an agent does best: handling the challenge of countless schedules, rate structures, and conditions.

The travel agent's income depends solely from commissions earned as a result of bookings on behalf of a client by handling its travel needs. Of course, we know that the airlines cut back on paying commissions on travel some years ago, but for groups, they are still offering commissions. Agents also represent hundreds of wholesalers and tour operators by booking their clients on special land packages all over the world. There are some local hoteliers that have been very successful in going after this kind of business. Lucille Yokel, Director of Sales and Marketing at The Wellington Hotel in New York City has recently been back from a very successful European trip visiting with agents, and is now preparing for trips to visit with travel agents and operators in China and Japan. A recent familiarization trip to her hotel by Chinese representatives proved to be very worthwhile, Hannah Zuleta, Director of Sales and Marketing for Tapestry Resorts, out of California, has been working with travel agents to help with the booking of their tour packages via tour operators and wholesalers for Tapestry's 17 properties they operate.

Popular approaches of going after the travel agent market include: Selling directly by phone, personal visits or phone calls; internet marketing via email flyers and messages; packaging a property with other travel components and marketing it via a wholesaler or tour operator; Connecting with wholesalers who are already selling packages within your area; Participating in trade shows geared to travel agents; participate in or create social travel networking groups. However, no matter how you approach this subject of going after travel agents, it certainly would be worthwhile for any property to consider spending a portion of the sales effort time in this market area.

Chapter 7

Sales Planning and Sales Calls

START THE NEW YEAR WITH A FRESH SALES PROGRAM

December 1981

The list is certainly not complete, but the following factors are a good "starter set" for small property owners, managers, and salespeople looking to revamp their sales programs. They offer in most instances a fresh new approach that could translate into higher profits for your organization.

- ▶ **Try new things:** Ask your 1staff for new ideas, even if it has been tried before; success could come the second time around. All personnel should contribute ideas from maids to maintenance men, bellmen, busboys, and desk personnel.
- ▶ **Business contacts**: Belong to local groups and be active in the community; it pays big dividends. The good business contact is your future sales. Some of your employees already belong to clubs. Ask for their support for business contacts.
- ▶ **Market knowledge**: What business do you need and when do you need it? What type are you trying to get and how do you get it? What market mix do you want? What's the best way to get it?
- ▶ **Analysis of your guests**: Can you identify your guests and their points of origin? Can you see a pattern of the types of guests you get (tourists, commercial, etc.)? Can you get more of the same?
- ▶ **What is your competitive edge**: How do you match up with the competition on · service, location, price and promotion of products? Are you selling your positive points? What do you have that is better?
- ▶ **Cost analysis:** How much waste is there? How much downtime on productivity? Is there wasteful promotional expense? Try to improve employee productivity, particularly from salespeople.
- ▶ **Selling techniques:** Is there an aggressive selling program: front desk, dining room, lounge, sales department? Is there a lead program with a follow-up strategy?
- ▶ **Quality of people employed:** How are they selected, trained and developed?

Is the end result in high quality? Sometimes it doesn't cost more to get better-qualified people. The selective process just has to be better.

▶ **Efficiency:** Are all departments and staff technically efficient? Always measure the effectiveness of people as well as the strength of your programs.

▶ **Good delegation:** All the way down the line, no one person can do it all. A delegation of duties must be done-and done well-to ensure proper performance.

▶ **Work smarter**: Some people work just to be busy. Plan better: organize, direct, control, motivate, communicate, train and develop.

▶ **Make things happen**: Let's get it done now. Don't spend too much time evaluating and discussing. Procrastination is the thief of time.

▶ **Measure results**: Know what pays off and what doesn't. If something doesn't work, stop and start something else.

▶ **Finally have pride**: Be proud of where you work, so that where you work is proud of you.

INCENTIVES WILL HELP BOOST YOUR WEEKEND OCCUPANCY

October 1982

The approaching fall and winter periods will find many lodging operators looking for additional sources of room business to help boost occupancy during what could very easily become a slow period. Hotels, motels, and resorts that are not already involved in the development of weekend business should certainly start now. Besides selling the weekend package directly to consumers and meeting planners, the incentive market is another avenue that could easily be explored. Here is an opportunity for you to develop a simple, uncomplicated weekend package to sell to companies for sales awards as well as dealer and customer incentives. This gives the company to which you are selling the package a low-cost motivational tool that can be applied almost immediately for reward purposes.

An incentive of this type appeals to many people (employees or customers) who would otherwise be reluctant to take money out of their own pocket for a weekend "escape" locally. In many cases, it is more memorable than an award of an appliance or housewares or clothing.

The package can be created to include a variety of saleable products such as king-sized beds, suites, cocktails, meals, sightseeing, as well as shopping and local attractions, etc. All components are priced appropriately and included in the one package price for two persons.

Logical prospects for weekend incentives include:

▶ **Radio stations:** Looking for high audience rating to use as contest award.
▶ **Automotive parts manufacturers:** Use for sales awards.
▶ **Retail Stores:** Looking for more customers. Use along with high purchases or sweepstakes.
▶ **Local insurance company:** Use for immediate incentives for the weekly

or monthly attainment of goals for salespeople.
- **Car dealerships:** For employee sales incentives and to customers who make purchases during special sales drives
- **Dairy companies:** To route men for sales or signing up new accounts.
- **Manufacturing plants:** Used for attendance and safety awards as well as production goals incentive.
- **Banks, savings, and loans:** Used to generate new accounts.
- **Condominium sales organizations:** Used for employee sales incentives as well as added gift for purchasers.

Lodging operators of small hotels, motels and resorts can be creative in coming up with ideas on how to market the week end package as an incentive. Of course, the same package may be marketed' directly to the consumer through local advertising or distribution of flyers.

A SUCCESSFUL SALESPERSON ALWAYS MAKES THE RIGHT "CALL"

July 1986

A salesperson's job is, of course, to bring in sales. A good salesperson creates a sales situation-a sales call-to develop prospects and encourage them to buy something. There are many different kinds of sales calls, and hotel salespeople need to distinguish one kind from another. "Cold calls," P.R. calls, appointment calls, presentation calls, telemarketing calls and "inside" sales calls are among the most effective methods used by the truly creative salesperson. Let's take a closer look at them:

- **Cold Calls:** This technique is best characterized as an exploratory call. The objective is to maximize the salesperson's time and develop leads for various types of business. Generally, these calls are made in a small geographical area within a short period of time.

  Again, the idea here is not so much to sell, but to gather information so that a follow-through strategy can be developed to get business as a result of the contact. It sometimes happens that a salesperson will run into a "hot prospect" who is interested in doing business on the spot. At that point, the good salesperson grabs the opportunity to turn the cold call into a sales call.
- **P.R. Calls:** The public relations call is made to a company that may already be doing business with the hotel. The salesperson drops by to say "hello," to ask if everything is going well, or even to bring a gift during the holiday season.

  The problem with a P.R. call is that it's generally made by a non-aggressive-type salesperson. This is a "comfort zone "call—the salesperson feels comfortable visiting these people because they are already, in most cases, doing business with the hotel. There's no real pressure to make a pitch for more business.

  But remember: There's always an opportunity to sell more, even to a

company that's already providing business to a property.
- **Appointment Calls:** Obviously, this is a sales call for which an appointment is made. Planning is of the utmost importance in making such calls. The salesperson should know as much as possible about the person he is scheduled to meet, and about the company's needs, so that business can be booked or information gathered for a follow-up call.
- **Presentation Calls:** These are usually made after previous calls have resulted in acquiring and developing information. An audio-visual presentation should be prepared (handouts are also a good idea) so that the salesperson can introduce as much information as possible to the person or committee being courted.

  This is strictly a sales situation in which the salesperson is putting the hotel's "best foot forward," with the objective of having the committee, or individual, make an immediate decision. During a presentation call, it's important to know who the decision makers are as well as what their "hot buttons" might be. The salesperson must have the confidence to make a strong sales pitch, answer questions and, of course, ask for the order.
- **Telemarketing Calls:** There is a place for soliciting business by telephone however, it should not take the place of other kinds of sales calls. Telemarketing is sometimes used by hotels/motels to gather information and qualify leads. Then, salespeople may follow-up with a presentation to secure business.

  Telephone calls should be used only for prospecting-not for a presentation-since the best sales pitch is always the face-to face visit.
- **Inside Sales Calls:** Many lodging properties have what they call "inside" salespeople. As you might guess, these are people, for the most part, don't go out on sales calls. They feel comfortable staying in their office waiting for people to come by or call in for information.

  Inside sales calls also include those calls from people who have been invited by for breakfast, lunch, dinner or cocktails so that an on-site sales pitch can be made. Remember: No matter what kind of sales call is used, the ultimate goal is to close a piece of business.

TIME AGAIN FOR *SALES CLINIC'S* NEW YEAR'S RESOLUTIONS

January 1987

Once again, it's time for our New Year's resolutions. If hotel/motel salespeople resolve to do the following, 1987 could be a very good year indeed:

- **Sell price last:** It's really the least important selling point of all.
- **Cut down on giveaways:** Comp rooms and various other "freebies" are not necessary ingredients in closing a deal.
- **Close as soon as possible:** The prospect you leave behind may be closed by the competition.
- **Increase your sales calls:** If you make more calls, you meet more prospects; if you meet more prospects, you do more

selling; and if you do more selling, you close more sales. (The law of averages works.)

▶ **Have pride in your product:** Be proud of where you work; then, where you work will be proud of you.

▶ **Be Persistent:** Most sales are closed after the fourth or fifth call.

▶ **Be loyal:** One of the greatest virtues of selling is loyalty to your hotel, your manager and your owner.

▶ **Ask for the order:** Too many salespeople are on P.R. calls; they forget to ask for the business.

▶ **Communicate more:** Let the staff (especially the switchboard and front office) know where you are and where you're going to be.

▶ **Have confidence:** The seller's confidence in the product results in the potential buyer's confidence-and that, more often than not, results in a sale.

▶ **Be nice:** Training people to be nice is tough, so hire only nice people.

▶ **Don't be too aggressive:** Selling isn't a fistfight. Find a happy medium between high- and low-pressure sales techniques.

▶ **Keep in touch:** Don't lose contact with a prospect. Phone calls, notes, even birthday cards, are good ways to remind people that you're still interested in them.

▶ **Don't criticize:** Sell your features against the competition's, but don't stoop to criticism.

▶ **Be a self-starter:** Don't wait for your manager or your home office to "wind you up" every day.

▶ **Be honest:** Don't sell what you don't have, and be honest about your hotel's facilities.

▶ **Keep your word:** Whatever you promise, you must deliver.

▶ **Talk quality:** The quality is what people want these days. Whether it's the quality of your hotel's services, its food or its staff—quality sells!

▶ **Write it down:** Reconfirm, in writing, everything upon which you and your prospect have agreed.

▶ **Be enthusiastic:** Nothing is as contagious as enthusiasm, whether it's with your staff or a client.

▶ **Educate yourself:** Keep learning as much as you can about our business. There are too many salespeople out there who don't know enough about hotel operations.

▶ **Don't gripe about work:** If you're unhappy with your job or your boss, straighten it out. Or quit.

▶ **No canned pitches:** Canned pitches are out, so don't use them to sell your hotel. Know your product and be convincing in a normal, conversational manner.

▶ **Forget the word "I":** You can "I" a person to death; say "we" instead. You're part of a team, so keep 'I' out of your conversation and correspondence.

▶ **Use action calendars:** They needed to plan the work ahead, and they allow you to adjust your schedule as necessary.

▶ **Dress the part:** Don't go overboard on clothing styles. The classic look is still the best for sales.

▶ **Be a joiner:** Be active in your community; join professional groups like MPI and HSMAI.

▶ **Talk to your manager:** Every day, keep him or her advised of what's going

on; it'll improve communications and productivity.
- **Don't be too sensitive:** Some salespeople are notoriously thin-skinned. If you get your feelings hurt, bounce right back.

From all indications, 1987 is looking like a banner year for hotel/motel salespeople. It will be especially good if we remember to keep our New Year's resolutions.

MORE SALES CALLS OUTSIDE MEANS MORE BUSINESS INSIDE

February 1987

It stands to reason that the more people you meet and the more contacts you make, the more prospects you can develop. Prospects lead to presentations, which lead to property visits and, ultimately, to more closings. And, of course, more closings mean more business for the hotel or motel. It's not that difficult to understand: The more sales call you to make, the more business you book-it's the law of averages. But it is difficult to understand why so many industry salespeople won't (not don't-won't) leave their offices to make contacts, develop prospects and look for deals to make for their properties.

EXCUSES AND EXCUSES

The old excuses are still around: "I've got too much paperwork.""I've got too many other jobs to do." There'd be no one to take care of walk-in business." There are all kinds of excuses like these. Of course, there are properties where people just walk in or call in to book business—and someone does have to be around to close these deals. But usually walk-up or call-in business winds up being short-term bookings of relatively small groups.

If this is the way salespeople are using their time, then the General Manager needs to sit down with the sales staff and review their job functions with them, It's only logical that part of a salesperson's job would be to develop new group business, as well as to book long-term, future business. Perhaps there are other employees who can answer the phone and take care of walk-ins.

TEN NEW CONTACTS

If each salesperson can make 10 new contacts a day—every day—it could add up to a lot of new business for the property (even if only 10% of the contacts actually book some business). There's still a place for telemarketing, direct mail, directory advertising, media advertising, sales promotions and what ever other methods salespeople use to avoid making outside calls. But there's nothing like the basic, face-to-face visit to make friends for the property, close sales, create referrals and develop long-lasting relationships for the hotel or motel.

Outside calls don't necessarily have to be presentation calls. For the most part, in fact, they probably should be prospecting calls; that is, calls for the purpose of finding out if the company or organization is a prospect for future functions at your hotel or motel. It's not hard to do, and it's not even time consuming. Using a brief questionnaire, a salesperson can ask key questions (writing down the answers) to determine if the organization could make use of the hotel's facilities for meetings, parties, room business or banquets. The entire interview need not take more than five or 10 minutes. The selling starts (and the closing as well) as soon as it's determined that a qualified prospect has been developed.

SATISFYING RESULTS

It gets most exciting when a "cold call" results in such a prospect. Then, the selling techniques start to unfold, the prospect gets sold on the spot and another piece of business is closed. It does happen. Just ask some experienced hotel salespeople-they have some interesting "war stories" to tell. They'll also say that when this kind of sale hap pens, it n be very exciting and very motivational. However, it'll only happen if salespeople go out and make it happen. More outside sales calls are the secret to more bookings. Ten new contacts a day is not too much to ask of a salesperson. In fact, 15 or 20 calls a day would even be better!

BACKYARD BOOKINGS HELP IN BOOSTING GROUP BUSINESS

February 1987

How necessary is it to take those sales trips all over the United States or, for that matter, to foreign countries? I'm not saying that trips such as these are a waste of time especially if they're for trade shows, where sales time can be very productive. The fact is, however, that for the typical hotel or motel, 80% of all room business (corporate travelers as well as groups) probably could be generated right in its own "backyard."

Unless a property is located in a very rural area, the sales staff (or manager, if no sales staff exists) should be able to generate most of its business through local contacts. Constantly calling on local businesses-which includes asking the right questions, qualifying, following up, inviting them to see the property—will transform cold calls into leads, leads into prospects, and prospects into closed sales. Each local company that generates overnight, individual travelers is also a prospect for meetings business. Even local companies that don't generate overnight stays to the property are prospects for some type of group movement into the area.

Most business people belong to some type of organization, and usually belong to several: trade and professional associations, civic organizations, fraternal or service clubs, religious or social groups. Persuading local business people to bring their organizations to the property for meetings is an approach that often is overlooked by sales staffs.

BUSINESS ASSOCIATES

A hotel's suppliers, distributors, and other business associates are often overlooked as sources of group bookings. You'd be surprised at the amount of business a sales staff could uncover by calling on these firms. Try this: Make a list of all the companies and individuals with whom your property does business. The list will include banks, the post office, legal firms, advertising agencies, the dry cleaner, plumbers and electricians, the air-conditioning service company, contract cleaners, landscapers and pool-maintenance people, food and beverage distributors—the list goes on and on.

Then there are local institutions: hospital, clinics, libraries, nursing homes, high schools and colleges (incidentally, teachers in all departments usually belong to at least one educational association—yet more sources for group business). Again, the list is a long one.

Community Organizations

There are civic and community clubs, too. A list of such organizations, with contact names, usually can be obtained through the local chamber of commerce. And keep in mind that such organizations usually have some kind of relationship with the larger state, regional and even national associations, which can generate additional group business.

Other local sources of business? Check out the local music stores. Usually, they can provide a list of clubs made up of musicians that play particular instruments: pianist groups, for instance, or accordion clubs, or associations for trumpet players and trombonists. These clubs, like other organizations mentioned above, are often associated with larger groups that hold statewide or regional meetings.

Military Units

Local military reserve units also have meetings, especially active military recruiting departments. Veterans' groups meet statewide at least once a year, and lists of these groups are usually available through local reserve units. Just think of all the potential meeting business that can be generated by local business and community organizations; think of all the companies and groups in your vicinity that you or your sales staff have not contacted recently, or have never contacted. Then think twice about buying that airline ticket to wherever.

COLD-CALL SALES APPROACH WILL RESULT IN HOT PROSPECTS

January 1988

There has been much discussion over the years about cold calls, and it appears that more and more marketing types pooh-poohing the practice. They have many arguments against cold-call selling, and some may be valid, but it still beats most other methods of developing prospects for new business.

As the old saying goes, "Ya gotta make calls if ya wanna get results." The face-to-face method of gathering facts, establishing relationships and putting the business on the books may seem old-fashioned to some people—but it works. Some salespeople are fearful of cold calling because there may be some rejection; they prefer to work within their own "comfort zone," calling only on existing accounts and people they know. While this is important, a salesperson's real function is to develop new business. Perhaps it would be easier if we didn't think of cold calls as sales calls. Actually, the cold call is a prospecting process, the objective of which is to determine whether that particular company, association or group has any potential business for the salesperson.

The process is very simple: The salesperson asks a series of questions, the answers to which will immediately indicate whether a prospect has been developed. Then, the selling can begin. Cold calling is an efficient way to develop business. Because it only takes about 15 minutes to gather the necessary information, a salesperson can make many calls in a given area. This helps cut down on wasted travel time. And while it is most unlikely that a sale will be made on a cold call, it has happened; many a salesperson can relate interesting stories about such sales.

Some salespeople will say that cold calls are a waste of time, or not professional, or that making calls without an appointment doesn't "look good." Such arguments are not valid. It is doubtful that a hotel ever lost business because a salesman made a cold call. It would be most unusual if any one ever became angry with a salesperson because of a cold call. The worst that can happen is that someone refuses to see the salesperson, or that he's asked to make an appointment.

ASK ANYBODY

Actually, it doesn't matter whether the information is developed through questioning the boss or questioning the secretary. Anyone at the place of business who can furnish the information is the person who should be interviewed. One of the best ways to train new salespeople is to have them make cold calls every day. It gives them practice in getting out and seeing people (and when salespeople see people, more selling is done); it gives them practice in penetrating through tough situations; and it helps them develop good attitudes about selling.

With cold-call training, new salespeople will begin developing their own interviewing techniques and learn how to influence people (convincing a secretary to answer questions is very good training indeed). Meeting new people and new situations are always good practice for salespeople just starting out.

DEVELOP POSITIVE ATTITUDES

With successful salespeople, a proper mental attitude often outweighs technical skills or ability. When it come& to selling, the first step in developing that attitude is to eliminate fear and negativism toward cold calls. Successful salespeople have developed a positive attitude toward the cold call because of it is probably the most efficient means by which to develop new leads.

If cold calls are not being practiced at your property, conduct this test: Every day for a week, have your salespeople call on area businesses. This should result in 30 calls a day per person. A minimum of one third of those calls will turn out to be new business prospects. Give it a try. Then let me know how it works.

SELLING IS NOT A SPECTATOR SPORT

April 1990

When we go out to see a football or baseball sport we sit back in the stands and watch the game. In hotel sales ... you can't sit back and watch ... you gotta play the game. The problem these days is that there are too many loading sales personnel really sitting back and watching, and waiting, and hoping someone will call or come by and ask them to play the game. If we are honest with ourselves we would admit, myself included (at some points in my career), that about 80% of the business being out on the books got there as a result of someone calling or coming by and wanting to do business with us. From there on out we do a great job in showing the property and even closing the deal. The fact still remains, however, that the prospect came to us and "we did our job" and made the sale. There's nothing wrong with this as long as we put business "on the book." If we can continue to keep generating the prospects to come in the door, or solicit us in selling to them. Then great . . . as long as we know we can make goals and maintain profitable sales so that a decent bottom line may be achieved.

The big problem is that many of our properties are not achieving sales goals or planned profit . . . more competition out there is making it tougher to closed deals, even with the prospects coming in to see us. They're learning who to negotiate better; how to play one property against another; and learning more about our business. So, what does this mean we have to do? We have to start doing a better job by playing the game. Of course, part of the game is selling; provided there is someone to whom to sell. The biggest job we've got in playing the game is getting out there looking for the prospects to whom we can sell.

What are some of the common ways of prospecting? By the way, just so that we're all thinking along the same line, let's define prospecting as "looking for potential customers." One common way of getting potential customers to come to us or call us or write to us is to advertise. Our industry spends millions of dollars doing this each year. We do this for a variety of reasons, but mostly we want "leads" . . . so now is our chance to sit back and wait for responses (are we back to the spectator scenario?). Well, we also try to contact potential customers by mail . . . direct mail . . . and again, we have to sit back and wait for responses. No question about it . . . we are "prospecting," but it only works if there are enough responses to make sure we do enough sales to achieve goals and profits.

Try to get the job done, it seems to me, it boils down to "one on one" . . . remember the old one on one basketball games? We can get the one on one done by making cold calls, in person and on the phone; attending conference and trade shows is another way. In order to maximized time in playing this game of prospecting a very number of contacts needs to be made. If for example, we use the formula that for every 100 contacts, thirty may be potential buyers, and from the thirty prospects we actually close ten . . . then it becomes a numbers game. The more calls made (or contacts made) the better chance we have of closing more deals.

Maybe this formula doesn't work for everyone, but let me suggest you check it out yourself and see with what figures you come up. The time has come for all of us in lodging sales to get into this numbers game . . . the more outside prospecting done (one on one), the more leads developed, the more sales presentations made, the more closing accomplished, the more business on the books and the more bottom line profit for the company.

Try it, it works.

MANAGING SALES CALLS

March 1991

Not the ones you make, but the one you receive!

Owners, operations, managers, sales directors, and trainers always seem to be preaching about "productive sales calls." Getting out and making calls seems to be the top priority these days. This is all well and god. It is important to get out to make those calls; however, has anyone ever wondered how much business we miss because we are not paying attached to the sales potential of calls coming into the property? Leads are all around us, but the hottest lead of all, and the easiest to book is the *inquiry*. This is when someone has an expressed desire to do business with a facility, or at least inquire about doing business. Why else would the person call or write or walk into the property? In many cases, before providing a consulting service for a client or prior to developing a sales training program I would make some phone calls to some properties inquiring about individual room reservation or about a meeting. The results always appear to be the same regardless of where in the country r to what size of property the call is made. 7 out of 10 usually do not try to get dates or make an attempt to close a deal, but generally, all will take an address to mail a brochure or a letter with information about the property.

There is no question about it. More training is necessary for areas of reservation offices at the property level (reservation centers op an outstanding job of trying to get bookings) as well as in sales and catering offices. Recently phone calls were made to seven hotels to try to make arrangements for a family reunion for 50 people over a weekend. In some cases, there was confusion when the phone was answered in determining whether the call should go to sales or to catering. Everyone responding to the inquiry was pleasant, answered questions, offered information. Two immediately asked for name, address and dates of the event. All offered to send information or have someone else return the call. Only one salesperson did offer to hold dates and send a proposal.

What happens when a letter of inquiry for a meeting is received at a hotel? How long does it take to respond? At a recent sales training session, I conducted I asked what the typical response to such a mailing would be and just about all responded that the first thing to do would be to call the person sending in the inquiry. Perfect response! However, in that particular group, we mailed inquiry postcards to 16 hotels of the 80 people in the audience a month prior to the training session. Although all 16 properties responded by mail, only one went to the trouble of calling to make contact with the person requesting information.

Here is what needs to be done to be more effective in managing sales calls—that come in.

▶ Telephone inquiries: Standards have to be established at the property level to make sure that telephones are answered promptly and that calls are directed to the right department. After the pleasant greeting in the sales department, it needs to be determined who is calling, getting the name, company identification, and

purpose of the inquiry—securing dates and number, whenever possible. It may have to be that someone else will have to return the call from the sales department, in the event, there is no one around who is capable of responding to the caller. In any event, an attempt should be made to hold space, and/or rooms on a tentative basis and a letter should be sent. Of course, someone will have to check to make sure of availability.
▶ Inquiries by mail: These should always be answered by phone, first. Gather all information over the phone, offer to hold rooms and space and follow-up with a proposal or confirmation.
▶ Walk-ins: The procedures are the same except that it is much easier to show by selling and usually much easier to close the deal when selling face-to-face.

The idea is that the offer should be made to hold space, on a tentative basis. This obligates the prospect to some extent. Of course, ultimately the sale may not be made. There still has to be selling, negotiation, overcoming objections and the closing of the sale.

LET'S GET THE FACTS STRAIGHT RIGHT UP FRONT—THEN SELL

November 1991

A good deal of the business is our lodging operations comes from inquiries made by telephone. Usually, the first question asked is "what do you charge for." We probably make the mistakes, and probably lose some business by providing the answer to that question. The idea in selling over the phone to people calling in is to get as much information as possible and get them to buy into coming to the property, if possible, before discussing prices. Of course, it doesn't work all the time and sometimes people may be offended if we do not come right out and answer their questions. However, it does work a good portion of the time when you show a real interest in their having a successful event.

Many times people only provide the vaguest information because they do not realize how important it is for you to know "the whole story" before you can help them plan their meeting or event. Sometimes they don't even have all the answers themselves and appreciate the fact that you are asking them about things they need to go back to come one to ask themselves. You really provide a very important service to them by doing this.

When professional meeting planners call it is one thing. Usually, there is no problem since they will know exactly what is wanted. However, when we get phone calls from staff people, secretaries, even presidents of companies who have never done a meeting before we can be a big help by asking key questions for them to answer before we even discuss prices.

Our objective in getting questions answered before we discuss price is twofold. First, we need to know who profitable the whole piece of business is worth to our property. Secondly, we want to be to be able to "sell" the prospect on the facilities, quality, service, etc., so that they are pretty much "sold" on the property and are ready to say "yes" before we even get around to price.

By asking appropriate questions we get a chance to do both. Most people would understand that you would have to get back to them with the information because rooms and space availability may not be readily available to you. This provides a good chance to put your information together for a proper presentation without having to be caught "off guard" with a quick answer on the phone. Of course, we all understand that we do have a better shot at the business if we could get the prospect to visit the property for the presentation. Sometimes this is not possible due to distance and time consideration. So we have to settle with phone and fax machines.

Sales department personnel need to have information sheets readily available right by each phone so that the questions that need to be asked are readily available on a form that can be filled in while the conversation takes place. Sometimes, those of us that are experienced think we can remember all the questions that need to be asked and we end up using a yellow lined pad. It really is embarrassing when we have to go back sometimes, with another phone call, to ask the prospect some basic question that we forgot about asking.

Always get: name, company, address, phone number, fax number, name of other people involved in the decision making, when the meeting or event is planned, number of days (dates and times), number and types of sleeping rooms to be blocked, types and usage of public rooms (meeting, general sessions, break out rooms, meals functions, special events), are any events sponsored by others, who pays the bill, where they have met before and when, budget considerations (granted sometimes it is tough to get, but you'd be surprised how much good financial information you can get, if you ask). And, you can continue to add to this list. The idea, of course, is to get as much information as possible, then always consider getting the best rates for rooms as well as functions . . . only after you know that they are "sold" on your property. In most cases, there will be some negation necessary, but as long as you know how profitable the total package is to your property, you'll know how far you can go on rates.

THE IMPORTANCE OF KNOWING YOUR CUSTOMER

So many times I have been asked "What's the latest in hotel sales techniques?," "What are some new strategies?," "How about some new approaches." Is everyone out there selling hotel products tired of doing the basics? What are we looking for—shortcuts? Harvey Mackay has it down pat, and that's why his business has been so successful. His key to selling has been and still is—what we've been preaching for a long time—KNOW YOUR CUSTOMER. Knowledge is the key to our success in selling, and it's one of the very basic sales philosophies. Too often we try to become so sophisticated in our selling approaches that we forgot the basics. After all, you really don't have to be a rocket scientist or brain surgeon to be successful in selling.

Have some of us lost sight of humanizing selling strategies: Let's think of putting little more of ourselves into selling rather than the

PC. You do this by knowing the prospect and showing that you care. Harvey Mackay says you can know all about a prospect during one lunch without having it come off like an interrogation. You're looking for needs, wants, exercises, feelings, "hot buttons," typically, information shown in the "Mackay 66." Try to get as much information that will help you know as much about the prospect and his or her organization as he knows.

Now, just because you have all this information doesn't guarantee a sale. There are many other things involved that have a lot to do with closing a deal on a meeting or rooms contact. Certainly, there has to be a possibility of the prospect of using your property over the competition. You also need to have the type of product, location, service, reputation, and rate range that meet the certain basic need. Then, it's up to you to make the deal. There is where your knowledge comes into play. And, you may not even sell the prospect the first time around, for any number of reasons, but you must remember there's always next time.

You use the information you gathered to stay in touch, show that you care (remembering to send a birthday card or anniversary card, or sending a note relating to a hubby) and most importantly, establishing a relationship. We've said it so often in this column, many hospitality salespeople will tell you that they credit their success in selling with having created a good, healthy, close working relationship with their customers. In fact, the well-known industry speaker, Jim Cathcart, also tells us about this in his book and audio tape series entitled "Relationship Selling."

If you haven't been inundated with comments and quotes on this philosophy yet, you will be. Although there's nothing really new about and it is as basic as you can get, you'll be reading and hearing all about it, all around you. To sum up this business of relationship, here's a Harvey Mackay quote: "the customer relationship is like a marriage—small shows of sensitivity and awareness maintain spice."

A REMEDY FOR HARD TIMES? HOW ABOUT LOOKING AT THE BASICS AGAIN?

February 1992

It's tough to sell even in good times, but in recessionary times, like right now, it is even tougher. But whoever said that sales work was supposed to be easy at any time? In the 1980s it was relatively easier to close the sales, at a good rate, than it is today. But this is the case typically of almost every industry, not just in our hospitality segment. There are several different approaches these days, with different philosophies about sales due to the recession, overbuild situation, more competition, several discounting, better-educated buyers or whatever the reasons that surface for lower occupancy at lower rates and unprofitable lodging operations.

There have to be some silver linings somewhere out there, but we have to go after them differently than the way we did it in the '80s. We all have to sell a lot more just to stay even . . . and everyone seems to work harder at it. The rate cutting that is going on is a real shame . . . and too many people see this as a way of doing more business . . . and in some cases, it does seem to work, but it can't for long with the number of lodging operations losing money and even going into receivership. The successful operations probably have been the ones that advocated taking positive actions in these troubled economic times rather than assuming a wait—and—see attitude. Most successful marketing and salespeople are those that are willing to innovate, even to take some risks, but are getting out doing things to create more business.

Here are some basic thoughts to help beef up the selling situation for lodging facilities:

▶ Make sure that all employees are taught to think SALES. We give a lot of lip service to this, but are we really doing this? How about getting staff people to generate leads for the sales department? Just about everyone that works at a property knows a neighbor, friend, relative, business connection that may be in a position to send people who are visiting the area to stay at your property. Think of employees' relatives who work elsewhere for organizations that have meetings that should be solicited by your sales department. Think of the local clubs to which some of your employees belong (or their spouse belongs) ... could they get a meeting to the property?

Don't stop advertising and promoting. In previous recessionary times, the onset hat pulled in their advertising horns were the properties that had the most difficulties. Advertising dollars may not bring in the same return now that they did in better times. But, when the curve swings up you'll be a great deal better off than those that did not advertise.

Do the best you can to stay away from a price war with your competitors. Anyone can lower prices, but sooner or later there will always be someone that will still manage to come in with a lower price. Think of how you are one cut above the competition and your property is worth the extra cost for someone to stay or for a group to have a meeting. Be sure the service provided at the property makes you stand out above the competition. Take another look at your features and relate them property to benefits for the prospective buyer of your property.

All department heads should get involved in local organizations and be active, accept leadership roles and become Mr. or Miss Holiday Inn, Sheraton, Hilton, Quality or whatever in your own community. This is the easiest, fastest and least costly way of bringing business to the facility.

We've discussed the very basics here ... but these are the things that do work ... along with all the other basic sales activities like making sales calls, following up on leads, etc. It really doesn't take a PhD or genius to get sales into a hotel. Just doing the basics better than the next guy, working harder, smarter and probably longer hours than the competition will always work.

THINK ABOUT WAYS TO MAXIMIZE SALES

March 1992

One of the biggest concerns we have these days, of doing business in a tough economy, with heavy competitive discounting, eroding bottom line profits and cutbacks in corporate spending, is in the area of putting more business on the books. We all try so hard, particularly those who are more concerned with the more "sophisticated" way of doing business, that we sometimes forget about the "old fashioned," simple ways of selling hotel rooms, food & beverage, and space. Perhaps it is time to think about how it was done in "olden times" and use some of the techniques we may have forgotten . . . or never learned.

Back then, when there were tough times for hotels, and there were, but we did manage to survive . . . when we needed more room business for a property, we sat down and thought about ways of being creative to uncover new uses and new customers for the property. The General Manager, food & beverage director and sales department sat around for about an hour or two and we did come up with some new ideas. This is still possible today. Look for some product applications that are hidden . . . there are still markets out there that have not yet been trapped by your property. List the types of business you are not doing and figure out why you are not then come up with strategies that will get that business for you (perhaps even without discounting, hopefully).

How about going through those lost business reports. There are probably many properties that do not even use these, but should. Just because a piece of business was lost at one time doesn't mean that it is not possible to get it back. In fact, you should just go through all the files and get back to people that were not followed up in the first place. We seem to lose a lot of contacts just because we fail to follow-up when appropriate.

I have been found, time and time again, that the more face-to-face sales calls made the more business one gets. One old report once reflected that if a salesperson spent as much as 5% more time in the face-to-face selling that sales would increase by 10%. Try it and see if it works. This gets to the time management situation of figuring out how salespeople spend their time. My guess would be that most industry salespeople property sends less than 50% of their time in real selling situations.

Sales goals is another area of concern. If management sets the goals too high, salespeople lose interest. If goals are too low there is no challenge; so what is "just right"? Most industry salespeople these days do have goals, either for room nights, average group rate, group revenue, food/beverage, etc. Do realistic goals create more sales/ Do the incentives we are now providing make more sale? Has anyone thought of daily goals for sales? Those readers who may have worked with me over the years may recall our "magic drawer" system of measuring daily sales generated by the salespeople.

Although in the "olden days" we really did not have a marketing plan do believe that one is urgently needed these days and it should include an action calendar. Coming up with strategies for new, creative sales without the benefit of some marketing intelligence doe tent to lead us astray. So before anyone comes up with a new idea it is not a bad idea to do a little research.

Many of us coming up through the ranks did not have the benefit of much sales training . . . it was kind of "shooting from the hip," but somehow we got the job done. Today there are great opportunities for learning, but unfortunately, very few hotel companies take advantage of what is available for salespeople to learn to do a better job of putting the business on the books. There's a lot of stuff out there for just a little investment. Take the Educational Institute of the American Hotel and Motel Association as an example. This outfit is loaded with books, study programs, audio and video tapes, training manuals as well as a department that produces that produces customized training programs for companies.

Maybe we just need to go out and do the basics . . . again!

THERE'S NO SUBSTITUTE FOR MAKING MORE SALES CALLS

July 1996

More and more salespeople, in my opinion, are making less and less sales calls. It seems that I keep running into property sales managers who really don't like the idea of making sales calls; even the appointment ones, much less the cold calls. Many will not admit that they do not like the idea, but have come up with a variety of reasons (excuses) for not getting out to make the calls. It wouldn't even be so bad if they would just make some prospecting calls by phone, but it isn't being done in many cases. Most of the reasons given for not getting out on sales calls have a lot to do with "being too busy." Perhaps we need to take a look at job descriptions and see what it is we really want a sales manager or sales rep to do. It seems to me that sales personnel need to develop new business accounts for a property. If this would be their sole responsibility perhaps they could get more new business. But, "being busy" involves a myriad of tasks that salespeople probably should not be doing. They "help out" with reservations, front office, and dining room, run errands for General Managers, look after catering events and they just feel they have to be around during a conference that they happened to book.

If you want sales folks to bring in more new business then why not just let them do the job of selling—which of course, involves making sales calls. In all probability, right now, most of the business coming in at the property level originates with leads. Yes, most salespeople will follow-up on leads—provided to them by responses from advertisements, convention bureaus, people calling in to make inquiries, direct response cards and "walk-ins." And, that's good! An, leads need to be followed up immediately and of course, leads need to be converted to prospects who are then converted to customers by good sales efforts. We don't want to sacrifice following up on these leads for the sake of making someone go out make cold calls –and, in many cases, this is the cause of our dilemma. "Do I go out and make sales calls while there is no one in the office to take care of people calling in?" What does happen when the sales staff is out? Who takes the calls? Salespeople probably have very little confidence in other people handling the business of sales while they are out.

Now, this becomes a management situation. The solution is having someone available and well trained to take telephone inquiries. Perhaps it may even be solved by adding staff. We hardly ever want to do that because it increases the cost of labor. However, would the increase in sales activity—resulting in additional sales, make up for the extra investment of another body? Many times when we make test calls to properties we find that when a salesperson is not available we do get a secretary or receptionist. Some do a good job in asking appropriate questions about needs, but most do not. It seems that they are there solely for the purpose of taking a name and telephone number—and sometimes they are not even good at that. If we already have that "extra" person why not take the time to train them so that they can do more than just take messages.

Selling is really the easiest job in the world. It is just a question of finding out the needs of people or groups and doing something to satisfy those needs—provided we have the facilities to meet the needs. Everyone in the sales office (and even other departments) should be trained to use a "Prospect Survey Sheet." This is just a basic survey looking type of form that prompts the user to ask specific questions to determine needs. Once you get the information it is a question of being able to meet the needs and having the space available. Then, of course, some selling has to be done to close the sale.

Prospecting is an Ongoing Job for Successful Sales Development

Unless your property is running full all the time you should come up with a program to make sure that someone at the location is doing some prospecting every day. Under normal circumstances, it may very well be the people assigned to the sales department who are responsible for bringing in new business. However, at limited service properties where there may not be any staff assigned to the sales function, then it should be someone. The manager, front office staff, night auditor (yes, why not?) or just about anyone who can spend a couple of hours a day looking for prospects. Why not some part-time students from a local college?

Prospects are people or organizations that may be in a position to provide you with rooms or meetings or restaurant business, but the problem is we do not know that they are prospects until we make a contact and ask specific questions. So we have to find these contacts first and here's how we do it.

Your own files: Somewhere on the property you've got files on people or groups that have had business with you previously, but they may not have been contacted again. There may even be some names of those that made inquiries, but never booked anything. Search out these files and information and start calling.

Your own staff: You'll probably be surprised when you start asking questions of your staff about who they know, the organizations to which they belong, their spouse connections, etc. Just your housekeeping staff will probably belong to more groups than you would ever expect. Religious groups, school organizations, garden clubs. A variety of special interest activities. And don't forget. Everyone has friends, neighbors, and relatives.

Your regulars: They stay with you on a regular basis and are in a great position to refer individuals and companies to you who may be prospects. You've just got to ask.

Current guests: Just "eyeball" the registration car-do each day and see who's there. Information on the card may be able to tip you off if your current guests are prospects for future business or have connections with companies that might be prospects for other business opportunities.

Your vendors: If you will take a look at your own accounts payable list you'11 probably be surprised at the length, and shouldn't some of these be prospects for business? After all, you are doing business with them why wouldn't they not only do business with you, but refer their contacts to you?

Meeting Participants: There's a pretty good chance that one out of three people attending a meeting or an event at your property may be a prospect. Whether it is a local civic club event; or a company meeting or social activity. Find out who is in attendance to make the contact.

The cold call: Don't be afraid of it. It's still probably a pretty good way to find prospects. Even if you can only devote an hour a day. You can do this while on the way to the post office, or bank or insurance office or even when shopping.

The newspaper: Probably the easiest way to find contacts that may be prospects and certainly the local paper is the best source. In any daily paper you will find many opportunities for prospecting. In just about every section of the paper from sports, to business, to social, the classified, even in advertising you will find tips on whom to follow-up. You should learn to read a paper with an "eye" towards looking for prospects. With a little practice, these will pop out at you.

Anyone of these opportunities for prospecting may be accomplished by just about anyone on the staff . . . in fact, let the salespeople do the selling (if you have anyone assigned to sales) and get you and your staff to do the prospecting. Once you've found a likely prospect all you need to do is the typical six questions of: WHO, WHAT, WHEN, WHERE, WHY and HOW to find out if, in fact, these are people or organization who can do business with your property.

HAVE A DIFFERENT PAIR OF EYES TAKE A LOOK AT YOUR SALES EFFORT

October 1998

Business is still great, hotel profits are still up as are average rates, occupancy is down a little in some areas, but demand is still there and most salespeople are still heroes. Are things so good in the area of sales that there is no room for improvement? No matter how well we are doing, from a sales standpoint, there is always an opportunity to doing things better. Sometimes when things are going well for us we neglect to step back a little to take an objective look at what's what. As long as our way of doing things work for us we tend to hesitate to look at other ways of accomplishing our mission.

Perhaps it may even be best to have someone come in with a "fresh" pair of eyes to take a look at the sales operation just to analyze the system; who is doing what, how do they do it, etc. What we're mentioning here is an opportunity for a property to have a sales audit conducted. The idea is to find someone with heavy hotel sales and operations experience to visit with you for a couple of days to look over what's going on in your sales department. Of course, there are always some people who may be very protective of their turf and would not appreciate anyone coming in to look over their shoulder. To do this successfully, a property would have to have some very open-minded people on the staff who understand that there may be a way to improve sales activity and productivity which would lead to more profitable sales. If an independent sales audit is conducted and there are no recommendations for improvement, then, at least that is some reinforcement for the staff, knowing that what they are already doing cannot be improved upon.

A sales audit is pretty basic—it may include, but wouldn't be limited to the following: A review of the property marketing plan, contracts or letters of agreement, rate management system, job descriptions, sales activity reports, lost business reports, sales account file and trace system; how the phones are answered when people call to make inquiries (on individual reservations as well as for groups and meetings); prospecting activity; how accounts are managed; interview with all sales staff personnel to determine actual job function, and individual sales productivity; ratio of actual sales to bookings; market segment analysis; advance bookings; and of course, there are other things to include based upon the type and size of property. There may also be some specific aspects of the business that property management may want to be analyzed during the sales audit visit.

Not only is the sales audit a good idea to try to improve sales performance now, but for the future as well. Thinking ahead a year or two, there may be a shift in the business cycle, there may be a drop in demand, and there very well could be a much over the built situation in our industry. Would it not be to every owner's and operator's advantage to be better prepared with a personalized report on how to improve sales activity, performance, production, and profitability?

The question is: "Who is going to do this sales audit"? Of course, it would have to be someone with hotel industry experience in sales as well as operations. There are many hotel consulting firms operating around the country with very qualified people on the staff who are very capable of conducting such audits. In addition to these companies, there are many very well qualified individuals who have already retired from the industry and would love to be involved in some part-time work. There are probably some already located in your own backyard. If some of those "old timers" read this article and are interested in doing some of this work just contact me and we'll make up a list; just in case we get some inquiries.

PREPARING FOR THE YEAR 2000

October 1998

Anyone out there a little worried yet about where we are going with our new rooms construction, rate increases, high demand, our seller's market and about our economy? Maybe we should start thinking about where we are going to be by the year 2000 as far as our occupancy and profitability is concerned. A good many owners, operators, General Managers and salespeople may have been taken by surprise in the mid-eighties. Are we in for a repeat performance? Hopefully, not! But, one never knows. Even the economists can't tell us too much about the country's future economic situation.

Right now the figures look great, for the most part. Smith Travel Research's latest report for the first half of 1998 looks pretty good. The report shows that everything is "up," around the country, over the same period last year, except for occupancy; at 63.7%, is down almost half a point. Supply of rooms is up 3.9% (compared to 3.3%); ADR is up to $78.90 (it was $75.14 last year); RevPAR is up to $50.26 (last year it was $48.20). Even demand is up 3.2%, which is stronger than last year when it was 2.4%. So everything is still pretty rosy. Even with a little drop in occupancy, room business is still hot in most parts of the country, the demand is there and certainly, profits are the best ever.

So now the question is: "how long can this continue? Who knows? What we need to do is think about what to do if the business starts falling off. How do we prepare ourselves? Even if this business climate continues forever wouldn't we do much better if we position ourselves to prepare for worse times? At least we may do better than our competitors that are not thinking about this and do not do anything to prepare. Let's take a look at the expected marketing environment for the year 2000:

Overbuilt Situation? In all probability, we will be overbuilt in the limited service area. Some areas of the country are already feeling the pinch. And, with all the various announcements being made by brands (the old and the new) it looks like new builds will continue to come on the market for at least the next two years. We may not be so hurting in the upscale, resort and meeting type facilities, but little by little we see more construction going on in those segments.

Strong Competition? Most of the time when we consider competition we look at the product itself to see how much better it is than ours (or how much worse). We also look at rates being charged by other properties so we can see where we fit in and it helps us decide how much we can charge. And, some look at market share to see where we stand in the community. What we really need to look at is the quality of the sales effort of competing properties. To what extent are they better than we are; are their sales folks spending more time on sales than our sales folks. Can we do a better job in sales than they are?

Market Segment Analysis? Are we taking a good look at the market segments to which we are selling? Which are our best markets from an ADR standpoint and what percentage of total business is being derived from each of our markets? We need to think about creating a sales plan which develops strategies to go after more of the business that is the best business for us. Who will do what, by when, and how much will it cost to go after the better types of accounts? This needs to be an important part of your marketing plan.

Selling and Confirming Opportunities: Take a look at the amount of time being spent on your sales effort. Whether it be by full time, part time salespeople, or even the General Manager. In talking with salespeople around the country attending our workshops we find that most salespeople are spending a whole lot less than 50% of their time in direct sales. If someone is good at sales, it just makes good sense for them to spend more time in direct selling than doing other things at the property.

LET'S NOT JUMP THE GUN ON LOWERING THOSE ROOM RATES

September 2001

Too many owners, operators, managers and sales folks seem to be getting nervous about the economy and in some areas, an overbuilt situation. As a result, we see some drastic cuts in rooms rates being offered to build additional occupancy. Some of this just doesn't seem to make sense. Has anyone looked at the mathematics of this? Just how much additional occupancy is needed to justify the rate cuts? If you lower rates, across the board by 10%, how many more rooms do you need to sell each night to make up the revenue difference? In order to maintain RevPAR rates need to be increased to make up for a drop in occupancy. So, what's the answer? It is really very simple: Let's do a better job of selling! Maintain the rate (maybe even increase it?) and build additional sales through the on-property reservations office, front desk staff as well as the sales staff.

Perhaps the phones have stopped ringing after the past seven years or so of being in the sellers' market. Now it's time to start getting serious about the sales effort. How productive is the sales staff? How much time is spent on real pro-active sales activity? From what I have gathered in my travels and discussions with salespeople, just about all spend less than half their time in active selling. Most sales folks also have not had any formal training in hospitality sales. It is amazing what many do not know about prospecting, selling techniques, working a trade show, writing sales letters, negotiations, as well as how to confirm business for a property. So many sales folks are spending too much time making cold calls, which is nothing more than survey work. Survey takers could very well be paid much less than salespeople. Perhaps minimum wage, part-time, college students can do this for you so that the sales staff can be more productive by spending their time only calling on qualified prospects.

At limited service properties, where there may not be a dedicated sales staff, how about and the General Manager taking over those duties? Even just a couple of hours a day, calling on qualified prospects, would generate a good amount of additional business for a property. Of course, there is always the opportunity of other staff members to help out in sales. During slow periods get some members of the staff together to go out to make cold calls, gathering information in your own community about who may be prospects for room business. They should use a Prospect Survey Sheet which lists the questions to be asked to determine if the contact could be a prospect. The completed forms would then be turned over to the General Manager for follow-up with the contact to try to book some business.

Above all just be sure that you don't miss any prospective business from people calling into your property. So much revenue is being lost daily because we don't answer the phone right away or we don't answer it properly, or we do not know how to ask the right questions, or use voice mail improperly, etc. Just a little training for our reservation folks and front desk staff would go a long way in improving sales. Just think again before you start cutting rates. Maybe the answer to the economy, overbuilt situation, and higher cost of operating a hotel, boils down to doing a better sales job.

WHY DON'T WE GET BUSINESS FROM OUR VENDORS?

March 2002

In these tough times, with a down economy, drop in occupancy, and every property manager scrambling for any kind of business, we just seem to overlook the obvious opportunities. It would seem logical; if we buy products from a company, why wouldn't that company buy our products? If any size property salesperson would take a look at the accounts payable print out from the accounting department, there would be a big surprise. The listing would seem to go on and on. It is really amazing the number of vendors with whom we do business. It wonders me that so many salespeople and managers are not aware that most of these companies could very well be prospects for some kind of business at our properties. These vendors probably have manufacture representatives or folks from their own corporate offices visiting from out of town from time to time. Shouldn't we find out where these visitors are staying? Most likely a competitor is getting that room business. Then, there very well may be some small meetings going on every once in a while. The opportunity for some social event business also exists. We really have to look upon all the people from who we buy something as being a prospect for some sort of service we are offering.

Besides going after these vendors for business they may be able to give your property, how about asking them for referrals? After all, some may not have people coming in from out of town, or have meetings, or social events. The question is, can they provide you with referrals of people they know who can give your business? It would seem very plausible that every vendor could very well be in the position to provide names of other organizations with which they do business that may be prospects for you. It is just a question of asking. A good idea would be to have someone in the accounting department, preferably the controller or the accounts payable clerk to make the phone calls. Now, we are not asking them to do any selling. What we want would be for them to ask specific questions to determine if the vendor is a prospect for business at the property. You should design a Prospect Survey sheet with questions that need to be answered to determine if the vendor could be a prospect for business. The completed sheet of questions is then turned over to the sales department at the property, or the General Manager, for follow-up to do the selling. A sample Prospect Survey Sheet has been offered previously through this column.

When selling the vendor, who is a prospect for you, be sure to make it a face-to-face meeting. Try to get to the top of the organization, like the President or the chief in charge of the local operation. The idea is to create a strategic alliance; "I buy from you, and you buy from me. There is a big value for both by working together. We help each other by making referrals." Remember that you have to ask for the business. NHUSSS (Nothing Happens Until Someone Sells Something).

HOW DO SALESPEOPLE SPEND THEIR TIME?

April 2002

It sure would be interesting if we could get a study done on how salespeople spend their time. Wouldn't it be good to make comparisons with other folks who are in the same job function as us, with regard to what we do and how much time we spend doing it? Maybe we can get the HSMAI (Hospitality Sales & Marketing Association International) Foundation involved in doing something like that.

Every once in a while, during my sales workshops, I do a little survey of the participants, to find out just that. Of course, the study is not very reliable since the sample is very small and I am not a research type, but the results are kind of interesting. Recently, we asked workshop participants to complete a survey form, that asked how time was spent in a typical day. The results reflected that sales managers spend 40–50 hours at work, in a five day week. It was observed, that in a typical day somewhere between 35–50% of the time involved direct sales effort; however, a large portion of that time was spent in making "cold calls," either in person or via telephone. Some of that time, about 10%, involved luncheon meetings with prospects or site inspections. It appeared that about 20–25% of the time took up answering email and telephone messages. Meetings took 10%, while 20–25% involved "checking on details of conferences taking place on site," paperwork (getting rooming lists, guarantees for meals, checking on contracts, proposals, trace files). 70% of the survey takers indicated that they start a day somewhere between 8:00–9:00 am and end at about 6:00 pm. The other 30% started at the same time as others but ended at 5:00 pm. 10% of those surveyed stayed inside all the time, and 10% worked outside all the time.

Please remember that this was a very unorthodox study, and doesn't follow any standard, authoritative research methods. Then, again, it may very well be typical of what is going in the industry. I wonder if any of our readers have come across a survey of this type that we may share with the other readers of this column?

Overall, the study made it look like all the participants had been very productive with their time spent in sales activity, whether it be making sales calls or responding to phone calls, emails, RSV's, tracing files, making proposals, etc. One of the big things with which we need to be concerned is that sales folks continue to strive for improved productivity. What we mean is being more productive in bringing in revenue, especially new business. It stands to reason, pretty much, that if the service and product meet the needs, and exceeds the expectation of our guests, then they will return. However, not everyone returns, for any number of reasons beyond our control. This, then means, that it is imperative that salespeople spend as much time as possible in going after new business, to replace the business that did not return. After all, in taking a look at occupancies at our properties, these days, we do not have very many full house nights. The answer to how we can bring in more new business is to spend more time in pro-active sales. Maybe we need to show our sales folks that they can spend more time calling on qualified prospects to confirm more business, and maybe less time doing other things. Cold calling is okay at certain times, but not as productive as taking that same time to call only on *qualified* prospects.

WHY ARE SALESPEOPLE STILL MAKING SALES CALLS?

May 2002

WOW! What a waste of a sales professional's productive time.

We need to start taking a better look at this business of cold calls to see how productive it may be, and who should be doing it. What we're talking about here is someone making a call on a contact, not knowing whether or not that contact can become a prospect for business for a property. The call is made either in person or by telephone. For the most part, these days, sales personnel are spending a heavy portion of their "selling" time doing this. Many properties include this requirement in the job description of sales personnel.

The whole idea of a cold call is to determine if a contact, or the contact's organization has needs that could be met by the property. The key to good cold calling is to determine the "who," "what," "when", "where," and "how" of a contact's hotel needs. Once it has been determined that the contact does have needs which can be met by the property, then that contact becomes a "prospect," and then the sales process can begin. Hopefully, the prospect will then become a customer.

My question is: "should salespeople be doing this"? The way to determine this is by evaluating the return on investment. Is it really worth the time of a professional salesperson to go around "knocking on doors," looking for prospects? How many of these cold calls need to be made before a qualified prospect is uncovered? How much unproductive time is being spent by a salesperson in doing this function?

What does make sense is to have salespeople spend their time selling only to qualified prospects. This means that the "cold calling" or prospecting, needs to be done by others. Once a prospect has been determined then the information is turned over to the sales folks to go after the business. This way our professional sales staff will be much more productive in conducting pro-active sales activities. They will then be talking only with contacts who they know are in a position to use the property, should the prospect decide to do so.

So, now the question arises: "Who can do the "cold calling" (prospecting) if it shouldn't be the sales staff? Good question! Would you believe that just in one day, with some training, property staff members in reservations, front office, and accounting, each person spending less than one-hour, can develop more prospects for a business that could possibly be handled by the sales staff? It is just a matter of training these other in-house staff folks to ask appropriate questions of people with whom they are already in contact. This is not difficult, in fact, it is very easy to do, and it takes just a little bit of training. Other than contacting people they already know (or with whom they are already doing business) they can also check out the numerous leads they can find in the local paper. Of course, the idea is to get this "cold calling" accomplished during some "down time" during the day. Leads and contacts are all over the place. We just need to use other staff personnel to do the prospecting and turn over qualified leads, using an appropriate form (like my Prospect Survey Sheet), to the sales department for follow-up.

Having professional sales staff personnel spending their time only on qualified prospects is a better return on investment.

PLAN NOW FOR YOUR 2003 SALES ACTIVITY

October 2002

Most of the time, in planning for our next year, we produce a marketing plan that includes a good deal of information along with some sort of action plan. This is good and every property should have a marketing plan. Many times the action plan is nothing more than a calendar of activity for each month of the year. We indicate, very generally, the things we are going to do to implement the marketing plan. The big question is: Do we really follow-up to make sure we do all the things we indicated we were going to do? Probably not! From what I have seen over the years, these plans end up on someone's shelf after the first couple of months of the New Year. We do the plans because the corporate office required them. However, what would more productive is a Sales Activity Plan that is written, not by month by month activity, but by action planned for specific improvement in sales.

Since most sales folks should be responsible for bringing in new business, we want to be looking to create a Sales Action Plan, with a step by step approach that is very specific, and that would produce more profitable sales. Here is the suggested plan:

▶ We start by looking at all the property business, for this year, from a market segment standpoint.
▶ The segments need to be listed them from top to bottom, with the one producing the best average daily rate at the top, then down to the segment that produces the lowest ADR.
▶ Indicate what percentage of the total room business, year to date, is generated by each of the market segments. In most cases, the top ADR segments would be producing the least percentage. We know that at the top of the listing would always be those segments that bring in your highest rate, the Rack Rate; however, they would in all probability represent no more than 5–8% of your total business.
▶ This gives you an idea of where the sales activity needs to be directed. The

idea, of course, is to generate more sales from those segments that are the most profitable.
▶ Take, for example, one of your rack-rated segments. You probably have that listed as "transient." However, there could be several different sub-segments under that category. They could be rooms generated by travel agents, central reservations, walk-ins, etc. So, pick one of the segments, like travel agents. Let us say, that rooms generated by travel agents at rack rate, represents 5% of our total room business. It surely could be more beneficial to the property to try to increase this particular segment.
▶ We now have an opportunity to create the first objective of our Sales Activity Plan. We will plan to increase this segment to 10% of our total room business.
▶ At this time we convert that 10% figure to an amount in actual dollars. If your total room sales, year to date, is $1,000,000 and you are now doing 5% ($50,000), your new goal is $100,000, double what you are doing this year.
▶ We continue to create more objectives of increasing business from other market segments. It could be an individual corporate business, or corporate group business, government, association, etc.
▶ Once goals are set the idea would be to come up with strategies for each market segment we want to increase.
▶ Taking the travel agent sub-segment of transient, as an example, our strategies may include:
   – Send out three direct mail pieces during the year
   – Participate in three different travel agent related trade shows
   – Create an ongoing internet travel agent promotion
▶ For each of the three strategies we then create an action form which includes the following sections that need to be completed:
   – Listing of action steps in sequence
   – Person responsible for each step
   – Beginning date
   – Targeted completion date
   – Actual completion date
   – Estimated cost
▶ A separate form is necessary for each specific strategy.

How many objectives and strategies have to do with the number of staff personnel available to assist in the plan. Consider assistance from staff other than sales staff to be helpful. We need to be concerned with return on investment, so costs should not exceed 10–15% of expected additional revenue generated by the plan.

TRENDS IN SALES: COLD CALLS ARE OUT, CONSULTING IS IN!

March 2003

Just take a look at the job descriptions for sales personnel at some of our hotel properties. So many of them require that a specific amount of "sales calls" have to be made on a weekly basis. What does that mean? It seems that for the most part, this refers to "cold calls" being made to find prospects for new business. Or, in some cases, probably calls on existing accounts to "maintain relationships." Whatever it means, it just doesn't make sense anymore unless we specifically describe what we mean by "sales calls." These days we see so many salespeople just going up and down the street "knocking on doors" trying to sell their properties to just about anyone without even trying to determine if there is a need.

For about eight years, prior to 2001, when the hotel business was doing just great, just about all we had to do in sales offices was to answer the phone. At the end of 2000, we had the most profitable year in our history. As things slacked off, starting in 2001, we had to start to scramble for business. Apparently, there were many sales folks who did not have the experience or training in finding new business to replace what was lost. So the answer, in many cases, was to get our folks "out on the street."

What we need to do is look at this business of "cold calling" to see how productive it is in actually booking business. It is only productive if we can find enough qualified prospects that can then be sold on using our properties. The whole idea is to get salespeople to create more sales. So what we need to do is get sales folks to spend more time on pro-active sales calls. That is, calling only on qualified prospects, rather than just making "cold calls." There is no doubt about it. It makes sense! Spending more time on making calls on people with companies and organizations that could use hotel facilities will ultimately create more sales for the properties. So, we need to get salespeople to properly prepare themselves with information on qualified prospects, then make appointments to make a sale. Where are these qualified prospects? Right in your own existing files. Most properties have an abundance of sales files that have not been followed-up in a long time. These files need to be revisited since all of them have been prospects in the past, and probably are still prospects for a variety of types of possible bookings.

With the information already available from our files, we can turn to a consulting position rather than a sales position and try to find ways to help those prospects in solving situations that require hotel facilities. The consulting process then begins with property salespeople, making appointments, and rather than just trying to make a sale, start consulting by asking specific questions to show an interest in how they may be helpful.

WATCH FOR HIGHER ROOM RATES IN THE NEXT COUPLE OF YEARS!

April 2003

Room rates have almost always seemed to be a major consideration when selecting a meeting and convention sites and locations. In fact, a recent survey by the HSMAI (Hospitality Sales & Marketing Association International) Foundation, to seek out demand generators in the business of selecting hotels by business travelers and leisure travelers, reflected that number one and two always seems to berate and location. After that, it has to do with features and amenities.

No doubt about it, corporate meeting and incentive planners have been well aware that for the past two years we have been in somewhat of a "buyers'" market with lots of negotiation opportunities to get lower room rates. Hotels have suffered a sizable loss in profits in 2001 and 2002 compared with 2000, when the industry made its biggest profit ever. Property owners have not been happy about that, but with the drop-off in business, creating a lower occupancy, along with an overabundance of available rooms on the market, just about all over the country, sales folks found themselves out begging for business and making deals. Still, with rate cutting, a softening of occupancy, hotel profitability is not back where the owners like to see it. Little by little, however, there is an improvement in hotel business on the horizon. Already we are seeing reports of leisure travel picking up, with more folks going on the road, taking trips, short vacations, buying more travel packages, and hotels are going after that business. Further, the short lead-time, smaller meeting market is also being seen as growing. We see a great deal more marketing going on with associations; where they are trying to beef up attendance at their conventions, and it has been working. And, on top of these indications of hotel occupancy starting to build back up, we find that investors, lending institutions, owners and operators are becoming shy in adding more rooms or building more properties because of the profit and occupancy decline in this industry.

So, what does all this mean to you, the corporate meeting planner? It means beware of what's coming in the next year or two. In fact, we will probably see a tightening up on negotiations from property sales folks, certainly by the last quarter of this year. With a big push on by property owners to get salespeople trained in selling better, providing more incentives for increased revenues in sales, and insisting that General Managers get more involved in sales, room rates will be on the rise again. Attendance at hotel sales training workshops has been on the rise these last few months, along with hotel management companies including sales training segments in their general management meetings. This means we will start seeing hotels increasing their occupancies as well as revenues. Sales folks are learning more and more about how to negotiate, write contracts, where to prospect for new business and seek out market segments which they were not getting previously. They are going through old files to re-establish accounts they have had previously. Good hotel salespeople are in demand with higher pay scales that have been paid previously. So maybe now we may have to use the term "Buyer Beware."

Rates will not be raised severely, but we are already seeing hotel room rates coming up, little by little. One of the big problems hotels have these days is in the area of the third-party Internet provider, who are being used by some properties to unload rooms that are not expected to be rented, at a lower rate than quoted by the hotel. We see more and more of the larger, meeting type hotels, deciding not to use these reservation systems since it interferes a great deal with contracted rates on meetings, and ends up with an attrition problem both for the property and the planner.

At any rate, you should expect to see room rates rise in the very near future and you should be able to prepare yourself for this by brushing up on your own negotiation techniques. Just be sure you always have options for locations and sites. Never lock yourself into one location. Having flexibility is definitely an asset when negotiating rates with a facility. Many times if you can move arrival and departure dates, to suit a hotel's availability, you sure would be able to make a pretty good deal. All you have to do is ask: "When could you use my business."

If you are able to lock in some great rates for your 2004 and 2005 meetings or incentives, then good for you. And, there is still time to do this before word gets out to the properties that they will have to raise their rates for those years. It might not be a bad idea to check on your contract just to make sure those rates are locked-in, without any fancy clause about rate increases for the cause. And, it doesn't hurt either, to stay in touch with the properties and your sales contacts. It would not surprise me if some hotels start looking their contracts for 2004 and thereafter, just to see what's on the books, and then try to improve their rates. There could be some turnover in salespeople, and even General Managers, at some of the hotels.

Sometimes, new people want to be heroes, and will want to get better rates out of you. You may even be living with a contract that was agreed upon by someone other than yourself, and perhaps you were not even involved in the negotiations. If this is the case, be sure to review the agreement very carefully, and be certain to live up to your end of the bargain so that there is no excuse for the other party not to live up to their end.

Just remember that the hospitality and meeting industry lives on cycles.

What goes up must come down. The buyer vs. seller market goes back and forth. This time the buyers' market may be very short lived.

SALESPEOPLE NEED TO SPEND MORE TIME SELLING

September 2003

During the past two months, we have conducted a bunch of Hotel & Motel Management sponsored sales workshops ("Selling In Uncertain Times") in different cities around the country through HSA International. We had a few General Managers participating, but mostly sales personnel were in attendance. At each workshop, we asked the question: "How much time do you spend in pro-active sales?" We defined pro-active sales as being a discussion with someone representing an organization that is a qualified prospect, that is, someone who could book business with the property, should they chose to do so. Making "cold calls" did not count. The results reflected that only a few, at each workshop, spent 50% or somewhat more, on pro-active sales; and those represented the larger properties with multiple sales personnel. Most in attendance spent 35% or less of their time on sales calls.

It just seems logical, and just about everyone agreed, that if you are good at sales, and can spend more time on pro-active sales, more business would be booked at the properties. Apparently, sales folks get hung up on doing so many things other than sales that their sales times are limited. A good many of them get caught in the battle of getting involved in "making sure" the events that they booked take place properly. This means they feel they have to be around when "their group" checks in just to be sure they are handled properly. These sales folks readily admit they there is a lack of confidence in their conference service department or catering department properly taking care of the group. Others advise that their job description, or General Managers, require that they make a certain amount of "cold calls" daily or weekly. All admit that making cold calls are not productive. Another area of concern among the sales folks was the amount of time necessary to complete reports for management. It appears that many management companies and property managers measure the performance of sales personnel on activity; reports are required to reflect all activities on a daily basis.

All this boiled down to everyone agreeing that they could be more productive in booking more business if they spent a good deal more time calling only on qualified prospects. So where do we find these qualified prospects? Right in our own files at every property. Our files (either the old paper files, or those found right in our computer) are loaded with prospects that have not been called upon in a long time. Most sales folks will admit that there are many prospect files have not been traced properly over a period of time. Just by going back over these files, making the phone calls, and asking appropriate re-prospecting questions, will develop new prospects for business. Just using a basic prospect survey sheet would be helpful. I'll send you one, just let me know if interested (howardf@vt.edu). Get some of your own property staff to help do this during the downtime. Or get some local students to work on a part-time basis, or even some retired hotel sales folks would be very good at this. This would be a minimum of an investment to be able to develop additional qualified prospects that could bring more business to the property.

SELLING OPPORTUNITIES ARE ALL AROUND US—MAKE THE MOST OF THEM

January 2004

Times are changing fast and we are already into an economic recovery period. Business is getting better, more people are traveling, occupancy seems to be picking up, and the latest information from the companies that measure our business indicates that we can expect about a 3.5% increase in demand. That's great news, so how are we handling our sales opportunities? Of course, we can sit back and wait for the business to come to us, and continue to do more business with the third party independent planners as well as those online travel companies. Or, better yet, we should be more aggressive and start hustling to bring in more new business to our properties.

Selling opportunities are all around us, right in our own backyard and in our own files. A big problem that exists in our business is that we have so many very qualified prospects in our old files that have not been contacted in a long time. Of course, we would like to feel that we trace all files regularly; however, it just doesn't work that way. We need to admit that it is virtually impossible to be able to trace all files on a regular basis, unless we have extra staff available to do this for us. Most sales folks just get too busy with so many things to do that we just get behind in the tracing. We need to understand that every piece of business we have ever had at a property is an opportunity for additional business from that same prospect.

Somehow, we need to find the time, or find some extra part-time help to review all of our old files to seek out more business. This is easier, costs less, and is much more productive than prospecting on the Internet or making cold calls in an area, or buying lists of companies and organizations, and then calling them to inquire about their meetings. It is easier and cheaper to work with and retain your old customers that look for new ones.

We also need to think in numbers. Isn't it much easier to make one group sale of $10,000, than to try to sell 100 individual rooms at $100 each? Let's face it, even if it takes a long time, the single $10,000 sale is quicker, and certainly lots more profitable than trying to sell the 100 individual rooms at $100.00 each. Everything costs less per unit when sold in bulk than when sold individually. So we need to start thinking in terms of group sales. Even with a limited service property, which may not have meeting facilities, there are groups that will gather without the need of meeting space. Think of wedding guests that need to spend the night, sports teams, entertainment groups, family reunions, military reunions, retreats. These types of groups can possibly be generated through leads provided by your own property staff. After all, everyone has friends, relatives, and neighbors; alert the entire property staff with how they can help bring in more group business.

Opportunities are all around us for group business, we just gotta make the most of it.

THE EASIEST WAY TO IMPROVE HOSPITALITY SALES

February 2005

Of course, increasing sales is not always easy, but perhaps we need to think of ways to make it easier for us. When it comes to hospitality sales, there are some areas of concern that directly affect our ability to do more business. There is the economy with which to be concerned; these days, however, it is in pretty good shape, and is working well for us in our industry. Improved employment always works well for us which in the long run creates more leisure travel. So in good times we usually think of building a business is somewhat easier. However, along with the improved occupancy and revenues that are generated by good times, we know that we will be looking forward to growth in a new room which will have a good deal to do with an over-supply. This, to some extent, will create more competition through a possible over-built situation, which in turn could decrease occupancy at some properties.

Regardless of the economy, whether it is good, not so good, or even bad we need to continually be involved, on a regular basis of going after new business. After all, most properties do not run 100% occupancy all the time. Travelers who stay with us at one time do not necessarily always stay with us again when they return to the same city. The same goes for the group market. Companies and associations do not always put the business in at the same properties all the time. This means that we must continually seek business to replace the business that is "lost" or has not returned. We need to think of ways that make it easier for us to get more business. All too often we find hospitality salespeople trying to book business by making contact with a wide variety of individuals, companies, and groups who really are not in a position to do any business with them. To avoid that kind of a situation it is much easier to do business with people and organizations that are already doing business us or have done business with us in the past. It should be easy for us to do this. It is just a question of staying in touch with accounts with which we have done business. Sometimes we get so concerned with booking a single piece of business that we neglect to ask our appropriate contact questions to determine if there is other business that might be available to us. This could through be through other employees of a company or members of organizations. We need to find out is who are other contacts that could be in a position to do business with us. In many cases, corporations have several different departments directing corporate business travel or making arrangements for meetings. Then, of course, there very well may be social events that are held throughout the year which could provide additional opportunities for a property.

Another area where we could make the job of sales much easier is to make sure we do not lose any telephone inquiries. All too often we hear complaints from qualified prospects about the "telephone ringing off the hook;" or "nobody returns my calls," and of course, no one enjoys listening to all those prompts on phone mail. There is no telling how much business is lost daily because of poor telephone practices.

There is a lot more to this business of making the job of sales easier by getting into the subjects of: product knowledge, selling the difference, building relationships, public relations, E-distribution, and the list could go on.

COLD CALLS BY SUPPLIERS ARE ONE OF MEETING PLANNER'S BIGGEST COMPLAINT

July 2007

There is nothing new about this; it seems that year after year, in all kinds of surveys of meeting professionals, in the corporate market as well as the association market, have been complaining about this. Generally, what is happening is that sales personnel from hotels and other meeting venues, get on the phone (and even email as well) to get in touch with planners and ask if they have meetings, and then try to make a sales pitch to use their facilities. Meeting planners say this is a big waste of their time, and that salespeople should be in a position to find out more about them to find out about their needs before they make calls or send emails to make those inquiries.

In a recent survey conducted by Meeting News magazine, of those planning meetings for corporations and associations, results reflected a pretty poor score of how salespersons conduct their business. The following table shows the percentage of respondents answering this question in reference to the issues presented:

What makes this all very interesting is that it seems as though nothing has improved in this area over the years. Surveys like this have been taken by this publication, as well

"Which of the following aspects of dealing with supplier salespeople do you most dislike?"

	Corporate Planner	Association Planner
Your calls/emails not returned in a timely fashion	48.3	44.6
Cold calls/emails with no knowledge of your meeting profile	39.9	56.1
Poor communication/follow-up by the salesperson	38.9	39.6
Lack of creative thinking/alternatives offered during negotiations	37.4	34.5
Often dealing with a new contact at a property due to high turnover	36.5	34.5
Proposals that don't properly address all your RFP items	24.6	25.9
Contracts that contain terms different than what was discussed	21.7	26.6
No post-meeting follow-up	22.2	18.0
Not empowered to make decisions	17.2	18.7
Don't understand some contract clauses	9.9	15.8

as others time and time again. The message here is very clear that sales personnel, at all types and sizes of properties, need better training opportunities.

Most major brands do have sales training programs for their personnel at their corporate operated properties as well as those franchised and managed by others. Where are we missing the boat? It would seem that the number one complaint, that salespeople do not respond to messages, wouldn't even exist in modern times with email, faster telecommunications with a wide variety of systems such as cell phones, and blackberries on hand with almost everyone.

Salespeople right now are as busy as ever with the high percentage of demand for meetings and convention space in just about all parts of the country, and it is not that salespeople are not doing their job. Most put in well above 40–50 hours per week, as well as working on weekends, to keep up with taking care of business. Part of the problem could be how the salespeople spend their time. When we raise this question during sales workshops, concerning how much of a salesperson's time is spent in pro-active sales, we get a response reflecting 35–50%. The remainder of their time is spent on doing a wide variety of other functions in connection with customer service or operations. Sales personnel need to come up with a Sales Action plan in connection with their Marketing Plan so they can be spending more time on communication aspects with business connections.

CHAPTER 8

HOSPITALITY SALES TRAINING

A BLUEPRINT FOR BIGGER PAYOFFS FROM YOUR SALES EFFORTS

February 1984

Want to make your sales effort more successful? The follow these guidelines when soliciting and booking business and servicing accounts:

▶ **Acquire and Use Good Marketing Knowledge**: Finding out who your customers are, where they are coming from, how they came to select your Property, how long they stay and how much they spend—among other information—will help focus your marketing effort.

▶ **Plan a Good Market Mix**: The mix of business reflects how much of what type of business you're doing. Ask yourself: What percentage of my total room sales comes from meetings, What mix of business would be most profitable?

▶ **Know Your Competition**: Having a thorough knowledge of the other properties near you can help you size up your property and determine the areas in which you can compete for the best-whether it is location, price, size, product, service or amenities. The idea, of course, is to sell your positives.

▶ **Question Your Multiple Price Policy**: There's nothing wrong with selective discounting. Hotels have been doing it for the years-with special off-season, corporate, group, senior citizen and military rates, among others. But which special rates are generating business for you? A periodic review of all your rates will help you establish the multiple-rate policy that's right for you.

▶ **Make Good Business Contacts and Make Them Work for You**: Getting good business contacts is the first step, and making sure they're bringing business into your property the second. Make sure your contacts are frequent users of your property-then ask them to provide you leads.

▶ **Try New Things**: Remember: most successful entrepreneurs would not be where they are today if they didn't take a chance and try new things. Come up with new ideas to promote business and don't be afraid to put them into action. Develop new sales techniques to book more rooms and new proposals to land more group business. If only half of your new schemes works, you'll be ahead of the game.

▶ **Be Attentive to Costs**: In any business, spending more than you take in, of course, is dangerous. Cost-effectiveness in selling for a hotel is very important. As total sales expenditures start to creep up, you must continue to expect a greater return from your sales effort. Budgeting for sales and monitoring the sales budget against results are two big musts.

▶ **Follow Good Management Procedures**: In general, try to do a better job communicating, developing, training, motivating, planning, organizing, directing and controlling. It's not enough these days to hire a salesperson and say: "Get out there and sell."

▶ **Recruit High-Quality People**: You want salespeople who are sincere, believable, down-to-earth, friendly, well dressed and well mannered-folks who will represent your property well.

▶ **Set Realistic Growth Plans**: Assess where you are today and decide where you want to be next year, the year after, the year after, and so on.

▶ **Sell Aggressively**: Aggressive salespeople are the ones who book the business. You can be aggressive while still being friendly, credible and sincere. Being tenacious, Following up, ensuring customer confidence-all these add up to aggressiveness.

MASTERING SALES METHODS—A 'DO-IT-YOURSELF' PROPOSITION

January 1986

No matter how many times we try to establish sales training programs for entry-level salespeople-or even to provide advanced training programs for seasoned salespeople-it really all boils down to a self-training situation.

There are, of course, company policies and sales training manuals that set standards for salespeople at a property or corporate headquarters. And there are certain forms and records that must be kept and systems for setting up and maintaining files. These are standard requirements established by companies for their people. When it comes down to basic sales techniques, however, it's a "do-it-yourself" thing; the more you do it, the better you get.

We need to challenge our salespeople to learn more about our customers and their businesses. It's very important that we learn what's going on in the customer's world so we can be better prepared to do business with that customer. By making face-to-face sales calls, we are able to learn more about the prospect and what his or her needs might be. This is really what selling is all about and it's so easy to do. It's merely a matter of asking; most people are more than willing to tell you about themselves, their businesses and what it is they're looking for. Then it comes down to knowing our product.

Once we know what the customer wants, and once we have a thorough knowledge of our product, then it's just a question of relating the customer's needs to the services our hotel or motel has to offer. An important selling technique developed in the field is finding out what we do right that our customers appreciate; what makes them want to come back to have more meetings and conferences with us. We should ask our existing customers why they like our property and its services.

Learning why present customers use what we're selling is very valuable to us is seeking new customers. What serves well for one group can also serve well for another group. This is where the salesperson must know the benefits a client realizes from coming to a property, Most of the time, we already know (or think we know) the benefits to a group using our property. However, by asking repeat customers what they enjoyed, we'll no doubt discover additional benefits we may not have been aware of earlier.

There's no question about it: the more calls we make, the more sales we make, Becoming an effective salesperson requires more calls, more inquiries, more leads followed up, more rejections, more digging, more effort in finding out what the client's needs might be, and more analysis of previous successes which can be applied to present selling situations-in short, a continuation of hard work and the selling process.

There's no real selling unless the seller and the prospect and up face-to-face, which means going out and making the calls or having prospects come in to look at the property. Before a sale is made, multiple calls must be made. It's unlikely a sale will be closed with the first "cold call" to a prospect. The object of a "cold call" is to gather information about the client, the company, the needs of the property can fulfill those needs.

It probably will take four, five or six calls on a prospective customer before an actual sale is made. So it's a question of making as many calls as needed to get the job done. Salespeople can learn as they go by asking prospects certain questions and listening to their responses. What works for one customer may not work for another. The idea is to be comfortable in making sales calls, comfortable in asking questions and comfortable in the relationship with the prospect. Becoming a professional requires an extended learning challenge. There are no shortcuts to knowing your customers. Developing selling skills is a self-training process that's accomplished in the field.

WHEN THE GOING GETS TOUGH, THE TOUGH GET POSITIVE

September 1988

There's no question about it: Times are tough in the hotel/motel business-not necessarily for the buyers, but for owners, operators, managers, sales personnel and others in our industry. For now; we're stuck with overbuilt (sorry, underutilized) situations, severe discounting by the competition, high employee turnover, and in some cases, poor products and service.

The list, of course, could go on. But don't despair-help is on the way. It comes in the form of a property's sales personnel. These are the people who can make a hotel hang tough (provided they themselves are able to hang tough). The old expression, "tough times don't last; tough people do "still holds true. There are too many negatives being thrown around nowadays, too many excuses for salespeople not being able to close more deals to help make the property more profitable.

We need to be more positive if we are to survive. Here are some suggestions that truly can help salespeople do a better job:

- **Get Back to Basics:** Our most important function, as Phil Wexler says, is "the acquisition and maintenance of guests." We need to be better than the competition in helping prospects solve their problems. Learn what the customer wants and needs, then deliver it to him. He'll react by perceiving your property as better than the others.
- **Repeat Business is a Must:** Provide continuous feedback to management as to ways to improve service. Get feedback from previous users. Get them to rebook. Retaining an old guest is much less costly than getting a new one.
- **Continue to Prospect:** "There's gold in them that hills." Really, there is! Personal calls, telephone calls, letters-whatever works, keep doing it. We must continue to find new accounts to replace the ones that are lost. Even in the smallest of towns, we can never run out of prospects to solicit.
- **Be More Aggressive:** This doesn't mean we have to be overbearing or obnoxious. It simply means we keep making the calls and keep asking for the business-always with a positive attitude.
- **Be Market-Driven:** This means the decision-making process needs to get as close to the customer as possible. The front-line salespeople are the ones closest to the customer. Everything we do must be geared toward meeting the customer's needs. Remember: the customer is still the boss.
- **Watch the Competition:** Don't take anything for granted-they're probably watching you. Keep in touch, know what's going on. (And by the way, it's not old-fashioned to scout the reader boards.)
- **Don't Let Tough Times Show:** Rumors spread like wildfire in our business, and they do get back to our customers. Always emphasize good news; be an actor if you have to, but accentuate the positive. Don't discuss bad times with anyone-especially, not with suppliers.
- **Be Accountable:** Establishing goals for all salespeople is a must. They should be reachable and measurable. Break goals down to a weekly or daily basis. There can be goals for actual sales, booked business, number of outside calls, etc.
- **Fish Where the Fish Are Biting:** Wherever our most profitable business comes from is where we should concentrate our efforts. If motor-coach tours represent only 2% of room revenue, why spend big bucks on that market segment?
- **Keep Salespeople Selling:** Managers make a big mistake when they use salespeople for other duties. Never take away their productive selling hours. If salespeople attend staff meetings, perhaps the

meetings should be scheduled for before 9 a.m. or after 4 p.m.

Forget all the reasons and excuses for lagging sales. Be positive, be enthusiastic-and sell profitably.

IDLE CHATTER MAY RESULT IN SILENCE WHEN CLOSING A SALE

April 1989

Recently, I was working with a salesperson for a particular product in which I was interested. This had nothing to do with hotel sales. Somehow, though, I ended with hotel sales. Somehow, though, I ended up getting a ride with him in his car which lasted about an hour.

From a salesperson's point of view, I certainly would not have objected at all had we discussed a particular product in which I was interested and if there were some questions pertaining to my needs or wants of the certain product. That Would have been very interesting to me, because I would have seen the man was doing his job and was trying to find out how much of a prospect I was and what I wanted out of the product.

However, somewhere along the line, this person must have gotten the idea that you must "talk your prospect to death." There was a content line of chatter which became most boring to me. And then a line of questions in which I was not interested at all and had absolutely no intention of answering-since most were of a personal nature.

I was dropping hints that I was not interested in his idle chatter and I was not interested in answering personal questions and didn't mind discussing his product. But these were to no avail. You needed a sledgehammer over the head to convince this fellow of my interests-what I wanted to talk about and what I didn't want to talk about. But without any indication that he understood my "gentle" hints he continued to harangue me with discussions of his trips, his hobbies, his personal life-everything except my need for the product he was selling.

Needless to say, I was ready to scram. I guess I could have been rude and just told him outright to shut up and let's talk about the product. But I was not comfortable doing that, So, you can imagine that by the end of our trip I was no longer a prospect and whatever sale might have occurred was lost.

What lessons can be learned from this? First of all, salespeople have to size up their prospect right away to find out if the prospect, indeed, is interested in any idle chatter; is interested in discussing his needs and wants pertaining to your hotel facility; if he's interested in talking about his own personal life or other areas of interest to him.

GENTLE PROBING NEEDED

This needs to be done by the hotel salesperson almost immediately. Then there's got to be some gentle probing on the part of the salesperson to find out what it is that the person or the person's organization needs from the hotel.

Through gentle probing and good listening techniques, a lot of information can come out that would be useful to the hotel salesperson in making a presentation and ultimately closing the deal. We should never be inquisitive to the point of getting involved and asking personal questions. This really has no place in our business.

All conversation should be business driven unless, of course, the prospect brings up some subjects which he would prefer to discuss. Then it's okay to proceed, but only upon his initiative. There's no need for idle chatter just because there is dead silence. Too much silence, of course, is not good either and sometimes the salesperson needs to take the initiative to have some discussion.

Open-Ended Questions

The discussion should pertain to the needs of the prospect. The conversation can take the form of open-ended questions of the client as they relate to needs or any discussion on the part of the salesperson in presenting the benefits of a particular feature of the hotel.

However, it is not necessary at all for the prospect to know anything about the salesperson's background or personal life. There is a time, however, that background and experience must be related. If any questions come up or there are any objections with regard to experience, then background information certainly can surface.

People selling hotel services need to be good listeners; they need to talk less, probe gently and sell more by learning more.

Asking Key Questions Can Help Boost Your Group Sales

September 1989

Sometimes, premature presentations kill sales instead of enhancing them.

All too often, salespeople at meetings, conferences, trade shows, sales calls, and on-site inspections jump at the chance to start selling. In their haste, they begin describing features that may hold no interest whatsoever for the contract. Frequently, this is done even before the salesperson finds out whether the person to whom they are selling is a prospect.

Not only is this a waste of time if the person is not a prospect; it often has the effect of killing a sale if, in fact, the contact is a prospect.

Focus on Client's Needs

In sales, we should divide our time between asking about our client's needs and presenting our product. But most hotel salespeople seem to spend too much time selling the hotel and pushing for a close, without taking into consideration that the majority of prospective buyers are more interested in their own needs than in what salespeople are trying to sell.

So it's only logical to spend more time discussing prospects' interests and needs rather than selling and closing. Remember: more time spent on questioning and listening to the client means less time spent on closing the sale. It is important to control the conversation by asking key, open-ended questions and writing down information gleaned from the answers. Find out what is important to the prospect and ask a lot of "why" will offer insight into providing alternate solutions to problems, particularly if the hotel can't address a specific need. Writing down the information makes the prospect feel that what is being said is important-which it is.

Also, the written word serves to remind both parties of what was said. Proper questioning persuades the prospect to list the features and benefits of the facility; these automatically will become advantages in the prospect's mind.

TIME FOR REVIEW

Even after all the questions have been answered, it still is not the time to sell. Rather, it is time to review all the items that were written down to see if anything else is needed. During this review, all of the items are related in terms of desired features, advantages and benefits so the prospect understands that the hotel can accommodate all of his needs.

There will, of course, be some items that will have to be negotiated and some objections to be overcome, but asking questions, writing down answers, reviewing and relating will save selling and closing time. In some cases, the responses made to the "whys" of the need may hint at alternative curses.

START-UP QUESTIONS

Let's say a deal is being negotiated to hold a company meeting at your property. Here are the kinds of questions that a good salesperson will ask the company representative:

- ▶ What's the goal of the meeting?
- ▶ What are you trying to accomplish by holding the meeting?
- ▶ What kinds of sleeping- and meeting-room space do you require?
- ▶ What are your entertaining and meal requirements?
- ▶ What dates and alternate dates do you need?
- ▶ Where was your last meeting held? What were the problems?
- ▶ Can you give me a detailed breakdown of the program, day-by-day, and hour-by-hour?
- ▶ Who will be attending? (Get demographics of attendees.)

These are some starter questions, and they aren't in any particular order. Remember to get a 'why' to an answer if it applies.

A good salesperson can't push a product without genuine concern for a prospect's needs. Spend 80% of the time asking questions and listening to the answers, 20% selling, and closing, and see what a difference it can make.

GOOD SALESPEOPLE HAVE OPPORTUNITIES TO BECOME GREAT

September 1989

There are many things that go into making a hotel salesperson great. Sometimes we'll hear somebody say, "That person is a born salesperson." I really don't believe that's true. Energy, looks, motivation, personality—all of this helps (and perhaps genes have a little to do with it).

Whether or not you have the right genes, there are many opportunities to make yourself great in the hotel/motel sales business. Here are 10 areas that are very important in improving a person's ability to become great in the business of selling for hotels and motels. A little practice in each of these areas will significantly increase a person's ability to sell:

▶ *Listen:* Listening more and talking less make you appear intelligent. Just ask some good, key questions and keep quiet until it's your turn to talk again.

And don't oversell. Silence sometimes can be a powerful tool. Learn to let "them" do the talking.

▶ *Features/benefits:* it's OK to discuss the features of your property-provided they can be discussed as a benefit to a prospect. There's no point in bragging about parking for 500 automobiles if everyone in the group is flying in.

Knowing a potential client's need and applying benefits and advantages as a result of your features is what works.

▶ *Looks and answers:* You can create a good or bad image by your looks, your voice and your ability to communicate effectively.

People make instant deductions from what they see. How you look upon contact with a prospect should be a prime consideration. An energetic and enthusiastic tone of voice is an important attribute of good salespeople-and knowing what to say and how to say it is another skill that they acquire.

▶ *Motivation:* Great salespeople are usually great motivators. In our business of selling hotel/motel space, it's important that we take on the job of internal motivation.

Service is such an important part of selling that salespeople could and should be the driving force to get service improved at the properties. Be the self-appointed morale-builder of the troops—it does help in the selling process.

▶ *See and be seen:* Be sure to get out of the property and be visible to the business community.

See as many people as possible, and be sure that you've seen at important local business events. People like to do business with people who are well-known.

Be believable—Do what you say you'll do; meet deadlines on time; deliver what was promised. Get a reputation for being honest, reliable and credible. Don't make promises that you can't keep or are not authorized to make.

Be a good communicator—Learn to be articulate and use proper grammar. Avoid using slang expressions. Get rid of the bad habits of repeatedly saying, "ummmm," "you know," "like," etc.

Good speaking habits add strength to your sales position.

Know your stuff—You must have knowledge not only of your own property, but of your competition's properties and the business of your prospect.

Read trade magazines; go to seminars; get active in as many industry organizations and associations as possible. It helps in selling if you're knowledgeable about the business.

Be determined—Persistence usually wins out for hotel and motel salespeople.

Don't let a "no" keep you from trying again and again and again. Be careful; a pest will never be a good salesperson . . . but persistence does pay off in the end.

Have confidence—sellers' confidence translates into buyers' confidence. Without confidence in yourself or your hotel unless you're a good actor-you can't be successful in sales.

If you practice all of these suggestions, your confidence will improve and so will your hotel/motel sales performance.

TIPS FOR SELLING IN TOUGH TIMES

There's no question about it. Times are tough in the hotel/motel business-not for the buyers (lucky for them), but tough for the owners, operators, managers, sales personnel and all other connections to our business. We're stuck, for now, with the overbuilt (sorry, underutilized) situation, severe discounting by competition, high turnover, poor products (in some cases), poor service (in some cases), etc. The list, of course, can go on.

DON'T DESPAIR! HELP IS ON THE WAY!

Help comes in the form of property sales personnel. These are the people that can make a hotel/motel hang tough-provided, of course, they themselves are able to hang tough. The old and probably worn expression, "Tough times don't last. Tough people do"-still works. There are too many negatives being thrown around because of the overbuilt situation-all kinds of excuses for salespeople not being able to close more deals to help make the property more profitable. We need to be more can really help salespeople do a better job.

1. *Let's get back to basics.* Our most important function is the acquisition and maintenance of guests (thanks to Phil Wexler for that great line). We need to help our prospects solve their problems *better than our competition can do.* Learn the wants and needs of the customer-then, deliver it so that it is perceived as "better" than the other property can provide.
2. *Repeat business is a must.* Provide continuous feedback to management in ways to improve service. Get feedback from previous user-get them to rebook. Retaining an old guest is much less costly than getting a new one.
3. *Continue to prospect.* "There's gold in them that hills." Really, there is! Personal calls, telephone calls, letters—whatever works—keep doing it. We must continue to find new accounts to replace the ones that are lost. Even in the smallest of towns, we can never run out of prospects to solicit.
4. *Can we be more aggressive?* Sure we can. This doesn't mean we have to be overbearing or obnoxious-just keep making the calls-keep asking for business—always with a very positive attitude.

5. *The whole property needs to be market driven.* This means the decision-making process needs to get as close to the customer as possible. The front line salespeople are the ones closest to the customer. Everything we do must be adapted to meet the needs and wants of the customer. The customer is still boss!
6. *Keep an eye on the competition.* Don't take anything for granted -they're probably watching you. Keep in touch, know what's going on and by all means -it's not "old fashioned" to scout the reader boards.
7. *Don't let the tough times show.* Rumors spread like wildfire in the hotel/motel business and bad rumors do get back to our customers. Always emphasize good news-be an actor, if you have to, always accentuate the positive. Don't discuss bad times with anyone-customers, employees and especially suppliers.
8. *Be accountable.* Established goals for all salespeople is a definite must. Goals need to be reachable and need to be measured. Break goals down to a weekly or even daily basis. There can be goals for actual sales, booked business, number of outside calls, etc.
9. *Fish where the fish are biting.* Wherever our most profitable business comes from is where we need to concentrate our efforts. If motor coach tours represent only 1–2% of room revenue, why would we want to spend big bucks on that market segment?
10. *Keep salespeople selling.* Managers make a big mistake when they use salespeople for other duties. Never take away their productive selling hours. If salespeople attend staff meetings, then perhaps the meetings need to be held before 9:00 AM or after 4:00 PM or maybe weekends.

Forget all the excuses (or reasons) why sales are behind. Be positive, be enthusiastic, sell profitably and be proud of where you work so that where you work will be proud of you.

"I WAS A CAPTIVE PROSPECT"

Recently I was working with a salesperson for a particular product in which I was interested. This had absolutely nothing to do with hotel sales. Somehow, though, I ended up by getting a ride with him in his car which lasted about an hour. From a salesperson's point of view, I certainly would not have objected at all had we discussed a particular product in which I was interested and if there were some questions of me pertaining to my needs or wants of that certain product. That would have been very interesting to me, because I would have seen that the man was doing his job and he was trying to find out how much of a prospect I was and what I really wanted out of the product.

However, somewhere along the line, in "Salesmanship 101"- this person must have gotten the idea that you must "talk your prospect to death." There was a constant line of chatter which became most boring to me. And then a line of questions in which I was not interested at all and had absolutely no intention of answering—since most of them were of a personal nature. I was dropping hints all over the place that I was not interested in his idle chatter and I was not interested in answering personal questions and didn't mind discussing his product. But all these were to no avail. You needed a "sledgehammer over the head" to convince this fellow of my interests—what I wanted to talk about and what I didn't want to talk about.

But without any indication that he understood my "gentle" hints—he continued to harangue me with discussions of his trip to Alaska, to India—of his hobbies, his personal life—everything except my need for the product which he was selling.

Needless to say, I was ready t "scream." I guess I could have been rude and just told him outright to "shut up" and let's talk about the product. But I was not comfortable doing that. So, you can imagine that by the end of our one-hour trip I was no longer a prospect and whatever sale might have occurred as a result of being together was certainly lost by this what I would call an obnoxious and very boring salesperson.

What lessons can be learned from this? First of all—salespeople have to size up their prospect right away in just a few minutes to find out if the prospect, indeed, is interested in any idle chatter; is interested in discussing his/her needs and wants pertaining to your hotel facility: if they 're interested in talking about their own personal life—or their areas of interest to them. This needs to be done by the hotel sales almost immediately.

Then there's got to be some gentle probing on the part of the salesperson to find out what it is that the person or the person's organization needs from the hotel. Through gentle probing and good listening techniques—a lot of information can come out that would be useful to the hotel salesperson in making a presentation and ultimately closing the deal.

We should never be inquisitive to the point of getting involved and asking personal questions. This really has no place in our business and absolutely has nothing to do with closing a deal for your hotel. All conversation should be business driven unless, of course, the prospect brings up some subjects which he/she would prefer to discuss—then, it's okay to proceed—but only upon their initiative.

There's no need for idle chatter just because there is "dead air"—there's nothing wrong with a time of silence. Too much silence, of course, is not good either and sometimes the salesperson needs to take the initiative to have some discussion. The discussion should pertain to the needs of the prospect. The conversation can take the form of open-ended questions of the client as they relate to needs or any discussion on the part of the salesperson in presenting the benefits of a particular feature of the hotel as an advantage to the individual or the group.

However, it is not necessary at all for the prospect to know anything about the salesman's background or personal life. There is a time, however, that background and experience must be related. If any questions come up or there are any objections with regard to experience—then background information certainly can surface. People selling hotel services need to be good listeners: they need to talk less: probe gently and sell more by learning more.

Yield Management: A Bad Rap?

More and more travel and meetings trade magazines are getting on the bandwagon of alarming meetings managers and corporate travel manager of this so-called "new" process of yield management. Of course, we all know there's really nothing new about this and we've been practicing this management tool for many years. Only recently have we been able to take advantage of technology to help us do a better job in maximizing our revenue.

Salespeople need to alert themselves to the negative reaction they will be receiving from good accounts because of these articles being circulated. Before long yield management will be a topic for discussion (it has already started) at conferences of meeting planners, corporate travel managers, incentive buyers, and travel agents. Let's be prepared.

Learning the workings of the process is important for all salespeople so that discussions on the subject can be handled positively. Sales personnel have the responsibility of generating profitable sales. In 1987, U.S. hotels and motels lost a median of $748.00 per room. The year before, 1986, the loss was in the $500.00 range, per room. When the figures come out for 1988 it probably won't look any better. The most probable reason for the loss was an oversupply of the room. However, besides that room rates across the board, did not rise sufficiently to eat up the additional 8% of costs hotels and motels incurred. It's just not easy making a profit in the hotel business today. That's why there is such a thing called a sales department, with the responsibility of bringing in additional, profitable sales.

A yield management process is a great tool for a sales department. If worked properly, the system provides a rate for a particular room, on a particular day, month and year. The rate prescribed is based on a number of factors: projected occupancy and average rate for any particular day, for any particular type or location of room: anticipated pick up of room blocks, anticipated walk-ins at rack rate, past experience for that particular time of year and, of course, other factors that would impact occupancy and rate.

The whole thing has to do with supply and demand. During low occupancy periods, we need to bring in business and we're willing to sharpen our pencil a bit. During heavy occupancy periods, we'll need to stick to the rate to maximize revenue. Other considerations also need to be acknowledged. For example, in dealing with groups, the expected revenue for food and beverage would certainly impact the decision on the rate.

Then, of course, we're still in the people business and there are other factors that come into play. We can't rely solely on a computer-generated decision as far as the rate is concerned. There's the question of "how important is this piece of business to the hotel or motel"? It could very well be that Doris Sklar at General Electric should get a better rate for her group since she can book many more meetings at your hotel—volume buyers certainly need to be looked at differently than a one-time buyer. There is such a thing as cheaper by the dozen.

At times there will be a group that wants your property during high occupancy times, but can't or won't pay the high rate. If this organization can place another piece of business with your property during a low occupancy period—then you'd certainly want to consider making a deal.

Jim Jones, formerly of Connecticut General fame, and now a meeting planner, sales consultant, and a great speaker as well as sales trainer for hotel companies, used to have a "thing" about negotiating rates. It used to be his philosophy that his business was so strong and powerful for a hotel that the contract would have to state that no other group in the house at the same time during his meeting would get a rate lower than CG. Many other companies picked up on this and are getting away with it during the negotiating process. To make yield management work, hotels should not "give-in" to this thinking. We have to explain to meetings managers that something could very well come up that would require you to bring in a group at a rate lower than theirs. It happens all the time you book a group during a high rate period at what the group would consider a favorable rate. Somewhere along the line, another group booked during the same period cancels or drops a chunk of the room block. Now—short time out—you're stuck with a hole. Sometimes we'll want to fill it any price—it happened to all of us.

As the subject of yield management comes up for discussion with buyers or at meetings let's meet it head on and explain our position. After all, reasonable people understand basic business practices. All we want to do is to be profitable. Salespeople need to have, not only a positive attitude about yield management, but an enthusiastic approach to dealing with the issue when it pops up and it well!

DON'T START SELLING TOO SOON

Sometimes premature presentations kill sales rather than enhance them. Too many salespeople at meetings, conference, trade shows, on sales calls and on-site inspections jump at the chance to start selling. They start describing features which may have absolutely no interest to the contact. This is all done even before finding out if the person being "sold" is a prospect. This not only is a waste of time if the person is not a prospect, but sometimes kills a sale, in fact, if the contact was a good prospect.

Generally, in sales, we divide our time between questioning for needs and presenting our hotel product. Most hotel salespeople seem to be spending too much time selling the hotel and pushing for a close. It would appear that most prospective buyers are more interested in their own needs rather than what we are trying to sell. It would be very logical then to spend more time discussing prospects' interests and needs rather than selling and closing. The more time spent questioning and listening will require less time in closing.

It is important to control the conversation by asking key, open-ended questions- and writing down information. Find out what is important to the prospect. Also, ask a lot of "whys." The "whys" will give insight to providing alternate solutions to problems if the hotel cannot provide a particular need. By writing down the information it makes the prospect feel that what is being said is important—and it is! Also, things written don't get changed easily—it serves as a reminder to both parties of what was said. Actually, during the questioning (if it is done properly) we are getting the prospect to list the features and benefits of the facility and they become advantages automatically in the prospect's mind. Once all the questions are answered it is still not the time to sell. Now is the time to review all the items that were written and to see if there is anything else that was needed. During the review, all the items are related in terms of desired features, advantages, and benefits; so that the prospect clearly understands that the hotel is able to accommodate the needs.

There will, of course, be some items that will have to be negotiated and some objections to overcome. However, the process of asking questions, writing down the answers, reviewing and relating will save the selling and closing time. In some cases where particular needs cannot be met the response made to the "whys" of the need may provide a hint to an alternate cure for the need. Some questions to ask which will help in getting the process started for a company meeting as an example:

- What's the purpose of the meeting?
- What are you trying to accomplish?
- Explain sleeping and meeting rooms space required.
- What are the entertainment and meal requirements?
- When are the dates and alternate dates needed?
- Where was this held last time and what were the problems?
- Please give me a detailed breakdown of the program day by day, hour by hour.
- Who will be attending (get demographics of attendees)

These are some starter questions, not in any particular order, and certainly, questions should not be limited to these alone. Where it would be helpful, always try to get a "why" to an answer, if it applies. A good salesperson can't simply just "push" a product without a genuine concern for the prospects" needs. Spend 80% of the time asking and listening—20% selling and closing—and see what a difference it makes.

TEN CLUES TO BEING A GREAT HOTEL SALESPERSON

There are many things that go into a good formula for making a hotel salesperson great. I really don't believe the old adage that salespeople are born, not made. Sometimes we hear people talk about "that person is a born salesperson." Energy, looks, and personality—it all helps and perhaps the genes have a little to do with it. Whether you have the right genes or not there are opportunities to make yourself great in hotel/motel sales. Here are ten areas that are very important in impacting a person's ability to become great in our business of selling for hotels and motels. Practice in each of these areas will appreciably increase a person's ability to sell.

1. *Listen*: The more you listen and less you talk makes you appear intelligent. Just ask some good key questions and keep quiet until it is time to talk again. And don't oversell. Silence, sometimes could be a powerful tool. Learn to let "them" do the talking.
2. *Features/Benefits*: It sure is OK to discuss features of your property—provided

it could be discussed as a benefit to a prospect. There's no point in bragging about parking for 500 cars if everyone in a group will be flying in. Knowing needs and applying benefits and advantages as a result of your features is what works.

3. *Looks and Demeanor*: You can create a good or bad image by your looks, your voice or your ability to communicate. People make instant deductions by what they see. How you look upon contact with a prospect should be a prime consideration. An energetic or enthusiastic tone of voice can also impress someone—and then knowing what to say and how to say it is another necessary skill.

4. *Motivation*: Great salespeople are usually great motivators. In our business of selling hotel/motel space it is important that we take on the job of internal motivation. Service is such an important part of selling that salespeople could and should be the driving force to get service improved at the properties. Be the self-appointed morale builder and motivator of the troops—it does help in the selling process.

5. *See and be Seen*: Be sure to get out of the hotel/motel and **GET VISIBLE**. See as many people as possible and be sure that you are seen at important local business events. People like to do business with people who are well known.

6. *Be Believable*: Do what you say you'll do: meet deadlines on time: deliver what was promised. Get a reputation for being honest, reliable, and credible. Don't make promises you can't keep or are not authorized to make.

7. *Be a Good Communicator*: Learn to be articulate with using good grammar. Avoid using slang expressions. Get rid of the bad habits of repeatedly using "ahhhs," "you know," and "like." Good speaking habits adds strength to your sales position.

8. *Know Your Stuff*: You gotta have knowledge—not only of your own property but of your competition and the business of your prospect. Read trade magazines, go to seminars, get active in HSMAI, MPI, ASTA. It helps in selling when you are knowledgeable.

9. *Be Determined*: Persistence usually wins out for hotel/motel salespeople. Don't let a "no" keep you from trying again and again and again. Be careful not to be a pest or overbearing—PIA will never be a good salesperson but persistence does pay off.

10. *Have Confidence*: Sellers confidence = buyers confidence. Without confidence in yourself or your hotel—unless you're a good actor you can't be successful in sales. Practice all the previous nine suggestions and your confidence will stand out.

TRAINING, TRAINING, TRAINING—IT'S STILL IMPORTANT

February 1990

In the food and lodging industry, it seems that we just don't pay enough attention to the importance of training and retraining. Of course, we see and hear a lot about this from some of the major chains and independents. And that's good, but not enough for the industry. We 're so far behind, and away from where we should be in technical skills and people skills.

One of the biggest industry problems we face today if turnover. Based on a variety of surveys we've seen lately, generally in the top five of the reasons for turnover has been: "not properly trained for the job." we hear a lot about the importance of training and development, but see so little of it. Most of the business is "on the job." This is where an "experienced" employee shows the new employee "how it's done." This could be good, but too often it is bad.

To ensure success in our industry owners, operators, and managers need to make a commitment to training. Those who have done it have been rewarded by improving service and reducing turnover. Ask the people at Hyatt, Marriott, Disney World, McDonald's and, of course, much other successful food and lodging operators.

A new study was completed recently by the American Society of Training and Development (with a grant from the U.S. Department of Labor) which concluded "job-related training is critically important for both individual opportunity and business competitiveness" and that training is both insufficient and unevenly distributed. The report includes recommendations to employers, educators, and the government with the principal challenge being to the employers. One of the report's recommendations was for employers to spend 2–4% of payroll for training.

These days with sales and marketing becoming more and more important in the struggle for market share and fighting the so-called "overbuilding" situation, owners and operators will be looking for a good return on investment from the sales expenditures. This is an area where we really need to take a strong look at the training of our salespeople. The number of people who have come into the lodging sales ranks with a formal education in sales is rather small. And the number of people who have joined lodging companies in sales capacities with experience in sales is also rather small. Therefore, it looks like we really should be doing something about doing a better job of providing sales education and training.

If we accept the recommendations of the report mentioned above, then let's consider spending 2–4% of sales payroll to provided education and training for the sales staff. If a lodging operator is investing $100,000 a year on a sales department payroll, it would seem reasonable to protect that investment with an additional $2–$4,000 to help the troops do a better job. The education and training dollars could be invested in a variety of ways:

Arranging through the Educational Institute of AHMA to prepare and conduct sales training programs for the staff.

Purchase and use the Educational Institute's video training tapes, books, and manual for self-training programs.

Pay membership dues for sales personnel to join and become active in organizations that are dedicated to education such as: Hotel Sales and Marketing Executives International: Meeting Planners International: American Society of Travel Agents: National Tour Association. And there are others, of course.

Encourage salespeople to attend national, regional and local educational conferences related to our business and make them company expense paid trips.

Pay for subscriptions to trade publications that have educational articles (most are free, anyway).

Support your salespeople in their own, Personal educational and training endeavors.

Have periodic sales training programs, in-house, using your own sales staff to conduct training sessions.

Continuous education and training are vital to anyone's career. Salespeople, especially, need to improve with technical and people skills. If a property doesn't have a formal, structured approach to training. It's OK. The idea is for salespeople to work towards that end little by little with their General Managers, and owners to strive for a training budget—no matter how small—just make a start.

YOU'RE WANTED ON STAGE

April 1990

Last November we wrote about how a video of your sales presentation would a great tool in getting you to improve your technique. By reviewing your own performance on tape you could easily see how you can do better. I mentioned that I experienced this sometime back and found it to be most helpful. I further stated that I would want to do it again to see how I can still improve my presentations when I conduct seminars and workshops for hotel companies and groups.

It didn't take long for the offer to come from Paul Cahill of Cahill Associates in Westport, Connecticut (see letters to the editor, February 26, 1990 issue). After some telephone discussions, we agreed that I would be caught on film while conducting a seminar in Dallas for the western region of the American Society of Travel Agents. Two ninety minute videos were made while we were in session. We then set a date to meet in Roanoke, Virginia, after Paul and his wife, Ellen, had a chance to review the tapes. There I was, in full color and sound! What an enlightening experience.

Most people who have gone through this probably will agree that anticipating viewing yourself on tape is a very uncomfortable feeling, but very rewarding. Paul handled it beautifully; nothing intimidating about it, he made me feel comfortable; he was energetic, enthused, and very likable; I felt "at home." Paul walked me through the tape, stopping it after every few minutes to discuss what we saw on the monitor. We looked at my introduction, my approach (how I accepted the applause and greeted the audience); the distribution of handouts; how I got people involved in the discussion; handling of questions. We discussed how to make better use of the overhead projection equipment and where to stand when using one; about my hand in my pocket (a no, no) about an hour and a half later, with four pages of notes, we wrapped up the viewing and critique. And, I felt great what great ideas I picked up from Paul. I always thought I was pretty good, but now I know how to be 100% better in my training seminar and workshop presentations.

Since my visit with Paul I've already used my "new" techniques while conducting a full day educational program for the Belz Hotels General Managers conference held at the Peabody Hotel in Orlando and it went just great. Paul's stuff really works here are samples of some of his comments:

▶ Expressing of ideas is more important than delivering words.
▶ Eye contact is more important anything else.
▶ Don't let audiovisuals (overheads) compete with you.
▶ Use hands to energize the audience.
▶ Let the audience see you listening when you speak.
▶ Don't spend too much time on issues that are not relevant.
▶ Deliver brief answers to questions.

Presentation skills training is only one of the programs that Cahill Associates offers. Since all hotel employees are actually "selling" their property in one way or another, Cahill Associates has developed many programs for the service sector. The service programs use unique training methods to help line and staff employees more effectively present themselves and their hotels.

YOU'LL NEVER KNOW HOW MUCH BUSINESS YOU'VE LOST BECAUSE . . .

June 1990

Your telephone can't tell you. Of course, we can't get along without it—we love it, we'd be lost without it so much business is conducted over it. And yet, there's information how many millions of dollars in business are lost to motel, hotels, resorts, and conference centers because we don't know how to use it.

It's the incoming calls that are hurting us—not the outgoing—although we probably need help there as well. We all are pretty well aware of the problems involved with incoming inquiries:

▶ Takes too long for switchboard to answer.
▶ Operator answers and puts the caller on hold.
▶ Operator doesn't come back after putting the caller on hold.
▶ Call relayed to the appropriate department, but again, takes too long for someone to answer.
▶ Call relayed to the wrong department.
▶ Call screened two, three or more times before getting to the right party or department.

And we can all add to this list, but you get the idea. Perhaps interested General Managers can get the local phone company to install a device that measures disconnect calls. This is a system that keeps a county of all calls where the caller hangs up before the call is answered. It would also be good if they could measure the number of disconnects after the main number is answered, and the callers are unable to reach the desired party. How many times do we get through to the main number, but upon being transferred, no one answers, but the switchboard operator doesn't come back on the line? Anyway, I don't believe there is a system to track this.

The solution to getting the phone answered at the switchboard (or reception) is to make sure that the person responsible for that job is not encumbered with taking messages or providing information or gets "tied up" with something else. This is generally the major reason for delays in getting the telephone answered. The function of the person answering the phone is to make sure *THE PHONE GETS ANSWERED QUICKLY* and relay the call to the appropriate party or department.

Information calls and message taking needs to be passed over to someone else. Too often at many lodging operations, the telephone is answered by front desk personnel who are busy doing a number of things in addition to taking telephoned room reservations, answering inquiries and handling guests incoming calls.

This looks great for controlling labor costs, but this needs to be related to the number of room nights lost, not only for individual reservations. But for groups as well, or the number of unhappy prospects who hang up after waiting too long. It's a good guess that more than half of the prospective business lost has to do with the initial telephone contact. Sometimes it may not have anything to do with the delayed answering. The reason could be because of the way the phone is answered, how it is relayed, being put on hold. The person on the answering end not knowing what to do, etc. We've all experienced this, of course, but with other business, *not our own.*

General Managers and salespeople need to check out their own properties by having someone call to try to do some business and see what happens. Callers should try different days of the week and different hours of a day. You should be wondering what happens after 5:00 pm during the week, on weekends and holidays.

You'll find it to be most interesting.

TIME MANAGEMENT FOR HOSPITALITY SALESPEOPLE

October 1991

Most of us never seem to have enough time in a typical workday to get the job done. We all get busy involved in doing things we never planned to do and we get our normal, everyday interruptions. There are just so many hours in a day and it is how we use our time that determines whether we can complete the jobs that we want to get done. In hospitality sales, time is real money. What we need to do is work on some kind of time management program to ensure that our time is spent more productively to bring profitable sales into a property. We all seem to get hung up on spending time on things that are not sales related. If more time was spent in selling there is no question about the results creating more bottom line profit.

If we want to get serious about this business of good time management then we have to challenge ourselves to get rid of all the things that waste our time. What are some of the typical time wasters in a lodging sales operation?

Meetings: We probably have to attend too many that are non-productive. Sales meetings or on property meetings involving salespeople should be scheduled any time that would not interfere with selling time. Think about 8:00—9:00 am or late afternoon or even Saturdays. Meetings should have a purpose a reason for the meeting to be held with anticipated results. Persons who are to make reports at the meeting need to be notified well in advance. The number of people in attendance should be kept to a minimum only those that need to be there. There should be an agenda, with a time slot for items to be discussed and a time for when the meeting should be adjourned.

Visitors: We probably get loaded down daily with our own property staff people visiting just for the purpose of schmoozing and wasting our time. Salespeople always want to be friendly and pleasant, so we put up with these kinds of visits. Try not to let these visitors get too comfortable. Stand when they come and don't give them a chance to sit down. Then walk casually towards the door and when a break in the conversation takes place, thank them for coming and apologize for having to get back to "get something out right away." This really does work. Remember, visitors who are customers or prospects, or staff members who help you with your work should always be welcomed.

Being imposed upon: this is a delegation in reverse. Where someone else gives you a job to do that takes away from productive sales time. We've all experienced being front desk substitute during lunch hour; hostess or cashier in the dining room; running errands to the bank or post offices; picking up items from the florist or bakery. Talk this over with your supervisor, General Manager, or whoever, and get it understood that in an emergency, yes, we all have to help out, but let's not make it a habit and expect the salespeople to be readily available to help out frequently.

Each of us can probably do pretty well in constructing a list of typical daily interruptions to our work real time wasters. You may want to make such a listing and see how you can eliminate some of them thereby making more time available for sales calls.

Think of time as an investment or think of it as an investment portfolio. You see, when you invest your money you place it in areas where you feel it will give you the most profit. You do the same thing with time. You invest your time for the best return.

ARE OUR SALE PEOPLE BEING TRAINED?

May 1992

What's going on in the hospitality industry today has caused a good deal of concern among management and all employees. In the last two years, we've seen properties fail, close down, go into bankruptcy, and being taken over by others. We've seen overbuilt locations, intense competition, severe discounting, lower occupancies, increased personnel turnover and corporate restructuring which put a lot of people out of work. All this with little, if any, the bottom line has really put the stress on management, especially when owners are calling for more cash flow.

After you've cut all you can what's left to do? It looks like the only way out is to produce higher revenues-in all profit centers of a property. The more you take in at the top the more will fall to the bottom line. So, now where do we start? Logically, it's with General Managers' Involvement with Sales. It starts with a property manager taking over the supervision of sales. And, the first question to answer is: "who is doing the selling." Of course, we have always said: "everyone sells," and that's true. But, for now, let's discuss the sales department of a property and its effort. Here we are dealing with people and do we have the right person in the right slot? I'm fearful that we do not in about 25% of the cases. Many people in our industry got into sales by accident (I know I did). A matter of expediency, perhaps. Generally, it is the case of finding the first warm body who appeared "interested"—or perhaps promoting from within—moving up someone who was very "nice to people." Where we may have failed in these instances was in the selection and hiring process. Many properties have not produced sales job descriptions; most managers do not interview correctly and in many instances, references are not checked. This group of salespeople probably represent about 25% of all those in hospitality sales.

Let's now say we actually do have the right people in the right slot. These people look the part, act the part and seem to like selling, and in many cases have some sales experience. This is a group of salespeople who work hard, have great interest, will do anything they are told to do, but probably lack the necessary sales skills training to produce the sales needed, no fault of their but the fault of management, not being able to provide the training, the direction, the supervision and perhaps motivation. This group probably represents about 55% of those now employed in hospitality sales. The remaining 20% of the sale force population is the really productive ones. They've been well selected have been properly trained, directed, supervised and motivated to a point where they are successful and "holding their own" pretty well. Where these salespeople may have lacked the support of management, in some cases, they've gone out on their own and learned the business of selling and become good at it. These are now our Vice Presidents of Sales and/or Marketing, Sales Directors and Sales Managers who are in a great position to assist management in the development of new salespeople coming on board.

Other than sales skills being passed down from one to another. How do we get around to provide professional sales skills training? From time to time there is a seminar to attend. Some management companies and franchise groups provide sales training seminars for their own people. There are HSMAI (Hospitality Sales and Marketing Association International) educational programs. Associations such as Meeting Planners International and American Society of Association Executives, and others, provide suppliers seminars during their national conference. But there needs to be more of these available. And training funds need to be made available so that salespeople may participate in the programs offered.

There are many opportunities for independent learning of sales skills through books, tapes, and videos made available through HSMAI, MPI and the Educational Institute of American Hotel and Motel Association.

DO WE HAVE TO LIVE WITH DISCOUNTING?

October 1992

We sure have had problems lately, being plugged with an ongoing mix of unfortunate situations-the overbuilding, the Gulf War, the recession, unemployment situation, lowered purchasing power of the retired group, hurricane disasters, and even the uncertainty caused by the presidential elections. One of our most unfortunate situations is the one created by us the fact that we have trained the traveling public to shop for discounts and bargains. Yes, we have done it ourselves and we need to do something about it. WHY, even the high-end market buyers-those that have been used to paying the five-star price are looking for bargains at five-star hotels and getting them.

Individual travelers, corporate travel managers, meeting planners, association executives, motor coach tour operators and even travel agent (who work on commission only) are looking for bargains for their clients. It is interesting to observe these types of third parties bookers, the middlemen of the industry, look for better room rates, comp rooms, free or lower cost food/beverage functions, free meeting space. For the most part, they all know that what they are getting when you quote them a rate is a good value for the price. Just about on one, these days is gouging, or trying to get more than a fair price for the rooms, food beverage, etc. So even with the knowledge that the price is right-they just don't feel good about themselves until they negotiate something better. We've probably taught them that they could get a better price by asking for it. We are scared to death that a possible piece of business will go elsewhere for a little less cost so we are eager to give in and quote lower rates

This business of our worrying about market share, getting more than the next guy-regardless of price, is what's driving us into bankruptcy. And we see this all around us. Hotel losses the past few years has run into the billions. The minimal sale of rack rated rooms reflects what's going on in the area of discounting in order to fill rooms. Just about everyone has special rates for all persons and all seasons. Cut-rate prices have never been a cure for anything accept may be spilled fruit. We are in the midst of a cut-rate war and the beneficiary is the traveling public (individuals as well as groups), and they are not buying rooms unless it's offered at a special price and even then, they want it for less.

Now there is a place for discounting, if it is done properly and effectively, but we have to get out of this mass and get back to selling value for a deserving price. It is more of the question of doing a better job in selling and having a better understanding of a return on investment philosophy and from the General Managers all the way up to the front line hourly wage personnel. From the sales end, we need to get involved in more training of our customer contact personnel. These in sales, front office, catering, reservations, food/beverage wait staff and of course, all the department heads, really require more training in the area of selling values as well as providing value through customer service.

It is not going to be easy and it won't happen overnight. Some property will continue the discount to get business; they need to cash flow! And, there is nothing wrong with discounting as long as it is done selectively approach and started with the establishment of a few discounted special rate available under certain specific condition. And I sure do hope it works,

Certainly one of the places where we could start is in the area of group sales for the meeting, motor coach tours, reunions, catered functions, etc. It seems that just about every property starts out selling to the group market by giving free marketing space, a "special" rate, comp rooms and other extras. Isn't it logical that any special rate, any comp room, or free meeting space or lower meal prices would be determined only after we know how important that piece of business is to the hotel? It is during the negotiation process that special deals could be made once it has been determined how profitable the whole piece of business would be to property.

Why wholesale discounted prices to everyone? Let's get back to improved selling skills and techniques so that we don't have to give too much away to get the business.

DISCOUNTING: THE DOWNWARD SPIRAL!

December 1992

"The unsold room tonight is revenue lost forever." How often have we heard this? Sometimes our salespeople have been told to book business at any price, just to build the occupancy or to get a better market share. When you see five-star hotels with regular daily rates of well over $150 per night competing with three-star hotels for the $50.00 group rate we gotta be in trouble. It is happening! But it can't go on for long with that philosophy hotels will have to fold.

If there is no tomorrow, then the unsold room tonight is really revenue lost forever. However, there is a tomorrow, and next week and next month, etc. You see, what we do today does affect tomorrow. If we keep reducing rates to protect occupancy, then our competitors, to protect market share, do the same, and on a continuing basis, we have caused the downward spiral effect. Therefore, over a period of time, the marketplace has established rates much lower than the real value of the product. The customers then begin to accept the lower rates and we have trained the prospective buyers to become good price negotiators. So what happens tomorrow? We are expected to continue to sell our rooms at discounted rates, the competition does the same; sometimes we even discount lower and lower just to get the business and we end up going out of business.

There has been continuous great pressure on salespeople to produce sales. This comes from the owners, operators, and General Manager, who of course, are also under pressure to produce profits. Where we have probably missed the boat is the training of the salespeople to produce profitable sales. In today's marketplace, it is not only the occupancy which we must achieve, but we must be tuned in to the "bottom line" sales philosophy. If we can't make money on a sale let it go. Too often salespeople turn back to the General Manager to approve a "special" rate for a piece of group business because the competition is offering something lower in rate. There may very well be what we think is a good reason for this: it's offseason, there is a "hole" to fill. It's a slow weekend, or even just to beat out the competition. For one thing, we sure lose a lot of credibilities when we offer a group one rate, then agree to lower it to meet competition. The first thought that comes to the meeting planner is: "well, if they can come down even lower in price now, then the first price offered was really too high, to begin with. Maybe they should come down even lower."

What we are doing is telling our salespeople and front office manager to book the business- "filler up." What we really need to do is improve our selling techniques. We're doing too much selling on price alone. Let's start selling value, not just worry about beating out the competition for occupancy at any price. Some operators feel they have to "buy" every piece of business.

We must also always keep in mind that we are doing to the value of the property. Sometimes discounting today could be in the best economic interest of the property. Sometimes discounting today could be in the best economic interest of the property, for today. But on a continuing basis, with the downward spiral effect, it will end up diminishing earnings as well as the overall value of the property.

Granted, there is a time for discounting selectively. There are special promotions and packages as well as off-season rates and from time to time, other special deals created. We are not saying that every room must be sold at rack rate. But we must be careful not to indiscriminately reduce rates just to fill rooms and not produce a bottom line profit on a piece of business. We must learn to sell value. Each type of product, whether it be budget, limited service, mid-price, upscale, etc., has a value for its rate. Certainly, people will expect to pay more for a five-star product compared to a three-star product we have to sell the value.

Right now our industry is suffering more, not by the lower occupancy, but by the lowering of rates. When you are making rate and pricing decisions consider the implications of lowering rates and how it would affect earning and ultimately the overall value of the property.

Sometimes it is better to walk away from a piece of business.

YOU REALLY SHOULD GET INTO CONSULTATIVE SELLING

March 1993

If, in fact, you are good at what you do if, you have the experience, knowledge and are confident regarding your ability to be a good salesperson for your property then you need to get into this business of consultative selling. What we're discussing here is really nothing new. You have probably been providing advice to prospects and clients every time you made a sale, but perhaps did not consider yourself a consultant. A consultant is someone who proves professional or technical advice. Every salesperson is really a consultant, but perhaps we do not go about our business as such. Perhaps we are too intent on selling something that we forget to provide the advice and counsel.

One of the biggest complaints of meeting planners is that some salespeople are so aggressive about selling a lodging product that they forget about asking the pertinent questions to find out even if the meeting planner is really a prospect or if there are needs to be met. There really is no point in trying to hit someone over the head with product knowledge unless the person is a prospect for the property.

A characteristic of a good consultant is being able to listen intelligently. Unless we know *and understand* the real need of the individual or group we cannot really figure out how it helps them. The consultant asks specific question determine the need or not, then listens intelligently to look for clues and "hot buttons." Then relates to the response by providing specific information that meets those needs or wants. When you think about it, providing professional or technical information or advice is really doing the job of selling. Before we get into trying to make a presentation to sell a lodging facility we need to find out as much as we can about the individual or group the usual who, what, when, where, how is fine, but we need to get more deeply involved if we want to provide the consultative process. In the case of a meeting planner looking for an appropriate location and site for a meeting think about using questions that make the person, think, analyze, speculate and evaluate even some questions that stimulate emotion will sometimes help draw out feeling that will provide you with clues as to what is needed to help you show the benefits of using your property.

We do this by asking a question that will need to be answered by more than a "yes" or "no." Such as:" Tell me what's most important to your group in the selections of a hotel ." Or," Please tell me what you liked best about your last meeting." During a site inspection, once you know you do meet the needs of a group, a good question would be: "Describe to me how your group would fit into our space here at the property." This is where you walk the property with the prospect and they describe where the registration area will be; how the meeting rooms are set up; how you will arrange for the meals, breakout sessions, etc. As you go through this process you are providing a little professional and technical advice here and there to enhance their own specification.

The big thing about being a consultant is solving people's problems; helping them make decisions; getting them to feel comfortable by having you around. Building long-term relationships are very important in this business of consultative selling. Maintaining good contacts with key prospects over a long period of time, and, at the same time having them know that they can call on you anytime for advice even if they are not booking your property, makes you more of a professional and good consultant.

Probably, building your network of contacts and being reliable, trustworthy, honest, informative, helpful, and loyal (to your property and to your contacts) and getting into the consultative process will do more for you in closing sales than anything else.

BE SURE TO GET AS MUCH DETAIL AS POSSIBLE WHEN BOOKING MEETINGS

May 1993

When inquiries are made for group meetings, it is always in everyone's best interest to get a lot of questions answered right up front during the booking process. Sales personnel should always be tuned in to gathering as much information as possible so that they fully understand customers' need. This helps us to close a sale by being able to cater to those needs and provide the facilities at a price and value that will meet both the prospect's and the property's objectives.

So, let's look at some of the things we need to know to make sure the property is prepared to take good care of meeting the group's needs and that the salesperson is in a good position to be able to negotiate in the property's best interest.

▶ **Dates:** It's not enough to know when the meeting starts. When people will be checking into the property is what's important. Sometimes people will need to get their rooms in advance of a meeting, depending on travel time from point of origin. For resort meetings, there's a good chance many will want to check in well in advance of the meeting date. Same for check out. Some may depart well after the meeting is over. Another very important factor is to understand the flexibility of the dates. The prospect may be able to work out different days of arrival and departure that would better suit the property's needs.

Sometimes, moving the date a day ahead or back will help you tremendously in the profitability of the business.

▶ **Number and types of rooms needed:** It is not enough to know for a group to want 50 rooms to come in all sizes, types, locations, and prices. It is good to know what the group really needs. Some of the people may want or need suites; some who bring spouses may want two double beds or king-size beds; individuals may want king size beds. Get the idea? You really need to know the preferences of the people who attend the meetings. It just makes for better planning so that the front office is not burdened with problems at the time of reservations being made or even at check-in time.

▶ **What's the program schedule:** This is very critical for a property since it will be important to coordinate the use of the various meeting and function rooms. Today we cannot afford to go along with "hold all space" until the group decides to make up its mind about how many and which meeting rooms they will use. You need to maximize the use of all your space. Find out about specific needs for meetings, breakouts, breaks, receptions, planned meals, entertainment, and any other activities they plan on or off the property. The program schedule also helps prepare the hotel operations team for meal hours when they are not having planned group meals. For example, it the meeting needs to start at 8:00 am and there is no planned breakfast, it probably would be best to open the coffee shop at 6:00 am on the meeting days instead of the usual 7:00 am; just to make sure all the people get a chance to have breakfast well enough in time to get to the meeting by 8:00 am. All this information will also get you to realize just how profitable the piece of business may be to help you with your negotiations. Also, if any meals are taken off property this needs to be known as well, for staffing purposes in the dining rooms.

- **Schedule of planned meal and bar functions:** Get a schedule along with approximate attendance at each. You should not assume that because the meeting is for 50 people that the property should expect to feed or serve only 50 people. Sometimes the prospect will invite local people to the functions; or even break up the group where half go elsewhere to eat. If there is any planned entertainment or dancing involved with any of the meal functions the property should know this well enough in advance since the size of the room reserved for the function will be critical.
- **Handicapped Attendance:** This is extremely important these days with the ADA requirements. Be sure to find out about what the group needs are in this respect.

By having all this information in advance, you will be better prepared to discuss the group's needs in relation to the hotel facilities. This will, at the same time, help maximize the productivity of your time in planning the meeting with the prospect, help you in the negotiations and make it much easier for the operations people at the properties to meet the needs of the group.

THINGS ARE LOOKING UP BECAUSE OF IMPROVED SALES EFFORTS

April 1994

All reports indicate that our business is getting better. The last six months of 1993 and the first quarter of this year reflect improved occupancies, improved average daily rates and an overall improvement in revenues and profits, Owners are probably crediting the decrease in new construction and increases in demand. Operators and managers are probably crediting better-operating methods and cost reductions. But, we all know it's because we're doing a better job in sales. Oh yes, there's still a lot of discounting going on, but we're learning real fast that it can be overcome with better selling techniques.

Although most property budgets do not reflect a line item for training, we do see more and more General Managers and salespeople showing more interest in, and participating in training programs going on around the country. Just about all the major chains as well as many management groups are getting into the sales training business. The Hospitality Sales and Marketing Association International (HSMAI) is promoting an entire series of sales educational programs and they are getting great enrollments. Tom McCarthy's professional Seminars on Sales is very popular as are the programs put on by The Educational Institute of the American Hotel and Motel Association.

TRAINING IMPROVES PROFITABILITY OF A PROPERTY

August 1994

"When a property spends money on training employees, it makes itself more profitable"... those are the words of E. Ray Swan, President of the Educational Institute (EI) of the American Hotel and Motel Association. And, Mr. Swan knows what he is talking about. Since his start at the Educational Institute in 1978, he has grown the non-profit organization from $400,000 to $12 million in sales of educational products. It is now the premier organization for educational and training materials in our industry with products being shipped all over the world (96 countries involved in EI purchases) to hotels, hotel companies, franchise groups, motels, conference centers, universities, community colleges, high schools, as well as private school. EI textbooks are currently being used by 150 universities and 850 junior colleges or vocational schools. There are up to 5,000 independently learning students at any one time enrolled in self-study courses provided through the organization. A separate department headed by James Davidson arranges seminar programs and speakers for all types of organizations.

One of the very important programs provided by EI is certification opportunities for employees of the industry at all management levels. Through self-training programs department managers can prepare themselves for certification examinations in all areas of hotel management. The latest program, which is just one year old, is the new Certified Hospitality Sales Professional (CHSP), designed for persons contributing at least 50% of their time to the sales effort. Many General Managers have been certified in this area. Dr. George Wolfe is the person heading up this department for EI.

CONTINUING EDUCATION IN SALES IS ONGOING AND VITAL TO SUCCESS

March 1995

A colleague related to me recently that thirty years ago we had so few salespeople in our business and that they were located only in large, convention or resort type locations. They were most knowledgeable and spent their time developing group business for their properties. There was very little negotiations going on and the visiting public wasn't too savvy about what they wanted or how to make deals. These salespeople dealt mostly with association executives of associations or presidents of companies and the term "meeting planner" was relatively non-existent. Time has changed since them. We have educated the public to look for deals; meeting planners are all over the place and very knowledgeable about negotiations; all types of hotels with a wide variety of segments have been developed. And where there used to be four gas stations on every intersection we now see four limited service properties. Now there are thousands of salespeople looking for the same business for their respective properties. Not only are there salespeople for the large convention or resort properties, but just about every roadside, suburban, downtown, airport hotel, motel, conference center, the facility has someone doing the selling. If there is not a person dedicated to sales then the General Manager of the property is probably doing the seeking out of business and making the sales calls.

TRAINING IS STILL VITAL TO INCREASING PROFITABLE SALES

June 1995

There is a big cry out there in the hospitality workplace from people who want to do a better job for their employers. We hear it all the time from employees in a wide variety of types of jobs in hotels, motels, resorts, conference centers, restaurants, car rental companies, attractions, airlines, cruise ships and just about any entity involved in this business. It seems that we either don't have the funds for training, or we don't have the time, or we don't know how to train. If we want our employees to be successful in their jobs (and they surely want to be) then we have to provide something more or even better than the typical on the job training—which is probably what's happening now in most of our properties.

We hear it all the time, particularly when it comes to sales. People are hired into a sales job because they look good or talk good, or know somebody and maybe have had some exposure to sales such as a sales secretary or front office experience. Or we hire someone away from another property and think they will be able to steal some business, perhaps. And, these new hires also probably learned the business while on the job.

Now with the proliferation of limited-service properties, we find those General Managers, who have to get the sales job done without a sales department, are crying for help. Most of these GMs have never been involved with sales and now they find

Even with the industry has made great gains in occupancy and average rate these past two years, and with 1995 looking great, we still have to do some selling. The competition will never go away. How do salespeople spend their time? Are they spending more time servicing than selling? Do they spend enough time prospecting for new business since every piece of business doesn't repeat all the time? Are there better ways to manage sales time so that we are more productive? Is the relationship between sales and operations favorable so both are working towards the same end of profitable sales? What about this business of technology? How does it help sales? Does it make sense to automate the sales effort? How does the salesperson become more of a consultant and build relationships? What are the characteristics of successful salespeople and how do these traits help close the sale? Don't we want our salespeople to be more knowledgeable about the customers so that they can do a better job in bringing in the "right" kind of business? There are ways to negotiate with groups so that it works to your best advantage do our salespeople know the techniques? Which market segments are the best and most profitable to your property and how do you go after them? Do you know how to change the mix of business to improve the average rate and make more profit?

One of the biggest problems in our industry is our inability to handle inquiries effectively. Inquiries come to us every day by phone, mail, fax, and walk-ins, so what's the best way to handle these "hot" opportunities? The inquiry is the hottest lead, but we seem to fail to handle it properly.

themselves grasping for any piece of help they can find.

Probably the best source of training materials and training programs today in the hospitality industry is the Educational Institute of the American Hotel and Motel Association. By the end of this year, that organization will have served over one million people, around the world, through courses, books, video tapes, audio tapes, seminars and certification programs. Most state hospitality associations conduct training seminars in a variety of topics, but they also must do more in the area of education and training for their membership.

CLOSING THE SALE IS CRITICAL FOR INDIVIDUAL RESERVATION INQUIRIES

September 1995

We're still not getting the job done properly at the property level in getting reservations made for those people making telephone inquiries. Oh, some locations think they are doing well because of the recent high occupancies and revenues we have been seeing; however, we still do not know how much room business we are losing over the phone. Ask around and you'll find out that only a few hotels, motels, and resorts are measuring the ratio of inquiries to reservations made. It just seems funny that in an industry which spends millions of dollars annually in advertising and promotion, hopefully, to direct the consumer to individual properties, no measurement is being made of the amount of business not being booked. We're not talking about the business we don't get because we don't have space available or we can't "sell-through" because of closed out dates. It is simply a question of how well do we do the job of selling on the phone when people call to inquire about availability, rates, etc.

We could do a much better job of closing sales over the phone if we would take the time to train our people to do some things better that would encourage people to buy. For example, how about speaking more clearly, distinctly and slowly so the inquirer wouldn't have to say "what did I say "! How about a little more energy so that the person calling wouldn't have to think he or she should apologize for awakening, someone!

Then there is the question of having some enthusiasm for what it is we are selling we need to show more interest in the product when inquiries are made. Most responses to inquiries are rather "blah." This is still a people business—we're not yet into robotics. Then there is the question of not taking enough time to find out the real needs of the person making the inquiry. This means showing some interest in the Inquirer and coming across like "we really want to help you." All too often we spout off a long list of the hotel features, when, in fact, many may not at all appeal to the person calling. We need to express only those features that would be of interest to the prospective buyer and then we have to relate the features as a benefit to them. For example, if the property provides airport courtesy vehicles it doesn't even make sense to mention it if the people are driving. However, if the employee knew that the prospect was arriving by plane, then, of course, it would make sense to mention the free airport courtesy transportation. Yes, the free continental breakfast served at some properties would probably be a great benefit for all inquiries.

Then there is the big item of not asking for the sale. Let's say we've done everything right—talked slowly, distinctly, showed a lot of energy and enthusiasm, found out some of the needs of the person calling, related appropriate features as benefits now what? We seem to wait for the caller to say something about making the reservation. Why is it that in most cases we do not ask for the reservation? Or even better than asking why not just assume the close and be positive by saying "I'll just go ahead and hold that room for you—what credit card will you be using"? Or some words to that effect.

Douglas Kennedy, President of Hospitality Services of America (HSA), agrees that generally a good enough job is not being done when inquiries are made. Kennedy, who conducts reservations and front office sales training programs around the country, states those improvements could be made to get a better ratio of reservations to inquiries through training. Reservationists and front office salespeople could be trained to: listen to caller's request; ask probing questions; provide benefit statements; offer personal recommendations; attempt a primary close and then a secondary close; personalize the conversation; convey confidence and pride in the product. And Kennedy ought to know, he has personally trained thousands over the last several years and the proof was in the increased sales and revenues for their properties.

TAKE A LOOK AT YOUR GROUP SALES POLICIES AND MAYBE MAKE A CHANGE

October 1995

We keep getting surprises at what's going on at local hotel properties around the country. At some recent Hospitality Sales Workshops sponsored by HMM, we found out about a number of things that really need to change with regard to policies in our sales offices if we are concerned with continually improving profitable sales. As owners, operators, managers, directors of sales, we should all be attentive to what procedures are being followed in connection with quoting rates, answering inquiries and attempting to close sales.

Here are some examples: At some properties, sales directors have to get proposals approved by the General Manager, and in some cases by a management committee, before any quotes may be given to a prospect. Never mind that someone may be in a hurry to make a decision. Whatever happened to empowerment? The prospect might just as well deal directly with the General Manager if the sales director can't make decisions.

Then there are a large number of properties whereby anyone trying to book a certain number of rooms have to deal with a "specialist" in the sales department—if that person is not available then, though, there is no else around to work with a prospect. Here's what happens you call a property (and we have done this a number of times) to inquire about 25 rooms for a group. You are told that group sales will handle it and the call gets transferred to someone. A secretary responds explaining that you will have to talk to so and so, but she is out and she would be happy to return your call when she returns. Had you asked for 24 rooms or less then you would have been transferred to someone in an "express" type of service. This person handles only the "small" groups. In either case, if the appropriate party is not available to believe it or not there is not anyone available that could quote you a rate not the secretary, not a General Manager, and unbelievable as it may sound, not even the reservations department. On a recent call, we asked the group sales secretary to transfer the call to the reservations department since we had to get a rate quote right away. The response was sorry, but they can't handle groups a company policy.

And, then there is another large number of properties that due to policy, will not let the catering department book rooms. Try to book a family reunion you'll get catering for the space for socials and meals, but you need to talk to group sales for rooms. And, then there are the sales departments that would not or could not talk to you about meals and receptions.

How about the policy on comp rooms? It's all across the board when it comes to how many comps for how many rooms sold. Why have a policy at all? Let's leave it to the negotiation table to decide if any comps are going to be given at all. Here's a good one! One property had it printed in all their contracts: "Complimentary room policy 1 for 50." When the sales director was asked about this we were advised that if comps are not provided then they just cross it out (and of course, ask the prospect to initial the change?)

How about your policy on handling telephone inquiries, whether they be for individual rooms or group rooms? Who gets the call? What questions need to be asked? What's the attitude of the employee who answers the phone? When we tested some properties during the workshop period the sales directors participating just could not believe what they were hearing even from their own hotels. All agreed—we're leaving a lot of money on the table. We just do not know how much business and revenue and profit we are losing because of company policies that just don't get the job done. Take a look!

KNOWLEDGE IS POWER WHEN IT COMES TO SELLING AND CLOSING SALES

November 1995

We've heard it time and time again that "knowledge is power" in sales work in any kind of business. In our business of hospitality sales, we need to know about our own product the hotel, motel, conference center, resort, and even the bed/breakfast and we need to know all there is to know about it. Then, we need to know about our company and all of our competitors. As much as we know our own property, we need to know about them. After all, we can serve our guests better and we need to be able to make comparisons, if necessary, and, probably most important knowledge of all is what it is that our prospects need or want. There is no point trying to sell selling. Sure, every now and then we can talk someone into something he or she does not want, but it is not typical in our business.

When it comes to product knowledge (our own property) it is amazing how much salespeople, front office and yes, even General Managers do not know about what it is they have to sell. Every once in a while every property should break down the areas of interest (rooms, food/beverage, meeting/function space, recreation/area information) and have each key staff person write down all the things they know in each of interest. When its what's the difference between the $79 come and $59 rooms (and please don't tell them it's only very some guests will be asking this question more and more people are interested in space, and this is a comparison they will be making among competitors, room sizes in newer properties don't vary as much as the room sizes in the older hotels) What's the difference in size of the double, queen and king? We need to know the difference between adjoining rooms and connecting rooms sounds basic, but every time we ask the question of how many "adjoining" rooms we get the answer to connecting" rooms. Then, of come, how many connecting rooms are there? The "extra person rate" could this apply to the single rate plus another person, or is it different than the double rate (two to a room)? For a third person in a room does the extra person rate including the use of a roll-a-way or is that extra also? How about "incidental charges" something as the company will pay off someone's room and tax, but incidental charges are paid by the guest, what does this mean? Answers that are the long distance our charges? This probably should be posted in each room near the telephone, but we do not see them too often, When we call down to the desk to ask it is difficult to get a straight answer.

In the area of food and average the big question about hours of operation get confusing, depending on who you risk, the night auditor always knows when the restaurant opens for breakfast salespeople are not sure is are the afternoon front office people, Same good for room service, if there is any, and some don't know that as well. The range for prices for breakfast, lunch, and dinner? These vary depending on who is asked, "What's the restaurant' specialty of the house"?

When it comes to meeting and function space-capacities are all over the board. Each property has some sort of a sheet that lists each room and the capacity for meetings and banquets. When was the last time anyone every set up the room for each configuration to determine exactly how many people would fit? When asked we usually get answers that include the word "about."

In the area of recreation and area attractions, it's just a question of knowing what is available in the locality and on the properly. Providing direction get them properly and see what happens.

SALESPEOPLE: IT'S TIME TO GET ON THE INFORMATION HIGHWAY

January 1995

Now, through your computer keyboard, you can interact electronically with your meeting planner colleagues around the world. Everything from general questions about destinations, sites, meeting requirements, or any topics for and about suppliers to the meeting industry may be answered by anyone on a new network called "MPI Net" without generating expensive long distance telephone charges.

MPI (Meeting Professionals International), a membership organization of over 12,000 with meeting planners and suppliers in 50 chapters in North America and around the world, launched MPI Net in June 1994, as a service for its membership. MPI Net was developed to meet the demands of meeting planners and suppliers, for now, and in the future and it is being continuously updated to make certain that is beneficial for its users.

From a hospitality sales standpoint we have always preached the value of prospecting and lead development, networking, developing relationships, continuous communications with customers, and getting proposals out as fast as possible. Now MPI makes these things easier for you to do and at a much lower cost than would otherwise be possible, just a local phone call. With MPI Net you get email, which according to Coleman (he goes only by the one name) past President of his own company called The Conference Works: "email is the most efficient means of exchanging correspondence or files anywhere in the world. I avoid steep hotel fax charges when keeping in touch with clients when I am on the road."

With MPI Net Message Boards you'll always find meeting planners searching for answers to program ideas, speakers, as well as meeting sites. Ed Simeone, CMP, Sr. Conference Manager for Digital Equipment Corporation states: "An online service's greatest benefit is giving people the chance to exchange information and ideas quickly and more efficiently than any other form of communication. MPI Net message boards are easy to use and addicting." And, what a great opportunity for salespeople to network with existing clients as well as with new prospects.

Matt Volk, Director of Marketing of the Lisles, Illinois Convention and Visitors Bureau sent an MPI Net message while I was attending MPI's Professional Education Conference held in Vancouver in December. He just raves about the opportunities offered him to network and develop a relationship with prospects for getting meetings to his area. Within the past few months, Matt has taken four convention leads from the "Locales and Facilities" section of the Message Board. By putting messages on the "Business Exchange" section of the Message Board he has received over 25 leads from interested meeting planner.

Here are some of the features of MPI Net for hospitality sales executives: Data Libraries—there are hundreds of documents, forms, templates created and used by meeting industry experts all over the world. Conference Room—This allows you to plug into live conversations with one or more people all over the world and eliminates long distance phone charges. Message Boards—electronic bulletin boards connect you with experts in a wide range of fields that can help you with ideas and solutions to your most difficult sales or management problems. On the Business Exchange section of this Board, you can tell about your property and anyone "listening" it can respond to you immediately if there is any interest. Electronic Mail—Private mailboxes offer the most economical way to deliver correspondence, proposals, contracts, etc. Update documents right on the screen and send messages back and forth 24 hours a day.

With MPI Net, subscribers may also have access via CompuServe to more than 1700 services in major news sources, travel information, databases, and online computer support. Subscribe you do not have to be a member of Meeting Professional International, but the software does cost less if you are a member.

IF YOU HAVE A PHONE ON YOUR DESK, YOU NEED TRAINING

January 1996

One of Nancy Friedman's opening lines in her training session during the recent New York Hotel show was "If you have a phone on your need training . . ." And, how right she is! We've reached this subject many times in his column, most of the time discussing how poorly we handle telephone inquiries. But Nancy covered just about all of us from desk clerks, to salespeople to managers, and even to corporate offices and what it is we do when talking on the phone and what it is we should do when it all comes to life. And we are sure to get to recognize our shortcomings when it comes to telephone conversations.

Anyway, the Telephone Doctor conducts seminars around the world helping companies improve their employees' telephone skills. Her company also produce a variety of training videos; some titles are: "How To Treat Every Call as a Welcome Guest" and "How to Handle the Irate, Angry, Rude, Unhappy and Sometimes Abrasive Caller," as well as "Voicemail—Curse or Cure?" If AT&T uses her stuff for training and they do (over 2000 of their workers have seen her videos) then there has to be something worthwhile to it.

Just stop a think how much of our business is handled by phone. From a sales standpoint, we just can't do without it. From prospecting handling inquiries, selling, closing a sale, servicing the business it's just about all done by phone. And, you know what? When have you ever really taken the time to train your people on telephone courtesy and techniques as well as telephone selling? There's no telling how much business we love because we are not familiar with Telephone "Doctor's Five Forbidden Phrases which Nancy covers exceptionally well during her seminar. It all has to do with how we respond to callers—making every call a genuinely satisfying experience for the caller especially if it involves a problem. There are certain phrases the Telephone "Doctor" tells us we should not use and what we should say instead, and how to say it. For example, how often have you heard one of your staff telling someone on the phone: "You'll have to talk to someone in catering" (or whomever)? The customer doesn't have to do anything. We shouldn't be giving orders. Instead, the Telephone "Doctor" tells us to soften the request by saying something like "here's how we can help you with that. I'll connect you with our catering department." We should not put people on "hold" without asking permission. Too often you'll hear someone saying "hold on a second' well, that's a big lie. According to Nancy, the word is "no' at the rejection." Do you have any rooms less expensive that?" Why not just answer the question without starting off with a "no"?

There isn't anything we can't find out for someone, even if we have to transfer the call or make a return call with some information. Too often we find our employees telling people on the phone: "I don't know." Why not respond with "Gee that's a good question. Let me check and find out"? That's the way to handle it according to Nancy, the Telephone "Doctor."

LET'S START WRITING BETTER SALES LETTERS

May 1996

Owners, operators, managers, salespeople grab the sales office trader file and take a look at the last batch of sales letters written. How do they look to you? Do they all have about the same appearance, sound alike, anything creative about any of them? They probably, for the most part, look computer-driven. Perhaps we've gotten into a cut-technology has taken over our ability to be creative and interesting and personable when it comes to writing letters. By now, most properties have personal computers with word processing software so it is pretty easy to draft a few, rather standard type letters. So, when it comes to sending a letter to prospects, or solicit business, or thank someone, or respond to an inquiry, or make a proposal we just "pop in" the appropriate name, address and a line or two here and there and BINGO the computers does the rest and now have a "beautiful" letter to send.

Most of us can recognize a computer-driven letter, especially those direct mail types where a salesperson is sending a batch of letters to various people met at a conference. It starts off very nice with something like: "How good it was, Howard, meeting you at a such conference." Then about halfway down, when it comes sometimes have you gotten a letter like that?

We've gotten so used to word processing that now it becomes difficult for some people to writs individually prepared letters with some creativity. We practice this at our HMM Hospitality Sales Workshops and it becomes pretty obvious that we can write better letters, with just a little effort. What's wrong with most correspondence coming out of sales offices? Here are a few thoughts:

▶ They don't pertain to need (we write standard solicitation stuff, without considering the needs of the person to whom we are writing).
▶ They are not creative (most sounds like a portion of the brochure was copied).
▶ Many are mysterious (sometimes we have to read the letter more than once to find out what it is the writer wants).
▶ Most are too long (how about trying to keep the info to one page?)

The best sales letters seem to be those that are well planned in advance of being written ask yourself: why am I writing this letter: what is it I want to get across: is there something I want the reader to do: will the addressee really read this?

It just seems logical, first of all, to want to write a letter and making it sound I like you. Why don't we write the way we talk? In conversation, we just don't use phrases like "pursuant to our conversation" or "enclosed herewith." And if you think about it, you'll find a whole lot more such phrases we include in letters, but never use verbally.

Most opening paragraphs are very standard and pretty boring. We need to do something in the first paragraph to grab someone's attention. What's wrong with coming on strong like: "We sure do have some good news for you, Mr. Brown"? There are many more opening statements that could be made to get someone to read on. Following your opening paragraph will be the paragraph that provides the information you want to get across. This is where you come up with something that pertains to the particular needs of the reader. The last paragraph should be a "call to action."

It is something that the reader should do or something that you will do like: "I'll go ahead and hold that space for you and will call you next Thursday to confirm all the details."

Since people buy with feelings (or even emotion) why not include something in the correspondence that makes the reader feel good? Refer to an award they received or acknowledge having met a family member—anything that will appeal to ego should work well. They'll remember you better.

Just remember to make sure your letters sound personable, pertain to needs, create an interest, get to the point, are not mysterious, have a call to action and make the reader feel good sometimes a handwritten "P S." works very well.

WHEN SOMEONE DOESN'T RETURN YOUR CALL, WHOSE FAULT IS IT?

June 1996

I've been feeling pretty guilty the last couple of weeks about not returning a call from someone I had not heard from in about three years. We used to work together and now he was looking for employment and wondered if I knew of anyone he should call. This was a great salesperson and I felt there were a few opportunities for him where he would fit in very well. My problem is not knowing how to get in touch with him. You see, the message was left on voice mail on a day that I was out of town. I really do want to get back to him because I believe I can be helpful, but as with most of us salespeople, he ran off at the mouth just too fast and the telephone number he left was just not understandable. Several "sound-alike" numbers were called, but to no avail. Other people listened for me—just in case it was me not hearing too well, but they too could not understand. Oh well, maybe he'll call back.

One of the big complaints salespeople have these days is that people do not return calls. We have to wonder if perhaps that fault lies with the messages being left-either on voice mail or with secretaries. Of the 50 or so calls a week that I receive through voice mail(out of town, gone to the bathroom, out to meetings, on the phone, teaching class, etc.) a good 10% come through with unintelligible names or phone numbers. And, it's not me just because I do not hear too well it's just that some people talk too fast and don't talk distinctly.

Here are some suggestions when leaving messages on voice mail. Speak slowly and start out with your name (spell it if unusual) follow with a phone number (with area code). Make a brief statement regarding the purpose of your call. We should not try to make it a mystery as to why you are calling. Then make a brief statement regarding the action to be taken. This could be the action you want the called person to take or the action you will take. Sometimes you don't even need people to call you back, maybe it is something you want them to fax to you or put in the mail to you: or perhaps you want to advise them that you are sending them something. You close by restating your name and phone number.

I get a good number of voice mail messages from hospitality salespeople that leave me great messages like "no need to call me back, unless you want to, but please fax me information about your upcoming workshop, my fax number is." Then, I also receive some very long, too long, messages where the caller is not prepared or organized and just rambles on repeating stuff over and over again. The whole idea is to be brief, to point and indicate what action is requested.

Sometimes voice mail is not involved when you are calling but the phone is answered by a secretary or an assistant to the person you want to reach. In many cases where you don't explain the purpose of your call, but just leave a name and number, there is a good chance the call will not be returned. There should be no secret to why you are calling -especially if you are in the position of being able to help the party. If it is prospecting you are doing and you just need to gather information, perhaps there are others available to talk with you to give you the information you seek you just need to explain to the person answering the phone what you are trying to do and ask if there contact anyway; and be in a better position to tell you what you need to know.

If you are getting a large number of unreturned calls just take a look at how you are going about leaving messages. Whether it be leaving messages with other persons or with voicemail you should enthusiastic and full of energy and explain how you can be helpful speak slowly and distinctly, get to the point, but be brief.

COMMUNICATION IS THE KEY TO STAYING OUT OF TROUBLE WITH CUSTOMERS

November 1996

If anything ever goes wrong at your property between a customer (individual, travel agent, meeting planner, tour operator, wholesaler, or any type of group) and any staff member or department (front office, catering, sales, housekeeping, maintenance, food service etc.) you can pretty bet that in 95% of the time it will result in better and more successful events, less problems happier customers and more repeat business.

Frequent communication in writing and by phone with group contacts will keep you on top of what's going on and if any changes in plans are anticipated. It is the professional way of doing things and a lot will be learned to help avoid troublesome situations that may arise at the time of the group arrival. How often do we hold maximum space for a group and find out at the last minute that it is OK to release some of the meeting spaces and even part of the room block? Could we have lost some other business had we known earlier? It has happened to me in the past and it was a good lesson to learn. This is particularly important when you are holding tentative space for the group through your sales or catering departments and waiting for "them" to tell you if it is confirmed. I would hope that whenever tentative space is being held that three is a cut-off date to confirm or release. However, even with release dates, I have seen function books with "hold all space" written across the pages (it is also being done on computer sales account file systems), and upon checking we find out that the group had decided "some time ago," not to use the hotel. Staying in touch can avoid this.

Even when everything is and in writing someone needs to stay in touch with the contact. So often people change jobs and with new people taking over a special event or a meeting, they want to make changes or do things differently. If there will be problems let's catch them as early as possible. Critical areas for misunderstanding generally may have to do with:

▶ Room reservations and out off-dates: Be sure to stay in touch to remind the contact of the actual pick-up periodically in advance of the cut-off date. Here's an opportunity to release rooms for public sale even before the cutoff date.

▶ Meeting space: Particularly if you have agreed to "hold all space" until a certain date. Keep checking to make sure that the group will actually use everything and try to get the contact to specify how each meeting/function room will be used. You may be able to recover some of this space for other groups.

▶ Deposits: If a deposit has been agreed upon, to be received by a certain date, stay in touch to make sure you get it. If they do not have the cash available, get a bond or letter of credit from a bank, but get the deposit somehow. Then often we forget to trace a file for the due date of deposit.

▶ Master Account: keep checking to see if there any changes necessary form the master account standpoint. Anyone else added to the list for those authorized to log in? Any other else Added to the list for those authorized to sign? Any other charges to picked up by the group for special guests, speakers etc.? Changes in billing information?

▶ Meeting Room Set-ups: As time drawers nearer for most large meeting group we find that planners are lining up their speakers. Even though you may have received information about how the meeting rooms need to be set, experience has told us that many times the speakers want different arrangements made. All too often those changes are not known until the day of arrival of a speaker. Staying in touch and asking about this will certainly save you a lot of trouble. Even offer to have someone call the speakers in advance to find out what is needed.

This past summer I was engaged to do a program for Management Company and about two weeks before the date I received a call from the catering department of the hotel where the event was to take place. They just wanted to confirm the arrangements that had been made for the meeting room set up. Needless to say, I was impressed.

If you want to do anything right how about this business of staying in touch I would suggest you go right to the function book or automated sales file system and take a look at all your "tentative" bookings. You'll probably be able to release a good deal of space that won't be confirmed. This will give you an opportunity to book more business.

WHAT DO WE MEAN BY SELLING "VALUE"?

November 1997

All too often we hear that in order to get people to stay in our properties or eat in our restaurants we have to offer "Value." Do we really understand what it is we are talking about? Value is just what we think something is worth, and it is a perception of worth related to what we pay for a room, a meal, a meeting, a package, a vacation, etc. Value has different meanings to different people . . . is it something in the eye of the beholder? Once a guest gets a bad feeling about something (a telephone that doesn't get answered; an unfriendly desk clerk; a dirty table in a restaurant; no hot water for a bath; a too-busy salesperson; or whatever) the value in using that property diminishes greatly. So many of us think that by reducing room rates, offering discounts' giving things away is the way to create value. Someone mentioned to me the other day when talking about a local restaurant "If they gave me meals for free I wouldn't go there." You have heard something like that yourself from time to time. Sometimes there is no value at all regardless of the price. Price alone does not indicate value.

Many years ago after opening up Frenchman's Reef in St. Thomas I invited Jim Jones for a site inspection. Jimmy was President o Meeting Planners International and I served with him on the executive committee. He was a very well-known meeting planner for Connecticut General Life Insurance Company and ran some 200 meetings a year. Of course, we wanted his business. He spent three days with us at the property, giving us a look-over. Nick Pourzal our General Manager (he is still there, by the way) gave Jimmy the extra special treatment. The big thing that impressed him during the visit was the way our guests were smiling and having a good time while there. Most of his time was spent visiting with guests and getting their reaction to the place . . . the big question in his mind was there a "value" for people staying at the resort? We ended up booking 8 groups with Jimmy and price was never a question . . . he saw the value in having his incentive and meeting programs at Frenchman's Reef because of a number of things that provided value to the guest. I've never forgotten that.

On a recent trip to Hilton Head, I had the same feeling that Jimmy Jones experienced. It was a very busy weekend at the Crowne Plaza and I saw nothing but smiles from the guests all over the property. Where I witnessed value was the cleanliness, everything was spotless and shiny, especially the staff. Clean, crisp uniforms; desk clerks wore white shirts, nautical designed ties with double-breasted navy blazers. (So many resorts think that golf shirts or sports shirts look better, but I feel coats and ties present a more professional appearance). And, probably one of the best things was that the staff was so pleasant, friendly, helpful; their smiles and greetings were contagious . . . resulting in all the guests smiling and greeting each other. It was just a wonderful experience . . . this is what I call "Value" Ron Antonucci, the General Manager of the property, keeps talking to his staff about this business of value. His feeling is that whatever people pay for a room or a meal there has to be something there that creates value for the price. The way the staff looks and acts and the way the products are presented are the things that create value. Having lunch at the property and breakfast the next day certainly reflected value . . . from the product standpoint and service, there was no question that the guests appreciated the value presented.

In fact, I suggested he raise the price of his breakfast buffet because it was such a terrific value. No wonder the property was named 1997 Crowne Plaza resort of the year . . . and Ron received two other awards for the property this year. I am sure we all have good stories to tell about how guests appreciate the value and what makes people return to a property . . . and it is not priced alone. Of course, the price is relative, but doesn't value have more to do with the perception of what you are getting for the price paid?

Salespeople can bring the business in, but how the guests are treated, what they see and how they relate to what they got for the price really determines the success of the operation.

PRICE/VALUE RELATIONSHIP IS ALMOST ALWAYS A CONSIDERATION IN BUYING

November 1998

No question about it! Before we make a purchase, any purchase, a house, car, dress, tie, even a meal in a restaurant, we just about always consider the value of what we are getting for what we are paying. So why shouldn't this question of price vs value be raised by prospects who are looking to rent a sleeping room, buy a meal at your restaurant, book a wedding reception or have a conference at your property? Your customers are evaluating all the time: "what can I get for what I am paying; is it worth the price for what I am getting"?

It stands to reason, in our business of selling rooms, food, beverage, space, recreation, and all the other outlets at a facility that we need to understand how to fit value to the needs of our prospects. Frankly, we do not do such a great job at this. There has got to be a value to everything we sell. It may be a real value or even a perceived value, but it only works when it can fit a customer's needs. People just don't buy the same product for the same reason . . . even a room for the night. The same product at your property (room, banquet room, food, etc) may be bought by different people for many different reasons. The key in selling is to find out the reason for a need. The value of a room at a certain price will be different because the prospective guest has different needs. A room is worth more when a prospective buyer urgently needs to stay at your property because of you a near where he or she may want to be. If it is more important to someone to be at a property with a restaurant, then there is more value to staying at such a property than another without a restaurant. If there is a meeting going on at a hotel where a prospective guest needs to stay, then the value is greater than the value of a property a couple of miles away.

The answer to a better understanding value related to price has a lot to do with why the prospect needs to visit your area. Understanding real needs for a person to buy a room, meal, space, etc., is critical in being able to sell with the price/value relationship in mind. And, before people buy anything they need to know the benefit of making the purchase-which, in itself, is getting to know the value. Once salespeople understand the benefits of all the property features which are available they are in a better position to prove value to the prospective buyer, provided they also understand the buyers' needs. Here's an idea:

▶ *Learn your benefit list*: The benefit list is the most valuable tool a hotel salesperson can carry. Actually, when selling, your benefit list is your product. When someone has to pay more for a product than he has been paying in the past, the decision to pay more will be made primarily on the basis of benefits the property can deliver (price/value consideration). Your hotel product features create benefits, but it is the benefits, not me features that sell your property. What's the point of bragging about your 500 seat ballroom, if a meeting planner is looking for space for 50 people? Therefore, all the "wonderful" features of your hotel don't mean much if there is no benefit to the buyer.

▶ *Meet with your department heads*: Have a meeting with all of them. Work with them in a workshop setting to discover *all* the features your property has to offer, and then describe them as benefits. Actually, make a list. Salespeople, as well as General Managers, will probably be surprised as these department heads come up with features about which they didn't have any previous knowledge. For example, what features in your bathrooms may be described as benefits to a guest? Anything in the lobby or front desk area? How about security devices? What about bedding, square feet in a typical sleeping room, phone system, etc.? You get the idea?

WHAT'S WRONG WITH OUR SALESPEOPLE THESE DAYS?

January 1999

There will be a lot of hate mail coming my way because of this topic, but here goes . . .

It is time to speak out on the topic of what's wrong with our sales folks these days. From my point o view, and from what I have seen over the past few years, we need some help when it comes to the basics of selling. Business has been so good with high demand, increases, and phones ringing off the hook. Perhaps we have not paid too much attention to whether the business was coming to us because we had such great salespeople or because business was just there for the taking. When times get tough (and it will happen) we'll find out.

Letters of solicitation from hotels seem to find their way to me. I guess I'm on a lot of lists, but I am not a buyer . . . so why should I be getting these letters anyway? Some of the mail includes some pretty expensive collateral and heavy duty postage. Why not find out first if I am a prospect? There is a place for prospecting type letters for direct response rather than soliciting.

From time to time, in preparation for one of my workshops or management company training sessions, I will make test calls to properties to try to book a meeting or reunion or something. It is really almost unbelievable how many of those calls are handled by hotel Sales offices. And, here we mean all types of hotels, everything from one-star to five-stars. We just seem to put people to work in sales without any training in how to handle telephone inquiries. Most everyone is very good at answering questions of the caller, but the very first thing is to get a name, organization and phone number, then ask specific questions to find out the need. What happens if no one is in the sales office? After all, we do have to eat, go to the bathroom, go out on sales calls, etc. This brings up the question of how and when to use voice mail. It is nice to have this type of technology and it is really very helpful, but a good many people will agree that it is misused. In sales, offices, a real live person should answer the telephone, even if it is call-forwarded to another department where someone can take a message. If for any reason there is no one available, the front desk or switchboard should be notified so they can take messages before even transferring the call. Sometimes, of course, you have no choice if all the lines in the sales office are tied up with other calls. But the message should be something other than "I'm either on the phone or out of the office." Just about all voice mail systems give you an opportunity to record a message when your line is busy and another when there is no one there to answer the phone. Check your options and leave specific messages, so a caller will know for sure what's what.

Then there is the business of making sales calls, in person or even on the phone. So often we find salespeople trying to sell their hotel to folks without even finding out what needs there might be. Do we even understand the need for prospecting? This is usually the case with new salespeople who are told: "get out and make sales calls." There is no sense in trying to convenience people to use a hotel when there is no actual need. Rather than asking questions to determine if the contact is actually a prospect, we find salespeople telling about all the wonders of the property without first determining if there is a need.

Before times start getting tough for business thinks about getting your sales staff trained to improve their productivity and your profitability.

HOW IS YOUR TELEPHONE IQ?

February 1999

We've written about telephone skills several times over the years, but it sure is worthwhile to mention again. With all the new property development that has been going the past five years or so we probably have a bunch of new employees that have never had any telephone skill training. Just call some competing hotels in your own backyard and see how many do a poor job on handling telephone calls. Try calling for room reservations, the General Manager's office, the sales office, the restaurant, etc., and see for yourself. It seems we do not take the time to help employees do a better job in delivering customer service on the phone. We think it is just a natural thing, something simple; after all, it is "just answering the phone," or is it? Do you think our employees are properly trained in telephone skills? Wanna find out?

If you have not yet heard of The Telephone "Doctor," then you should certainly know that one does exist. The Telephone "Doctor," an international training company, is offering an online quiz that measures your property's awareness of customer service. This is a 15-question quiz that can be taken on the Telephone "Doctor" website at *www.telephonedoctor.com* The Customer Service IQ test is designed to evaluate the working knowledge and basic skills of customer service professionals (and it's free). The quiz consists of questions covering customer service and telephone skills. At the conclusion of the quiz, you can access your results by clicking the "grade my quiz" icon. The results come back with: the original question, your answer, the correct answer, and the Telephone "Doctor" (r) explanation. All results are confidential. I took the quiz and got 12 out of 15 correct. So, I need some help myself. Nancy Friedman, President of the Telephone "doctor" (r) says that the test is more of a barometer to let you know how effective you and your staff may be in delivering excellent customer service.

"It also can be used for pre-employment testing and is taken privately on our website. We don't ask for your name and you get graded right away," according to Friedman. (And again it is free)

In our own business, in my opinion, our other big problem with telephone skills has to do with handling telephone inquiries from people who want information about rates and space. There is no telling how much business is lost because we just do not take these inquiries serious enough. Just about nobody is able to measure the amount of lost business as a result of not handling these calls properly. In addition to proper telephone handling skills which we mentioned earlier, we still need to provide training in the questions to ask to determine to what extent the caller is a prospect, and how we can help in fulfilling needs. One way would be to provide everyone who may have the opportunity to answer the phone with a form which prompts them in asking specific, important questions. You can create your own, but if you would like a sample, I will send you one to use or to help you design your own. In the past, some readers of my column have been good enough to send me forms they are using for this and I have taken a little from each to create one.

"HIGH TOUCH" STILL BEATS "HIGH TECH"

March 1999

There's been a tremendous growth in conducting business in our industry via technological innovations. We see it online with interactive websites, forums, email, and other internet and intranet methods, and all kinds of computer software, as well as with voice mail, fax machines, and lots of new technology stuff coming to us, for our use, almost daily. But, no matter how much technology is available for us, it will never replace the need for hospitality salespeople. In fact, right now, with all the advancement made in technology, my guess is that we have more salespeople on the road than ever before in the hotel industry. It really all boils down to people buying from people. Yes, we are getting individual room reservations via internet, and that will continue. But the bulk of the buying of rooms, meals, and services will always be through our people-reservationists, front office people, catering-banquet staff and the sales staff. Actually, the whole hotel really does the selling-every employee regardless of the type of job or function is involved in selling-whether directly or indirectly through service. Technology will never replace High Touch. Technology will never be a substitute for the hands-on approach to sales. Although our customers may use technology to do research and gather information, when it comes to making the purchase, they would prefer the credibility and value that they receive from "working" with a real live person.

Building loyalty of customers is critical in sales. If there is no customer loyalty to your brand or your specific property, you'll be out of business in a very short period of time. If there is a way to build loyalty with technology I do not know of it. This seems to me to be a people thing. And, the other part of this is that if you do not build loyalty among your managers, associates, owners, and franchisees, you are not going to get the customer loyalty you need to stay in business.

Last January, during a break in a sales workshop I was conducting at Marriott's Westfield Conference Center in Chantilly, Virginia, I ran into Kurt Krause, Marriott's General Manager. We chatted for a while about how impressed I was, not only with the property, but the way the staff worked so well together, how they all respected each other and how friendly everyone was, not only to the guests, but to each other. Kurt said: "we preach loyalty," with that he pulled out a small, folded, wallet-sized, piece of paper upon which was imprinted "Westfield's Marriott-our Vision: To be the worldwide LOYALTY leader for customers, associates, owners and franchisees in the quality tier lodging segment." Inside the folded piece of paper were 20 statements (one each for each working day of the month). Each associate at the property has a copy and is encouraged to read, remember and practice the standard set for each day he or she is at work. An example: "we practice teamwork and treat each other with the same respect we afford our family and guests. We adhere to Mr. Marriott's belief that if we take care of each other, we will be able to take better care of our guests.'"

Now, doesn't this sound like good sales stuff that can't be done with technology? Here's another: "Genuine care and comfort of our guests to ensure they return is our highest mission. Display genuine and enthusiastic interest in the guest, and always pay complete attention."

In sales, it really all boils down to High Touch more than High Tech.

POOR VOICEMAIL TECHNIQUE CAN HURT YOUR SALES

September 2000

Our good friend, The Telephone Doctor, Nancy Friedman, was singing my song when she came up with some voicemail pet peeves in one of her recent newsletters. Hospitality Salespeople just need to get on board, and do a lot better in this area if we want to sound more professional, and not turn-off sales prospects. It really bugs me when I call a property and get a voice mail message like: "Your call is very important to me, etc.,." ok, if my call is so important to you, why aren't you there? Then, the message goes on to say: "I'm either on the phone or out of the office." Now, that doesn't tell me a thing more than I already know. The Telephone Doctor says, in her newsletter, that people want "good, in-depth information, like where you are, when you'll return, and who they go to for further information. I think most voice mail systems will allow two messages which can be recorded with your own voice. One, when you are really on the phone, and the message indicates so; and the other when you are really away. Check it out!

And, speaking about using your own voice, I agree with The Telephone Doctor; she says: "It's a good idea to have your own voice on your voicemail greeting. Not recording it yourself gives the caller the feeling that you can't be bothered with them." What's the big deal about changing your message daily? Tell us what day it is, when you'll be back, and how we can reach you, if necessary.

Now, on the other side of the issue with voice mail. When you call someone and you get their voice mail, What kind of a message do you leave? It bothers me very much when I get a message from someone who is not clear in leaving a name or telephone number, or what the call is about. So often I get ill-prepared, rambling messages that tell very little. Granted, I have a little hearing problem, but sometimes I ask an associate to listen for me to a phone number or name, and they can't understand it either. So it's not only me. In leaving voice mail messages, speak slowly, distinctively, be brief, get to the heart of the message, then get off.

The Telephone Doctor's quarterly newsletter, The Friendly Voice, is chock-full of great tips, especially for salespeople. See if you can get on her mail list (info@telephone-doctor.com). She also has some great telephone training videos available, especially for people who handle your call-in reservations. I just reviewed two which were terrific. And, if you have not yet taken the Telephone

Doctor IQ Quiz on their website, try it. It's free, fun, and you will learn a lot about your own telephone techniques. You even get an immediate, online report on well you did.

More on how poorly we handle the telephone. Just got back from doing a sales workshop where I made several test calls to properties to demonstrate to the group just how we don't handle phone inquiries properly. Every hotel we called during a 45 minute period "bombed." If these calls were real inquiries a lot of business would have surely been lost. We still need a good deal of training in how to handle inquiries for group business (and that goes for transient business, as well).

DISTANCE LEARNING—THE LATEST IN TECHNOLOGY FOR SELF-EDUCATION

October 2000

Yes, it looks like you don't have to leave home, or your office to get some education these days. Just turn on the computer, tap into a website, make an 800 telephone call and you are in business. If you haven't tried it yet you are bound to do so pretty soon. Kind of confusing at first, but I asked Mike Hampton, Ed. D., Chief Executive officer of HSA International what this is all about. He described distance learning as Synchronous technologies which involve real-time delivery, incorporating live events during which the facilitator and the participant can interact with each other at the same time, such as net conferencing, audio conferencing and video conferencing. Then there is the Asynchronous technologies which provide learning experiences on demand without "live" facilitator/participant interaction, such as webcasting, web-based course, electronic performance support systems, and CD-ROM.

When! Sounds kinda too high tech for me, but Mike Hampton says: "The overriding consideration with distance learning is that learning experiences can now be made available 24 hours per day/7 days per week to employees in virtually any location worldwide. This enables learning to take place when it's needed instead of only when it can be delivered. The implications for employee behavior modification are profound."

To see how this thing works I tapped into our industry's premier live, online reservations sales seminar on August 17th, presented by HSA International. This was the first in a series of various sales and service seminars that will be offered to specifically focus on topics and issues that are distinctly pertinent to the hospitality industry. It was a snap for me, a non-techno. An email message from HSA gave me instructions on the website to access; I signed on to the website to get the visual presentation, and then dialed-in to a toll-free teleconference line to get the audio portion. The 90-minute seminar was hosted and facilitated by HSA International's president, Doug Kennedy, who conducted the session which involved participants from 35 hotel and resort properties, as well as several hotel management companies and brands. Well over 100 people, from all over, including Hawaii and Canada participated. Many of the properties had assembled their reservations staff around a speakerphone in a meeting room and hooked-up a computer to an LCD projector in order to take advantage of the opportunity to orient new employees to the reservations sales process, and to reinforce prior training activities.

As Kennedy went through the program, participants could ask questions. Rather than raising their hands, as we do in a classroom, they simply clicked on a microphone icon that was on the website screen. This alerted Kennedy, and he was able to respond to them by having a technician who was monitoring the event open their microphone for everyone to hear. Participants were also able to communicate with Kennedy by typing in questions, comments or other contributions to the presentation. And, just as students often pass notes to one another in a class, the participants could send private text messages to one another or to the group as a whole.

The whole thing was just amazing to me. Doug Kennedy, who does many on-site, live presentations, said: "it was almost like doing a live workshop, except that we were all in different locations. Although we couldn't see each other, the telephone conference feature seemed to make everyone an integral part of the program. It was especially encouraging to have gotten so many good questions as we went along, which appeared to show that people were really thinking about and absorbing the content."

APEX IS WORKING FOR YOU, MAKING YOUR CHALLENGES EASIER

September 2004

As owners, operators, General Manager, sales personnel as well as all department heads, you need to be aware of APEX (Accepted Practices Exchange), a meeting industry-wide initiative be ringing together all stakeholders in the development and implementation of meeting industry accepted practices, to create and enhance efficiencies throughout the meetings, convention, exhibition, and hospitality industry.

We wrote about this a few years ago, in this column, when the Convention Industry Council (CIC) started on this project by having industry professionals joining together to accomplish the goals of this initiative to ultimately improve the industry. CIC, an umbrella organization of 31 industry organizations involved in meetings, conventions, exhibition and hospitality industry, in order to achieve its objectives of standard forms, reports, terminology, systems, etc., through APEX. To accomplish this it required a buy-in from the various organizations that make up the industry. The participating groups also include convention and visitor bureaus, educational institutions, major hotel chains, convention centers, associations, and planners.

The whole idea of accepted practices is that the industry will see such direct advantages as higher service levels, improved productivity, and clear terms for relationships between supplier and buyer. APEX will also derive the industry to keep pace with technology. The streamlined process resulting from APEX will significantly reduce duplication of efforts and ensure consistency from city to city. New tools and shared systems will provide instant access to information that is credible, consistent and available. This way, all participants will rely on the same data and enjoy greater accountability when it comes to maintaining industry data. Hundreds of professionals have been at work over the past few years on committees to analyze and put together their thoughts regarding the standardization in addressing seven key areas: terminology; history/past event reports; resume and work orders; requests for proposals (RFPs); housing and registration; meeting and site profiles, and contracts.

This is ongoing in all areas with continuing updates as changes take place in the industry and technology. What works well for the project is that CIC provides an impartial and inclusive forum for APEX and the development of accepted practices by being represented via all facts of the meetings, and hospitality industry.

Our industry is really getting overwhelmed with technology and communication. What is challenging is that a hotel or meeting facility could have as many as 20 different computer systems integrated at a property that needs to "speak" to one another to process a credit card transaction or process a housing reservation, receive an internet an Internet-based RFP, etc. This combined with the variety of technology being used at the meeting buyer level or travel agency could, and does cause problems. There is an ever-widening technology gap with different approaches to managing technology. The amount of time, labor, a cost that is wasted could be enormous. While the APEX initiatives are not intended to solve all of these industry-wide problems, it can make a solid impact in communicating best practices throughout the industry.

HOW BORING WAS THE LAST MEETING YOU ATTENDED?

November 2004

It is happening so often, and people have been complaining about it for a long time, but now it is even getting much worse. Meetings are just getting to be so boring. We can tell it is getting worse when you take a look around at many seminars, workshops and other types of meeting, and participants are doodling, checking their raspberries (or is blackberries?), or palm pilots. Most would rather run out of the meeting room and check their cell phones; of course, everyone has been notified at the start of the meeting to turn off the cell phones. At some recently attended meetings, a lot of this was going on. Why do meetings have to be boring? Well, they don't have to be. It is just a question of the people planning these meetings getting some help on how to do it successfully.

What we are discussing here are not the 2–3-days conferences or conventions planned by the professionals, but the one-day meetings planned within our own companies and organizations, by people who are not trained on how to put on successful meetings. Of course, the whole idea is to find out why do we have to have these meetings in the first place? Sometimes companies have meetings just for sake of having meetings with no real purpose or objective.

To make a meeting successful the first step needs to be for the person who calls the meeting to lay out its purpose. That is, to answer the question of why are we having the meeting in the first place; what is going to take place at the meeting. Following that we need to lay out an objective such as what is it we want to happen after the meeting; what kind of a goal can we set to get something accomplished within a certain period of time, as a result of the meeting. Of course, the objective needs to be something that is measurable. This will help us to determine if the time and money spent on the meeting was worth it.

For those who have to attend we need to create the meeting content in such a format that will keep them interested. The monotone reading of complex reports and studies, of course, doesn't help the matter and attendees end up "checking out" of the meeting mentally, if not physically, using wireless technology to check email or surf the Internet, and even play games. A big problem also is that so many people facilitating the meeting can see this and know it is going on, but do nothing to correct the situation. At a recent meeting in a major city, with about 100 participants, all this was observed to be taking place. Speaker after speaker (including a "motivational" speaker), and panel after panel (except for the last panel) kept the audience sitting on-edge. On-edge, of course, waiting for it to be over so they can dash out the door.

Good meetings, no matter how uninteresting the subject may be can be made exciting and interesting to the audience. It is a question of reformatting the presentation, and making it highly interactive. That is, do something to get the participants to participate. It helps to seat the attendees at rounds table and offer items for them to discuss, even for just a few minutes, then reports their findings. The idea is to involve the audience in the content to some extent. Most people in business today just cannot sit still for an hour or more listening to others make speeches. We are all used to being active in our jobs, so sitting still, without any participation does become boring.

BETTER IMPROVE YOUR COMMUNICATION SKILLS FOR SELLING IN 2005

November 2004

A good deal of people who plan meetings have been complaining a lot about communication skills of folks with whom they are doing business; and rightfully so. We covered that issue a few months ago in this column, whereby we mentioned a survey of meeting planners rating this as the number one complaint. With economic and our industry recovery on its way, it just seems that salespeople need to start doing a better job with telephone messaging, voicemail, letter writing as well as Internet communications.

Let us start with the telephone. Help is really needed when calls are made and voice messages are left for the person being called. It is amazing how little we are prepared to leave messages. For the most part, leaving a message for someone to return a call needs to be brief and to the point. There are so many of us who just go on and on, repeating information without any idea of when to close the message. What we need to do is explain briefly, the purpose of the call, who is calling and how to return the call as well as when we will be available. When someone makes a return call then all that is necessary, if the first caller is not available at that time, is to respond to the original message relating to the purpose.

In many cases that will end the issue without the necessity of phone-tag. Next, don't we all get upset when we try to call someone and we get voice mail: "I'm either away from my desk or on the phone."? That doesn't tell us anything. The person we are calling could be off on vacation or out of the office for a long time, or maybe they will be back in five minutes. The idea of a good voice mail message is to make sure you change it every day or sometimes more than once a day. What the message we leave needs to do is advice the caller: who we are, what number to leave for a return call, when we will return, where we could be reached, and how we could be contacted if needed right away. Of course, above all, we really need to return calls as soon as possible, certainly within 24 hours.

Letter writing is another area that could use some help. We all seem to use letters that are already available in our computer systems. These probably were originally written by salespeople who no longer with the organizations and do not reflect our own styles. Why can't we take the time to write personalized letters? Why can't we write the way we talk, so that people can really be related to our correspondence? Why do all of our letters have to appear as though they are computer drive? Because they are! Why can't we do something with our letters that add feelings, something personal? Let's throw out all the letters stored and start over.

The big thing about email is that so many salespeople, as well as General Managers, and others do not have a signature line at the end of their messages. Why not? This way everyone receiving a message knows the complete contact information of the sender. Makes sense!

PRESENTATION SKILL TRAINING IS CRITICAL IN SELLING TO GROUPS

April 2004

Generally, in the business of making a sale, we think of a sales presentation as something a salesperson does when addressing a group, or presenting a product to a large audience. Most of the time it is a well-planned program presentation, including some visuals which help explain how a product or service would be a value to prospective buyers. However, in our business of hospitality, when we go after group business for a property, for the most part, we are presenting to a prospect. The prospect is generally one or two people who represent an association, or a company, or any organization as its meeting planner. It is usually a one-on-one conversation involving the gathering of information which reflects the needs of the meeting group. The property salesperson asks questions, the answer to which, reflect can deliver to meet those needs, This process is also considered "presenting." There are certain skills involved in becoming successful in the sales presentation process.

Presenting to one person is also presenting to an audience, according to Public Speaking Coach, Victoria Chorbajian, who is founder of Chorbajian Speaking Enterprises of Basking Ridge, New Jersey (*www.VictoriaTheCoach.com*) Victoria states that the key to effective presentations, in dealing with one-on-one selling, is to excite the audience; the same as if you are talking to a large group. Using a series of appropriate questions that pertain to the group meeting needs is most important in getting the prospect interested and participating in the process. The idea is to use common sense questions, relative to needs, that get a "yes" response. Let us use as an example, in discussing a family reunion gathering where there will be a good deal of family members with young children, and the property has a large number of connecting rooms. The property salesperson could very easily ask a questions such as: "Wouldn't your people feel great to know that with the large number of connecting rooms at our hotel they would be able to room their children right next to them, with a door that connects to the children's room, for safety and security, what more could they ask?" This would have to get a very positive response from the planner. What we want to do is captivate the audience, even an audience of one, with questions, the answer to which make them feel good, according to Coach Victoria.

In presenting to prospects, Coach Victoria, states that sales staffers need to be careful in how they use their voice. Sometimes we have a tendency to talk too much about our property, and mention things that do not directly relate to the prospect's needs what they want to know about is their own issues, and how their needs may be met.

HOTEL SALES MANAGERS FAIL TO RESPOND TO CUSTOMER INQUIRIES

November 2007

It is a wonder that we do not know how much group business is lost, on a daily basis, because of sales personnel not responding promptly to telephone inquiries. We have known that this problem has existed for some time with electronic Request for Proposals because we reported on this several months ago. Meeting News magazine did a study and found out that the number one complaint from meeting planners was that hotel sales departments were not responding promptly to RFPs. Since we reported on this we have received a number of responses from our readership with information on how some companies have installed new high-tech devices to take care of that problem. We will cover this new information in a future column. Now we have found some, not so new, information regarding telephone inquiries getting similar poor attention.

Group sales are a significant source of revenue for most hotels, yet according to a test completed by Quality Track International during the month of September, hotel sales departments fail terribly at the most basic function: responding to initial customer telephone inquiries. Fully 25.5% of initial customer contact inquiries go unanswered. Quality Track International, a leading provider of group sales mystery shopping services exclusively to the hospitality industry, placed in excess of 3,00 test calls to more than 750 on-property sales departments in North America, and tracked responses. Across all tiers and brands evaluated, a sales manager was available to handle customer inquiries just 49.4% of the time. If a sales manager was not available, a detailed inquiry message was left. Sales managers failed to respond to these messages before the end of the next business day more than 51.7% of the time. So, when we look at this from a standpoint of almost half of the 3,000 calls did not get a call-back, how much business could that possibly represent? IT sure looks like a lot of money left on the table.

Results were tracked by brand and tier (deluxe, upscale, and mid-price). Surprisingly, little or no difference was found in the performance of sales managers across tiers. From reviewing the results of the study it appears that in most cases where messages were left, it was accomplished via voice mail. In some cases, messages for sales managers were left with a person. In cases where messages were left with a person, call-backs occurred at a higher rate: 54% vs. 49% via voice mail.

In our business of hotel sales, we know that it is not unreasonable for a sales manager to be unavailable maybe even more than 50% of the time. After all, the job function entails a lot of being out of the office, or even busy elsewhere on site with other department functions, and visiting prospects. However, each property sales department's staff needs to involve themselves in making sure that all inquiries are handled in a proper manner. Further, there is no excuse for failing to return calls from prospects 100% of the time.

SALESPEOPLE JUST NEED TO START DOING A BETTER JOB

September 2008

Sales staffers will need to get around to doing a better job in working with those people who are responsible for booking groups. We have written about his a number of times in this column. The subject just keeps coming up, and maybe because we have more and more new salespeople working at properties who have not had any sales training prior to getting their jobs. Perhaps, have not attended any sales workshop programs offered around the country during conventions and conference.

Anyway, of course, sales are better all around, and have been pretty good for the past few years with higher occupancy and rates. However, has it all really been so good because our salespeople are doing a better job or are there other factors? Now, we really do not know what the future holds for our hospitality industry. Next year and the year after could very well be OK as far as our business goes: however, there is a good chance that some properties, in some locations will become overbuilt. In addition to that, the economy may worsen, and the cost of fuel may still be high, thereby affecting travel. We have to prepare ourselves to improve productivity, and that means to get some training on how to work with your prospective group buyers.

It just never seems to stop: you just can't please everyone: and it seems that everyone wants to complain: at least that's what we hear from salespeople. But, let's stop and listen for a while to our prospective buyers and learn what they say about salespeople. Sometime, last year Meeting News magazine conducted a survey of some of their readership which included 342 meeting planners working for companies as well as for associations. A lot of them had complaints about how salespeople do their work: obviously, they selected the wrong meeting planners to survey.

Here are the top six complaints:

▶ Calls/emails not returned in a timely fashion

- Cold calls/emails with no knowledge of meeting profiles
- Poor communication/follow-up by a salesperson with conference services person
- Lack of creative thinking/alternatives offered during negotiations
- Often dealing with a new contact at a property due to high turnover
- Proposals that don't properly address all the RFP items

These six areas of sales involvement with prospective buyers would be an excellent start-up kit for properties to use as an outline for an in-house training program. There is just not enough of ongoing training at properties. It is easy to create a "do-it-yourself" two-hour session once a week. First, put together your sales team, plus even a couple of department heads to meet once a week for a couple of hours. Alternate department heads every week so that you get different departments covered. Start with a brainstorm session on just one of the six issues listed above. Discuss why the group thinks each issue came up by the buyers, and then all should participate in how to remedy the situation. Each week handle a different issue. When you run out of the six, contact me and I'll send another half dozen. Just keep the training going.

HOTEL SALESPEOPLE TALK TOO MUCH

May 2012

Quite often, when we run into folks who have small children who do a lot of talking, we hear someone make the comment: "Oh, they are going to grow up to be a salesperson." It just seems that there is a common notion that salespeople are "good talkers." In reality, the most successful salespeople are those that talk very little. The whole idea behind making a good sale is not in the talking, but in the asking. Just think a little about how people make a buy decision: there are two things that are considered: Needs and value. Rather than spending a good deal of time in telling all about the hotel's attributes, what a salesperson needs to do is spend the time in asking good questions to find out specific needs, and maybe even "wants."

The whole thing behind someone to make the "buy" decision is not in "telling" all about a hotel property, but getting the prospect talking. The selling process really involves in asking open-ended questions that put a prospective buyer in a position of talking all about the reasoning behind them seeking a hotel location for their needs. It is hardly ever productive to ask a question which ends up with a "yes" or "no" reply. It is the "who," "what," "where" type questions that will get the prospect of talking. Then, of course, the salesperson needs to be a great listener. Listening well to the response will provide opportunities for the salesperson to be able to relate something positive to the conversation. This would be the time to show understanding to needs (or wants), and be able to relate something of value, or a benefit to the needs of the individual or group.

When a prospect visits a site to make an inspection to determine if the property is appropriate for a group event most salespeople start with a walk around the property showing off all the main discussion to determine specific needs (or wants), then do the "site inspection." This way there is the opportunity to address specifically the needs of the prospect, by showing how specific features reflect a benefit in meeting specific needs, thereby reflecting a true value. It is not enough to show and say: "this is our indoor pool." It would be better to present that feature by saying something to the effect of: "and this is where we will be able to create the setting you want for the welcome reception, when." it is never enough just to tell what the feature is, it needs to be explained as a benefit. Further in prospect is interested in having a meeting for only 50 people. The idea is to stay away from showing features for which there is no need; it takes away from a favorable buy decision.

Just keep listening and stop talking too much.

THE ONE-HOUR PROPERTY SALES MEETING

June 2012

Hopefully, property sales meetings are still going on. With reports of business for hotels beginning to return, along with some rates getting back up, and an increase in demand, local sales departments need to start getting busy to bring in its share of the group market. A weekly one-hour sales meeting would lend to being extremely helpful to the sales staffers as well as the various department heads at a property (they need to be invited). After all, aren't all employees in the business of sales? Even at a small property, with a one-person sales staff, there should still be a weekly meeting held, and conducted by the General Manager. All too often weekly sales meetings are held on an on-and-off basis, with the General Manager not even involved.

The purpose of a sales meeting should be to improve sales productivity, not only from those directly involved in personal selling to prospects, but for all employees to guarantee all guests' return. Therefore, the meetings should be conducted by the General Manager, on a regularly scheduled day, and time each week, with members of the sales team, along with department heads to be invited. Following are some thoughts to be considered in items to be covered in the meeting held for only one hour.

An opening of the meeting with some mirth by a different salesperson selected each week; maybe something funny that happened the previous week; or even a joke. This should lighten things up since so many employees get "uptight" when they have to attend a meeting. Then, a success announcement-anyone can contribute a success story on a situation of the previous week; maybe some nice comments from guests, or a catered event rebooked; or a housekeeper telling how someone helped a guest in a situation.

Administrative details for the week could then be announced by the General Manager or the Director of Sales. A different department head each time could then provide something on Product Knowledge (about their department operations) for about 10 minutes, or less. It is amazing how much sales staffers do not know about their own product, the property. This could be followed by someone from the sales team providing a sales subject lesson of the week (salespeople can take turns each week, for about 10 minutes). Here is where other department heads could learn something about sales and how it works.

Frustrations sharing for just a couple of minutes is always a good idea for the purpose of "blood-letting." Let the participants tell about things that bother them, with solutions offered by anyone; just another 10 minutes for this segment. Each salesperson should then report on networking accomplished the prior week; new contacts made who may become prospects. This, then, is followed by the top five prospects review (providing names, status, and expected results). Expectations of the week-each salesperson affirm what will be accomplished this week (sales calls, bookings, entertaining, etc.).

Two minutes of something motivational by the General Manager could wrap up the meeting.

CHAPTER 9

THE NEGOTIATION PROCESS IN SALES

DON'T LET SALES TALK ABOUT "FREEBIES" UNTIL THEY HAVE TO

January 1984

Naturally a hotel or motel's sales staff are very eager to book business at any cost. However, being swept up in the enthusiasm of trying to close a deal more often than not leads them to give too much away just to get it.

Right at the beginning—before serious sales negotiations even get underway—salespeople will typically offer discounted rates, complimentary rooms, free cocktail parties, rent-free meeting and exhibit space, and a host of other "goodies." And no one thinks to ask if they're even necessary to get the business!

Sometimes, of course, flexibility is necessary for landing accounts, especially during off or shoulder seasons. Then, perhaps through trade magazine advertising or direct mail to group prospects, it could be very worthwhile to entice business through such deals. But the offers should be limited, both in terms of the time available and the amount of business an interested party has to book in order to qualify.

But during normal business periods, salespeople should steer clear of "special deals" at least until the negotiations stage. Rather than starting off the sales pitch by calling attention to possible "freebies," salespeople should initially present themselves as a "consulting service," especially when dealing with meeting planners: A meeting planner has a job to do; the sales staff is staff should present itself and the property it represents as resources to make that job easier.

When this is done, all the "giveaway goodies" become secondary. Then, during the negotiations period, when a deal is about to be closed-and only then-should those goodies enter into the dealings. It may be necessary to do so given the buyer's budget or special meeting needs so as not to lose the order.

In general, the most important things to be aware of in selling to the meetings market are:

▶ Knowing your own product, its facilities and services, and how they meet a prospective client's needs.

- Knowing how you compare to the competition in terms of facilities, service, product quality, and price.
- Knowing your prospect's business.
- Knowing the needs of the group being solicited.
- Knowing the primary concerns of the person responsible for doing the booking.

By being a "problem solver," a salesperson helps the meeting planner or group buyer resolve problems and helps create an atmosphere of credibility, reliability, and friendship. This is an atmosphere for better selling without the need to lure prospects by giving things away.

In short, save the goodies for the closing.

GIVEAWAYS ARE NOT ESSENTIAL WHEN NEGOTIATING GROUP SALES

February 1986

All too often, salespeople tend to start "giving away" too much too soon when negotiating group sales such as meetings, conferences, and conventions.

GIVEAWAYS ARE OFTEN UNNECESSARY TO SUCCESSFUL GROUP SALES

Many corporate and association meeting planners are cancer only with being at a particular location and receiving an excellent product and quality of service for their money. The question of getting better rates, complimentary rooms or other extras often doesn't even occur—unless the salesperson brings it up.

Generally speaking, salespeople need to learn to listen more and speak less; they need to learn when and how to negotiate in the best interest of the property. An important part of effective negotiation is knowing when and how to make proper use of giveaways.

When negotiating group sales, the sales staff should pay special attention to the following areas:

Room Rates: Salespeople often make the mistake of bringing up room rates as one of the first topics covered in group sales negotiations-or even when a planner merely indicates an interest in using a hotel or motel. The salesperson's first objective should be to satisfy the needs of the meeting planner and the group. Selling the features and benefits of the property and convincing the prospect that it's his best choice for the meeting should come before the negotiating process. Hotel salespeople should learn to put off discussing rates until the last possible moment. Furthermore, it is not always necessary to sell at discounted rates. Many times, the regular rack rate will be quite acceptable if the property otherwise meets the planner's needs within his budget. It is also important for salespeople to learn something of the planner's experience with other properties: how much he was charged for rooms, what was the room-block pickup, etc. When negotiating room rates, the salesperson should consider how profitable the meeting will be for the property, how badly the property needs the business, and the extent to which the meeting or conference will generate other business such as food and beverage sales and meeting room rental.

Complimentary Rooms: This is another area where salespeople often make unnecessary or premature giveaway offers. It seems that one "comp" room for fifty sold has become a standard expected by meeting planners—but why? This should not be a standard practice. More often than not, meeting planners do not ask for complimentary rooms—salespeople offer them. Complimentary rooms should be a part of the negotiating process, but only as a "trade-off" for something else. It is not necessary to be consistent—the idea is to give something away only if it's necessary for finalizing the deal.

Letter of Agreement: All negotiations that are agreed upon by both the salesperson and the meeting planner should be put in writing, and salespeople should be certain that the planner is authorized to make the agreement. Generally, this should be established in advance when information is gathered during the prospecting and development of the account. Such details as meeting dates, food and beverage functions, meeting space, cutoff dates, guarantees, who pays for what (if billings are separate), when payments are to be made, cancellation clause, penalties on cancellation-all should be spelled out in writing. Salespeople will do much better in closing a group sales deal by knowing all that is possible about the group.

And all that information is available to them simply by asking the meeting planner.

Salespeople should listen a lot, talk little, and should not commit to any extras without there being some kind of "trade-off."

IN NEGOTIATING GROUP BUSINESS, SELL VALUE BEFORE PRICE

July 1988

As hotel/motel salespeople try to close business in the era of underutilization (or, if you will, overbuilt markets), they are finding that buyers are looking for all kinds of deals.

By looking at lodging advertisements and direct-mail solicitations to meeting planners, one can easily see that hotel operators are playing into their hands. Many of these ads and mailings are price-oriented as well as a giveaway-oriented. Some, of course, are value-oriented. Knowledgeable meeting planners will always try to negotiate downward from any price offered by a hotel salesperson. For this reason, advertisements aimed at drawing group business should create an interest on the part of the prospective buyer—enough of an interest so that an inquiry will be made.

TALK PRICE LAST

Although many meeting planners may start off an inquiry by requesting information on price, it's not always in the hotel's best interest to reveal price early in the negotiations. In my experience, it has always seemed that the best approach is to concentrate on keeping the prospect interested. The point is to refrain from talking price until the meeting planner is ready to buy. There must be an understanding of value on the planner's part—if there isn't the salesperson will probably have to overcome a series of objections.

Defer all price questions until the very last, when a "buy" decision is ready to be made. That's when the buyer is convinced that the needs of the group can be met, that the product is what is needed and that he's getting a good value. In short, the planner must be sold on all aspects of the deal before the subject of price is discussed.

Value, of course, comes in many forms. Hotel salespeople should not take the term "price-value" to mean "let's give a good discount to get this piece of business."

"Value" Defined

There are several aspects of value that should be emphasized: There's great value for the buyer if the salesperson is reliable, credible and easy to work with; there's great value if the hotel is in good shape product-wise and facility-wise; there's great value if the hotel or motel can meet all the needs of the group; and there's great value when the planner is satisfied that the food and beverage service is of high quality.

Rule of thumb: The value of a product-staff, service, facility-means more to a professional meeting planner that its price. Of course, the price may not really be a problem when a meeting planner has other important considerations. Rather than price, there may be a concern for meeting space without rental costs; complimentary rooms may be more important; or there may be other items that need to be negotiated.

Determining Needs

The successful salesperson always listens carefully to determine what the meeting planner's *real* needs are in terms of effectively serving his client. There are many areas that will create value for the group that may very well be more important than price.

Those items, once identified and satisfied, will help make the sale. Salespeople must get the planner to understand what it takes financially to put on a meeting, and that the money spent should be viewed as an investment rather than a cost. And selling value first will go a long way, toward helping the planner understand what is really important to the organization he represents. Price will always be a consideration, but let's leave it for last-after the prospect is convinced that the hotel will satisfy his client's needs.

Remember: A successful meeting will be remembered long after the price has been forgotten.

TRAVEL MANAGERS FRET OVER YIELD MANAGEMENT'S IMPACT

July 1989

More and more travel and meetings trade magazine are alerting meetings managers and corporate travel managers to the "new" process of yield management.

Of course, we all know there's nothing new about yield management-we've been using this management tool for years. Salespeople need to be aware of the negative reaction they'll be getting from good accounts because of these magazine articles. Yield management has already become a hot topic of discussion at conferences of meeting planners, corporate travel managers, incentive buyers, and travel agents.

Let's be prepared. Learning the workings of yield management is important so that these discussions can be handled in a positive way.

TOUGH TIMES

It's tough to make a profit in the hotel business today. In 1987, U.S. hotels lost a median of $748 per room. The year before, the loss was about $500. When the figures come out for 1988, they probably will tell a similar story.

That's why we have sales departments whose task it is to bring in profitable sales. Yield management is a great tool for doing that. Used properly, the system provides a rate for a particular room on a particular day, month and year. That rate is based on a number of factors: projected occupancy and the average rate for any particular day for any type or location of the room; anticipated pickup of room blocks; anticipated walk-ins at rack rate; past experience for that particular time of year; and other factors.

It all has to do with supply and demand. During low occupancy periods, we need to bring in business and we're willing to sharpen our pencil. During heavy occupancy periods, we must stick to the rate to maximize profit.

But remember: We're still in the people business and we can't rely solely on the computer-generated rate decisions. There's the question of how important a piece of business is to your hotel. It could very well be that General Electric, for example, should get a better group rate since it can book many more meetings at your hotel-volume buyers certainly need to be looked at differently than one-time buyers.

LET'S MAKE A DEAL

There is such a thing as cheaper by the dozen. At times there will be a group that wants your property during high occupancy times, but can't or won't pay the high rate. If this organization can place another piece of business with your property during a low occupancy period, then you'd certainly want to consider making a deal.

Jim Jones, formerly of Connecticut General and now a meeting planner, sales consultant and sales trainer for hotel companies, used to have a "thing" about negotiating rates. He followed the philosophy that his business was so strong and powerful for a hotel that the contract would have to state that no other group in the house at the same time during his meeting would get a rate lower than his. Other companies picked up on this philosophy and are using it during negotiations.

BE HONEST WITH CLIENTS

To make yield management work, hotels should not succumb to this philosophy. You have to explain to meetings managers that something could very well come up that would require you to bring in a group at a rate lower than theirs.

It happens all the time. You book a group during a high-rate period at what the group would consider a favorable price. Somewhere along the line, another group booked for the same period either cancels or drops a chunk of the room block. At the last minute, you're stuck with a hole-and sometimes you'll want to fill it at any price. As the subject of yield management comes up for discussion with buyers or at meetings, let's meet it head-on and explain our position. After all, reasonable people understand basic business practices. All we want is to be profitable.

Salespeople need to have a positive attitude when dealing with the issue of yield management.

SUCCESSFUL LODGING SALESPEOPLE ARE GOOD NEGOTIATORS

March 1990

The key to success as a salesperson in the lodging industry is being able to bring in new business at the proper price. There is no question about it, this will be another tough year in our industry and owners will be looking for a better bottom line. The sales function at properties is becoming more and more important as owners and operators look to the sales staff for increased profitable sales.

Going out to get the business is tough enough, but bringing it in at the right price is tougher. The key to success in this area is in the selling and negotiations skills of lodging salespeople. The best negotiators in our business have these things in common:

▶ They are well prepared
▶ They clearly understand the prospect's needs
▶ They establish relationships
▶ They do not rush
▶ They practice good negotiation strategies

Preparation: Never try to negotiate a deal without being prepared. This means having superior knowledge of your own lodging establishment. Everything from floor load capacity per square foot in your ballroom to ceiling heights, to the number of king, bedded rooms, to size of the pour at cocktail receptions, etc. Knowledge of your rate structures; what you'll be able to "give" during what periods of the year and days of the weeks, certainly is important, and on and on. The salesperson will need to have superior knowledge of all the competitions' facilities so comparisons may be made, if necessary. Buyers will be looking at other properties, as well as yours. Of course, being prepared also means knowing the history of the prospect and the value of each proposed piece of business being negotiated.

Buyers' Needs: Understanding the prospects' needs is a high priority in successful negotiations. The selling technique here is to probe and relate. The more we listen and less we talk the better off we care. Ask a key open-ended question—then relate to the answers with good benefits of the features you are selling. The objective is to sell to fulfill the needs of the buyer. The buyer doesn't care for your "wonderful," spacious, lot for 500 cars if the meeting attendees fly in.

Developing Relationships: It is no secret that people buy from people they like, they respect that are reliable and dependable. Successful lodging salespeople establish healthy, long-lasting relationships with people who are in a position to buy. We do this by living up to and delivering our promises and through contacts via associations. Meeting Planners International is a good example of where an organization was the vehicle for excellent long-term relationships were developed between buyers and seller during this 1-year in existence.

Never Rush: The prospect should never fell that he or she is being "pushed" into a deal during the negotiation process. There is nothing wrong with being aggressive and always trying for a close when closing signals are observed. Give the prospect whatever time is necessary in the discussion process. Buyers should not feel that they are an interruption of your work—give them full attention.

Negotiation Strategies:

- Look for mutual benefits—areas of shared interest and areas of agreement. This will establish a feeling of goodwill.
- Separate people from the problem. Personalities should never enter into the discussion when differences or areas of disagreement come in to play. Avoid "I," "You," "We," etc.
- See the prospects" point of view (understanding it doesn't have to mean that you agree).
- Keep control. You set up the agenda; what is to be discussed and when. When you control the agenda you control the meeting. And, make sure the discussion on rate, if it is an issue, is kept for last.

IMPROVE PROFITABLE SALES WITH GOOD NEGOTIATING PRACTICES

April 1991

Good negotiating techniques will almost always produce increases in profitable sales for property sales personnel. Entering into a negotiation session to close a sale need to be a planned event. It should not "just happen."

Some good, hard research needs to be done during the planning stage. Gathering information is an important part of the process. There are two things that the best negotiators have in common:

1. They are well prepared.
2. They clearly understand the other party's needs.

By being well prepared we mean knowing certain things in order to evaluate the value of the piece of business. For example, some of the basic things to know would include:

- When? Not only the specific dates, but how those dates relate to when the property needs the business . . . not only on or offseason, but specific days of the week. Even during the season or during any busy period there may be a "hole" as a result of a cancellation or other reason, and by filling it with a new group it can be most profitable.
- Length of stay? Sometimes a longer stay group may be more profitable than a shorter stay depending on projected revenue from all outlets. Sometimes, it may not be so profitable if it knocks out other business that could be more valuable to the property.
- Ratio of meeting rooms to sleeping rooms? If a social or incentive group can be booked, using only a minimum of meeting space—or not space—then there is the opportunity to use the space for other groups or local catered events.
- Other income from the group/ the more food and beverage functions and space sold—the more revenue produced. This can mean a more profitable sale.

▶ Exhibit space? Many times having a conference with exhibits, a property could be selling more suites, more food/beverage and sometimes linger stays for exhibitors. Along with the sales of the space, this means additional revenue for the property.
▶ Repeat business? Multiple meetings from the same group should prove to be more profitable, over a period of time.
▶ Group history? Very basic, but a key to knowing the group experience is just a phone call away. It helps present a good picture of the budget, spading habits, rooms pick up history and problems which may have surfaced previously.

Clearly understanding the other party's needs makes a sale person a real professional. Serving as a consultant to the buyer shows interest, concern, and a real desire to help make the meeting a success. Questions such as: Why is the meeting, conference or event being planned? What is the purpose and objective? What are the results expected to be? With this information, as a starter, the salesperson will get to better understand the needs of the organization.

Good negotiators always look for mutual benefit—areas of agreement and shared interest. This establishes a feeling of goodwill between the parties. It sets the tone and mood of the session. Sales personnel should also pick up on something that is wanted or needed by the property, but will also work well for the group. As an example, getting a Sunday night arrival could be most important for a hotel. If a group is expected to arrive on a Monday for a meeting starting on that day it could very well be in the best interest of both, the meeting planner and the hotel to work out some arrangement so that the group arrives on a Sunday.

Above all, salespeople should not feel intimidated by a negotiation session—it should not turn out to be an adversarial encounter, but a friendly session for mutual benefit.

WHAT'S NEGOTIABLE WITH LODGING FACILITIES . . .

June 1991

Corporate executives planning meetings and travel incentives are always concerned with what is negotiables a lodging facility and to what extent they can negotiate a "better" deal. We have all heard from time to time that "everything is negotiable," well, perhaps . . . sometime. To what extent one can negotiate has a lot to do with clout. Before any executive gets to the negotiating table some planning needs to be done. The key to successful negotiations is to determine clout. This involves knowing how important your business is to the property and how badly they need you. Clout can mean of several things. You may have clout if you can:

▶ Bring business to the property when they need it (offseason or specific days of the week).
▶ Arrange a heavy budget for food and beverage functions.
▶ Provide multiple occupancy.
▶ Bring multiple meetings to a property.
▶ Have pre- and post-meeting stay overs.
▶ Produce a tradeshow or have exhibited with meetings.

- Book a site while it is under construction.
- Select an advance date when the property has nothing on the books.

Those executives that can be flexible when it comes to dates and can "fill a hole" when a property needs it has the best shot at negotiations. Even if you have only a little clout or feel you have no clout at all—it does pay to try to negotiate anyway. You never know! And, it certainly doesn't hurt to ask. It is amazing how much lodging salespeople giveaway just because a prospective buyer asks.

Corporate executives looking for deals from lodging facilities can certainly find them these days. Most people are already aware that the hospitality business has taken a pretty bad hit these past few years. Especially, just recently, between the gulf war and the recession, the first quarter of 199 was truly a disaster for the industry. In 1990 64% of the hotels lost money. The ones hit the hardest were the mid-price establishments. Resorts seemed to hold its own, but the ones that did better than most, from a profit standpoint were the budget or limited service category. From a buyers' standpoint, because the hotel industry is still in the overbuilt situation and there is a strong need to recover business after the first quarter, there are great bargains out there and good opportunities for negotiations.

It would be well for a prospective buyer to be aware of where money is made and not made in the lodging business. Most hotels and resorts do make money from the rental of meeting or exhibit space and the rental of sleeping rooms. Generally, there is no money to be made on food and a little may be on bar operations of there is a large volume of business in that area. Therefore, the best opportunity for negotiations in connection with company meetings or travel incentives is with room rates and space for meetings. Successful hotel and resort sales personnel are always interested in knowing how profitable any piece of business may be prior to their getting involved in negotiations with a prospective buyer. It would be important to impress salespeople with the volume of business available. Although there is no profit in food and little in beverage, it is most important for lodging facilities to get as much of this business as possible to offset costs. The more profitable the whole piece of business appears to be the better opportunity for the buyer to negotiate.

In the area of rooms, there are a variety of items to negotiate other than room rate. Based on the number of rooms being picked up for the meeting or travel incentive movement the facility would probably make available a certain number of rooms on a complimentary basis. Although there are no industry standards for a formula by which the number of free rooms are computed, many operations would offer one complimentary room for every 50 paid. Some would require more and some may require less, depending of course, on how badly the property needs the business and how important it is to them.

Negotiable areas to consider:

1. **SLEEPING ROOM**
   - Rates and rate structures
   - Complimentary Rooms
   - Extra Housekeeping Service

- ▶ In-room amenities
- ▶ Room locations
- ▶ Early check-in; late check-out
- ▶ Deposits
- ▶ Overflow housing
- ▶ Room gratuities

2. **FOOD FUNCTIONS**
- ▶ Meal and meal plan prices
- ▶ Service changes
- ▶ Guarantees
- ▶ Extra labor
- ▶ Table decorations
- ▶ Menu items
- ▶ Surcharges

3. **WINES AND LIQUOR**
- ▶ Brands used on receptions
- ▶ Dry foods and butler service
- ▶ Bartender fees; gratuities
- ▶ Cost of house bulk wines vs. bottled brand wines
- ▶ Price structure (per drink, per bottle, per person)

4. **MEETING AND EXHIBIT SPACE**
- ▶ Price structure, labor, and technicians
- ▶ Audiovisual
- ▶ Receiving and storage of supplies
- ▶ Conference service manager

5. **MISCELLANEOUS**
- ▶ Parking fees
- ▶ Recreational activities
- ▶ Signage
- ▶ Collateral for attendance promotion
- ▶ Airport pickup
- ▶ Local phone calls rate
- ▶ Chairman office and phone

UNDERSTAND THE NEGOTIATING PROCESS BEFORE MAKING THE DEAL

June 1992

In today's tough selling environment, salespeople really need to understand the process of negotiating when working with the group market. It is always in property's best interest for you to know what and how to negotiate. Although you don't have to negotiate everything. Generally, everything is negotiable when it comes to trying to book group business in tough times. The problem we have is that there are too much of us trying to "give away the house" before we even get into the negotiation process. The whole idea of getting what you need for a property is not to give anything away, at first. Always try to get what's good for you with giving nothing—until it gets to the point where negotiation is necessary. You always what to leave the door open so that there is a possibility for a "trade-off," making concessions, giving something here and taking something there. After all, negotiation should be an opportunity for both parties to do some selling. The prospect wants to sell you on the idea that the piece of business is good for the hotel and the hotel really needs that piece of business because of any number of reasons. On the other hand, the salesperson needs to convince the buyer that the property meets the needs of the group, it is comparable or better than the competitions and that there is a price value relationship that is favorable to the planner.

Salespeople should never get into a negotiation situation until there is complete knowledge available. That is, knowledge of your own property, competition, and the prospect.

- *Your Own Property*: How badly do you need the business? Do the required days of the week, month, time of the year already look good for you because of projections? Then, perhaps you do not need the business so badly and will not have to "give" too much during the negotiation session. On the other hand, if there is a "hold" that needs to be filled, then you probably could "give" a little more to book the deal. How far can you go on rates, camp rooms, amenities, food and beverage prices, etc.? There has to be a cutoff point on how far you can reduce prices and give stuff away and still make profit on the deal.
- *Your Competition:* You already need to know with what other property the prospect may be dealing. Find out before, while you are gathering information, which cities, which properties are being considered. Know all there us to know about them to see how you "stack up" against them. How far do you think those competitions will go to get the business? What is their reputation for making deals? You probably will have to make comparisons with them if it comes up during the negotiation processes. Don't be fooled with the prospect of making statements about what the competition will do.
- *The Prospect*: It is not difficult lining up information about the prospect before getting into the selling and negotiation situation. Always, during the prospecting period, we need to find out all we can about the people with whom we want to do business. If we do a good job of asking questions and listening we can get just about all the information we need. Except, we must also always check up on the prospect's reputation at the previously used hotel location. A phone call to the sales department of properties where the prospect has done business previously will generally get the information you need. What is their reputation for: picking up the block of rooms; providing meal functions for the group; use of profit center outlets (bar, recreations, sports, etc.); providing deposits upon booking; stay overs; creating damage to the property; early arrivals; late check-outs; trying to re-negotiate after the contract is signed; real need for comp rooms; paying the bill and of course, other things that come to mind because of the particular nature of the type of business. Probably the biggest and most important thing to know about the prospect is: HOW BADLY DO THEY NEED YOU? Sometimes you may be the only "act" in town because of the location, type of property, facilities you have available. This would give you an indication of how far you need to go during the negotiation process.

Don't let the term "negotiation" throw you; don't be intimidated. It sounds so legalistic, but all it means is: Let's Make a Deal . . . Partner.

WHEN NEGOTIATING WITH GROUPS, HOLD OFF ON RATE QUOTE

May 1993

In prospecting for and selling to group planners or arrangers we really need to consider how well we do our job when it comes to negotiating. All too often many salespeople are so enthusiastic about closing a deal that we probably give too much away when it comes to rates. In the present climate of pressure for revenue and bottom line orientation, we need to improve our negotiation skills to the point of being able to book the business and be sure that it is as profitable as possible. Often we are too eager to quote rates for a group booking before we even know how profitable the piece of business may be to the property. There are certain steps to go through before a salesperson should get into negotiations with a group prospect.

We should all learn to prepare ourselves well before we even attempt to negotiate rates or "give-a-ways." Knowledge is the key to being prepared. Of course, it stands to reason that we should be completely knowledgeable of the property. The more we know about the product we are selling, the better job we can do in selling it. You need to be able to answer any question that comes up about the facility as well as offer suggestions about how the property could best be used for the prospect.

The next area with which we need to be concerned is knowledge of the competition. Hopefully, based on the type of group movement, people, dates, requirements of the planner, etc., we have a pretty good idea with which properties we may be in competition. The more we know about our competition the better position we are in for negotiating. After all, there are some features of our own property, compared to the competition, which creates a much better benefit for the prospect. Or, the competing property even may have features which may appeal more as benefits to the prospect. We need to know everything there is to know about those facilities we are negotiating against.

Then, of course, there is knowledge of the prospect. Whether it be a company, association, leisure group, whatever, we must be sure to know as much about them as possible. The business they are in; something about their operation, the purpose, and objective of the meeting or group movement as well as their needs with regard to the particular event. History is of major importance; where they have met before, what was good or bad about that experience, what they paid, an outline of the complete program, room block pick up, etc., how will the decision for site and location be made, who will make the decision.

During the selling process, we have to make sure that we just don't jump into negotiations. First, we must ask questions, listen actively (intently), and continue to ask questions about needs and relate our specific features which represent benefits to the prospect's group. There's no point presenting features that don't relate. For example, if your property has airport transfers why even mention it if everyone will be driving in for a meeting. The same holds true for the "very large parking lot," if coming in by air. The continuing process of asking questions is the probing we have to do to try to uncover areas where can present benefits and put us in a favorable light when it comes to negotiating.

One of the biggest problems we have is offering a rate (which may be discounted) before we even know how profitable the business will be to the property. Or, if even providing a discounted rate is necessary. Save the rate quotation for the last in the negotiation process.

If objections to price comes-up, then try to put it off until all other objections have been handled favorably to you.

Successful salespeople know that the best time to get into the rate quotation, if there is any sensitivity in this area, is to do it only after we are convinced that the prospect is sold on the property and has a real desire to make the deal.

SALES AND NEGOTIATION TIPS FOR WORKING WITH GROUPS

December 1993

When it comes to working with groups, whether leisure or business, we need to be careful in the way we go about "offering" some things that may not be necessary. In these times of tough competition and discounting running rampant, be on the lookout for closing group sales without giving away too much. If we start out without giving anything away and quoting higher rates we may be surprised about how much better we'll end up in closing the deal.

▶ *On Comp Rooms*: Do we really need a "policy" of 1 for 50 (in some cases 1 for 25)? Somehow we have trained group buyers to think there is some sort of universal policy in our business to provide comp rooms. We do not have to be consistent in this area. Sometimes you won't give any, sometimes more, sometimes less. It all has to do with how profitable a piece of business may be and how badly you need the business, but more importantly, how good a negotiator you are.

▶ *Let's Get Paid as Quickly As Possible*: Even meeting planners complain that hotels take too much time in getting bills to them. Then there may be a hassle about incorrect charges which delays payment further. Work with your controller or billing department to set up a system whereby the person in charge of the group meets daily to go over the previous day's charges and again just before departure. This gets to resolve any differences with regard to guarantees, specific charges, or whatever. Then impress upon them the value in getting the billings out immediately after the departure of the group. Your property's account receivables will be helped immeasurably.

▶ *Advance Payments*: Don't forget to think about getting deposits or advance payments in connection with groups. This could work very well for you in the negotiation process. More and more groups already have the money available and may be willing to pay as much as 60%–70% up front (at time of arrival) for something in return like comp rooms, upgraded cocktail reception, reduced meeting room rentals, etc.

▶ *Computer-driven Letters*: We're probably getting a little too technology-oriented when it comes to word processing. More and more sales departments have long lists of programmed letters being used for all purposes. Are we getting a little bit too lazy to personalize our correspondence? Customers view these letters as too impersonal and in some cases downright discourteous. Even the proposals should be individually written since there is nothing standard about any kind of a group meeting or function. At least, the buyer doesn't think so.

▶ *On Sales Calls*: For the most part, we are told that salespeople generally do a lousy job. Many salespeople are not prepared and just seem to waste the prospect's time. There's a difference between a prospecting (fact gathering) call and a sales call. Before a sales call is made just be sure you know about the prospect, their needs and what you have that meets those needs, know all there is to know about your property and just about as much about the competition.

▶ *On Site-Inspections*: Meeting planners say that property salespeople waste too much of their time on site-inspections. Mainly, we are not prepared with an agenda. Spend 100% or your time with the prospect, arrange for no interruptions which keep the prospect waiting. Pre-arrange sleeping rooms to be checked (show the best and the worst). Get a copy of the group's previous conference schedule and design the visit to walk through the property showing how you meet their needs with respect to what they will actually be using according to their program.

▶ *When to Come to Contracts*: Be specific about everything. There needs to be no misunderstanding. It may not be a bad idea to have an industry-related attorney (those that specialize in meetings contracts) to help you with designing a contract that is fair and that protects your interests.

IN SELLING, WHEN IT COMES TO NEGOTIATIONS, KNOWLEDGE IS POWER

August 1997

For those properties that are involved in selling to the group, it is important to be prepared to properly negotiation before making the deal. Groups are not always large, but also come in small, profitable packages, that are available to the smaller, even limited service type properties. There are opportunities to book group business with companies that have small (or large conferences); association meetings; motor coach tours; travel agents; there are also military reunions, family reunions, school reunions; and variety of small to large groups. To a certain extent, negotiation will just about always come into play when trying to book this type of business.

In a high demand period, like these days, for hotels, managers and salespeople need to make the best deal possible for the property, but also leaving the prospect as satisfied as well. This mean negotiating to your best advantage. When it comes to selling you need to be prepared by having as much knowledge as possible about the prospect with whom you will be dealing. Knowledge is power. You will be able to do a much better job in negotiating if you had a feel for the Prospect's Position in advance of negotiating: Here we mean—to what extent are they ready to book? Are they very close to agreeing to meet at your property or not?

You can get a feel for this if you take the time to get information about:

- *History of the group:* How often do they meet; where have they met before (were previous properties used similar to yours); what was good about the site and location, what was not so good; do your facilities have exactly what they need?
- *Budget:* What have they paid previously for the room, food, space rental, etc.? Are those rates similar to yours/does it look like they have the budget for what you want to change?
- *Competition:* What other properties are being considered by the group? How do you stand against these other properties—are you in a better position to satisfy the prospect or nor?
- *How badly doe the group need you:* This is really the key! If not other property has the space available and the group needs to meet on specific dates which you can accommodate, then they need you pretty badly.

In addition to understanding the prospect's position you really need to understand your position:

- How badly do you need the business; If you expect to run full, at rack rates, during the period the prospect needs you, then you can "hold tough" during negotiations without offering any considerations for rate or extra amenities. On the other hand, if you could use the business during the period because of a slow season or projects reflect a low occupancy, then you will want to treat your negotiations differently.
- How important could the business be to you: Sometimes, you may not need the business so badly, but taking the business could be important to you. Here you want to examine:

Will this business bring in heavy food and beverage sales? Will it bring in people who may be in a position in the future to bring in other business? Is this going to be an important event that will bring in the news media and provide favorable public relation for you? Does this company (or group) have other business that you could get at other times when you need it? Is there any opportunity for a multiple year contract?

Also, in preparing for negotiating you should have a very clear understanding of how far you can go in discounting a rate for rooms or space or "giving in" on other considerations. If you know you can use the business then decide in advance how far you will go in negotiations.

TIPS FOR NEGOTIATING SUCCESSFULLY WITH BUSINESS PROSPECTS

October 1999

Negotiations, in our business, has a lot to do with developing a partnership, according to Jonathan Howe, President of Howe & Hutton, a leading Chicago law firm specializing in contract relationships between hotels and groups planning meetings. So often we talk about negotiations being a "win-win" situation, which sounds pretty good. But what we are really trying to do is establish and maintain an ongoing, favorable relationship with a prospect for continued business. Therefore, in our negotiations process, we want to make sure that we do not create an adversarial or confrontational situation where both parties fight over a series of demands and concessions. We want the outcome to become a true partnership; therefore, the negotiations should be approached as a problem-solving process through which both the hotel and the prospect reach a mutually agreeable solution, ending up with a contract acceptable to both parties. This does not mean we have to give away the whole store. An objective, of course, has got to be a favorable bottom line.

Being prepared for negotiations is critical to a favorable outcome for both parties. From the hotel standpoint, it is necessary to understand the property's position. How badly do you need the business (does it fall during a slow period; is occupancy projections soft during the time of the event)? Sometimes we do not need the business if projections reflect that the property will be busy, at good rates, without this event. However, taking the business could be important to you. For example, there may be other business from the same organization during another period where you could use it. It could mean very favorable publicity for the property. Or, the make-up of the group (chief executives of top companies) could be very influential for other group business. Heavy food and beverage business during the event could also make the group important to you. Some preparatory work needs to be done to analyze the business. Part of your planning will need to be how far you can go in negotiating. Would you be willing to reduce rates to get the business, offer comp rooms, provide free meeting space or other comp amenities, etc?

Sometimes salespeople talk too much during the selling process. Instead of trying to "hit people over the head" with talking about all the wonderful features of the hotel, how about letting the prospect do the talking? You can do this by asking open-ended questions which prompt the prospect to provide you with additional information. "What's the purpose and objective of the meeting (or event)?;" "Tell me about the make-up of the group;" "What did you like best about your last meeting?" These types of questions will bring information to you, to which you should be able to relate some sort of benefit of having the event at your property.

You should be able to identify key issues (price, location, service, etc.), but put off discussing (negotiating) any key issues until you have a good feeling that it looks like the prospect is "sold" on your property. This provides you with more time to do a better job of selling. It is not always in your best interest to start giving things away reduced rate, comp rooms, free space, etc., unless there is a trade-off. That's what negotiations are all about."I'll do this if you do this." All too often we start giving things away too soon in the negotiating process.

OVERCOMING OBJECTIONS IN HOTEL SALES IS A TOUGH JOB

October 2002

Without resistance to buying there would be no need for salespeople.

Resistance is the most important hurdle every salesperson in our industry has to get over. It is the toughest part of our job, and we don't always win. When people resist buying anything, it is really that they have an objection to something. So, we need to know what is keeping someone from buying (booking a room, a meeting, a social event, etc.). Once we are aware of the right reason then we can bring up the right benefit (value) to minimize resistance.

One of the least effective means of overcoming resistance, that we have observed over the years, within the ranks of hotel sales folks, is taking a defensive position. This is where the salesperson starts coming up with reasons that defend the hotel with regard to whatever an objection might be. A buyer can care less about reasons. One manager once told me at one of our workshops that when someone complains about the price he tells the prospect that he cut his rate right after 9/11, and now he has to go back up to "get even." Another poor defensive tactic that we hear quite often, when an objection comes up over price is: "Well, how much do you want to pay?" or "what's your budget?" In selling, we need to be in control by taking the offense rather than be defensive.

The first real order of business is to find out the reason for an objection.

After all, when people object we should take it for granted that what we hear is the real reason. Here are some reasons why prospects resist buying, but they do not always tell us; they may give us other reasons:

- Lack of Knowledge (maybe we did not express enough information on benefits)
- Skepticism (uneasy about being unfamiliar)
- Need to Compare (they are looking at other properties)
- Trading Advantage (wants some advantage, something extra)
- Prejudice (against the salesperson, or the hotel brand or owner)
- Bad Reputation of Property
- Poor Presentation (we just didn't do a good enough selling job)

A prospect may be too uneasy about bringing up any of these real reasons so they object to price, location, facilities, etc. The whole idea is to try to bring up the real reason. And, we do this by asking "why" type questions like: "What makes you say that?;" "Please explain what you mean by that?;" "How come you feel like that?" What we want to do is keep the prospect talking, with more questions about their reason regarding the objection. Rather than the salesperson feeling that he/she has to do the explaining when a person objects, we should try to get the prospective buyer to explain. This way we may be able to find out if there is another reason for the resistance. Once we are satisfied that we have identified why the prospect is not buying then is the time to come up with the advantages of doing business with us.

Here is where we want to show what value there is in doing business at the property. Perhaps we may even want to offer additional benefits or value to get them to make the buy decision.

Sometimes the objection may be obvious, so we have to be careful and not ask the "why" questions. For example, if someone objects to making a buy because the property is 50 miles from the airport, and they are correct, then we surely can't ask the question "What makes you say that?" There is nothing wrong with admitting the obvious and showing the advantage of the location. We may even be able to capitalize on the objection as many resort sales folks do in explaining the advantages of their location.

Remember, overcoming objections is not easy. It takes lots of practice, and with some experience, it will work well for you.

BE BETTER PREPARED FOR NEGOTIATIONS WITH GROUPS

May 2004

In this current period of business recovery for all lodging properties seeking group business we just need to do a better job in negotiating with prospective buyers. We should not have to be in a position of giving away so much as we have been doing in the past couple of years. All indications from various companies that measure our business outlook are that we can expect a good increase in demand while growth in new rooms is down. When demand increase is greater than supply increase that means we should be able to increase rates, occupancy, and sales productivity.

From the standpoint of group business, sales personnel need to be sure to start spending more time going after qualified prospects, and being better prepared for negotiations. All too often we find ourselves in a position of negotiating for group bookings without having enough information to do a good job, and end up not maximizing a sale. Being good as a negotiator means being better prepared by learning as much as possible about the buyer's position. That is: "what is it that you already know concerning the prospect's position with regard to entering into an agreement with you." Or, "how badly do they need you"? It is not enough just to know the other party may be interested in having a meeting at your property. You need to do the research before you put yourself into a negotiating position with them. Here are some tips on what information you should gather prior to setting up a negotiating meeting or even a site inspection:

▶ Where have they met previously (at least for the last two years or the last two meetings)? Knowing just what cities is not good enough; find out the names of specific properties used so you can compare what you have comparable to the others.
▶ The make-up of the group. What are the demographics? Is this a group of all men, or women, or are there families with children? Find out about the point of origins, age groups, economic status, types of businesses represented. Having a good picture of the group makeup can

give a feel of how you can accommodate certain needs and show the value of having the event at your property.
- What is the past booking pattern of the group? You should know arrival and departure days to see how well that would fit in with what you may already booked at the property or expected occupancy during that period.
- Will, you be dealing through a third party independent meeting planner and is there any commission expected on their part. This probably would affect the rate negotiations.
- What is the expected spending habits with regard to what would appear on the master account folio. You also need to know the spending habits of the meeting participants. What has been the history at previous hotels with regard to the use of various hotel profit centers (bars, recreation, restaurants, shops, Internet time purchases, etc.)?
- The amount of meeting space used in relation to a number of rooms actually used by the group. Does history reflect that the prospect paid for meeting space, or special set-ups, or audio-visuals?
- Other locations and sites which the prospective buyer may be seeking.

Of course, there are a good many more areas into which a property salesperson must investigate in order to be properly prepared to negotiate so that business may be booked generating the best revenue possible.

NEGOTIATIONS ARE STILL THE HOTTEST ISSUE WHEN BOOKING GROUPS

June 2004

Hospitality salespeople, as well as those who plan for group bookings, still get nervous when it comes to the point of having to negotiate. Most of the time it gets frustrating for both parties and often time-consuming. Hopefully, it doesn't get to be adversarial, where you fight over demands and concessions. The outcome of negotiations should be a win-win situation for the prospect and the property. The outcome is more likely if negotiations are approached as a problem-solving process through which both parties reach a mutually agreeable solution.

Here are some tips that may prove helpful in negotiating successfully with prospects

- Prepare for negotiations by carefully analyzing the prospect's position as well as your property's position. Here we mean how badly they need you compared to how badly you need their business.

This involves organizing information gathered earlier in the sales effort, and researching additional information that may prove useful when negotiating key issues with the prospect. Although all the information may not be available, try to research the following items:

- *Previous Budget*: While the prospect's current budget constraints may differ from those in the past, knowing the prices paid at earlier events provides

insight into important aspects of the prospect's negotiating position.
- *Competitors*: Find out if the prospect is negotiating with other properties. Analyze the competition, and be prepared to make a comparison and offer reasonable concessions.
- *Special Needs*: Determine the concessions and special services provided by previous hosts. These may be important items on the prospect's bargaining list.
- *Past Problems*: Knowing what went wrong in the past may help anticipate demands.
- *Ultimate Decision Maker*: If the prospect doing the negotiations is not the ultimate decision maker, make sure the prospect leaves the negotiating session with the conviction, and tools to sell the hotel to the boss, board of directions or whomever.
- *Decision Deadline*: Knowing when a decision must be made affects the hotels' negotiating position. Negotiating on the day a decision must be made calls for a different strategy than negotiating with a prospect that will not make a decision until sometime in the future.
- *Key Issues*: Being aware of what the prospect considers to be the most important negotiating issues provides the opportunity to prepare specific strategies for trade-offs, upgrades, comps, etc., and yet maintain a required level of profitability.
- *Interest Level in the Property*: Having a feel for the level of interest the prospect has in booking the property tells a lot about their negotiating position. If there are no other options for the prospect, or if your facility will best meet their needs, they may have a weak negotiating position. However, if they have other options for locations or sites, that can meet their needs as well as your property, they probably have a strong negotiating position.

How badly the hotel needs a prospect's business depends on the amount and type of demand for the hotel's facilities and services on the dates in question. If the prospect's event is planned for a month or day of the week when demand is typically low, the hotel may need the business badly enough to lower profitability requirements. This allows the salesperson to develop generous bargaining strategies before negotiating with the prospect. However, if the prospect's event is planned for typically high demand dates, higher profitability requirements may allow for different negotiating leverage.

KNOWLEDGE IS POWER WHEN IT COMES TO NEGOTIATING

May 2005

So many hospitality salespeople get a little nervous when, in the selling process they need to get involved in negotiating with a prospect. Just about everyone wants to get a "deal;" whether it is a lower room rate, complimentary accommodations, a free something or other, etc. We have to expect this, and sometimes, more often, than not, salespeople feel they have to "give-in" to get the business. All too often we find ourselves in a situation where we start off the selling process by negotiating something before we even have enough information.

The key to successful negotiations is not to get involved in this part of the selling process until the salesperson has enough knowledge about the whole situation. There are three areas of knowledge that is a MUST in all negotiation situations: Knowledge of your product; knowledge of the competitors and knowledge of the specific needs and wants of the prospect.

Product Knowledge: Using a typical hotel, as an example, how much do the salesperson know about the whole hotel operation? Sure, we know a number of rooms, rates, restaurant hours of operations, number and size of meeting rooms and a whole lot of other stuff. But how about the number of connecting rooms? Could this a good piece of knowledge in selling to a specific market segment? Or, how many parking spaces do you have? Just in case the prospect's group is driving cars to get to the property. What is the size of the sleeping rooms? Sometimes the size of the room is important in comparing with the competition. The idea here is for salespeople to conduct a complete product analysis of a property and make up a fact sheet covering every possible aspect of the hotel's facilities.

*Competitor Knowledge***:** Every salesperson should do his or her own competitive analysis; making a comparison of all the features of his own property compared to all competitors. Here coverage of rooms, food/beverage outlets, meeting and banquet facilities, all services have to be covered. We just need to know where we "beat" a competitor or where the competitor "beats" us.

Knowledge of the Prospect: In considering "Knowledge is Power" when it comes to negotiations, this is the critical part of the process. There is no point in entering into a negotiation session without knowing all there is to know about the prospect's needs and wants. The key here is the questioning of a prospect, with open-ended questions that get him/her to talking. What you are seeking are "hot-buttons." In responding to a response to one of your questions you then have the opportunity to show that you understand the group's needs, and then you can express something that relates to the benefit or value of using your property. From there you can then proceed to another question. Ask questions that get the prospect to think, analyze, or even speculate.

Remember that negotiating has to do with "making a deal" that is good for both parties. However, the idea is not to give anything away without getting something in return—a give and take!

UNDERSTANDING A PROSPECT'S POSITION IN NEGOTIATIONS

April 2006

An important part of preparing for negotiations, during the process of making a sale, is analyzing the prospect's negotiating position. That is, how close is the prospect of booking the business with you, or how far away that prospect may be. As discussed in the "Negotiations Are Still the Hottest Issue When Booking Groups" column, the negotiation involves organizing information gathered earlier in the sales effort, and researching additional information that may prove useful when negotiating key issues with the prospect. Although all the information may not be available, try to research the following areas: previous budget, competitors, special needs, past problems, ultimate decision maker, decision deadline, key issues, and how far away they are from booking (or interest level in the property).

An important part of preparing for negotiations is also the analyzing the hotels' negotiating position. Profitability requirements determine limits the kind of bargaining strategies a salesperson can develop before negotiating with a prospect. Profitability requirements are established by determining how badly the hotel needs the business as well as how important the prospect's business could be to the hotel. This depends on the amount and type of demand for the hotel's facilities and services on the dates in question. For example, if the event is planned for a month or day of the week when demand is typically low, it means that the hotel may need the business badly enough to lower profitability requirements. Therefore, this allows the salesperson to develop generous bargaining strategies before negotiating with the prospects. However, if the prospect's event is planned for typically high demand dates, then the higher profitability requirements may allow less negotiating leverage.

Another factor that affects the hotel's negotiating position is whether doing business with the prospect may lead to opportunities for future business. This may provide the opportunity to develop generous bargaining strategies for negotiating with prospects that could bring additional business to the property, particularly during low demand periods.

BE PREPARED FOR NEGOTIATIONS WHEN SELLING TO GROUPS

May 2008

In past years, we have seen, in many cases, when hotel sales staffers have sold to groups, that no negotiations had taken place at all. Quite often, in cases of small groups seeking conference space as well as sleeping rooms, and meals, whatever was offered by a property salesperson had been accepted, and a contract signed without any negotiations taking place. This has happened as well with corporate travel managers accepting an annual contract for a guaranteed rate on sleeping rooms for their traveling executives. However, now with continual room rate increases of the past few years, we see more and more group meeting planners, as well as corporate travel managers, seeking to make the best deal possible by asking for better rates and other considerations. This, of course, puts hotel sales personnel into the position of having to negotiate to make the sale.

Being a good negotiator take practice; the more one does it, the better one becomes at the art. The whole idea behind negotiations is to end up with a deal, and the deal, of course, has to be satisfactory to both parties. From the hotel standpoint, the main thing to consider is: "How badly do we need the business?" Naturally, a salesperson has to be well prepared, well in advance of the planned discussions with a prospect. With proper forecasting, the details will be obvious regarding what business is already on the books or expected, and what revenue is expected to be generated already on the dates being requested by the prospect. If business coming in during that period is good, then, the property does not need this business so badly, and the salesperson would not be willing even to offer anything in negotiations to book the business. However, if expected occupancy is not so good, then the property may very well need the business very badly, and the salesperson would want to make some offers in negotiations to get the business. What salespeople then need to consider how far they may have to go to get the business by giving-in, and providing more of what the prospect will want? This may come in the form of better rates on rooms, free or lower rates on space rentals, or other considerations.

If, during the dates that are requested by the prospect, the hotel does not need the business so badly, salespeople need to further consider: "How important could this business be to the property?" Even though the prospect's business is not needed so badly there could be some things that could make the business important enough, such as:

▶ The make-up of the group expected would bring in high profile executives that could be bringing other business to the property.
▶ This group would be heavy users of the property's catering services.
▶ Attendees of a group would be heavy users of hotel's profit centers such as lounge, recreational activities, spa, shops, etc.
▶ Nature of the group could expect early arrivals and stay-overs, using rooms pre- and post-event dates.
▶ Prospect may offer a multi-year contract for this same event.
▶ Future booking on future dates for other events by the prospect.
▶ Heavy up-front deposit on the expected event, or a pre-pay of a portion of the expected costs of the event.
▶ This could be discussions with a long-term account doing business with the property.

Obviously, it is very important for a property salesperson to be very well prepared in advance of visiting with a prospect in connection with the possible booking of any type of event. Knowing how badly a property needs the business, how important the business could be, and far one can go in negotiations to try to get the business.

UNDERSTANDING PRODUCT KNOWLEDGE... THINK COMPETITIVE ADVANTAGE

To be effective in sales, all staff personnel, at a property need to be thoroughly knowledgeable about everything available in a facility, be it a hotel, conference center, convention hall, or any meeting hall. This means knowledgeable about not only the physical property, but services that are available at the site, as well as off the site.

Sales department staff members need to have available to them, a property fact sheet, created by themselves, physically and individually, by conducting a personal physical inspection of the facility, and asking questions of key staff personnel in all departments. Pertinent information would include (but would not be limited to) the following:

- General Property Description—location, age, anything unusual, or appealing
- Guest Rooms—number, type, size, rates, taxes, in-room amenities, security, ADA accommodations, parking available, number connecting rooms
- Restaurants and Lounges—number, and types, sizes and capacities, hours, menu types, sample prices, taxes, room service, house specialties
- Meeting/Banquet Facilities—number, sizes, capacities for different types of set-ups, banquet menu information, audiovisual services available and prices, ADA services; have floor plans available
- Transportation—distance to the airport, transfers available with cost, any local transportation provided by the facility
- Recreational Facilities—types, locations, rates, transfers
- Outside Services—secretarial services, babysitting, special vendors, shopping nearby, hair salons, entertainment activities
- Local attractions
- Guest Mix—types of travelers, groups, market segments come to the property

Different things appeal to different people and groups. Understanding everything there is to know about a property, and matching facilities to needs of a prospect is what helps make a sale.

TRANSLATING FEATURES INTO BENEFITS

Successful salespeople realize that they do not sell products or services; they sell the benefits that prospects receive when they purchase products or services. Most prospects that are potential guests become interested in the features of a property only when they perceive how those features will benefit them directly. Translating property features into prospect benefits takes practice. One good exercise is to prepare a features/benefits worksheet for each type of visitor or market segment group. For example, the following features/benefits worksheet format might be prepared to help sell corporate groups:

Property Features:
- Large size guest room
- Wi-Fi available in the room
- Electronic door locks

Prospect Benefits:
- Plenty of room to also do work
- Easy access without cost
- The feeling of safety and security

Note that the benefits of a property or service are often intangible. What prospects really expect to buy is the satisfaction of their needs. Satisfaction often takes the form of feelings, pleasures, and images.

The following sales phrases indicate that, when selling to prospects, an experienced salesperson can smoothly translate property features into prospect benefits:

Features:
▶ "I'm glad you asked about security in our guest rooms. We have the latest in keyless lock technology.

Benefits:
▶ Our electronic keycards are simple to use, and provide the kind of security that will make everyone in your group feel safe."

The benefit list is the most valuable tool a hotel salesperson can carry. Actually, when selling, your benefit list is your product. When someone has to pay more for a product than having paid in the past, the decision to pay more will be made primarily on the basis of benefits you can deliver (see, "Giveaways Not Essential When Negotiating Group Sales" section for additional information).

OVERCOMING OBJECTIONS IS THE CRITICAL PHASE OF THE SELLING PROCESS

August 2012

Most salespeople would probably agree that the toughest part of selling something to someone is getting over the hurdle of resistance. Of course, the easiest way of making a sale is to get a prospect to buy rather than have to "sell." However, in our business of hospitality, there always seems to be some resistance when it comes to the group market. With group markets, there always seems to be a wide variety of objections, for a wide variety of reasons. Some objections would be very real, and some are just brought up to be able to negotiate something extra, like a better rate, or something without additional cost.

The whole idea, for a salesperson, is to really understand what is keeping the prospect from buying; what is the real reason? Objections that prospects raise during the course of a sales presentation should be handled immediately. If not, credibility in the eyes of the prospect is lost by their thinking that the salesperson is ignoring or avoiding the objections. The one exception is an objection concerning price.

If the price is discussed too early, the prospect may become preoccupied about rates throughout the sales discussion. This would provide a salesperson with little opportunity to bring out other positive points. What needs to be done is to defer discussions about price, and get into the asking of open-end questions ("putting that aside for the moment, please tell me something about your last meeting").

Develop a frame of mind that sees objections as a prospect's way of asking for more information. Address objections with empathy, and without arguing. Never become defensive or create an adversarial relationship with a prospect. Never interrupt the prospect, wait until the objection is completely voiced; give a sympathetic response, and restate the objection in your own words. Always try to clarify a prospect's objection by asking open-ended questions. The prospect's response provides clues to how an objection may be overcome. For example,

- "What makes you say that?"
- "Why do you think?"
- "Please explain what you mean by?"
- "How come you think?"

These type of "why" questions work very well; however, we must be careful in using them when an objection arises that is very obvious, such as a property is pretty far away from an airport. We really cannot use one of the above open-ended questions, when the objection is obvious. Instead, we need to show some positive advantages of the location, and point out other favorable aspects of the property.

Many sales objections are predictable. Experienced salespeople prepare answers to common objections well in advance of a sales presentation. If an objection can be anticipated there is a much better chance of overcoming it and confirming a sale.

NOW IS THE TIME TO BE GREAT NEGOTIATORS IN BOOKING GROUP BUSINESS

May 2013

OK, all the reports seem to be in. We are on the upswing. Business is great; occupancy is up, rates are up, in most cases, and we can look forward to a terrific year for profitable sales this year, and in all probably 2014, as well. OK, what are we doing about it? We must remember that hotel business is pretty much cyclical. We can just about tell that when business is good, like it is now, the developers of hotels, investors, management companies all look to expand, and build more properties. It makes sense, for them. So, as long as demand keeps up, and the supply of hotel rooms does not increase, we will be in hog heaven in our business. However, we know for sure that the pipeline is already increasing, and the move is on to build more properties around the country. So we sure can expect to start seeing some building of more somewhere in our backyard. As the supply of new rooms keeps coming on board, the demand for business must keep up, and stay ahead of supply for us to continue having great occupancy and revenues. Once the supply exceeds demand, we are in trouble.

For the remainder of this year, and even next year, the reports indicate that the group market will continue to be good. Companies, clubs, organizations, seem to reflect that they will have more meetings, conferences, and conventions for this year and next. Great for us!

Now the sales staffers at properties need to understand that if the market is good, and we can expect the meetings business to continue to be good, how good are we in negotiating the deals? We all know that to be good in sales we have to place the RIGHT BUSINESS, in the RIGHT PLACE, at the RIGHT TIME, at the RIGHT PRICE. All this boils down to be good at negotiating with prospects. It really all gets back to the basics of knowing your product and how it meets the need of a prospect, so that you can show value for the quoted price. This is not the time for "giving away" anything, unless it is done via negotiation. As an example "I'll do this for you, if you do this for me." Sales staffers should not get into a negotiating situation until they are ready. That means no specific needs of the prospect, gather the history of their previous meetings like where they met previously, how much did they pay, what other location sites are they checking, understand how well you can meet their needs, and how you can show value.

Revenue managers will tell you that when business is in high demand that is the time when good negotiators know how to make the bookings more profitable.

OVERCOMING OBJECTIONS IS A KEY TO CLOSING SALES ON GROUP BUSINESS

June 2014

Without resistance to buying there would be no need for salespeople. Resistance is the most important hurdle every salesperson has got to get over. The whole idea is to know what's keeping a prospect from buying. Once the real reason is known then, perhaps a salesperson can figure how to get around that particular issue. We need to take a look at the various reasons why people might have an objection to making a buy decision. Some of these reasons to resist buying might be: Lack of knowledge (needs to know more on value/benefits); Skepticism (uneasy about being unfamiliar with brand, location, etc.); Need to compare (what other competing hotels offer); Trading advantage (price, trying to get something extra); Prejudice (personal, company); Bad reputation of property ("so and so said" or "I heard").

A salesperson needs to figure out: WHAT DO PROSPECTS REALLY MEAN. Perhaps, it is just "I do not want it so I'll complain about," or "I don't like this salesperson;" or "Maybe the price is more than budget;" or just "Wants to look good with the boss;" or perhaps it really does cost more than another property, or just that perhaps desire is too low. Once the salespersons get a pretty good idea of what the issue might be, it is time to come up with the process of attempting to overcome the objection. The first rule in this business is never to start defending or explaining your position once someone rejects an opportunity. If someone indicates that "the price is too high," as an example, we do not want to defend the position and explain why that rate is quoted. What you have to do is try to find out if that is a real reason, or maybe it is something else. To do this, questions need to be asked of the prospect. Every time an objection comes up a salesperson need to ask questions to try to analyze the objection. Use a series of "why" questions to help define what the prospect really means.

Sometimes we have to admit the obvious. If a prospect says: "Well, you are 40 miles from the airport." Since you already know that, you cannot say "why do you say that?" In a situation like that, it may be that the salesperson needs to point out additional benefits that might make up for that issue.

It is better to anticipate objections so they may be handled during the presentation Managers and sales staff should make a list of most typical objections they have experienced. Following this, practice sessions should be planned on how to prove benefits during a presentation to overcome the objections before they are even presented. Set up strategies to overcome those objections. Remember to ask questions to analyze the objections answers help define the problem.

NEGOTIATE TO YOUR BEST ADVANTAGE WHEN WORKING GROUP PROSPECTS

June 2015

Yes, we are in a Sellers' Market, and will continue to do so the rest of the year, next year, and in all probability even in the following year. However, don't forget the cyclical type of industry we are in. Once the owners, developers, lenders get around to it, the pipeline will start to open, then, in all probability, we will see a load of new rooms in the marketplace, like in 2008. So, sales folks, let's make hay while the sun shines.

All too often, now in the business of sales, prospects usually hit us right in the beginning of wanting to know what we are going to charge them for the room, food/beverage, etc. Here is where we miss the boat. We seem to get into a negotiating situation right off the bat, before we are even ready. We need to get away from that. The idea is not to get into a negotiating session until we are very ready. We are never really "ready" until we know the needs and wants of a prospect, and to what extent that business would be profitable for the property. It is a situation of us needing to ask the appropriate questions so that the responses we get would help us determine to what extent we need or even want the business. The key lies in the answers we get to WHO, WHAT, WHEN, WHERE, WHY AND HOW, before we are ready to get into a negotiating situation. While being a good listener to all the responses to our questions, we should be able to relate to how our property may add value to the prospect's needs, thereby putting them in a "buy" position. Getting the prospect in a "buy" decision makes it easier when negotiating. The key questions above relate to the following: WHO (we need to know who is the decision maker); WHAT (get details on specific needs of the prospect and the group involved); WHEN (find out when the decision will be made; the shorter amount of time, you will be in the better position); WHERE (location and property where they have met previously); WHY (their reason for having the meeting); HOW (your best opportunity for booking this business).

Of course, a key in being a good negotiator is Knowledge, i.e., Knowledge of your property, Knowledge of the competition, and Knowledge of the prospects' needs. In all probability, in many cases, there is the issue of overcoming objections during the negotiating process. Having accomplished a complete product analysis of your own property, as well as a competitive analysis, and show how you can better meet the needs of the group, and provide value makes you a better negotiator in overcoming objections.

MORE OF WHAT YOU NEED TO KNOW WHEN NEGOTIATING WITH THE GROUP MARKET

July 2015

The whole idea behind the business of Negotiating with a Prospect is to end up assisting the prospect in making a "VALUED BUY" decision, while at the same time making a successful sale for the property. Of course, we all understand that the process of negotiating should never be an adversarial situation where both parties argue over demands and concessions. The outcome of negotiations is more likely if it is approached as a problem-solving process by which both parties reach a mutually beneficial solution.

In preparation for the process, it is important for the property sales representative to carefully analyze a prospect's position as well as the property's position. This has much to do with how far away the prospect might be in making a "buy" decision. The closer they are (time-wise) to making a decision the better negotiating opportunity for the property. However, if it will be some time, down-the-road, when a decision is made, the prospect may be looking for more concessions because they very well have the opportunity to seek out some other locations. The whole idea is for the property sales representative to gather important information, early in the sales effort that will prove useful in negotiations; such as:

- *Ultimate Decision Maker*: Does the one negotiating make the decision for the group?
- *Decision Date*: Will it be soon, or sometimes down the road?
- *Key Issues*: What the prospects think is most important?
- *Budgets*: Prices paid at earlier events.
- *Competitors*: What other properties are being considered?
- *Special Needs*: What concessions were provided by previous hosts?
- *Past Problems*: Know what went wrong at the last event.

Although all this information may not be available in advance of the discussion time with a prospect it is important to be able to secure this information so that the sales representative is prepared for negotiations. Be careful about not getting into a negotiation session too early in the sales effort. Sometimes it develops into concessions being made too soon. Also, it is important to understand that concessions are a "give-and-take" situation, such as "if you do this, I will do that." Be prepared to make appropriate concessions or suggest reasonable alternatives. Do not do too much talking during the process. Ask good open-ended questions and be a good listener. You never know what information one can pick up by listening well. Tips garnered in the discussion may be very useful in negotiations.

Understanding the property's position is the other important part of the process. The big questions is: How important is this business to the property? Will the group provide additional revenue via food/beverage, shopping at the property, Spa or recreational facilities be used, early arrivals, stay-overs, use of suites, favorable public relations for the property, multiple year contracts.

HOLD OFF ON NEGOTIATING RATES WITH GROUP MARKET PROSPECTS

October 2015

No question about it, we are in a Sellers' Market with improvement in occupancy, average daily rate, and demand, and will no doubt continue in 2016 and even most of 2017. All too often in this business of hospitality sales, we find that we tend to offer rates for groups too far in advance of our securing critical information about the prospect's needs. In a seller's market, the idea is to maximize revenue. Of course, in many situations, revenue management folks advise the sales department the rates that need to be used in selling to groups. However, there are other reasons for coming up with rates that may be higher, or even lower, based on a number of things.

The whole idea behind rate quoting has to do with two things: How badly does the prospect need your product, and how badly you need the business. The first rule in selling has to be not to quote rates too early in the discussion with a prospect. What has to be accomplished is first to find out the needs of the group, and this has to do with the needs of the planner, the organization and of course, needs of the attendees. From this, we determine where the revenue will be coming from. Will it be in rooms only, or is there additional revenue on space, food, beverage, recreation, or other activities. From this, we may be able to determine what overall revenue may be generated. We should not be quoting rates, or getting into a negotiating session, until we understand the whole value of the package. Then, in discussion with a prospect, how well do we meet their specific needs as well as determining how badly they need your product. Are there competitors that may also have the opportunity of booking their business? If not, then, of course, you may be "in the driver's seat."

Of course, if competition does come into play, there are other considerations from a selling standpoint. The big issue would be: How Important is this business to your property? Do you need the business that badly during that period? Does this group use multiple occupancies? Would they be providing "heavy" type food and beverage events? Are attendees the type that may bring additional business to the property on future dates? Would the attendees be providing additional revenue via other outlets (spa, gift shops, recreation, etc.)? Because of the nature of the group would there be favorable publicity via media outlets?

Is there an opportunity for multiple year contracts for this business? Could the group (company) generate other individual or even other group business in the future for the property? Thinking about all this there may be a negotiating opportunity in proving value to the property.

Chapter 10

Attributes of Successful Salespersons

AGGRESSIVENESS CAN CURE EVEN THE WORST SALES ILLS

July 1984

We all know there's nothing wrong with being aggressive in selling the services of our property, whatever those services may be and to whomever we sell them.

Aggressive doesn't mean "obnoxious" or "pushy," though; it means "tenacious." We simply need to hold onto a prospect until we've sold him something. Let's face it: almost no one is waiting in line to buy something from us. We need to be aggressive by constant follow-up on prospects. It's never appropriate for a salesperson, when queried about progress on a prospect, to reply: "He said he'd let us know." It's the sales department's job to be on top and always in touch with prospects, and be in constant command of any and all sales techniques possible to close a sale.

To be aggressive, a good salesperson needs to answer yes to the following questions:

▶ Did I tell my prospect how important it was for me to get his or her business?
▶ Did I find out why the prospect couldn't say yes right now to the deal?
▶ Did I inquire just what I needed to do to get the prospect's business?
▶ Do I take the time out to learn how to be more aggressive, to learn such things as selling and closing techniques and body language?

Of course, being aggressive doesn't stop with the sales department. Everyone at the property can become more aggressive. There are great opportunities being missed at the time reservations are made and again when the guest arrives at the front desktop check-in. These are two perfectly good times to merchandise and "upsell" our rooms. We can certainly be aggressive in a pleasant fashion, of course, and improve our average rate by being so.

And how about being aggressive in our restaurants, particularly in the area of selling and dinner? Does anyone, for instance, ever try to get the second cocktail order? Or sell the after-dinner cordial?

And when was the last time the General Manager, in person, called on an account to ask why the account's room volume was starting to fall off a little? And when was the last time someone in the catering department sold the "extras" to increase the meeting planner's budget?

Clearly, all hands need to think about being more aggressive. Being "laid back" and merely "taking orders" just doesn't work anymore. There's too much competition out there and only the aggressive types will get the business. Customers aren't waiting to buy-they're waiting to be sold.

CHARACTERISTICS OF SUCCESSFUL SALESPEOPLE

October 1990

There are certain characteristics that outstanding, successful salespeople have in common, and that distinguish them from all the rest. Some of these include: being a good listener; having enthusiasm; being persistent; willingness to improve and learn; knows how to build relationships; having a sense of humor, and the listing could go on and on. It is important for owners, operators, General Managers, corporate sales staffers and directors of sales to be on the lookout for people with these kinds of characteristics during the selection process for new sales personnel. We'll review just a few of these:

Having Enthusiasm: This is an attitude -a healthy attitude because enthusiasm reflects a person's excitement about a lot of things. Being excited about the prospect, the prospect's organization, the prospect's needs; being excited about the property being represented, and excited about the property being able to fill the prospect's needs. Successful salespeople are the ones we notice that seem to be "up" all the time, excited about the job, full of life, always positive. Doesn't it feel good to be around someone like that? Enthusiasm is not something that a person is born with—it gets learned and has a great deal to do with personality development at an early age. Of course, there are some people that have difficulty showing enthusiasm—they may not be cut out for sales.

Being Persistent: This just means not giving up. Even if some groups just won't book with us—remember there are "other fish to fry." Too many salespeople get discouraged and fell "down" and lose enthusiasm when a piece of business is lost. The idea is to know that we are not going to book them all—and sometimes, we even have to learn to walk away from a piece of business that is not profitable. Persistence is keeping the spirit up and to keep going after business. It is important not to confuse the meaning. It is not to be confused with being overbearing or obnoxious. Persistence is not badgering a prospect. Staying after an account with appropriate follow-up by phone, letter, personal call, and networking properly (with that prospect) when the occasion arises at a public function or conference.

The Need to Build Relationships: We don't know how true this is, but we've all heard the statement: "Many deals are made on the golf course." If you don't play golf, don't despair. It is not the golf game that helps close the deal (unless maybe losing to the prospect), but the relationship that has been established. People, generally like to do business with people they know. Think of your own situation. Don't you like to buy at a shop where you are known and greeted by name? Don't you feel more confident in a product if you are buying it from someone with whom you have established a good relationship? This is how it is in hospitality sales.

People will deal with you more readily if you have established a great ongoing relationship. Think of how many people you know, who can give you business, that you can call by the first name and who will readily see you in their offices, without an appointment and will take your phone call without it being screened. PLEASE STOP READING NOW AND MAKE THAT LIST. Where do you stand? Do you have 25? 50? 100? Maybe 1000? If you have 1,000 you pretty much have it made. You can probably stay home and do all your business by just answering the phone. BUT let's not stop there, keep building more such relationship it all leads to profitable sales.

You don't build these kinds of relationships by only making sales calls. It takes a lot more than that. The key is meeting and getting involved with these business contacts in a different, non-selling environment. This could mean joining appropriate organizations and getting involved in committee work and attending conferences where you can be with these prospects.

IN TOUGH TIMES GOOD SALESPEOPLE PREVAIL

March 1991

It is tough all over in these recession days. Lodging operators are feeling the pinch. And, every buyer wants a "deal." Corporate travel managers, association executives, contract room buyers, meeting planners, tour operators, travel agents, social groups, military, and government agencies, and just about anyone looking for a room or group of rooms wants to buy "wholesale." We can't much blame them, for aren't we all trying to do the same when are out buying something?

Although many times the price of a room doesn't surface as a real objection to a purchase, It does come up as a concern. All too often lodging salespeople, when confronted with a price concern, treat it as an objection. And jump to the conclusion that if the price isn't reduced the sale will be lost. Salespeople should know that there is always time to reduce the price, if people should know that there is always time to reduce the price, if necessary, and if possible. Sometimes we really do have to say "no," even at the risk of losing business, if it is not profitable. The point is, however, many times the question of price is not really an objection to the purchase, but the buyer is just trying to get a better deal. Explaining value for the price would get the job done most of the time. If price, in fact, is a real objection then there are two really successful ways of handling it.

The first, and best way to overcome any such objection is not to get the objection in the first place. The idea is to develop a strong and trusting relationship with the prospect so that he or she is absolutely sure you are always looking out for mutual benefit, not just your own. Another way is to do such a great sales presentation, making benefits so clear that price doesn't become a dominant factor in the discussion. The whole idea is to convince the prospect to want to do business with you because there is such a need for your property (for a variety of because, i.e., location, accessibility, facilities, service, reputation, etc) and that there is a good value so that price need not become an issue. The lodging salespersons need to always sell something other than price to get a piece of business. Fulfilling a need, selling benefits is the approach to take.

When the question of price comes up before the convincing, enticing, sales presentation takes place the issue should be sidelined in some manner and saved until the prospect is convinced to buy (except for the price, perhaps). In any sales, presentation price should generally not be discussed until the prospect is sold. Then, during the negotiation session, certainly, price or other negotiable areas will surface. Even during the negotiation session price discussion should be reserved as the last item to cover. We need to overcome all objections, one at a time, and save what appears to be the most serious objection for the last item.

If the price seems to be the biggest objection, we keep putting it off to last by saying something like: "beside your concern of the cost are there other things we need to discuss?" Or "If we could resolve your question of price do we have a deal?" From a sales closing standpoint, we need to know that once we have resolved the major issue of objection that the deal is made and we do not need to go back and renegotiate anything else. The prospect needs to keep focused on value and has, in fact, made a good deal based on needs and your ability to fulfill those needs.

CONSULTIVE SALESPEOPLE BRING IN MORE BUSINESS

June 1991

Before any sale is ever made there has to be some sort of interaction between someone with a need and someone who can provide a satisfaction to that need. In every case of a sale being made the process involves getting information, giving information and making the sale. There needs to be a time for listening and a time for talking. Making a sale in this industry sometimes becomes more a matter of getting information rather than giving it

Consulting selling means getting to know needs and wants, then responding with appropriate solutions. The process involves: probing for information (the interview), listening, and then responding with benefits. The whole process starts with the interview, or getting information from the prospect. Let the prospect do the talking. This is where the successful salesperson identifies and qualifies the prospect. We get the "who," "who," "where," "when" and "why" through the asking of open-ended questions. By listening intently we can determine needs as well as problems that need to be resolved. Before any solutions are offered it is best to have a thorough understanding of the prospect's situation. Too many lodging salespeople jump into the conversation too soon and start relating features of a property before getting the whole picture.

What is probably most important during the consulting selling process is to lay the groundwork for the development of a long-term relationship. Establish Rapport! People buy from people they like, they can trust, who are dependable, who understand their needs. The key to consulting selling also has so much to do with providing solutions, not offering services and facilities. We need to sell the ability to provide solutions before we even try to sell service or the lodging product.

During the probing process, we should be able to determine if the prospect is the decision maker and what is the decision process. Many times we get involved with someone who may influence the decision rather than make the decision. Once it is determined who and how the decision is made, the salesperson can then determine how and to whom to offer the solutions.

The whole idea of "probing" should not suggest to the lodging salesperson to perform an interrogation session. A series of fast questions may make the prospect uneasy and will probably "turn off" people. Learn to use a 4–5-seconds pause after the question is answered. Even a little relating works well. This is where the salesperson shows empathy or feeling to the response such as: "uh huh," "I understand," "Oh, I see," etc.

Here are some tips that will help with the probe and improve listening techniques:

- Show empathy
- Use pauses between questions
- Be attentive
- Show concern
- Use humor, if appropriate
- Observe non-verbal communication
- Do not make assumptions
- Concentrate only on the prospect
- Do not interrupt
- Take notes

Remember, think and act more like a consultant than a salesperson.

BE SEEN AND BE HEARD TO SELL BETTER

January 1993

What it really boils down to in making sales is getting people to want to do business with you. It's not only the advertising, the public relations, the market research, the direct sales, the promotions that work. Of course, all these marketing efforts, overall, do help to some extent. Just making the sales call and asking for the business doesn't really close deals. Having the best product on the market doesn't necessarily do it either. Advertising will create some visibility, but do people really care about well you tout your own product? In order to create customers, we have to find people who know about us and need what we have to sell and want to do business with us. After that, it should be an easy sale.

To find people who know about us, first, we have to create visibility. Locally, for those properties where most of the business is generated through a local contact, the best shot, besides making sales calls, is to be active in the community. Be seen and be heard. General Managers, department heads and salespeople may be able to do this best by getting invited to speak at local civic club meetings. This is wonderful exposure and it gets very high visibility. Most local organizations would welcome a hotelier as guest speakers there are so many things to talk about and we are only talking about a

20-minute presentation. There's the economy, ADA, employment opportunities, the romance of the business, etc., that come up with a "hot" subject and call the presidents of the clubs. You can get the listing from your local chamber of commerce. For resorts, out of the way locations, conference centers and offshore properties visibility will be created by getting involved with organizations that represent your customers. ASAE, MPI, SCMP, ASTA, NTA, ACTE, and others and be sure to be active in order to get the visibility. It does work!

Next, you need to determine who needs what you have to sell? This can be accomplished by making inquiry calls or prospecting for new business. Your files should already give you most of the information on those individuals and groups with whom you have already done business. But, the key to staying in business is getting new customers. So, we have to make the calls. Doing in a short period of time. Use your own staff or get local college students from hotel schools or from marketing courses.

Once we have created the visibility and found out who needs the products we are selling we have to make sure the prospects want to do business with us. Generally, people will do business with people they know, like and trust and there needs to be a price/value consideration. Getting the visibility by making personal contacts is one thing, but getting people to like you is another and then there is the question of being trustworthy. It really is a question of how you treat people and the service you provide. It is not difficult being liked being a good listener, smiling, offering to help, being considerate of individual needs, staying in touch, following up immediately with phone calls, all add up to getting people to like you. Living up to promises, telling the truth, not being too pushy or obnoxious will add up to being trustworthy. The question of price/value has to do with your product and how you sell it. When we start discounting and everyone gets to know that you will cut your price because your competition does, certainly does not put you in good standing from a price/value viewpoint. The idea is to get your price because your product is worth it and you stand behind it. Sometimes you have to do a better job of selling benefits. And having a better understanding of your prospects' needs will put you in a better position for this. If you've got a five-star product don't try to steal the business from the three-star product at a lower price. You've got the five-star value, so get the price. The same holds true for that three-star, full-service properties trying to steal business from the economy line. Leave them alone. Go get your own business at the price people should pay for your full-service product. You've got the value get the price.

We all just need to do a better job of selling what we've got. Create visibility, solve peoples' problems by selling them what they need and get them to want to do business with you.

YOU CAN CAPTURE ACCOUNTS BY CAPTURING DATA

March 1993

In today's sales game information is power. Successful salespeople have always talked about the power of knowledge. Knowledge of your property: knowledge of your competition: and knowledge of the prospect or client. Just about everything we do in sales today is driven by information. Think about it! When someone calls into the property to inquire about it! When someone calls into the property to inquire about a meeting or social event we always have to ask certain questions (information): when we are out making sales calls we are always asking questions (information): during a conference or trade show we know we have to find out certain things before we can make sale (information). We just cannot get along in this business of hospitality sales without getting information.

In the "olden days," we always thought good salespeople were born, not made. A good smile, friendliness, a slap on the back and of course, being in the right place at the right time often made a sale. Those days are really gone forever. Now it is information that is needed to make a sale. "Knowledge is power" is more applicable today than ever before. In the 90s, to be successful in the group sales are the answer is in targeted marketing—knowing who your target audiences are, getting information to determine who may be a qualified prospect and then going after the business.

Closing sales have more to do with professional research than most anything else. Getting to know who the real prospects are will improve sales productivity—why waste time trying to sell your property to people who may not be qualified prospects. Sales personnel can save untold hours of wasted time by gaining access to a source of information that will provide details of prospective groups that would "fit" the needs of the property. What you need to know is what groups have a history of booking rooms in your type of hotel, at the time when you need the business. Once you are able to find that out you can spend your time developing relationships with the people that make decisions for those groups. This certainly saves a great deal of time in prospecting and looking for the right groups.

These days more and more successful hotel sales executives are getting into database marketing—a technology for gathering information about all types of groups and putting the information into a computer program so that the data may be withdrawn based on certain qualifications. What if you are interested in locating group business during July and August, midweek, to all 200 rooms and sell a good deal of food and beverage functions. Where would you look? What you really want to do is target certain types of groups that will fill your need; those groups that specifically need what you have to offer.

There is now a company called Database Research Group which is an information research center for the hospitality industry. Their purpose is to gather valuable, strategic and current information on the meetings industry. Every month members of their staff talk directly to more than 3,000 meeting planners for the purpose of researching their needs and buying patterns. This information goes into their database and is made available to their clients so that the property salespeople do not have to spend their time cold calling and searching for appropriate accounts. With this information, salespeople are able to spend all their time selling to the right groups. Once a search is made, a client profile is provided reflecting a wide variety of information that qualifies the prospect for the property. A perfect match is made so there is no question that the prospect could use the property requesting the search. Histories are provided, reflecting where they have met previously, a number of rooms picked up, rate range paid, facilities used, etc.

RELATIONSHIP SELLING: A GOOD SALE BENEFITS THE BUYER AND SELLER

June 1994

There is nothing new about this business of relationship selling. It is a practical strategy involving ongoing relationships built between the buyer and the seller. Many of us in hospitality sales have made our successes through applying this strategy. You probably have been practicing it from time to time and think of it as "good working relationships" with customers or "consultative selling." We're really in the business of providing service to others and we do this best when we establish a healthy, ongoing relationship with the people with whom we are doing business and provide a valuable service through this relationship. Jim Cathcart, a good friend, explains it all in his book, *Relationship Selling: How to Get and Keep Customers*. He also has a great video on the subject.

For years I have been preaching to salespeople that the whole idea in our business is to let people buy; they do not have to be sold to. Selling is service oriented. There needs to be a need or a want for people to buy something, and we provide the service that fulfills that want or need. There has to be trust. People need to trust you before they will buy anything people buy with emotion. There has to be a good feeling about something before someone will buy. Doesn't it really boil down to people buying from people they like? None of us will do business, in any form, with people we do not like and/or do not trust.

So relationship selling is the process of establishing a positive feeling between the buyer and seller, and it has to be ongoing. We just don't want to make the sale and forget the customer. After all, we do want and need repeat business, but it is even more than that. The long-term relationship developed enhances the climate to the extent that the buyer even begins to help the seller. A friendship develops that goes on and on over the years. Wherever you end up working, the relationship will continue.

When we talk about long-term relationships and relationship selling we are just not addressing those of us in the job of "salesperson." What we are discussing here applies to everyone in our business. Owners, operators, General Managers, front office salespeople, anyone in contact with customers, after all— aren't we all involved in sales? And, we're just not discussing sales from a standpoint of the large convention hotel or resort property. All this applies equally well to the small, limited service property, with a small staff and no one in the specific job title of the salesperson. In fact, the successes of many of the smaller properties, in cities as well as resorts, can boast about their excellent occupancies and revenues as a result of relationship selling.

Staying in touch is critical to this strategy. We don't call or write only to book business. We stay in touch to find out how things are going, what's new, may we do something for you, etc. We try to do favors when we can. It becomes more of a personal thing, rather than just looking to book a piece of business. This results in the customer then calling you when there is a need for a reservation or a group meeting or function. When people need to buy they will come to you.

So how about now getting to your Rolodex or business card file or your sales files and browse through them to see who it is that you have not been in touch with lately? Going through some of your old cards it may feel like a stroll down memory lane. You'll probably find some good leads (opportunities for business) based on contacts you have made over time, but have just not been in touch lately. It's just a question of following up and practicing this business of Relationship Selling.

LISTENING IS STILL THE KEY TO BETTER SELLING AND CLOSING OF SALES

October 1995

During one of my recent HMM Hospitality Sales Workshops, we placed some phone calls to hotels, using a speakerphone so the participant in the workshop could listen-in. The purpose of the calls was to demonstrate what a poor job is generally done at properties when people call to make inquiries for sleeping rooms, meetings, and catered events. As was expected, most of the handling of the calls were rated very poor. We just do not know how much business is lost daily at any one property because we do not handle the inquiries properly. It just seems that the people answering the phone at hotels are not trained in how to answer the call, what questions to ask to further identify needs, and especially how to listen. In most cases when we made an inquiry about having a family reunion, and if indeed, we were lucky to get to speak to someone in sales or catering, we found they did more talking than listening. It just seems that only in just a couple of instances did we get to speak to someone that was really interested in what it was we needed or wanted.

The best way to sell-anything-is to ask the key questions to determine needs and then relate to how your property can meet those needs. But, we have to be sure that we listen carefully to understand what the person inquiring really means. Listening is an art and we need lots of practice to be a good listener. It is a communication skill used most but taught least.

Here are some tips to help you improve your listening skills:

▶ Ask questions-the answers tell you what you need to know in order to make the sale. You'll never know until you ask.
▶ Don't interrupt-sometimes we think we know what someone is going to say, but don't be tempted to finish a person's sentence. We all do it don't we?
▶ Show interest-use body language, your expression, your eyes, nodding the head and even interjections Like: "Oh really?" or "Yes, I see."

- ▶ Limit your own talking-even when you can relate to a particular feature of your property and how it meets a particular need-use it when appropriate.
- ▶ Don't let your mind wander-keep in tune with what the person is saying-you'll always learn something.
- ▶ Take notes-this really important. These will remind you of how to position property to meet specific needs and how to present your features as benefits.
- ▶ Don't let biases and prejudices interfere with what the person is saying.
- ▶ Show enthusiasm for what the person is saying.
- ▶ Observe the person's nonverbal language. Their body language also speaks to you. Note their expressions and movements.
- ▶ Look for "hot buttons" and buying signals. This is an important key in closing a sale.

What we need to do in this business of being a good listener is to understand what it is we need to learn from the prospect. You want to be able to meet needs and wants so listen for specifics in regards what the prospect is looking for. When you start relating to needs of an individual or a group you don't want to try to sell something that doesn't pertain. If a group is planning a family reunion for 50 people do they really care that you have a ballroom to accommodate 500?

You should be asking specific questions the answers to which will help you close the sale. There is no point trying to make a sales pitch when you don't know what it is someone wants to buy. Ask one question at a time, let them respond-listen very carefully. When people call in to make an inquiry they are looking for answers—let's not try to sell them the whole hotel until we know for sure what they need.

WHAT IS IT ABOUT SALESPEOPLE THAT MAKE THEM GREAT?

February 1996

In recruiting, interviewing, and selecting sales personnel do we take the time to try to figure out what it is we are looking for and what it is we want in a salesperson? Oh yes, we want someone who looks "good," has a great personality, meets people well, is a good networker, great in public relations, knows all about product knowledge and competitive analysis, can write a great marketing plan, gets along with everyone, etc. Perhaps we should start looking a little further. We need to take a look at what top salespeople want, what drives them to be the best, what is it about the makeup of great salespeople that we should be looking for in bringing salespeople on board at the property. What makes them different than the other salespeople who are good, but not great?

According to Harvey Mackay, Chairman of Macky Envelope Corporation, and author of the bestselling books, titled *Swim with The Sharks Without Being Eaten Alive* and *Beware of the Naked Man Who Offers You His Shirt*: Different things drive different people pride, happiness, and money. But all great salespeople have one thing in common: an unbeatable drive to excel. It's a drive that probably goes back to grade school or kindergarten. There's a hunger in them to continuously improve. And, that's what we need to look for when hiring salespeople. We need to understand our salespeople needs and help them to satisfy those needs. Let's help them and be great.

There's almost an unending list of characteristics or traits common to most good salespeople such as: good communication, friendly, honest, reliable, great personality, listens well, sense of humor, persuasive, professional, and you get the idea, the list could go on. However, as a result of many years of research, while interviewing prospective salespeople for their corporate accounts, the Gallup Management. Consulting Group found out that the best salespeople bad four key talents: the ability to close a sale; motivation: disciplined work habits: building relationships. The "big out," so to speak, is this business of knowing how to close a sale. Having many characteristics as listed above will not do us any good if cannot get the prospect to buy. During the many *hotel & motel management* magazine sponsored Hospitality sales workshops that we have conducted this problem surfaces every time. Salespeople generally do well getting up to the point of closing-it's just the question of asking for the sale or even assuming the sale has been made if there are no objections. Salespeople should understand that to be great and complete a high percentage of closings there is a process to follow. It's a great deal more than just making a presentation. By the way, we do too much of this and then expect the prospect to say: "Wonderful presentation, sign me up." Closing is more than just presenting. Follow the process!— the probe, relate; features/benefits; negotiate; overcome objections; and close.

When we probe we are looking for more than just the basic details which should have been discovered during the prospecting process. Here we are looking for real needs relating to the buyer. Be sure to ask open-ended questions to get the prospect to do the talking (and you be the good listener). Then you need to relate to the needs at this time expressing the different features of the property that will be a benefit to the buyer, and be sure to talk only about those features are benefits. Then there may be a matter of negotiations: a trade-off, perhaps. Like I'll do this if you do this. It takes training to be a good negotiator. You need to have knowledge of the buyer, history, budget, and specific-important needs, how hard they need your property, how badly you need the business and how far you can go in giving anything away. When a prospective buyer has some objections-what do you do? How do you respond? Again, here the great salesperson is one who has been trained in handling objections. What has always worked for me is to respond with a question. And be surprised at the objection. After the response to the question come up with another question. It usually ends up with the buyer making the sale for you. When it comes to closing, why not assume the person is ready to buy and close the sale without even having to ask the question?

SELLING MAY NOT BE SO EASY—IF YOU'RE IN IT YOU GOTTA LOVE IT

October 1996

In a recent column, I made the comment that "selling is the easiest job in the world."

A reader took me to task on the comment and went into detail with regard to how difficult it is to be a salesperson in the hotel business. After thinking about it, I realized that the reader had a point, and I am willing to back down. Probably because I have been in love with the business of selling for so many years that. I may have forgotten about how tough the job could be. Of course, there were pressures, lost sales, irritable bosses (and customers), heavy competition, rejections, bad days, service problems, short staff, etc. But even considering all those difficulties. I thoroughly enjoyed my work. I guess that's what it is all about-perhaps, after time, we only remember the good things and successes.

People working in hospitality sales should get out of the business if they find the work too hard, too tough, not enjoyable or if it takes the joy out of life. We only go around one time in this world, so let's enjoy it to the fullest. Here are some things to look at to help you determine if you can be successful in this business of sales:

▶ No matter how hard you have to work, if you look forward to going to work every day, you are in the right business.
▶ Can you take rejection and keep coming back for more?
▶ Are you seeking additional specialized knowledge about the hospitality business, always?
▶ You are focusing on building strength where it will do you the most good in sales.
▶ Always work to improve presentation skills.
▶ Understand your property inside and out by conducting your own product knowledge experience. Have you spent a night at your own property recently?
▶ Know as much about your competition as you do your own property.
▶ Make sure you understand how your property makes money or does not make money. It sure will help you sell better.
▶ Study sales skills, even if you think you know it all. It could mean survival in tough times. There is always more to learn.
▶ Having experience is not the same as having skills.
▶ Work for a company that is willing to help you learn more about your job and help you with training tools or encourages you to employ your own initiative in developing yourself.
▶ Become proficient in all the components of sales building; develops relationships, networking prospecting, qualifying prospects, presenting, overcoming objections, negotiating and closing the sale.
▶ Practice on every contact you make. Practice make perfect.
▶ Get computer literate from email to the internet.
▶ Get yourself organized. Keep appointments, carry business cards (and a writing instrument). Make call-backs promptly; respond to letters, faxes, etc., within a day. Work phone mail property and let people know where you are or can be located if you are out of town.
▶ Are you up to date on tracing of files? We lose so much business because we are neglectful in following up on accounts at a specified time.

- ▶ Don't forget to look in your own "backyard" for business. In many cases, 80% of all our business comes from just around the corner.
- ▶ Communicate all the time—internally and externally. That is, with your own property staff (they can make you or break you) as well as with your contacts outside the hotel.

Yes, I guess after reading all this you can say being successful in hotel sales really takes a lot of hard work. But that doesn't mean it doesn't have to be enjoyable.

- ▶ Are you really sincere with people—or putting on an act? Being falsely excited or impressed is usually sensed by the recipient and it fosters distract and ill will.
- ▶ Good salespeople don't make excuses or blame other people.
- ▶ Every salesperson need to conduct his/her own product analytics. That is, understand everything about the product you are selling; from sleeping rooms, meeting space, food service, and all guest services.
- ▶ You can't make a comparison with competing properties unless you have done the competitive analysis yourself; understanding everything about the competition and how they relate to your own facility.

Of course, we are just scratching the surface in discussing the basics, but you get the idea. There are so many basics involved in the selling process. Every salesperson needs to rethink what it is supposed to do and how it should be done.

THE BASICS STILL WORK WHEN IT COMES TO HOSPITALITY SALES

June 1997

Yes, life in this business of hospitality sales has been made a lot easier for us because of improved and new technology. We now have fax machines, word processing, phone mail, cellular phones, automated sales file systems, email, websites, desktop publishing, and a whole bunch of other stuff that makes marketing a lot easier, faster and less expensive. But when it gets down to selling, whether it be face-to-face or over the phone, it's the basics that will help make the sell. All too often salespeople keep looking for shortcuts in the sales process. We still have to ask questions, listen to answer, understand what it is we are selling, and make presentations, overcome objections, negotiate and ask for the business. Sure there are shortcuts. In all probability, a very high percentage of business that we close comes to us over the phone from a customer who has been with us before, knows what is wanted and just books the business. It is in the area of bringing in new accounts, new customers, a new business where we need to understand the basics of selling. So let's take a look, again, at some of the basics:

- ▶ Understand the business of prospecting. Not everyone is a prospect. It doesn't pay to take the time to try to talk to people and explain all there is to know about your property if there is no interest or desire or need for them to use your facility. So often I have seen salespeople making calls on non-prospects, trying to explain how wonderful their hotel is, before even finding out if the person is a prospect. We see this entirely too much at trade shows. Not everyone coming by your booth is perfect. All a salesperson needs to do is get answers to the questions of who, what, when, where, why and How to determine if someone is a prospect.

▶ We talk too much. Many salespeople are so "happed-up" about their property that they can't help but tell someone, even a prospect, all about it. A prospect doesn't need to know everything there is to know about a facility. They only want to know how you are going to meet their needs. So when you get a prospect you just need to ask more questions to determine specific needs, then relate how you can meet those needs.

▶ Many of us are poor listeners. We are impatient, we don't give the prospect time to finish a sentence before we start responding. We anticipate what someone is going to say so we jump right in. Sometimes, we lose eye contact because we may be looking around (especially at cocktail parties and trade shows) for the next contact. Show respect, be a better listener.

▶ Not responding promptly to phone call messages. Will lose you some business. "A customer is never an interruption to your work; he is the reason for it" (Jim Cathcart in his book "Relationship Selling").

▶ Follow-up with prospects when you say you will and don't turn it over to the secretary or an assistant unless you explain.

SALES CERTIFICATION PROVES YOUR PROFESSIONALISM

July 1999

We see more and more certifications being made available today in all industries and professions. And, this is good! If your good at something, and proud of it, and operate in a very professional manner, most people's will want to do business with you. Becoming certified in a specialized field says a lot about a person. In our business of sales and marketing in the hospitality industry, there are many who are very professional, experienced, and do a most worthy job for their companies, but do not take the time and effort to show how good they are through certification. That's OK, it is an individual choice. Frankly. I feel certification is a darned good idea, if someone can qualify, then he or she should go for it. There is no question about it, being certified in a field says a lot about a person when it comes to doing business, getting a job, and even a promotion.

Being one of the first to get certified (CHSE) by HSMAI, way back in 1974, did a great deal for me in my employment, promotions, Salary and a lot of respect in the industry. The same was true when I was in the first batch of Certified Meeting Professionals (CMP). It will do the same for you, and open many doors.

Recently the Educational Institute of the American Hotel and Motel Association

(EI-AHMA) joined forces with the Hospitality Sales and Marketing Association (HSMAI) to help get sales and marketing folks recognized through two certifications. The Certified Hospitality Sales Professional (CHSP) through the Educational Institute, and the more advanced Certified Hospitality Marketing Executive (CHME) through HSMAI. The CHSP and CHME represent a dedication to our industry, and demonstrate an in-depth understanding of hospitality sales and marketing. For those sales folks who have been in the industry for over a year, EI's CHSP certifies knowledge of the sales discipline, with an examination which is geared towards an applicant's knowledge of core sales competencies, such as sales forecasting, sales strategies, account management, prospecting, relationship building, and basic business practice. HSMAI's CHME goes to the next level, for those who have been in the business for five years or longer. It verifies an applicant's knowledge of the hospitality marketing discipline. Examples of core competencies include market segmentation and penetration, marketing strategies, positioning, promotional activities, incentives, distribution channels, as well as research.

Remember, there are rewards for being certified in your field of endeavor: Tangible recognition for your high level of skill and experience; valuable opportunities for professional development and personal growth; a strong competitive edge in the global job market; self-satisfaction; the respect of your peers. Go for it!

THE SUCCESSFUL SALESPERSON OF THE FUTURE

July 2001

"The Successful salesperson of the Future" is the title of an article in the spring issue of the HSMAI Marketing Review, a quarterly publication of HSMAI (Hospitality Sales and Marketing Association International). If you are a member of HSMAI you have already reviewed it. The article, written by Dr. Richard McNeil, CHME, a faculty member in the School of Hotel and Restaurant Management at Northern Arizona University, not only tells it like it is, but how it is gonna be (or gotta be) in our business of hospitality sales. No doubt about it, the successful salesperson of the future will be the one who changes to become the consultant salesperson.

According to Dr. McNeil: "As more and more business is conducted on the Internet, it becomes clear, that in the future, a successful salesperson will be one who is able to add value by playing a consultative role." His theory of salespeople, in our business, must make the transition from the transactional to consultative sales model. He is right on target. We keep "spinning our wheels" at properties doing the same thing over and over again, year after year, targeting the one-time piece of business. Spending hours knocking on doors, making cold calls, with most being nonproductive, is how many local salespeople spend their time. And, most of the time all this being directed by a General Manager who, for the most part, never had any experience in sales. Once a piece of business in booked by a company, association or any organization, there is little or no follow-up until it is time trace that accounts for the same piece of business the following year. What we have missed is what Dr. McNeil discusses in his article; the continual follow-up with the account to maintain and build the relationship, and do more business as a result. Many salespeople are just satisfied with booking the one piece of business that they neglect to develop the account further to determine other types of events that would be productive for the property.

To operate in a consultative manner we need to gather a good deal more information than is typically called for a simple Prospect Survey Sheet. All that from does is help determine who may be a prospect. Once we have a prospect, we should have one for life. That means getting more involved with the business of the prospect, what meetings and events do they sponsor so that we can find ways to help with the details for them to accomplish their objective. We need to get close to the account so that we understand their buying habits. It is better to allow people to buy from us than for us to sell to them. Having a consultative attitude rather than a sales attitude means developing a long-term relationship with an account so that there is repeat business (automatically, we hope), business from other departments within the account (whether it be lodging, food/beverage, space, recreation, etc.), and of course, referral business.

More productive salespeople can bring in more profitable revenue which can add greatly to the bottom line of an operating statement. The whole idea is that we can make sales folks more productive through the consultative process. We just have to look at the lifetime value of an account. Rather than salespeople spending their time running around getting single sales customers, we need to be on accounts that can produce profitable repeat, long-term business, and referrals.

BARRIERS TO SUCCESSFUL SALES COMMUNICATIONS

April 2003

From time to time we will hear folks comment about small children who do a lot of talking. Sometimes, people will say "That child is a good talker, he/she will grow up to be a good salesperson." Perhaps that could be true in some cases; however these days our best sales folks are those that do a better job of listening than talking. The successful, professional salesperson asks good questions, listens well and relates to the prospect's response. Then, of course, in the process of good sales communications, must be aware of certain verbal and body language signals that may deter from the sales process. Following are some tips that could be helpful in understanding some of the "turn-offs" when it comes to selling.

- ▶ *Sunglasses*: Indoors and even outdoors in bright sunlight, you cannot do a good job of selling if a prospect cannot see your eyes.
- ▶ *Eye contact*: Good eye contact is essential. The roving eyes show a lack of interest if what a prospect may be saying. We

see a lot of this at cocktail parties and trade shows.
- *Fidgeting*: Most of the time, we do this unconsciously, but we have to be aware of this. It is distracting to someone else who wants your full attention.
- *Crossed Arms*: This is something we also do unconsciously. Sometimes salespeople don't know what to do with their hands, so crossing arms is comfortable. However, it creates a barrier to good communications.
- *Crossing Legs*: Another barrier being created when sitting opposite a prospect during discussions. A psychologist once said to watch the other person, if they cross their legs then you should do the same. So maybe that does work too.
- *Interrupting*: This is so very common with sales folks. We want to show that we understand what they "are about to say" so we interrupt and say if for them. Or we interrupt to reinforce what a prospect may be saying, just to make a point about a feature of our product that would work well for them. Many eager salespeople
- *Chewing*: OK, we don't see this too often, but have to mention it because it is something that sometimes we do not realize that we have a piece of gum or mint in the mouth.
- *Tables*: We see this quite often at trade shows, where a table is set up in a booth and a salesperson talks to prospects from behind the table. Restructure the booth so that table could be set up against a wall, and work from the front of the table so there is no barrier between the salesperson and a prospect.
- *Conference Tables*: Sometimes a salesperson may have discussions with a group in a conference room, with everyone sitting around a table. Of course, this could create a barrier; however, there is no choice. Except, a good idea would be for the salesperson to locate a seat next to the prime decision-maker of the group. This provides close access rather than having the table as a barrier.
- *Office Desks*: Not a very good idea to have a selling opportunity in an office, sitting behind a desk. Your office, with a desk barrier, creates an intimidating situation. Move to a barrier-free environment, like a couple of easy-chairs in the office or to another location.

I'll bet many readers can come up with some more barriers to good sales communications.

BUILD SALES BY HARD WORK AND BEING A GOOD LISTENER

July 2003

Lots of times we hear about people having "good luck" when they book a big piece of business for a property. Or someone might say: "they were in the right place at the right time." Maybe that could be true, but I believe more in the fact that the sale was made as a result of hard work and building a good relationship with a prospect.

In the business of selling it is not the talking and telling, but the asking and listening. It takes a lot of real, hard work to get to understand the specific needs of a prospective buyer, build relationships, asking critical questions, listening to the answers, and then responding with something that would create a value in order to make a sale. We do not always win, and make a sale, but those sales folks who know how to follow-up, are persistent, work on the relationship, and ask for the business are the ones who are successful.

It is so much more of having the determination to work harder on the relationship to make a sale that creates the successes. Too often many salespeople give up when they cannot make a sale on the first or second try. We all know that building business for our properties has more to do with staying after a qualified prospect, over and over again with good relationship building techniques that make the sale for us.

What helps in building relationships is the business of asking questions. The more information sharing there is the better the understanding between parties. Answers to critical, fact-finding questions, lead to getting to know the specific needs and wants of a prospect. By asking questions you are leading into a more detailed conversation which is most valuable in determining what you have to offer at your property which can be of value to the prospect. Many people do not know what they are buying when having need of a hotel, resort, conference center. They know the number of rooms they need, what space, and or meals they would like to have and little else that may be of value to them. What we really want to do with these prospects is to help them with some problem solving, and provide some consulting services to better meet their specific needs. This is the value in our ability to ask some specific, open-ended questions, so we can get them to talk. Getting information is always more important than giving it. Here is where we need to be a good listener, because we have to be able to evaluate the answers to our questions so that we now have a chance to relate to something that could be of a value to the prospect. Once we learn of something that is critical, or important to the prospect we need to say something that shows we can accommodate that specific need with something of real value. Someone may indicate, for an example, that at their last meeting at the XYZ hotel, they were not able to get all the 100 rooms in the one hotel. It would not be enough for our response to be: "Oh, no problem we have plenty of rooms to make sure all the people are together here at this hotel." They probably already know you have more than 100 rooms, so what is it you can say that adds value to the comment? Maybe your rooms are larger than XYZ hotel, maybe your rooms have a better view, and whatever extra value you can show about your rooms is what you have to relate.

EFFECTIVE CONSULTATIVE SELLING IN HOSPITALITY SALES

October 2004

You have probably read it here, in this column, before, or have heard me say from time to time that we need to do more consulting than selling. After all, it is a question of working with prospects to help the solving of issues like where to stay or where to bring a group of people. Recently in a conversation with a good friend, Catherine Alfus, who is employed with Medicinal Aesthetics/Manhattan, the issue of "selling" was discussed. Catherine agreed with me that concern of bringing in more business, in any field, has more to do with consulting than anything else. Her theory in the consulting process has to do with seven phases; it does make great sense, and it looks like it would work perfectly well in hospitality sales. With Catherine's permission here are the seven phases I learned from her:

▶ **Planning:** Her goal is to prevent practicing on the customer; have a dress rehearsal before you see the prospect. The difference between success and failure in a consulting situation is slight, planning separates the two.
▶ **Credentialing:** To build an optimal relationship with a customer you must have credibility and trust. If a prospect does not believe you are capable or knowledgeable enough to be helpful the sale could fall apart.
▶ **Confirming:** Never bulldoze forward without asking a commitment question. This results in positioning yourself as a typical salesperson. The idea is not to talk AT the prospect, but WITH the prospect.
▶ **Discovery:** Here, the goal is to gain an understanding of the prospect's mindset BEFORE you proceed with trying to make a sale.

   Here are three essential components of Discovery:

   – *Smart Questions:* Part of the discovery process is the ability to ask good questions. These need to be open-ended and get the prospect to think, analyze or even speculate. The idea here is to understand specific needs.
   – *Listening:* Another part of the discovery process, getting information is more important than giving information. Don't talk too much, but listen to responses to your questions to determine needs, and particularly "hot buttons."
   – *Objection Handling:* Consider objections as opportunities; a request for additional information. You need to listen, demonstrate to the prospect that you heard them, then explore by asking a "smart" question to gain a better understanding of their concern ("How come you say that"), and respond with an opportunity to provide additional information to address the concern.

▶ **Design:** This next phase is to develop solutions to address the prospect's concerns. Make sure you design the solution specifically to the concerns; don't over-do it with a long list of features. Here is where you need to show value by presenting the benefits of specific features of the property.
▶ **Delivery:** Now you are ready to get the prospect to take action. Making the sale needs to take place by assuming the prospect will do business with you.
▶ **Post-Call Evaluation:** Most of us miss this phase, but it is critical. An evaluation at the conclusion of each consult is critical in enhancing our consulting skills.

SALESPEOPLE SHOULD BE HEARD AND SEEN!

January 2010

We have to wonder, these days, if salespeople are being seen by their prospects rather being contacted electronically via email, and the various social networks out there. Do we do a better job face-to-face in making a sale or do we think we can do better by just being in touch by other means? Hey, maybe not even face-to-face, but maybe even voice to voice could work better; like using the old-fashioned telephone, perhaps?

Anyway, new electronic devices keep coming up that try to do the work for salespeople, in making contact with prospects for business. How much difference is there between personal contact and electronic contact, in the business of effective selling? Some research has actually been done in this area by Albert Mehrabian, a pioneer research of communication in the mid 20th century. He found that in any face-to-face communication the components of the communicated message are:

▶ 7% words
▶ 38% tone of voice
▶ 55% body language

In the email-driven society of today, tone and body language are entirely stripped from communication. Using video to communicate adds back the ability to use intonation, expressions, and gestures to communicate a personalized message. Professor Frank Bernieri, while at the University of Toledo studied first impressions and how it can affect a candidate applying for a job. The results of his study showed that observing a short video of a candidate provides a view to the candidate akin to interviewing them in person. His study reflected that a 15-second video being observed was nearly identical to 20 minutes, in person, interview. This has got to tell us something about the value, in sales, of getting a prospect to see and hear a salesperson rather than reading an email or website message.

With all this research behind it, a few years ago, a Stanford University graduate put together a program for the recruiting industry to help recruiters do a better job in the interviewing process. The product that was developed is called "Vipe," and of course, looked like something that should work as well for the hotel industry, and it does. From a sales standpoint, Vipe which is a video presentation, allows a salesperson to interact with prospects as if the person was there in front of them. It works on an email message to a prospect once the email is opened, the prospect sees the salesperson delivering a short message who asks for some interaction, the recipient can then respond via email or link to more information and, or a website. To see the science behind using Vipe, or a demo of vipe, readers can go to *www.vipepower.com*.

WHAT MAKES A SALESPERSON SELL WELL?

February 2012

What we have to think about when considering what makes a salesperson sell well, is really how do we get a prospect to "buy" what it is we are selling. So maybe it, not a question of being a "good" salesperson, but being a person who could make a person "buy." in general, we know that people need to "buy" what they need, especially when it comes to hotel products. No one is going to buy 100 rooms for a specific date if there is no reason to do so: of course that goes for a single room as well. So, we need to have sales staffers think about "how do we get someone to buy."

For the most part, people buy from people they know, like, and those who understand what is needed. Perhaps it even starts off with the first impression a sales representative makes when calling on a prospect, personal appearance is critical, and should be compatible with normally acceptable business attire in the geographic area where the call is being made if in doubt, be conservative.

Too often sales staffers are too eager to start off with telling everything about what it is they are selling. This does not help the prospect in making a "buy" decision. In getting to the "meat" of the process what is needed are answers to questions where the prospect's response will permit insight into what he or she considers important in making that "buy" decision. It is just a question to finding out what is needed by the prospect. As the relationship is established, we need to ask questions that allow prospects to express thoughts and feelings about what they consider to be the most important aspects of their need. Ask an open-ended question like "To what extent was last year's meeting successful": "What did you personally like about last year's meeting?": "Explain some of the challenges you had with previous meetings." The whole idea would be to get the prospect to do the talking, and the sales staffer needs to be a good listener. Do not interrupt. Listening shows interest in the prospect, and their needs, as well as creates rapport.

Once it is known what is important to the prospect, focus the discussion on the specific needs, and how your property would meet those needs, and the specific benefits the prospect will enjoy when they make the decision to use the property.

Careful listening, and artful questioning also helps anticipate prospect objections that may have to be overcome for them to make the buy decision.

BUILDING YOUR CONFIDENCE WILL BUILD YOUR HOTEL SALES PRODUCTIVITY

September 2013

Right now we are seeing a very large growth of hotel sales staffers around the country, working in a very in a tremendously large competitive environment. We see experienced hotel sales managers moving up the ladder or out of that job function, and are being replaced, for the most part by new limited or inexperienced, and mostly untrained staffers. Here is the situation: those moving up are becoming General Managers or getting involved in some corporate capacity. Those moving out are going into software companies, some third party provider or into other industries, probably for higher pay.

So, what can we tell our new salespeople? Of course, there are many things they need to be told. In addition to all the training they may be given, and whatever things they may have been told about how to make a sale, the big thing that they need to know is to be great in hotel sales they need to have confidence. The key to having confidence in the business of hotel sales is knowledge, and that is three-fold: knowledge of the product, knowledge of the competition, and knowledge of the Needs of the prospect.

Knowledge of the Product is where we should really start with a salesperson, new or already (supposedly) trained or experienced. Even every veteran salesperson needs to gain knowledge of what it is that they are selling. The product is the whole hotel, and salespeople need to be thorough about knowing everything that is within the product. It should be the need of every sales staffer to through the hotel property creating a product Analysis: making a checklist in the area of rooms, meeting space, public space, food outlets, front desk, health club, all the recreational outlets, and all the outlets that are mentioned in a brochure or website. There are just so many things that are needed to know. The list goes on. Thorough knowledge of a property is the start of having confidence.

Knowledge of the Competition is next on the building block of Confidence. Sales staffers need to know how their property stands in relation to other properties with which they compete. In view of this, it is logical that a product analysis also is conducted on competitors. It is necessary for salespeople to make a comparison of the products in order to know where to concentrate the attention of the prospect. How properties compare is critical. This is a key step in having confidence.

Knowledge of the Needs of the Prospect is the third step in the building block of confidence. Once a salesperson has a product and competitive knowledge, then they can get involved in making a sale. This third step comes only when there is a prospect at hand. A prospect is someone who is in a position to do business with a property should they chose to do so.

CHAPTER 11

HELPFUL TIPS FOR HOTEL SALES STAFF

UNDERSTANDING BODY LANGUAGE CAN ASSIST SALES EFFORTS

June 1985

Successful personal selling depends on many things including displaying positive attitudes such as sincere interest, confidence, knowledge, and enthusiasm. Expressing these attitudes and feelings is accomplished not only through words, but also with the tone of voice as well as with action. In fact, more than half are expressed nonverbally through body language. Certain body language signals can help defeat a salesman's ability to make a sale. Sometimes the words may be right, but signals sent through body language tell the prospect to be cautious or to beware. Salespeople should learn to read their own body language and learn to use it to help sell.

Some things to look for include:

▶ Obstacles between people that obstruct communication and thus create a wedge. Holding a package, a presentation, a briefcase or anything else obstructing free space is not healthy. Don't create barriers.
▶ Leaning forward into the prospect's face. This is infringing on the other person's territory and is never considered to be a good idea when selling.
▶ Using your hands for expression. When using your hands to emphasize a point, do not use the motion against the prospect.

OPEN POSTURE BEST

Salespeople need to learn how to use relaxed and open postures that communicate confidence. Most buyers are persuaded more by what they see than by what they hear. Feelings of trust are communicated through body actions.

Learning to read a prospect's body language will help the salesperson to better understand that person's true feelings during a presentation, and if the salesperson reacts properly it may improve the sales opportunity.

Naturally, we tend to feel we know how a person reacts to things by watching facial expressions. This does not always tell true feelings. Facial muscles work very well and people can be misled by a false smile, a nod, or an understanding gaze. True feelings can easily be hidden in a facial expression. There are several other non-verbal communication channels that are not so easily controlled and will alert the salesman to a prospect's feelings or attitude.

Hands, for example, may tell a lot. Fists can obviously reflect aggression, clasped hands reflect guardedness, and self-touching may reflect tension. A positive selling signal is relaxed and open hands, while involuntary hand gestures can reflect guardedness. When negative hand signals are apparent it is important to let the prospect air his feelings. Use open-end questions to get the prospect to talk.

The crossing of legs also tells a lot. If a prospect's legs are crossed toward the salesperson or uncrossed it reflects an open and favorable attitude. If the legs are crossed away from the salesperson and there are other negative nonverbal communications, something must be done to regain a positive attitude.

Get Prospects to Talk

The crossing of the arms reflects a "show-me" attitude or signals caution. Steps need to be taken to get the prospect into the conversation and to "open up" to ask questions. Body angles also help tell the story. When someone leans away, caution is reflected. However, leaning forward is a good signal.

It is important for a salesperson to observe as many body language signals as possible before determining what course of action is best to take.

EMPLOYING FORMULA SHOULD GIVE BOOST TO SALES ACTIVITY

July 1985

Many areas of the country have been hit with over-development of hotels creating more of a supply of rooms than what is needed. Some areas have been affected with an economic situation that has created a decline in room demand. Both of these types of situations have created volume problems for hotels and motels that adversely affect the bottom line of a profit and loss statement. There is a desperate need for additional sales at locations that have this problem.

Some hoteliers, in the interest of showing good performance, start by making cuts in service and sales, when the answer to the problem really lies in improving sales performance—getting a better share of the available business in the market area. It may even mean an increase in the sales budget—adding more salespeople—if that will work to improve productivity.

Feiertag's 10-percent formula is a very basic system that we have tested at selected hotel properties and have found to work. The theory is that 10% of selected sales activity will result in a definite business immediately. If you are serious about booking business we ask you to put this theory to a test and to send us a letter detailing the results.

Here's how it works.

▶ **Cold Calls:** Take two full days devoted to cold calling only, making at least 30 calls (more is better). Call only on companies whose business you've never had before. Don't make pre-arranged appointments—no phone calls—all cold calls outside and in person. We'll bet 10% of the calls will result in definite bookings, plus at least another 10% will be excellent leads for future bookings.

▶ **Files Review:** Take two full days devoted to examining and following up on old existing files. Review and work at least 30 files. Call the contact on each old file and attempt to book some kind of business. Invite the contact in for a tour of your facilities. Just start at the beginning of any file drawer and pull the first 30 that are not due to come up on the trace system in a short period of time. Here the focus should be on files that have not been worked for a long period of time.

Again, 10% of all files worked should result in some type of confirmed business.

▶ **Wedding Receptions:** Using the newspaper society section or other sources for wedding reception leads, call 30 prospects for a wedding reception for wedding receptions and invite them (bride-to-be, parents, prospective groom) over to show your property and sample your food. At least 10% should book with you.

▶ **Existing Room Accounts:** Many times, companies that are already supplying overnight room business are overlooked as meeting-business prospects. Telephone or visit 30 of the current room business accounts. Find out who makes decisions on meeting locations and sell them on using your meeting facilities. At least 10% of the calls will result in definite bookings.

▶ **Legitimate Proposals:** Send out at least 30 proposals to prospects (not covered in the four categories listed) with whom you are already working and who have shown interest in your property. Much more than 10% of the proposals will result in definite bookings.

Having a positive attitude about doing business in a distressed situation will result in productive sales activity. Just following the basics will result in business. We challenge readers to take this test and report the results. Let us know the number of contacts in each of the five categories and the number of definite bookings—so we can compute a percentage—to see how the formula works for you.

COURTEOUS TELEPHONE MANNERS WILL HELP MAKE THE SALE

November 1985

It's happening at switchboards and in sales offices—telephones are not answered properly. People who call to do business should not expect to be kept waiting while the phone "rings off the hook." Friendly and courteous people handling phone calls at a lodging operation help make the sale. No one really wants to do business with a grouch.

From time to time, salespeople should place calls to their own hotel or motel just to see how long it takes for the switchboard to be answered and then how long it takes for someone in the sales office to answer once the call is forwarded. Do the people who answer the phone sound as if they are people with whom to do business?

How long does it take to return telephone calls? Salespeople need to get in the habit of responding to every telephone message immediately. Salespeople who have the reputation of promptly returning calls seem to do more sales than those who delay responding to messages. Consider every phone message a prospect for business. The name or the company on the message may not look like a prospect; it may even appear to be someone trying to sell something rather than buy. But one never knows. A telephone inquiry is probably the hottest lead any salesperson can get. It shouldn't be ignored and a response shouldn't be delayed.

The same holds true for letters of inquiry. The best response to an inquiry letter is an immediate phone call—especially a long-distance call. Most people who write to hotels/motels inquiring about meeting space or room blocks are probably writing to several locations. A quick response by phone is generally appreciated, gets questions answered and results in bookings.

"P" FACTOR

One of the most important prospects for group business is someone who is already a guest in a hotel or motel. Salespeople should establish an alert system with the front-office manager. This should work out so that the sales department is advised of the "P" factor (prospective prospects) people who are expected arrivals each day.

When reservations are made, certain information is generally obtained which could be a good "tip-off" to a "P" factor person. A meeting of the General Manager, sales manager, and front-office manager could result in a list of factors which might help determine if an expected arrival could be a prospect for future business. Certainly, corporate executives, association executives, company salespeople, travel agents, tour company representatives, and airline representatives would be good starters for the list. Salespeople should explore this possibility with General Managers.

Of course, once the guest is identified as a "P" factor person, then contact has to be made to explore further whether or not that person is, in fact, a good prospect for future business. This may be accomplished with a nice note waiting for the guest upon arrival, indicating a warm welcome and a request to call the sales manager (or General Manager) for a brief visit. A phone call or visit could help identify whether that guest is a prospect.

Sometimes, a phone call to the guest's room immediately after arrival will work just as well. Does he or she:

▶ Influence decisions regarding hotel accommodations for other people?
▶ Help determine locations for meetings?
▶ Play a leadership role in professional, trade or social organizations?

A "yes" to any of these three questions means that this is indeed a person that the sales manager or General Manager should meet personally. An invitation for cocktails or dinner, or even breakfast the next morning, would be in order. More details may be developed at that meeting, and perhaps a tour of the property's facilities would be conducted.

LISTENING CAREFULLY IS OFTEN UNHEARD OF IN SALES OFFICES

April 1986

All of us in a hotel/motel sales should wonder how much business we lose due to improperly handled telephone inquiries. A good percentage of meetings and catering business is done by telephone. Even though initial contact may have been initiated through a personal sales call or advertisement, prospects generally will call by phone to check on availability of space and dates. But all too often, the person who answers the phone in the sales office is not prepared to take inquiries or answer questions. In many cases, they are not trained to ask specific, pertinent questions—or even to listen properly.

See if this scenario looks familiar:

- **Prospect:** "Hi. My name is Howard Feiertag with Servico, and I'd like to know if you can handle a meeting for me on June 25th for fifty people."
- **Sales Office Person:** "I'm sorry, I didn't get your name."
- **Prospect:** "That's Feiertag. Howard Feiertag."
- **Sales Office Person:** "OK. When did you say you wanted a meeting?"
- **Prospect:** "June 25th."
- **Sales Office Person:** "Thanks. How many people was that for?"

You get the idea. It can go on and on this way, and usually does.

The answer, of course, is for everyone in the sales office to be trained to listen carefully, ask the right questions, be knowledgeable and exude confidence. The prospect calling in must get the feeling that the person at the other end of the phone is a professional. By using a banquet function sheet or other types of the worksheet that covers all the questions that need to be answered, the sales office person will not only get all the questions answered, but will also sound professional. Avoid using writing tablets and pads of blank paper—use only worksheets.

Develop a worksheet form with questions on it that cover all situations. This way, almost anyone can get the information on an inquiry.

An effective worksheet will include:

- Prospect's name, title, company, address, and phone number;
- Name of the group for which the inquiry is being made;
- Dates of meeting or function, number of sleeping rooms needed and number of nights of stay;
- Number of people attending and their breakdown (men, women, age groups, spouses, children, special dietary needs);
- Daily schedule of activities;

- Group functions needed (breakfast, lunch, dinner, coffee breaks, receptions, etc.);
- Social activities;
- Types of setups needed for each meeting; the space needed for exhibits, entertainment or demonstrations;
- Budget available;
- Who will make the decision on the location of the meeting, and when they can come to visit the property.

Obviously, such a worksheet is elaborate, but it can be simplified. Arrangements for a local luncheon meeting, for example, would not require all these questions. The idea is to listen carefully to determine the prospect's needs. If answers to the prospect's questions are not readily available, then the prospect should be told that someone will return the call as soon as possible. The worksheet is then passed on to someone who can handle the sale. Before the call is returned, all the information should be gathered so the prospect doesn't have to answer all the questions yet another time.

All available information about the prospect and the group should be recorded. Also, it is important not to interrupt when prospects are talking—very often the things they say and how they say can provide good tips that may help close a sale. Never feel uncomfortable about asking questions—the more information available, the easier it is to close a deal.

Listening properly is hard work. It is one of the most important skills in closing a sale, but seldom is emphasized in our industry. Listening, unlike talking, never lost a sale.

PAYING ATTENTION TO MEETING DETAILS CAN PAY OFF IN SALES

April 1986

Professional meeting planners want to do business with hotel/motel salespeople who understand a planner's role and needs. Sales personnel need to impress upon their General Managers and department heads the importance of taking proper care of meeting details. Attention to detail can help a property develop a reputation for professionalism in handling meetings and conferences.

In the areas of guest registration and housing alone, there are a number of situations that should be discussed with the meeting planner and then properly handled by the hotel/motel staff. These items can apply regardless of group size.

- Rooming List: Ask the planner to provide a rooming list two to three weeks in advance of the group's arrival. (Make sure the planner identifies VIPs on the list.)

  Make every attempt to have all rooms ready for check-in. However, if rooms can't be readied because of heavy occupancy the night before, arrangements should be made to check luggage and to have some type of activity available for early arrivals. In any case, salespeople should work with the staff to ensure room availability at least for VIPs.

  It is also a good idea to provide the meeting planner with two copies of the room list, one by number and one alphabetically.

▶ Check-In/Out Areas: With large groups especially, it is in the best interest of all concerned to provide a special check-in/out area away from the front desk. If extra personnel is required, meeting planners may be willing to pay for the extra service. Discuss this with the planner well in advance of the meeting date.
▶ Bell Staff: It is important to be aware of group participants' arrival times so that enough bell staff is available. The staff should make sure that luggage waiting to be delivered is in a relatively conspicuous area.

Also, advance details should be worked out with the meeting planners for luggage tags printed clearly with the names of the owners and the group. To save time, large numbers corresponding to the rooming list can be put on the tags. Also, a "blanket" gratuity system for luggage should be discussed when the system is explained to the planner.
▶ In-Room Amenities: A good opportunity for creating additional revenue for the property is to provide a printed list of in-room amenities that the meeting planner may purchase for the group. This could include fruit, cheese trays, wine, nuts, liquor bars for VIPs, turn-down service, candy, flowers, etc. Prices for each item should be listed. This will turn out to be a very good selling tool.
▶ Accounting: Suggest to the meeting planner that each day at a mutually convenient time, the accounting department will make available for inspection all master account charges as well as a few individual folios. This is to ensure that posting is being handled according to agreed-upon instructions. When the meeting is over, there should be no discrepancies in the billing and it will encourage quick settlement of the account.

Most meeting planners need to pay the account as quickly as possible so they can close their books and get busy planning other meetings. They do not want to delay payment due to billing inaccuracies.
▶ Post-Meeting Summary: The sales department should be provided with some sort of internal worksheet. It should show actual room pickup by day and by category compared to the room block, gross food, and beverage revenue and meeting room usage. This provides a good record to use for future booking of the same group.

The sales department may also send a copy of this to the meeting planner with the appropriate "thank you" letter. Meeting planners appreciate details like this, and they'll likely remember your property when planning a future meeting or conference.

SALESPEOPLE SHOULD LEARN ABOUT HOTEL'S F&B FACILITIES

May 1986

Hotel and motel salespeople need to know all they can about their property's functions facilities and F&B operations. Those who have not had exposure to these areas should inform the General Manager so that a program can be developed to give the sales staff a working knowledge of them.

Function Space—Often, printed materials are available that describe meeting and banquet facilities. They usually include charts showing dimensions and capacities for receptions, banquets, and meetings. Frequently not included in such material, but what salespeople need to know, are:

▶ Ceiling heights (could low-hanging chandeliers interfere with movie or slide projection?);
▶ Audio control locations for all meeting rooms, particularly for rooms that divide into separate sections;
▶ What are the dimensions, including the height of risers, of stages used for head tables or presentations? How many are available?
▶ What size tables are used for banquet setups? For the classroom, theater, or conference style setups?
▶ Are housemen available to freshen meeting rooms during breaks?

Also, meeting planners should be asked to provide set-up details, including diagrams, three weeks prior to the meeting date. Planners should then be given a copy of the property's function sheet so there is no misunderstanding of what needs to be done at the time of the meeting.

Bar Functions—Salespeople should have a complete understanding of such details as:

▶ Which method of beverage sales best meets the needs of the meeting planner and the hotel (by the drink, by the bottle, by the person, etc.)?
▶ How many bartenders per hundred guests?
▶ Are single or double bar setups to be used?
▶ What is the per-bar sales figure used to determine whether the planner pays for the bartenders?
▶ What constitutes an "open" bar?
▶ What are price differences between premium and standard brands of liquor?
▶ What size bottles are provided (fifth, quart, liter)?
▶ What size drinks are poured at an open bar? At a cash bar? Are drinks free-poured or measured?

By the way, tip glasses should never be used at hosted bars.

Food Functions—While salespeople don't have to be experts in the area of food preparation, a basic knowledge is important in selling to groups—especially when it comes to banquet setup and feeding. Generally, menu details are handled by the catering department. Salespeople, however, do need to know:

▶ Menu choices and estimated counts should be available two to three weeks ahead of time;
▶ What is the property's policy on guarantees?
▶ Over what percentage the guarantee can arrange the property to feed? Can the meeting planner increase or decrease the guarantee up to 24 hours in advance?
▶ What table setups are available? What table sizes are available? How many does each seat?
▶ How many guests are to be seated at each table? Does this create a problem

for the banquet department? Does it create service—or union—problems?
- How many covers does a server handle for breakfast, lunch, and dinner? For buffets?
- What are room capacities for banquets that require a head table? A dance floor with bandstand? Special entertainment? Awards ceremony? Visual presentations (especially rear-screen projection)?
- If wine is to be sold, how do prices compare with those on the dining-room wine list?
- Can drink orders be handled without disturbing banquet service?

If a salesperson is prepared to answer these questions ahead of time, then he or she is a true professional—and probably does quite well-booking business.

BREAKFAST MAY BE THE BEST TIME FOR BOOKING GROUP BUSINESS

April 1987

Most hotel and motel salespeople will agree that group-business and meetings sales are usually accomplished when the prospect has an opportunity to see the property. It just makes sense that a hotel, as a product, is easier to sell in a face-to-face situation. The time for salespeople to get their best shot at a close is when a prospect is on the premises. This is when he can see if the property will meet the needs of his group, and this is when the features of a property can be presented to their best advantage.

Sometimes the site inspection involves a prospect staying overnight at a property, with at least one meal scheduled with the salesperson handling the account. In many instances, a decision to book a group function is made by a local person and a subsequent invitation is extended to "come over for lunch or dinner." In any case, it's not uncommon for deals to be made during lunch or dinner.

- But a trend is starting to take shape: More and more, salespeople are starting to view breakfast as the best time to do business—and that makes a lot of sense.

BOTH PARTIES BENEFIT

Inviting prospects for breakfast and a tour of the property is becoming very popular and appears to be most appealing to buyers. But they're not the only ones who benefit from breakfast business-deals: the hotel does, too.

Here are a few reasons why the business breakfast is gaining so much favor:

- There are fewer cancellations for breakfast appointments, mostly because the invitee hasn't yet had the chance to get "tied up" at his office.
- Doing business at breakfast allows the prospect to get back to his office early and get in a relatively full day's work.
- The salesperson's time becomes more productive because he or she is spared the time-consuming process of interrupting daily duties to wait for a lunch or dinner appointment.

- Time spent at breakfast is more productive than at any other meal; there is less wasted "chitchat" because participants understand that the purpose of a business breakfast is to create more productive work time for both parties—you can get right down to business.
- The question of "To drink or not to drink" doesn't arise. This avoids the touchy subject of whether to have drinks before the meal.
- No matter how elaborate or how much is ordered, breakfast meals are the least costly to the hotel or motel.
- At breakfast, the salesperson will usually get the prospect's full attention, and vice versa. That's because both parties are likely to have clearer minds in the morning. With midday or evening presentations, there's always the strong possibility that a prospect's mind—or salesperson's—is occupied with other matters that have come up previous to the lunch or dinner meeting.
- At many locations, there are more people eating breakfast in the dining room than there are during lunch or dinner, and it's always good to show off a busy dining establishment. What better way to show a prospect that your F&B facilities are popular?
- Staff personnel throughout the hotel will probably look and act more alert and fresh during the early-morning hours than at other times of the day.
- The salesperson handling the account will look fresh, alert and attentive. He or she won't be worn out from making sales calls during the day, as is sometimes the case at lunch or dinner meetings.
- Each of us can probably add a few more good reasons for being able to close better over breakfast than at lunch or dinner.

The "power breakfast" has gained a lot of publicity lately for its popularity among executives in all areas of business.

Why not make breakfast the "in" time to do business in the hospitality industry?

TODAY'S SALESPERSON NEEDS TO "SEE THE PEOPLE" MORE

September 1987

One of the biggest problems that w have to face in our industry today is the falling-off of sales—or the lack of increased profitable sales. As a result of the recent overbuilding in certain markets, we are finding that the supply of rooms is exceeding the demand. Obviously, this is not a very healthy situation for our industry, and it means that we have to do a better job of selling our products—rooms, meeting facilities, and food and beverage.

Perhaps we are not doing the selling job that needs to be done. More and more, we see too many business cards bearing titles like Marketing Director, Marketing Manager, Sales Administrator or Director of Sales and Marketing. But whatever happened to the honorable title of Salesman (or perhaps should we say Salesperson)?

Hung Up On Titles

New people entering the sales force these days are too hung up on titles. It seems that many are concerned with how long it will be before they can become a "director of sales and marketing." There is nothing wrong with looking forward to growth and development, but there is nothing wrong with being a professional salesperson, either.

Putting business "on the books" is critical to success in our industry, and it's selling, pure and simple, that does the job. Unfortunately, the productive selling time of hotel salespeople is diminishing. More and more, hotel salespeople are getting involved in activities that take them away from selling.

The three basic rules of selling have always been:

1. See the people.
2. See the people.
3. See the people.

Today we salespeople spending too much productive selling time at the hotel. For any number of reasons, there seems to be some difficulty in getting out to "see the people." Salespeople get involved in advertising, creating brochures, coming up with packages, servicing parties, helping in other departments, running errands and a wide variety of other duties the General Manager assigns them.

Law Of Averages

Let's get our salespeople out to meet the people. The more people they meet, the more prospects they develop, the more leads they create; and the more leads they create, the more bookings they produce. It's simply the law of averages—and it never fails.

So while the salespeople are out, who takes care of all the marketing-type jobs that need to do be done? Who talks to the media reps? Who lays out advertising plans? Who makes sure the meetings at the hotel are being serviced properly? Who does the direct-mail programs? Who arranges publicity?

The answer to all those questions is the General Manager. All these activities certainly help the sales effort, and it's the General Manager who is responsible for the overall operation of the hotel and for getting all the marketing aspects of the hotel accomplished—either by himself or through other people.

Honorable Profession

However, the people who are hired to do sales *must sell*. Selling is a most honorable profession, and people who like sales and are good at it are probably a hotel's most important asset.

It is the challenge of selling—the close of a sale—that provides one of the most satisfying feeling of accomplishment a person can have.

Good salespeople love to sell—so let them do what they do best.

Salespeople just have to have the opportunity to get out more, make more calls, see more people, develop more prospects, turn the prospects into buyers and close more sales.

It really works, and given today's highly competitive marketplace, the salespeople who get out to "see the people" more will sell more.

TITLES OF CERTIFICATION ARE SIGNS OF TRUE PROFESSIONALISM

October 1987

Hospitality industry people are realizing the importance of becoming more professional in their specific duties. Among other things, it takes experience, love of work, peer respect, involvement in associations, sharing with others and developing relationships to become a professional. It is well worth the effort: To be recognized as a professional is truly an honor in any industry.

There are several ways to achieve recognition through certification programs sponsored by specific organizations. For example:

▶ Qualified travel agents can be recognized as a Certified Travel Counselor (CTC) by the Institute of Certified Travel Agents (ICTA).
▶ In the hotel industry, a professional can be designated a Certified Hotel Administrator (CHA) by the American Hotel & Motel Association.
▶ The Certified Association Executive (CAE) is the professional designation awarded by the American Society of Association Executives (ASAE).
▶ Meeting planners can qualify for the Certified Meeting Professional (CMP) designation awarded by the Convention Liaison Council (CLC).
▶ Hotel/motel salespeople have the opportunity to qualify for the coveted title of Certified Hotel Sales Executive (CHSE) awarded by the Hotel Sales and Marketing Association International.

Any organization's certification program is designed to provide recognition for qualified persons in a particular industry. In order for one to qualify, he or she usually accumulates points based on experience and contributions made within the specific job category or industry. An application of qualification, an examination, and a published article are usually part of the requirements.

CHSE IS A VALUABLE ASSET

In the hotel industry, it has become increasingly important for salespeople to achieve the CHSE designation. The buyers of hotel services who themselves are professional in their own areas, recognize what the CHSE designation means.

Buyers really want to do business with industry salespeople who are recognized as true professionals—and more likely than not, properties that have a CHSE on staff have a better opportunity to do business. As mentioned above, CHSE designates professionalism in the sales and marketing area of the lodging business. It serves to identify to both industry peers and customers that the individual entitled to this special designation has achieved a high level of competence in sales and marketing experience, education and service.

To qualify as a candidate, CHSE applicants must be actively employed in the lodging industry. They must complete a pre-examination form and then pass an examination component. The pre-examination component includes industry related experience, association service, and education requirements, with a required total of 250 points attainable from any combination of factors related to experience, association service or education. The examination component contains two parts: a written examination and an original research paper prepared specifically for CHSE consideration.

Although the certification program is sponsored by the Hotel Sales and Marketing Association International, it is not necessary to be an HSMAI member.

SALESPEOPLE: USE THESE TIPS TO HELP IMPROVE YOUR SKILLS

March 1988

Hotel/motel salespeople should always be concerned with selling more profitably and improving their sales skills. Owners and managers should share these concerns and provide necessary support.

Here are some tips on how to become a high achiever in hotel/motel sales:

▶ **Establish goals:** This is one of the first steps successful salespeople take. Goals can be set on an annual, quarterly or monthly basis—even weekly or daily—and should be in the form of room nights booked, revenue booked or any standard by which sales can be measured.

Goals can also be established to reflect action is taken, such as sales calls made, proposals written, site inspections conducted, etc. The object of goal setting is to help measure production and performance. The more frequently goals are monitored, the easier it is to determine performance and take any necessary corrective action.

▶ **Self-feedback:** Salespeople should provide themselves with their own feedback. Successful salespeople always know the quality of work they're doing. It is good to feel free about providing self-criticism and then laying out a plan of corrective action. Salespeople need to catch themselves doing something wrong—before someone else does it for them.

▶ Look for responsibility—Seeking and accepting additional responsibility is the sign of someone who wants to be successful. In hotel sales, the job is to bring in new business. Looking for additional responsibility within the job description can be a big help.

But don't be overly ambitious: Accepting responsibilities that aren't sales-related can be burdensome, and are often detrimental to achieving sales goals.

▶ **Seek successful methods:** In hotel sales, there are many opportunities for improving oneself. Reading, listening to tapes, attending seminars, practicing improved sales skills—these are all health outlets that can lead to success.

Salespeople should always be on the lookout for information and material that will help make their jobs—and their careers—more successful. If you're lucky enough to have a mentor, it can mean a great deal in achieving success.

▶ **Expand the comfort zone:** Many hotel/motel salespeople continue to work n their own "comfort zones;" that is, they call only on people they know, and they do only the things they're used to doing.

To be successful, salespeople need to reach out and do things that are more productive—even if they feel uncomfortable doing them. To overcome this problem, salespeople need to "bite the bullet" and practice doing those tasks that may not appeal to them. For instance, if a salesperson doesn't like to make cold calls, he must overcome the problem by making cold calls until it feels comfortable.

▶ **Take moderate risks:** Sometimes it becomes necessary for a hotel/motel salesperson to make a business decision in order to accomplish a particular mission. It may be a matter of quoting a rate, offering a complimentary item or agreeing to provide a particular service.

To be successful in sales, such risks sometimes have to be taken. Frequently, occasions arise when an OK from management might create enough of a delay to jeopardize the closing of a deal. Therefore, managers should encourage salespeople to make on-the-spot business decisions—even if they may not agree with the decision. It may mean the difference between closing a deal and losing it.

Of course, this list is just a beginning—there are many more sales tips that can help pave the way to more sales and bigger profits. Hotel/motel owners and operators should work with their sales teams to expand this list, and should take steps to support them as they attempt to put these tips to use.

USE AN 'ACTION CALENDAR' TO BRING HOTEL SALES UP-TO-DATE

May 1988

It's just too easy for hotel owners and operators to accept the often-heard reasons—or excuses—for plummeting sales. We're all very familiar with the litany: "The supply is more than the demand" . . . "Too much competition" . . . "We're in a distressed economy" . . . "The market is soft" . . . "We're overbuilt" . . . and so on and so forth.

In getting to the real heart of the problem of declining sales, hotel owners and General Managers need to think of the situation as an "under-utilized" condition. This means, of course, that business is out there somewhere—it only has to be brought in to create better utilization of the lodging property. Granted, that's easy to say—but not so easy to do.

AGGRESSIVE SUPERVISION

Still, if owners and operators take a more aggressive approach to supervising their sales departments, there will be better utilization of their facilities. It just requires a strong commitment to getting involved, on a daily basis, with the people in the sales department.

First, there has to be a plan. Most properties work from a marketing plan. But even if that isn't being done, there has to be an "action calendar." An action calendar is a guide that is prepared for each month of the coming year. The calendar specifically describes the type of sales action that will be taken during a given month (even by a specific date), and indicates which member of the sales staff is assigned to get it done. For example, if a local sales blitz is to take place in September, the planning can start in July. The entry in the action calendar would include the name of the person responsible for getting the meeting organized. Sales trips, meetings, special projects, and trade-show participation should be entered in the action calendar.

REVIEWS AND UPDATES

The overall program for the year needs to be reviewed by the General Manager, and changed or enhanced as necessary. Once approved, of course, the plan not only needs to be followed, but should be updated when appropriate. Circumstances may change over time, and that would require an adjustment in the action calendar. The General Manager must accept the responsibility of reviewing the calendar on a regular basis and making sure appropriate action is taken.

The plan may even include the number of sales calls, presentations or proposals to be made that month by each member of the sales staff. From the standpoint of involvement, the General Manager (or owner) should visit the sales office at some point during each day. The visit is merely to find out "what's going on;" discuss who's doing what; review the reader file; report on what's being booked; review daily sales calls. Taking an hour a day for this kind of involvement is definitely a wise investment of time.

GOING ON SALES CALLS

Periodically, the General Manager should join salespeople on their outside calls to get a feel for whether some direction needs to be provided. Getting feedback from accounts during sales calls can be very enlightening. From clients, a General Manager can learn firsthand the operational aspects of the hotel that need to be improved.

Calling five local accounts a day could end up being very productive. A "Thank you" or a "How are we doing?" goes a long way toward reassuring an account that the hotel is interested in doing business. Lodging owners and operators should always understand that the sales effort at a property doesn't start and end with the sales department. Rather, it starts and ends with the "chief in charge"—the G.M., the owner or the operator.

'PROBE' METHOD MAXIMIZES SALES, MINIMIZES OBJECTIONS

July 1988

Most people in the lodging industry are very eager to close a sale. In the process, however, they frequently feel that the only way to close is to tell the prospect all there is to know about the property.

This might work sometimes, but nowadays there's more to selling than just expounding on a hotel's features; anyone can do that. The truly professional salesperson will relate the product—be it a hotel, a motel, a conference center or whatever—to the prospect's needs. And the truly professional salesperson will show how the property's benefits and advantages can help the prospect solve specific problems and meet specific requirements.

Traditional selling should begin with the "probe"—that is, the questioning and qualifying of the prospect. For instance, a large parking facility provides a group buyer no benefit if attendees are flying to the meeting; but it's a real benefit and advantage if the attendees are driving. In the latter case, then, the salesperson should emphasize the parking facility; in the former case, the parking facilities are irrelevant.

Successful salespeople realize that making a presentation without first determining the buyer's needs doesn't help close a deal. Presentations should only be made after the salesperson fully understands what the prospect wants. This concept applies to the company that is considering a rooms-only contract; the association meeting planner; the tour wholesaler; the motor-coach tour operator. It applies to sell for class reunions, bridge tournaments, or wedding receptions. In short, the concept applies to all segments and all markets.

Furthermore, all hotel departments should get into the act. Anyone who's in a position to sell should remember not to present a product (rooms, food service, space, etc.) before finding out what is needed. Catering managers, front-office salespeople, reservation personnel—every person on the staff should practice the probe technique.

Asking the Right Questions

Even if it is just a single room that is being sold, questions need to be asked. What type of room does the guest-to-be prefer: king bed, double, suite, studio? Does the guest want a room with a view? Does he want a room on a higher floor or on a lower floor? Does he want to be on the concierge level?

In all cases and in every situation, the price should not be discussed until all the information is developed through probing. Once all the needs, wants and problems have been brought out into the open, it is time to start relating the features in a way that solves all the problems meets all the needs and wants of the buyer.

This is the point at which all the hotel's relevant features are expressed in terms of their benefits and advantages. And this is the point at which the selling phase comes into play.

Minimizing Objections

Using this technique, most objections can be overcome (often before they even can be raised as objections). The question of price should never be brought up unless the buyer does so. And if that happens, salespeople should try to delay discussion about price until the end of the presentation and negotiations. The idea, of course, is to design the presentation in such a way as to precisely fit the needs and solve the problems for the buyer—so that the buyer is sold even before the price is mentioned. What counts is getting the buyer's agreement that the property is exactly what is needed.

In addition to the usual who, what, when, where and how questions during the probe, salespeople will find it helpful to find out about good and bad experiences the prospect may have had at another hotel, motel or conference center.

Making a presentation too soon in the selling—without a complete probe—can lead to objections that need to be overcome.

Remember the order: probe, relate, present, sell and close.

POSITIVE SALES ATTITUDE LEADS TO POSITIVE PERFORMANCE

September 1988

Is your attitude showing? If it's a positive attitude, it should be.

It seems that people who are successful in hotel sales always display a positive attitude. They are enthusiastic, friendly, happy, never negative about anything—or at least they appear that way.

Many negative things affect our attitude: The ballroom needs a new carpet . . . the General Manager doesn't understand our problems . . . the controller wants us to collect an old bill . . . a prospect complains about a bad meal—the negatives go on and on. In spite of all this, we in sales must continually exude a positive attitude, it is essential to selling effectively.

When the temperature is high, after a couple of rejections on sales calls, it's tough to be very enthusiastic on the last call of the day, knowing full well that the General Manager will ask, "What did you book today?"

BE CONSISTENTLY POSITIVE

The last call may very well give you the response your General Manager wants. It is important that—in spite of rejections, unproductive sales calls, the heat, humidity, whatever—salespeople "keep cool" and present a positive attitude on that last call.

Professional salespeople maintain the positive attitude they have for the first call of the day all the way to the last call. Train yourself to react consistently, as if you had successes throughout the day.

Salespeople must also have the proper attitude about training. Ongoing training is a must in the hotel-sales business. At the same time, self-training through reading or listening to cassette tapes and attending workshops is very important. Too many salespeople have the wrong attitude. They tell themselves, "I've heard this before," or "This stuff is too basic." That's negative thinking. The attitude has to be positive, geared toward self-improvement.

Learning the "basics" over-and-over again may very-well reinforce positive approaches and methods of doing business.

SEEK MEANINGFUL CRITICISM

There also has to be a positive attitude with regard to owners, managers, supervisors, and others responsible for making sure the salesperson is successful.

There will be times when criticism is necessary. The successful salesperson welcomes criticism with the attitude, "I need these comments to help me improve." Many hotel salespeople can't take, or don't like, criticism from the boss. That inability to respond to criticism reflects a negative attitude. Sometimes we are too defensive when someone comments on how we can improve our sales technique. Salespeople should seek evaluation of performance and encourage employers, associates, friends, and family to criticize them in a meaningful way.

It is most likely in our business to encounter some rejections, such as tentative responses to sales pitches that didn't ultimately result in definite sales, or site inspections that didn't turn into bookings. It happens all the time and your boss knows it. However, if a manager considers such an incident a failure—which is not—salespeople must learn to cope with that perception.

LEARN TO "FAIL" SUCCESSFULLY

The ability to "fail" successfully must be developed. Ask yourself these questions: Where did we lose out? Why? What could we have done better? What's our chance of getting the deal the next time?

These questions are good ones because they make your probe your performance. A positive attitude—even about "failing"—makes hotel salespeople successful.

Let's try to control our attitudes at all times.

And remember, a positive attitude brings positive results.

INTERVIEWS WITH SATISFIED CUSTOMERS CAN HELP SPUR SALES

October 1988

The best sales pitch in any business is the favorable referral from a satisfied customer. Somehow, though, I feel there's only a small percentage of hotel/motel salespeople who use this wonderful, cost-saving sales tool.

Many properties keep a complimentary letter file to use as a referral tool. Generally, the letters are unsolicited; copies go to the boss, the home office, parents—even the staff gets to read them. Then they go into the file and are rarely retrieved to show to prospects. Would it hurt to ask every group account to send a favorable letter that you can use as a reference for other prospects? Of course not. There's nothing wrong with soliciting complimentary letters. When they're received, they should go into a binder that can be carried around on sales calls to show to prospects who raise questions about the hotel's ability to perform.

VIDEO INTERVIEWS

A taped interview with a satisfied customer is another great sales tool. It's a terrific way to show someone why you're proud of your property.

Here's how it works:

▶ Locate some accounts that have been doing business with you over a period of time. These would be people with whom you're in touch regularly and with whom you've developed a long-term relationship. They should represent different segments of the business, such as corporate, meetings, motor-coach tours, travel agencies, etc.

▶ If you've done your job in a professional manner, if they're happy with you (and if you've made them feel important), then there should be no problem getting them to agree to a taped interview.

▶ Set up the interview location (at your hotel, if possible, as opposed to the interviewee's office). Explain the purpose behind the interview and how it would be used. Ask the interviewee to explain his reasons for selecting your property; ask him if your services met his needs; ask him what he especially liked about his experience at your hotel (friendliness, food, location, rooms, etc.).

▶ Now it's time to record. Start the tape and proceed with an introduction, such as: "My name is Howard Feiertag, and I'm here with Joe Dokes, President of Dokes Inc. Joe, can you tell us why you're so satisfied with your decision to hold your meeting at our hotel?"

The idea is to make sure the interviewee makes some positive comments about the property and its service. During the course of the interview, it's appropriate to bring up experiences that the interviewee might not have mentioned. For example: "I remember, Joe, that after your last meeting you mentioned how especially pleased you were with our banquet set up staff. Tell us what happened that pleased you so much."

Of course, your remarks have to relate to an actual situation.

▶ Following the close of the taping session, ask the interviewee for a reference letter relating to the comments made in the interview. It's appropriate to send a small gift and a thank-you note to the interviewee to show your appreciation.
▶ The tapes should be collected and used in future sales negotiations. (For instance, if a situation were to arise in which a prospect questions your banquet staff's ability to handle unusual setups, that would be a perfect time to bring out the Joe Dokes interview.)

Even before a presentation is made, a salesperson can show a taped interview in order to reinforce his position and the hotel's features. A tape can also be played at the close of a presentation to reinforce the salesperson's credibility. A variety of interviews with different clients who had different experiences can be a big help in overcoming other kinds of objections. There's nothing you can say that's as powerful as a video interview or a letter of recommendation. Taped interviews are just another low-cost method of sales support that can be used in developing professional sales skills.

START THE YEAR RIGHT BY REVIEWING TIPS FOR BETTER SALES

December 1988

In terms of improvements in occupancies and room rates, 1989 doesn't look like it will be much better than 1988 was. New construction hasn't slowed significantly, and demand probably won't be all that much greater. This, of course, makes the sales staff's job all the more important. It is their responsibility to make sure that the market share is improved and that there is a good return on investment in the department's sales efforts.

With this in mind, let's review some methods by which profitable sales can be improved:

- Create an action calendar for every month of the coming year. This is a reminder list of things that need to be done each month.
- See the people. Regardless of how it's done, see and meet as many people as possible. Make new friends continuously—they often develop into prospects.
- Check on your attitude. Are you always positive, cheerful and in a good mood?
- Be a good communicator both internally and externally.
- Confidence is contagious. The best way for a salesperson to make a buyer confident about a property is for the salesperson to be confident.
- Make more "cold calls." Time after time, it's been proved that cold calls work.
- Handle telephone inquiries more professionally. All too frequently, inquiries are lost because of poor telephone technique.
- Keep in touch with key accounts. Make sure the ones that give you business know they're appreciated.
- Do a better job with sales letters. (And watch out for those computerized letters—they don't apply to all situations.)
- Be a better listener. You can do more selling and closing by listening than by talking.
- Look at the total picture on a piece of business before you give anything away.
- Don't offer reduced rates or complimentary rooms at the start of meeting contract negotiations.
- When negotiating, put off discussions of rates until all other objections or problems are resolved.
- Concentrate on that segment of the business that will be most profitable for your hotel.
- Make sure all phone calls are answered within two rings. Whenever possible, calls should be returned the same day. (Consider every call a prospect for business.)
- Solicit "backyard" business. Use neighborhood contacts as sources for future business.
- Act as a consultant to business prospects. They often need advice on how to put on meetings, plan menus and handle the many details of a conference.
- Review job descriptions of all salespeople. Do the descriptions exactly define the jobs you want your salespeople to do?
- Aim for more time spent on direct sales and less time spent on relatively unimportant tasks.
- Up-to-date reader files need to be kept. Managers should review them regularly so that they know what's going on.
- Make a good first impression. As the old saying goes, you never get a second chance to make a good first impression.
- Encourage your staff to turn in sales leads. Every employee can provide leads to the sales department. Like most people, your staff members belong to organizations and have friends, relatives, or colleagues that may be prospects for future business.
- Know your competition upside down and inside out.
- Check reader boards—it's not old-fashioned. Reader boards still provide leads

on business that is being placed at competitors' hotels.
- ▶ Measure results. Know what pays off and what doesn't.
- ▶ And here, perhaps, is the best advice for this—or any other—New Year: Be proud of where you work, so that where you work can be proud of you.

SALES OPPORTUNITIES DEPEND ON YOU, NOT ON THE MARKET

November 1989

There still are many opportunities for sharp hotel/motel salespeople to sell in depressed markets, though that sounds almost impossible when we hear about why "we can't get business because . . ."

Don't Sit Back And Wait

Of course, the reasons (or excuses) are endless. The funny thing is that General Managers and corporate staff accept these stories, sit back and wait for the "turn-around."

Successful salespeople don't have stories. They go out and generate sales—and at good rates. They don't believe the stories justifying why people won't book their hotel.

You might hear someone say a city is overbuilt. How many times have we heard that? Of course, it's overbuilt. There are more hotels now than there was a year or two years ago. So what? More business people are traveling, more meetings are being held and people are taking more frequent vacations.

If you know your hotel/motel and city well enough, you could probably list features that would make people want to stay at your property.

Whoever said a hotel/motel sales job was supposed to be easy? It takes hard work, networking, prospecting, soliciting, knowledge, and experience.

Not Supposed To Be Easy

Sometimes we get a "no-brainer" where a piece of business is dumped in our laps. Someone calls and inquiries about availability and we book a nice group at rack rate with no comps or giveaways. This probably happens more often than we let on. We feel better when we think we sold a deal like this. In reality, we only took an order.

What do we mean when we use the term, "depressed market"? We probably use it too freely. In many cases, we're talking about more competition in the area. When people stop traveling—or businesses stop having meetings—those could very well create a depressed market. It's more important, however, not to worry about it but to create a marketing plan, an action calendar, to come up with ways to improve your sales.

Review Your Past Business

Our Best shot for this is to review all business of the past few years. The easiest sale is for people who have had previous satisfactory experiences. You may think you trace-file system works so well it takes care of this, but it probably doesn't. Someone should go through the files one by one and go after the repeat-group business.

A Systematic Approach

The successful hotel/motel salespeople are the ones with a systematic approach to prospecting. Goals need to be created. How many prospects did you meet last week? Is it possible to establish a goal of 20 new prospects per week? Maybe more? Maybe less?

If you really want to sell successfully you've got to make the calls to get the prospects. Fore very three good calls, you may develop one good prospect. Say you get three good prospects out of 10 cold calls (people, companies, and associations that have not been called on before). Make 50 calls a week to get those 15 prospects.

If you are really good, one out of three good prospects will book something with you within six months, or most likely in less time. Nowadays, we find more and more short-term (within 90-day) bookings. The more calls you make, the more prospects developed and the more business booked. So don't worry about a "depressed" market. There's business out there. You can't sit back and wait for "good times." You may get closed down before that happens.

BECOME A STAR ON VIDEOTAPE TO REFINE SALES TECHNIQUE

November 1989

Not too many of us get the opportunity to watch ourselves in a selling situation. Not just role-playing, but a real sales pitch with a live prospect. If it could ever be worked out, get yourself videotaped while you're in action. It will probably be hard to concentrate on the sales part knowing full well you're on camera. It may even be difficult lining up a prospect who would be willing to participate. Maybe it could be done with a hidden camera, sometime when you least expect it.

Everyone in hotel and motel sales needs to get into this kind of stuff. The idea, of course, is to see yourself in action—as the prospect sees you. It will be very, very revealing, and most of us will cringe when we see ourselves fumbling through the situation.

Here are some areas where we may think we're great until we see ourselves "live" on tape:

- ▶ Do you come across with sincerity?
- ▶ Are you believable?
- ▶ Would you buy from you?
- ▶ How enthusiastic are you?
- ▶ Does the prospect appear motivated by you?
- ▶ Take a good look at your methods and style.
- ▶ Spot your weaknesses.
- ▶ Spot your strengths.
- ▶ Figure out how to build on your strengths and get rid of your weaknesses.
- ▶ Look at the prospect's actions.
- ▶ Check prospect's body language and verbal reactions.
- ▶ Are you being persuasive?
- ▶ How about your own body language? Is it positive?
- ▶ Check your posture, gestures, mannerisms.
- ▶ Do you have good eye contact?
- ▶ Do you like your tone of voice? Your speech?

- ▶ How many "ers" or "ahs" or "likes" can you count that you use in 30 seconds of conversation?
- ▶ Do you appear to be in control of the conversation?
- ▶ How are the questions answered?
- ▶ How good are you at overcoming objections?
- ▶ Did you present a good value and bring in benefits as advantages?
- ▶ Are you pushing for a close?

Many times after having lost a sale we think, "I should have said this" or "I should have mentioned this." It happens. Perhaps we even think we covered an important area but failed to do so. A couple of years ago I was doing a sales training program for a group in Toronto. The sponsor wanted to videotape the presentation. At first, I thought, "How can I do a good job knowing I'm on camera—especially doing a full-day program?"

LIGHTS, CAMERA, ACTION

I agreed to do it anyway. To my surprise, it didn't bother me a bit (I'm a ham anyway). After the first few minutes of getting into my training program, I even forgot about the camera. I probably drove the cameraman crazy with all my walking up and down into the audience. They sent me a copy of the tape. I was afraid to look at myself in action. It wasn't until about six months or so later that I was brave enough to take a look at it. And boy, was it rewarding.

I picked up on the areas where I could use improvement—things that stood out like a sore thumb to me and probably to the audience. I noticed that some of my words were not understandable. Some of my jumping around could make people nervous. It was a great lesson. In a short time, I corrected what I thought were deficiencies. I still watch the tape about once a month or so to remind me where I need improvement. Now I'll have to arrange to get taped again. It's a great experience. Try it.

STOP SELLING . . .

May 1990

And start solving problems. These days, with competition being what it is, everyone is so hung up on selling, selling, selling. Perhaps we are too uptight about our people getting out and "selling" something. Managers keep asking the salespersons, upon return to the office, "Well, what did you sell today?" Let's think about asking: "Whose problems did you solve today?" From a psychological point of view, it appears that professional lodging salespeople could do very well for themselves if they would consider being problem solvers rather than salespeople. After all, in many instances, what really determines if someone buys has a lot to do with the presentation rather than the physical facilities of the property.

If we really do not serve as problem solvers we can truly screw up the sale. We talk about being good listeners and some salespeople are great at this; however, we have those that ask good questions, appear to be listening, but really do not listen to the answers. What we have to do is consider ourselves consultants—whereby we find out the problems (or needs) of the prospect and proceed to take care of them.

There are various steps to becoming this problem solver: Generally, people go to people they like to help them solve problems. Building relationships with prospects, getting the prospect to like and respect you as a hospitality expert is the first step. You can accomplish this by first finding a common ground, that is, something similar between you, representing the lodging property and the prospect. It could be any of many things: You both have children of the same age; you went to the same school; you served in the same branch of the military service; you both play golf, tennis or whatever; etc. It appears that people like to do business with people with whom they have something in common.

Another way to establish that rapport that is important, is to find something of which to be complimentary, but this has to be genuine or it looks phony. We're all tired by hearing about the nice tie we are wearing today or the pretty dress, etc. If there is something praiseworthy, then the compliment could be sincere. To make this really work, you, the problem solver, have to sincerely like the prospect. Once the prospect recognizes that he/she is liked, that's when the trust and respect come into play. Once the prospect and "problem solver" establish this rapport, the prospect ceases to be suspicious, recognizes the relationship, is not afraid of exploitation and there is mutual trust.

Lodging salespeople need to be concerned with how influential they can be with buyers to create profitable sales. It's no longer a question in our business just to be available and to take orders from people calling in to book a meeting or rooms catered affairs. In these days of underutilization of our product we need to do better than just being an order taker. Of course, some business does come in this way, but for the most part, we really have to hustle to achieve our sales and profit goals. For the long haul, it is building strong relationships over a period of time, with people that can bring in the business that will make the difference between the great, the good, the mediocre or the poor lodging sales professional in our industry.

One sure way of getting into this relationship stuff is to belong and become active in organizations where your customers belong, such as Meeting Planners International, American Society of Association Executives, Society of Incentive Travel Executives, American Society of Travel Agents, National Tour Association, and, of course, there are others.

DON'T LET THE TIME WASTERS GET IN THE WAY OF SELLING

May 1990

Lodging salespeople could really do a lot better in spending more productive hours in selling. We need to take look at how many hours a day, and week are spent in getting out and prospecting and selling and closing deals. Of course, there are a good many necessary activities that do eat up the hours such as travel time to sales calls, waiting to see people, doing paperwork. However, the idea to increase sales is to find more time to sell. This means maximizing the productive selling times and minimizing the time lost doing things that may not be necessary.

Let's take a look at some time wasters:

- **Meetings:** Here we are talking about the in-house stuff with department heads, General Manager, etc. Yes, meetings are important and a lot of good direction and motivation can come from them. The problem with meetings is that many times they are held during productive selling hours. Try to work out arrangements with other staff members or department heads to arrange for the meetings during non-productive selling time. This means sometime in the morning before 9:00 am or in the afternoon, after 5:00 pm, or even on Saturday morning. Meetings also run too long, sometimes.

  Discuss the possibility of getting the agenda through within an hour. To help save time during a meeting there should be an agenda and all parties in attendance need to know what is on the agenda and what their individual participation might be so they may be prepared to report or discuss items. If there are no agenda items then there is probably no reason for the meeting.
- **Luncheons:** Eating alone is not much fun and not very productive, unless of course, there is some reason for eating alone (relaxing, getting thoughts together, unwinding, getting away from people). We see salespeople lunching with other salespeople, secretaries, other department heads and others just for the sake of having company. It would definitely be more productive for salespeople to invite clients and prospects to the property for lunch. It gives the guest a pleasant setting away from the office and does help turn a prospect into a customer, sometimes. It is also a nice way to thank an existing customer. A lot of business can be done this way. Also, try to bring them in for breakfast or cocktails or dinner. It presents a good selling opportunity.
- **Excessive Conversations:** We're all really guilty of talking too much. Generally, salespeople are very outgoing, friendly, and like to talk too much. Although we do a lot of talking (when we should be listening) when we are out making sales calls, the big problem is all the unnecessary conversation taking place in the sales office or elsewhere on the property. A lot of it is gossip, a lot is just friendly chit-chat and most of it is going on with other salespeople, secretaries, other department staff personnel and of course, a lot of it over the phone with friends, family, and even customers. This will be a hard one to measure and improve upon since many times "it's someone else who is at fault." If you have a business-like approach and shy away from these people who just like to talk you'll notice that little by little the chit chatters will use up less of your selling time
- **Paperwork:** Let's not use a cluttered desk as an excuse for not to make sales calls. There are reports to make, files to follow-up, letters to write, proposals to get out, etc. If you have to start earlier in the morning or work late . . . then do it. If you have to come in on a Saturday to catch up, then OK. Just don't use valuable selling hours to do the paperwork.

Successful salespeople just never let anything or anyone get in the way of the precious selling time. Start today.

MOVE FASTER AND SUCCEED FASTER

July 1990

"The early bird catches the worm." If you believe it then move faster and win faster than your competition. Sometimes "slow and steady wins the race," but I don't have time to mess around while those competing lodging properties beat me to the punch.

I'm continually amazed by salespeople (in all industries) that do not move fast enough to suit me. It's happened with real estate people, several auto salesmen, media reps and I guess several others over the years. And all have lost a sale—because I was a "hot" prospect in each case.

We should really wonder how much lodging business, particularly group or contract sales we lose because we don't move faster and act fast enough on a lead. Just returning phone calls promptly is something about which lodging salespeople (and General Managers) are getting sloppy. If you are into sales then be a successful salesperson and get the reputation for returning phone calls right away. This should not only apply to calls from prospects or clients, but anyone.

Even a media rep wanting to sell you an ad or someone you know wants a comp room or a personnel manager who wants a reference on someone. It doesn't matter—courtesy and good business practices dictate that calls be returned promptly. The non-prospect now may be a prospect someday. Word also does get around and if calls are not returned promptly it hurts a salesperson's reputation.

What about all those cards you collect at a conference or trade show? When will you "get back" to all those people you promised to send information? They'll probably get a form letter from you (that your secretary will sign) in about two to four weeks. Because of all the cards you have you've decided to use a computer formatted letter that will merge with the names and addresses on the cards . . . and this is sure to take a couple of weeks. Is this working faster?

Also, what's the big deal in getting proposals out? Years ago, before word processing and fax machines, it seemed that we got them in the mail the next day after discussing a deal with a prospect (sometimes it went "special delivery"). Now, with computers, fax machines and next day guaranteed delivery services we can be faster than ever.

THINK FAST, ACT FAST, GET THE JOB DONE . . . DO IT NOW!

Next time you are visiting a prospect and you get the interest in a site inspection—get them to do it right then and there. Other people besides myself have done it many times. And it does work! Ask Jimmy Jones, who now teaches selling skills to hotel salespeople, formerly meeting professional with Connecticut General Life Insurance Company. I made a follow-up prospecting phone call to him one day trying to get him to see our Frenchman's Reef Property in St. Thomas (after turning me down several times previously). He finally agreed—he said: "OK sometime." I said: "How about right now?" He laughed, but did agree to go the next day, if I would agree to go with him to San Juan after that to help him with a site inspection of a property there. We spent a day and a half and we ended up getting eight meetings for the next two or three years. If you'll just get "action now" oriented—move faster and don't put things off, you gotta become more successful in building lodging sales.

Bridge the gap faster between need and satisfaction.

CANCEL YOUR NEXT SALES MEETING... UNLESS...

July 1991

... you can make it worthwhile. Companies really miss the boat so many times by calling sales meetings that are not worth the expense or time. Sometimes these conferences are called without much preparation time at the whim of a CEO who sees the need for "jacking up" the sales force.

If a sales meeting is to be productive and cost-effective there are certain steps to follow. The person responsible for the meeting must first establish the purpose and objective of the meeting. The purpose reflects what will be going on during the meeting, what will happen. Is the purpose to inform, educate, train, motivate, solve problems or what? Or maybe even, all of these. The objective reflects what is supposed to take place after the participants leave the meeting. Have goals been established for them to attain? Are they to close more deals, make more sales calls, complete more reports, do more or less of anything? Once the purpose and objectives have been established the planning may begin. The purpose tells the planner a good deal about where and when to have the meeting: what needs to be covered, who needs to be brought in to speak or train; what types of entertainment is needed, as well as what the budget might be. The planning for an awards program, which should be motivational, could be absolutely different from the planning of a training session, which could be entirely different from a review of the profit and loss statement.

Sales meetings usually work well around a theme, especially if the purpose is to motivate, educate and train. With a well-focused theme, participants usually get a good idea of what is expected. A theme, if used, should run through the entire meeting experience—from invitation through a return to home base.

Active participation should be encouraged throughout the meeting process. It is very hard to expect salespeople to sit through any session without having something to offer. Corporate officers and others who are leading session need to be aware that the more authoritarian the behavior, the less salespeople will engage in give-and-take discussion. To really get the participants to speak up and share information sometimes the leader needs to give up on the role of the teacher and step into the role of the student who wants to learn from the group. Skilled discussion leaders often conduct sessions by asking questions of the group. It seems to work best not to try to offer solutions to problems (sales are down), but to ask questions that lead group members to solve their own problems.

Productivity during a meeting improves when the pace of the meeting changes ever so often. The planner should prepare an agenda that forces the attendees to alternate four different mental activities: listen, think, speak and act. The idea is not to leave the group in any one of these modes too long. To get best results, twenty minutes in each mode works best, most of the time. Having a speaker for one full hour . . . regardless of speech content, is just too long. Presentations for twenty minutes work well with a feedback period of twenty minutes and followed with case study work or role-playing. Professional presenters who conduct 90-minute training sessions usually manage to work into their presentations all four modes of mental activity.

The same philosophy that has been presented here should also apply to the property weekly sales meeting. In fact, a change of pace from time to time will probably add to the productivity of the meeting. Change locations of the meeting . . . it would be best not to use the GM Office . . . in fact, why not take it off property to a local diner for breakfast? Have the meeting early, even 7:00 am sometimes. Be sure to be over by 9:00 am so that productive selling time is not wasted. Invite the head of another department to participate, who probably will have something to contribute.

Sales meetings should never be routine or boring whether they are the usual weekly meetings or a three-day conference. The meetings need to solve problems, produce results, increase profits or be canceled.

VISUAL ENTHUSIASM HELPS IN THE SELLING PROCESS

July 1992

Do we really get excited when we are trying to sell someone on using our property facilities? Sometimes we may feel that we are excited and enthused about what we are doing, but seldom show it. It helps to let people know that we are enthusiastic—excited about doing business with them. After all, enthusiasm is catching. There is nothing wrong or bad about being excited about something . . . especially your own hotel, motel, resort, conference center, cruise ship or whatever. What we are really discussing here is a question of pride. Enthusiasm reflects pride. If you do not have pride in your product, your -boss, your fellow workers, your owner and in yourself it is probably going to be very difficult to get excited . . . unless, of course, you are a very good actor. My philosophy has always been: "Be Proud of Where You Work, so That Where You Work Is Proud of YOU."

When you talk with meeting planners and others who bring business to your property they will tell you that it is much easier for them to do business with salespeople who show enthusiasm. Not only when they are in a selling mode, but all the time when it comes to discussing their particular property or destination, or even just schmoozing. It stands to reason that people just generally like to be around people who are "up" all the time, not faking it, but are really in a good mood, talking favorably about things, not being negative, loving life, seeing the good things around them. It sure creates a much better selling environment than being around someone who is always talking gloom and doom and wears a long face.

How do you look at people in your own working environment? Do you wear one face and attitude when it comes to dealing with prospective buyers and another when you are working with your own staff at the property? If you do you are creating a very serious problem. The way you work with prospects should be the same way you work with fellow employees. By being visibly enthusiastic around the property you create a very healthy sales environment for the whole hotel. You really should want to see all the employees acting the way you do. You, the salespeople set the mood for the facility with the way you carry yourself and act.

You would be amazed at how much you affect everyone else's job—the desk clerk, housekeepers, bell staff, dining room staff, kitchen employees, even all the administrative staff... everyone looks upon the sales staff as the local motivator. YOU set the tone, the pace, the look of a property. You see, it is not only an external matter of showing enthusiasm and excitement. It is also very much an internal situation. People, in all jobs at a property, put their best foot forward when they feel good about the facility and themselves.

So how do we go about setting the right mood, internally? You start with being very visibly excited. Your appearance and greeting, the first thing in the morning, when you report to work tell a lot about how you feel. Look good! Walk at a swift pace! Hold your head up high! Smile! Be friendly! Greet everyone—employees and guests! This would be a great start for you. Then, during the course of the day—show your interest in your job and be visibly enthused with whatever you are doing. How do you answer the phone? Can the people at the other end of the line get the idea that you are excited about being where you are? There are too many salespeople that sound "down and out" over the phone—this is not good for business. Make sure that the people that work with you also take on these visible enthusiastic ways. You will find that the way you look and the act becomes catching... people will copy what you are doing. It will all make for a happy environment in which to work.

By the very nature of your job, in sales, you will be dealing with difficulties and problems every day. What you need to do when there are bad feelings, as a result of something going wrong, be sure to keep it private, try not to show it, stay away from people until you overcome the feeling. Your feelings will always show in your face; it is something you can't help. When people know you well they can tell when something is wrong. There is no point passing along bad vibes; it doesn't help anyone or anything.

Of course, walking around smiling all the time, looking like a "goof-ball" is not good either because it may cause you to come out looking like a phony. Having enthusiasm and being excited about the job, the property, your fellow workers is not the whole answer to being a great salesperson. There are other steps to the selling process, but this would be a good start.

HOW IMPORTANT IS IT TO KNOW YOUR COMPETITION?

February 1993

Without question, knowing about your competitors' products, services, staffing, new developments, market mix, major accounts, and pricing strategies is most essential in improving your own marketing strategies.

All too often we really do not know all we can or should know about our competitors. Oh, we call them once in a while to shop them for rates. And, of course, salespeople take turns copying from their function board to see what groups there are in the house . . . All this is fine; however, there is a lot more to good competitor intelligence which would help you determine your market direction and who knows, you may find out something that may even create a change in your own organization. There is nothing like a good competitor from whom to learn. The good competition provides good challenges and makes you do a better job.

What is that you want to look for besides rates being charged by the competition and what groups they have in the house in any particular week? The first competitor you look at is the one that is doing a better job than you are. How about taking a look at their salespeople. What is the makeup? Who are they? How do they work? How good are they? What are they doing that you are not doing? Are there any innovative approaches to sales that they are using? What are some of the weaknesses of their salespeople; their shortcomings? What don't they do that you do that works so well? Well, how do you get this kind of information? You do it by visiting the competition. You may be surprised with how open people will be if you just ask the questions. Then, you can also visit their customers who are providing them business, but not to you. All you have to do is ask, who, what, why, etc.

Another area of competitive intelligence is to seek out weaknesses in their product, compared to yours. This just takes a visit from time to time to the other properties and walk around. Even observe service during a conference so you can make comparisons. The idea is to incorporate all competition intelligence into your selling strategies. Use the information to your advantage. Be careful, it is important to know what your competition is doing, but you don't want to tell your customers what they are doing. In other words, we do not want to "knock" the competition. Sometimes in the area of overcoming objections, we may have to bring out how we compare with a competitor in a particular situation (location, size, accessibility, product condition, price, etc.).

Networking with salespeople who work for competitors is also a good way to gather intelligence. Just in casual conversation, you may be able to pick up some helpful information. Whenever possible try to observe a good competitor in action. For instance when you are attending a trade show where they are exhibiting observe how they make contact, what they say and what type of sales presentation is made. Are they doing something that works well that you don't do? What do your competitors do when they attend an association meeting, like Meeting Planners International, National Tour Association or American Society of Association Executives (particularly the state association meetings)? How do they approach prospects; what do they say; how do they establish relationships. It is amazing what can be learned from a good competitor.

Gathering competition intelligence is everyone's job. Salespeople need to be aware of the value of this and should keep an eye out for this information all the time. Don't let it consume all your time. We need to be careful to continue our job of selling and bringing in new business. Don't try to be agent 007 or work "undercover" to get the information.

LET'S MAKE HAY WHILE THE SUN SHINES (1995)

February 1995

There's no question about it, just about everyone agrees that 1995 will be a banner year for improved occupancies and rates. At least so much better than what's been experienced over the last several years. Already, 1993 and 1994 have produced better figures and 1995 should be even a whole lot better. It seems that demand has increased considerably and the development of new properties has slowed. So what does this mean from a sales standpoint? Just don't sit back and wait. Just because things are getting better you shouldn't wait for the business to come to you. It also means that you need to be somewhat more aggressive in getting the type of business you want that would be more profitable to you than some other type of business. And, it also means it is OK to walk away from business that is not profitable. It also means that you need not to be ashamed of the rates you need to get; you need not be so "discounting" conscious. What we are discussing here is let's take advantage of the good times to bring in more business by being more aggressive in our sales activities and building high occupancies at high rates.

Now is the time to be more positive about the properties you are representing . . . sell the value . . . sell the benefits . . . whether it be over the front desk in getting the top rate for your rooms. Or in the dining rooms getting the best dollars for your food and beverage services. And, of course, in the group market. Getting meetings and conference business should not have to be as a result of giving things away. Now is the time to get meeting room rentals, advanced deposits, guarantees, charges for your in-house audio-visual equipment, even set-up charges. The positive thinkers who are proud of the properties they are selling and who know how to meet customer needs (and wants) and can show value for the dollar spent will be the winners in the 1995 race for better sales.

Now is the time to take a look at your market mix, if you haven't done it yet . . . (and this should have been looked at months ago in preparation for your 1995 marketing plan) do it right now. You should think about going after more of the types of markets that are bringing in the best business. Look at the three segments that are getting you the best room rates. What percentage of the total amount of room revenue is coming in from each of these? Take a look at the bottom two segments that are giving you the least room rates. The objective is to increase the percentage of your total room business in favorable market segments. This means you may have to reduce the amount of business coming from the bottom end. The idea is to spend more time going after the better business and less time on the least profitable business. Be careful, this does not necessarily mean you have to get rid of the lower end. Sometimes this is a good fill-in business for periods when you need the business. So what you need to look at are the better occupancy periods and that is the time to move the business from the lower end to the higher end.

Just because business is better do not let down on the prospecting. However, you do it ... by database marketing, by telemarketing, lead development companies, reading newspapers, doing student sales blitzes ... however ... keep it going. Just figure that all the business you're getting now just isn't forever. You'll always need new business—even in the best of times. Prospecting should be never-ending.

Get back to basics; look after the customers for repeat business; continue to be aggressive in sales activity; watch the competition (they are probably watching you); don't slow down; have goals that are measurable ... and above all be sure to keep the salespeople selling.

SOME TIPS PROVIDED BY MEETING PLANNERS FOR ALL TYPES OF HOTELS

May 1995

Over a period of time, people who plan meetings and book rooms at hotels, have raised issues concerning how salespeople, General Managers, catering folks, reservationists, etc., can do a better job of working with prospects and closing sales. Here are some excerpts from comments made by prospects about how we handle our business and where we need to improve if we want to book business.

- ▶ Return phone calls, promptly. This seems to be the biggest problem with people who want to do business with us. Sometimes our customers feel that we overuse our phone mail systems. If we do not get back to a prospect right away they certainly will look elsewhere to book business. This applies not only to salespeople, catering staff, but to General Managers as well when there may be a problem to solve. Sometimes when a General Manager is called the message is turned over to sales to return the call ... not a very courteous thing to do.
- ▶ When it comes to working on menus with people in catering, not very often are suggestions for meal items made. Most of the time it's a question of "watch a want" instead of "let me suggest ... "
- ▶ Unannounced sales call. There is a difference between a prospecting call and a sales call. Some meeting planners and corporate travel managers do understand that cold calls for prospecting purposes are being made and for a short few minutes interview to determine needs there is no problem. However, if we intend to spend some time trying to close a sale, then make the appointment.
- ▶ When visitors are invited for a site inspection do introduce them to key staff people who may be in a position to be of assistance before and during their meeting. Many site visits are very unorganized. Salespeople should spend more time planning for the visit.
- ▶ Many hotel employees do not know the "big picture" of hotel operations. They know and understand the operations of their own departments, but need to know more details of the entire hotel operation.
- ▶ Salespeople need to be more enthusiastic about their properties when making sales calls. It is important to get excited

about trying to close a deal. Enthusiasm is catching.
- First impressions of a property are most important. People at the site need to be courteous, the facilities need to look clean and organized.
- When sending proposals to prospects use an outline form. Too many are going out with long narratives with pages and pages of paragraphs.
- Group business prospects want to meet the General Manager when looking over a hotel. It has a lot to do with having confidence in making the booking.
- Overbearing, obnoxious, "pushy" salespeople turn people off. It is OK to be eager, enthusiastic and persistent.
- Suggest a post-conference meeting where the planner can sit with the property key staff and go over the details of the meeting that just ended. This is also the time for the planner to review with the accounting department all the bills for approval.
- Brochures usually are terrible. They do not provide the detail that is important to people who plan meetings. Floor plans of meeting facilities are critical.
- Price is important; however, meeting needs, facilities, the attitude of staff and service are more important.
- It is not important for the sales personnel to be available at the time of the meeting . . . as long as there is someone who will be in charge, who knows the details and who can make decisions on behalf of the hotel.
- Personal, handwritten notes from salespeople go a long way. They work.
- Many prospects would indeed prefer invitations for a breakfast rather than for lunch or dinner. It provides less time away from their other duties.
- Get prospects to feel more at home during site inspection with something in the room, a personal welcome note from the General Manager.
- Formulas for meeting room capacities usually not correct. Properties need to improve in this area.
- Pre-conference meetings with all key members of staff are critical.

SALESPEOPLE: DISCOUNTING DOESN'T BUILD CUSTOMER LOYALTY!

February 1996

"Gee, we love you and your property it's just right for our group and we would like to do business with you, but your price is too high. Can't you do something for us so we can bring our group to your hotel?" How many times have you gone through this scenario?

Do we get a little nervous and worry about losing this piece of business; particularly during a period when the property really needs it? Most of us recognize the technique of trying to get the prospect to focus on the price/value relationship and get away from the price issue. However, what usually happens we try to find out how much the prospect would be willing to pay and then negotiate the price down to almost that level. When this question of overcoming price objection comes up at our Hotel & Motel Management sales workshops most salespeople in the group respond to it by trying to find out how much the group is willing to pay and then working out an equitable rate. This means they are willing to come down off the quoted price.

Don't you become suspicious when you want to buy something and you are offered a discount without much reason? How can a company offer to sell something at a discounted price when a higher price was asked for in the beginning? Maybe they could have lowered the price even more. "Maybe I should have asked for even a lower price." Instead of building customer loyalty, when you think of it offering discounts tends to reduce it.

So many salespeople, afraid of losing a piece of business, think that they can get the business if they lower the price. When people object to booking a group, they may indicate that it is the price that is keeping them from buying, but it could very well be something else. Or, they may even be willing to pay the price, but by objecting they feel they may get a better deal. And, many times they get it—even though they weren't serious about it. It's hard to tell the difference. Salespeople who are good negotiators understand this and have trained themselves to be better prepared for negotiations. The idea is to be armed with information, well in advance, of the history and needs of the prospect. You need to learn where the group has stayed before, what is their history of room block pickup, room rates, other profit center use. What's the total value of the business and specifically how your property can best meet the needs of the group, better than any other competing hotel? And, above all, you need to understand how badly the group needs your property. Prepared with all this information you should be able to close the sale without having to discount the price. When the question of price becomes an objection—you need to show surprise and start asking questions like: "What do you say that." It is time for the prospect to do the explaining—not you. You should not have to defend your price. After one question is answered, ask another. Keep the prospect defending—rather than have them take the offensive position.

If you feel you have to come down in price to make the deal then there should be a very good reason to do so. We do not reduce price just to satisfy the prospect—there has to be a trade-off. Reducing the price for no reason does not get you respect or builds loyalty. Perhaps you can get another night's stay, or another group reception or dinner—or maybe even take back the comp rooms you offered in the beginning. There may even be the chance to get additional meetings at a later date or even a multiple year contract for that piece of business.

IF YOU WANNA BE GOOD IN HOTEL SALES, THEN BE A GOOD LISTENER

March 1997

It's not easy to be a good listener, particularly if you are in the business of selling. It just seems that many people in sales want to (and like to) do a lot of talking. For years a salesperson has been portrayed as the fast-talking, glad-hand, the fellow who always had a good joke to tell. Well, if that's the way it really was in years gone by, it isn't gonna work anymore. These days in hospitality sales it's more a matter of getting information rather giving it in order to make the sale. Maybe we can even think of it as exchanging information. Selling has to start off with finding out whether someone or a group, is in need of what you have to sell. There is no point in trying to convince someone—or sell something to someone—that has no need or is not even interested in what you have to sell. Once we can get the people to tell us what they want we need to listen carefully to what is being said so that we can understand what they really want. Sometimes we need to ask more questions to get more information before we even try to relate back to the people how we can meet those needs.

It seems that most of us would rather talk than listen—somewhere along the line I read that the average listener hears only 2.5 minutes of every 10 minutes spoken. If that is true then we really are missing a lot. Probably the best way to listen is to keep silent. This should be easy since it doesn't require you to do anything but listen. However, keeping quiet is generally not our cup of tea, for those of us in sales. Here are some of the things we do that show we are not good listeners, because we can't keep quiet:

▶ We interrupt—we don't let the other person finish
▶ We want to defend something—if we don't do it now we may forget to make a point of it
▶ We think we know what they are going to say and want to finish the sentence for them
▶ We know what they are alluding to so we already know the answer and want to respond before they are finished

So, let's say you may not be a good talker and would never do any of the above. Then, there are other signs of people who are not good listeners, for these other reasons:

▶ We're thinking of something else because we have a lot on our minds because of family, job, or whatever
▶ There are distractions: other people around, noises, interruptions, entertainment going on
▶ We don't like what they are saying so we "turn 'em off"
▶ We don't respect the person who is talking
▶ We have prejudices against the person talking for any one of a variety of reasons
▶ Our minds race ahead of the speaker since we can think about five times faster than the normal person talking
▶ Sometimes we have a tendency to let our eyes wander away from the speaker (are we looking for our next contact perhaps?)
▶ We fidget with paper or pencil or rings

In the prospecting or selling situation, we need to concentrate on the other person and make sure we understand what is being said. We need information—knowledge is power when it comes to selling. It sure is OK to relate from time to time to show that you are listening and that you are interested with nods of the head or a "oh yes" or "I see." Eye contact with the speaker is good, but it should not be a stare which could be intimidating. The whole idea is to let the other party do the talking –you ask key, open-ended questions which will encourage them to talk. Don't interrupt! It works!

RELATIONSHIPS SELLING: THE KEY TO GETTING AND KEEPING CUSTOMERS

March 1997

We talk a lot about it, we hear about it, but I wonder to what extent we are practicing it. That is, this business of relationship selling. Relationship Selling—The Key to Getting and Keeping Customers is the title of a book by Jim Cathcart who travels the world teaching salespeople how to sell. This book is a "must" for all of us. It is very easy to read with many tips of how to do what and when and how. In paperback, it's all about ten bucks (Perigee Books—Putnam Publishing Group).

When we practice relationship selling we are really involving ourselves in good communication. The idea is to get and keep customers by building strong business relationships. Here's a great example of doing just that. Many years ago in Meriden, CT, we had a two hundred room Holiday Inn. Neil Salerno was a sales trainee at that time and I recall his getting a call from a person he had met somewhere along the line, in New Haven, CT. This was an attorney who was looking for a meeting site for a Bar Association meeting. Apparently, the group had to have specific dates which were not available at our property and although Neil tried his best to move the group to another set of dates, the group could not change. Neil, by the way, a rookie salesperson at the time advised the attorney that he would go ahead and try to locate them at another property that would meet their needs. After several calls, Neil located a hotel in Hartford that would accommodate the group. The sales manager at the other hotel then followed up, called the attorney, and made the booking. about two or three days later Neil received a call back from the attorney who indicated that since Neil had gone out of his way in locating a competing property for them, that the committee met, discussed the situation and decided that they wanted to have the meeting at Neil's place even if they had to change the dates. Now, this is what I call building solid relationships. And, that group booked monthly meetings and a state-wide convention year after year, I think until Neil left to go elsewhere.

As Jim Cathcart points out in his book, "In selling you need skills in three areas; technical knowledge skills, interpersonal skills, and self-management skills." When it boils down to doing business in our hospitality industry I think we will all agree that the most important of all would be the interpersonal skills. Or, how do we get along with people? After all, people buy from people. People buy from people they know, from people they like, from people they respect and from people they can trust. Building healthy relationships reflect all this. Have you ever bought anything from someone you didn't like, respect or trust? Probably not.

What relation selling will do for you will to keep them coming back. We not only want to confirm business for one time only and forget about it. We need to continually stay in touch, not only to try to book the business, but to let the customers know that we are thinking of them. Newspaper and magazine articles that may be of interest to them should be clipped and sent with a note. Birthday, anniversary cards, congratulation cards or notes when someone gets promoted or something nice happens to them. Even just a phone call now and then to say "Hi." Caring, being concerned, keeping them informed, inquiring about family health, sending special event cards, and think of some other things you can do to help build the relationship. Of course, remember you also need to be sincere about his. Anyone can spot a phony!

SEEK OUT BUSINESS FOR WHEN IT IS MOST PROFITABLE TO YOU

September 1997

When demand is high, as it is these days, it takes very little selling to book business in a property. We see more and more corporate group business being booked within a short period of time. We find very nervous executives or secretaries calling hotels for the small, last minute, a corporate meeting that needs to be held "next week." You quote a rate, they pay it, no rooms, just meeting space and maybe a break and lunch. We're filling up our meeting and banquet space with this kind of business. We are so busy answering the phone that perhaps we're not looking at the long haul. Yes, we need this kind of business and not suggesting we ignore it. However, shouldn't salespeople be spending more time looking for more profitable types of business when we need it?

Sales personnel, as well as General Managers, should be taking a look at the mix of business being booked into a property. What different market segments are buying rooms, space and food functions? Which of the segments, in the mix, provide you with the best profitable business? Once you have taken a good look, make a listing with the best profitable segment on top and work your way down to the least profitable type of business being booked into the property. Your next step would be to figure what percentage of your total business comes from each segment. The idea here is to work on getting your best-producing segments to the top three or four on the list.

In order to increase sales from the top segments we need to take a look at our resources and how much time, effort, money is spent to bring the different types of business. It may very well be that you may be putting more effort and time into developing accounts that may not be as profitable as some others. Maybe at some locations, we need to rethink our marketing and sales plans. What we do here is select the market segments which need to be improved and come up with objectives (goals) for each. Here is an example: You find that the best type of business, from a profit standpoint, is the two or three-day corporate executive retreat (or meeting). Usually, this type of group business is not rate resistant and pays for all expenses of their people, including, rooms, room service, bar bills, phone calls and has many planned meal and cocktail functions. A goal could be to increase this segment of business by 25% (or even create a dollar figure). Then you will come up with different strategies to accomplish that goal including who on the staff will do what, by when and what will it cost to do it. This will be a very important part of your marketing plan for next year. Then come up with another profitable market segment and do the same thing. Again, the idea is to increase the resources necessary to improve the business in the market segments that are more profitable.

Remember, you need a mix of business. Although it may look good now, but it surely is not healthy to get all your business from one market segment. Right now business is terrific at airports, downtown and some suburban locations with the short term corporate meeting market. But having all your eggs in one basket could lead to trouble if something happens to that market and it falls apart because of any number of reasons as has happened in the past. The same holds true for the properties that are enjoying good profits from the transient leisure market segment. Some properties need to take a look at the long haul, as well. What do you have on the books for a year from now or even two years out? If there is nothing there, already contracted, someone needs to get busy to find some good, long-term, profitable business to book.

TIPS FOR IMPROVING PROFITABLE SALES IN 1993

January 1998

Of course, we are still in a "sellers' market" ... the demand is still there, rates keep going up and even the occupancy is still pretty much up, but getting flat and even dropping a little here and there. And, how about those competing properties grow up in your own backyard. Except for the gateway cities, large meeting hotels and the upscale properties, everyone is pretty much getting the feel that the good times may not a role for much longer. If the economy falters a little, if the demand doesn't keep up and the dark clouds start looming over us, what will we do? The answer is not to wait until we feel a crunch, but prepare ourselves now for what may come later this year and probably next year. Here are some tips that may be helpful from a sales standpoint:

▶ Get back to basics ... to be better than the competition you need to help prospects solve problems. Learn what the customer wants and needs, then deliver (for a proper price of course). What is it you have that could be perceived as being better than your competitor?

▶ Keep an eye on the competition ... keep the competitive analysis up to date, know what they got and know how you compare.

▶ Salespeople should have sales goals ... we must all be accountable. Goals need to be reachable and measurable.

▶ Keep on prospecting ... new accounts are always necessary in our business ... not everyone returns all the time. Before you can make a sale you need to find a prospect (someone that can use your property and is in a position to make the buy decision).

▶ Keep making the calls and keep asking for business, always with a positive attitude.

▶ Keep salespeople selling ... it seems that nothing has changed over the years. We still see salespeople doing things other than sales related. Never take away productive selling hours from a salesperson.

- If you need to have meetings get them done before 9:00 a.m. or after 4:00 p.m.
- Trace your files when appropriate . . . it is amazing the amount of business that is lost because we don't get around to following up when we should.
- How much business do you lose over the phone . . . because of not getting a name and phone number before checking space?
- Fish where the fish are biting . . . wherever your best business comes from is where you need to concentrate your efforts.
- Take a look at your market segments. Plan to go after more of the most profitable segments you got last year.
- Stay in touch with travel agencies that have provided business to you previously and ask for more.
- Group business is for all types of properties even if you do not have meeting rooms. Go after social groups.

There are so many opportunities out there for business for everyone, all types of properties. It is just a question of going after it. Even with an increase in demand we still have competition; so sitting back and waiting for the phone to ring won't last forever . . . prepare now for the future.

HOW MUCH BUSINESS DID YOU LOSE TODAY?

March 1998

It really continues to amaze me that in our hotel industry we spend millions of dollars in advertising our products, employing salespeople, conducting promotions, merchandising what we have to sell, and conducting public awareness campaigns. And, no one seems to know how much business is lost on a daily basis at any one property. Do we lose business every day? Of course, we do, but we can't get a handle on it. And, don't tell me your sales department's lost business report tells you the story. In fact, it only tells you a very small part of the whole story. Most of the business you lose on a daily basis is a result of very poor handling of the telephone inquiry. Will a reservation person or front desk employee or a salesperson tell you that we lost a piece of business because we screwed up when someone called to make an inquiry? I doubt it!

At many of our HMM Hospitality Sales Workshops, we do some test calling around the country to get a feel of how hotel employees handle the telephone inquiry. We call to try to book a room, or a meeting or a catered affair, etc. And those folks who have attended these workshops will verify my claim that in just about 9 out of 10 calls the business would have been lost had the inquiry been real. So what's the problem? There's a long list!

Sometimes the phone "rings off the hook" before someone answers. When it does get answered, we get a very fast spoken, long message telling us something about the property which is indistinguishable . . . and that's just for starters. Very seldom do we ever get someone who shows interest in what it is we called about . . . seems that everyone is always too busy (people don't buy from people who don't show interest). Often we talk with someone at a property in reservations or the front desk and ask about a room rate for one night. We get the information and then indicate the rate is too high and ask for referrals to another nearby property which may cost less. Without any other discussion we find that the person is very helpful and does offer, not only names of other hotels, but phone numbers as well. Whatever happened to trying to book the room by doing a little selling? Sometimes, after getting the rate, we offer buying signals by showing interest and saying something like: "hey, that sounds pretty good for that area, isn't it?" We get an affirmative response, but still not selling ... not even ... "let me go ahead and hold that room for you."

Then there is the problem of trying to book a meeting ... we get passed around from catering to front office, since at some properties one department books the meeting space and the other books the rooms ... whatever happened to dealing with one person? Most of the time very little questioning goes on when we try to book a meeting, except for how many people and what type of meeting room set-up is needed. Hardly anyone ever asks for a name and phone number. Knowledge is power, the more information we can get the easier it is to do some selling to confirm a sale.

Also, there is the problem of voicemail. Technology is wonderful, but we sure do misuse it. At some large, convention type properties we get voicemail right away and have to go through a long list of prompts to get to the right department. Then there is another list of prompts to get to the right person, and then the party we are calling is on ... you guessed it ... voicemail.

Let's make it easy for people to do business with us and we will do more business. What we need to do is take advantage of some of the training being offered to help your staff do a better job in handling inquiries and booking business. Hotel & Motel Management and Hospitality Services of America have scheduled a series of front office and sales workshops around the country this year.

LET'S TALK PHONE MAIL ... A VITAL COMMUNICATIONS LINK

April 1998

There probably has never been a time like now, when just about everyone is as busy as can be. And, how wonderful it is to have such technology available today which can help us with our ability to communicate with others ... we've got cellular phones, fax machines, and even email (my favorite) which provides the written communication which may be sent and retrieved from just about anywhere in the world. But of all the tools we have available to send and receive messages probably the most abused is phone mail.

Here's the problem. Most of us have not been trained in its proper use to make it work effectively and reduce our frustration and stress level. Particularly when it comes to hospitality sales. Telemarketing is a wonderful opportunity for us to do business with prospective buyers of our hotel services. Whether it be for cold call prospecting, returning calls, confirming information, networking, or just schmoozing to stay in touch, we really screw it up. How? Well, for the most part, we do not know how to leave messages when we are calling someone and we get voicemail indicating the party is not available. So we are prompted to leave a message. I'm out of town a lot, so my voicemail is good for about ten to twelve messages a day. It is so frustrating when I retrieve the calls because of a couple of things: Sometimes a caller talks too fast when indicating who is calling and just as fast when a telephone number is left for me to return the call. Then I play the message two or three times and still cannot understand the name or number . . . sometimes I will be able to pick up the name of the company and maybe the city . . . so that helps but it does make me upset for people not to leave distinct messages. The idea here is when you leave a phone mail message (especially if you are trying to sell something) you need to speak slowly and carefully, leaving a name, company identification, and a return phone call number. I like the idea of getting a brief message regarding the purpose of the call so that when I return the call I may even have a response. It's a good idea to start off the message with telling the called party your name, company, and phone number. Then leave a very brief message of why you called, and repeat your name and phone number. No big deal, but why not just make it easier for me, and others to return the call?

What prompted this subject was the three (out of twelve) phone messages waiting for me when I got back from an out of town trip. Three people were asking for information about previous HMM columns I have written and they will probably get mad at me for not responding, but I can't. Maybe they'll call back. Another big problem has to do with people who do not know how to end a message. There are folks who go on and on and on, even repeating what was previously said just do not know how to finish a message. It is just not a good sales technique to go on rambling. We just need to be better organized in preparation for making a call.

OK, now what about the message you will leave when you are not available to answer the phone? I hate the one that most people leave: "I'm either on the phone or away from my desk . . . blah, blah.etc." That doesn't tell me much . . . can I expect a return call pretty soon, or will I have to wait until you get back from vacation? Why just be specific? It is no big deal to change your message as frequently as you can. Call me and try it out and see what my message is like. Sometimes I even say: "I've gone to the bathroom and will be right back, etc." When I am out of town I always, yes always, will leave a message advising where I may be reached and when I will return to the office.

If it is in the selling business we are in, and, aren't we all? Let's impress the hell out of people with the proper use of phone mail.

LET'S NOT UNLOAD UNPROFITABLE BUSINESS ON OUR STAFF

April 1998

Yesterday I had an interesting chat with Dennis Flannery the Director of Sales at the Richmond (Virginia) Marriott and that's what prompted this column. Dennis takes pride in not taking every piece of business that comes his way . . . and sometimes it is tough telling his salespeople to say "no" to a prospect. He feels that it is most important to focus resources and efforts in the most profitable direction and turn down business that does not meet certain criteria which would bring profits to the hotel.

We've seen it so often in these "busy" times of high demand, when a good many salespeople need only answer the phone to book some business. Sometimes it is very difficult for salespeople to tell a prospect that we cannot take the business when it looks like we've got the space available. The whole idea is to thoroughly understand that "filling up the house" does not mean, necessarily that we are doing good business. Sometimes we just have to say "no" to opportunities that will just keep the staff busy and return very little, if any, profits.

When demand is high, as it is these days, we just need to do a better job of selling. This means looking for, searching for the most profitable business for the property. Of course, when the phone does ring and we can sell something at a profitable rate, then we want to do that as well. However, if we take a look at the amount of business we do at rack rate, we probably see it running at about 10% or less. Is there business out there that we can get at a higher rate . . . if we look for it. I think there is! So many properties rent most rooms at some sort of a discounted rate. At many hotels, when you call for a rate, the first thing you hear is: "Are you a member of AAA or AARP?" Take a look at meeting space. How much of it is given away without a charge? Sometimes we run our catering and banquet staff to death (not to mention the paperwork) just trying to keep up with the small, and/or limited profit building meal functions.

Dennis mentioned his secret of how he moves some prospects, with unprofitable business, to the competition. This reminded me of my early days in hotel sales in Charlotte, NC. I really took pride, and had a lot of fun out of telling prospects, if I couldn't close at the rate I wanted, how I can get them a better deal at another hotel. Some people were forever surprised at the way I did business. The whole idea was to have space available for my most profitable business, which I knew I could get to fill that space I "gave away" to some other property.

Salespeople need to be trained and motivated to seek out the best possible business from a profit standpoint and to be able to reject business because it wasn't in the best interests of the hotel. Sometimes it is tough for people to understand that having a full house does not necessarily mean it is good for the property. And, I am not suggesting that low occupancy at high rates is the answer or even good for the property. Here we are talking about doing a better job of selling . . . yes, a full house is great, but let's do it at better rates. The whole idea is to concentrate on: prospects that bring in the most profit; the best long-term accounts; the service that provides the best competitive advantage; motivating the sales force based on profit rather than gross sales; and increasing repeat business through customer loyalty.

Maybe we're talking about quality instead of quantity. And, that will result in increased profits and a motivated workforce.

WRITING THE WAY YOU TALK MAY IMPROVE YOUR SALES EFFORTS

June 1998

Why can't we write letters that are "me to you"? That is, using words that sound like we are talking to the people whom we are writing. Sometimes we get carried away with the correspondence we create, so much that we tend to use words or phrases that we would never (or hardly ever) use in verbal communications. Just take a look at some of your recent letters. Particularly those that have been computer programmed and are rather "standard," which are used to respond to inquiries, solicit business, transmit proposals/contracts, and especially those that say "thank you." Are they "me to you"?

How often have you used words and phrases such as: "for your perusal," "as per," "enclosed herewith," or "don't hesitate to call"? And, I am sure you can find a bunch of other such phrases which we would not use in normal conversation. Have we forgotten how to write creative, individual letters or are we just lazy and figure it is just easier and quicker to use our computer-based stuff?

I get loads of letters from hotels soliciting my business. I don't have any business to give to anyone, but I must be on a zillion mailing lists. All the letters sound the same . . . they can just about all be from the same property. Hardly are any creative, most are boring, none respond to any of my needs (I'm not a meeting planner, travel agent, corporate travel manager or tour operator). But, even if I was a prospect, hardly any of the letters impress me or make me feel I want to do business with the property or salesperson sending the letter. The letters lack personal appeal.

In writing a good sales letter it would seem logical that the writer would know something of the needs of the person who is to receive the letter. Why write at all if there is no need? Although it doesn't bother me, but it would many people . . . how about making sure you have the correct spelling and title of the person to whom the correspondence is addressed? The opening paragraph of any sales letter should certainly try to grab the reader's attention. Say something that makes the person read further. A "Let me introduce myself . . ." opener is not very exciting. The second paragraph should include the information—say what you are going to say. Make sure it relates to needs and it certainly is OK to make some sort of a sales pitch. The last paragraph has to do with action . . . something you, the writer, will do, or what you expect the reader to do. Just one page should be able to do the trick. Of course, if you are putting together a proposal, or contract, or providing detailed information, it could be included within an enclosure or attachment.

Here's another helpful thought in writing good sales letters. A decision for someone to buy something has a lot to do with emotion—a good feeling. Think of a comment you can make in the letter to get the reader to feel good. And, how about being a little more creative than indicating "Cordially Yours," or "Sincerely Yours"?

HOW NEAR-BY COLLEGE STUDENTS CAN HELP YOUR SALES EFFORTS

June 1998

Need some help in prospecting for new business? It may be available at a minimum of cost and is probably right in your own backyard. There are probably over 150 universities in the U.S. and about twice that number of community colleges with some sort of hospitality curriculum available for students. Many of these institutions would welcome the opportunity of providing students to your property to make prospecting sales calls just for the experience. We have been doing this in our Hospitality and Tourism Management curriculum here at Virginia Tech for many years. Students do not get paid for their work, but are graded on their performance by the faculty. They are reimbursed for use of their vehicles and are provided housing and meals during the two or three day period of making calls on area businesses. In most cases, properties within a reasonable driving distance from a campus could benefit from this program. For example, this past semester students from the Blacksburg, Virginia campus of Virginia Tech conducted prospecting sales calls for hotels in Virginia, North Carolina, Tennessee, Maryland, and Washington, DC.

The idea would be for you to search out universities and community colleges which are reasonably close to your property and see if they have a hospitality curriculum. If they do, then talk with some of the faculty members who may teach the courses and develop their interest in allowing students to gain some first-hand, practical experience in hospitality sales. This is a great way for students to get a "feel" for our business. Our graduates who are now employed in sales positions around the country have told us how this program has benefited them in many ways in their career development.

Here's another most worthwhile program with which hotels may want to get involved. There is a new industry program that assists college students to gain first-hand work experience with top hotel and resort executives during summer months. The Hospitality Mentors Fund was started with a grant of $25,000 from travel industry veteran Melinda Bush. The fund will enable universities with hospitality programs to send selected students to an eight-week summer work studies program. Students will work directly with a hotel or resort executive on a specific project or group of projects involving sales and marketing. Students selected for the program will receive credit hours from the school and $5,000 grant from the Hospitality Mentors Fund to cover living expenses for the eight-week assignment. The fund will be administered by the American Hotel Foundation, an affiliate of the American Hotel & Motel Association. According to Ms. Bush, the program provides a simple, easy mechanism to allow talented students to gain experience and knowledge that would not only benefit them, but help the participating hotel companies as well.

This new program is now underway for this summer with five students selected from around the country, each to be awarded the $5,000 grant.

MAKE BETTER USE OF THE BASIC TECHNOLOGIES AND IMPROVE YOUR SALES

September 1998

Just about every day I get at least a dozen messages from readers of this column concerning content, questions about the industry, improving performance, career advice, our sales workshops, etc. These messages arrive via email, fax, voice phone and sometimes even by U.S. mail. Based on my observations on how we use these basic technologies I would like to offer some suggestions on how we can improve sales via better use of these tools.

Email: I do not claim to be an expert on computers... being a novice, I am still learning. A very high percentage of email messages I get come from property managers and sales staff requesting information (brochures, copies of the form, etc.) to be sent to them. Most of the time I need to email back requesting U.S. mail addresses and/or fax numbers... since everything I have cannot be sent via email (or at least I haven't figured out yet how to do it). But, if all your email messages have a signature line at the bottom of the messages it could reflect your name, title, company, address, phone number, fax number, website, and anything else you can to have printed at the end of each message. And it will do it automatically. All you need to do locate the toolbar icon that will open the "signature" line, then include any information and it will automatically appear after each email message you send to someone. This is great for promoting just anything, especially when you are sending messages to prospects.

Fax: Great for fast communication of information and documents for most of us that do not yet have access to email. However, the problem I have is with getting fax cover sheets that look terrible. Most of the time, when we run out of fax cover sheets we take the last copy and make another bunch ... when we run out again, we repeat the procedure. So, we end up with these forms that are third, fourth, fifth, etc., generation, and they really look lousy, and are hard to read. The fax cover sheet should be crisp, easy to read and contain enough information to be able to identify the source of the message, address, phone and fax numbers and even some promotion information (but don't clutter it up) and some space for a brief message... of course, always indicate the name and title of the sender. When prospects and clients receive information from us we could very well be evaluated by the product we are sending. And, we want to make a good impression. Just take a minute right now and look at your fax cover sheet and find ways to make improvements.

Phone mail: Not too long ago I dedicated an entire column to this, but it sure is worth mentioning again. There are too many poor impressions made upon people with whom we want to do business by the way we leave phone mail (voicemail) messages. I've been hearing this a lot from meeting planners, travel agents, tour operators, and others over the past few years. Some salespeople have a tendency to speak too rapidly which makes the message being left hardly distinguishable. Messages should be as brief as possible; some people just do not know how to end a message. Your name should be clear and even spelled if it is a difficult name like mine (however, over the phone, I just spell it like it sounds: "Firetag," just to make it easier for the listener. Take your time in leaving phone numbers. With ten digits to record, just make it easier for people to be able to transcribe the number.

All this sounds rather basic, but it is important in selling and closing sales to make good impressions upon the people with whom we want to do business. There is no second chance to make a good first impression.

THE ON-SITE VISIT—TURN IT INTO A SUCCESSFUL SALE

September 1998

Here is your best shot to close a sale—the on-site visit. People visit a property to take a look, and gather information to help determine if the site will meet their needs for overnight stays, or for a meeting or catered event. They come because they are definitely interested in using the facility for an event. Or they want to make a comparison with another facility they may be interested in booking. Either way, it sure is a great opportunity for a "show & tell" for you. You are in control; you have their undivided attention, and it is the best opportunity for you to close a sale.

There are two types of on-site visitations. One is the unexpected walk-in. For the most part, this involves local social groups and wedding prospects. The other is the planned visit. This is where you invite a meeting planner, corporate travel manager, group organizer, tour operator, etc., for a property inspection during the process of your trying to make a sale for group business. Both types of visits should be of the greatest importance to you. This is the best chance you'll get to "show your stuff."

All too often, whether it be a planned visit, or a walk-in, we tend to start by taking the people for a "walk" through the house, showing everything we've got to show. In either case, we'll have a much better chance for closing a sale if we do not start off with the tour. One of the most important steps in the selling process is to gather information. Even if you already have a file on the account from previous business or from having prospected the account earlier, now is the time to talk (and listen) some more. Certainly, on the walk-in, you will want to know the who, what, when, where, why how, etc., of the prospective piece of business. So it is better to have a chat first to find out all the needs. No matter how much data you already have—you can always use more knowledge of the prospect. Once you think you know all their needs and know you can meet those needs you can start asking some pin-pointed questions to get them to do some talking. The idea here in selling is to ask some open-ended questions to get the prospect to do some evaluating or speculating or getting them to show some feeling about something. Here are some questions that would work very well for you: "Tell me about some of the wedding receptions you've been to that you liked very well." "What did your members like most about the last meeting you had." "Explain the different room set-ups you've had previously that worked well." "How did you work out room reservations at the hotels you have used previously."

The whole idea behinds these types of questions is to get the feeling of what went well, or what was liked or not liked at other facilities. This will give you the opportunity to relate how you handle things at your property, especially when you start taking the tour—giving you the chance to explain benefits as you show the site. You can ask these types of questions while you are taking the prospect around the hotel to show how you will meet their needs.

Too often we like to brag about our property and show things that do not meet the specific needs of the prospect. It is OK to do that, but first, let's take care of the business at hand. There is no need to show the ballroom which seats 500, if we are concerned with a meeting of 100. Or, no need to show off the "largest parking lot" in the city, if everyone in the group is flying in. Hold these things off for later. Let's get the event booked first.

WEBSITES THAT MAY BE HELPFUL TO YOU IN SALES AND MARKETING

May 1999

Websites galore, all over the place, and you can keep yourself busy 8 hours a day, or longer, just checking out websites, and many people probably do. Sometimes we just don't get around to finding those that really could be helpful in our business. The idea is to seek out locations which can provide some education and training, and even offer tips or provide services which may be able to build business for our properties. Let's take a look at some:

▶ **http://ei-ahma.com.** Of course, we all know about the American Hotel and Motel Association, but are you aware that the Educational Institute of AHMA is the premier supplier of books, guides, and videos to educators and the hospitality industry? Just get to the website and try the links to Hospitality Training, Distance Learning, Industry Certification, Research Reports, and of course, their listening of books, guides, seminars, and videos.

▶ **http://hospitality-1st.com.** Here's a pretty new one which is really great. This site serves hotels, restaurants, and the travel industry, and offers "targeted solutions and proactive programs for your business." Lots of good stuff here on sales, marketing, public relations, and advertising. Lou Taverna offers a weekly, electronic newsletter (you can subscribe for free) with ideas and tips, easy to read and pretty brief, right to the point. He also provides a "positions available" edition for newsletter subscribers.

▶ **http://hsatraining.com.** HSA has a great site for training, not only salespeople, but front office staff as well. Hospitality Services of America (HSA) provides on-site training, regional workshops, training products, property sales analysis reports, and customized training and monitoring services. HSA conducts regional workshops all over the country with one-day and two-days programs on Guest Service Techniques, Reservations, Yield Management, Marketing in Cyberspace as well as Hospitality Sales.

▶ **http://MADSearch.com.** This is another new site which is directed mainly to people who plan meetings; however, a really helpful site for hotel salespeople. It is a search tool for meeting planners to locate facilities for meetings. There are thousands of listings of hotels all over the world, and the site provides links to those locations. Through the site, one can get updates on the industry, career opportunities (for all kinds of positions, including hotel jobs), and you can even make announcements of your property. There is an interesting section which provides information on hotels that have short-term space available with rates being quoted. I've seen announcements where there has been a group canceled, and the property was offering blocks of rooms for the dates at a lower rate than is published. The site provides meeting and destination news ... so if you have something new to announce you can probably get it in their electronic news bulletin, and for free.

▶ **http://mpiweb.com.** Even though you may not be a member of Meeting Professionals International you can still get a lot of good stuff out of this site. Of particular interest would be the discussion group section. I've seen many hotel salespeople making announcements about their properties. It is a good place to ask questions about working relationships with meeting planners. Lots of good meeting planning tips are provided as well.

HOW ABOUT JOINING AN ONLINE REVERSE AUCTION TO FILL ROOMS & SPACE?

February 2000

We're hearing a good deal lately about technology, all over the place. So here's the latest! This is where the seller (the hotel) becomes the buyer—it's a reverse auction. Meeting planners become the seller—putting up their meetings for bid. Got a hole to fill on short notice? Did a block of rooms cancel out on you? Did you just wake up and noticed that business is not what was projected for a certain period? You may find an answer to your prayers via the Internet by checking with www.EventSource.com to see if there are any meeting planners out there looking for a "deal." There are already probably over 10,000 member properties online with Event Source waiting in the wings for auction invitations.

Here's how it works: This is really an RFP (Request For Proposal) situation where meeting planners are looking for space, submit an RFP indicating their requirements, and member hotels submit their "bids." Event Source has certain criteria that need to be met before the process begins. A meeting must be 180 days out, it must involve only one city; and the meeting must have a minimum of 50 room nights on its peak night. The meeting planner decides on an auction for the business, advises Event Source, who then distributes the specs to whichever of its member properties that can meet the specifications. Properties interested in the business then complete the RFP online. The planner can then take a week or two to make site visits or gather additional information on the hotels. Now the auction may begin. Each participating hotel is assigned a number, which is used to make a bid (similar to the basic auction process). Although properties which participate know which other hotels will be involved in the auction they do not know which one makes which bid. The auction itself takes place in a "chat room," accessed via passwords. So far, Event Source has set it up so that between four and eight hotels participate at one time. There can be as many as four rounds of bidding taking place. The meeting planner can see each hotel's rate bid; and hotels can see other bids, but would not know which hotel bid what. Based on what they can see, hotels can decide whether or not to make a lower bid during the next round. This still doesn't mean that the lowest bid will be accepted. It is the meeting planners choice since other factors may be considered. There is a time limit for the auction set by Event Source which probably would run under 20 minutes. It appears to be a very fast-pasted situation.

Hotels winning the business pay Event Source 10% commission. So far, nothing but favorable reports for participating meeting planners and hotels, since the program started in October 1999. Another company getting into this e-auction business last November was www.StarCite.Com. Hundreds of meetings are expected, by both companies, to be auctioned off in this manner the first part of this year. Meeting planners, of course, want to drive competitive rates in this sellers' market which still exists in major cities. Since it is all short-term business, looks like a pretty good deal all around. Depending on how badly a hotel needs the business, during a certain period, could very well dictate what rates would be offered to make the hotel attractive to a buyer. And, it may not only be lower rates. There are always other consideration in the selection process for meeting planners.

WE'LL BE DOING MORE AND MORE BUSINESS ONLINE, SO LET'S GET TO IT

June 2000

No question about it! Technology is here to stay, so we'd better get to it, understand it, and use it. We gotta realize that just about everyone wants things done faster. We used fax machines to replace regular mail service; now we have email replacing fax machines. No need to fax copies of documents . . . just scan them into the computer and email the document. We can create documents and send them in something called PDF files, and they can get printed looking just exactly as the original document. But, now you don't have to even do that. Just create the document in Microsoft Word and send it as an attachment. The receiving party just prints it, and gets the actual document.

When it comes to technology, it looks like the big problem we've got in our hospitality industry, particularly when it comes to operating sales departments, is that we are "behind the times." We're just not up-to-snuff when it comes to communicating and trying to do business with meeting planners via the Internet, according to what some planners are saying. The big complaint I hear is that so many sales offices are not Internet-connected, that is, no email. It seems that many computers in sales offices are old, not up to date; sales folks don't have their own email address, and are still using faxes or playing telephone tag.

Here are some tips for hotel salespeople (to help them work online), which appeared in a recent issue of Meeting & Conference Executives Alert (mcea@mediaone.net). They were provided by Doug Fox, publisher, Event Web newsletter (doug-fox@eventweb.com) and T. Scott Falk, Director of Sales, Arrowhead Conferences and Events (tscottfalk@msn.com):

▶ Have immediate Internet access from their own computers through a modem or corporate network.
▶ Have their own email account, which they can access from their offices and from the road.
▶ Receive proper training in how to use the Internet, email, and popular software programs.
▶ Stop sending expensive printed brochures (unless requested) and start directing planners to the appropriate web pages on hotel websites or online customized presentations.
▶ Know on which meeting and convention industry websites their hotel information is listed, and they should make sure that their property descriptions are accurate and up to date.
▶ Learn how to use the top meeting planning websites so that they can respond quickly and effectively to requests for proposals and other communications from planners who use these services.
▶ Be very knowledgeable about their own hotel websites and be able to walk planners through the appropriate information and web pages by telephone.
▶ Know how to send and receive email attachments, such as proposals sent in Microsoft Word format.
▶ Have access to an instant messaging program during contract negotiations so that they, and their customers can discuss changes to an agreement without playing telephone tag.

Doug and Scott suggest that all salespeople get up to speed on using the Internet, so everyone can do their jobs more effectively.

All this makes great sense. Besides getting things done much faster, and being able to communicate much more easily via the Internet, I have found I am writing much less letters and using the fax much less frequently. I am also getting much value out of subscribing to Doug's EventWeb newsletter. You may want to try it (www.eventweb.com). I've learned a bunch of good stuff from him. You may even want to subscribe to MCEA Newsletter. The website is (www.mcea@mediaone.net). It's also loaded with great stuff for planners as well as suppliers.

LOADS OF SALES TIPS AVAILABLE IN NEWSLETTERS

July 2000

It is amazing the number of excellent quality sales tips that are made available to us in a variety of newsletters covering our industry. Regardless of the type of property you represent, whether it be limited service, full service, large or small, five-star or one-star, there seems to be something in these newsletters to help everyone.

Even though there is a lack of sales training available around the country, for all types of properties, subscribing to newsletters, electronic or otherwise, could in itself be a very worthy method of self-training. There are so many new salespeople being employed these days, due mostly to the development of new hotel properties around the country, and the increased competition, that so many have not had any proper sales training. Having access to these tip-filled newsletters is a great way of learning some good sales techniques that really pay off in improving sales productivity and increasing hotel sales.

Although there may be several electronic newsletters available devoted to hospitality sales and marketing, the one I am most familiar with is www.hospitality-1st.com which provides a pretty brief, easy to read the weekly message. There is always something in there which someone can use to increase sales productivity.

Just out, there is newsletter entitled "Sold Out" which is "The Motel/Hotel Newsletter That Shows You How to Rent More Rooms." It is written by Bill Flor who is a co-author of the book: "Sorry! No Vacancy." I just received the inaugural issue which is absolutely loaded with wonderful stuff directed mainly towards the limited service market, but has good tips for everyone on how to build room business.

There are also a good number of hospitality consulting firms who produce newsletters. One that also just came in the mail was from Grantham, Orilio and Associates, Inc., out of San Diego. They put out a monthly Hospitality Industry Report which covers a variety of issues concerned with operating hotels. In this last issue, they covered Proper Telephone Training, How Bartenders Become Your Partners, and Modern Master-Chefs of the 21st century.

LET'S NOT FORGET ALL THE "... INGS" WHEN IT COMES TO SELLING

June 2001

Thanks to John Hogan, Director of Training for Best Western International for his contribution to this column which will be appearing in our new book to be published this fall. There is a good deal involved in the selling process when it comes to getting it done. So here is what is involved:

▶ **Managing**: Someone has to be in charge and handle the sales effort to make sure it gets done. No question about it ... responsibility is still in the hands of the General Managers, whether there is a sales staff or not.
▶ **Planning**: We need to have a scheme for getting something done. So we have a business plan, a marketing plan, and a sales action plan. All are needed to make sure we go in the right direction to improve profitable sales.
▶ **Organizing**: Of course, we have to put it all together, to make it a functional sales effort, and creating a balanced sales/

operational team to make it all work; and providing the best service so that our guests return time and time again.

- ▶ **Directing**: Surely, there has to be someone in charge to direct the affairs of the selling effort. And, this (telling) needs to be done on a daily basis whether it is by a Director of a sales team or the General Manager.
- ▶ **Controlling**: Here we need to be concerned with the checking and measuring of results. Yield management comes in to play here to be sure we are booking the right kind of business. Inventory control is critical, as is control of the budget.
- ▶ **Motivating**: This is a key activity for General Managers to provide the stimulus for sales, as well as other staff folks, to continue to do their best. Inspiring and encouraging through support and constructive criticism is the way to go.
- ▶ **Communicating**: Providing information via two-way discussions among staff is what prevents "break-downs," keeps guests happy, and ensures a productive sales effort.
- ▶ **Training**: Everyone needs self-improvement. This needs to be an ongoing program. Too often we bring new folks onboard without taking the time to properly train them. Salespeople, especially, need help in this area. We should take advantage of opportunities all around us through organizations like HSMAI, MPI, the Educational Institute of AHLA, as well as a variety of sales workshops available.
- ▶ **Listening**: The most difficult for most of us, especially in sales, as we tend to want to be telling our story. We just need to be better listeners. It is the real key to doing a good job in selling.
- ▶ **Meetings**: Do we have too many, and maybe too often? Those weekly sales meetings should not be held during productive selling times. Try having them after 4:00 pm or before 9:00 am.

I think there is a whole load of these "... ings," as they relate to sales. Got any ideas? Let's see how many we can collect from the readership of this column.

SELL... ING HAS SO MUCH TO DO WITH A LOT OF "... INGS"

August 2001

When you think about it, just about everything that goes on in a hotel has to do with sell ... ing. Of course, everyone is involved, from the General Manager on down through the staff ranks. In a recent issue of this column, we came up with a bunch of words ending with "... ing" that were related to sales or sales activity. We were bombarded with reader responses adding a whole bunch of other words ending in "ing." They were so good we want to share them with you.

Here's a big batch which came from the students enrolled in a Convention Sales & Promotion class at Paul Smith's College, Saranac Lake, New York. These were submitted, on behalf of the class instructor, Tracey Beyer:

- ▶ Appreciating your staff
- ▶ Cooperating with clients, vendors, and other departments

- Evaluating your incoming business and competition
- Caring for your internal and external customers
- Complimenting all employees for a job well done
- Trying just a little harder than the day before
- Educating ourselves and our guests
- Empowering our frontline employees—without limits
- Celebrating the accomplishments of your sales team
- Preparing for the future—keep up with trends
- Comparing your features/benefits/pricing to your competitors
- Learning continually
- Competing fairly, but fiercely!
- Developing our selling and negotiating skills
- Sharing our vision with each department
- Serving our guests to the best of our abilities
- Questioning any failures to determine their cause
- Thinking creatively and outside the box
- Practicing new sales techniques
- Delivering above and beyond service consistently
- Comprehending the true needs of our guests
- Probing for the need behind the need
- Timing our sales calls perfectly
- Experiencing the service we offer from the guest perspective
- Maintaining quality in the products and services we offer
- Negotiating fairly to maximize our revenues
- Thanks, Paul Smith's College students . . . ya done good!

And, then we heard from Susan LeMaster, Director of Sales, Holiday Inn Airport, Englewood, Ohio. Here is her batch of words:

- Producing—showing a profit for the hotel
- Profiting—more $ brought into the hotel = more profit
- Networking—working in the local community, offices, etc.
- Collecting—all information from prospects leads to a sale
- Monitoring—your services to customers
- Learning—always learning to newest techniques in selling
- Closing—the actual sale
- Dedicating—service to the customer; making sure each customer is treated the same and the customer service equals the same no matter the dollar amount of the client.

Thanks, Susan, appreciate the input.

And, we also heard from Randy Stuart from Holiday Inn Harrisburg/Hershey, PA

- Closing—asking for the business
- Suggesting—providing solutions and options to customers' questions
- Providing—giving the customers what they want
- Creating—to offer new ideas
- Negotiating—arriving at an agreement where both parties are satisfied with the results
- Prospecting—looking for new customers through various resources

- Networking—exchanging information and services between peers and customers
- Learning—gaining knowledge of all the customers' needs and requirements
- Fulfilling—to deliver and meet the expectations you promised
- Understanding—the customers' needs
- Writing—taking good notes

Thanks, Randy . . .

And only room for one more . . . this time from Wayne Kearney of the Pinehurst Company, Group of Resort Hotels.

- Follow-Through—and we all know what Wayne means by that.

HI-TOUCH IS STILL BETTER THAN HI-TECH IN SALES AND SERVICE

September 2001

We keep reading about it all the time, all over the place: service stinks! But, we are really not as bad as the public or the press, make us out to be. Frankly, I strongly believe we are doing as well or better than we have done in the past, as far as taking care of our guests. In my over 100,000 air miles a year, visiting a good many hotels, and as a veteran of the industry, I see a lot of things that can use improvement from a service standpoint.

However, as a whole, our hotel and restaurant industry is holding its own compared to all other types of businesses. Of course, in many cases, it is the little things that bug our guests and prospects. Things like a front desk associate who will stare at an arriving guest, waiting for the person to be the first to say something, rather than offering a nice smile and warm greeting. Or the toll call I made to a property trying to reconfirm a reservation and had to go through a voicemail answering device as well as a series of prompts and never did get a live voice. Then, after having called back two more times (long distance), I never did connect with a real live voice, and am still trying to get a reconfirmation . . . ain't nobody there to talk with me. And the list goes on. Are we getting too high tech with so many things that we forget the opportunity we have in building business through a good high touch?

At one of my recent HMM sponsored Hospitality Sales Workshops, one of the participants insisted that is was most important to prospect for new business over the Internet. OK, maybe so, but think how time-consuming that could be. Have we forgotten about all the prospecting we can do just by going through the old files that have not been traced over a long period of time? Your own files contain such valuable information that can give you a much better return on your investment of time. Just getting back to the folks who have done business with you in the past will yield much better, and faster results. Your local newspaper will also reflect numerous leads that can be qualified by phone. Good leads are all over the place. Think about your own property staff and how they can provide leads for new business through their own business contacts. Some department heads make purchases locally and your accounts payable department has a listing of companies with which your property is doing business. Why wouldn't they do business with you? Bartenders and restaurant servers know many local folks who do eat and drink at your place whose companies may very well be prospects for out of town guests visiting your city, or for meetings, or local social business. The same holds true for your staff working reservations or the front desk. The point is we just don't take advantage of the prospecting opportunities around us that just require a little high touch (personal contact), rather than spending too much time concerned with using technology to get the job done.

Another thing we found out through an unofficial survey of our workshop participants is that salespeople are spending only about 25–30% of their time in pro-active sales. That is, talking with qualified prospects with the idea of making a sale. In our business, people buy from people; whether it is a person working reservations, or the front desk, or catering, or sales. The more contacts we make equals more prospecting, equals more business we will bring in to our properties. A new challenge we have in sales is the amount of time we are spending in responding to RFPs (Request for Proposals). There are so many new websites out there encouraging meeting and event planners to make use of their sites to locate possible facilities, and they are being used more and more, creating a real burden on our sales forces around the country. We know we have to respond because we do not know which one out of maybe a hundred would be the one organization that would book with us. However, why do salespeople have to be the one to do this administrative task? Some properties that receive many of these RFPs have employed an administrative or clerical person to handle that function. This frees up a salesperson to do the job of being more proactive in selling.

SALES STRATEGIES FOR SURVIVAL AND RECOVERY

December 2001

We've been hearing a good deal lately from sales folks, General Managers as well as owners of properties that "these are the worst times than ever." They are referring, of course, to the lower occupancy, average rate, and RevPAR. Perhaps we are "jumping the gun" a little by making such remarks, since there really have been much worse times for our industry. No question about it though, these are pretty bad times from another standpoint for our country because of the terrorism situation. But, getting back to where we stand in the hotel business, we all surely can remember, or at least, heard about the very "dry" periods of the mid-70s during the oil embargo. Those were the days when all types of properties suffered for lack of business because gas was just not available. Then we had the period of 1986 to 1992 when the hotel business, as a whole, throughout the U.S. suffered tremendous losses those eight straight eight years. As a group, hotels did not break into any profit until 1992. During that period it was reported that 64% of properties in this country could not make debt service. But, with some suffering, hard work, an economic recovery, then an increase in demand, and after having learned a good lesson, we managed to survive. Then it was followed by eight years of increased profitability for our industry. So here we are today, with an economic slowdown, some people who are fearful of travel, and a drop in occupancy, a lower average rate as well as reduced RevPAR. But this is not the worst time ever for our business.

We have learned some good lessons from the past, and we know that we have to look forward to doing some things perhaps a little differently from which we were used to over the past eight years. From a sales standpoint, we are not talking about rock science or shortcuts to making business happen. If you want to bring it in at the bottom line (profit), then you have to bring it in at the top (more sales). We've heard it over and over again throughout our industry: "Let's get back to the basics." Here are some tips for survival and recovery:

Let's start off by just answering the phone. We've written about this many times in this column. Test your own property! How long does it take for the phone to be answered? We'll never know how many people who want to do business with us get turned off, and hang up to call another property when we don't answer the phone right away. What's the attitude of the person answering the phone (or is voicemail answering it for you)? Is the voice friendly, enthusiastic, do we ask the right questions? It is easy to turn off an inquirer just by the way someone answers the phone.

Why do so many property owners and managers feel that they have to cut rates because the competition down the street is doing it? Be proud of your property, and the value it can present for the rate you want to charge. Sell the difference! We have heard that so many times, but it does make sense. There is something about your property that is different, and you have something that is of more value than what your competitor may have, for the price. Rate cutting can hardly ever build enough occupancy to make up for the difference in RevPAR.

Have every staff member work on building relationships with existing guests and accounts. It's easy and a very friendly thing to do. It builds loyalty! It makes sense! A little public relations goes a long way. Make contact, and get friendly with the folks at the print and broadcast media. They will help, and work for you, and it won't even cost a cent. Get better known through community projects; speak at local civic club luncheons; be visible.

Let's get back to basics folks. It works!

SALESPEOPLE GOTTA MAKE CONTACTS TO MAKE SALES

February 2002

Of course, we teach our children never to talk to strangers. That's good and makes sense. However, at some point in life, especially, if we are working in sales, we have to make contact with strangers if we want to be successful. We can surely understand that there are many people who are shy (someone once said that 93% of adults have indicated, in a survey, that they are shy), but do these people belong in a sales position? Over the years I've attended hundreds of conferences and trade shows involving potential buyers of hospitality services. At these functions, I have always been surprised at the number of sales folks who stand around waiting for someone to talk to them. The most successful of our sales teams have always been the ones who can walk up to anyone, offer a handshake and make introductions. By the way, there are also a good number of General Managers who seem to be "wall flowers" at events where there are potential prospects for business.

One of the big things we always talk about in training sales folks is that building the relationship is the key to doing business. So, the idea, at a business event, when mixing with prospective customers (and almost always at industry events we have prospects), don't feel that you have to put yourself in a position of having to make a sale. The idea is to be friendly, introduce yourself, explaining where you work and what you do. Provide a good handshake and look the person you are greeting in the eye. Speak up, don't mumble, and pronounce your name clear enough so there is no misunderstanding. At most events attendees will be wearing name badges, so that will make it pretty easy for you to pick and choose to whom you will want to meet. Certainly, you should be looking for people you do not know, and who may be a prospective buyer. But, remember that this is not the time to try to do any selling. The idea is to just make contact and offer a business card. In all probability, the other person will also offer a card, if not, then ask for one. If the other party does not pick up the conversation by asking you a question or telling you about what they do, then you should be able to make a comment or ask a question to keep the conversation moving. There is nothing wrong with asking a couple of key questions which would give you an idea if this person could be a prospective prospect. You could even write some quick notes on the back of the card to remind you of some detail about the person or business for follow-up at a later time. You want to make the conversation comfortable for the other person. A series of questions sounding like an interrogation may be threatening which results in losing the relationship building aspect.

All too often some salespeople think that they have to kind of "hit people over the head" with all sorts of information about their property to try to make a sale. This is not the time to do any selling, all we need to do is to start building relationships with new contacts, getting some information which would indicate if the contact could be a prospect, and make the follow-up at some later date. It is not necessary to spend too much time with the new contact. After all, you and they have other things to do. You need to make more contacts. The more contacts you make, the more prospects you can develop, and more prospects lead to more customers, which then leads to more sales.

Here are a few tips in making contacts at business events:

▶ Be the one to initiate conversation.
▶ Have a brief introduction of yourself (tell them what you do).
▶ Exchange business cards.
▶ Avoid any semblance of selling—establish the relationship.
▶ Get the correct name of the contact and use it often in your conversation.

SALES TIPS ARE AVAILABLE ALL OVER THE PLACE

April 2002

With the soft economy, drop in occupancy, lots of rate cutting, and some overbuilt destinations, it is getting pretty tough out there to build profitable sales. Owners, managers, and sales folks could use all the help they can get to improve sales productivity. Well, sales help is available almost all over the place if we just know where to go for it. It is amazing to see all the emails and phone calls I get from salespeople, and General Managers who are looking for tips to bring in new business. Of course, we help with some ideas wherever we can. Here are some websites, as well as some hard copy newsletters that provide a good deal of great sales tips for all sizes and types of properties.

The leader and top organization for those in sales is the Hospitality Sales and Marketing Association International (HSMAI). Property owners and managers should insist that their salespeople become members, and attend their training and educational programs all over the country. However, even for non-members, there is a great deal of sales educational material offered on their website, along with their online bookstore (www.hsmai.org).

Meeting Professionals International is another favorite for training and educational opportunities through memberships and attendance at their conferences. They have about 20,000 members (meeting planners and suppliers) all over the world. Many tips for members (and non-members) are offered on their website, along with links to other organizations as well as an online bookstore (www.mpiweb.org).

The Educational Institute of American Hotel & Lodging Association is the leader in hospitality training all over the world. A variety of sales related books, videos, and audiotapes are available through their online bookstore (www.ei-ahla.com). For training in sales, front office, guest relations, yield management, marketing, customer service, as well as customer shopping evaluations there is HSA (Hospitality Services of America). They provide on-site training as well as public seminars, and provide "train-the-trainer" tools, books, videos, etc., via their online bookstore (www.hsa.com).

Some online newsletters are also providing excellent sales tips and even leads for business. One of the most active is www.hospitality-1st.com. This is a free weekly newsletter, easy to read, pretty brief, but full of great sales and marketing tips. Lou Taverna, its editor, also has a great job listing that goes out weekly, as well as a weekly reference, with links, to important news items related to sales.

Equally as active, and mainly for meeting planners, but loaded with terrific ideas for salespeople involved in group meetings solicitation is www.MADSearch.com, Another free weekly online newsletter. Madeline Drysdale, its owner, has some very creative stuff going on with this site. You can even click on a link to get a phone call from the staff in case you have a question. Take a look!

Here is some wonderful, hard copy, newsletters that are absolutely loaded with sales tips. "Sold Out" by Bill Flor, is a newsletter "That Shows You How to Rent More Rooms." It has been around for quite some time. Bill also is the author of "Sorry! No Vacancy," conducts sales workshops, and does on-site consulting around the country. Contact him at info@no-vacancy.com. Another one is "Increase ... Sales and Service Issues for Hoteliers," A Due West company publication. Janie Wiltshire always has some great and creative checklists in all the issues. She also conducts sales workshops for properties and management companies. Contact her at sell@janiewiltshire.com.

The whole idea is that there is a load of information, education, hot tips, and training available around the country. There are a good number of brand new sales folks out there that do need some help and there is no reason why they can't find it.

WE NEED TO START WRITING BETTER SALES LETTERS

October 2002

No question about it! We all get plenty of solicitation letters from companies trying to sell us some products or services. Don't they all almost sound just about the same? Well, I get loads of letters from hotels, soliciting my group and meeting business. However, I don't have any such business to provide to those soliciting me. Somehow, or other, I seem to get on mailing lists as a result of belonging to organizations like Meeting Professionals International; however, I am listed as a Supplier Member. This means I am really not a prospect for any meeting business for any hotel. Also, I speak at a good many conferences for Travel Agents, Association Executives, Meeting Planners, Tourism Organizations, etc. And, of course, I get on the attendee list. So, what usually happens, some hotel salespeople get that listing and do a mailing to everyone on this list. I guess that's how I get all these mailings.

And, would you believe that almost all those letters, asking for my business, just about read-alike? I'll get the same letter that some real prospects, with lots of potential business, will get.

Most of the letters are very blah, not interesting, not creative, read like a brochure, don't appeal, don't relate to needs, and appear to be computer written. Everyone should take a look at the word processed letters that contain all those standard type paragraphs that we have been using for too long a time. Throw them out and start all over. A really good letter should, first of all, make sure that the name to whom the correspondence is addressed is spelled correctly. Then, the salutation needs to reflect how well we know the person. If we are on a first name basis, then, of course, the first name could be used. However, if we do not know the person, personality, or well enough to use first names, then the last name should be used.

Before we even start writing the body of the letter we need to think of writing that reads like "from me to you." We probably spend too much time trying to act sophisticated by using words that may not be necessary. For example, we have seen words or expressions in letters as: "Per our discussion," or "Pursuant to your request;" or how about: "enclosed herewith" or "don't hesitate to call." And the list goes on. We don't use words like that when we talk to people. Just write the way you talk and people will remember you better.

The body of the letter should include something in the first paragraph that gets the attention of the reader. Make it short, but get someone's attention so that they are inclined to read the rest of the letter. This should be followed with an information paragraph which relates to the needs of the individual or the group. And, it should be specific, rather than general. There is no point trying to solicit someone's business if we do not know their needs. In that paragraph is the place to make a "sales pitch" explaining how you can meet those needs. The last paragraph should be an "action" method. Either it is something you want them to do or something you will do. This is important so that we keep the selling process on the move all the time. And, then maybe a "P.S." with something that deals with emotions. Something personal as mentioning a promotion, an award, their hobby, sport, a pet; anything that will get them feeling good about you.

THOSE SALES EMAILS, HOW ARE WE HANDLING THEM?

March 2003

Just from the flood of an email message I receive as a result of this column, I would suggest that, generally, so many folks are handling them poorly. To start with we should consider that emails are evidence of one's level of professionalism. In the early days of email, use was casual, and we gave little attention to spelling, grammar, spacing, or punctuation. Today, in our business of doing an improved job of selling, and communications, we need to be more professional and treat emails as another form of business correspondence. Don't you agree on those with whom we want to do business, evaluate us in the way we write messages?

Let's begin with how we should be using the subject line. Make it descriptive! Let the reader know what is to follow. I get many messages with mysterious subject lines. Next, use a greeting. Most of the time I say "Hi Joe" or "Hello Lois" and sometimes I put the greeting in the first line of my message; like: "It was great getting your message, Sam." Some form of greeting is polite and courteous. And, how about a "thank you" when someone sends information, or a "please" when requesting information? Sometimes I get very lengthy messages. Email should be short, and to the point. If the message should be lengthy then send it as an attached document. Try to send messages with only one subject at a time. If you have several subjects to discuss, then send separate emails. This way there can be a separate response to each one, which helps in an event separate files are kept. When I get a long email covering several topics I print it and refer to it each time I respond to each item, this could be frustrating. Sometimes I go into the original email and respond by scrolling down to the end of each topic to make my response. Many times, when using the reply function I change the subject line to make it clearer about what it is I am responding. Quite often, with a good deal of responses back and forth, the original subject line no longer refers to the content. We probably should be starting a new email message if the subject has shifted during the multiple responses.

We should always make sure of our spelling and grammar by using our spell check tools. Also, we need to follow the standard business format for memos, such as using appropriate upper and lower case letters. One big problem I seem to have with emails I am receiving is that so many sales folks do not use a signature line on their messages. This should include full name, job title, company, and address, phone number and even a website for your organization. This is the most helpful in the event someone wants to make voice contact or send something via snail mail.

One last tip. Be sure to review your message before you hit the send button. I try to do this all the time, and I have caught many errors which could have been embarrassing to me had I not corrected them.

SELLING VALUE IS THE KEY TO IMPROVING PROFITABLE SALES

February 2003

OK, we all know that business is getting better for us as a result of the economy, increase in demand (about 3.5% increase is projected), more rooms not coming online right away, improvement in the employment situation, and of course, hospitality sales folks are being trained. But, we cannot just sit back and wait for the phone to start ringing. We've got prospects all over the place as we've indicated in previous HMM issues, but the big question is: "How do we approach these prospects?" No question about it, we have to start off with knowledge. First, we need to know what it is that we are selling. Complete knowledge of the property is the place to start. That is, understanding all the features of the property and how each feature can be presented as a value to someone who in need, or could use that particular feature. It is never enough just to know what you've got, but how to present it to someone is the key. Not every feature at your property is a value to everyone, so there is no point in "pushing" something that is not needed or wanted by someone or some group. This leads to the next piece of knowledge you must have in doing a great sales job; knowledge of the needs (or wants) of the prospect. All too often salespeople will start right off in a sales pitch telling everything about the property, but who cares? People only care about what they need or want. Before we start to do anything, we need to gather more information than what you may already know. When you think about it, we are putting ourselves in the position of a consultant and trying to assist with solutions for the prospect. During your questioning and conversation, as a need or a desire pops up, then is the time to do some selling of a particular feature by presenting it as a value, or benefit. You do not want to speak in generalities, but you need to be specific. For example, if the issue of security comes up it is not enough to say something about how "the place is very secure." You need to express the security with specific information about the secured features of the rooms with viewports, deadbolts, electronic door locks, keychains, or even in-house security folks who patrol the property, the lighting in the hallways, and parking lot, or even the record of the property not having any security issues raised over a certain period of time.

All too often we try to sell by bragging about how wonderful we are, with generalities. Sometimes we make a lot of claims, like being the "best" or the "biggest." If you do make a claim, then it needs to be backed up with some verification. Remember, it is the value in what you are selling that will get people to buy from you.

WHAT MEETING PLANNERS SAY ABOUT HOTEL SALES COMMUNICATIONS

September 2003

It seems that nothing much has changed over the years when it comes to what meeting planners expect from hotel salespeople with regard to how they preferred being contacted. Of course, technology has changed and probably helped a good deal in some areas, and then, in some areas probably made it more cumbersome (like phone mail, for instance). A recent survey, conducted by Northstar Travel Media LLC (parent company of Meetings & Conventions magazine), came up with some interesting facts that should be very helpful to hotel sales personnel.

According to the study, most meeting planners do not want to be called on the telephone, but prefer face-to-face time. If they do get a call from a salesperson what they do not like about it is: "they call and begin a pitch without asking if I have the time to talk;" "they don't research to learn if their services are suitable for my needs;" "they contact me repeatedly." Even I have received calls like this, and I am not a meeting planner. Of all the meeting planners surveyed only 5% indicated they would prefer being called. Technology really helps here since 57% of planners indicated that they would prefer being contact via email. And 50% prefer direct mail (snail mail). It also seems that planners complain about unsolicited sales pitches from suppliers (68%) and that they receive an overwhelming amount of material with details that they do not need (49%). We just send out too much stuff. What they really want is "brief, at-a-glance information," by direct mail, according to 67% of those surveyed.

Interestingly enough, only 6% want to get together over dinner, but most (over a third) prefer to meet over lunch, and 44% indicated they would like to meet face-to-face in their office. What is interesting about these figures, these are probably the preferred method by most sales folks anyway, face-to-face contact gives sales folks the opportunity for relationship development. There isn't anything like face-to-face selling.

We always seem to get back to basics. Relationship selling has been the key ingredient to sales success as long as I can remember. Jim Cathcart, speaker, and author, has been preaching this for years. His bestselling book, Relationship Selling, is really a must for every salesperson. Now there is something brand new that is available to all of us, the Relationship Selling Email Course. He sends out weekly emails with specific sales strategies, exercise, and calls to action. And, it only costs a buck a week (wish I had thought of the first). The whole idea is to keep Relationship Selling in the forefront of salespeople's minds. I'm thinking that this would be a great way to brands, as well as management companies to do some great training for sales personnel at their properties.

TIPS FOR SALES FOLKS ON EMAIL

March 2004

In a recent issue of Lou Taverna's online newsletter (www.hospitality-1st.com), there were some very worthwhile tips on using email. We all get a good amount of stuff (especially Spam) on our computer every day, but so much of the legitimate messages that I get really bother me. It is not the messages itself, but the manner in which some salespeople write them. Right from the start, we should certainly do a better job on the subject line, before we even get into what we want to say. Two things tell me a lot and help me determine if I want to bother reading the message; first, I want to know who is the mail from, then, what is the subject. We need to do a better job of making a statement on the subject line about the purpose of the message. Just indicating "hello," or "hi there" or "a special message" doesn't say anything. Be pretty specific, but brief, about what will be included in the content of what you will be writing. The receiver needs to know what to expect. With all the spam going around, without a good subject line your messages just might get deleted and not read. Sometimes, as messages go back and forth between writers, the content changes. If that is the case you should be sure to change the subject line to reflect the new content.

The body of the email message should be brief; get to the point of the message quickly. Plan on what you want to say before you start writing. All too often some message gets bogged down with repetitive statements and unnecessary "garbage." If what you need to say is rather lengthy then record it in word format and send it as an attachment. All too often some salespeople are sending proposals within an email. I feel it would be more professional to write a formal proposal and send it as an attachment. However, if you want to send an announcement of some sort, or a promotion piece, that would work in an email format by making it appear as a one-page flyer. That way, the recipient can decide to read it all or not. Be careful with spelling, grammar, and sentence structure. What you say and how you say it represents you and your property. Always read over what you have written, and use your spell-check tool before hitting the send key; and of course, be sure you attach whatever you indicate you are going to attach (that's one of my big problems; I count it as a senior moment).

There is also the matter of signature lines on email messages. I see no reason why anyone in business would not want to have an automatic signature reference that would include name, title, company, address, phone number, website, and even a promotional message. A very high percentage of what I receive do not reflect this information. Very often the message is just signed off with a first name, and from people, I do not know and have never heard from them previously. And, even though you do have a signature line, it is still rather polite to sign off each message.

I've also got a big problem about we handle voicemail, but we'll tackle that one in the next issue.

VOICEMAIL PROCEDURES REFLECT SALES PROFESSIONALISM

March 2004

You can really tell a lot about a salesperson's ability in relationship building just by listening to messages left by the person on voicemail. For example, when a salesperson is not available to answer the phone, the message left for a caller needs to be very clear about providing some information. And, we do not mean a commercial for the property. The message needs to reflect where one can be reached, if necessary, and when there will be a return to the office; of course, the caller can also leave a message upon a prompt. Sometimes, the call may be forwarded to another party for a response by providing instructions. What we are noticing is that more and more salespeople are leaving voicemail messages that really don't say anything that could be helpful. What we hear is something like: "I'm either out of the office or on the phone . . ." This is not at all helpful to the caller since it is rather obvious that no one is there to answer the phone. A caller needs to know how to reach you, if it is important and when you will be back to the office. Most phone systems allow you to program at least two automatic responses. One for when you are on the phone and another call comes in. In this case, you can have an automatic, standard voice message that tells the caller you are on the phone, for them to leave a message and you will call them right back. The other program should be one that is changed frequently, at least once a day. On this one, your greeting could mention the date and day of the week, and where you can be reached (if you are on a trip or expect to be out of the office for what period of time). You may even expect to be in the office all day, but just out for a little while, could be another message, along with advising the caller that you will return shortly. The whole point is to provide good communication that reflects your ability in excellent relationship building with everyone who calls.

There is something else that a lot of salespeople do not do. And, that is to program the telephone system so that voicemail kicks in automatically when you are not there. Why do we have to sit and wait for a phone to ring four times, or more when someone is not there to answer the phone or the person we are calling in on another call? When you leave the office it is a matter of just hitting a few of the phone keys to program the equipment for voicemail.

Here is another suggestion that will help you be more professional in relationship building: When you leave a phone message for someone you are calling, please be brief and to the point. Indicate why you are calling, and what you want in return. That way, when that person calls you back and you are not available to be on the phone, they can leave a message that responds to your request, and that should end the phone-tag.

LET'S STOP THE WHINING AND GET OUT AND SELL

April 2004

Hospitality Sales and Marketing Association International (HSMAI), just released a new survey of its members, listing the top then issues facing sales and marketing staff in 2004. The survey reflected these in the following order:

1. Current Economic Situation:
2. Increased Competition:
3. Finding New Contacts & Qualified Leads:
4. The Whole Issue of Pricing:
5. Reduced Budgets:
6. Impact of the Internet:
7. Time Poverty:
8. Building Awareness:
9. Making Money:
10. Recruiting and Motivating Salespeople:

To review more detail on the survey, in each of these categories, please go to the organization's website: www.hsmai.org.

Congratulations to HSMAI for doing the study; it is most important for all of us to take a look at what is bothering salespeople in our industry. Bob Gilbert, President and CEO of the organization indicated that we expect to see some of these concerns and issues eased over the next few months as recent reports indicate a steadily improving economy and environment for the hospitality and travel industry and in particular for business travel.

We hear it all around us and all over the place; the economy, fear of travel, cancellation of groups, need new prospects, airline problems, overbuilding, third party internet providers, and on and on. Not only salespeople are complaining about all the reasons why business if "off," and how hard it is to put the business on the books, but so are managers, and owners are well. What's going on here? Have we forgotten all about how to sell, where to look for the best business, and how to maintain rates?

In reviewing these ten major issues it may just look like a lot of whining, and to some extent, it really could be. We are not doing the business we should, so let's find all the reasons for this and decide what we can do to fix it. First of all, the business is out there. Latest figures from the companies that monitor our business have reflected that in 2004 and 2005 we can expect an increase in demand of about 3.5% which is considerably greater than the 1.5% increase in new rooms. So when it comes to supply and demand we are certainly better when improvement in demand is better than supply. Maybe we cannot really complain about overbuilding.

Latest facts reflect that more people are traveling and taking vacations, as reported by AAA and TIA (Travel Industry Association). Occupancy is projected to show an increase in 2004 over 2003. So with all this, what is there to cry about? Need more and better prospects? What we have to do is to take a look at all our old files and we will see that the majority of files have not been traced or worked in a long time. We worry about getting new business, what's wrong with going after old business that has been a prospect for a long period of time, but we just have not gone back to them to develop relationships or ask for the business.

There is a lot more to be said about the other issues reflecting what salespeople can do to improve the selling situation. But, the answer is just doing more selling in the right manner of doing it. General Managers and Owners need to get more involved in the business of sales by providing Attention, Direction, Supervision, Training, and Motivation.

MEETING PLANNERS COMPLAIN LOUDLY ABOUT HOTEL SALESPEOPLE

May 2004

"When meeting planners have a bad experience with a salesperson, they usually don't take their complaint to the top. They're more apt to switch suppliers, or even just to tell their peers about their experience, than to contact the sale person's boss." That is a quote from an article written in the March 15, 2004 issue of Meeting News magazine. The story was written as a result of a survey conducted by that publication with 235 random meeting buyers.

The results of the survey, in general, were not too favorable about how hotel sales personnel conduct their business when it comes to communicating in with prospective buyers. On the top of the list, with over half of the survey respondents, was the complaint on "Receiving Cold Calls." We have covered this several times in early Sales Clinic columns, where we have mentioned that cold calls are a big waste of time for salespeople. It is more productive to call only on qualified prospects, and there are loads of these in old files at all hotels. Cold calls are so very much unproductive, and it gets buyers very much upset by getting those calls. Just about as many meeting planners complained: "Your calls not returned in timely fashion." It is hard to believe that we are not getting back to people right away when they want to do business with us.

The other items on the list of complaints were as follows, indicating the percentage of respondents complaining:

▶ Won't take "no" for an answer (41.1)
▶ Receiving contracts with terms other than what was discussed (36.9)
▶ Act like their needs outweigh yours (29.9)
▶ Proposals that don't address RFP items (28.6)
▶ Having to repeat specs for a meeting because of supplier staff changes (26.6)
▶ Salespeople who refuse to negotiate (25.3)
▶ Constantly having to work with new salespeople for key suppliers (22.4)
▶ Receiving material from suppliers you're not interested in (21.6)
▶ Poor communication between sales staff and CSMs (19.1)
▶ Not empowered to make decisions (18.7)
▶ Lack of creative thinking (14.9)
▶ Don't understand contract clauses (12.4)

Respondents on the survey were able to select up to four responses as indicated above. It seems that nothing much has changed over the years when it comes to meeting planner complaints. In discussing this with some "old timers" in the meetings industry, they just shake their heads and say: "Nothing's changed."

All this has to tell us something. Two big things come to mind. First, salespeople need improved training opportunities, either at the local level or through their management companies. Secondly, salespeople are not getting direction and supervision from their management team. At so many properties salespeople make cold calls because they are told that they have to do a certain amount of calls a week. We can be so much more productive if we spent our time working with, and developing relationships with only qualified prospects, which are those meeting planners who are in a position to use our facilities if they choose to do so. Further, we've got to be great communicators if we want to improve profitable sales.

However, here's a good note. In spite of all this, when asked how the level of professionalism among industry salespeople today compared to that of five years ago, almost half of those surveyed, 49% said sales operations are more professional today. Only 26% said salespeople are less professional, with the remaining respondents citing no difference.

IMPROVE SALES WITH BETTER VOICEMAIL TECHNIQUES

July 2005

We would all agree that first impressions are important in the business of selling. You never get a second chance to make a first impression. When someone calls a property sales office, for the most part, there is some level of doing business with the sales department, whether they call directly or through the hotel operator. It then seems logical that we surely would want to make a good impression in the way we answer the phone.

There is probably no problem when someone is in the sales office to answer the phone with a nice, cheerful, welcoming voice. The problem exists when we are not available and we use our voicemail system to leave a message for a caller.

So, here's the problem; too many salespeople leave the same old trite message which tells the caller nothing new. Most of the time the message is: "I'm either on the phone or out of the office." No kidding? Really? We are telling the caller that we are not there which they already know. People want to know where you can be reached, not where you are not. This type of response, along with "Your call is important to me." Wow! Isn't that informative? Or "I'm sorry I missed your call." Well, of course, you are! What we need to do is start recording personal messages whenever we are not available to answer the phone. We need to tell a caller where we can be reached, or when we will return to answer the call. Of course, we record all of this with a cheery, friendly voice. Or even give the caller an opportunity to speak with someone else that may be taking your calls, and providing instructions for transferring the call by dialing the operator. Most voice messaging equipment provides an opportunity for a message to indicate that you are on the phone, and this automatically drops in when your phone is actually busy. So if you are in the office and on the phone the caller will automatically be aware. If you are not in the office, but out for coffee, or lunch or a meeting, then you should be leaving a message to so indicate and advise when you will get back to return the call.

At some sales offices, we use the automated attendant on voicemail when we are not available. That is that automatic electronic voice that says something to the effect as: "the party you are now calling (and you have inserted your name here) is out of the office right now . . . etc., etc." According to Nancy Friedman, the Telephone Doctor, who does telephone training for companies all over the world: ". . . The first voice you hear when you call a company sets the mood, sets the tone for all future interactions. Then why on earth would you leave a robotic, monotone, dull voice to greet your callers?"

Another thing that Nancy reminds us of is the outdated greeting on voicemail. In one of her company's surveys this was high on the list of voicemail "no, no's." Greetings need to be updated every day, and it really isn't necessary to provide a date and time. All too often we slip up and do not record each day, making your previous message outdated when you provide a date.

50 LITTLE TIPS THAT MAKE A BIG DIFFERENCE

September 2005

There is a terrific little book, just published, with a wonderful collection of tips that are truly most worthwhile for people involved in the sales effort. Actually, the tips are worthy of just about everyone in hospitality whether it be sales, front office, food/beverage, conference services, etc. It was written by a lady who is dedicated in getting all of us to do a better job in communications, particularly with the telephone; she is Nancy Friedman, exceptionally well-known all over the world as "The Telephone Doctor."

Following is a random sampling of these "50 Little Tips That Make A Difference" with some little editing for this column:

▶ When getting on the phone to do some prospecting, or follow-up, ask the person you are calling if they have the time to talk before you start rushing into what you want to say. Without this permission, you are barging in and may very well be an interruption.

▶ How about putting a nice little hand-written note on letters going out to prospects or customers? Just a "Hi" or "Thanks" after your signature makes it more personal, and it's a little thing that could very well make a big difference.

▶ When leaving your phone number on voicemail be sure to speak very clearly and leave the number twice. Rushing through the number makes the person rewind or play back the message twice. (Frankly, this is a very big pet peeve of mine, personally; I have a tough time with this, quite often trying to understand number when the voice message is rushed)

▶ Give full attention to the person with whom you are speaking, on the phone or in person. You cannot do two things well at once. Make eye contact if you are face-to-face or ear contact if you are on the phone. Next time you attend a trade show watch some salespeople making contact and conversation with a passer-by and watch their eyes roaming the room while they are speaking or listening; not a good idea!

▶ On email: Respond to most non-spam messages unless the message clearly

closes the exchange. A few words would suffice. It assures the message was received.
▶ Screening telephone calls are at its best intimidating; at its worst, humiliating. Why screen call at all? This really bugs me when I am calling someone, and a secretary or receptionist wants to know the reason for the call. Why would anyone in sales want to have their calls screened? What's the big deal about having to know the reason someone is calling? Just accept all calls and talk with everyone.
▶ On your voicemail greeting tell the callers where you are, not where you are not. "Hi, I'm not here right now." Duh ... just say something of value like when you will be back or where you can be reached.
▶ Show appreciation constantly. Even for something, you take for granted. Particularly in the area of sales when there are so many other employees at the property that are helpful to salespeople most of the time; show appreciation to them.

APEX TIME-SAVING REPORTS HELPS YOUR SALES EFFORT

June 2006

Finally, after a number of years working on APEX (Accepted Practices Exchange), the Convention Industry Council (CIC) has come up with a user-friendly package of tools designed specifically for meeting planners, and suppliers. The idea is to assist the meetings industry with easy to use, standardized forms and systems that would improve communications between buyer and seller. These ready-made tools can be customized for individual events and include the following:

The APEX Industry Glossary has over 3900 industry-approved terms, acronyms, and abbreviations, and their respective definitions.

The APEX Event Specifications Guide which is designed to help meeting planners and suppliers prepare detailed instructions for events more quickly and efficiently.

The APEX Housing and Registration Practices Guide, which includes event registration, housing, and rooming lists forms and templates.

These tools are now available in CIC's APEX toolbox, which also includes completed samples of all hard-copy APEX templates that can be used for reference, as well as all checklists from the CIC Manual. However, there is a lot more going on with APEX. When the project was started there were seven initiatives with which volunteer committees from the meetings and hospitality industry became involved.

The following is an update of the ongoing work of the committees:

▶ Terminology—Completed, but ongoing with revisions as necessary
▶ History/Post Event Report—Completed
▶ Resumes & Work Orders—Completed
▶ Housing and Registration—Completed

An asset for hospitality sales staff is some new work that CIC has in development, through APEX, in two key areas of accepted practices. The first one is called

"Action," which would be an accepted practice addressing the "how" of doing business and exchanging information ... for example "The most recent Post-Event Report for an event should accompany any request for proposal (RFP) sent to solicit proposals for future occurrences of that event."

The other is on "Information"—accepted practices outline specific information that should be shared, transmitted, stored, etc., about events. This would be information from the Post Event Report such as Published Event Start Date, Event End Date, Event Type, Event Frequency, Original Expected Attendance, Total Pre-Registered Attendance, Total On-Site Registrations, No-Shows, Number of Exhibitors Attending, Actual Attendance.

These two new initiatives in the APEX program certainly would be a big help to hotels sales staffers in assisting them in the creation of proposals in response to inquiries.

Through APEX, the CIC will be able to yield a wide variety of tools that could be used throughout the meetings/hospitality industry. Although the work is still going on in connection with the seven initiatives, CIC is making possible the use of the tools already available in the APEX Meetings & Event Toolbox, a computer software package, which includes the three areas of Terminology, Meeting Specifications, and Registration/Housing. This is a user-friendly toolset designed to interface with the Microsoft Office Suite of software, specifically Word, Excel, and PowerPoint. The package is being sold by CIC for $99.95. The toolbox has the ability to create an Adobe-compatible PDF file, so that users can distribute secure files via email or post them to the website.

All of the already CIC approved practices are available for free download from their website (www.conventionindustry.org).

The Convention Industry Council is an umbrella organization made up of 32 meeting and hospitality related national organizations.

HOW ABOUT USING YOUR BUSINESS CARD AS A SALES TOOL?

November 2006

One of the most underutilized and misdirected sales tools in the hospitality industry is the business calling card. Every salesperson, as well as managers on a property, has this low-cost, easy to carry, easy to distribute sales piece readily available. However, in many cases it is not used as a sales tool, or—not even used at all.

The business calling card should always be on hand (keep them in your pockets or purse, and forget about those fancy little meal card holders which are difficult to open, and only hold about 4 or 5 cards). Cards should be distributed at every opportunity with the ultimate objective being to develop more business. Thank of it as a sales tool, like an advertising piece, a brochure, or a reminder letter. Use it as an introduction to people you have just met or as a reminder to people to whom you are already known. Hand it out with little notes written on the card, such as a "thank you" note, or "great seeing you again" note, etc. Use it in all solicitations, direct mail programs, payment of bills, as well as account receivables. Yes, the property controller should also have a card. The reservation staff folks should also have cards to send along with confirmations.

How about furnishing business cards to all the folks who work the front desk? What's wrong with providing a card to guests when they check in? The same goes for restaurant guests. When they pay the guest check the server leaves a business card with a "thank you" note. Property managers should make a habit of giving out cards as they greet guests all over the property. Guests love getting a card given to them by management or staff. It makes them feel important, and it shows that property staff does care.

Primarily the card should be used to identify the person presenting it; therefore, the name needs to be very prominent. It needs to stand out more than the name of the property. A title on the card helps to identify the job function, and the property name, of course, needs to be included, along with a logo. There should be street address, phone number, as well as an email address, and maybe even a cell phone number for sales personnel.

We need to think about providing business calling cards for just about every staff member at a property. It does a lot to make the employees feel good about themselves, the property and the brand or company name. Everyone sells in our business, so everyone should have a calling card.

Some very creative people have developed business cards that look more like an advertisement, and there is nothing wrong with that. The more creative you are the more attention the card will draw. Some people have put colored photographs of themselves on the card, and some have photos of the property. Whatever the format or copy, the idea is to use them, a lot of them. Distribution of cards should be a habit, not an afterthought. Anytime a meeting is attended, or any gathering, almost everyone in attendance should get a card.

Set a goal of how many cards will be distributed at a property within a year. Try for 2000 a year which would be 500 cards each quarter. Just think of all that low-cost advertising.

MOST HOSPITALITY SALES LETTERS STILL NEED A LOT OF WORK

February 2007

Have you taken a real good look lately at some of the letters coming out of the sales office going to customers or prospects? Here is a good idea: pull copies of all the form letters stored in the word processing equipment. These are probably letters that are used to send a proposal, or say "thanks," or to confirm some booked business, or to solicit business, or maybe just to network. These letters should be reviewed, just pulled apart, word by word, to see if this is a letter that truly represents the property or the person writing it. So many of these letters are being used over and over again, and were probably composed by a former sales staffer who has been gone for quite some time. Does the computer-driven form letter, once read, sound like it is really coming from the person who signed it? We have gotten in the habit, over a period of time since word processing came into our lives, to do so much with computers that we may have lost the personal touch in doing things, like writing letters. When was the last time anyone in sales actually wrote a completely fresh letter?

Why don't letters sound like they are coming from the person writing the letter? Somehow or other, we must have been taught that in writing letters we must impress the reader with our intelligence. What's wrong with writing letters the way we talk? The people reading the information should get a picture of the writer because that is who it should "sound." There are many phrases used in those kinds of "standard" letters that people just don't generally use when speaking with someone.

For example, people just don't say things like: "I am handling this to you, herewith." However, in a letter, we see it all the time" "Enclosed herewith please find . . ." Also, from time to time there is the written phrase: "Pursuant to our recent discussion . . ." People just do not talk like that. How about "please feel free to call me?" Then there is the all too frequent familiar first paragraph of a letter stating: "Pleased be advised . . . "

Now let us take a look at a typical structure for a letter. We have the salutation, which is the usual greeting: "Dear _____;" of course, we should always use the last names unless the relationships have been developed on a first name basis. The first paragraph should say something that gets the readers' attention. Try to be creative in this area. What is it that can be said to make the reader continue? It should be brief, not more than two to three lines. The next paragraph contains the information, the meat of the letter, but get to the point quickly. Beating around the bush makes for a boring letter. The last paragraph is the call-to-action section. This is where the reader is asked to do something that will create some action on their part, or something that the writer will do to keep the communication going.

Keep the letter to one page, but include other information as an enclosure (but don't say: "enclosed herewith").

HANDLING THE ISSUE OF SALESPEOPLE NOT RESPONDING TO CALL AND EMAILS IN A TIMELY MANNER

July 2007

It seems that the issue of sales folks not responding to meeting planners' emails and phone calls have more to do with the RFPs. In a recent report by Meeting News, almost half of all corporate and association planners surveyed indicated that this was a major problem. In discussions with some hotel sales staffs, it is apparent that the big problem may very well have to do with the quantity of Request For Proposals (RFPs) coming into hotel properties on a daily basis. More and more corporate and association planners, as well as many independent planners, have been using electronic RPP in seeking information as well as proposals from properties. Handling the normal flow of business in sales offices, as well as sales folks being out on the road developing business leads may very well be the cause in the slowdown of responding to those RFPs.

In response to this issue, some hotel operations have geared up their system to be able to improve timely responses to electronic and phone call leads. Hilton Hotels Corporation updated its standards relative to responding to leads last year as a result of customers' increasing feedback on the need for a quick response, according to Lisa Bertomeu, Southeast Regional Director of Sales & Marketing for Hilton Hotels. All their properties are held accountable for responding to queries within 4 hours of receipt. Systems are in place to identify those hotels who are not meeting this standard so that they can follow-up and get the properties the tools they need to meet their customers' expectations. According to Ms. Bertomeu, they are evaluating tools that will allow their Sales and Catering Managers to respond with professional, thorough proposals that answer all the questions outlined in a customer's RFP. It looks like Hilton Hotel properties are complying very well with the new standard for fast response.

A new online quote device is already in the works with Cendyn's MAD-Marketing's MAD-Quick Quote which claims to have solved the hotel industry's major online RFP issue, by providing an immediate answer to a meeting planner's first question: "Is space available, and what is the cost?" Within seconds, the MADQuick Quote presents a planner with an instant estimated cost of their meeting, breaking down each day, including the cost of rooms, food, and beverage, audiovisual, recreation and service charges, from any hotel that is online with this service. The Jackson Lake Lodge, a Rock Resort, was one of the first properties to install the system and recognizes the benefits of presenting an instant answer to an online RFP. Cinda Culton, Director of Sales for the property said that "... it has been used by numerous visitors to our site with wonderful results for them. The system has already paid for itself."

MAD-Marketing is a division of Cendyn (www.cendyn.com), a full-service interactive marketing firm established in 1996, which is a provider of e-solutions for hotels worldwide, in the area of software services that will expedite and enhance the group venue search and booking process.

HOW MANY ELECTRONIC RFPS HAVE YOU BEEN RECEIVING WEEKLY?

September 2007

Yes, I would like to get a response to this question from our readership. Just how many do you receive weekly from the variety of meeting planners who work for corporations, associations, SMERF groups, government agencies, any organizations as well as from Independent Meeting Planners? In recent columns, we have tried to address the issue of meeting planners via a survey conducted by Meeting News complaining about hotel sales staff not responding in queries in a timely manner. There seems to be a big issue about a large number of electronic RFPs coming into properties with meeting space, and following a response, no other word comes back the prospective buyer who sent the request in the first place.

Please send me an email to howardf@vt.edu with the information about the total amount of these requests you get in a week. If you have the time, and the information, and can advise me of the breakdown by market segment, it would be helpful, but not necessary. I do not want to create another load of work for you. This will help me with a follow-up in response to prospective buyers about the workload involved in responding to them. People may not realize the research that needs to be done with some of the request received in order to make a response. Of course, all this does create an additional burden on the sales staff.

Several companies seem to be working on software programs to help curtail the time it takes for sales professionals to respond to electronic RFP's. One of the new solutions is The Direct Book from Newmarket International which allows meeting planners to actually book online an entire meeting in two to three minutes compared to the usual 24–48 hour turnaround time it takes to manually schedule a meeting. Hotels and conference centers that join New market's Direct Book program would have a link from their website directly to the program. From there it allows the inquiry to get access to a consolidated database that will bring up all inventories available to them, based on their user password. Properties that participate in the program load information, and have total control over which function rooms, menus, and resources it offers. This allows the hotel to have control over the booking process. However, the program is set up to handle small day meetings only. A hotel that might benefit would be a property like the 147 room, The Inn at Virginia Tech and Skelton Conference Center in Blacksburg, which has 10 conference rooms, and a 700-seat ballroom. The on-campus property, managed by Hilton, already runs New market's Delphi System. If The Inn creates a basic package including a meeting room, lunch, breakfast, and a morning and afternoon snack, it would satisfy an estimated 90% of a planner's requirements, and a la carte items can be added if desired. There is also space on the screen for a planner to enter special requirements.

DON'T CUT RATES EVEN AT THE EXPENSE OF OCCUPANCY

October 2007

Most property sales & marketing staffers may be getting ready to prepare a marketing plan for 2008, and with that are checking out the local marketing environment. In many areas of the country, we are seeing some growth of new rooms, and even some new brands. All this means more competition in someone's backyard. Sometimes, once a new competitor comes on the market, and opens the property with some very competitive rates, many owners and operators start getting nervous. Maybe they will be thinking about cutting rates to keep up with the competitor and not lose business to the new property, or some other property in the area that may have already cut rates.

Reading what is being projected in some of the hospitality trade journals, we are being told that because of new construction picking up that we are to expect somewhat of a drop in occupancy around the country. This may even happen with this last quarter, and then followed into 2008. It will all probably be due to a switchover in supply and demand. For the last few years, the industry has done exceptionally well in profitability. With a considerable increase in demand for business overgrowth of supply of new rooms, hoteliers have been able to continually raise rates over the past few years. History tells us when this happens, watch out! It is inevitable that there will always be growth in new product following good years of profitability . . . and that happens in most any business.

Just looking back to the mid-70s, mid-80s, and again in the mid-90s we have experienced just that. Right after a period of great increases in occupancy, and high profitability, we see a growth in new rooms. It seems that right after a couple of great years we see the franchise brand salespeople, investors, landowners, lenders, developers, management companies, and various entrepreneurs coming out of the woodwork, and new properties start coming online. Of course, we are seeing that right now. It just all means brand new competition for most everyone in the business.

One of the approaches taken in earlier years by many owners and managers was to cut rates to try to keep the business they had, and not lose their good customers to competitors who are charging less. They have all regretted that move. History tells us that it doesn't work. Back in the mid-80s, a chart was published in HMM reflecting that a drop in the rate of 10% requires that a property increase its occupancy by about 7% in order to break even in profitability, and that still applies today.

Managers should think about this while planning for next year. It just does not make sense to cut rates to keep or to build a business. It is better to take the drop in occupancy. Another point to consider is how difficult it has been in earlier years to come back up in rate once the rate has been cut. We've seen this right after 9/11 when people have stopped traveling for a while, and many properties got nervous and started to cut rates. Within a rather short period of time, people were traveling again, companies and associations were continuing with their meetings. It sure wasn't an easy thing to get rates back up.

BOOSTING SALES ACTIVITY IN TROUBLED TIMES

August 2009

Many areas of the country have been hit with some over-development of lodging properties, last year, and to some extent early this year. So what we see now, in this recent economic cycle is not only an additional supply of rooms, but troubling economic times. Occupancy is down, RevPAR is down, as well as profit, in most cases. While we are waiting for the turn-around, which may very well already be on its way, we need to boost sales activity. All we need is a positive attitude about doing business in any distressed situation, and it should result in some productive sales activity. Just following some of the old basics of sales will result in some solid business.

Some hoteliers, in the interest of showing good performance, may start by making cuts in the service, as well as sales and marketing efforts. When the answer really lies in improving sales performance. The idea is to start getting a better share of the available business in the market area. It may even mean an increase in the sales budget, adding some or more salespeople.

The 10% formula is an old, tried and true, basic system, that has been tested, and it really works. The idea is that 10% of selected sales activity will result in a definite business almost immediately. Here is how it works:

▶ Send out at least 30 proposals to prospects with whom you are already working, and who have shown interest in your property. Probably more than 10% will result in bookings.
▶ Many times companies that are already supplying overnight room business are over-looked as meeting business prospects. Telephone or visit 30 of your current room business accounts; find out who makes decisions on meeting locations, and sell them on using your meeting facilities. At least 10% of the calls will result in bookings.
▶ Take two full days devoted to examining and following up on 30 or more, old existing files. Select files that have not been traced or have not come up on the trace system in a couple of years. Make contact and inquire with the old Who, What, When, Where, Why and How. Just remember: "Once a Prospect, Always a Prospect." At least 10% of those followed-up should result in bookings.
▶ Using the newspaper, or other sources for engagement announcements, call 30 of those a week. Ask the appropriate questions, and arrange visits to your property, in preparations for their wedding receptions, other related parties, as well as rooms for guests attending from out of town. At least 10% of the contacts made will result in some business.
▶ Any other pro-active activity, as reflected above, should work on the 10 formula. Of course, whatever is done, needs to be done well. It is the positive attitude that makes it work.

Get the salespeople to take some time away from the computer, and their social networking to do some old-fashioned sales type of activities, and just watch this 10% formula work.

CUTTING RATES MAY LOWER PERCEPTION OF VALUE, JUST SELL BETTER

June 2010

Over the past couple of years, we've heard so much about discounting rates because "the other guy" was doing it. Many sales staffers say that price is the key to our guests' decision making. Of course, that could be true in some cases, but not in all.

Many prospective buyers may want to give us the impression that price is the only critical thing with which they are concerned, but isn't that just a way to get a lower price—even if they are really satisfied with the price being quoted? Trying to buy cheaper is a natural thing to do.

In many opportunities to work with a person who plans group meetings, conferences or events, we find that most figure that "it wouldn't hurt" to negotiate a little further, to get a better rate. In most cases, they find out that it does work, and the hotel salesperson does arrange for something "off" the rate quoted. Could the salesperson have held out for the original rate quoted by going a little further by showing the advantages and value of having the event held at the property? The chances are that the prospect would have realized the value and paid the price quoted.

What's the point? Don't give in too readily to people who push for discounted room rates. Just do a better job of selling, and find out the prospect's real needs and how you can meet those needs. Then ask yourself these important questions:

- What is the prospect's position with regard to booking with you?
- Where do you stand vs. the competition from a product and services viewpoint?
- How badly does the prospect need you?
- What do you have that the competition does not?
- How badly do you need to business—badly enough not to make money on the sale?

Find out what other objections the prospect may have with regard to using your property. Think why people want to stay with you in the first place; why have they shown interest? There must have been some thought of benefiting from booking with you. Could it be location, service, reputation, space, quality of the facility? Play up whatever the reason might have been.

To overcome this business of cutting rates (or discounting), we have to change our mindset. We have to be proud of what we have to sell. We have to believe in our product and believe that there is great value for the price we are offering. Then we must do a better job of selling the value.

Someone once said: "There is less to fear from outside competition than from inside inefficiency, discourtesy, and bad service."

WE ARE IN A "SELLER'S" MARKET; IT IS TIME TO START THINKING ABOUT SALES FOR 2016

August 2014

With all the data coming in from a variety of sources that measure U.S. hotel occupation, demand, supply, ADR, etc., it sure does look like we are in for a banner year right now, and to continue in 2015. So, what is it we should expect for 2016, and even beyond? It just never changes in our industry, we are so very cyclical. This happens not only because of the economy (and we just never know what will happen in that area), but mainly to the change in supply in demand. When profits start looking good in our business, we start seeing developers, lenders, owners, brands, consultants, all coming out of the woodwork and talking about: "let's get into the act." And, that is pretty natural. When business is good, and there is profit to be made, the idea is to get into whatever business where that happens; as it is happening right now in our industry. We will continue to see a good picture by the end of this year, and likely in 2015, but be aware of what could happen in 2016, as it happened in 2008. There, most likely will be a surge of new properties coming on board throughout the country.

The 12-month outlook (May 2014—April 2015) by Travel Click, recently released, reflected that for that 12 month period, the committed occupancy will be up 4.9% over last year. ADR is projected to be up 3.4%, and that was based on reservations already on the books, so it could even be greater than that. Here is something very interesting from a sales standpoint: Transient bookings are up 5.2% over the year, and ADR is up 4.8%. Travel Click has indicated that for the second quarter this year transient ADR was up 4.1%, and even though the group segment occupancy was up 4.0%, the ADR for the group was up only 0.1%. Now, that tells us a great deal. What has happened to the group rate? It seems that we really have not recovered enough since we started dropping rates as a result of the 2008 figures. As business falls off, for a variety of reasons, sales staffers and General Managers get nervous and start cutting rates. The problem with cutting group rates is that when we see a competitor cut-rate, we feel we have to do the same. Instead, we should be holding rates and selling Value. We have already proven, over the years, that once we cut group rates it becomes very difficult to get back up to where we were when the business gets together.

So where are we going to be when new builds start coming on board in the hotel business, in your own backyard? This could be in 2016 and even after that. Are we going to have another 2008? What are the sales staffers doing right now to get group business on the books for 2016 and thereafter, at the rate that is needed? Now is the time to get busy going after the various group markets for those future years. Remember we are in a "Seller's Market."

MEASURE SALES STAFF GROUP BUSINESS ROI BY PRODUCTIVITY, NOT SALES ACTIVITY

September 2014

In evaluating the Sales Department at a property, we certainly do want to understand the activity with which the staff associates are involved. The idea, of course, is to make sure that the activity leads to productivity that is measurable. So, with that in mind, we need to look at how sales staffers spend their time. All too often they may very well get involved in doing things, and "helping out" in other departments which take away from their sales productivity time.

When it comes to measuring productivity we need to look at the actual revenue generated for the property by each salesperson. Salespeople do get involved in a good deal of activity which may lead to booked business, but can it be measured in terms of revenue generated? Those involved in sales should be examining their own activities, and measure their own productivity based on performance, keeping in mind that activity should lead to productivity. There are probably many activities performed, but which will lead to measurable business? One of the big areas where sales staffers get involved is the marketing aspect of a property. Granted that Marketing is the business of informing the public of a product, but sales are delivering the product for a price. Of course, Direct Sales is a function of Marketing, but so many marketing activities are difficult to measure in ROI of booked business. Advertising, Public Relations, Promotion, Social Media, are all important marketing tools, but it is Direct Sales that put the business on the books. Sales Associates just need to spend more time on their direct sales activities, as in following up on qualified prospects.

Here are some sure-fire ideas to improve measurable business:

▶ Follow-up on Prospects (make the phone call). Once a Prospect, always a prospect, so don't give up.
▶ Make sure traces are accomplished on a daily basis (via Delphi or other software programs for Sales Accounts).
▶ Find prospects via networking with business organizations (remember that prospects are those who are in a position to do business with you, should they chose to do so).
▶ When exhibiting at trade shows find real prospects. (See our HM Sales Clinic column on "Working a Trade Show").
▶ Search old files which may not have been traced regularly.
▶ Seek qualified prospects reflected in local daily newspapers.
▶ Respond to leads immediately.
▶ Follow-up on RFPs with a phone call (Yes, sometimes we are notified not to call. Call anyway!)

Owners, operators, and General Managers should be able to measure productivity on each sales associate, assigning a specific goal of revenues collected for a period of time. Here we are being specific about revenue collected, not just "booked" business. Some form of incentive compensation works very well in this area.

ARE REQUEST FOR PROPOSALS (RFP) DRIVING YOU CRAZY?

October 2014

So many hotel sales associates are going nuts trying to handle all the RFPs coming into a property every day. Of course, they all do not get reviewed or answered immediately. Many meeting planners have been complaining about their requests not getting responded to (according to Meeting Professionals International). These requests keep coming in, not only directly from meeting planners needing the information to help them make a decision about a meeting location, but also from third-party independent planners, convention bureaus, and a wide variety of software companies trying to assist planners who have their software programs. This does create a deluge of requests daily for hotels, conference centers and resorts to answer. How much time does it take to review all these inquiries, and then to determine if this is business in which the property may be interested?

In most cases, as the RFPs are reviewed, if they do not meet the needs of a property, in all probability they do not get answered. This makes the inquirer very unhappy, and creates a poor image of the property, which may very well reflect very badly in case there is other, future business that may be available. Also of great concern is how much time is spent by sales associates in reviewing and responding to all these Requests For Proposals? So much time is spent in responding to RFPs that are never booked. Is there an opportunity to reduce the amount of time spent by sales folks, by having someone else do the reviews to determine which may very well be the best prospects for a property? This would help greatly so that a salesperson may spend more time in actually following up on a qualified prospect. Further, management should try to determine what percentages of the RFPs coming in represent qualified prospects for a property; it may very well be a pretty low amount.

There are many of these requests, mainly from CVBs, third-party planners, and software companies, indicate not to respond with phone calls. Once reviews of incoming requests are made, the ones which represent the best prospects for the property need to be contacted. The best bet would be to actually go ahead and make the phone call to the meeting planner for an organization or company, not necessarily to the person sending the RFP. Requests coming in directly from the sponsor of a group whether it is a company or an organization should be handled immediately. This should be a priority, even though this particular request may not qualify for the property. We never know when that sponsor may have another opportunity for a group to be placed at that location.

With the limited number of sales staffs located at properties these days, it is important for management to find other staff associates to assist, to some extent, on the reviews of RFPs, and have sales staffers work those which would qualify for the property.

HELP! WHERE TO FIND SALES HELP WHEN YOU NEED IT

February 1981

Everyone, from the most experienced salesman to managers who don't know where to start a sales program to employees who "sell" while performing other functions at the property, calls out for sales help from time to time. Fortunately for one and all, there are quite a few ways to get it.

▶ **Membership in national trade associations:** Hotel Sales Management Assn. International (HSMA), Meeting Planners International (MPI), the American Hotel & Motel Assn. (AH&MA) and many others are very reliable sources of information. All provide their members with a wide variety of material pertaining to sales and related considerations. Most also have local chapters that hold educational meetings on a monthly basis.

There are also a number of local associations affiliated with national organizations that are readily available in smaller cities. These include the Chamber of Commerce, merchants' associations, sales and marketing executive clubs, innkeepers' associations, restaurant associations, and chapter to the American Society for Training and Development.

▶ **Schools:** Many colleges, universities and community colleges conduct hotel and motel sales courses at local branches. Further help is available from extension programs. In addition, groups like the

AH&MA and the HSMA offer correspondence courses.
- **Public libraries:** Never discount this source when it comes to research on almost any topic. Extensive bibliographies listing books on hotel and motel sales, as well as sales in general, may be right there on the shelf. Also, keep in mind that the AH&MA and the HSMA, the National Tour Brokers Association and MPI all maintains resource libraries. The latter offers a sourcebook listing relevant books, pamphlets, and magazines, complete with the publishers' name and address.
- **Conventions and trade shows:** Attendance at meetings of this type often offers workshops or seminars on sales. Among the shows offering this type of help are the International Hotel, Motel, and Restaurant show in New York; the National Restaurant Association convention; and the American Society of Association Executives.
- **Publications:** Magazines like this one often focus on sales problems and how to cope with them. You'll find other tips in trade association publications, like the HSMA WORLD.

Occasionally, an opportunity will present itself in the form of a one-day workshop held locally. This could be an inexpensive way of getting new sales ideas or having nagging questions answered. Readers who are interested in attending such a workshop in their locales should communicate directly to this writer at P.O. Box 14100, Roanoke, VA, 24022.

Chapter 12

Action Plans for Marketing and Sales

Letter Writing Hints for the Small Lodging Operator

May 1982

The key to writing letters, whether they are solicitation letters or confirmation letters, is to keep it simple. Personalized, individually typed solicitation, sales, or confirmation letters will get more attention and responses than a printed form letter. Try to avoid the use of mimeographed or printed form letters with a name typed in at the top. They look horrible, and do not get much attention.

It is most important that conversations pertaining to booking information on facilities or price should be confirmed in writing to the prospect. Too much confusion and bad feelings can result from not recording the details on paper. Of course, keeping it simple is the key; therefore, the letter should not be too long, and neither should the sentences. The letters should be easy to read, easy to look at, and not be crowded. Consequently, it is necessary to leave wide margins on either of the letters.

Language

Don't try to impress your reader with your knowledge of the English language or the use of long words. Don't worship correct grammar. The emphasis should always be on smooth reading. Most of the words should be five letters or less, excluding proper nouns. This helps keep it simple.

The shorter the letter, the better. Just get the points across, and don't try to be cute or write a novel with a lot of adjectives. The sentence structure should be such that it would not exceed 15 to 20 words—anything longer becomes very wordy and confusing.

In a sales letter, the opening paragraph should be an important message or a benefit. The last paragraph should conclude with some direction or action that should be taken by the reader. State the message clearly without overstating your case.

The use of an appropriate P.S. (Postscript) could be very helpful, provided it contains something important. But don't repeat what has already been said in the opening paragraph.

Probably one of the biggest mistakes that a letter writer can make is to send out a letter addressed with the person's name spelled incorrectly or even the wrong title. Usually, people do not appreciate that. If it's worthwhile sending a letter to a person, it is certainly worth the time to get the name and title correct.

The most important letters that can be written are those confirming a piece of business and responding to an inquiry. Questions from the inquiry must be responded to in writing accurately. The confirmation of a booking, for example, reflects all the details that were agreed upon for the prospect to conduct business at your property. This avoids confusion when the business arrives.

For direct mail programs in connection with the solicitation of business, try to use a flyer effect rather than an individual letter style. The flyer-type mailing will certainly attract more attention, look neater and will probably be more believable. This should only be used if you're considering a large amount of direct mailings.

USING YOUR BUSINESS CALLING CARD AS A SALES TOOL

November 1982

One of the most under-used and misdirected sales tools in the hospitality industry is the business calling card. Every lodging operator, manager, and salesperson have this low-cost, easy-to-carry, easy to-distribute sales piece readily available. But, in many cases, it is not used as a sales tool or not even used at all.

The business calling card should be always on hand and distributed at every chance with the ultimate objective being to develop more business. Think of it as a sales tool—almost like an advertising piece, a brochure, or a reminder letter. Use it as an introduction to people you have just met or as a reminder to people to whom you are already known. Hand it out with little notes written on the card, such as a "thank you" note. Use it in all correspondence (direct mail programs, payment of bills, as well as account receivable mailings, etc.). If you want to give a complimentary room, meal or drink, use the calling card with an appropriate message.

PASS OUT CARDS

Lodging operators circulating among the guests should pass out the cards when greeting people. Check the dining room during meal hours-meet the customers and hand out your card. Pass out cards when checking guests out in the morning. Guests' love it because it makes it makes them feel important—and it's a good reminder for them to return.

What information should be on the card? Primarily, the card is used to identify the person presenting it; therefore the name should be prominent. A title needs to be included to identify the job function, as does the property or company being represented along with the logo, if there is one. A street address is necessary, and a telephone number is a must. If the lodging operator, manager or salesperson really wants to do his or her job thoroughly and enthusiastically, then a home telephone number should be listed.

"Always avail able" means you are really serious about doing business and servicing your guests.

Dieter Schulz, Vice President and General Manager of Holiday Inn Conference Resort in Decatur, IL, has a fold-over card which gives the pertinent information on the front. When the card is opened, there's a listing of features including a number of rooms and suites, capacities of meeting rooms, and a listing of lounges and restaurants. This makes for good selling.

Distribution of business calling cards should be a habit, not an afterthought. Lodging operators should set a personal goal of distributing at least 2,500 a year. That's only about seven per day. Just think of all that inexpensive advertising.

SUCCESS AT TRADE SHOWS STARTS WITH MARKETING STRATEGY

March 1985

Success at trade shows and conferences depend on the marketing strategy you develop to sell your property and the tactics you use to turn those leads into sales calls and, eventually, bookings.

The astute salesperson will not only increase the number of sales calls at shows and conferences but also expand his prospects and personal growth over the long term.

From the standpoint of productivity, more sales calls can be made at a meeting, conference or trade show in two days than in the same amount of time on the street knocking on doors.

Selecting the show or conference to attend is easy to do once you have targeted the markets you wish to reach and have completed a marketing plan.

Which groups do you want to reach? There are gatherings for association executives, meeting planners, corporate travel managers, travel agents, wholesalers, incentive travel buyers, and planners, training directors, religious conference managers, insurance conference planners and group travel producers.

Besides being able to attend a trade show, where contact may be made by several hundred potential buyers, there is also the chance to develop yourself, to experience new ways to conduct business and to start long-term client relationships.

Your performance at a trade show or conference is measured by how much business you are able to do at the event. You can count the number of qualified leads and actual bookings. It turns into a numbers game: the more contacts that are made, the more leads that can be qualified. Establish goals for yourself. "I plan to make 80 new prospect contacts the first day." Plan to leave the show with a specified number of qualified leads to follow-up.

Making productive use of your time may mean that you will have to stay out of the "comfort zone." You may not be able to mix only with people you know because the strategy may not be productive. The object is to make new contacts that can provide business for your property. Too often salespeople spend too much time mixing with other salespeople from other properties.

When you select a table for group meals, look for tables occupied by potential customers. During coffee breaks keep moving to maximize the time for new contacts.

DON'T WASTE TIME

Spend time with prospects who can buy or are in a position to recommend or influence the buying decision. Too much time is wasted on sales presentations to persons who are not in a position to buy or have no need for a particular hotel or motel.

All it takes is five or 10 minutes and a few key questions to determine if you are talking to the right individual. Ask a contact to explain how meeting sites are selected. Or you could find out how the prospective client recommends hotels to corporate accounts. Or you could ask: "How do you propose I go about getting business for my hotel from your association?"

By working diligently, using only five to 10 minutes per contact, a good salesperson could generate six to eight "cold calls" per hour or 40 new contacts in a day. Even if only 25 qualify as good leads you are still doing well.

Using attendance listings, target those people who really need to be contacted, and, if they are not seen readily during the day, find out where they are staying and contact them there. In order to follow-up on a prospect be sure to obtain the individual's name, title, company or association, address and phone number. For best results, get a business card. It saves time and provides correct spellings. But besides these vital statistics you should also find out:

- ▶ Who else is involved in decision making?
- ▶ What does the contact need?
- ▶ What feature of the property particularly appealed to the prospect?
- ▶ What dates are being considered?

The more information you get, the easier it will be to follow-up. Then, don't forget to follow-up.

TRADE DIRECTORIES OFFER VALUABLE ADVERTISING EXPOSURE

October 1985

Hotel and motel operators are really missing the boat many times over by not being certain to take advantage of free listings that are offered by trade directories.

Generally, information forms are sent to the properties-some with an introductory letter and some without. Of course, the most logical approach would be for the General Manager of the property to personally see to it the requested information is supplied to the trade directory. It should then be put back into the mail right away.

However, what usually will happen instead is the information form gets passed on to the sales department in order to be filled out. Depending on how busy the salesperson might be—or how short—staffed the department is—will usually determine when the form is finally completed. There are many of these information forms that appear to be too complicated to they are stuffed somewhere never to take the time and energy to answer, so found again. All this results in either the forms getting mailed back to the publisher of the directory too late for inclusion or never being completed at all.

When it comes to paid advertising in directories, the problem usually is that not enough information is provided by the property in order for the advertisement to generate reservations. One of the most popular and important trade directories is the Hotel & Travel Index (HTI). This is probably the directory that is most widely used by travel agents. It is also the world's largest hotel directory and is published four times each year.

National representatives of HTI a real ways working with hotels and motels to help in the design of the advertisement (if requested) and offering advice regarding what information should be included.

ADS GENERATE RESERVATIONS

The objective of advertising in such a directory is to generate reservations. Too often advertisers come up with good, attractive graphics but fail to include basic information that travel agents will need to know in order to properly serve their clients.

The idea, of course, is for the lodging property to include as much information as possible so travel agents will not have to waste a telephone call in order to ask questions about the property that could have been answered through the advertisement.

Hotel Travel Index recommends the following information be included in advertisements:

- Good photographs that clearly show what the property looks like;
- Address and location of the property with regard to distance from downtown, airport or major attractions;
- Specific room rates by dates the rates are in effect;
- Special package information with rates;
- Meeting and convention facilities, restaurants and lounges with hours of operation;
- Percentage of taxes that apply to rates;
- Listings of nearby facilities such as hospitals, ballparks, civic centers, other attractions, and major company headquarters.

Other descriptive information about the property that would further describe its features should also be a part of the advertisement. That includes attractions such as a health club, an exercise room, indoor/outdoor swimming pool, secretarial services, gift shops, barber/beauty salon; listing of credit cards accepted; travel agent commission policy; and how to make reservations—listing an 800 number or local phone number.

If the property is affiliated with a national chain, then be certain to include the airline access code of the chain.

GIVEAWAYS AREN'T NEEDED FOR EFFECTIVE HOTEL MARKETING

September 1986

When it comes to giving things away, the hotel industry ranks as one of the leaders in the field.

We give away rooms, meals, liquor, flowers, candy, fruit baskets, turn-down service, newspapers, meeting space, transportation. Who knows what we'll be giving away next? We have a good deal more to give away than do car-rental companies, airlines and ground operators and we give it on a daily basis. We give to meeting planners, association executives, travel agents, motor coach operators, wholesalers, corporate travel managers, airline personnel as well as to an assortment of relatives, friends, and friends of friends.

Now, a normal extension of our sales efforts is to provide certain complimentary amenities when we are entertaining people who may be in a position to give us business. And we extend similar courtesies to our colleagues within the industry. There's nothing wrong with this.

Nor is there anything wrong with providing a "comp" room complete with a fruit basket, some wine and cheese, and perhaps a dinner invitation when we are dealing with prospects who are on-site inspections or familiarization trips.

But when hotels start offering personal incentives as favors in return for providing business, then a question of ethics arises. Some hotel/motel salespeople must feel that it helps business when they give away individual prizes in return for business. Of course, there are always buyers who really go for these gimmicks and are all too ready, to accept them. They are probably the same ones who continually look for comp rooms (though often, they're not looking to do business) and complain that the fruit basket is too small.

And we've all experienced the so-called professional meeting planner who always seems to need a suite for three or four nights (in season, of course) so he can bring his family along on the trip while he "looks over" your property.

DISAPPROVAL VOICED

Many members of the Meeting Planners International have been very vocal about this questionable method of doing business. Professional meetings managers, travel agents or corporate travel planners consider this a bribe, and are turned off by incentive offers. Whether in the form of gifts or trips, this sort of thing comes off as a poor way to conduct business.

When will we learn that this isn't the way professionals work? To paraphrase the commercial, let's do business in the old-fashioned way—let's earn it. And the way you do that is by *effectively marketing your product*.

If your property meets the needs of the buyer, if you do a good job of selling, if you have confidence in your property and can deliver what the buyer needs at a price that's fair for the value then you'll make plenty of sales. And you won't need free trips, furs or diamonds to do it.

Let's be professional in the way we conduct business. Forget the gimmickry, just sell!

SMALL PROPERTIES HAVE SPECIAL NEEDS FOR MARKETING PLAN

October 1986

There's no question about it—marketing a hotel or motel is costly, especially if it's a small property. In addition to the dollar costs of certain marketing activities, the small property operator must consider time and effort as a cost of doing business.

There are several things to consider in marketing a small operation. Because many small hotels do not have a full-time (or even part-time) salesperson, the main consideration is that the General Manager or owner must undertake the marketing of the property. That task can include market research, advertising, public relations, sales promotion, merchandising and direct sales.

In order to do the job right, there must be some sort of plan. The manager of a small property can produce a relatively simple plan that focuses on the kinds of business (corporate, leisure, meetings, etc.) the hotel is getting, and which kind would be the most profitable.

DEVELOP STRATEGY

The plan should reflect what types of business need to be secured for the property to be more profitable. Once that's determined, the manager can create a strategy by which to develop and promote those specific market segments that are more meaningful from an occupancy and profit standpoint.

Perhaps the best ways to get more business at a small property is for the manager or owner to become involved in community activities, and to make sales calls in person and by phone every day. If a manager can devote enough time to make 15 or 20 personal sales calls and the same number of phone calls every week, it will generate a good amount of additional business. And it is certainly the least expensive method of marketing.

Some small operations may be able to hire a part-time salesperson or secure the assistance of a student from a local college in order to help with its marketing plan.

Making the personal sales calls really does work. And, of course, the more calls you make, the more business you'll develop. It's just a matter of making the effort.

COMMUNITY INVOLVEMENT

When a manager gets involved in community activities, he or she immediately becomes known as the local hotelier. Working with people from the community, a small-property owner or manager can develop additional business just through personal exposure.

Marketing must be considered a major part of a hotel's operation. By providing the best service possible, knowing who the guests are, making them feel at home and providing a quality product, a small hotel or motel will be able to ensure that its guests will return.

But considering that every guest does not repeat all the time, it becomes necessary to create new sales for a property. This is the reason that outside sales-call solicitation, as well as telephone solicitation, is so important to a small-hotel operation.

Unfortunately, the management of many small operations gets so involved in the operation of the property that no time is left for marketing. This is a big mistake, and many small-property owners/operators are guilty of making it.

MARKETING IS CRUCIAL

In planning the operation of a hotel, 20 to 25% of the manager's time could easily be devoted to the marketing efforts discussed above. A hotelier who doesn't make the effort to market his property dooms it to failure (unless the hotel is located in a popular geographical area to which guests flock automatically-and there aren't too many of those properties around).

Today's hospitality business is very competitive. Therefore, a strong marketing effort-especially for a small operation-is crucial for hotel's survival.

THE BEST-LAID HOTEL MARKETING PLANS NEVER GO ASTRAY

May 1988

One of the hardest things to do in a hotel or motel operation is to turn goals into action. Too often, goals are set to accomplish certain objectives, but a reasonable plan to reach those goals is not established.

Property operators may have ambitious plans for improved sales and profits but they fail to devise strategies for successful completion of their mission. Without a workable plan, The property probably will fail to achieve its true potential.

The reason for creating a marketing plan is just that: to plan. All too often, a marketing plan is created and approved, but then sits on a shelf gathering dust until someone decides to look at it to "see where we are." Sometimes, it doesn't leave the shelf until it's time to prepare the next year's plan.

In order for a marketing plan to work, certain steps must be taken. Here are the most important ones:

▶ Those responsible for making the marketing plan work should be involved in creating it. And because everyone on the staff will be involved in making it work, they should all have some input when the plan is being developed. Of course, owners and managers will have final approval on the plan's goals.

▶ Goals should be attainable. A plan made up of unrealistic goals will make people want to give up without even trying. The other side of the coin is that if the goals are too easy, it will defeat the purpose of the plan.

▶ Goals should be specific, and there should be at least one goal for each department. Also, time periods should be established within which each department will be expected to reach its goals.

▶ Setting a goal without devising a plan to reach it won't work. Specific goals must have specific methods for reaching them.

▶ Action plans for each department are needed. This means that on a month-to month basis, each department has the responsibility of mapping out its plans for the 12-month period. These action plans should include the date a particular action is to begin, the date it is expected to be completed, and the person to whom the job is assigned. As with the overall marketing plan, all members of each department should have a hand in creating the action plan.

▶ Once the action plans are reviewed and approved by General Managers and owners, a follow-up system should be implemented. Each month, someone should be assigned to review what has been accomplished in relation to action plans. If an individual or departmental goals are not being met, perhaps a change in strategy is necessary. This may mean revising a department's operations so that it is better able to meet its goals.

▶ Goals should be measurable. While goals concerned with sales or profits are easy to track, measuring success in other areas such as service or quality standards can be difficult. However, it must be done. Improving customer satisfaction definitely needs to be an objective, and the extent to which goals are met in customer-satisfaction areas can be measured. Guest comment cards, letters and comment books are effective tools for tracking these results.

▶ Employee performance appraisals should be done regularly. The extent to which individual goals are fulfilled should be part of the evaluation. Salary increases can certainly be based on reaching-or exceeding-objectives. Connecting the achievement of goals to salary levels and promotion has a definite influence on employee performance.

The marketing plan—its goals, objectives, and strategies—will only work well if the property's owners and managers support it. The plan must be taken seriously and reviewed frequently.

The most important point is this: Management's job is to create an environment in which all employees want to work toward achieving the property's goals.

MARKETING PLAN MUST BE MORE THAN A COLLECTOR OF DUST

July 1989

If you've looked at your marketing plan within the past two weeks, you do not need to continue reading this column. For the rest of you, do yourself a favor and review your marketing plan.

You should pay particular attention to the section that contains the action calendar. Did the sales department accomplish all it had set out to do at the beginning of the year? Has the plan been updated? Has the General Manager reviewed the plan and discussed its workings with the sales department?

It seems as though we knock heads two or three months before the new business year begins to make sure our staff starts work on the new marketing plan. Then, when complete, it probably sits on someone's desk for a while, waiting to be reviewed. Then, perhaps, it goes to a corporate office for more review. Someone will probably review it partially and ask for adjustments in the average daily rate, occupancy or market mix. Then it goes back for adjustments. If no one asks for it—and in all probability, no one will—it will get shelved to gather dust for eternity.

One of the greatest tools we have in sales is the marketing plan and action calendar, but we generally treat this helper poorly. The plan needs to be continually reviewed and updated. After all, it was prepared to be used as a guide for the whole year in the interest of improving sales. The action calendar itself belongs in front of the hotel's General

Manager, who should be checking it against the activities of the sales department.

For the most part, the action calendar is really a timetable of various tasks that need to be done by a particular person at a particular time. If the assigned responsibility is not given to a specific individual, it will create confusion about who was to do what by when. In addition to being as signed, the action needs to be given a specified completion date. It is not sufficient to make a list of things that need to be done within a given month. If you do it that way, it just doesn't seem to get done-it usually gets held over until the following month.

Changes Are Required

From time to time and for a variety of reasons, certain activities will not be completed on schedule. Some items may have to be delayed or even eliminated. Due to changes in market conditions, manpower or budget considerations-or even a change in management or ownership there may need to be a deletion of items from the action calendar. A series of scheduled trade shows or attendance at conferences may have to be canceled. Whatever the reason, the calendar must be changed and updated, not ignored.

Other areas of the marketing plan need to be checked. In many cases, the competition analysis is completed in advance of the new year and never looked at again. It stands to reason that competitors, like ourselves, do change rates and features during the year. We need to keep up with what's going on with the competition if we want to compete well.

Reviewing Marketing Mix

The marketing-mix projection is another area that needs to be reviewed periodically. The objective is to create a mix of business that will generate the best revenue and produce the best profit. It may be, for instance, that during the course of the year it becomes advisable to go after more contract business or more group or transient business.

At airport hotels, an airline strike would create havoc for us. In anticipation of something like this; the market mix must be changed. New hotels in the area that would siphon occupancy surely need to be anticipated, and a change in strategy could be necessary. Highway rerouting, economic conditions, strikes, weather, remodeling projects, and other situations may all call for a change in the mix of business, and adjustment in the marketing plan and action calendars must be made.

The marketing plan is something that shouldn't be relegated to a shelf or hidden in a closet. It's a tool for management and a guide not only for the sales department, but for the whole organization.

THE RIGHT APPROACH TO DIRECT MAIL

January 1990

We do a great deal of direct mail in our industry, and most of it, unfortunately, is a waste of money and time. Many salespeople use direct mail as a "cop out"- something to do to show the boss that we are keeping busy trying to increase sales. This doesn't mean that direct mail isn't a good idea, in fact, it is a great idea if implemented properly, followed up and measured for a return on investment. We need to related dollars spent to dollars brought in.

Direct mail (also referred to as direct response) should never be an "afterthought" or the last minute, rushed deal to show we are doing something to stimulate sales. It works best when it is planned well enough in advance; a mission established (purpose and objective of the mailing); goals projected for response; mailing list reviewed, reviewed and reviewed again so that recipients are a specific targeted audience; a plan for immediate fulfillment to respondents; message and copy well thought out, checked and rechecked; project is budgeted completely; the program is included in the marketing plan action calendar; and the entire staff is made aware of the project. Sometimes it is best to seek outside help—using an agency that is experienced in direct response programs for the hospitality industry such as a Bob Stein with Gardner, Stein, and Frank in Chicago.

Effective direct mail program is to generate inquiries or responses when the mailing involves a letter, brochure (or information sheet) and a direct response machine magnesium such as a prepaid postage reply card or envelope. In preparing the copy for the letter the first paragraph must be brief. It should tell the reader why this letter is important to him or her and should make them want to read more. Being cutesy doesn't work well all the time—the reader needs to know "what's in it for me."

The second paragraph should tell more of the story, but should also be brief. Copy needs to be specific in what we don't need to know. The last paragraph is the "Call to action." It tells the reader what to do and how to do it and again, needs to be specific. In most industry-related direct mail the purpose would be to generate leads or to qualify a prospect; therefore, a single page letter should do the trick-all we really want is for the reader to respond, indicating an interest in the property. Another type of direct mail where products are being sold, and the response indicates a purchase (we get a lot this directed to our residents), direct mail gurus says that two and the four-page letter is work better, particularly where sweepstakes are involved. Ours are business letters from which we hope to get responses which will require proper follow-ups to do the selling.

Standard 8-1/2 × 11 letterhead Works well for a lead generation, computer-driven letters are OK, provided they are personalized and produced So that they do not appear to be off the computer. Also, if a computer it is to be used, the letter content must apply to all addresses. We should not be telling everyone to "stop by for lunch someday" when some of the addresses are hundreds of miles away. Information in the letter should be applicable to all persons receiving the mailing. Literally should be signed by only one person that carries is an authoritative title and should be signed in the blue link. The use of a P.S. (Postscript) does work well, sometimes, if it says something worthwhile.

Make it easy for the reader to respond. But in the reader's name, title, organization, and address on the response mechanism usually help get a bigger response. Using addressed labels are acceptable—usually, someone other than addressee opens the envelopes anyway. First class mail isn't that important, so money can be saved by using third class mail. The envelope does not have to be stamped, metered mail washed as well.

What should the response be on direct mail programs? It is hard to see. However, we need to be concerned with how much business can ultimately be booked through leads generated by this type of program.

LOOK FOR ROI ON TRADE SHOW EXHIBITING

July 1991

It is a game of "show and tell," in which we participate when we exhibit at a trade show? Hopefully, it is more than that. It should be an opportunity for maximizing time, effort and expenditures for developing Business prospects. There is a real power in trade shows for drawing people who may be in a position to buy or recommend to buy or call me before your property to others. It is just a question of how well sales executives take advantage of their time spent at these events. Exhibiting at trade shows present great occasions for building an image, exposing a property and for getting prospects As well as selling and yes, even closing deals. The cost of exhibiting, along with Allied expenses for travel, Hotel, meals, entertaining and manpower may even seem prohibited for some. However, when the expenses are pro-rated on the number of exposures and contact it usually works out to reflect that the cost per prospect could be a worse investment.

One area where sales executives can improve the ratio of cost per contact is to make sure there is enough manpower available to cover the show. All too often we see a great exhibit with many attendees, but only one representative at the booth. Considering the cost of additional people working in the exhibit to the additional number of possible contacts to be made, we wonder why sometimes we fall short in this area of manpower.

To help maximize the effort in the trade show exhibiting think about:

▶ Having one representative per 400–500 expected show attendance.
▶ Avoiding food and drinks in the booth.
▶ Asking you questions of visitors to determine prospects.
▶ Not asking "Can I help you?"
▶ Keeping business cards only on those people who may be prospects or could help on networking.
▶ Taking notes (on the back of business cards) in Prospecting.
▶ Getting information as opposed to only giving.
▶ Not working from behind the display table. Get close to the people.
▶ Networking during hours other than exhibit time.
▶ Having an upbeat, and Enthusiastic Approach—all the time.
▶ Being attentive to each prospect—use good eye contact.
▶ And an immediate follow-up to contacts.
▶ Writing "Thank you" notes while on the plane home.

The whole idea in making exhibiting at trade shows worthwhile is to ensure a return on investment (ROI). How do you determine if the total cost of exhibiting (including the time of the salespeople) provided a good return on investment? If you were able to actually close some sales it would be easy. Every once in while this will happen, but most of us aren't that lucky. We usually have to follow-up, follow-up and follow-up. And, if my experience holds true for other people we will find that at least one out of 10 good Prospects will close within six months, if we do our job right.

What is an average piece of a group business worth, in revenue, to your property? Let us say, for example, that an average booking brings in $20,000 (or whatever). If it takes 10 good prospects to bring in one closing, then each of these prospects Leads is worth $2000. If, during one show, you develop 40 good prospects then exhibiting created a return on $80,000 (40 times $2000). If the cost of the attending the show was $4000—you have a 5% cost of sales.

It is important to continually Track your trade shows to see how many good prospects are developed at each. You follow this to see what results in Bookings you get from these prospects to find out if my "One booking per 10 prospects" Works for you. Then you need to find out how much each piece of group business is the worst by averaging out all your group business for a period of the year. From this, you can create your own formula for return on investment. My opinion is anything up to 10% cost of sales (a dollar spent exhibit divided by projected ROI) is well worth the investment.

WHAT'S MARKETING ALL ABOUT?

January 1992

In the 60s, we heard a little about this thing called "marketing." Then in the 70s, we became more aware of it since some of the larger hotels and chains were getting into it. In the BO's we were a lot more aware of its importance and just about all hotels, motels, resorts, conference centers, cruise ships were really into it—to some extent. In the 90s we found out we can't do without it. However, even today there are still so many misconceptions of what it is all about.

Very simply put, marketing is the process of getting and keeping customers. Of course, it's a lot more than just making a sales call. In these tough economic times, with little or no bottom line profits, heavy competition is counting all over the place, we really have to do more than just making sales calls. We have got to do some planning. The answer is in producing a marketing plan for a lodging property, regardless of size, type or location. Large or small; downtown, airport or resort; chain affiliated or independent; we must put in the time and effort to have a plan.

If you have not done one yet for 1992—it is not too late. Having one for 9 or 10 months is better than not having one at all. It could be a simple, shortcut version or an elaborate 150-page document. There are many things to consider in helping to determine how extensive a plan should be. Obviously, there would be a significant difference in the extent of the plan between a 75 room limited service property and a 500 room resort, but some sort of marketing plan you should have.

Too often lodging operators (and some salespeople) cannot make the distinction between marketing and sales. It is just not the same—and to think that your sales team if your marketing team would be a grave error. There are separate roles for each. In marketing secrete the right conditions so that more sales may be made and salespeople do the selling. To "dump" the job of creating a marketing plan for the sales department would not be the best approach. For the sales team to help, is good, but the job itself needs to be accomplished by the chief operating person at a property. Of course, with the help of all the department heads.

With all the discounting going on around us, some operators may even feel that price is the most important thing to prospective customers. A good, well-documented, marketing plan, with proper research could very prove this to be wrong. When you start examining your market segments, occupancies, average rates by periods, you will probably see a way to improve your yield per room by setting goals for higher rates at particular times for particular segments. Then, by preparing specific action plans you can, in fact, start getting better rates.

Once you start working on a marketing plan you will begin to realize that much of the information you need to gather may not be readily available. The information is probably already there on the property, someplace, but has not been collected in a manner to provide you with some good research. But you learn as you go and next year you will have the data. Marketing plans are kind of a hit and miss, at first. But they get easier to do each year and they become more meaningful.

Be sure to include actions plans on things to do year round. Not only for the sales department, but for just about all the department heads. Front Office, Reservations, Dining Room, Catering, Room Service, Bars. Wherever there is a profit center there should be an action plan. The plan outlines specific market segments and strategies for getting business from that particular segment and a budget for each effort should be specified along with goals to be attained.

The job is not done once the plan is completed. It must be used and worked throughout the year. It should be updated when necessary as things change. It's tough the first time around, but it will become so meaningful to your revenue and bottom line.

WHAT WORKS AND WHAT DOESN'T WORK IN ADS FOR MEETINGS

June 1992

Recently, I've done a video of a focus group that made up of meeting planners who were discussing what they like and didn't like about hotel and resort promotions and advertising. Some results surprise me and others did not. Sometime last summer *Meeting News Magazine*, with the help of Dana Communications of Princeton, NJ conducted the focus group with the New York area meeting planners. The magazine wanted to find out what planners, and general throughout the hotel advertising. The group responded to companies, Associations and consulting firms. All were experienced and responsible for multiple Meetings of various sizes, Domestic and international.

The group was presented 20 different advertisements representing for different advertising concepts: Price, location, facilities, and guarantee of a successful meeting. Within each group, 5 ads demonstrated a different approach to executing the four concepts. The group's Reaction to each category of the advertisement was recorded. As a result, 15 specific observations were made by Dana Communications and the following Reflects some of the highlights of its report to MEETING NEWS:

PRIZE ADS ARE EFFECTIVE

There may be certain situations where this is not true; however, in general, price advertising is not related to meeting planners. They feel that it is part of their job to negotiate rates and that they can beat any price listed in advertising.

COMMON SENSE: THE KEY TO IMPROVED PROFITS AND SERVICE

September 1992

When you think about it, isn't common sense the main ingredient to any endeavor? And, isn't common sense the least expensive of all ingredients? Every once in a while just stop what you are doing and look around and you will probably find something going on or someone doing something that just doesn't make sense. Think about some of the things we do to create sales, or how we sell, or how we advertise, or how we do site inspections and work with groups during their meetings. Here's a good example: It would make good common sense for a General Manager or director of sales (GM is better) to take a few minutes to welcome a group of meeting planners or travel agents that are meeting at his or her property. Just recently at three conferences, I attended (two were strictly meeting planners and the third was all travel agents) no one from the hotel welcomed the group or even tried to mix with the participants. Common sense would dictate to me that if I was running a hotel I would want to make some sort of a welcome address and spend a little time rubbing elbows with these types of buyers of my product. In fact, I would create some visibility for my key management staff during the period of the meeting in the hotel.

WOW, what a missed opportunity for so many GMs and Sales Directors who get these types of meetings at their properties.

How about comp rooms? We just seem to volunteer to give them away, without any real reason, when we are trying to book groups. Somehow or other, someone, sometime ago made everyone believe that the standard policy is one comp room for every 50 rooms used during a meeting. Ask any meeting planner that has been around and he or she will tell you that. It would make more sense, from my point of view, that we do not have such a "give-a-way" policy. Comp rooms need to be earned based on profitable sales and should not be offered up front prior to the negotiations. If a prospect needs or wants comp rooms the subject will certainly come up. And, if it doesn't, you may not have to give any comp rooms. It's one of those things that's determined during the negotiation process. The same applies to group rates. We should learn to start out selling at rack rate and then come down during the negotiations, if we have to, provided there are tradeoffs. There is a value to everything. If there is a $100 rack rate on a room—then that is its value. If you start selling the room to a group for $80 the value is reduced to $80 and it is no longer worth $100. However, if you start selling at $100 and come down to $80 during the negotiations, then the prospect knows he or she is getting a $100 value for $80. It just makes common sense.

It doesn't sound like common sense for me to get a letter from a hotel sales representative following my participation at a meeting, thanking me for attending and trying to convince me to use that property. Although I may have been on the registration list of those who attended, I am not really a buyer of hotel rooms. So someone decided to take the registration list and use it as a prospect list. Along with the letter came beautiful brochures which probably cost about $6.00 apiece and the postage alone was $2.94. And I'm not even a prospect. Of course, to top it all off the letter was computerized.

It also makes a great deal of common sense and is a good business practice for all salespeople to have a sales goal. Whether it be in room nights or revenue there should be something to shoot for and be evaluated upon. Everyone has to be accountable for a job function. If selling is your thing, then you have to be judged on how well you do your job. The bottom line is really how much business has you brought into the property and how does it relate to what you are being paid. There has got to be a return on investment on every person employed at a property. You either measure up or you don't. And that's common sense too.

DO WE NEED TO SPEND MORE TIME SELLING?

October 1992

Just about everyone, especially General Managers, would agree that salespeople need to spend more time selling and less time doing other things. So many salespeople are loaded down with other than sales things to do. Some by their own choice, but mostly because of:

▶ Someone else is not doing his or her job.
▶ Having other assigned responsibilities.
▶ Paperwork requirements.
▶ Meetings.
▶ Lack of time management skills.

A 1990, Dartnell Corporation Survey on Sales Force Compensation reflected the following on how salespeople spent their time:

▶ 30% face-to-face selling.
▶ 20% selling over the phone.
▶ 14% administrative tasks.
▶ 13% service calls.
▶ 23% waiting/traveling.

We need to understand that this survey was not conducted by inquiring of hotel salespeople exclusively. In fact, I really doubt that any hospitality sales personnel were included at all. However, the figures look fairly reasonable—if we could just reduce the 23% of waiting and traveling time to about 10% and add the leftover 13% to face-to-face selling. If we were able to spend 60–65% of our time in active selling (in person and by phone} it should please any General Manager or operator.

Whatever amount of time is spent on selling there is usually room for improvement. If we are being paid as a salesperson then we should strive to maximize our time in prospecting, selling and closing sales. If you believe this, and want to improve in this area, here's a suggestion—and it is as basic as you can get. Most time management instructors would tell you to keep a time log for two weeks. Record whatever it is you are doing every fifteen minutes. That means everything, even going to the bathroom, taking a coffee break, etc. Following the two week period (or checking it at the end of a day or a week—which most of us will do just for curiosity sake} review the times spent doing things that need not be done at all or things that should be done by other people. The idea is to eliminate the times, as much as possible, that are not sales intensive.

We probably will find that a good deal of our "wasted time" comes from interruptions. A caution here: "Customers Are Never an Interruption to our Work. They are the Cause of Our Work," according to Jim Cathcart, a great sales trainer and author of Relationship Selling (How To Get And Keep Customers). Look over the log and identify these "interruptions" that could have been avoided; come up with a plan to eat away at these by having someone else take them or somehow get them eliminated by not encouraging them. Sometimes, we, by ourselves create the atmosphere to have these interruptions.

Then, there are all kinds of meetings being called by different people. Meetings are important, if there is something important to discuss. Try to get these meetings scheduled during periods that would not interfere with good selling hours. Then, of course, there are the usual office visits and telephone chit-chats with friends (and not necessarily business calls either). Are you spending too much time trying to figure out on whom to call; or too much time traveling between outside calls? When you are working the phone are you calling a prospect when it is the best time, from his standpoint, to call or when it is best for you? A good look at the log will surely give you some ideas of how to substitute "wasted" time with good selling time.

Of course, the old adage of: "If you are out selling then you wouldn't be in to get those interruptions, etc.," still works today.

ENVIRONMENTALLY CONSCIOUS HOTELS CAN CREATE NEW SALES MARKETS

February 1994

Four years ago the Saunders family, owners of three hotels in Boston, MA, made a strong operational and financial commitment to install a policy to in rate sound environmental management into all standard operating procedures. It started with a simple white paper recycling initiative and today has progressed into a campaign with more than 186 company-wide initiatives at The Boston Park Plaza, Copley Square, and the Lenox hotels.

During a recent trip to The Boston Park Plaza Hotel, we were able to observe a truly environmentally sound hotel in action. For example, we saw: the elimination of polystyrene foam, plastic tableware and aerosols throughout the hotel; printing of all stationery and forms on 100% dioxin free recycled paper (and printed on both sides); a new bathroom amenities system that eliminates the use of nearly million plastic containers annually (and uses the packaging cost savings to buy even higher quality amenities); new double-glazed windows throughout the property, faucet aerators and water-efficient showerheads in all their guest rooms; there are recycling programs for paper steel, glass, cardboard, shipping pallets, aluminum, plastic; an employee fund initiated with the proceeds from redeemable soda cans; and an ongoing program of donating linens, blankets, food and furniture to local veteran shelters.

According to Jose Campo, Managing Director of the Boston Park Plaza Hotel, each year the program saver over 16.5 million gallons of water; 300 trees (through the paper recycling program); 300,000 pieces of plastic dishware; 29,000 gallons of fuel oil and 200,000 pounds of "trash." What was really impressive was the communication and educational programs that have been initiated with the hotel staff. Everyone on the staff seems very high on the program. Many employees have indicated that because of the hotel's program they have initiated steps in their own home to be environmentally conscious.

So, what's all this going to do with sales? To start with, just the public relations aspect has been the most astounding. The program has generated interest all over the world, with recognition from the White House (the President's Environment and Conservation Challenge Gold Medal), coverage in USA Today, New York Times, and in other national daily newspaper, major consumer magazines, various trade journals as well as coverage on NBC, CNN, CBS and other broadcast media. There's nothing like good old PR to make people aware and bring business to a property. And, it works! The hotel's marketing plan includes a new market segment entitled: "Environmental," with operational as well as marketing strategies. According to Mr. Campo, as a result of this program, the hotel generated just over $1 million in additional convention business during 1992 and it is expected that this figure will jump to over $2.5 million in 1993.

The hotel sales team directed by Gary Mitchell plans to fully utilize the environmental policy as an additional sales tool in differentiating and communicating the advantages of The Boston Park Plaza. Other than convention and conference business that could be generated through many mainstream companies and organizations, there is a tremendous market out there for transient business. Major supporters of environmental action programs in the U.S. are the business and leisure travelers 25–45 years of age. Database Marketing Research, a division of The David Green Companies in Chicago, is creating a database for the hotel that will be used developing business in this new market.

The program was initiated, created and designed for the hotel by Tedd Saunders, President of Eco-Logical Solutions, Inc., whose firm's mission is to "consult business for environmental success." His company is pioneering a new business philosophy within the service sector. Saunders says: "The opportunity to create efficiencies has allowed leading companies to redirect savings into improving quality and therefore solidifying customer loyalty." He goes on to say: "It was never intended that these programs would form the sole basis of a marketing strategy, but it does look like going green pays . . . in more ways than one."

PUBLIC RELATIONS SHOULD PLAY A BIG ROLE IN YOUR MARKETING EFFORT

April 1995

Let's not get too uptight about this business of public relations. It's a tool that is important in getting your message out to the public without taking out an advertisement. What this is about is letting people know what's going on, what you are doing, your property's activities through publicity and trying, at the same time to create a positive image or favorable public opinion. Sometimes we neglect this aspect in trying to improve profitable sales. Of course, we think of advertising, direct mail, and direct sales and, sometimes we do think of PR (the thing that some salespeople think they do when they are making sales calls). But, the real meat behind public relations is the ability to get your property and/or staff remembered and recognized as the place where prospects for business ought to be doing business.

It was probably a hundred years or so ago when I first started in hotel sales in Charlotte, North Carolina. Being brand new to the business I sure didn't know any better so I made a call on the business editor of the Charlotte News, Emery Wister. That call was the start of a long time, endearing friendship with a man who became my PR mentor. He was so impressed that I went to him for advice. My only question was, Mr. Wister, what is it that I need to do to get favorable publicity for the Queen Charlotte Hotel? Since then I've been preaching that it is important to meet the people at the local paper, find out from them what to do and how to do it to get publicity. Try it, it's easy and the news folks will probably love you for it.

According to Robert Allegrini, Public Relations Director of the Swissotel in Chicago, if you really want to get into the PR game then orchestrate noteworthy publicity stunts. He states: "A hotel is an extremely versatile multifaceted place with many departments, services, and functions. As such, there simply exists more interesting possibilities for publicity than in most businesses." After all, according to Allegrini, "People still perceive hotels as a magnet for glamour, intrigue, and excitement. As such, they are more likely to be a source of interest to journalists who are eager to provide the public with information that will be well read."

The idea behind good, favorable publicity, is to come up with a thought, an idea, involving the hotel operation that would whet the appetite of your local news media (and national, if possible, of course). Anyway, if something is happening or has happened at your hotel that the local press feels would be of interest to the local folks then you'll get coverage. So put on your thinking cap and come ups with good ideas. To help you along, Allegrini has published a great little booklet called "The Hotel Publicity Cookbook: A Guide to the Top Ten Hotel Publicity Secrets. The 1994 publication was sponsored by the Hotel-Motel Association of Illinois. One of the ideas in the booklet has already been used many times, mostly by large hotels or hotel chains. It involves the polling of guests to find out any number of things. Allegrini suggests polling out of town guests; the results of which would then reflect how guests perceive the city or region where the property is located. This certainly would be an item of interest to your local media . . . I can see it now: "Donaldson Brown Hotel And Conference Center's Annual Poll of Guest Perceptions of Blacksburg, Virginia." Wow! What a headline! Ya know . . . stuff like this works!

LET'S MAKE SALES LETTERS WORK BETTER FOR YOU

June 1995

It's amazing the number of letters that are sent daily by hospitality sales executives that don't work well. You've gotten them, I'm sure. Most are dumped when they are direct mail solicitations for things that don't meet our needs. Before a letter is written we should take the time to examine what it is we are trying to do.

▶ *Why write it?* Generally, the reasons for writing would be to prospect for new business or solicit new business (even old business); respond to an inquiry; follow-up on an account or perhaps thank someone. None of these is the same type and each should be specifically designed to meet the needs of the correspondence. When prospecting we are trying to find out if the person or organization is in a position to provide business for a property. When soliciting, we know the addressee is a prospect and we want to secure their business. Then there is the letter in response to cm inquiry, whether the inquiry is by phone or letter—we know what they want and we want to be able to tell them how we can meet their needs. In following up, the basic information has already been exchanged and we are trying to either secure more information, confirming something, providing a proposal, confirming a booking or just staying in touch. Of course, the "thank you" letter is always a great one to write.

▶ *The format*: Generally the letter format will just about always be the same. It starts off with a salutation—that's the greeting. Most letters begin with Dear Mr. or Ms. To me that sounds a little old fashioned and kind of blah. Here's an opportunity for you to add some creativity to attention. Those of you that have received correspondence from me may recall that all of my letters leave off the general salutation and the first paragraph starts off with something like this: "Thank for the call, Mr. Jones, it was great hearing from you." After the salutation, you start the opening paragraph. This is where you should try to attract the attention of the reader with something that is or could be important to that person. "It was great seeing you, Charlie, at the MPI conference. We've got just the thing you need for your next meeting." The opening paragraph needs to be short, to the point and it is followed with the information paragraph. Be careful here. Do not start adding stuff that sounds like your brochure. Be specific and to the point. Add information that meets needs, state-specific features that are real benefits. The last paragraph is the close, which is the call to action—either they need to do something or you will do something—just don't leave things "hanging."

▶ *Most sales letters:* do not pertain to need—they include a lot of stuff about a property, but so what? If I don't need it why tell me about it? Many are just transmittal letters where we send along a brochure or meeting kit: "enclosed herewith is the package of information we discussed" Ugh! Most letters are not at all creative and are not appealing and surely look computer-driven—not too personal and they say "I'm not very interested in you and I have been busy so I asked my secretary to send you one of our #12B computer-driven letters."

▶ *Good sales letters:* Are "me" to "you"—something personal has to be included to show this and they have to sound like you are talking to the person. Forget using words that you usually don't use when you are taking: "Pursuant to your recent request . . ." "enclosed herewith . . ." You don't really talk like that, do you? Good letters should not be mysterious . . . we should not have to read it twice and they still don't understand what is meant or wanted. Good Letters need to get to the point right away and should, of course, identify the product that is needed or wanted. Good letters should also push what it is you are trying to sell—specifically make a point about a particular feature that would definitely be an advantage to the prospect. And, don't forget that people buy with emotion. Say something that would dig into feelings.

HOW TO MAKE YOUR DIRECT MAIL SALES PIECES WORK BETTER FOR YOU?

August 1995

According to a study conducted by the Travel Resources Institute of the Potomac ," 64% of the direct mail correspondence sent out by hotel industry professionals has proven to be ill-conceived, poorly executed and ineffective in accomplishing its original objectives." The study involved the review of sales letters and direct mail pieces directed to meeting planners from various hotels and resorts. It was conducted by Bob Spraker, Xerox Document University and Duncan Farrell, General Manager of STAG (Society of Travel Agents in Government).

▶ 84%—Had a lack of clear purpose and/or response. They were passive, in that they didn't clearly state the benefits of their product or service. They didn't answer the question, "Why should the customer buy now?"
▶ 75%—Had proofreading mistakes: spelling and grammar

- ▶ 70%—Didn't "write the way you speak," had overused adjectives, and were "too flowery"
- ▶ 68%—Lacked adequate quality/professional appearance
- ▶ 61%—Overstated the product or promises about service
- ▶ 59%—Had incomplete or outdated prospect name/address
- ▶ 54%—Didn't have the relevant timing of offer mailing and/or selling
- ▶ 35%—Failed to sign each letter personally

According to the authors of the study—if you are involved in doing direct mail be sure that you remember that you are Selling. State the benefits that the prospect will realize by doing business with you. You should always ask for the business, or appointment or whatever other desired response. Strive for the highest quality and professional appearance your budget will allow. It communicates the style and image of what you're selling. If you are sending a letter, sign each one personally. It also helps to add a brief handwritten note on the letters to persons you know well. Write the way you speak. Be brief and to the point. Check and double-check spelling and grammar. Be careful about the timing of your direct mail piece. Always get your stuff out well in advance of when the information will be important to the prospect (if you are promoting a fall meeting package don't wait until August to get it out). The property staff aware of what you are doing. Make everyone on All the elements of the direct mail program should be communicated through all departments. If an inquiry about the direct mail comes in through another department your staff members should know how to respond. And pass out copies of the mail piece.

All too often we attempt a direct mail program for the sake of showing we are doing something to attract business, but we do not take the time to plan the campaign. You should always start out with "What are we trying to accomplish?" . . . It should relate to your marketing plan. Then, you need to identify the message, what is the offer? What are we going to say that will be a benefit to the prospects, will be believable and will get a prospect to respond? How are you going to measure the effectiveness of the mailing—should be the next step. What are the criteria? Will the results of the mailing be measurable? Now, how about your target audience? Who are they—do you have an updated list that makes sense? Be sure your stuff is being directed to the right people. What will the mailing look like and do you have an adequate budget to take care of the "professional looking" piece along with the cost of mailing? There is always the question of whether to send the mailings by first class mail or bulk mail (bulk is less expensive). Check with your local post office for the rules covering this. There are some advantages and disadvantages to each and you have to make up your own mind which way to go. Doing direct mail is very time consuming if you do it properly. There's a lot to understand about this medium. Perhaps the job of direct mail should be left to the professionals. There are many direct mail houses and advertising agencies all over the country, that would do the job for you. It may really be in your best interest to find an agency that could do the job for you. And it even may cost you less and get you more response.

THE 10 TOP MARKETING TIPS FOR THE 10 BASIC PRINCIPLES OF MARKETING

November 1995

All too often there is a tendency to equate marketing with sales and throughout our industry, we find salespeople with marketing titles, when in fact, most are not really marketers. Simply put we can define marketing as the development and delivery of a product to a marketplace. Selling could be defined as delivery of a product for a price. Direct sales (or selling) is a facet of marketing as would be research, advertising, promotion, public relations, and merchandising. All these things that we do, in the area of marketing, hopefully, end up in assisting the profitable sales effort.

Let's take a look at the top ten tips of marketing:

1. *Have a plan:* Most properties do have a marketing plan. There are probably as many different types of plans as there are numbers of properties, but the type doesn't matter as long as a plan does exist. The extent of the plan depends on the size of the property, profit centers, the type of product itself, etc. Some plans run over 200 pages while others run much less. Some small properties may just have a sales plan. And that could work well too. Whether a marketing plan or just a sales plan the key is to include objectives (goals) and strategies for securing business from different market segments. The Hospitality Sales and Marketing Association International located in Washington, D.C. is an excellent resource for how to develop a marketing plan.

2. *Owners and managers involvement:* It has always been a wonder to me why owners, operators, and managers don't get more involved in the marketing effort. They are really very logical people to follow-up on specific types of leads, where their position alone would influence a decision to get people to stay at a particular property. During my employment with American Motor Inns, many years ago, I never hesitated to ask Adolph Krisch (Chairman) or Joel Krisch (President) to follow-up on a specific lead if I thought it would help get the business—they did—and we did get the business. It seems to me that there are always some key accounts that need these types to make the one important phone call. General Managers also need to do their share of marketing at the property level . . . most do not even want to get involved.

3. *Let salespeople sell:* Too many salespeople use the term "marketing" in their job title. Some may very well be involved in marketing; it has been our experience that most just there to "deliver the product for a profitable price" ie: make the sale. Too often salespeople are asked to do many things other than sales. Properties should use weekly sales activity reports which reflect the number of hours working in direct sales and number of

hours doing other things. This would tell us a lot about how to improve sales productivity.

4. *Everyone sells:* The butcher, the baker, the candlestick maker ... everyone on the payroll. How often dies a General Manager deliver that message to the staff? Not only from a service standpoint—and we do understand that excellent service does get a single guest or group to return—but do we explain how staff members can provide business leads to salespeople? Everyone has friends, neighbors, relatives, and even spouses who may very well be connected to the business somehow.

5. *Advertising does pay:* Yes, it does pay to advertise, but there are a lot of questions that need answering before you start spending your money on ads. Too many hotel operators can't answer all the questions alone. Get professional help. There are many advertising and marketing agencies out there specializing in the hospitality and tourism industry that can figure out what you need, how to present the message, create the advertisement and make placement in the right media. Seek help!

6. *Understand public relations:* When something bad happens on your property, you are usually in for it with the local media. However, if you do your job in media relations—that is, get to know the city editor and/or others at the local paper, get to know the news editors of local television stations—they'll probably call you first before they do anything. Be "up front" with these people; don't try to hide anything from them—they'll usually find out anyway. Favorable PR is easy to get once you know the people. Keep them informed about what's going on at the property (i.e., promotions, special events, special meetings taking place, special visitors to the property—anything of a human interest nature). Good public relations start with you (the manager, the staff) and will help position your property, build awareness and create business for you. Keep your name out front, with news releases sent to the media whenever you have something of interest to relate.

7. *Work on profitable market segments:* Look at the breakdown of types of business coming to your property. Rank each segment based on profitability. Increase the percentages of the best segments by devoting more time and sales effort to these, and less to segments that create the least amount of profit. Understandably, during slow periods, you may want those lower-profit segments to help build occupancy—so don't ignore them completely.

8. *Understand and practice RevPAR*: If you are not already into working with Revenue Per Available Room, consider it. With RevPAR, you compute the revenue taken in from all rooms available to rent. If you have a property with 100 rooms and rent 756 (75% occupancy) at $100, you have taken in $7,500 with an average daily rate of $100. However, with RevPAR, you divide the $7,500 by the total number of rooms available for rent (100). Now, your revenue per available room is only $75. However, had you sold an additional 15 rooms at a

reduced rate of $80; of course, it reduces your a.d.r., but look at what happens to your RevPAR: The additional 15 rooms sold created revenue of $1,200. Add this to the $7,500 for a total of $8,700. Divide this amount by the number of rooms sold, and you have an a.d.r. of $97. Divide the total revenue by total rooms available at the property (100), and you'll have a RevPAR of $87, which is a more realistic way to view your rooms revenue. It's only a simple way of combining your rate and occupancy and is a better measure.

9. *Watch your promotions:* Sometimes we can go crazy with a "good" idea and go overboard on costs without even thinking about return on investment. Some years ago, we had a manager Who created a weekend package deal. The package sold for $40. He ended up spending over $4,000 in advertising and promotions and giveaways. When I asked him how much business he expected out of the deal—how many packages he expected to sell his response was 100. You figure it out. (By the way, he only sold a dozen.) When you send salespeople to trade shows to promote and sell, have a formula for figuring out ROI. Between the cost of the exhibit space, materials, travel expense, entertainment and portion of salary, you could run up to quite a bill. What do you expect in return if the value of all this represents an expenditure of, let's say, $3,000? How many leads or prospects were developed that would result in business within six months or a year? Did it pay?

10. *Make a training investment:* One of the biggest things we've overlooked in our industry has been the dedication of dollars as an investment for training our profit-producing personnel (reservations, front desk, sales, catering, restaurant, bar, recreation outlets, etc.). Those responsible for bringing in business or handling inquiries need help to do their jobs better. When searching for companies or individuals to conduct training sessions, look to those with experience in the hospitality industry. While many "genetic" training programs exist, those conducted by people who "have been there" can better relate to your opportunities.

ARE YOU COMMUNICATING BY E-MAIL OR FAX?

September 1996

A little while back, while in Houston, a colleague from another university and I were discussing the wonders of email. Two days before my trip I sent him an email that I would be in Houston to conduct a workshop and suggested we meet day a returned message confirmed our get-to-gather. The whole process of sending and receiving the notes couldn't have taken more than two minutes—allowing, of course, for the time in between of being available and turning on the computers. Much easier than writing a letter, faxing it or even telephoning. If we had tried to call each other we would probably both end up getting voicemail messages (and that could have done the trick as well).

There is something about email though that is more specific and no chance of being misunderstood when it comes to times, dates and places, and a hard copy is also available.

Anyway, my friend, the professor told me that he is the site selector and planner for an association meeting to be held in San Francisco next year. He received an email message from a hotel salesperson soliciting the meeting. He was impressed—it was the first time someone has prospected him by email. He responded, of course, but we don't know yet if there will be a booking for the hotel. The fact is however, there is an opportunity to prospect and solicit by email. Think of the time and money you can save. Of course, you need to get email addresses to do this. However, more and more meeting planners are getting "online." As you do more networking and meet more potential buyers you will soon develop your own email listing (as I am now doing this myself). Right now most salespeople send out letters, dictated to, or written by, or pulled off a base of computer-driven letters by a secretary, reviewed by the salesperson and signed. Probably a pretty expensive collateral piece is included in the mailing and postage of anywhere from $2.50 to $5.00 is spent. And we send these out in droves. Many are "dumped" upon receipt since the mailings, in many cases, are not going to real prospects. With email you do it yourself—no secretary needed for this. And, it only takes a couple of minutes for a short message just to find out if there is a prospect on the other end and if there is any interest in your property. I feel the rate of response will probably be much greater, more immediate, less expensive for both parties compared to mail, fax or even a phone message. People using email, I feel, respond to their messages.

A few weeks ago I received an email message from Italy. There was an interest in having me conduct a sales workshop during a large hotel show being held in Rimini, Italy in November. An immediate email response reflected my interest. A fax arrived within a couple of days confirming the arrangements. I responded by fax and then email a contact there regarding some details. My message went directly through in less than a minute.

My email address log is growing as I am in touch with more and more people with whom I am networking. Now I can "talk" to my editor at Hotel and Motel Management and can even forward this column to him via email, instead of faxing it. When I fax it he has to copy it into his computer, by email, he just transfers the column.

BUILD MORE SALES THROUGH PUBLIC SPEAKING

February 1998

People would prefer doing business with people they know, people they like and people who are respected. One of the better ways to become known, liked and respected is to get involved in public speaking. We don't mean getting involved in the big stuff like addressing large audiences internationally, but doing things locally, in your own backyard. After all, in all probability, most of your business is generated from local sources. Once groups know that you are available to address their members on topics that may be of interest to them you probably will get called upon pretty frequently. And, what's so bad about that? Your title and the name of your property always goes along with you when you speak and creates wonderful awareness and visibility for you and your hotel.

Ask any program committee member for any local organization about the problems they have getting speakers for their monthly membership meeting and you'll see why there is no problem getting booked. Of course, you need a story to tell, a group to tell it to and not be afraid to do it. This is where most people have a problem—being afraid. One of the biggest fears people have is to speak in public; most people dread doing it. Even with professional speakers, there is always some level of anxiety, but the amount of anxiety gets reduced with experience. The more you get out and address audiences the easier it becomes and probably the more fun you'll get out of it. What causes most anxiety is a lack of confidence and experience. To reduce the anxiety you need to have confidence in what you are going to say; therefore, only select topics that you care about, in which you have a great deal of experience or knowledge and feel comfortable in addressing.

You should start off with small groups, particularly with those in which you already hold membership and know some of the people. In just about any city, if you take a look at the local telephone directory you will find a listing of clubs and associate ions that would welcome you as a speaker. And, there are loads of church groups, garden clubs, senior citizen groups, and the list goes on. Check with the Chamber of Commerce for a listing and even ask the Executive Vice-President for advice about which groups to approach.

Just about any topic involving the hotel business, travel and tourism, and food service, would be of interest to most groups. We are in a very glamorous business and people love to hear about it; there is so much they do not understand but find it most fascinating. And, here we are not talking about any long-winded speeches. Most groups only want 15–20 minutes for the monthly meeting. Certainly, any property manager or salesperson can come up with enough information on an interesting topic to cover that amount of time. Think of all the funny hotel stories or sales stories you can tell. Or tell about the hardships or changes in the industry or just telling about how you do your job could be very informative and exciting.

So where do you start? First, pick one or two topics with which you feel comfortable addressing. Make an outline to cover material for twenty minutes. Work in some support material (overhead transparencies, slides, posters, videos, etc.). Make sure you have an interest in getting an introduction to the topic (maybe even ask a question that would stir people to think). Practice the presentation with friends or family. Go through it a couple of times to make sure you are comfortable with it and at the same time, you can fine-tune it.

As you get more experience you will want to add topics and audiences and without any doubt add to your business. Just be sure to hand out your business cards to the people you meet.

HOW WELL IS DIRECT MAIL WORKING FOR YOU?

March 1999

Each week I get about a dozen or so pieces of direct mail from hotels soliciting my business. They come from all types of properties, from one-star to five-stars; from major cities to small towns; some contain very expensive collateral, and weigh a lot, creating costly postage. Most of these are asking for my meetings and convention business; some want my corporate travel account. I am not a meeting planner or a corporate travel manager and certainly not a travel agent. So, why would someone at a hotel want to spend money going after my "business" when I am not a prospect? Somehow, I am on some direct mail lists as a result of being a member of organizations such as Meeting Planners International or Professional Conference Management Association. Or, perhaps I was on a registration list for a conference I may have attended. Maybe I put my business card in a bowl, at some trade show, to be drawn for a prize. In any event, I am getting on a lot of different lists, somehow.

My point is this maybe we need to be more careful in the selection of names that go into our database for direct mail purposes. Sometimes sales personnel get hold of a list of names through a membership list, or attending a convention, or participating in a trade show, but perhaps do not consider that not all those names are prospects. We gotta do a better job of cleaning up our lists.

Another consideration in doing a direct mail program, is to try to understand how effective it is. Do all your direct mail pieces get opened and read? Who knows? Is anyone measuring results? I like to think that just about everything we do in marketing, and sales are measured, based on return on investment. If I plan to spend $3,000 on a direct mail campaign, I would also plan on what results I would expect in ROI. Sometimes, a good response mechanism could tell you who is reading your stuff.

Too often we try to do direct mail ourselves without the help of professionals. Many times a sales staff person, without any expertise in this area will be assigned to do the job. It really pays to engage the services of a marketing company or advertising agency to get this work done for you. They can design the campaign, create the collateral, get the appropriate mail lists, etc. Yes, it does cost something, but you will probably end up with the better-looking material, better lists, and a better response.

Want to know how to get direct mail opened? Pitney Bowes did a study last year and they found in order of importance, the following:

▶ First: Address your mail to a person and spell the name right. But even an envelope with a misspelled name has a better chance of being opened than one with a generic address.
▶ Second: Use a stamp. Metered mail is next best. But a preprinted permit will deter recipients from opening your envelopes.
▶ Third: Include a return address with a name. No return address is a reason why recipients don't open mail.

- Fourth: The standard No. 10 envelope won the "most opened" contest against 20 other sizes.
- Fifth: Choose red ink for teasers because red positively influences recipients to open envelopes.
- Sixth: Window envelopes are the most "openable," probably because bills usually come in them. A directed printed address is a close second. But address labels deter "open ability." (Source: MAIL: The Journal of Communication Distribution, Milford, PA 18337; cited in COMMUNICATION BRIEFINGS (January 1999)

EXHIBITING AT TRADE SHOWS IS A LOT MORE THAN JUST SHOWING UP

June 1999

Hotels around the country are probably spending millions of dollars participating in trade shows at conventions, and a variety of special events, all over the world. The objective, of course, is to bring more business to the properties. Exhibiting at these happenings is a good idea, and an important part of a marketing plan, but are we maximizing our potential? All too often owners, operators and General Managers expect the sales folks to return from such an event with business in hand. It doesn't exactly work like that most of the time. Exhibiting at a convention for a specific market segment (meeting planners, travel agents, tour operators, corporate travel managers, association executives, etc.) provides a great opportunity for developing prospects for business who, hopefully, will be sold at a later date.

The way to be more productive in this type of promotion is to start with a plan. One should start with the idea of researching the event to determine to what extent the attendees would be or could be business prospects for the property. Then, gather information relative to previous events' attendance. Here you are looking at numbers of buyers only, not combined figures for buyers, suppliers, speakers, guests, media, etc., and the ratio of buyers to exhibitors. Very recently I heard a report from a hotelier who exhibited at a conference where there were more exhibitors than buyers, and very little "action" on the floor. I learned my lesson many years ago with the same situation happening to me.

Once the research is done, and you have decided to participate, plan on the number of your hotel staff who should attend. This has a lot to do with the length of the "show" (exhibit hours, and days). You probably do not want anyone working a booth longer than three hours at a shot without a break (some of these go on all day long). One staff person for every 400 expected buyers (over a period of time) should work well. If expected attendance is 1200 then it is not unreasonable to have three people "working" the event.

Budgets need to be prepared to determine the entire cost of exhibiting. Productivity has all to do with ROI (return on investment). Come up with a formula to reflect how much business needs to be booked (eventually) as a result of your participation. Not everyone will agree on this, but a good starting figure would be to look at a minimum return of ten times the investment. Of course, don't expect it all to be booked at the time of the event. Look for the return on investment in booked business within a year of the show. The whole idea is to develop prospects from those buyers in attendance so that the sales staff may follow-up within the following days and weeks after return to the office.

A critical function in "working" a booth is to make as many contacts as possible so that the number of prospects who may be developed will be maximized. So, another part of the plan would be to estimate the number of new contacts to be made during the whole event. Then estimate the number of those contacts who will become prospects. With some experience, a formula for this will be reached. However, to get you started, just figure that for every ten contacts you make (within a specific market segment) three may become prospects for some future business. A good idea is to have a prospecting form ready to use on every contact. The form covers the "who, what, when, where, why, and how" questions; the answers to which help you determine if the contact is a prospect.

AN EXPERT'S ADVICE ON MARKETING VIA THE INTERNET

May 2000

We get a good deal of inquiries from managers and sales folks who want to know how they can use the internet to their advantage in improving their sales and marketing techniques. There's a whole lot of new stuff out there in the area of technology which we can use, if we just knew how. So, not being up to snuff, myself, I asked my good friend, Lou Taverna, editor of the popular Hospitality–1st newsletter to give us some of his thoughts on the subject.

Here are a few tips from Lou:

Lou says: "For many of us in the hotel business the name of the game has always been "location, location, location." As we join the Electronic Commerce World we had better be thinking "repetition, repetition, repetition." Give every one of your email messages the same tagline (or as many call it—'sig file'). That represents the easiest way for your recipient to recognize and/or contact you. Most recipients of your messages do not take the time to build a contact database, so make it as easy as possible for them to reach you, at a moment's notice have to agree with Lou. Of the over 75 emails I get in any one day, well over half do not have a signature line. Those without are missing a marketing opportunity. Take a look at my sig line. When you get an email message, or response from me it will always carry my full address, phone number, and advertisement of my sales workshop, along with the website.

Need an email address for someone at a particular organization? Lou says: "If you already have one name/email address, most other email addresses will follow the exact same pattern, i.e., if you know that John Doe is john.doe@xyz.com then Carol Smith's email address would most likely be carol.smith@xyz.com.

How about creating your own email newsletter, and make it available via your own website? Lou suggests, however, that you make it legitimate. "Be sure to always include a fast and efficient way for people to 'unsubscribe' without having to jump through hoops. Make your business email as professional as possible."

There will be a good many other ideas coming from Lou at our sales workshop, but in the meantime, you may want to latch on to his website (http://www.hospitality-lst.com), and sign up for his free newsletter. I've picked up a load of good ideas from him this way.

If you are interested in knowing what's going on in the minds of meeting planners with regard to their working with hotels, you may want to look into, and become part of a discussion group. Take a look at (www.mim.com) which has links galore to help meeting planners, as well as suppliers, in doing a better job at planning meetings. You can also sign on to their listserv which has participants from all over the world. You'll find lots of discussions going on regarding information requests on facilities; questions on contract clauses, meeting room set-up, receptions, banquets, and commissions. There also always seems to be some complaints about how hotels operate. Great stuff going on for conference service managers, catering managers, sales folks, etc. Take a look at the website, you may want to join in.

MARKETING TIPS FOR THE PROFESSIONAL HOSPITALITY SALES EXECUTIVE

July 2000

More often than not, there is some confusion within our ranks in our industry with regard to what is marketing and what is sales. Most often, we hire someone to be a Sales Manager or Director of Sales, and provide a job description which requires a certain number of sales calls or bookings. A marketing person would more than likely be involved a good deal more than that. Their responsibility would, in all probability, involve all or some of the facets of marketing to include: research, advertising, public relations, merchandising, and of course, direct sales. However, we should not forget that no matter what the specific job functions of salespeople might be, according to a job description, there is always a little bit of marketing involved to get the job done professionally and effectively.

With this in mind, we want to offer some thoughts on some basic marketing opportunities which exist for salespeople that need to be practiced on a regular basis, year-round, all the time, to help in the sales effort. Although most of these are very basic, you would be very much surprised, as I have been, that most of these tips are not practiced on a regular basis by many salespeople.

▶ *Business Cards*: How about using business cards of a much better quality than we usually find in the possession of hospitality sales executives? The five-star properties, and many independents have it down pat, but then why not all the others? Keeping a healthy supply of cards on hand is critical. Give them out very freely; it is the least expensive form of advertising.

▶ *Professional Organizations*: Join and be active. We are always learning, and what better way than from each other? Education and networking with other sales folks will pay off for anyone active within their specific trade organization. For us, in hospitality sales, belonging to the Hospitality Sales and Marketing Association International (HSMAI) is a must. If you do not belong, check it out (www.hsmai.org) or call me for information.

- *Sending Facsimiles*: We do this all the time, but why send any fax which would not represent your property properly. I just received a confirmation of a reservation from a Hyatt hotel in a major city. Hyatt's are great, quality products all over the place. Having stayed at this particular Hyatt, I know the fax I received, of the confirmation, was not indicative of the type of property. The form used looked like about a 20th generation reproduction of the original; could hardly read anything on it, much less the telephone number, since I needed to call them on something. I'm saving it to use as an example at my next Hospitality Sales Workshop. Salespeople send lots of faxes. Let's be sure that the transmittal sheet and other forms being sent are not multiple generations of the original. What you send should be clean, crisp, and contain the appropriate information.

- *News Media*: Take the time to get to know your local newspaper business editor, and other editors, as well as the folks in television and radio broadcasting. Stay in touch with them on a regular basis even you don't have any news to report. You will be the first they will call when an issue comes up regarding our industry. You should be their local contact anytime they are doing a hospitality story. It never hurts to get your name and the property name in front of the public.

- *Be a Speaker*: Local clubs always seem to be looking for someone to address one of their monthly luncheon or dinner meetings. You can be the local hospitality "specialist" and get invited to any number of civic clubs, garden clubs, church groups, etc. Hospitality is an exciting and glamorous topic to everyone (even to us). It doesn't take much to prepare an exciting 20-minute presentation on our industry. Loads of information is available through the American Hotel & Motel Associations (www.ahma.org). Or, just talk about some interesting stories of about what goes on in this business. The more contacts you make this way the more business will come your way.

BE CAREFUL OF YOUR E-MAIL MARKETING

Most of us pretty much resent getting email messages about things in which we have no interest. Like getting a bunch of fourth class "junk" in our daily U.S. mail. What we're talking about here is "spamming." This business of sending huge numbers of the same message to a wide variety of people who have not indicated any interest in receiving the stuff. It makes some people mad to get spammed, and creates negative feelings toward the sender.

No question about it, direct mail has been a very good form of promotion for many hotels. However, it is expensive, and getting more costly all the time. So now is the time to take advantage of technology. E-mail marketing is like direct mail in many ways, but we just have to be careful in using it. We start out by building a database of names and email addresses, by categories. The lists need to be separated by special interest. There could be lists by corporate travel, leisure travel, travel agents, corporate meeting planners, association planners, government meeting planners, SMERF market contacts, local restaurant customers, local meeting and banquet users, etc. This way you can send specific information pertaining to specific needs; that is, customizing the messages. But, the idea is to send information primarily to people who have already expressed some level of interest in what you may be offering. This way you won't be spamming.

OK, so how do you get these contacts? Create your own list of how to do this:

- From your own files of previous property users
- Telephone inquirers—people who call in for information
- General prospecting—cold calls, sales blitz
- Contacts made at trade shows—get the business card
- Visitors to your Website—be sure to be able to capture information
- Those who respond to your advertisements
- Respondents to your direct mail—make sure there is a response method
- Information secured by your reservations department
- Guests already in-house

In all cases encourage people to provide you with email addresses, and get an "OK" to be put on your mail list for special promotions and/or newsletter (should you chose to create one). On your Web site have a link with a form visitors can complete to be put on a list for "special promotions." It's a good idea to include in your email promotion messages, that you are sending, that the recipient has "subscribed" to this, but also give them a chance to unsubscribe from the list. Also be sure, on all your messages, you are linked to receive a direct email back in case someone has questions or needs information. You will find that this business of email marketing will prove that it is the lowest cost of direct marketing you can ever do. Remember to segment your email lists to target special interests. Just be sure your messages are interesting, but not lengthy. And, be careful of the subject line in the address. Don't put anything that is misleading. Maintain your list and keep it clean, up-to-date.

An important thing about sending messages on the Internet is that so many salespeople, and General Managers do not include a signature line. Every message you send should reflect your complete name, title, company, address, phone number, email address, and Web site. This is done automatically for you by setting it up through your toolbar. You can create a standard signature line with all that information, and even put in a brief advertising message, should you chose to do so. Every message you send will automatically reflect your standard information.

MARKETING VS. SALES OR MAYBE A HYBRID

February 2002

Sometime last fall we covered the subject of "Eliminating Marketing Tasks" in this Sales Clinic column. It was amazing to see the number of responses on this topic from General Managers, salespeople and from those in marketing. Most agreed that salespeople should sell and marketing people market the product. Of course, we had some hybrids—those folks who do both and are qualified to do both. Everyone agreed that both functions are critical in these tough times of an economic downturn. Business is off at most properties around the country. So, do we concentrate on more marketing or more sales? And, who does what? Or do we have someone to do both?

Many agreed that, unfortunately, there are people who have been hired to do sales for a property, but are given the job of doing some marketing as well. Most have not been schooled in marketing or have even done any of it previously. But still, they are told to make up brochures, makeup and place advertisements, do direct mail, create a marketing database, conduct promotions for a restaurant, produce packages for the travel market, and on, and on. Just about everyone agreed that all this marketing stuff does take away time from direct sales, which of course, is also important. So what's a salesperson to do? The answer lies in the hands of the General Manager. Do you want to have someone to sell or to market the property?" Most people can't do both. From time to time we do run into some hybrids, that is, those who have had experience, training and are knowledgeable about both functions.

One such hybrid is Jodi Cross, Director of Marketing at the Doral Golf Resort and Spa, in Miami, Florida. Jodi has been in our business over 15 years and came up through the sales ranks, at different properties, but now has a career path solely through marketing. Jodi agrees that most properties today see marketing and direct sales as one and the same, but in reality, the disciplines of sales and marketing are distinctly different. Most salespeople do not have the marketing skills, but still, not only have the title of Director of Sales and Marketing, but are required to fulfill the marketing function at a property.

According to Jodi (and I agree), "Marketing encompasses the areas of Public Relations, Direct Mail, Database build and capture, Advertising, Promotional/Partnership Alliances, Internal Marketing, and External Marketing as well as Internet Marketing." The difference between sales and marketing is that "marketing focuses on direct and indirect awareness, whereas sales focus more on touch techniques." Jodi also mentions all the internal things at a property that are marketing oriented and need to be accomplished such as: bounce back opportunities, birthday/loyalty programs, in-room signage, video advertising, on hold messaging and voicemail marketing. So, who is to do all this? Do we really want our local salesperson to get involved on these projects? If so, are they qualified to do so, then who will do the direct sales effort?

Over the years we have all seen the acronym NHUSSS. I don't know who originated this, but it does make sense. It stands for: Nothing Happens Until Someone Sells Something. We need to let our salespeople spend all their time selling.

There are some really good sales (and even marketing) tips available for you in Lou Taverna's Hospitality—1st Newsletter.

THE POWER OF THE PRESS CAN HELP BUILD YOUR BUSINESS

March 2002

It seems that owners and managers of properties around the country are, for the most part, fearful of the press. We probably look at reporters of print and broadcasters as our enemy, looking to dig up some dirt. Frankly, unless there really is some dirt to dig up, this theory just doesn't hold water. The fact is that reporters are generally looking for good stories to tell. It is amazing that we do not take advantage of using the press to our benefit.

Quite often, General Managers, and owners have some very good, favorable, stories to tell about their property, but either doesn't think of announcing it to the press, or do not know how to go about doing so, or are too intimidated to take the first step in making the approach. There is no question about it, but the press is very powerful by reaching tons of people who, of course, will believe whatever is written. It just seems plausible and realistic that people like to do business with people who are respected (getting your name in the paper or on the radio or TV, in a favorable light, will show respect for you). Building awareness and visibility is a key to bringing in more business. We keep sending salespeople out to make 15 or 35 calls a week, contacting local businesses, to do just that. Imagine how many business folks could be made aware of your property just by reading a daily paper.

So how do we go about accomplishing this business of getting some free publicity? The step would be to get to know the key people at radio stations, television stations, and local newspapers. This should be the job of the General Manager, not necessarily the sales department staff. You can't imagine how impressed media folks are when hotel General Managers personally make a call on them. It is really magic. Try it! The whole idea would be to let these people be aware of your availability in case they need a resource about the hospitality or tourism business. Inviting them to have lunch with you works especially well too. Also, you will need to stay in touch with them periodically to let them know how the business is doing in your area as well as other interesting stuff going on your property.

We need to understand the business of "news." These business editors, social editors, sports editors, etc., need to put out a newspaper or news broadcast every day. They NEED news to tell. Sometimes nothing new is happening around the world and these news people are really hungry for stuff to broadcast or write about. Learn how to write a news release. The staff folks at a local paper would love to help you learn how to do this. It makes their job easier. Don't try to figure out if something is newsworthy. Just keep sending out the news releases. If it doesn't fit, it will get dumped, but don't let it bother you. There probably is a lot more important things going on in the world that need to be reported more than your release.

Think about a local slant on a national news story. If you see something written about a national hotel brand, and you have that flag, think of a favorable incident at your property, which may be a tie-in to that story. All too often, a property will win some kind of an award from a hotel brand, but nothing appears in the local paper about it. If the owner or management did not think about telling the media, it won't get told. Sometimes, that could be a very hot story for the newspaper and broadcast stations.

Many times the staff at a property has some interesting facts about their current accomplishments or history. We once had a chef who had a cooking stint at the White House. It created a good deal of publicity once we got out the word. At one property we had a porter who once worked on the railroad in a Pullman car. When we noticed a series of stories about railroads not doing so well as in the past, we let them know about our employee. This resulted in news stories, pictures and television interviews. Great free publicity for our Holiday Inn!

Getting noticed by the media has a great deal more impact than paid advertising, It is a third-party endorsement of your business.

SALES NEWSLETTERS CAN PROVIDE SOME GOOD TIPS

June 2002

On one of my recent flights, I had plenty of time to catch up on reading a batch of trade publications. In the collection of magazines was the spring issue of "Inn Crease Sales & Service for Hoteliers." This is a quarterly hospitality sales newsletter produced by Janie Wiltshire. In a previous Sales Clinic column, I listed this newsletter, along with others as a good resource for our readers. Anyway, as I was reading Janie's Spring issue, I couldn't get over how interested, and excited I got by all the great ideas she was putting out to her subscribers. I felt I just had to devote this column to some of the stuff about which she was writing. The whole issue was devoted to how hotels, of any size and type, can do more business.

Here's a sample of some of the things she covers for a property manager or salesperson:

▶ How to score big with guests checking in—create loyalty with simple, easy ways to develop the relationship. Here is where the staff needs to understand how important it is to provide the smiling, warm greeting as guests arrive, as well as other little things that make guests feel good about the property.

▶ Your top accounts are your best customers—how to find new prospects from existing accounts. We are usually satisfied with whatever that one piece of business may be. However, we do not pursue other types of business from the same account.

▶ Nine ways to make better use of email and fax. We are just not maximizing the abilities we have in making better use of these technologies.

▶ Janie uses a chart to show you the positive emotions to keep in mind a goal for interaction with guests and prospects. This is a good one.

▶ How to make a big value in using a hand-written "P.S." note on sales letters. It is so easy; I don't know why we all don't take advantage of such an easy, worthwhile thing to do.

▶ She offers some clever ideas on how to get that "difficult" appointment.

- There are some suggestions on how to emotionally connect with guests to further build loyalty.
- How to get sports teams through their coaches and other sources is another opportunity for building this particular market segment.
- Jamie suggests these two words to make you a better listener: "Shut Up." How true!
- A section on "Spring Cleaning" gives us 26 tasks to do during the 13 weeks of spring. Do two a week. These are things like purging files, send solicitation letters with a seed packet (and what to say); hosting a realtor breakfast. Anyway, the list is super creative.
- Some very clever ways of getting a prospect's attention are also listed.

NHUSSS (Nothing Happens Until Someone Sells Something): The phones aren't ringing like they were during the better years of 1992–2000. Let's make sure we start selling again.

WE NEED TO START LOOKING AT NEW WAYS TO HANDLE OUR SALES AND MARKETING ACTIVITIES

July 2003

We have all seen the ever-changing face of the hospitality and tourism industry, especially since 9–11, threats of terrorism, the rise in unemployment, our economic situation, a drop in occupancy as well as revenue and RevPAR. We need to start thinking of perhaps not doing things the same way as before, particularly in the area of sales and marketing. The sales are out there, but what's the best way to get your share of the business? The answer? A good strategy and plan!

Minor adjustments in sales and marketing, as well as operations and delivery of services sometimes do nothing more than making improvements on the margins. Sometimes a real "look into the soul" of the business is needed a bottom to top, and back down review of the business strategy. Notice, I put the emphasis on the bottom-up, and we are talking about your property staff, all department managers, and maybe some others.

What actions can be taken to improve your Return On Investment? The traveling public (business and leisure) is still out there, but what can you do to capture more of that business? We spend a good deal of time around here at Virginia Tech, in our department of Hospitality & Tourism Management and Pamplin College of Business, talking about just that. According to our top faculty and staff experts one of the most unanimous answers is that an organization should periodically do a realistic strategic assessment of where they are, where they want to be, and how can they best there. Particularly when it comes to sales and marketing productivity.

There are many agencies and others in our industry who are experienced in helping an organization review its strategic plan, but a good many of them are still teaching methods that are becoming dates in meeting the dynamic needs of today's hospitality industry. So what's out there that can make a real impact upon the strategic plan of your organization? There is a new strategic planning system that's been uniquely developed just this past year by some people right here at Virginia Tech.

The Virginia Tech Group Cognitive Mapping System. How does this work? The process consists of a series of custom-designed activities using web-based software to brainstorm an organization's issues, identify categories that group the issues, rate their importance, and finally identify causal relationships. The whole idea is that all department heads at a property or management company are involved. This mentally demanding, but highly results-oriented process allows the group to understand the diversity of opinions held by team members, and provides a context for the group to move forward as a team in solving problems and setting new strategies.

YOUR BUSINESS CARD IS AN EFFECTIVE SALES TOOL

October 2003

Do you remember your first business card? It really was a big deal, perhaps the start of your career. The company probably printed about 500 of them with your name in big letters, along with your job title. And, you gave out your card to just about everyone you met. You used up a huge amount of cards the first several months on the job. The next batch of new cards lasted a much longer time. In the old days of American Motor Inns we printed 500 cards for each of our sales folks with instructions that they had to distribute at least 50 a month, and hopefully, they got rid of a whole lot more. We were selling the idea to our salespeople, as well as our General Managers, that the business card was an important sales tool.

An important sales tool it still is today. Back in those days we just had the name of the person on the card, along with a title, the hotel brand, and logo, address, phone number, and that was about it. But it still was an important sales piece, low-cost, and easy to give out. Today we have much more information available to put on the card to make it easier for anyone to be in touch with us. Now we have a fax number, email address, Web site, cell phone number and like I have, a home phone number. Business cards also come in all kinds of shapes, designs, and colors. Some folks have doubled fold cards because there is a lot of information to include.

Of course, the whole idea of a business card is to make sure it gets mass distribution so that people who get these cards will remember who we are, and are able to get in touch with us. Naturally, we want to use these cards effectively, and since they are of low-cost, we want to maximize distribution. The idea is to make sure that everyone we meet gets a card; even if only one out of ten people keep it, and use it, it's worth it. A card should be included in every piece of correspondence mailed. Have your accounting department include a card in all billings going out of the property, as well as with all checks being mailed. Cards need to be distributed at all networking opportunities such as the local Chamber of Commerce gatherings.

Why not have front office salespeople distribute their cards when people check in or check out of the property? By the way, it surely would be a good idea for properties to make sure that business cards are printed for each front office salesperson working the desk, on all shifts. In fact, why not have cards for all guest contact folks working the property? All staff personnel has friends, relatives, neighbors; they belong to organizations, local clubs, churches, etc. Ask them to give out their cards, as many as they can. Think about how proud your staff members will be when they get their first card, as you did some years ago.

GOOD PR CAN HELP INCREASE SALES AT LOW OR LITTLE COST

September 2004

No question about it. People generally want to do business with people and properties that are well-known and well thought of in the community. If this is the case, then there should be no reason why any lodging facility should not get involved in an ongoing community relations program.

Why not make your property the focal point for all this is going on in your city? It should be in the public eye all the time. Of course, the real "P.R." person at the location is the General Manager who needs to create some sort of an ongoing plan which should include features such as:

▶ membership in local organizations
▶ news releases to local media
▶ student project in cooperation with local schools
▶ speeches to local civic organizations
▶ property tours for local interested groups
▶ participation in community projects

Working through the plan, you need to decide who on the staff could be involved in the different phases of the plan. It is not intended that the General Manager would be the only person involved, but the idea is to get as many personnel as possible to participate in some phase of the program. For example, a desk clerk who may be pretty good at writing could do the news releases. There is always something happening in which the local papers would be interested. Human interest stories area always sought by local daily and weekly news publication.

Who on the staff has done something unusual or has an interesting hobby, or won an award? Have there been any special interesting people staying at the property? The first step on doing news releases is to get to know the news editors; tell them what you want to do and ask for advice on how the releases should be written. Find out what type of news items would interest them.

The General Manager should, of course, join one of the community civic clubs while other staff members should join others. And, it is not a question of just being a member, but you have to be active and get to be well known as a hard worker for the organization in order to get noticed and recognized as Mr. or Miss XYZ Hotel. The idea is to get visibility for the property. Once you and the property become well known in a local club, members will seek you out to make reservations for visitors.

Speaking about the hospitality industry is not difficult, and local organizations are always seeking guest speakers. All it takes is writing letters to the local clubs advising them of your availability as a guest speaker. Most clubs are looking for only 20 minutes of a presentation. For speech material, you can use news articles appearing in Hotel & Motel Management or other trade publications. Information is always readily available through our American Hotel & Lodging Association.

Then, there are school children tours through the property that you can make available with all the staff participating. Getting the property involved in fund drives, as a "collection" headquarters is also good public relations.

Getting involved in community affairs, creating an awareness of your property, exposing your facility to the public, is all part of the sales job. Getting out and asking for business is most important; however, the public relations aspect should at least play a role.

YOUR 2005 SALES ACTION PLAN

January 2005

If you have not yet completed your new Sales Action Plan for this year, it is still not too late. Whether you are the General Manager of a mid-level type property or the sales manager of a much larger facility, it would be worth your while to come up with some type of plan for going after the most profitable business for this year. What we want to do is maximize revenue so that more falls to the bottom line of a profit statement. Most properties probably already have some sort of marketing plan which gets into many things that are very helpful in running the marketing effort of a facility. Few, however, probably have a detailed sales action plan reflecting what needs to be done, from a sales standpoint, with an action calendar. The key is to start out in determining which market segments would bring in the most revenue, and how much more business can we get from those segments.

We start out by making an analysis of all the markets from which we have gotten business in the previous year. These should be listed in order, with the segment producing the highest ADR (Average Daily Revenue) on top. In just about all cases this would be the rack rated business. Right next to that figure, for each market segment, we indicate the percentage of our total room sales that is represented by that amount. For example, if the rack rate for a property is $100, and for the entire year the property took in $100,000 at rack rate, and the total room sales was $1,000,000 this would reflect the percentage as 10%. We do this for each market segment. This would provide an excellent picture of all the different types of business from the top down. Once it is determined what percentage of the total business is coming from each market segment, it is easy to take a look at all the figures and make some decision regarding from which market segments do we want more business.

The next step is to come up with an objective or goal of how much of an increase in revenue we want from certain market segments. Of course, we would be looking at the segments that are bringing the best ADR. It is likely that will find that the markets bringing in the highest ADR are those that reflect the lowest percentage of total room business. Also, we will probably find that those segments coming in with the highest percentage of total room business booked would be generating the lowest ADR.

Once we have determined to what extent we want to increase our sales in a particular market segment we then have to come up with an action plan. First, we need to come up with a strategy to create the increase. The strategies could be wide and varied. It could be doing more prospecting within that segment; making more personal calls on those prospects; attending trade shows covering that particular market segment; join and be active in associations that represent those segments; cultivate more site visits from key persons who book the business within those markets, etc. The idea here is to be very specific about who will be responsible for each action step; when will the action be started; when will it be expected to be completed and how much would it cost to do it. The cost factor here is very critical. We need to be careful to be able to relate cost versus how much more business we plan to bring in. If we set a goal of increasing business by $10,000 in a particular market segment, we surely need to be careful of how much we want to spend to bring in that amount of business. Everything is relative; however, if the cost of expected sales is figured at 10%, that should be considered a pretty safe amount.

For each market segment that is planned for an increase in business, we should come up with two or three strategies that need to be implemented. Create some sort of form to use that will show the name of the market segment, a particular strategy that will be used. Then create a table to reflect: the action steps to be accomplished, who will do each action step, when will the action start, when it is expected to finish, and how much will it cost. A separate sheet should be used for each strategy. Of course, it is reasonable to expect to have perhaps ten or more sheets if there are three or more market segments that are selected to show an increase in revenue.

CREATE A SALES PLAN FOR YOUR 2006 MARKETING PLAN

October 2005

If you have not yet started on your marketing plan for next year, now is the time to do it. One of the things to consider including is a sales plan for the marketing plan. This is a document that has to do with the strategies that are outlined with a plan of action for the sales department team, with details of who will do what, along with a timeline. This should end up being a guideline of activities to ensure that the sales activity for developing business for next year reflects that the sales team will be most productive in going after the best business possible.

Considering all the very positive figures coming out lately indicates an excellent 2006 year in our industry, we should be in a very good position to go after business that would generate the best bottom line for our properties. Generally, bringing in group business with the highest average room rates is what creates the best profit. With this in mind, the sales plan research should start with coming up with a listing of market segments that bring in the best dollar room revenue. Here is how you do it:

- Take a look at all of this year's group market segments and list them in order of highest to lowest average room rate for each segment. Your accounting department should have these figures on a year-to-date basis.
- We then want to find out what percentage of total room business is represented by each market segment. To get this figure all you need to is to get the total property room sales dollar amount on a year-to-date basis, and divide it by the total room sales revenue for each group segment. This will give you the percentage of total room sales for each group market segment. For example, if your total room sales amount for the year was $1,000,000.00 and your corporate market segment generated $100,000.00 in room sales, then that market segment represents 10% of your total business. If your motor coach tour business market brought in $50,000.00 in room sales, then that market segment represented only 5% of your total room revenue.
- Follow this process for each group market segment, and make a listing reflecting the group markets, with the segment averaging the highest room rate on top. The chances are that the segment having the highest room rate average will be producing the lowest percentage of total room revenue generated for the property.
- As you look down the listing of segments you should get an idea in which markets you want to increase the percentage of the total business. From this, you can come up with a sales plan objective reflecting a percentage of business increase in certain market segments that you would declare as a goal.
- Be sure to reflect a dollar amount based on the percentage of increase you are projecting for next year. For example, if in a particular market segment
- (pick those that reflect the best average daily rate) you are doing 10% of total revenue which amounts to $100,000, and you have an objective of increasing it by 10%, then use the dollar amount of $10,000 as your objective.
- The next step is to come up with a sales action plan listing the things that are expected to be accomplished to meet the goal. This would include the action steps: who will do what, when the action will start, when it is expected to be completed and what it will cost.
- You want to be sure that the total cost, of all the steps to the action plan, does not exceed an amount that would be unreasonable in achieving your goal. Somewhere in the neighborhood of 10–20% is reasonable.

WORKING A TRADE SHOW? HERE'S HOW TO MAKE IT MORE PRODUCTIVE

July 2006

One of the most productive ways to develop qualified prospects for the group market is participation, as an exhibitor at trade shows. All too often, however, sales staffer who has been assigned to "working" the show have not been trained adequately to maximize productivity. The ultimate objective for salespeople in hospitality is to increase profitable sales for the property being exhibited. Before any sale is made, it stands to reason that a prospective buyer must be found. Not everyone attending a trade show would be a buyer for every exhibitor; therefore, it is up to the staff working the trade show booth to determine who is, and who may not be a prospective buyer. Quite often, when visiting exhibits, it is noticed that exhibiting staff are making presentations to people who are not qualified, buyers. What it really comes down to is the opportunity for exhibiting staff to be able to make contact with people who provide information regarding their specific needs and buying practices. So, rather than trying to make a sales pitch to every person coming by the booth, it is better to just ask some critical questions to determine who is or who may not be in the position to either buy or recommend the property. If we can just use the trade show environment to seek out prospects rather than try to make a sale we can be more productive. This way we can surely make more contacts which would, hopefully, lead to more prospects, in less time.

It is in the follow-up where more information is developed, more needs are determined, going through the sales process, and confirming a sale. All any staffer needs is two minutes of conversation at the booth to determine if anyone is a prospect. What we need to do is talk to as many people as possible to maximize our productivity. It is not a question of making a sale when exhibiting, but getting as many prospects as possible with whom to follow-up at a later date.

Here are some tips to help make your next trade show exhibit time more productive:

▶ Always seek out new contacts, people you do not know.
▶ Working the exhibit booth is only part of the job; don't forget participating in all the meetings, meals and events. There are people there who do not get to the trade show floor.
▶ Collecting a pile of business cards really doesn't help. Keep only those cards that are true business leads that may be developed in prospects at a later date.
▶ When you take time to talk with other sales personnel it reduces the number of other people to whom you should be talking.
▶ Standing with arms folded intimidates people; they will be less likely to stop and talk to you.
▶ Avoid having chairs in your booth. You do not want to attract tired people who probably are not prospects anyway.

The return on investment in participating at a trade show is so much greater than in any other kind of prospecting situation, provided, of course, that you pick the right shows in which to participate as an exhibitor.

EXHIBITING AT TRADE SHOWS TO BUILD PROSPECTS AND BUSINESS

June 2007

Hotels and other meeting venues that do not participate in exhibiting at industry conferences and trade miss a great opportunity for the development of new prospects and a great return on investment. It would be very difficult for any salesperson to be able to meet as many new prospects in one central location within a short period of time in any place other than while exhibiting at a trade show, and yet there are many properties that do not send their sales staffers to industry conferences and conventions where trade shows are being held. Further, from a return on an investment opportunity, there is no greater way to show it.

Conferences and conventions are held all over the world and throughout the year involving a wide variety of market segments in which hotel should be interested. Search for any market segment association or organization and locating information on dates for these events taking place would be very easy. For example, HSMAI (Hospitality Sales and Marketing Association International) has a number of "Affordable Meetings" shows going on around the country all year (www.hsmai.org) or being a member of MPI (Meeting Professionals International) would get you into two very large exhibiting opportunities. These programs are attended by meeting planners from a wide variety of market segments seeking educational programming as well as contacts on locations for future meetings. The WEC (World Education Conference) in July, or the PEC (Professional Education Conference) in January (www.mpiweb.org). Attending either of these conferences would be over 3,000 professional meetings planners and suppliers from the corporate as well as association market segments, along with a wide variety of independent meeting professionals. If you are interested in prospecting and developing business in the association market then seek out your own state affiliation with American Society of Association Executives; since each state, the executive association has its own annual convention, with exhibits. If there is interest in making contacts with meeting planners specializing in military reunions The Reunion Network (TRN) has about a dozen ConFAMs a year in various cities around the country where these meeting planners seek new destinations and meeting venues (www.reunionfriendly.com) The Society of Government Travel Professionals (SGTP) have two conferences, with trade shows a year during February and September, held in Northern Virginia/Washington area with 200–300 meeting planners from various government agencies attending, (*www.sgtp.org*). There are four, three-day travel agent conferences, with trade shows, put on by Travel Trade Publications each year with about 1000–1500 travel agents in attendance (www.traveltrade.com).

All of these programs mentioned have workshops and seminars designed for all participants whether they are buyers or suppliers, so in addition to being able to prospect for future business, there is certainly an education value for hotel sales staffers.

And the list goes on and on . . . for just about any market segment, there will probably be an association or organization putting on an at least one annual conference along with a trade show. Of course, participation will not be without cost, such as attendance fee, space for a booth cost, travel, hotel, meals, cost of literature and special advertising items to be distributed, perhaps entertainment, etc. The work is hard with long hours each day. However, the opportunity for prospecting is exceptional, with great chances for follow-up and bookings.

THE SALES PLAN AS PART OF THE MARKETING PLAN

January 2008

With the start of the New Year, in all probability, most properties have its marketing plan for 2008 already in place, and ready to go. In fact, most plans for the coming year have been prepared months in advance during 2007. Most plans look great, but the big question is: "How do we get it going?" The best way to get the sales department going to implement the plan is to create a Sales Plan for the Marketing Plan. What we are talking about here is a document outlining the strategies necessary to accomplish the mission of the plan, along with the details of how it will be done.

For example: If an objective of the plan is to increase business in specific market segments, then what needs to be outlined in the Sales Plan is what is the specific objective? Here we need to mention the specific segment of the market to be increased, with the amount of increase expected in terms of percentage as well as the dollar amount. From there we proceed with outlining two or three, or more, very specific strategies that will accomplish the goal.

Now, for the next part of the Sales Plan, we need to get into the actual action that will take place within each strategy. The plan will next reflect a specific action such as: "having three site inspections, by meeting planners who are prospects for the business, take place within each month." Then we will need to identify, who, specifically, by name, of the sales staffer who will be responsible for this action, when will this part of the plan start, when will it be completed, and what will be the expected cost of taking this action. In outlining all the actions to take place for each segment, we have to be careful in figuring out what all this would cost the property. When adding up the costs we need to come up with the final figure representing dollar cost and how much this would come to in a percentage of the expected increase in business from the objective. If the expected increase in a specific market segment reflected in the objective is $100,000, then how much is the property willing to spend to accomplish that mission. If, for a property, a reasonable amount to spend to increase sales for a specific market segment would be 10%, then spending anything less than $10,000 to increase sales by $100,000.00 should be acceptable.

There very well may be a number of strategies to accomplish only one specific objective, with a variety of activities working within those strategies. Then, there would be a variety of objectives to reach in the sales action plan to meet the needs of the overall marketing plan. This document is then very critical in laying out the activities of the entire sales department for the year; which if followed, should lead to a most productive period for any property.

Considering the increase of over 135 new rooms to open in the U.S. in this coming year, and to at least maintain about the same occupancy period as this year, with a projected increase in average daily rates around the country, plan as indicated herein should prove to lead to a successful year for any sales department.

TO WHAT EXTENT DO ALL THESE SOCIAL NETWORKING SITES BRING IN NEW BUSINESS?

February 2009

Hopefully, we are all concentrating on bringing back repeat business to our properties. Chances are that if we have done our job on service we will get some repeat business without any sales activity with those prospects. However, some time needs to be directed towards getting some new business to replace those prospects who do not repeat. It appears that a good deal of time is being spent by sales staffers in this business of "social networking" via sites like LinkedIn, Naymz, Reunion, and many others, with the idea of a way to bring in new accounts vita the contacts made. Is this really doing the job of sales? Does it work? I am not too sure that all the time spent with sales staffers "working" these sites really does bring in new business.

Then again, maybe that is not the real purpose of social networking; if not, how does it really help sales? When we talk about networking, in our business of hospitality sales, isn't it supposed to be making contacts who can lead us via the network to other contacts with whom we may be able to do business? Has anyone measured results? So far we have not seen any firm results reflecting sales published on this.

Right now I, like most others, are pretty well-loaded with time-consuming "network" messages from people who "want me to be their friend," and join in on their network. As a courtesy, I do respond to people I know, not that I really want to be in their particular network, but if I did not respond, those requesting my being linked to them, may not look upon me with favor. In all cases I have responded with a note back, appreciating their request to be linked to them. So, now what happens" Now that I am linked, what can I do, via my network of contacts, to be of help? Only one person, so far, to whom I am linked, sends me articles which he thinks would be helpful to me. He even suggested an idea for this column. Now, that is a good contact; however, no one else has contacted me again for any kind of assistance. So why did they want me on their network? Think about this; couldn't any of these contacts made, contact me directly via my email address? If anyone wants me to be their friend all they have to do is send me a note, or make a phone call to me, that would be good networking, without having to be connected via any of these sites.

Where is the new business coming from via contacts made on those networks? Would it not be more productive for users of these sites to spend their time making direct contact with old and new prospects with an email? Also, a phone call to those existing accounts would also work (I know, no one answers their phones anymore, they get used voicemail), but a good percentage will get through.

Sales staffers and managers should start measuring time productivity on these networking sites. Further, measure the productivity of spending the same amount of time in developing relationships via phone calls and email messages. How about spending some time going through old files, and contacting existing accounts on a regular basis? Even reading the local paper, and contacting probable leads for new networking relationships, would perhaps work even better than those networking sites.

It would be interesting to get some readership response to this with information on the value of these social networking sites.

REVENUE MANAGEMENT IS THE KEY TO PROFITABLE SALES

August 2010

Now that business is picking for hotels in many areas of the country, so many owners, operators, managers, sales staffer are in a quandary about how to get their rate back up. It seemed so many hotel properties jumped the gun when business was falling off, and started with the rate-cutting business. We've been through this many times before in this cyclical business of hotel operations, and we just never seem to learn to "hold that rate." Once a property starts, many follow, just to get the occupancy up by reducing rate. Now, with occupancy increasing, we are in a hurry to get it back up. OK, so how do you do it?

There are many areas of concern with revenue management, but that is the key to being profitable. It looks like occupancy will be up a little this year, but much better next year. There is a little increase in demand, but supply will be down for a couple of years. So, that formula of supply vs. demand will be very helpful. One of the big areas when it comes to revenue management is getting the "right" kind of business, at the "right" rate, at the "right" time. There is no point in sending salespeople out all over the place to bring in any kind of business. There needs to be a plan. Part of a marketing plan is to determine the right market to go after. We do that by coming up with a Market Segment Analysis which is taking a look at all the market segments in the past year, listing them by ADR, and what percentage of total room business each segment contributes. The most valuable business is in the segments that have the highest ADR. Then the idea is to come up with a goal for how much more of a percentage of that kind of business the property needs to seek. Once that is determined strategies are planned to accomplish that goal. Of course, management then has to have the sales staff go after the most profitable market segments.

The objective should be reached through the plan being organized using a Sales Action Plan form. This would list all the steps that need to be taken, by whom, when, and how in order to reach the goal. Of course, what also needs to be included is the cost of each of the actions planned. The idea is not to spend more than what needs to be budgeted to be profitable.

EXHIBITING AT TRADE SHOWS A GOOD WAY TO FIND LEADS FOR NEW BUSINESS

January 2011

The objective of participating as an exhibitor at a trade show is to see as many people as possible to develop information that will locate prospects, and ultimately lead to increased sales. Bear in mind that not every attendee at any kind of event with a trade show would be someone who needs to use a hotel property for any kind of meeting, or visit for any special reason.

Just about all associations, global, national, statewide, and regional invite exhibitors to participate at their trade shows. Just checking the websites of different associations that involve meeting and event planners offers an idea of which to select. It is obvious that selecting state-wide associations could be a better possibility of meeting more lead potentials than attending a global or national event. However, there are national and global conferences that could very well provide superior opportunities for group business leads such as HSMAI's Affordable Meetings (www.hsmai.org), or Professional Conference Management Association's annual meeting (www.pcma.org). Then there is the American Society of Association Executives (www.asae.org). Most of all national associations have related St ate associations. Just searching for associations in any State would reveal a good number of conferences and events taking place each year. There are groups such as STAG (Society of Travel Agents in Government), and SGMP (Society of Government Meeting Planners). Then there is AIBTM, the very large annual trade show put on in June of each year that brings in over 2,000 qualified meeting planners, and will be held in Baltimore in June 2012, www.conworld.net).

The whole idea behind exhibiting at these types of shows is to be able to be productive in finding enough leads that could be followed-up and prospected to determine which lead would work best in bringing in group business to a property. Unfortunately, so many salespeople who "work" their booths do not do it professionally and productively.

Most exhibiting staff have never been trained in how to work a trade show booth. The main thing is to make sure that there is a return on investment for the property represented. This means getting many leads that can be developed into prospects which, hopefully, would develop into a meaningful business. Experience in working trade shows would provide an excellent formula for determining ROI. This may be just an example: At a conference of 400 qualified meeting planners for two days, an exhibitor may have contact with 200 or more attendees, of which 50 may turn out to lead, but only half could be qualified prospects. This then, requires to follow-up of the 25, immediately following the show. Maybe 10% may just turn out to book business. The estimated dollar value of the prospects, over a long-term, against the cost of attending the event could reflect the ROI. If the total dollar value of the prospects' new business is, for example, $100,000, and $5,000 was spent in attending, representing 5%, that would be a pretty good value. Keeping the cost at lower than 10% is a very good return.

Are Salespeople Still Selling or What?

Do we really know what salespeople are supposed to do? It seems that if some hospitality employees have the word "sales" in their title, then they should be selling. If not selling, then what else is it they should be doing? Well, we all know pretty well what else they are doing. It would be no surprise to find out that in all probability some employees in sales may very well be spending only 50% of their time in actual selling. It seems that to some extent salespeople do get involved in a variety of activities which come under a marketing effort. They will be doing public relations work, writing news releases or newsletters or "working" some events; arranging open house programs. Then, they could be handling the advertising for the property, or arranging to do direct mail, and taking care of any property promotions that need to be done. And, of course, the new thing on the block is doing Social Networking, and don't forget some will be taking care of the Website. The list goes on and on, all which really have to do with marketing. Now, we have to think about how much education, training or experience most salespeople have had in marketing? Some General Managers or owners of properties really think that their salesperson should be and could be doing the marketing stuff. Big Mistake!

Certainly, we do need to exclude the larger hotels, conference centers, and resorts from these comments since, for the most part, all those types of properties would have a marketing person on board to make sure those other facets of marketing are being handled properly. Yet, still, sales are a function of the marketing process. So, there is a big difference between sales and the other parts of a marketing effort. We can take a look at a very simple definition: Marketing encompasses all the things we do to bring a product to the attention of the public (or market). Selling is the delivery of the product for a PRICE. Simple enough? Salespeople get paid to bring business to a property; their efforts are measured by production. So, it stands to reason that salespeople need to spend their time selling, actually making the sale. It is a whole different process and effort than what is involved in the other marketing facets.

Yes, it is important to carry on a marketing effort to let the public, and more importantly, the prospective buyers, know about the property, but someone has to get them to buy, and that is the salesperson. More and more companies are offering incentives to salespeople based on their productivity. More sales equal more business for a property which equals more income for a salesperson. It seems that we would want salespeople to busy themselves only with doing things that benefit the direct sales process.

Social Networking And its Relationship to Hotel Sales Productivity

It sure does look like the "hottest" thing out there today for sales staffers is this business of Social Networking, through a wide variety of types of electronic messaging. No question about it, one sure does make a lot of connections around the world through these media opportunities. Making connections in hospitality sales is very critical to building a business, so the question remains: "what is it about making all these connections that really results in new business?" We need to take a look at all the connections and reflect on how many of these could be prospects for the business, and what kind of an approach needs to be made that would result in new sales for the hotel properties represented.

One of the important things to also consider is the relationship building with existing accounts to make sure we continue to do business with them. But don't we already do this via our tracing of files that exist on the properties? With a sound tracing and follow-up a system in the sales office so that staffers are consistently following up with accounts, without the necessity of searching for them or seeking them via any Social Networking process.

When we take a look at sales productivity, we are not looking at sales activity. For a simple definition we can look at Sales Activity as what is it we are doing, what method are we using to become more Productive. As far as Sales Productivity we should be looking at the result. What are we "putting on the books." Here we look at actual sales. In other words: What is it that a salesperson is booking. With this in mind, the issue is: "What is it about Social Networking that could make a salesperson more Productive?" Here we could be talking dollars and cents or even building of occupancy, but it needs to be something that is productive. Not just the ability of making new contacts.

Of course, some could look at Social Networking as a marketing tool. And, that is OK too. Yes, it would be a marketing effort which we can consider bringing a product to the attention of the public; something like advertising, but of course, it does not cost any dollars like advertising would do. However, it is not "delivering the product for a price" which is what sales productivity is all about. Hopefully, a sale may come about with the appropriate follow-up with someone on the network who could possibly be a "lead." Then the follow-up would have to consist of a series of questions to determine if this contact could be a prospect. Then, there needs to be a continued follow-up with more questions to find specific needs, and responding to those needs with the relating of benefits in meetings those needs. So, is all this done via the social network via their media? Of course not. Somewhere along the line real words need to be spoken between the salesperson and the contact. Then, hopefully, we may see some productivity when a sale is made.

This business of social networking can be very time intensive, with not much immediacy of revenue production. It could be considered a component of a marketing effort. There are many more things a salesperson can do that would create more productivity with less time activity.

IMPROVE YOUR PRODUCTIVITY WHEN EXHIBITING AT TRADE SHOWS

There is room for a great deal of improvement in productivity when exhibiting at a conference or convention. Too often we participate as exhibitors, but we do it without establishing a plan or goals.

We need to start off with a reason for participating, then, being more selective and attend only in those shows that will help us meet our sales objectives. After all, there are just too many opportunities to join in on trade shows, and we cannot go to all of them. Then, we need to do some planning in order to work the show more effectively, and to be more productive. Following the show there needs to be an evaluation: was it worthwhile? Would we do it again? Was there a decent return on investment?

Working a trade show effectively involves integrating a long list of elements: background, show selection, pre-show planning, working the booth, post-show evaluation and follow-up on leads developed. Think about the reason for attending and having a booth at the show. For most hotels there are different reasons; however, the top reason, and most productive is the opportunity of networking, finding leads for business and turning them into prospects.

Well in advance of the show opening identify prospective attendees through the sponsoring organization of the meeting. Here is a chance to make use of all those hundreds of "friends" you have on LinkedIn and Facebook. Start using your social media to let them all know that you are attending and want to see them come to your booth. Target the most probable important contacts by email, with a personal invitation to meet with them during the conference for breakfast or lunch, in case they cannot get by your booth.

Develop a pre-show action plan that specifies the number of contacts to be made, and the number of leads generated (which hopefully, later will be turned into prospects). The idea is to gather just enough information on each contact at the booth, as well as elsewhere at the conference, to allow you to follow-up following your return to your home base. That is when the real work is done by making contact and gathering additional information to determine if that lead can become a prospect for future business.

Since we want to make as many new contacts as possible while attending the conference, and working the trade show, we need to be careful not to spend too much time with each contact. We need just enough time to record some basic information that will allow us to follow-up later. This may not be the time to try to make a "sales pitch," which could be time-consuming. The more contacts made = more lead developed = a good percentage of real prospects = more customers, when sold. This is how you can measure your return on investment when you know the value of just one new group book.

AN EASY STRATEGIC PLAN TO MAXIMIZE ROI ON SALES PROGRAMS AND ACTIVITY

January 2006

When you think about it, it is really amazing how many different ways individual guests and groups get to visit, and enjoy a stay at your property. Just take a look at all your market segments that make arrangements to stay on any one day. Why are they there? How did they decide to make the reservation? Was there any one way, or more than one way that a decision was made to get to your property? Which of the market segments were providing you with the best ADR? Which of the market segments booked within the dates that were helpful to you in adding occupancy? As you can see, there are loads of information that could be valuable to you in developing more and better business, and, more importantly, what manner of marketing and sales activity helped provide you with the best business?

How can you keep track of all that available information coming from so many sources? Reservations keep coming into a property on a daily basis, but we really know what made that guest select your property? How did they know about it; where was the source of their information that helped them make that selection? At most properties, more than likely, do the reservation agents, or front desk associates really ask that question? It just seems most logical that in order to use a variety of sales and marketing methods to bring in more business, we would need a system that would perform the best for us. Why wouldn't properties install some sort of a reservation system that measures information to help us get improvement in ROI, and guides the sales and marketing effort in getting more of the more productive market segment?

A recent report from a large destination resort reflected that they have shown a remarkable increase in ADR, and improved direction in their sales and marketing effort by using such a system for their reservations. The system, called NAVIS, does it for their property. The way it works is that a unique toll-free number, for Inquiries and reservations, is assigned to each marketing or sales method used to promote the property. When a call is made to inquire, the phone number used automatically identifies the sources of the query or reservation. This then tells the property which marketing or sales tool was used to "capture" that call. This way property management has an idea which sales or marketing efforts should be continued or dropped. With the help of this NAVIS system that resort property reservation director said: "this lets us create more profitable marketing strategies that boost revenue. The system continues to make a huge difference in our revenue, and occupancy.

NAVIS does a good deal more with its method of gathering Information, and getting to confirm more inquiries to boost occupancy. It captures, guest and prospect data, tracks key revenue metrics, and provides 24/7 reservation call center services to help operators increase occupancy, and ADR, and close more leisure business.

CHAPTER 13

Working on Contracts for Groups

AGREEMENT LETTER IMPORTANT WHEN BOOKING HOTEL SPACE

September 1985

In a recent survey of hotel salespeople, about 90% of those that responded said they ask for a signed letter of agreement from a client when booking function space or guest rooms.

The survey was conducted by Cornell University on behalf of the Hotel Sales and Marketing Assn. International Foundation.

The survey also reflected that not cited with any frequency as appearing in letters of agreement were the following items:

- ▶ Room reservation cutoff dates
- ▶ Room block cancellations
- ▶ Billing procedure
- ▶ Contract return date
- ▶ Non-performance, arbitration and damage clauses

Other highlights of the survey:

- ▶ Nearly 50% in the event of a reduction of a room block—30% do so if less than 30 days notice is received.
- ▶ Nearly 37% of respondents do not assess a penalty in the event of a cancellation-approximately 45% do so if the cancellation is received less than six months out. When a penalty is assessed, the most common way to calculate the amount is based on one night's revenue per room held.

A fair amount of variation exists among respondents regarding the issue of group deposits. When required, the most common form of calculation is based on one night's revenue per room held. Of the respondents that require group deposits, 86.5% do so six months or less prior to arrival. About 90% of the respondents that require a group deposit do not pass along interest if the deposit is held longer than three months.

Rooming lists are required two to four weeks in advance of a meeting by 77% of the respondents. Individual reservations must be made two to four weeks in advance if part of a group booking, according to 70% of the respondents.

On for future of deposits:

- Twenty-two percent of the respondents refund deposits in the event of an individual cancellation.
- Fifty-seven percent will refund a deposit if notice of the individual cancellation is received five days or less prior to arrival. The percent require five to ten days notice.
- Eighty-one percent will forfeit an individual deposit in the event of a "no-show" reservation. If there is no individual deposit, 48% of respondents would charge the group master account.

On meal guarantees:

- Sixty-six percent of the respondents always require 48 hours and 13% require 4 hours notice if a weekend intervenes.
- Sixty-six percent do not permit a leeway if a group's meal account falls below the guarantee.
- Ninety-one percent of the respondents determine the number of complimentary guest rooms to be granted based on the total number of guest rooms occupied.
- Seventeen percent use one per 25 occupied as the guideline.
- Sixty-seven percent use one per 50 occupied as the guideline.

The information contained in the survey was obtained by collecting data in completed questionnaires sent in by nearly 1,000 members of the Hotels Sales and Marketing Association International.

The purpose of the survey was to publish a report which would offer valuable insights into the practices of the HSMA International memberships regarding areas such as contracts, group deposits, meal guarantees and familiarization trips. The completed survey is published in "Policies and Procedures of the Hotel Sales Office," which is one sale by HSMAI Foundation for $15.

HOTEL CONSTRUCTION PROJECTS CAUSING BUYERS TO BEWARE

November 1986

In today's lodging market, meeting planners and group buyers seem to be getting more and more cautious about making deals with properties that are either under construction or being renovated.

Many buyers have expressed their disappointment-even their outrage-with hotels that are under construction, mainly because certain promises made to them were not met, As a result, the industry in general and salespeople, in particular, are rapidly losing credibility.

It is very important to the success of a property that it has business on the books in advance of opening. Salespeople can do this and, for the most part, they do it quite well. They run into problems, however, when an owner, developer or manager does not come through with a finished product on time.

'CUSHION' BOOKING

It is getting to the point that salespeople themselves are booking business against a two-to-three-month "cushion" period prior to a new hotel's projected opening date. Construction schedules have to be realistic to ensure that the new property opens on time.

Group buyers, it seems, just don't trust opening dates. Granted, they understand they can probably negotiate a very good room-rate deal for booking a new hotel in advance. But so many of them have been burned so often that they tend to shy away from these "good deals." Salespeople for new properties often get this response from group buyers: "Forget it. Let me know when you are open, and I'll take a look."

Salespeople also have to deal with "partially finished" openings. This is getting to be a popular practice in the industry, and it causes problems for sales staffs.

PIECEMEAL APPROACH NOT HEALTHY

It's understandable that owners and developers, needing to see some cash-flow, insist on opening before the guestrooms are ready. But from a marketing standpoint, this piecemeal approach to opening a hotel is far from healthy-it just doesn't look good, especially to group prospects. It's safe to say that meeting planners on-site-inspection trips don't exactly fall all over each other to look at a property that is only partially finished.

There are similar problems with properties that are being renovated. Many such hotels make promises to meeting planners, then find they can't fulfill them, It's rapidly approaching the point that when a salesperson says, "We plan to renovate," it's more of a hindrance than a help in making the sale.

Such problems may well start to have an effect on the way business will be done between hotels and meeting planners.

Hoteliers should be on the lookout for meeting planners who will want certain protective clauses written into agreements. Hotel personnel and General Managers, in particular, should make a point of carefully reading all agreements in order to protect themselves. Some meeting planners already are giving themselves an "out" on cancellation if any renovation is going on in a hotel during their meeting. Still, others will be writing penalties into agreements to cover situations in which hotel construct interferes with a meeting.

HOTELS NEED PROTECTION

In most contracts, we agree to deliver rooms, meeting space, food functions, etc., over a period of dates for a group at specified prices. But when an agreement is written a year or two in advance, how is the hotel protected should the owner(who in many cases in new) decide to do some renovation that may interfere with the delivery of the product and/or services agreed to in the contract?

That's a question, it appears, that will have to be addressed by all of us. It might be a good idea to get our lawyers busy working on ways to protect us in such situations.

WE GOTTA GET TOUGHER ON GROUPS THAT CANCEL

March 1990

Lodging properties have just been too generous with groups that cancel or do not pick up properly on room blocks. It is time to look out for your property to ensure maximum sales and revenue. Losing a piece of business altogether on short notice where replacement is almost impossible is an injustice. And losing rooms at the last minute because a group doesn't pick up all of its blocks also really hurts a property's bottom line. We all know how tough it is out there with lots of properties fighting for the same business. In some areas, it is a real buyers' market, so there are salespeople who feel they have to go "easy" on deposits and cancellation clauses in order to get the business.

How tough does one have to be? That's hard to answer since each situation is different and operators have to ask themselves: "How badly do I need this piece of business?" Sometimes we actually do lose a booking if we get too tough on deposit and cancellation clauses-provided the buyer has another place to book the business. It is a risk. However, there is a plenty of opportunities for us to do better-particularly during the negotiation process.

The idea is to start high, with the best deposit cancellation clause for the property. Then, if necessary, it can be negotiated down to where the deal is livable. From a legal standpoint, if there is a contract without a cancellation clause-the property has it made if the buyer pulls out. If there is no "out" written into a contract and there is a cancellation the one who cancels must deliver whatever was agreed to in the contract or suffer the consequences of a lawsuit. If the contract is good, the court would no doubt find in favor of the harmed party. But, no one really wants to go to court; in our business, or cause a lawsuit, so we eliminate this type of a situation by writing in a termination clause and a cancellation clause that is acceptable to both parties.

There are always two reasons for breaking a contract according to attorney John Foster: "One is a good reason and the other is no reason." For good reasons, we have a termination clause which simply states when and why each party may terminate for certain reasons, without recourse. On the other hand, the cancellation clause allows a party to cancel under certain terms. The terms are usually monetary, with some sort of sliding scale; the further out a group cancels the lesser amount is paid; the closer in a group cancels, the more is paid. The whole idea behind this is to discourage groups from canceling. The higher the cancellation fee the less likely a group will cancel.

Property salespeople need to be very cautious about contracts prepared by buyers. The termination clause could be worded in such a way that the buyer could get off the "hook" too easily. The termination clause should be spelled out carefully and be simply stated. If it looks confusing-get help. Sometimes a meeting planner may use a loosely worded statement such as: "the hotel has to be in good shape." What does "good shape" mean? Look for statements that indicate causes for termination because of renovation, change in manager, weather conditions, etc. These "causes" should be carefully worded, if you are agreeable to them.

Regarding cancellation clauses, we must remember that our rooms are a very perishable commodity. We cannot afford to lose a group without replacing the bodies or getting compensated for the loss of revenue. If a group is sincere about using a property there is no reason they would not agree to a clause that protects the facility with a monetary consideration for cancellation.

Here's some good advice. If a group cancels start trying to resell those "lost" rooms the minute you know about the cancellation and keep documentation of what was done to resell the rooms.

In the event the breaking of the contract goes to court you will probably have to prove that you have tried to sell the rooms that were lost due to cancellation. The hotel has a duty to mitigate damages. Since there are no punitive damages in these types of cases the hotel may be awarded compensation for the lost revenue from the rooms not sold, but the hotel must try to sell lost rooms.

Actually, both properties and meeting planners benefit with cancellation clauses it's not a one-sided deal.

Be wary of pre-printed contracts provided by some companies and associations. To play safe, have an attorney review those types. It is always best for a property to initiate and write the contract; keeping it as simple as possible, but covering all that needs to be covered.

REMEMBER THE BASICS OF WRITTEN CONTRACTS WHEN BOOKING GROUPS

March 1998

Too often, in our business of booking groups into our properties, we run into the situation of "he said, she said.," when a situation arises regarding what was agreed upon between two parties. Verbal agreements, of course, are legal, and there are still being used, but to a very limited extent these days, and rightfully so. It only makes sense that when an agreement is reached between two parties, something needs to be put in writing so that there is no misunderstanding about who is supposed to do what.

When it comes to room blocks, rates, attrition, meeting space, food functions and any other item that is agreed upon, you need to put it all in writing, and in such a manner that there is no misinterpretation of what is meant. The agreement should be in the form of a legal document which confirms the obligations of the seller and buyer; and any non-performance of the agreement may be subject to legal redress. That is, if you don't do what you say you are going to do, a court may decide what damages there may be. This, of course, could end up with a long, very detailed document, depending on the size and complexity of the piece of business, but it is important these days. We hear about more and more court cases with hotels taking associations to court and vice versa, for non-performance of contracts. Without the piece of paper outlining what's what we would never see any favorable outcome for any party.

Here are some important things to consider when getting into a contractual arrangement with a group prospect: Everything in the written agreement needs to be very specific. Do not use terms that need interpretation or are ambiguous such as: "attendance is expected to be" or "meeting room rate will be determined by the number of meals served." Always specify numbers. Damages for non-performance should be written in the agreement with very specific numbers and dollar amounts. Include cut-off dates for room reservations or guarantees for meals. It is usually in the best interest of the buyer to have a cancellation clause. So, if a prospect crosses out such a section, it is really in your favor because they really have no legal "out" if they want to cancel. There are times when a prospect will make changes in a contract you send to him/her. Once there are any changes made in the document it will nullify the contract, unless both parties initial the changes. Sometimes a multi-page contract is signed and returned to a property, it goes in the file without checking to see if there are any changes. It's very dangerous! In line with this, always let the buyer to be the first to sign the contract. After it is reviewed, very carefully, you should sign it. There should be two originals prospect signs both and you sign both and send one back. The prospect does not have a legal agreement with you until he gets a copy with your signature.

You probably would want to include a clause reflecting that amendment. This often happens, and you want to make any chances of which you both agree and make it part of the contract.

There is a whole lot more to writing a good contract than what we have covered her and it is not as easy as it may appear to be. With all due respect to corporate lawyers, many do not understand the ramifications involved in our business of creating good contracts with buyers of hotel services. This is a unique business, and there are a number of attorneys around the country who specialize in this field.

CANCELLATION CLAUSES IN CONTRACTS WORK FOR BOTH PARTIES

August 1998

All too often we've been hearing comments from meeting planners and others booking groups into hotels that they do not sign contracts with cancellation clauses. Do these people think that because a contract does not have such a stipulation that they can cancel without redress? Well, we have news for them and for you. Without something written into an agreement that will allow a party to cancel (with or without payment of damages) they may very well be responsible, financially, for a whole lot more should they cancel and be taken to court. Every good contract should reflect under what conditions a party may get out of an agreement.

Here's how it can work! A termination clause should be included which will allow either party to get out of the contract without any payment of damages for specific reasons, such as an "Act of God" that would make it impossible to convene because of disasters. Either the property is damaged to the point where it cannot deliver the rooms and meetings space or the participants cannot get to the meeting location as a result of extreme weather conditions (flood, fires, hurricanes, etc.), of course, both cancellation clause. Here is where either party can get out of the contract for any reason. It doesn't matter why a party wants to cancel; however, the canceling party should be responsible for payment of damages, which must be specifically identified.

Generally, the further out from the date of the meeting the damages would be less than a cancellation right up close to the date of the meeting. Hotels should include a "Liquidated Damages" provision in amounts that would reimburse the property for any loss. This clause may include a statement regarding forfeiture of deposits, plus a dollar amount to be paid so many days after the cancellation notice. Courts have held that hotels are required to mitigate damages. That is, properties need to try to sell the rooms and/or space and only collect on what was not sold. Hotels have a duty to attempt to resell canceled rooms and must make a good-faith effort do so. However, with a "liquidated damage" clause, the hotel recovers liquidated damages, and can resell the rooms that were previously held by the breaching party. What this does is replace the common law duty to mitigate and makes it easy to establish what amount needs to be paid without the hotel having to show proof of having tried to resell the rooms. Educated meeting planners will want to add something to the liquidated damage clause indicating that the hotel should make reasonable efforts to resell canceled rooms and credit those revenues against the liquidated damages. In essence, this would require the hotel to mitigate damages. This should not be objectionable. Proving that a hotel has attempted to resell rooms has always been a tough thing to do, but properties should always make an effort anyway to resell canceled rooms.

Many contracts today have a cancellation clause providing an "out" for the buyer of the hotel services. But, very few provide an "out" for the hotel. Why shouldn't the hotel have an opportunity to cancel for any reason, providing, of course, there are liquidated damages to be paid to the other party?

The whole idea of a contract is to make it clear to both parties what mutual responsibilities are placed upon each. It is simply an agreement to do or not to do a particular thing and the performance of which the law recognizes as a duty, and for the breach of which the law will provide a remedy.

TAKE A GOOD LOOK AT THE MEETINGS & CONVENTION CONTRACT

December 1998

It seems that more and more associations, as well as companies, are getting interested in writing their own hotel contracts for meetings and conferences. Those that accept the standard contract provided by hotels are doing a lot of "Crossing out" and amending. They are getting some pretty good advice from attorneys who specialize in this field. And, nothing wrong with that, except you just want to be sure that the contract reads the way you want it to read.

In the November issue of Meetings & Conference Executives Alert, there were some "pointers" directed to meeting planners with regard to some significant wording which was recommended to be included in the hotel contract. Such as: "No later than one year prior to the Event and every three months thereafter, the Hotel shall mail to the Planning Organization a copy of all bulletins or mailings for the (name of the city convention and visitors bureau) that list groups scheduled to meet in the city during the month of (state month)." Now, why in the world should the hotel be responsible for this? It is understandable that meeting planner would want to know what groups are meeting at the same time, but it should be the planning organization's function to find out without putting the burden on the hotel.

In another "suggested" paragraph regarding the addition, there is an indication that if the organization falls show of its pick up of the contracted room block there would be a slimming scale of the amount the organization would pay. This is a good one for the meeting planner. It shows that if the pickup is 75% of the contracted block there is no fee due from the planning organization for the remaining rooms in the block that was not sold. This means that if there is a block that was not sold. This means that if there is a block of 100 room committed and only 75 are actually rented, then the organization has no responsibility for paying for the 25 rooms that were not picked up ridiculous. It further reflects that if only 50–74% of the guest rooms are sold the organization is responsible for paying only 50% of the guest room rate for each room not sold.

Here's another good one (for the planner): if the planning organization exceeds the Block, 50% of the revenue of those excessive block rooms will go to offset all charges to the Master Account and/or other fees and expenses." Can you believe this? I can't believe any hotel operator would go along with this kind of a contract.

What we are indicating here is that sales personnel involved in working our contracts with meeting planners just better be on the lookout for "different" things like those mentioned; of course, in negotiations, it is always a question f how badly you want the business. So you may even want to go along with these types of paragraphs in the contract, if you feel it is in your best interests in being able to make a deal. The Meeting and Conference Executive Alert is a newsletter which provides a good deal of information for meeting planners as well as for hotel executives. It is a good educational tool for sales personnel.

YOU NEED TO HAVE CONTRACTS FOR MEETINGS AND EVENTS AT YOUR PROPERTY

April 1999

Hard to believe, perhaps, but some properties are still doing business with groups for meetings and events without any written agreement. In these cases, I feel pretty sure that if there are any differences of opinion of what was agreed upon, the hotel ends up taking the shortfall.

If you have been reading any of the meetings type trade publications lately you will have noticed a good deal of material being written for meeting planners about contracts, and what to include to make sure there is an advantage to the buyer. It pays for anyone involved in agreements with buyers to be very careful these days with how the contract is written.

A contract is an agreement between parties to do or not to do something. And, if a party doesn't do what it has agreed to do, then the law may find a remedy. The agreement may verbal or written; of course, we all agree that it is best to have it in writing to avoid any misunderstandings. We don't want to have any "he said, she said," etc. There is still a lot of that going on these days with smaller meetings and catered events. Even if there is not something that looks very legal-like, you can still have a contract letter of agreement will work as will banquet event orders that call for two signatures.

Sometimes people think if they cross out something or make a change in the wording and initial the change, they have an agreement. Not so, at least not until the other party agrees to the change or deletion by also initialing. Many times this happens with banquet event orders or letters of agreement with regard to prices, dates, room blocks, menu items, guest counts, etc. Usually, we send out the agreement asking the buyer to sign and return by a certain date. Just be sure that You Do Not sign it until it is signed by the buyer first and returned to you. You should carefully review the agreement just to be sure there are no changes, then if you are still happy, you may sign it. If there are some changes (which have been initiated by the buyer) and you agree with the changes then you too may place your initials in the same area. Not until both parties have signed an agreement is there a legal document. You can also change your mind about anything in the contract and can revoke the contract anytime before both parties have signed. Just because you mailed the agreement to the prospect doesn't mean you have a contract: therefore, it stands to reason not to be the first to sign it. I've seen so many BEOS and letters of agreement signed by the hotel representative before being mailed to the buyer. Let the other party sign first.

There are many cancellation clauses in hotel contracts that favor the buyer and so few that provide equal protection to the property. These clauses should provide equal protection with the right for either party to cancel: however, the other party must be protected with some sort of damages to be paid.

WATCH OUT FOR CONTRACTS WITH "THIRD PARTY" PLANNERS

September 1999

Last year, about the same time, we discussed in this column the relationship hospitality salespeople have with the "third party planners." But, we did not discuss contractual situations. Here we're talking about those folks who contract with companies, associations, tour groups, etc., to take care of, and make arrangements for meetings, incentive trips, tours, etc. These independent individuals or companies work either on a fee basis with their client or collect a commission from the suppliers with whom they deal (hotels, ground operators, attractions, caterers, printers, etc.). Many collect a fee and commission. Whether you chose to deal with these folks, who work exclusively on a commission basis from you, has a lot to do with how badly you need the business, and how you want the commission structured. If you do choose to work with them you need to be particularly careful of how your contract with them is written and signed.

If you enter into a contract with a third party planner (independent planner, destination management company, travel agent or whatever designation someone may have) for a meeting or an event, and their client decides to cancel or perhaps does not live up to terms of the agreement, where do you stand? You can't go after damages from their client. You probably can go after the third party planner for non-compliance with the contract. But if they are a small-time operator (and most are) there probably would be nothing to collect. What is important is for your property to make sure that whatever agreement is made, you want the contract to be signed so that it protects you.

John Foster, an attorney in Atlanta, specializes in this area of contracts between meeting planners and hotels has a lot to say about this type of a situation. Here is an excerpt from his new book: "The Law of Meetings, Conventions, and Trade Shows."

"In the case of a third party, such as a travel agent or an independent meeting planner booking a meeting on behalf of an organization, the third party should have a letter from the sponsoring organization specifying that he or she is designated as its agent to perform duties of a meeting planner. That agreement needs to state that the sponsoring organization will back up the independent third party, spelling out the specific responsibilities, authority, and what the agent can and cannot do. The hotel or facility that is dealing with the independent should ask for and demand to keep a copy of that letter so that the parent organization cannot later come back to the facility saying that it terminated their relationship with that independent planner of agent and wants to cancel the meeting. If that independent was authorized when the contract was signed, the contract is still valid."

Be cautious of contracting with independents who want to book business with you, but do not want to tell you who their client may be. Either follow John Foster's advice, and get that letter by the sponsoring organization, or make sure the contract is signed by the sponsor or their authorized representative.

WHEN CONTRACTING WITH GROUPS, MAKE SURE YOU HAVE AN ATTRITION CLAUSE

November 1999

I have always believed (and preaching in my workshops) that a good contract is a binding agreement making each party responsible for living up to whatever was agreed upon. And, that the law (courts) may take action against the party who breaches the agreement. If I agree to use 100 rooms and I only use 75 then I should be responsible for paying for the rooms not used. Is an association, company, or any group responsible for the use or payment of all rooms that a hotel contractually agrees to hold, even if the contract does not contain language specifying that the meeting sponsor accepts this liability? Over the years, it has been my opinion, and that of many attorneys, that the buyer is truly responsible. This may be changing, so hotel operators need to be aware of what has happened very recently with regard to a suit where a hotel claimed payment for unused and unsold rooms during a convention held at the hotel.

Legal action was taken in an attempt to collect more than $100,000 in damages for unsold rooms during a tournament. The lawsuit was based on a 1993 contract, written by an association spelling out the terms for the property to serve as its headquarters hotel. After the organization failed to host enough attendees at its 1996 meeting to fill the room block, the hotel sought to recover the losses, stating that the contract expressed that the association was responsible for filling the rooms. The hotel claimed that the association was responsible for payment for the very rooms which it booked, which the hotel was contractually obligated to hold, and did hold, but was never used. In response, the association argued that without an attrition clause, it was not bound to pay for rooms in the block which were not occupied by the attendees.

The judge, in this case, agreed with the association that a specific attrition clause was needed in the contract for the group to be obligated, and that the agreement did not include any such stipulation. The result was that the judge dismissed the lawsuit brought by the hotel. Although this decision was made in a U.S. District Court in New York State it only applies to New York. The attorney who won the case for the association was quoted in a recent issue of USAE (U.S. Association Executive Weekly Newspaper) saying: "Technically and legally, the case only applied to New York State, but the ruling could be persuasive." This does not mean the decision is precedent-setting, but it could be. Other judges can very well go the other way in a similar case.

We have mentioned in previous columns the value of attrition clauses in contracts. Such a clause benefits the buyer as well as the seller. The whole idea is to make it very clear to all parties what was agreed upon when it comes to how the buyer will pay for rooms guaranteed, but not used. It is a question of just spelling out what the agreement was during the negotiation period. Every contract should have an attrition clause, not only for rooms, but for space, food & beverage functions, as well as any other revenue areas considered in the agreement. Be very specific.

According to attorney John Foster: "A well-written attrition clause will also give the hotel what it needs a guaranteed amount of revenue to offset lost profits if the group's attendance is less than what was agreed to by the parties. Meeting Producers and hotels should always take the necessary time up front to negotiate performance levels to include in the contract."

Also, have an attorney check over your contract format just to be sure you've got yourself protected.

LET'S DO A BETTER JOB ON DRAWING UP CONTRACTS ON GROUPS

November 2000

Group business is a great business for all type of properties. But we just need to do a better job in writing contracts so that there is no misunderstanding about what's what. Meeting planners are complaining more and more these days about undisclosed charges, which tend to come up during and after a meeting or convention at a hotel. The whole idea behind a contract is to make sure there is an understanding of who is responsible for what. It just does not seem right for any party to a contract to have to be responsible for any charges, which have not been disclosed and previously agreed upon by both parties. Of course, things will come up well after a contract is written, and even the last minute during a conference which may requires extra charges. This happens quite often, as we all know. However, there needs to be an understanding between the parties regarding policies and procedures of a property with regard it additional charges. To maintain a healthy relationship with accounts, which hopefully, will be bring more future business to the property, we really need to make sure that each party understands who is responsible for paying for what. And what anticipated charges there may be, based on additional needs, which may come up after the contract is signed. This means we just have to be more careful in our writing of contracts with group planners to make sure we disclose what additional charges may be imposed, and under what circumstances.

Recently I attended the VSAE (Virginia Society of Association Executives) Fall Marketplace in Richmond, VA and participated in a terrific contract negotiations seminar conducted by Mr. James Goldberg, a Washington, DC attorney. Jim specializes in meetings management issues, and consults with companies and associations in connection with hotel contracts on meetings and conventions. Among the many tips, he provided the meeting planners and hotel representatives in the audience was the importance of disclosing charges in the contract. It saves a lot of headaches, on both sides, avoids ruining healthy relationships, and helps get the bill paid sooner. Some planners in the audience indicated that they were already adding items to contracts reflecting that they would not be responsible for any charges, for any services not agreed to in the contract. Not being an attorney I hesitate to offer any advice in regard to this in this column, but would suggest that if there were any interest in this area to contact an attorney who specializes in this type of legal stuff. Most general legal counsels do not have this specialty expertise. If interested in being provided with a listing of some attorneys around the country that do this just let me know.

In many of my own sales workshops, I've been finding out from salespeople, when we discuss contracts, that they are signing the contract before they mail it out to a prospect. This just doesn't make good sense. It is best to send an unsigned contract (three copies), asking the other party to sign and return two. Then when they are returned, and there are no changes, go through it very carefully to make sure that there are absolutely no changes, then if you still agree, sign both and return one to the other party. This way both have contracts with original signatures.

RENEGOTIATING EXISTING CORPORATE AND GROUP CONTRACTS

May 2002

It seems that many market segments being concerned about the economy, and uncertain conditions, which may affect travel as well as meeting attendance, are seeking to renegotiate existing contracts for future dates.

Attorneys will tell us that a contract is a contract, and each party to the agreement needs to live up to its terms. However, as long as both parties to the contract agree, then certainly the agreement may be amended. From the buyer side, in all probability. What is really wanted is some relief on rate, easing up on an attrition clause, a relaxing on guarantees, and perhaps some "extras." All of this, or some of this, would be as a result in cost-cutting within the ranks of the sponsoring organization. It probably also is a fear in the anticipated drop of attendance when it comes to association or SMERF meetings. From the supplier side (hotels, resorts, conference centers, etc.), it becomes a question of whether it is or is not in your best interest to even consider a renegotiating of the contract. What concerns most General Managers, and sales personnel is what will happen to the buyer/seller relationship for the future if the property does not agree? Some things need to be considered if you chose not to renegotiate:

▶ To what extent does the relationship now hurt future business?
▶ Should the buyer fall short of guarantees how far will you go to collect attrition?
▶ If the buyer just refuses to pay or cannot pay would you follow with a lawsuit?
▶ If the buyer gets angered and decides to cancel the event would you follow through with a lawsuit?
▶ Would you agree to a lesser amount of attrition fees after the buyer falls short of guarantees?
▶ Is the lead time sufficient to book other business if the buyer cancels

It very well may be in the best interest of the property to consider renegotiating a contract, provided the new deal is a "two-way street." Helping the buyer with a new deal would surely put the property in a favorable light with the sponsoring organization (and make a hero out of the planner). However, it only makes sense if the property gains some advantage as a result of the renegotiations. Here are some thoughts to consider before agreeing to renegotiate:

▶ Is there other business available from the buyer for which they will contract now?
▶ They will tell you up front where their interest lies in renegotiating (i.e., what they want).
▶ Will the buyer provide additional events during the meeting which may bring more revenue to profit centers.
▶ Is there an opportunity to get additional years contracted for this same piece of business?
▶ If the group is made up of influential attendees who may be in a position to bring other business to the property.
▶ Perhaps there could be an increase in the rate agreed upon if the purpose of renegotiating is to relax the attrition or cancellation clause.
▶ A heavy deposit up front, or a prepayment of a large portion of the anticipated master account.

The whole idea in renegotiating a contract is that it needs to be worthwhile to both parties. If you give something in one hand, you then need to get something in return in the other hand.

BE SURE TO PROTECT YOURSELF ON ATTRITION CLAUSES IN YOUR GROUP CONTRACTS

July 2002

Lately, we've observed many meeting planners seeking opportunities to avoid payment of attrition fees by including additional "protection" clauses in group contracts with hotels. Attorneys, representing associations and corporations are providing plenty advice on how to do this. Generally, from their standpoint, regardless of how a hotel reservation is made, or how much is paid for a room, all rooms rented by attendees at the meeting should be counted towards the committed room block. In most hotel contracts, when a group does not live up to the room block commitment, a payment is required to make up for the rooms not rented. According to the terms of the attrition clause in the contract. Further, the number of rooms rented may also affect other consideration provided to the group such as complimentary rooms, reduced fees on meeting space etc.

Due to more and more Internet activity, by meeting attendance, in the booking of hotel rooms online at very special rates, we find that properties are not receiving the special group rate quoted in contracts for meeting and conventions. Because of this, some hotels are not crediting the group with rooms are booked with the group rate. This does make send; however, not with many meeting planners and their attorneys. They claim that rooms rented at a property, regardless of how it is booked or at what rate, should be contacted in the room block for attrition purpose. That belief was supported in an article written by attorney Mr. Raysner suggest that meeting planners include a definition of "attendance" to clarify how to determine how a property should count which rented rooms qualifies for attrition and other consideration. Here is what Mr. Raysner suggest for a special clause for meeting planners to include in a contract:

"For the purpose of this agreement 'attendee' means any individual associated with the(Meeting), including(Group's) and its affiliated organization directors, employees, representative, agents, speaker, exhibitors, members, delegates, guests, invitees, contractors, and subcontractors with guest room accommodation at(Hotel)."

Certainly, the author of the article. If he would be representing hotels, would never make this kind of a suggestion; it, of course, not being in the best interest of a property. All contracts are based on "consideration." A hotel by offering special group rates to an organization is the "consideration." in getting the group to meet at the property. In most situations, there may be a whole list of "considerations." Such as reduced rate, free meeting space, a formula for calculating complimentary rooms, upgrades, etc. If hotel agrees to count all rooms reserved by attendance as a part of a committed room block, regardless of how much less they may be paying for rooms via Internet bookings, we may as well consider the meeting a non-profit entity. As a result of negotiations for a meeting site, a planner accepts the responsibility of living up to the commitment of delivering a certain amount of rooms at specified group rates. Therefore, for attrition purpose, only those rooms rented at the contracted price need to be considered.

It may not be a bad deal at all to include an "attendee" definition clause, as suggested by attorney Raysner, but how about changing that last line to read: "guests, invitees, contractors and subcontractors with guest room accommodations at Hotel reserved at the contracted group rate."

UNDERSTANDING THE VALUE OF AN ATTRITION AND CANCELLATION CLAUSES IN GROUP CONTRACTS

September 2002

During a recent "meeting world" event held in New York City, I had the privilege of moderating a session with about 100 meeting planners in attendance. The topic was the hotel contract. The particular emphasis specifically leads to discussions on attrition and cancellation clause. It seems there is still a great deal of misunderstanding among meeting and event planners when it comes to contract clauses. Unbelievable, as it may sound, there are still group prospects that still want to cross out attrition and cancellation clauses. There are even some who absolutely refuse to use a hotel if they have to sign a contract with these clauses. Their thinking here is that if they avoid these issues in a contract they are not responsible for fulfilling their responsibility reflected in the contract Not so' the lack of an attrition clause does not relieve the group of meeting its obligation of sleeping rooms or food/beverage commitments. Also, the lack of a cancellation clause does not relieve the buyer of any responsibility of holding the event as committed in the agreement.

The purpose of an attrition clause is to spell out in what manner a buyer will honor the agreement to pay for the rooms (or food/beverage) that were committed but not used. For example, if the contract called for the property to hold 100 rooms for a group, and the group failed to use 25 of those rooms, the attrition clause would provide a mutually agreed upon sum to be paid if the room picks up was not met. Of course, the amount is determined through negotiation. Without an attrition clause, the failure of the group to fill the room block would be considered in court to be a breach of the contract for which damages must be paid. A court would then decide on the amount. The attrition clause makes it very clear as to the agreed amount to be paid. Rather than trying to negotiate a complicated formula based on percentages, I feel it makes it very clear when both parties agree to a set amount per night, per room or for each dollar not picked up for food/beverage guarantees.

With regard to cancellation clauses, the inclusion of such really does more to protect the buyer than the hotel, Without an opportunity for an event planner to get out a contract(even by paying an agreed-upon amount) the group may be responsible for paying a much larger amount for breach of contract, should the case go to court. By the way, I should mention, most contracts I have seen provide a cancellation clause for the buyer, but not the hotel. "What's good for the goose is good for the gander." A cancellation clause should work for both parties.

Beside a cancellation clause which should allow either party to cancel, for any reason, but for an agreed upon amount to be paid in damage, contracts should include a termination clause. This clause provides an opportunity for either party to get out of a contract for a specifically outlined reason, without having to pay any damages. Generally, this clause is referred to as an "Act of God" clause, and usually, converse conditions that create a situation making it impossible for either party to perform.

Of course, it is always best to have an attorney who specializes in meeting and event contracts to work with you in coming up with an agreement to protect your interests.

THERE'S A MOVE ON TO CHANGE YOUR CONTRACT CLAUSES FOR MEETINGS

June 2003

The hottest issue today in the sales, with regard to booking a group for meeting and convention is how you are wording some of your contract clauses. The three areas of particular concern will probably fall into clauses covering contract termination/cancellation, attrition, and liability. Right now we see a big move on among folks who are responsible for company meetings, and association conventions to change your contract clause, particularly in the area of event cancellations. These buyers are being egged on by their attorneys to seek changes in wording which would provide more of an "out" for many reasons without paying any damages.

All contracts should certainly have something in a contract which would allow either party to get out of the agreement. However, whether damages need to be paid or not is a big consideration. A termination clause in a contract should allow either party to cancel without damages for the specific reason. This has been commonly known as an "Act of God" clause, and it would list thing that could happen that would make it impossible to live up to the term of the agreement (Flood, Tornado. Hurricane, Earthquake, etc). Lawyers like to use the term "Force Majeure" clause. A cancellation clause in an agreement would allow either party to get out of the contract for any reason; however, there would have to be some damage paid, and this amount is negotiable, and once agreed upon, the details would be specified in the document.

Some lawyers and meeting planners have been pushing for a change in the termination clause. They state that since 9/11 things are happening in the world, creating more risks which should allow buyers to be able to get out of contracts, without damages, for a lot more reasons. Other than for allowances of the act of God and impossibility to perform they want to include threats of terrorism, fear of the personal safety of attendees, war, economic downturns, and probably a lot of other things to help get "off the hook" should they decide not to hold an event. And, at first glance of this may sound reasonable, but is it in the best interest of the hotel? The fact that we were at war in Iraq or there have been, and possibly will be more terrorist activities around the world, what would that have to do with people attending the meeting in the U.S? Or, if a company loses money one quarter, should they be allowed to get out of this contractual obligation without paying damages? Even if you agree, and want to add more reasons in a termination clause for not having an event, then you need to be sure that the reason is very, very specific.

The Whole idea is to be very careful in the wording of a termination clause to be sure that a buyer accepts the responsibility of holding an event unless something happens that makes it IMPOSSIBLE for either party to perform. It is just too easy to get out of a contract for any reason, by indicating the reason to be one of those loosely worded terms those lawyers want the meeting planners to use. Pretty soon we will see business terminated, without damages, and the event held at a competing site, if we are not careful. Of course, canceling an event for any reason is certainly OK, provided the details of what and how the damages should be paid are spelled out very specifically.

ATTRITION CLAUSES IN GROUP CONTRACTS NEED TO BE CLEARLY ADDRESSED

June 2003

In our last column (HMM, June 2, 2003) we addressed the issue of termination and cancellation clauses in group contracts which have been hot topics in these troublesome times. However, another most-talked about the issue among meeting planners is this business of how to handle attrition in connection with groups. An attrition clause in an agreement is simply a segment explaining what arrangements are to be made to resolve a situation whereby a buyer does not live up to its commitment as specified in the agreement. Generally, this would refer to a group committing a specific number of rooms to be rented, at a specific rate, and not being able to deliver those rooms. Or it may pertain to blocked meeting room space, a specific amount of dollar revenue to be generated in food and beverage, or a specific number of persons guaranteed for a meal, at a specific price.

The objective is for the seller (hotels, resorts, conference centers, etc.) to be able to recoup lost revenue as a result of the buyer not living up to its agreement. At the same time, the attrition clause will also protect the buyer by specifying in what manner the issue would be resolved. Without such a clause, should the buyer default on any portion of the commitment, they could very well be held responsible for a much larger amount of damages than would have been specified in the agreement. Having the attrition clause provides the opportunity for both parties to negotiate the terms of the damages to be paid. The whole idea is for both parties to have a clear understanding of what would happen should either party not live up to its agreement.

The latest wrinkle in attrition is the issue of a buyer committing a block of rooms to be used, at a particular rate, and now here comes the third party Internet provider. Prospective attendees at a meeting, searching the Internet, may be able to make reservations at the property, at a lower rate than specified in the contract. The buyer indicates that those rooms should be credited towards the block. However, since they were booked at a lower rate than committed by the group, should those rooms be credited towards the block? This is the issue now at hand in our business of group bookings. If the hotel accepts those rooms as counted in the block, although booked outside the block, at a lower rate, then there would be lost revenue for the property. So, it may only be five rooms today, but maybe 50 tomorrow, etc. In the long-run, there would be no point in coming up with a group rate for a block of rooms. Hotel operators need to protect potentially lost revenue when it comes to wording in the attrition clause. Of course, it will come to negotiations with the buyer to come to terms that would be acceptable to both parties. At the same time, there is also the issue of other benefits to the buyer by meeting its room block. In many cases, hotels are offering free meeting space or other added features based on the buyer "picking up" the room block, or picking up a specific amount of food and beverage revenue. We must be careful, and very clear in specifying how the revenue will be protected.

It is worthwhile to use an attorney who specializes in this field.

WE'RE INTO A SELLERS' MARKET AGAIN, SO WATCH THOSE CONTRACTS

November 2004

With our economy turning around, everything we've seen with regard to business travel and tourism seems to be very positive. For the first quarter this year it looks like hotels have had about a 4.4% increase in occupancy over last year. Also, rates have been gaining as well as RevPAR. So things do look good for the rest of 2004 and 2005. However, there are group room buyers out there that are still talking about us still being in a buyers' market and we still see a lot of hotel room rate discounting going on. We need to remember that profitability has more to do with improving revenue than occupancy, and there are salespeople out there that are still giving the thing away that they may not have to. It really boils down to doing a better job of selling value and negotiations.

When it comes to contracts on groups we should be aware that meeting planners are being advised at workshops being held around the country on how to get around termination, cancellation and attrition clauses. Of course, it is up to individual properties as to how these contracts are written; do not get fooled into using some "canned" clauses that you read about in most of the meeting magazines. Remember that attorneys representing the buyers (associations, companies, and a variety of SMERF groups) are writing these clauses in the best interest of their clients, the people who plan meetings and buy rooms, space, and meals from hotels.

The first thing to remember about signing contracts is not to be the first person to sign the contract. We still see sales personnel sending out contracts with their signature already on the document. By being the first to sign you do not have a chance to correct anything or change your mind about something once the buyer has signed. Always send three originals of the contract, asking the buyer to sign two and return them to you, and keep the third for their file. Then once you receive the two signed originals, if everything is in order, and there are no changes or corrections or additions (watch out for addendums being made part of the original contract) then you can sign both originals and send one back to the buyer.

It is always a good idea to include a "Termination" clause in the contract that should reflect under what conditions EITHER PARTY can get out of the contract with paying any damages. This clause is usually referred to as the "Act of God" clause. Either party should be able to get out with paying damages if anything happens that makes it IMPOSSIBLE to conduct the meeting. Be careful on any changing of the wording, which could reflect that a buyer can get out of a contract for any other reason. Of course, in writing a cancellation clause you should reflect that EITHER party can get out of the contract for any reason; however, you would indicate what damages would need to be paid should either party cancel.

When it comes to an Attrition clause you just want to be sure that the property is protected in the event the buyer does not deliver what has been agreed upon in the agreement. This should reflect on room pick-up, guarantees on food and beverage as well as any prices indicated for meeting space not used. One thing about which you should protect the property is the number of rooms that are booked at the property that, for any reason, were able to be booked at a rate lower than the group rate indicated in the contract. Those rooms should not be counted in the block of rooms being held and credit should not be given for any of those rooms rented at some lower rate, regardless of how it was booked.

WHAT'S IMPORTANT ABOUT CANCELLATION CLAUSES IN GROUP CONTRACTS?

May 2006

There is no point in entering into any kind of an agreement without including a clause that should be able to get either party out of the contract. Any lawyer will tell you that. There are a good number of group contracts being written today where the cancellation clause allows the buyer to get out, under a certain condition, but makes no mention of the hotel also being able to cancel. There should be a clear understanding in the agreement as to what happens if the prospect needs to cancel or if the hotel needs to cancel.

The terminology in the contract needs to be very specific as to what will happen if either party cancels by a certain date. The clause could reflect that a party could cancel up to a certain date without damages, and then go on to stipulate what the damages would be, period by period, as time drawers closer to the date of the event.

There also needs to be a termination clause in the agreement. This is where either party can terminate the contract for a specific cause without paying any damages. This is also called an Act of God clause. Any amount of specific causes could be included, provided of course, if both parties agree. This clause should specifically indicate that any of the causes listed would make it impossible for the event to be held. Sales staffs should be aware that many prospects for group business are attempting to include an addendum to contracts to cover a wide variety of items, some of which may make it impractical for them to hold the event, but that would not make it impossible.

Some prospects are under the assumption that if they cross out the cancellation clause in a contract, and the hotel agrees not to include it, that they are off the hook and can cancel at any time without paying damages. This is not the case, and sales staffers at hotels, resorts, conference centers, etc., need to advise these prospects that the damages would probably be worse if there is not an opportunity for them to cancel under certain condition. Technically, if there is a legal contract and no termination or cancellation clause allowing and "out" for the prospect's organization, the property is in a very favorable situation. In this case, the canceling party may very well be liable for picking up the tab for the entire expected revenue from the organization. The property, would, of course, have to mitigate damages (do the best they can to resell the rooms and space). Even though this may look good for the facility, it present problems not have a cancellation and termination clause in the agreement.

Neither the cancellation nor the termination clauses have anything to do with a prospect not filing the complete commitment to the agreement. That is, not picking up the complete block of rooms committed, or food service guarantee as agreed upon. An attrition clause would cover those details, listing specific amounts to be paid for sleeping rooms, meeting space, or drop in food/bar service guaranteed.

CONTRACTS ON GROUP BUSINESS GETTING MORE COMPLICATED

November 2006

There is no question about it; contracts on group business at hotels, conference centers, convention halls, etc., are getting longer and longer, with more and more clauses to cover everything possible. You hardly see any contracts with less than 3 pages and most are running 10–12 pages. It seems that some buyers from the corporate and association markets feel that hotels are using outdated contracts, and they need to be beefed up. This is according to a recent Meetings Today (Business Travel News) survey of buyers. As a result of this thinking, a good number of buyers have been creating standard addenda in their contracts with meeting sites. These addenda include clauses which are designed to further protect the buyer in a wide variety of areas; mainly in cancellation clauses, attrition and food/beverage functions. When this happens it is very critical for the vendor representative (in most cases a sales manager) take a close look before any acceptance of the addendum is made. More than likely there will be clauses in the document that conflict with clauses already included in the vendor's contract. It all really boils down to the negotiations process, and how flexible the vendor might be. The key in negotiating, from the seller's standpoint, has more to do with how well they know the value of the prospect's business and how badly the property needs the business Then, the critical point is in the reading, and understanding the terminology included in the addendum presented. If a rider to a contract is accepted, more than likely the information in the rider (addendum) will take precedence over the wording in the original contract.

This business of companies and associations creating addenda to contracts is nothing new; it has been going on for at least the past 20+ years. However, we see more and more prospective buyers using that tool, and they are getting longer, and longer, covering a myriad of items already covered in the original contract presented by a property. According to the survey the following elements, included in a standard meetings contract addendum, are listed in order of being most common clauses: cancellation clause, food & beverage, fees or surcharges, attrition, audio-visual, payments, rebooking renovations/construction, overbooking and reselling.

A good deal of credit needs to be given to our Convention Industry Council for the work they are doing on APEX (Accepted Practices Exchange). Included in a variety of meeting industry initiatives is the APEX panel on Meeting Contract. The panel is made up of representatives from all segments of the meetings industry, and headed by Tyra Hillard. The work of the panel has been completed and is being reviewed by CIC. The paper presented by the panel gives an overview of meetings contract basics, some general best practices, and explains 23 of the most common meetings contract clauses, and approaches to be considered.

It does not include any sample contracts clauses, since the document is intended to be for people from all aspects of the industry, according to Hilliard.

GROUP MEETING CONTRACTS KEEP GETTING MORE COMPLICATED

June 2007

The whole idea behind a contract with a hotel prospect for group events is to make it clear to both parties what was agreed upon during the negotiations. However, at the same as the organization; which means making a deal that is profitable to the property.

Over the years we have seen hotel/prospect deals go from a one-page letter of agreement to a multi-page contract, and it seems that they are still getting longer as well as complicated. The wording in contracts has changed so much over the years that now it is almost necessary to have attorneys review contracts before they are signed. There are some contract clauses that special attention to be sure that a facility's profits are protected.

▶ **Attrition:** This is the clause that explains in what manner the buyer will pay what charges that are due for agreed upon services that were not used. If 100 rooms nights were to be used at a specific rate, and only 75 rooms were used then what amount in dollars would the buyer have to pay the hotel. This amount would have been negotiated well in advance of the event, but it needs to specific amount that was not collected, for any reason on space rental, food/beverage, recreation or any other expected revenue, as a result of the group meeting at the property. It is critical to actually put in dollar amounts rather than percentages of expected revenue.

- **Termination:** The content of this clause spells out the opportunity either party has to cancel the holding of the event, without paying any damages to the other party. However, the reasons for cancellation need to be spelled out in very specific terms. Generally, this clause is used when the weather makes it impossible for the event to be conducted, by either party. Sometimes this is called the "Act of God" clause, but could be used in the event of other conditions that may exist which would make it impossible for the event to be held. Be sure that the word "impossible" is used.
- **Cancellation:** With this clause, either party should be able to cancel, for any reason, but damages would be paid, as spelled out in this clause. Usually, a scale is created to determine the amount of damages to be paid based on how far in advance the event is canceled. The further out the cancellation notice is given the less amount would be paid. If the event is canceled within short notice, the harder it would be for the property to rebook the space, therefore, the highest amount would be paid.

Of course, all clauses in a contract are important, and all need to be written in terms that are completely understandable. When sending a contract to a prospect to be signed, just be sure that the hotel representative sending the contract does not sign it until it is returned, and reviewed to be sure there are no changes. The Convention Industry Council (CIC) has recently completed a study and review of buyer/supplier contracts within their APEX initiatives. For complete details on the review and recommendations please go to www.conventionindustry.org and link to APEX.

UNDERSTANDING THE ATTRITION CLAUSE IN GROUP CONTRACTS

October 2007

We have covered this subject in a limited way in previous columns, but there is always something new to learn with respect in covering the issues to protect your property's assets. Attrition still seems to be the number one concern for meeting planners in trying to have the contract written in their own best interests. It is therefore important that hotel contracts written by the sales staff manage the wording of the attrition clause to make sure that the properties best interests are covered.

The idea is to be sure that all the revenue due the property, as a result of a group event, are collected, if the group defaults in fulfilling the terms as set for in the contract. The attrition clause in the contract addresses the issue of how any amounts due should be computed for collection. If the contract reflects that the group has a guaranteed a room block of 100 rooms to be used each day, at a certain rate, and only 75 rooms have been used, the attrition clause spells out how much would be due the property for the 25 rooms per day that was not used. Although some planners have been successful at negotiating fees due based on percentages of actual rate, or percentages of the block not used, make sure whatever method is used to come up with the amount due, it should be reflected in dollar amount.

When computing amounts due for other contracted items not used, such as meeting space, food, and beverage, audio-visual equipment, exhibit space, etc, the amounts due should be reflected in dollars. Most of the time, in group contracts, we see attrition fees calculated on the total number of room nights used during the period of the event. What we need to do today is make the computation on the number of rooms picked up from the block on a daily basis. If the agreement reflects that 100 rooms were blocked for each day of the meeting, and if on the first day only 80 rooms were used, then the hotel is due to the agreed amount for the 20 rooms not used. The fact that the group uses 120 rooms the next day does not mean they will get credit for the extra 20 rooms to cover what was not used the previous day. This could be a critical thing to cover when negotiating with the prospect regarding the manner in which the attrition clause is written in the contract.

Other areas to consider within the clause should be that any rooms booked by the members of the group, at any rate, lower than the specified group rate in the contract would not be counted as part of the block of rooms guaranteed by the group. Sometimes, group attendees, have the ability to seek out lower-rated rooms at the property via a third-party Internet provider. Further, there should always be a cut-off date which would allow the group to reduce, or even increase, the guaranteed block of rooms, by a certain date in advance of the event. In the event there is a change, agreed upon by both parties, then it should be in writing, signed by both parties, as an amendment to the contract.

CAUTION IS NEEDED WHEN SIGNING GROUP CONTRACTS

March 2008

In many cases, today, sales personnel who sign contracts for group business, booked at most properties, large and small, do not understand the ramifications of putting their name on contracts. The whole idea is for anyone to limit his or her potential liability. In putting your name on a contract, it reflects that you are personally responsible for the terms of the agreement. It is always best to sign as a representative of the property. For example:

ABC Hotel

(**Your signature here**)
As it's authorized Sales Director

Of course, it also applies to Catering Managers, General Managers and any other person who is authorized to sign contracts.

Then there is the question of who is authorized. Generally, the law recognizes people such as sales managers, catering managers, and others, who have "apparent authority," by nature of the job or duties, to sign the agreement on behalf of the property. This may very well be a minor issue since I have never heard of a case where a salesperson was held personally liable in a suit involving a hotel, even though they may have, personally signed a contract for a group. Some years ago I was called in as an expert witness on behalf of a hotel property being sued for nonperformance of a contract. The sales director of the property who signed the contract was also listed as a major party to also being held responsible. Although the hotel was found at fault in the case, the sales director was not considered liable.

This particular issue of signing contracts also applies to the group planner who signs a contract as a buyer. Right now there is a very large percentage of group business in hotels being booked by third-party independent planners. These independent planners seek locations. Enter into negotiations, and have been signing contracts on behalf of groups they represent. From the hotel standpoint, sales personnel involved in these bookings need to be cautious by making sure that the contract is signed properly by a truly authorized representative. It is not in the best interest of the hotel to accept a contract signed by an independent, making it look that they are a major party to the agreement. They, too, should be signing in the same manner as the property salesperson, as an authorized representative of the group. Further, in dealing with independents, it is always best to make sure they truly do represent a particular prospect for business. You can do this by asking them for a letter, from the group planner, reflecting their representation.

Today we are getting involved, to a certain extent, in entering into contracts electronically. It sure looks like paperless contracts are on the way. Some states have already adopted the Federal Electronic Signatures in Global and National Commerce Act (E-SIGN), as well as the Uniform Electronic Transaction Act (UETA). These laws state that electronic transactions cannot be challenged just because they are executed in an electronic manner. However, both parties to such an agreement have to agree to be bound by it. Better check with your attorney on how to get into this if you are interested.

NEW CONSTRUCTION AND RENOVATION PROJECTS MAY CAUSE GROUP BUYERS TO BEWARE

February 2016

Yes, no question about it. This year, 2016, will reflect some great improvements in average daily rate, occupancy, and demand. In all probability, 2017 will also be great. So what does this mean for the following years? History tells us that over the years, in our hotel industry, we have had our "ups" and "downs." When the industry shows improvement in business, developers, owners, and investors go to work on plants for new construction, as well as renovations. With that, we need to understand that group business buyers always seem to be cautious about making deals with properties that are either under construction or being renovated.

We all understand how important it is for the success of a property to have business on the books in advance of openings. Sales staffers can do this, and for the most part, do it very well. They do run into problems, however, when construction does not come through with a finished product on time. Sometimes, it gets to a point that salespeople, themselves, are booking business against a two to three-month "cushion" period prior to a new property's opening date. Construction schedules have to be realistic to ensure that the new property opens on time.

Some group business buyers, it seems, just will not trust opening dates. Granted, they understand they can probably negotiate a very good room rate deal for booking a new property in advance. However, in the past, some have been burned that they shy away from these "good deals." Salespeople also have had to deal with partially finished openings which certainly does cause problems for the sales staffs.

It is understandable that, perhaps, some owners and developers, needing to see some cash flow, insist on opening before the guestrooms are ready. From a marketing standpoint, this piecemeal approach to opening a property if far from healthy. It just does not look good, especially to business group buyers. It is safe to say that meeting planners, on-site inspection trips don't fall over each other to look at a property that is only partially finished.

There are also similar problems with properties that are being renovated. Many make promises to meeting planners, then find they cannot fulfill them. It gets to a point that when a salesperson says, "We plan to renovate," it is more of a hindrance than a help to making a sale.

Salespeople should be on the lookout for meeting planner group buyers who will want certain protective clauses written into contracts. It is important that sales staffers, and General Managers, in particular, should do some careful reading of all agreements to protect themselves. We are probably already seeing some amendments to property sales contracts where planners are giving themselves an "out" on cancellations if any renovating is expected to be going on at a property during their meeting.

We particularly have to be aware of how to protect ourselves on agreements written a year or two in advance of an opening or expected renovation. This is an issue that will need to be addressed, and we need to get our lawyers busy on way to protect us in such situations.

CHAPTER 14

EVERYONE AT A PROPERTY IS INVOLVED IN SALES

TURNING EVERY EMPLOYEE INTO A SALESMAN/MARKETER

October 1981

Who says that the promotion of hotel/motel services belongs only to the sales or marketing department? Growth responsibility of marketing and sales must be placed in the hands of the "front-line" people; front-desk staff, waiters, waitresses, housekeeping staff, bellmen, bartenders, and yes, even the maintenance man, yard man, and pool attendant. In short, all the people on the staff with whom our customers are in contact.

All hotel/motel employees must realize that selling all services of a property is a part of the job regardless of the specific duties assigned. Everyone must participate. In each of our jobs, we have to extend ourselves to our guests—not wait for them to ask us. It's like the waiter or waitress in your restaurant who doesn't wait for the customer to order a cocktail or wine or an appetizer. All these "extra" sale items should be "sold" to the customer by the server. And the server who understands this, who gets the extra sale, the larger dinner check, will, of course, also get the larger tip.

MORE GUEST SATISFACTION

Likewise, it's the front desk salespeople that have the opportunity to "sell up" on rooms—creating a better average rate for the property—and, in all cases, providing better satisfaction for the guest. Here also is the best place to start promoting your restaurant with a sincere recommendation from the front-desk sales clerk. A promotion poster near the desk advertising the lounge and/or restaurant is a good tool to use. Handouts can also be useful.

From the laundry room through housekeeping, all on the staff have the best opportunity to sell the customer on a return visit. The most important items in getting a guest to return are staff attitude and clean rooms. A clean-looking, happy and friendly housekeeping staff can make guests want to return. A friendly "hello" from a maid with a recommendation for a meal in your restaurant will go a long way in convincing your guests to eat with you.

The bellman, looking fresh and being friendly and helpful, is a strategic position from which to sell all the extras at the property, including the restaurant, lounge, game room, and vending machines. Your best salespeople for local banquet business are probably your own hourly employees. Don't forget, they too have all kinds of contacts, like neighbors, friends, relatives; and they too belong to veteran's groups, churches, garden clubs, school groups, etc. And it's not unusual for a bartender to be your best salesman for meetings and banquets, just by chatting with guests at the bar.

Meetings should be held regularly with sessions conducted by sales and management staffs on how each of the various employees who come in contact with guests can sell all the services of the hotel. In fact, why not just ask them? They'll tell you how they can help you with marketing and sales.

SALES INCOME OFTEN DEPENDS ON INCOMING PHONE CALLS

December 1985

Meeting planners are continuously amazed by the lack of professionalism displayed by hotel/motel salespeople in the performance of their jobs. From the handling of an inquiry to the solicitation effort to the booking stage and follow-up, salespeople need to be better performers.

It all starts with the switchboard. From the start, we tend to put people in a bad frame of mind when incoming calls are answered tersely or unpleasantly. Conversely, of course, a pleasant voice is a welcome sound to a caller, and can make the caller feel welcome and at ease. Well-operated hotels/motels sometimes are reflected in the promptness by which the switchboard is being answered as well as the attitude of the hotel being displayed through the manner in which the call is taken. As a starter, people calling in may be "turned on" or "turned off" by how the switchboard is answered.

Then, when the caller is connected to the sales department, the same opportunity exists to either impress or displease the caller. How many times does the sales department phone ring before it is answered? Once, ideally, and not more than twice, hopefully. Then, what is the attitude displayed by the secretary or salesperson answering the phone? A positive, friendly, eager-to-help voice is often the key to a successful sale.

QUESTIONS CAN BE A TURN-OFF

If the caller has to answer too many questions before he or she gets to the person wanted, the call (and possibly a lead or a piece of business) may be lost forever. Too often a caller has to answer questions such as: "Who may I say is calling?" "What is the name of your company?" or "What is the nature of your call?" These can all be real turn-offs. If the call is for a particular person in the sales office, it should be taken immediately—without questions. Salespeople should take every call without screening. Sometimes it is much easier getting through to the President of a large corporation than to the sales manager.

Sales personnel also should examine the number of calls they do not get right away, and how those calls are handled by secretaries. If salespeople are out on the street making sales calls that is something the caller is able to understand. If the salesperson being called is busy with a prospect—showing the hotel or helping him plan a conference—that, too, is understandable. However, when a prospect or client calls and the salesperson is "in a meeting" with other hotel people, the meeting should be interrupted so the salesperson can take the call. In fact, a good policy for salespeople to follow would be to have all in-house staff meetings before 9 a.m. or after 5 p.m. All other time in between needs to be available for selling.

Another area regarding the telephone is the length of time it takes to return phone calls. Some calls naturally have priority over others. This can usually be recognized by the name of the caller, the company name or the messages left. Many calls are made and messages left by what may appear to be "bothersome" callers. However, one never knows. The one call never responded to, which may have appeared to be a magazine sales representative trying to sell magazine advertising, could very well have been that person trying to set up a sales meeting for the people from the magazine. It's good to prioritize telephone messages, but at the same time, good salespeople always manage to respond to phone messages within 24 hours.

The sales personnel at a property need to look at themselves and the manner in which they answer the phone—enthusiasm shows up very easily as does the lack of it. Then, they need to look at the sales secretaries and other people in the department and the switchboard.

Let's all be enthusiastic and sell.

DON'T BLAME SALES STAFF IF CUSTOMERS AREN'T RETURNING

August 1986

With increasing frequency, it seems, guests are beginning to wonder whether we as hoteliers really do appreciate the business they give us.

I think that somewhere along the line, we're dropping the ball—and it's being noticed not only by our guests, but by the people who bring guests to us: travel agents, meeting planners, motor coach companies, incentive houses, association executives . . . they're all beginning to notice that hoteliers often aren't doing the job they should be doing.

Stop and consider the following:

▶ If every guest that ever stayed in your hotel came back to you each time he or she visited your city, you would probably be running at full occupancy every day of the year, and turning away crowds daily;
▶ If every meeting ever held at your hotel repeated with you whenever meeting time came around again, you'd never have to budget anything for marketing, and you wouldn't have space to handle all the groups;
▶ If every local social event repeatedly returned for a catered event, you wouldn't have enough banquet space to accommodate them all.

If you think about it, the real reason for placing advertisements hiring salespeople, creating sales promotions and spending marketing dollars is that, overall, we're doing a lousy job of motivating our customers to do business with us on a return basis. We all know that business returns when the customer is happy with a hotel's services, prices/value, treatment, and so forth. It stands to reason, then, that hoteliers who are running their properties right shouldn't have to worry about new hotels coming online, highways being moved, competitive pricing or frequent user prizes—if the hotel is operating properly, it will stay full.

Marketing Dollars

More and more money is being spent by hotels for marketing campaigns aimed at bringing in new business. The reason for going after new business is obvious: you have to replace the old business that's not coming back. Every year, the industry spends millions, perhaps billions, to compete for and bring in new accounts.

But if an individual or group has stayed at your hotel before, they already know what it offers, and they know whether they like it or not. No selling should be necessary—if, of course, the hotel is doing a good enough job for them to have reason to return. And they will return, if the price/value relationship is right; if the staff is helpful and friendly; if the facilities are right to meet the individual and group needs; if the property is clean and comfortable; if the food quality and service are better than expected—if all these jobs are done right, then there's no reason for customers not to return.

By not seeing to it that guests want to return, hotel/motel managers and owners are missing the opportunity to reduce their advertising and marketing costs. After all, the most important function of management and ownership is to make sure the hotel is being run so effectively that every guest or conventioneer that stays at the hotel wants to come back. You must exceed the guests' expectations.

Expensive Job

The job of acquiring new guests is getting more and more costly. Anything we lose is getting increasingly harder and more expensive to replace. Our most valuable asset is our customer base, but are we managing that customer base properly? What is our relationship with our top accounts? Are we showing them that we really care? Are we really doing the best possible job to ensure that our guests and meetings customers come back to our hotels?

Salespeople get blamed for not bringing in new business—but who's responsible for making sure that the new business comes back?

IN THE LODGING INDUSTRY, STAFF ENTHUSIASM SPAWNS SUCCESS

November 1986

Regardless of the size of a property—whether it's a 15-room motel or a 1,000-room mega-hotel—it's salesmanship, probably more than product, that determines the success of the facility.

Salesmanship is the real backbone of success.

What is this thing called salesmanship? It is something only salespeople are supposed to have? Salesmanship encompasses a variety of things, but the most important ingredient is enthusiasm.

Ideally, an enthusiastic attitude starts with the switchboard operator and is reflected in the attitude of the entire staff throughout each department. How the phone is answered; how a guest is received; how employees treat each other . . . A guest's perception of a property is positively influenced by a staff's enthusiasm—and negatively influenced if it's lacking.

ENTHUSIASM BREEDS CONFIDENCE

Generally, customers feel more confident when they deal with salespeople who show enthusiasm for what they're selling. The way a secretary answers the phone in the sales office, catering department, reservation desk or manager's office, is a key to salesmanship. An enthusiastic greeting with a "telephone smile" gets half the sales job accomplished right at the start.

Salesmanship is also knowledge of the product: confidence in the facility and its staff; warmth; empathy; courage; conviction; sharing information; asking the right questions; overcoming objections; credibility; and simply asking for the business.

No one ever buys anything—a car, house, suit or dress, and especially a meeting—without having confidence in the product being purchased. If the salesperson doesn't display confidence in his product, there's hardly any chance to make a sale. A salesperson's confidence generates confidence in the customer—and that usually results in a sale.

MOST IMPORTANT KEY

Enthusiasm is by far the most important key to displaying confidence. And we're talking about sincere enthusiasm—phonies are easily detected. Since enthusiasm is catching, it's relatively easy for every hotel or motel to create an "enthusiasm campaign"—I guarantee it will sell more of the property.

Since everyone sells, everyone on the staff needs to be enthusiastic about the property, its management, and their jobs. Naturally, the campaign starts with the property's owner or operator and its management team.

Try out this test: Count the number of employees you find displaying enthusiasm during the course of a given day. Check on how the phones are being answered; watch various people doing their jobs; observe the front-desk action; sit in the dining room and observe how the servers greet their customers; visit all departments, especially sales and catering—how do they rate on the enthusiasm scale?

Without sales, we cannot exist. And without enthusiasm, selling is hard to do.

Remember: Salesmanship is the key to success—and everyone on the staff is a salesperson.

ACCOUNTING DEPARTMENT CAN HELP SELL AND REBOOK MEETINGS

June 1987

It's amazing how much time and money is spent to generate sales—and how easy it is to mess up and not get a piece of business to rebook.

Let's not think about the obvious things that can happen to mess up a meeting; we would all agree the list is almost endless. Instead, let's think about the not-so-obvious mistakes that the sales department may hear about.

Let's look at bill collecting. In a majority of cases, properties are creating problems because of the manner in which billings are handled. Meeting planners continually mention that they are willing to pay their bills promptly—if they are billed promptly and accurately. Here are some of the complaints that are heard most often:

▶ Delayed bills (many as much as 30 days late);
▶ Bills sent without sufficient backup data;
▶ Unclear printing or overprinting on folios, making them unreadable;
▶ Charges for items that were not previously authorized;
▶ Prices on items that were not previously agreed upon;
▶ Incidental charges or individual folios added to the master account when not previously approved;
▶ Bills that aren't appropriately identified for specific meetings (in cases of planners with multiple, simultaneous meetings);
▶ Bills being mailed to the company or organization rather than to the specific person responsible;
▶ Counts on functions that differ from meeting-planners count or guarantee for functions;
▶ Extraneous items that the hotel agreed to pick up.

This list is far from complete, but it gives you an idea of the nagging problems that meeting planners often have with billing procedures. These kinds of problems can cause the hotel a loss of future business.

Preventing Problems

You can, however, prevent these problems and look very professional in the process. The key is to make use of your accounting department.

During the consulting period (the selling process), and certainly, during a site inspection, sales personnel should use the accounting department as a sales tool. As an example, tell the planner that as a part of the hotel's professional ability to handle his meeting, the accounting department will handle the account any way he sees fit.

Further, explain that the controller will meet daily with the planner to review all previous days' charges so that any inequalities can be resolved immediately. A final review of the complete billing with appropriate backup may be made before the planner's departure from the hotel. You can also suggest that 90% of the bill be paid at that time, reserving the remainder for possible future adjustments. (Salespeople might be surprised at the number of planners who will agree to this).

It should also be explained that the complete bill will be sent within the week and that all charges will be completely legible and will have appropriate backup data. And don't forget to introduce the planner to the controller as your partner in getting all this accomplished.

PROFESSIONAL APPROACH

Taking this approach will make the hotel look very professional in the eyes of most meeting planners, and could possibly help close the deal during the consulting period.

Of course, after these promises are made, they have to be kept. Sales personnel should consult with their General Managers and ask for a meeting with controllers in order to implement a program similar to the described above.

The accounting department can help the sales department sell and rebook. The controllers will be happy because such a program makes life simpler in the accounting department. And the sales department will be happy because they'll collect the money sooner!

LET'S MAKE MORE USE OF OUR 'FRONT OFFICE SALESPEOPLE'

April 1988

We don't pay enough attention to our front-line employees and their ability to sell for our hotels and motels. Desk clerks, cashiers, night auditors, bellmen, doormen—they all sell for us. They're the first employees the guest sees when he arrives, and the last ones he sees when he departs.

You don't get a second chance to make a first impression, and this fact should be emphasized to your front-line employees. Similarly, a guest's final impression of a property is made by the employees he sees upon departure—again, the front-line workers. The destiny of a lodging property often lies in the hands of these employees—and we should be paying more attention to them.

Adolph Krisch, Chairman of Krisch Hotels, once said that we should not call our check-in people "desk clerks;" instead, they should be called "front office salespeople." And rightly so. The first greeting to a guest and the check-in procedures set the tone for the guest's attitude during his stay.

A positive first impression can make other things that may go wrong somehow less important. Start a guest off on the wrong foot, and beware: Anything slightly out of kilter could create an avalanche of complaints. Creating a positive impression for arriving guests is the key ingredient to ensuring an enjoyable stay for them.

GETTING RETURN BUSINESS

Of course, the ultimate objective is to get the guest to return. So much time, effort and money are spent to get someone to stay at a property, but General Managers and owners also should concern themselves with making sure that repeat activity is very high. And remember: Guests are not only needed as potential repeat customers; they are needed as goodwill ambassadors to spread the word about their great experience at your hotel.

If anyone would undertake a bit of research, it probably would show that many people return to a lodging property as a result of the treatment they receive at the front desk. In addition, front office salespeople can be a great asset to the sales department. Who but front office employees have the chance to talk to each person checking in and out of a property? In all probability, many of these overnight guests are good potential prospects for booking groups at a later date—but they have to be asked.

Perhaps your property's official sales staff needs to talk more with these front-line salespeople. It could help the sales staff learn more about who the guests are, where they're from, what companies they represent. In fact, many front-office salespeople would probably enjoy talking to guests about the possibility of them bringing meetings and groups to the property. (They may even be able to do it better than some of the sales staff!).

Providing Leads

Sales personnel can prompt the front office salespeople on key questions to ask; the front office people can then provide leads to the sales office. The idea is to catch the prospect while he or she is still on the property. Believe me, it works—and it's a great way to make a sale.

It's also a good idea for the owner or General Manager to periodically call a sales meeting for all front-line salespeople. It doesn't have to be a lengthy meeting; it would simply involve the sales department making a pitch for how to develop leads, why the front-line people are really salespeople, how their attitude affects the stay of a guest, and so forth. And don't forget to tell them how much this positive interaction with guests is appreciated.

To paraphrase Phil Wexler, a renowned speaker, and sales trainer: "Everyone's function at a hotel or motel is the acquisition and maintenance of guests."

In other words, all employees are part of the sales department.

PROPERLY HANDLED PHONE CALLS CAN RING UP BIGGER PROFITS

November 1988

There's no telling how much business lodging properties lose because of poor telephone practices. Perhaps millions of dollars a year move into competitors' hands because of mishandled phone calls.

The inquiry is the hottest lead a lodging property can get. Most inquiries are made by telephone—and too frequently, we mess them up. The first shot we get at messing up a call is when we don't answer the switchboard right away. Five or six rings can feel like an eternity to the person making the call. If the phone isn't answered promptly, the call will probably go to a competitor.

The second shot we get at losing an inquiry is when the telephone is answered by someone who mumbles incoherently or who sounds as if you've interrupted his morning coffee break.

The third shot we get is when callers start to explain why they are calling and our operator does not know what to do with the calls (this often happens when we have trainees or a brand-new entry-level employee to answer the incoming calls).

The fourth shot is when the call is transferred to the sales or catering department, and it takes almost forever for someone to answer. The switchboard operator never comes back on the line, and the caller finally hangs up.

Even transferred calls that are answered promptly can get messed up. For instance, someone—a secretary, let's say—answers the phone. "There's no one here right now," says the secretary, who then proceeds to transfer the caller to yet another department. More times than not, this will result in the caller hanging up in frustration.

Daily Occurrences

Scenes like these happen every day in the hotel business. And the bad part about it is that no one seems to know that it's going on! How can we clean up our act and not lose inquiry calls? Well, we can start by not letting inexperienced people answer the telephone. Before an employee takes on that job, he or she should be knowledgeable about the property, know who works in which department, understand the working relationship between departments and have a pleasant, intelligible and enthusiastic voice.

Perhaps most importantly, a telephone receptionist should understand the importance of answering calls promptly –which means never letting the phone ring more than three times. The person responsible for taking incoming calls should only be responsible for directing the call to the proper department, and should never be given the responsibility of answering questions. This ties up the call and delays the answering of other calls.

If there is a department in the property that will not have anyone on hand to answer the telephone, the switchboard operator needs to be informed as to whom to direct those calls. (Of course, a call-forwarding system is an ideal way to take care of this.)In any event, if the department does not answer in three rings, then the call should go somewhere to someone who can answer, or at least ask questions.

Smiling, Happy Voice

Perhaps we also need to install a mirror next to the switchboard (maybe even next to all office extensions). The mirror could be placed in such a position that people can see their own expression as they answer the phone. Putting on a happy, smiling face makes one's voice come across as happy and smiling to the caller.

Let's answer our telephones promptly and with smiling faces. Be enthusiastic, speak clearly, transfer calls properly, make sure people are available to answer transferred calls, and take messages accurately. Don't lose any calls for any reason.

The more inquiries received, the more sales for your hotel.

It's just that easy.

BIGGEST KEY TO BETTER SALES IS IMPROVED GUEST SERVICE

November 1988

You've probably seen a situation like this happen at your property: A good account calls to complain about a reservation being messed up, lost or whatever. The very understanding, quite concerned director of sales says, "From now on, call my office and I'll handle your reservation personally."

Or how about this one: A meeting planner complains about a client's coffee break or lunch not being served on time. Our very understanding, quite concerned director of sales says, "From now on, call my office and I'll personally look after the next meeting for you."

Sound familiar? It should. Incidents like these happen at hundreds of hotels and motels every day. The result is that the sales department is faced with taking over reservations (maybe even the whole front office operation) and the catering department.

What's next ... the restaurant? The housekeeping or maintenance department? Maybe we should have salespeople schlep guests' baggage around in their spare time!

The point is, of course, that salespeople must continue to sell—and everyone must service. If there is a complaint from a client, each department head has to be responsible for correcting the situation under the direction of the General Manager. The whole unit works as a team—no one department should have to do the job for another department. (In an emergency, of course, everyone should pitch in to help "put out the fire.")

We see it everywhere: Poor customer service seems to be running rampant in all the service industries. In recent months, there have been numerous articles by nationally known columnists addressing this very topic.

More Pride Needed

In our industry, with the proliferation of rooms and high turnover, guest service needs to be greatly improved. Salespeople need to take pride in the property, as well as in the management of the property. Sales will improve when a salesperson can say with pride, "We guarantee the service." Sales will improve when the property's staff consistently delivers.

The quickest way to lose a customer is through bad service. It's absolutely infuriating to call a hotel and listen to the seemingly endless ringing of the phone. And how about the meeting planner who calls and gets switched around from catering to sales to reservations?

We never know how many prospects we lose because of poor service. Statistics tell us that more than 90% of all dissatisfied guests never tell us about their complaints. Instead, they'll stop purchasing our product and services—and probably bad-mouth us for years to come.

Staff department heads and General Managers—in fact, every member of the hotel's staff—must guarantee the delivery of the product as sold if we expect to increase profitable sales.

Educate Yourself

Jim Cathcart, one of the world's leading professional speakers and author of *Relationship Selling: How to Get and Keep Customers,* has come up with a new video training program called "Think Service." It would be well worth your while to check it out. Cathcart is a master at teaching "people" skills. One of his philosophies is "Customers can't be interruptions to your work, because they are the purpose for it."

Not a bad philosophy.

MAKE WORKERS FEEL IMPORTANT AND WATCH SALES ZOOM

March 1989

Well-informed employees work, think and relate to customers better than those who work in a vacuum. Too often, the most frequently overlooked audience in a hotel is the staff: Sell to them and we may just manage to be able to sell to the rest of the country.

It's not surprising that there are many cases of hourly hotel employees not really knowing what the people in the sales department do. There are many jokes about salesmen going to the movies—or saleswomen going shopping—on a workday. (How many times have you heard an hourly employee giggle when a salesperson says, "I'm going out to make sales calls"?)

IGNORING KEY ELEMENT

The message that we're spreading outside the hotel to bring business needs to be spread inside the hotel as well. We seem to ignore the most important element of our marketing strategy: the employees.

A typical hotel staff has no idea of what goes on in the sales office. They don't know what the sales goals are, they don't know how salespeople do their jobs, and they may not even know how important it is to bring in new guests.

And they can't possibly know unless we take the time to tell them. We should never take it for granted that hourly employees understand our sales-and-marketing strategy. Do managers ever explain to bellmen or housekeepers what happens when a prospective group buyer comes to the hotel for a site inspection? Who takes the time to explain to the switchboard operator why inquiries are important or what is supposed to happen when someone calls in and says, "I want to book a meeting."

Does anyone explain to the front desk clerk what has actually happened when a guest checks in and the reservation is stamped, "Travel agent?" Does anyone ever explain to the front office staff why contract business is charged much lower than rack rate, and why that particular type of business is important to the hotel's profitability?

Is the banquet staff ever clued-in as to the importance of meeting or banquet business in the hotel—and how important it is for that business to want to return?

Is the importance of meal preparation as it relates to return business ever discussed with the kitchen staff? Are they told how important well-prepared breakfasts are in making a guest want to come back to the hotel?

Has anyone ever told the bus boys how important it is to refill water glasses or clean tables promptly? Do they know that these simple tasks go a long way toward making a guest want to return?

FEELING OF ISOLATION

From a sales standpoint, the very people who are in a position to help the hotel get repeat business are the ones who are generally neglected. Many hotel employees perceive themselves as being isolated from anything or anyone else at the property. They have specific tasks to perform, they work for one department—and that's it.

Employees need to hear the same message that you want your customers to hear. Employees have to get beyond their own individual tasks and understand the extent to which those tasks and understand the extent to which those tasks help determine whether customers return.

When your various departments have their individual weekly meetings it would be beneficial if every so often the General Manager, assistant GM or sales manager made an appearance. They should talk about sales and explain to all employees how important their individual jobs are and how they relate to the hotel's sales strategy. They can explain what it means to the hotel from a business standpoint that they do their jobs well.

Employees have to think of themselves as being not just clerks, busboys or van drivers, but as part of the hotel's overall sales strategy.

Make sure your employees know this. If they do, you can expect the next time you take a guest into your hotel's dining room for a meal, the waitress, busboy, dining-room manager, and cook will all know how important their jobs are—and how important the guest is.

SELLING AND SERVICING OR JUST SELLING?

September 1990

One of the really hot questions today in lodging sales operations is: "Should salespeople just sell or should they sell and service, as well?"

The answer, of course, depends on a good many things which could include: size of property, availability of convention services staff, catering staff, quality of service, sales staff ability to service, size of sales staff, volume of group business, ratio of group business, need for additional business, attitudes of sales staff—F&B manager—General Manager, and many other considerations.

If a property is already staffed with a convention services department, then this question does not enter into the decision of who should service; the service and the sales department sells. These days we see more and more lodging operations getting into the business of having a convention services manager, even the smaller operations are doing this. From a meeting planners' standpoint, this appears to be the way they prefer it to work. One person, with authority to make decisions, should handle all the servicing of a group. Actually, even with a convention services manager, from time to time, the salespeople may get involved, just to "touch bases" with an account—which is not a bad idea.

The problem really arises when there is no clear-cut policy at a property regarding the responsibility of servicing groups. Do the salespeople get rooming lists and take care of all the changes with respect to arrivals? Does the sales department work on meeting room arrangements and meal functions and all those changes taking place over a period of time?

It would seem logical that once a piece of group business is confirmed (with contract)—the other departments would take over, if there was no convention services department. The front office manager would be in touch with a group to handle all sleeping room arrangements; the catering department would take care of all meeting and food service arrangements. Doing it this way, of course, frees up the time for the salesperson who booked the deal to go out and close some more groups.

There are very many salespeople who do not want to "let go" of handling arrangements for groups they have booked; and there are some good reasons for this. The biggest reason appears to be lack of confidence in service staff—there is a feeling that someone will "drop the ball." Then, there is a feeling that there is better communication when one person handles the whole thing. Some feel very responsible, making sure "it goes right," since "I booked it and they are looking to me to handle it." Or, even, in some cases: "If it wasn't for me the group would not have come here in the first place."

In this highly competitive business of lodging sales, we see more and more salespeople opting for letting someone else handle the details. With owners and operators putting more pressure on salespeople for more profitable sales—the salespeople are getting the message: "I'll book it—you take care of it."

The sophisticated professional salesperson will want it handled that way—but still keeping tabs on what's going on and not divorcing herself/himself from the account. After all, we want the rebooking too.

ALL EMPLOYEES REALLY DO HELP SELL

June 1991

We heard it somewhere before: "The sales department is not the whole hotel, but the whole hotel is the sales department."

So what does this really mean? If each person does his or her job well and treats the guest properly and helps make the hotel a nice place to visit and does things that encourage guests to return, then the whole hotel is the sales department. Well, does this really work? We each like to think that our property does the best job of satisfying guests, making them happy and getting them to return. There are probably hundreds of hotels, motels, resorts, and conference centers that can qualify. But every once in a while, one would really stand out in its performance with guest services. It was a pleasure for me recently to stay at the Hotel Maclowe in New York City. The hotel and conference center is less than a year old . . . the place has great décor, rooms, bedding, bathrooms, food service . . . it was all just superior in all respects. But what was truly outstanding, what really impressed me, as it did other guests, was the make-ya-wanna-return attitude of the entire staff. And I mean the entire staff. During the two night stay, I had the opportunity to observe employees in just about every department: doorman, bellman, desk clerk, telephone operator, housekeeping; dining room, banquet department, engineering, as well as the fitness center. Mind you, I did not go out of my way to seek contact with these people. Contact was made during the normal course of my comings and goings in and out of the hotel.

Every employee spoke to every guest whenever there was a contact . . . just about the friendliest, sincere greeting you can imagine. While walking through the conference center I was greeted by a most friendly banquet staff member who was working in a refreshment break. She had been there since opening and appeared to be so excited about working at The Macklowe. She explained that the group which she was serving had been there eight times for meetings; they remember her and she remembers most of them, by name. Another happy employee helping to make meeting groups return.

When you call the hotel from outside, or even from your room and get a voice that is so clear, warm, distinct and helpful it is hard to believe you are in a hotel. Two particular telephone operators Frank and Edward just make you feel that they are there just for you. Someone has trained them well. Housekeeping staff members also make the guests feel right at home . . . they all smile, are friendly, are concerned about your welfare and just overall make one feel comfortable and secure being at the hotel. The staff in the fitness center were so friendly as were the staff members in the lobby cocktail lounge. The dining room hostess, waiters, and bus people were also warm and friendly, as was the lady chef that prepared outstanding dishes (she came out to the dining room to greet guests—what a great touch). Front Office personnel upon checking in and checking out were also tremendous.

One of the most unusual things I noticed one day was the food and beverage director (he didn't know me or was even aware I noticed) while walking through the lobby, stopped to shake hands and greet each bellman on duty. I noticed bellmen, coming on duty, greeting each other with a handshake. Apparently, the secret to the success of getting the staff to be good greeters is to get them to greet each other as well. I was so impressed by this I mentioned it to Michael Little, the General Manager. He indicated that this was a general policy of the hotel, that staff members greeted each other at the start of the day with a handshake. WOW, what a great idea.

It was very difficult departing the property . . . it is just a hard place to leave; they just made me feel so much at home. I talked to a few other guests in the hotel, while I was in the lobby lounge, waiting for friends. They too felt the same way I did. Even a guest from Germany who escorted an incentive travel group that was staying at The Macklowe commented on the friendliness of the staff.

So, what's the point of this experience? The whole idea, in order to get repeat business, not only from daily guests, but from groups, as well, is to make sure they feel at home. The ultimate responsibility for getting the repeat business lies with the various department heads and the key people—who are the hourly employees. Salespeople should thank these heroes who take care of the business that is booked by the salespeople. Sales personnel need to help management in motivating the hourly employees to perform in such a manner that guests find it hard to leave the hotel. This is where the sales effort really is.

INTERNAL COMMUNICATIONS IS A KEY TO SUCCESSFUL SALES

October 1991

If we believe that "everyone sells" and that the "whole hotel is the sales department" then let's take advantage of it and make it really work. Just about any action taken by any employee at a property affects the guests and future business. All we need is a housekeeper, accountant, dishwasher, cook or whoever else on the staff to screw something up and we could have a potential loss of business due to a dissatisfied guest. Of course, we have many employees who are in direct contact with guests such as bell persons, front office personnel, salespeople, food servers, etc., whose actions could lose business as well. We really need to impress upon all our employees—especially those working behind the scenes, who indirectly affect a guest's satisfaction with a property, on the importance of indirect salesmanship.

Most hotels, motels, resorts, conference centers are already involved with customer service programs. I just wonder how many of these programs involve all employees, not just the direct customer contact people.

Perhaps it is time for salespeople to help in this area. The part sales personnel could play would be to impress upon the property staff what it is salespeople do to bring in business to discuss what is happening in the industry today: to bring everyone up-to-date with the cost of sales; the selling process; the difference in cost and effort between bringing in new business and getting existing guests and groups to return. Indirect customer contact employees need to be impressed with the facts of how important their jobs are as it relates to guest satisfaction—and how they perform determines whether guests have a good experience—and, if they return to the property at a future date. We may have to explain how a clean bathroom, a clean water glass, a clean lobby, a television that works properly, a friendly voice on the phone, a warm greeting at the front desk, a tasty meal, are just as important as the work of the sales department.

We're not talking about a lot of time here. A sales manager or director of sales could be scheduled on the department head weekly meeting agenda for 10–15 minutes for the purpose of "selling" sales to a property's department heads. about once a month a sales manager or director of sales could be invited to attend a departmental meeting to discuss sales with employees in each department. You will probably be surprised to learn that many hourly employees do not know what salespeople do—how salespeople spend their time or how they bring business to a property. This is the forum in which to explain how important the particular jobs of the various employees are in relation to the sales jobs.

The internal selling should be an ongoing situation. Doing it only once just doesn't get the job done. Attending these meetings on a regular basis puts the sales department in closer touch and creates a wonderful working relationship with all employees. This will also help the property sales effort from a feedback standpoint. Employees will feel comfortable in providing information which would be very helpful in sales. Employees will also start providing leads for business from their friends, relatives, neighbors, and organizations to which they belong. It does work that way.

This mission could be accomplished without any loss of productive external selling time. Work out arrangements with property management so that the meetings are held during a time other than your productive selling time.

CAN WE IMPROVE OUR ON PROPERTY RESERVATIONS OPERATIONS?

January 1993

Generally, our on-property reservations department are staffed with information services oriented people; which is good. But, most of them on not trained in sales. Shouldn't our reservations department be operated as a sales function? Isn't that department given the responsibility of selling rooms? In testing reservations departments, on properties, from time to time, I find the people are nice, friendly, answer questions and really want to help the caller. But, for the most part, they do not try to close sales and get a reservation.

In our industry, reservations are usually a part of the front office operation (which, by the way, is another important sales outlet for a property). Most front office managers are great rooms divisions people, great in service orientation; they can train and supervise their personnel well, but have little sales training themselves.

We find very few properties that do a good job in measuring conversions in the reservations department. That is, the ratio of reservations made to the number of inquiries coming in. The central reservations offices for hotel chains do it. We should take a lesson from them for on property use. You would probably find out that once a measurement takes place the conversion rates go up.

To help with our on property marketing effort, how well do we gather information from reservation requests that didn't get converted (where a sale was not made)? What a great database could be developed if we would ask a series of questions, the answers to which could help us market our property better. Why do inquirers fail to make reservations? From where do the inquiries originate? Are we keeping a list of the top cities of origin? Who is the person inquiring (company travel department, secretary to the traveler, the traveler, family member of the traveler, etc.)? How many reservation requests were denied each day because of non-availability of rooms? Then, is someone really checking to see if we ran 100% on those days when we had denials? Do we ask for and keep names and addresses of denials for future sales promotion?

Of course, there are many other questions with regard to the demographics that could be asked. The answers to all these questions, when compiled, can tell you a great deal that will help with your marketing effort. After all, we are in the information age. Your quest for knowledge and the proper utilization of the information will make you successful.

It just seems logical that we should be rethinking our organizational structure with regard to the point of sales and point of purchase locations in our hotels and motels and the job descriptions of persons working in those departments. Why wouldn't those departments be considered part of the sales and marketing department at a property? Why shouldn't they be supervised by a sales professional who has been educated in and is experienced in the business? We've been doing things the same old way for many years. Why not a change?

We give a good deal of lip service to training, but very few operations include a line item budget for it. Very few hotels and motels ever consider reservations and front office staffs as salespeople. But aren't they? When was the last time a sales manager met with front office and reservations staff to discuss telephone sales skills or how to get an upgraded price at the time of check-in and, generally, how to improve revenue through these departments? And, how guilty are we of putting a new employee or trainee in the job of answering incoming calls? Ugh!

The information age is here. Marketing at the property level means more than just advertising and selling. I don't expect that every property is ready to install a computer system for database marketing, however, information could be gathered and utilized manually, even if only on a small scale. And, how about some sales training for reservations and front office people? Now is the time to get a little more sophisticated in these areas—if you don't your competition probably will.

POOR CUSTOMER SERVICE WILL RUIN YOUR SALES EFFORT

June 1999

We read about it quite frequently in the daily newspapers and magazines that consumers and consumer organizations are up in arms about the poor customer service in most industries around the country. And, our hospitality industry certainly has not been excluded. Even some of our own salespeople have been complaining about the service "their" customers have been getting. In fact, at many properties sales, folks use that as an excuse to "be around" while "their" group is in the house just to make sure everything goes well. Does this mean our salespeople do not trust our service people enough to let them take care of groups without someone looking over their shoulders? What it does mean is that sales folks are probably spending less time selling because they are "servicing" more.

There should be no question about it. Poor customer service kills our sales effort. Sales folks work hard putting the business on the books, and it gets most discouraging when there are complaints about service. Individuals and groups will go elsewhere when there is a service problem at a property. Quite often we've mentioned in this column: "the sales department is not the whole hotel, but the whole hotel is the sales department." Every staff member is an adjust member of the sales department because through the service we provide the customers we are involved in either creating additional sales, getting repeat business or creating advocates for our cause (which is new customers).

We are all customers of other businesses, so it is easy for us to recognize poor customer service in others. However, in dealing with our own properties, it is never quite so simple because sometimes we do not recognize the poor service we are providing. It takes months and years, and a good deal of effort on the part of our sales team to cultivate a reputation for quality customer service. It is tough to influence customers' opinions of us. Customers generally remember negative experiences twice as often as positive ones. It only takes a couple of negative situations to wipe out years of sales work into creating a good reputation. Even with a sellers' market that we have today, dissatisfied customers can ruin a property faster than all the new competition coming into an area. When we allow poor customer treatment and service to exist we start a whole series of unfavorable reactions that have a long-lasting impact.

One of the problems we have in our industry is that there is a lack of customer service training in our hotels. Yes, we all think we are doing a good job in this area. Maybe so, but couldn't we do better with some ongoing training for our service personnel? Here are some "Steps for Improving Customer Service" which were presented at a recent seminar on "Customer Service Excellence" conducted in Roanoke, VA by the Better Business Bureau and Management Association of Western Virginia:

- ▶ Hire Nice People—and teach them job skills. Hiring based on skill, and trying to teach niceness later, is more difficult.
- ▶ Avoid Saying No—at all costs.
- ▶ Empathize with Upset Customers. "I Don't understand how we could have let this happen to you" is a good way to begin after a complaint.
- ▶ Solicit Complaints: Treat complaints as opportunities to win customers. Studies have shown that 95% of customers with small problems, and 82% of customers with big problems stay with the company that upset them if their complaint is fixed quickly.
- ▶ Empower Frontline Employees—to rectify customer problems.
- ▶ Salespeople—communicate with customers after the sale.

A handwritten "thank you" note works well.

GOOD COMMUNICATIONS LEAD TO IMPROVED PROFITABLE SALES

July 2001

OK, so maybe it's not a big deal to some of you, but we generally do a poor job when it comes to wearing name tags. Where is it written that a badge identifying the wearer should be worn on the left side of a garment? It just doesn't make sense! Think about it! How often have you strained your neck trying to see a person's name on a badge as you shake a hand or just greet a guest relations person at a property? From General Manager, desk clerk, food service workers, salespeople, to the hotel and restaurant staff personnel, let's start seeing these badges on the right side, and make it easier for guests, and all of us, to communicate.

Now about food prices at hotels! I know my good friend and co-columnist at HMM, Tony Marshal, has written about this, but the subject is a room-service menu. Recently I stayed at a resort property, and felt like having only a toasted bagel with cream cheese. Looking at the menu in the room, the price for the item was $2.50, plus $2.50 service charge and 20% gratuity. OK, no complaint. Really not bad for a resort property. However, I opted to go to the restaurant on the property where a breakfast buffet was being served. Upon entering a menu was offered, so knew I had an option. There was no bagel listed on the menu, but I ordered one anyway, no problem. The check came and the bagel with cream cheese charge was $6.95. Coffee was $2.95. Anyone care to figure that one out? Where is the good communication to guests regarding equity in the pricing of bagels when you compare room service to the restaurant?

Then, there are the staff folks at many properties who just refuse to smile and/or say: "hello" to guests (or each other). At the $6.95 bagel restaurant, the server was nice, not rude, but did not smile or offer a warm "good morning" greeting. She even seemed disturbed that I only wanted a bagel and coffee. So where is the good communication which makes guests return? Sorry, no tip from me!

And, how about sales personnel who over-communicate? Being a "good talker" doesn't necessarily make a good salesperson. So often sales staffers are told to "go out and make sales calls" without being provided good direction or any training. So they visit offices in the community and talk "their heads off" about their wonderful property. They talk to anyone who would listen. Then they leave a packet of information, with a business card, and leave the message: "Call me if I can help ya!." Most of the time no questions are asked by the salesperson regarding the need or to determine if the contact is a prospect for business at the property.

Even with proposals to prospects, some sales folks over communicate. Proposals should be fact-driven, and as brief as possible, making it easy to be read and understood. There should be no reason to include adjectives or flowering brochure copy. Just the facts ma'am, just the facts! Be careful that there are no "hidden" charges that may show up in the contract which may be written later. That would be poor communication leading to no business.

So, OK, I'm hard of hearing, but I do hear pretty well on the phone. Why is it I have so much trouble understanding what people say when they answer the phone when I call a property? Do they really have to say all that stuff, and say it so fast that sounds garbled? Many times I have to ask: "Is this so and so property?" Greet the phone callers in a way which makes them feel comfortable with the property. Good communications equal good sales.

There's a lot more, but you get the idea!

FRONT OFFICE STAFF ATTITUDE TRAINING IS VERY CRITICAL IN THESE TOUGH TIMES

November 2001

Training of our personnel, in all departments, has always been important in our business of taking care of guests and making sure they have a wonderful stay at our properties. However, it is even more critical now, at the front desk, from a sales standpoint, than ever before, considering our industry has had severe occupancy and rate challenges all over the country, at all types and sizes of properties.

It seems that just about all of our front office training of new staff has to do with the mechanics of the job, rather than attitude, motivation, how to deal with people, being positive, etc. Some properties, however, are into positive attitude training and it shows up in the results. For example, Craig Smith at the Monterey Downtown Travelodge in Monterey, California says: "Just about when business is dropping, a friendly, smiling face is welcomed at the front desk, and it brings in business. Underestimating this will cost many property managers dollars they never thought they could lose." Craig has worked 17 years at or near the front desk, and he knows that there are several things that work in appealing to guests, but are not commonly emphasized in training. They are, however, easy to learn and helps make his business more profitable. Fifty percent of his business is "walk-in" guests, and he keeps 100% of them because of the front desk training being accomplished at his Travelodge property.

All new desk personnel starts off their training with role-play situations. Eye contact is the first thing they are taught. The idea is to recognize a person just prior to, and after the guest opens the front door and enters the lobby. To do this, staff members have to be alert and aware of what is going on around them. What they are taught to do is recognize the people by providing eye contact, and a smiling nod, just to let them know that they have been acknowledged. Too often desk folks are busy with other guests checking in or checking out, or on the phone or in the middle of a conversation. An "eye-greeting," a smile, and a nod is sufficient until time opens up so they may be greeted with a "Welcome to . . . ," in a very upbeat, cheerful voice. We need to take a look at what type of greetings we are providing to people coming up to the front desk. My own personal experience reflects that not too often go I get a "Welcome to . . ." What I usually get is "Hi" or "Can I help you" or "Are you checking in." Sometimes I don't ever hear a voice just a peculiar look that says "What are you doing here."

Role-play is critical, according to Craig; it gives the trainee the chance to practice with other members of the staff, and to critique each other on how to greet guests. They also call around to other hotels to see how their competition answers the phone. Just experiencing how other properties handle the telephone inquiry is a good lesson on what not to do, in most cases. We do this just about all the time at our HMM Sponsored Hospitality Sales Workshops. Those who have attended know that it is an eye-opening experience. It is amazing that we do any business on the phone by the way most properties answer the telephone nowadays.

It seems that training budgets get cut every time our industry has crisis situations. At times when properties need more business with improved profitable sales, it would seem logical that training of staff would be a priority, especially at the point of sale, the front desk. And, it doesn't have to be costly to do the training that's what is so funny about this. The type of training we are discussing is what Craig Smith is doing at Travelodge. Do it yourself.

PROPER TELEPHONE HANDLING WILL IMPROVE YOUR SALES

September 2002

How much individual room and group business did you lose today because of poor telephone handling skills at your property? Of course, we don't know. There is no way of telling the dollar amount of revenue we missed because someone at the property did not answer the phone within a few rings. Or, the person answering the phone showed little interest, was too busy, did not transfer the call properly or perhaps your automatic voicemail system answered the call for you. These, and other reasons, cause prospective buyers of hotel services to lose interest and turn to our competitors. And, the competition has not been doing any better either.

It is amazing that over the years we have not done more to correct the way we handle telephone communications so that we can do more business, especially in these tough times when we all need all the business we can get. Even with all the new technologies available, we are not doing the job properly. We misuse and even abuse the new stuff, like voicemail. How the telephone call is handled in every department of our properties is the key to success in doing more business.

The hospitality industry is being challenged today. It is a lot more difficult to get that individual customer or group than it has been over the past eight-year period through 2000. It just seems so unbelievable that with all the time, effort, and resources we spend on marketing and sales that we would not take great care in how we respond to telephone inquiries when someone wants to do business with us. Have we stopped training people in all departments on how to handle telephone calls? Seems that General Managers, owners, and operators of properties would be more interested in getting this type of training accomplished before anything else. This should be the "hot spot" for property training.

At many of my hospitality sales workshops, we conduct an exercise in making telephone calls to competitors of those folks who are in attendance. What we want to demonstrate is how poorly property staff members answer telephone inquiries. We use a speaker-phone so that attendees can clearly hear the conversation between me and the person at the property being called. In just about every time we do this we get about a 99% failure rate. Here are the things we find wrong:

▶ The phone rings more than four times before it is answered.
▶ The greeting is said so fast that it is indistinguishable.
▶ The voice answering the phone is rushed or shows no interest.
▶ Proper questions are not asked to determine where the call should be transferred.
▶ Sometimes there is a voicemail greeting with prompts to try to get the caller to the proper department, if possible.
▶ Upon the transfer of the call to the sales department, reservations, or catering department, we end up getting a voicemail from that department.
▶ If we get to a live person in sales or catering, proper questions are not asked to determine specific needs.
▶ Most of the time no one ever asks for the name, address or phone number of the caller.
▶ Many times the person taking the call cannot offer information, but can only take information and refer to others, who may not be available.

Wanna do more business? Let's just do some training in this area of how to handle telephone inquiries. We cover a lot of this in

Chapters 35 and 36 of my new book *Lessons From the Field*.

Nancy Friedman, President of *"Telephone Doctor—Customer Service Training"* does a good deal of training and speaking to hospitality groups on this subject. The company has a free newsletter with many tips, and offers a free TEL DOC 3 minute Self-Assessment quiz on its website: www.telephonedoctor.com.

CUSTOMER SERVICE WEEK SHOULD RUN FOR THE WHOLE YEAR

November 2003

By the time you read this column Customer Service Week will be over. It was the week of October 6, but we should be celebrating it all year long. There is some really basic customer service "things" of which we need to be reminded. And Nancy Friedman, *President of Telephone Doctor,* has six wonderful Cardinal Rules of Customer Service which I would like to adapt to our business of Hospitality Sales and Service:

- ▶ People Before Paperwork: How often, has it happened in any hotel where we have guests waiting while someone was doing some paperwork. It happens at the front desk, restaurant hostess or cashier stand or someone "fiddled" with something? Nancy says: "Paper can wait, People should not." Paper won't walk away, but customers might. Drop what you are doing and pay immediate attention to the guest.

- ▶ Don't Be Too Busy To Be Nice: How often do folks call a property to make a reservation or inquire about a meeting, and get an employee who sounds rushed or too busy? It happens quite often when I call a hotel . . . and I feel like saying "I'm sorry I interrupted you." We gotta have people answer the phone in a nice way, and not act like we are too busy to even answer the phone.

- ▶ Rushing Threatens Customers: Whether it's on the phone or in person, avoid rushing people. Make each experience a great one. And, remember, one-word answers make you sound cold and unfriendly. Slow down, stop rushing people. It threatens them.

- ▶ Don't Use Hotel Jargon On Guests: Keep our lingo and terms internal. We are all familiar with the terms we use, but how about our guests? Let's not confuse the public, use words everyone will understand. Be a better communicator. Many times when I call a property to book a meeting, I get: "How do you want the room configured?"

- ▶ Be Friendly Before You Know Who It Is: You never know who might be a customer or future customer in our business. For sales folks, especially, you want to be friendly wherever you go. I cannot begin to count how much business I've developed over years by running into folks on planes, hotels, restaurants, trade shows, stores, yes, even in bathrooms, who have been customers and friends. Why wait to be friendly. If you are friendly before you even identify who a contact may be, you can make a whole lot of extra points.

▶ "There ya go" Is Not "Thank You" "Uh Huh" Is Not "You're Welcome": Count, starting today, just for one day, how many times people forget to say "thank you" and "you're welcome" to you. When guests at our properties spend their money with us, they want to hear a big smiling "thank you, we appreciate your business." Speak clearly, don't be grunter. When you are thanked by someone, give a great big, smiling "You're Welcome" right back at them.

Telephone Doctor, Inc. is a Twenty-year old training company which has helped over 20,000 organizations improve the way they communicate with customers. Take a look at some clips from their video training library www.telephonedoctor.com/keynote.asp.

YOU NEVER GET A SECOND CHANCE TO MAKE A FIRST IMPRESSION

October 2004

Voice, dress, demeanor, look, handshake any of these initial components of a guest contact can turn a person on or off. The idea in operating a property is to improve profitable sales for the facility, and at the same time, provide superior service to the guests. If we want to improve our repeat business, and get more people to stay at our properties, then it stands to reason we want to put our "best foot forward" when greeting people who may be in a position to give us business. The initial impression made by a property is seen through a guest's eyes at the time of contact by a "sales representative." This term does not imply only those persons assigned to the sales department; everyone employed at a property is really in "sales" regardless of job assignment.

When a guest checks in, his or her initial, and most important impression of the property is most often left up to the front office salesperson. The greeting needs to be cheerful, audible, refreshing and sincere. Is this happening at your place? Helpful, and as important also, is the appearance of the outside of the building, its lights (any bulbs burned out?); an attractive landscaping adds to its appeal; clean driveways, and parking lots do leave a good or bad impression, and the building itself says something to the arriving guest. Also, a clean and neat lobby, and front office area could make an arrival feel welcomed. The dress and manner of the front office salesperson also play an important role in how a guest perceives the property. What is appropriate attire? Some properties have uniforms, which is fine, but not entirely necessary. Clean, wrinkle-free business dress would do fine, even jackets are not necessary all the time.

When you get out making sales calls you also have the opportunity to make a great first impression. Here, we are not only talking about full time or even part time salespeople, but property managers as well. For those called upon, the prospect's first impression of the property has more to do with the person making the call than the appearance of the property itself. Favorable first impression goes a long way in booking business. Here are some things to be careful about when out making sales calls:

- **Dress**: Shoes that shine, and clothes that fit, and are within local style limits; as wrinkle free as possible and neat are important.
- **Gait**: The way one walks and carries oneself means a great deal when it comes to first impressions. The slow, shuffly pace, as well as the overconfident "know-it-all" strut, can both turn off a prospect very easily.
- **Voice**: A loud, obnoxious greeting is as bad as a mumbly, timid greeting. Prospects like to hear self-assured, audible and enthusiastic hellos. The voice of the salesperson throughout the sales call should always be natural and never affected. Observe the listener carefully to see if the volume needs to be raised (some people are hard of hearing). Always smile during the "hello," speak clearly and be expressive.
- **Handshake**: It has been said that a handshake can tell a lot about a person. It is a good practice to offer the hand when making sales calls (for both women and men). A clammy, soft handshake will not make points on first impressions. A brief, firm handshake, works well; and maintain eye contact. The handshake is the only physical contact you have with a prospect, and it does help to break down barriers and create warmth.

Learning to improve that first impression should be self-imposed part of a person's training. Good first impressions are needed in all phases of hotel work to ensure repeat business as well as develop new business.

THE WEARING OF THE NAME BADGE

July 2005

Where is it written that we must wear a name badge (or name tag) on the left side of our body? Wherever we go these days, when it comes to services personnel, in just about all industries, the name badge seems always to be located on the left shoulder area (or breast pocket area) of a service person. We, in hospitality, should know better. The only appropriate place for an identification name badge to be located would be high up, close to the right shoulder area of the body.

Just because a jacket or shirt pocket on a garment is located on the left side is not a good reason for a personal name for identification item to be placed in that area. The main reason for a name badge or tag is for anyone to be able to identify the wearer, and the easiest place for someone to see the name is at the right shoulder area. Just do the brief exercise of facing someone and shaking that person's hand. A quick glimpse, without moving the eyes or the head will lead anyone to the right side of the other person to be able to recognize a name for identification, if the badge is in that location. If the identification is on the left side of the other person your eyes and possibly your head would have to glance over, noticeably to find the person's name.

When attending any conference, convention, or trade shows you will see that mostly, the personnel wearing a name badge on the right side of a garment would be experienced hospitality sales personnel and meeting professionals. Most members of the professional societies like Meeting Professionals

International, Hospitality Sales & Marketing International, American Society of Association Executives, as well as members of many other associations, understand very well wear the name identification item needs to be worn.

So, why do we see so many hotel property staff members in almost all types of hotels, as well as restaurant employees, in all types of restaurants, wearing the name identification pierce on the left side of the uniform? Some answers we have received when making such inquiries are: "Well we have badges that clip on to the pocket of the jacket, so we have to wear the badge there." Or, "Well, I guess since most of us are right-handed it is easier to pin it on the left side." Or, "I don't know, that is where we are supposed to wear it." None of these responses really work. After all, if we are in the service business we have to provide service to our customers. If the purpose of the badge is to let the customer know who we are, why don't we wear it so that the customer can see it easily?

Our management people really need to get this right. Let's encourage our management to get all those badges moved over to the right side of the body and make it easier for guests and customers to see who's who.

QUALITY CUSTOMER SERVICE SUPPORTS THE SALES EFFORT

June 2006

We would all probably admit that the process of making sales comes easy to some people, but for the most part, being a salesperson is a pretty tough job. There is a good deal to it. Before we can make a sales we have to find a prospect, someone who is in the position of using a property, should they chose to do so. But, even before locating a prospect we need to be able to develop leads, and then question those to see, if, in fact, there is a prospect behind the leads. This preliminary work, before we even do any selling, sometimes is tough and very time consuming, but it needs to be done. Most any hospitality salesperson will admit, that after all the preliminary work is done, followed by the sales process, which includes developing the relationship with a prospect, we lose the deal. And, it all has to do with the service delivery process.

It has often said, in the column, that *The Whole Hotel Is the Sales Department.* What this means is that the service effort of everyone on the property staff has helped determine how good the facility performs in bringing in and keeping the business. Quality customer service has more to do with the sales effort than most anything else. If we do not meet customer expectations when it comes to service we lose business. Yes, we are now having a great improvement in occupancy and rate with a demand for business exceeding the supply. All indications are that this will continue at least through 2007. But, there will be time, as history has indicated, that the business will turn around. There are peaks and valleys in our industry, so what we do now, during good times, when it comes to customer service, has a great deal to do with what our business will be during the valley period. If we provide quality service now this will no doubt create loyalty with our customers, and hopefully, they will remain, good customers, when the business climate turns around.

Where does quality service begin? It starts with the first contact someone makes with a property, by either a telephone inquiry or a walk-in inquiry. How these prospects for business are treated at the time helps determine whether a sale can be made How we handle the initial contact by phone or in person is determined by the people we hire, and how we train them. In the April issue of Nancy Friedman's (*The Telephone Doctor*) Monthly Communication Article, she covered the topic of *The Seven Lingering Effects of Bad Customer Service*. This is something that is very worthwhile to read, especially for our industry.

The key to improved sales at any property has a good deal to do with how we train and treat the entire staff. We need to look at employees as customers of the manager, operator, Management Company, or brand. How well the entire staff gets treated reflects on how well they treat our guests. Once the sales department books business then it is up to the entire staff to provide the quality service that is expected to make sure the guests are pleased, and happy, and that they return at some later date.

IS CUSTOMER SERVICE PART OF THE SELLING PROCESS?

October 2010

So much has been written, and discussed about customer service in the hospitality industry, that it is almost too redundant to write a column about it. Just recently CBS covered it pretty well on its Sunday Morning News Broadcast, with demonstrations on how customers are being turned off by how companies are handling their telephone inquiries. The same subject has been covered in this column a number of times of the years.

When it comes to improving sales in the business of sales productivity, how does customer service play a role? Of course, generally, it does not come in to play much when a salesperson is involved during interaction with a prospect; hopefully, that is providing excellent customer service. The issue of negative customer service occurs during a time when a salesperson is not involved, but when other staff associates play a role. We can truly say that most of the issues of poor customer service have to do with the telephone. This was demonstrated very well in the CBS broadcast.

What's our problem in the hospitality business? How we answer the phone (or don't answer the phone)! Automated voice answering devices and the "hold" button are the big part of the problem. It costs a company much less when a call is answered and handled via an automated system than when handled by live people. There does not seem to be a problem when a call is answered by automation, and handled that way to satisfaction of a customer. However, when someone calls and really needs to speak to a live person to handle an issue that is when the poor customer service problem really begins. Once we get through the ten-button response system, and feel we are finally getting to talk with someone, we are put on hold forever and ever, with continual messaging telling us how important we are.

It is not only the telephone, but in so many other ways we can turn-off people, that it directly affects sales. After all, the whole hotel is the sales department. Everyone, somehow or other is involved in improving sales. Maybe we seem to have forgotten that hotel employees, in all departments, need continual training in customer service. There are so many little things that are important in making guests, and prospects for business satisfied.

There is a new book out called: "*54 Golden Nuggets* . . . the best of *The Telephone Doctor*," written by Nancy Friedman, The Telephone Doctor. Nancy has a Customer Service Training Company, has trained corporate executives and all types of employees on this subject, and has appeared on many television shows speaking on how customer service relates to business and improving customer sales. She would say that Customer Service IS part of the selling process.

Take a look at Nancy's *54 Golden Nuggets* for easy reading and for quick tips on how to cure your business communication skills.

SALES AND OPERATIONS: A PERFECT TEAM FOR BUILDING PROFITABLE SALES

February 2013

There is nothing better than a partnership in accomplishing anything. It works so much better in the hotel business. Marketing gets to let people know about our product; sales bring in the business, and operations deliver the product. That's how it works. So, let's think about how to make it work better to improve profitable sales.

There should be more working together between sales and operations to make it happen. To start, sales folks need to know more about the product, it is amazing how much a salesperson can learn about the property features from the various department heads, who could be invited from time to time to address the sales staffers at their meetings. A lot could be learned that could help in producing the sales effort. Sales staffers need to know a lot more about front office operations than they think they know. Just having salespeople get a familiarization visit to the front desk from time to time would be a grand learning experience in actually checking in and checking out guests. Understanding needs and wants are critical in sales, and a good deal about this can be learned just by listening to guests coming and going. A wide range of information can be picked up when there are complaints as well as compliments. All this provides a wonderful experience for making additional future sales.

Housekeeping is another area of operations where salespeople can learn about beds, linen, in-room amenities, and what guests expect. It certainly would not hurt to have salespeople spend a night in one of the rooms, to sample a guest experience. Then, there is the engineering department for salespeople to get to understand that department's interaction with guests. Learning about in-room security should be helpful to salespeople, as well as any sustainable activities that are going on at the property.

When it comes to food and beverage there is a whole new world of knowledge that is critical for sales production, particularly when it comes to sales activities for group business. Also, what goes on in Conference Services that could be helpful? Salespeople working a couple of days there should be most helpful to learn of the different problems encountered with groups.

We should also remember that in a partnership, the other party needs to know what is going on in sales. So opportunities need to be provided so that all operations folks need to know and understand what salespeople do, and how operations knowledge helps the sales effort. So much of all this could easily be accomplished by having department heads visit sales meetings from time to time to talk about their operations. At the same time, these departments could learn a good deal of what is going in sales so that they all can work together to improve profitable sales production.

WHO ELSE AT A PROPERTY MAY BE HELPFUL TO THE SALES EFFORT?

April 2013

How often, in the hospitality business, have we heard the expression: "Everyone Sells"? It seems that we say it, but in actuality, does it really happen? Well, it can, if we make it happen. For example, when a salesperson is not available on property, which Associates are in the best position to be helpful, in place of the salesperson? Would we not want to think that perhaps a front office associate would be the most probable? After all, in most cases this would be someone who is meeting guests upon arrival, they smile, they greet, and they are friendly and have great knowledge of the property's features. So, why not use them to that effect?

The following is a good example, quoting Mr. George Johnson, President and CEO of Johnson Development company, with respect to something that took place at one of his properties: "I recently asked this question of the Sales director at one of my properties: If something were to happen to you next week, that would preclude you from doing your regular job, what would happen to our sales effort? Her answer was immediate; it would be a disaster! Who would make outside calls, who would make and deliver proposals, who would prospect? As I sat with a smile on my face, she suddenly began to realize that it did not need to be a disaster, and that she had some really competent people that with a little more training could do so much more. We ended our conversation with an agreement that we would identify those people, and start the training immediately. That was one year ago. Today that hotel has made budget 11 out of 12 months, and exceeded its annual budget by almost 20%. Was it worth the small investment we made in some of our GSA's? You bet!"

Not only does a little sales training assist when the salesperson is not available, but think about the personal development being provided for the good of the person involved. It is rather easy to show some of the associates what to do when, for example, it is known that the salesperson is out, and an inquiry for business is called into the property. Rather than just advise the caller that the salesperson is out, and passing the call on to the voicemail system, why not ask the appropriate questions the salesperson would ask? "How may I help you" would work. Then record information which would reflect the needs of the inquirer. A standard Phone Lead Questionnaire sheet could be provided at the front desk for this type of information to be recorded.

HOW MUCH REVENUE DO YOU LOSE DAILY, UNKNOWINGLY?

June 2013

If there ever was a way to measure it would all be amazingly surprised. Are we measuring the number of real prospects we lose every day only because they tend to hang up the phone when they call, because there is a big delay in someone answering the call? In this fast-moving world we all work in we all want our call answered immediately. There are always other options for a call who gets disturbed waiting for someone to pick up the phone. If we could only measure the number of folks who really do hang up we would be surprised. Another area where we lose business from callers is the manner in which the phone is answered. What is the attitude of the person answering the call? Does the answering voice deliver a message about the property? Then, there is the question of voicemail. Are there people who will not leave voice messages, so they hang up on voicemail too? Most voice messages say something like: "I'm not here right now, leave a message, etc." The caller knows you are not there, so why make a comment like that. We need to leave a message indicating where we can be reached, and/or when we will be back to respond to the call.

Another area of concern in losing business unknowingly is with our current guests, or customers. The best new business we can ever get is from the people who have already been to your property and had a great experience. Not a "good" experience, but a "great" experience. There is such a thing such as a remembrance. What is it about your property that guests remember best, and take home memories of a great experience? The one big thing is how they were treated by the staff. That is what people remember best. It could just be some little thing about how they were served or how they were greeted, or how they feel when they leave the property. It is not the General Manager providing this experience, but any or all of the staff, in and around the property. It is the basic things that make it work. Everyone should be smiling and greeting guests anytime they pass by. What is it the staff could be doing to be extra helpful to visitors? They need to do the extra things that make people remember the experience. Most everyone has certain expectations of the service they will receive at a property, so the answer is to things that exceed their expectations.

It may be that the training opportunities at properties are failing. Training needs to be conducted on a regular basis. Doing a one-time training session is not sufficient; it needs to be ongoing, and needs to include how to provide "personal" service that is memorable.

SERVICE RECOVERY HAS AN IMPORTANT ROLE IN SALES PRODUCTIVITY

August 2013

In the business of sales, it is more than just "Get out and bring 'em in." We have to be sure that those we "get in" come back for another visit. We will all agree that the repeat guest cost us less than getting a new guest. So, what's the problem? Do they all come back? Of course, not! But, we do want to be sure that we do not lose them because of a "service recovery" situation. That is, did something happen during the visit that made the guest unhappy, and did we do something about it at the time it happened, right away? How good are we really in "service recovery"?

All hotels have seen these comments: "my air conditioner was not cooling, I contacted the front desk but nothing was done'; or "my room did not have any towels, I called to the desk, and was told I would have to come to the desk to pick up towels', or "we could not take a shower because the shower did not work. The Front Desk said they would send someone up to fix it, but no one ever arrived." These type of guests' complaints showed the root cause to be were failing in the "response and recovery" side of our guest service.

These issues were brought up by Ray Hobbs, Vice President of Operations for Kelco Hotels. Hobbs also commented: "Guests who share their concerns with us provide us with an opportunity to respond and resolve situations. When we do this right, we can retain our guest loyalty. When we do this wrong, our guests will take their business to our competitors. And, worse yet, disgruntled guests will air their negative impressions, and experiences for the world to see in one of the online review sites."

Kelley Slay, President of Kelco Hotels, made everyone (both GM & DOS) aware of Kelco's goal to be recognized by their guests for "outstanding" service levels. When she realized this goal was not being achieved, they researched further and found their "disconnect" to be attributable to the responses of our Associates when things did not go right. Kelco has now changed the line-up of their training programs to provide each Associate a "Service Recovery" guideline/checklist at the very beginning of training. Hobbs remarked that with many hotels seeing high turnover among staff, there needs to be a continual training effort at all properties to be sure that all staff members are properly trained before they start interaction with guests. At Kelco hotels their "service recovery" training provides written step by step instructions to each associate for responding to guests. This training instills confidence and empowers the associates to respond properly, quickly resolve guest issues or offer options for resolution, and follow through to initiate actions needed from others.

Sales staffers at all hotels work hard to bring in new business as well as repeat business, and they count on property staff to provide appropriate service to guests. However, those in sales often are also responsible for handling issues that come up from guests, and there also have to be Sales Recovery at that end.

PERFORMANCE ENHANCEMENT HELPS IMPROVE SALES PRODUCTIVITY

July 2014

Sales staffers know too well how important staff performance at a property will enhance the sales operation. The experience that meeting planners, as well as all guests, receive when visiting a hotel is a big key in assisting with the confirmation of a booking, whether it be for individuals or groups. All too often, just a little misstep on the part of any service employee will make the difference in the decision-making process of a reservation. There are so many opportunities for "things to go wrong" during guests' visits that need to be avoided. In all probability, any salesperson at a hotel property would be able to compile a listing of such situations.

Foodservice employees taking "smoke breaks" near the hotel's front entrance; the hotel manager who didn't say "hello" as a guest walked by; an associate who, while serving a guest, griped to a colleague about not having a day off. These encounters are cited, from his own personal experience, by Vince Magnini, in his new book: "Performance Enhancement: Twenty Essential Habits for Service Businesses." Magnini, associate professor of Hospitality and Tourism Management in the Pamplin College of Business at Virginia Tech, indicates that, we as consumers witness employee actions that are inconsistent with the experience that a hotel, or any service business is trying to create.

Academic research exists that, if properly communicated, service employees' actions could significantly enhance the opportunity for improving sales productivity. The book represents making such managerially useful studies more readily accessible. It presents 20 habits that are important for anyone working in any hospitality or service business to cultivate. The suggested habits range from the behavioral (using drama, laughter, and verbal and nonverbal cues) to the technical (analyzing and measuring return on quality). Service employee habits actually can impact and shape customers attitudes and behavior, which of course, may result in future bookings for a hotel property.

You never get a second chance to make a first impression, so of course, those first impressions of a hotel or resort property are so very, very important. We need to think about what is it about the impression a guest (particularly a meeting planner) gets upon arriving at a property? Probably, even before the arrival, the first and most important, could very well be when the phone call is made to make a reservation or an appointment with the sales department. What is the caller's impression when the call is answered (or not answered), by a live voice? Then, there is the physical appearance upon arrival; the check-in at the front desk, etc. First impressions really do count when it comes to booking profitable sales.

GREAT CUSTOMER SERVICE ENSURES FUTURE GROUP BOOKINGS

August 2015

There is no question about it. It is the wonderful customer service that is provided at any property which would ensure a group booking as long as it can meet the needs of a prospect. In today's world of communications, a favorable comment about a property will no doubt reach millions of people, and maybe more-so with an unfavorable comment. Each guest, and especially a meeting or conference attendee, staying at a property, becomes a "walking mouth-piece," according to an article by David Eisen, in a recent issue of Hotel Management. Guest reviews, in a variety of methods, will have a tremendous effect on Millennials who are becoming the group booking prospects of the future.

It is interesting to note that one of the issues on top of a list of things meeting planners would like to see improved at a property, in regard to customer service, has to with the General Manager. M & C (Meetings & Convention magazine), did a meeting planner survey to find out what they think could be improved at a property for group meetings. Of all the items mentioned by group planners, the one item that stood out was that they thought General Managers need to be more visible. Suggestions included for the GM to attend the pre-conference meeting of an event; meeting and greet planners when they visit for site inspections; be around, and visible sometime during their meetings.

Customers Service really begins when in advance of any group meeting event. It even begins well in advance of any personal contact made with a prospect. For example, one of the biggest items, every year, on a list of things meeting planners complain about is the slow response they get from sales staffers in replying to RFPs (Requests for Proposals), according to MPI (Meetings Professionals International). This also was reflected in the study made by M & C. Not responding within a day or two to requests for information whether via an RFP or phone call represents poor customer service.

The issue really is what do we do to improve our customer service? There are so many things involved when we talk about this subject. It starts with answering the phone, how we address guests, how our employees look, act and demonstrate good customer service. It really all comes down to employing the "right" people, providing appropriate job descriptions for all employees, and training, not just one-time training, but on a continual basis. In the business of bringing in more group business, a salesperson needs to rely more on the entire staff to perform appropriately, and to provide the service necessary to assure the success of the event for the planner who made the arrangements.

Sales staffers know that booking a group should not be a one-time thing, but getting them back again is what is critical. It is superior customer service that will do the trick. Customer Service has to do with Basic Human Needs.

STAFF ENTHUSIASM STARTS THE SUCCESS IN BUILDING SALES

February 2016

Regardless of the size of a property, it is a salesmanship, probably more than a product that determines the success in building business for the property. So, what is this thing called "salesmanship?" Is it only something salespeople are supposed to have? This term encompasses a variety of things, but the most important ingredient is enthusiasm.

Ideally, an enthusiastic attitude starts with the first person with whom a guest makes contact, whether it is by phone, van driver, bell person, front desk, food service server, etc. It is reflected in the attitude of the entire staff, throughout each department. How guests are received, as well as how employees treat others makes a big difference of a guest's perception of a property. No question about it, guests are positively influenced by staff members' enthusiasm—and negatively influenced of it lacking.

Generally, in the business of selling, prospects feel more confident when they deal with salespeople who show enthusiasm for what they are selling. The way someone answers the phone in any department is a key to salesmanship. An enthusiastic greeting with a "telephone smile" gets half the sales job accomplished right at the start.

Salesmanship is also knowledge of the product: confidence in the facility and its staff; warmth, empathy, courage, conviction, sharing information, asking the right questions, it all adds to credibility, and helps in making a sale. No one ever really buys anything, especially a business meeting, or conference location, without confidence in the product. There is hardly any chance in making a sale if confidence is not displayed by the staff. Confidence is by far the most important key to displaying confidence; and it is important to be sincere about it. Phonies are easily detected.

Since everyone sells, everyone on the staff needs to be enthusiastic about the property and its makeup, as well as its management of their jobs. Of course, all this starts with the General Manager and the management team. Here is a little test for department heads to initiate: count the number of employees in the department you find displaying enthusiasm during the course of a given day. Check on how the phones are being answered; watch various staff members doing their jobs, particularly with guest contact at the front desk, dining room, and housekeeping; observe how guests are being greeted. How do the employee's rate on the enthusiastic scale?

Without sales, we cannot exist, and without enthusiasm, selling is hard to do. Remember that salesmanship is the key to success, and everyone on the staff is a salesperson.

TURNING EVERY EMPLOYEE INTO A SALESPERSON VIA EXCEPTIONAL GUEST SATISFACTION

March 2016

Who says that the promotion of hotel services belong only to the sales department? The growth of the business through improvement must be placed in the hands of all the employees at a property, especially the frontline staff, i.e.: bell staff, van drivers, front office food and bar servers, housekeepers, maintenance, yard and pool staff, etc. In short, all the people on the staff who come in daily contact with guests.

All employees must realize that selling all services of a property is part of the job regardless of the specific duties assigned.

Everyone must participate. From the laundry room through housekeeping, all on the staff have the best opportunity to sell the guest on a return visit. One of the most important items in getting a guest to return are staff attitude and clean rooms. A clean-looking, happy, and smiling friendly housekeeping staff can make guests want to return. A friendly "hello" from a housekeeper goes a long way in creating guest satisfaction. A bell person, or van driver, looking fresh, being friendly and helpful is in a strategic position from which to sell for the property. It is not unusual for a bartender to be a great salesperson by the manner in which a guest is greeted, and engaged with chatting when visiting the property lounge. Yes, of course, a real key in providing excellent guest satisfaction is the front desk employee, who in most cases is the real official greeter for the property. When checking-in a guest do they really look at the guest, with a smile and a nice greeting, or do we find them looking down at the computer? Even in a very busy restaurant, there needs to be pleasantries and smiles by the greeters and servers to help make the guests really enjoy being there for a meal. It really means so much to a diner to make them feel "good" while they are there for the meal, and it is up to the staff to make that happen by the manner in which they provide service.

To make all this happen, and to be sure that all staff personnel are on the sales team. There need to be regularly held training sessions conducted by sales and management staff to explain and show how all employees are responsible for the continued growth in sales.

Chapter 15

A Professional Approach to Hospitality Sales Along with Networking

DO BUSINESS BY SOLVING OTHER PEOPLE'S PROBLEMS

August 1983

Think about it. When a guest decides to use your lodging facility, it is because a problem needs to be solved. The guest needs a place to sleep, eat, conduct a meeting or be entertained. If the problem is solved to the guest's satisfaction, then the opportunity exists to get repeat business.

What helps determine if that guest will return depends on how well he or she identifies with people with whom contact is made at the property. If the guest is made to feel comfortable or feel important or feel good, and if the property staff is perceived as having understood the guest's needs, then identification has been established. People generally like to do business with people they know or like or with whom they feel comfortable.

We can create magic—just by treating guests as human beings with problems that we can help solve. Starting with the bellman and front office salesperson, magic can be created. Behaving decently in trying to help others is what our jobs are about.

In order to solve other people's problems, we need to start with solving our own.

Personnel selection is a major problem in the lodging industry. In many cases we select and employ personnel as a matter of expediency—a vacancy exists and we have an "urgency" to fill a hole. This happens when an owner is looking for a manager as often as it happens when a manager is looking for a night auditor or waitress or salesperson. Poor techniques in recruiting, interviewing, selection, hiring, training, supervising, and developing abound in the food and lodging industry; and this poor handling of the process is directly reflected in the amount of business coming into a hotel or motel property.

If we expect to help solve other people's problems (which means doing business), then we need to employ personnel who have the attributes of solving problems—not creating our own problems with the wrong selection of people.

To be a problem solver in the lodging business, one must be understanding of people's needs and pay attention to people's needs, so that the needs of the guest may be satisfied. Smiles, of course, friendly greetings, warm welcomes, "How may I help you?," "It's a pleasure to serve you," "My pleasure," are phrases that should be used sincerely.

The background check and interviews with applicants will help determine if an applicant is a proper candidate for the job (whether it be manager or bellman), and then it must be clearly understood by the candidate prior to selection that the primary objective of anyone being employed in the industry is to solve other people's problems, and that this is done by being friendly, helpful, smiling, and taking care of the needs of guests—always with a positive attitude.

Another problem that needs to be solved in our own industry is the lack of follow-up attitude training. Even though we are sure that we have employed the right people, a continual follow-up program is needed to ensure proper attitude exists throughout a property. This is accomplished with periodic meetings conducted with staff members. Communication is the ultimate weapon against poor attitudes. When employees are kept "in the know" and there is a good communication back and forth, then a property probably will experience good staff attitude. The communication sessions need to be held with all staff members, not just department heads; and it really starts with the property top management. If the proper attitude is displayed at the top, it will automatically filter down through the staff. As a smile is catching, so is a good attitude.

Unfortunately, when it comes to not keeping guests happy and not satisfying their needs and loss of business results, it could very well lead to the culprit being the man at the top. There are many excellent, well-qualified managers who just don't have the right attitude. Too many managers act stuffy, feel too important, don't mix with the guests, don't speak or visit with employees, don't smile, are too negative, appear hurried and harried and even look unhappy. These spreads and ends up with a property having many of its own problems, and no time to help solve the guests' problems or satisfying the guests' needs. Too bad!

But hurrah for the manager who comes to work peppy and happy, full of enthusiasm, with a healthy, smiling "Hello!" to each employee as the rounds of the property is made. Hurrah for the manager who helps check out guests at the front desk in the morning and pours coffee in the dining room and conducts staff meetings periodically and enjoys a happy, healthy relationship with employees. This is the property that is satisfying its guests' needs—and, by the way, probably does good business.

LET'S NOT GET TOO TECHNICAL

December 1983

As long as you are dealing with people and plan to get along with them, never get too technical. When we get too technical, we lose sight of the human aspect of personal touch, personal understanding and that people need and benefit more from personal involvement. In the art of touching or in the day to day contact with one another—if we allow ourselves to become too technical—then life and everyday association with others become too stiff—too cold—too impersonal.

We lose sight of how important a simple touch of human understanding can outvalue all the rules. The technical "how to get a job done" becomes too one-sided and the scales are tipped. There is too much impersonal attitude today in dealing with people. Have we forgotten how to touch a shoulder, shake a hand with sincerity, smile a smile that can be felt as being human rather than mechanical?

In Hotel/Motel sales we're finding too many mechanical robots—going around doing a job—"according to the book." Between the lawyers and accountants and all the other "rule writers" hotel, salespeople better learn to be careful. To be careful never to lose the personal touch. We're hearing an awful lot lately about negotiating, contracts, deposits, cancellation clauses, guarantees—etc. It's almost impossible to try to do a good job and try to take a non-adversarial approach. The pressure is on for advance bookings with definite space blocks—with agreements signed in blood (almost). Salespeople are under great pressure these days—but that's OK—as long as we can get the job done without losing the personal touch. That is, internally as well as externally -it's just as important for managers/owners of properties not to get too technical with salespeople as it is for salespeople not to get too technical with prospects.

Have owners and managers gotten too wrapped up in technicalities themselves to forget how to say "hello" or "good morning" to sales staffers? What happened to the old slap on the back routine we used to see years ago? Who shakes hands these days among the staffers (salespeople) when they greet each other in the morning—and with a smile, hopefully. The handshake and smile are not only good for the guest or prospect—it works internally, as well.

Making the sale is not only getting the contract signed—it is everything that goes into establishing a credible relationship with a prospect. Starting with helping to satisfy a need to developing a sincere interest in being able to assist the buyer with his problems -all accomplished by being personable, friendly, sincere—the technicalities ultimately have to follow—but let's not be too technical.

NETWORKING—IT WORKS!

May 1989

One of the best marketing tools we have in our lodging industry is relatively inexpensive, readily available, and takes very little advance planning; but, many hotels, motel salespeople don't know how to maximize its use—and, that tool is Networking.

The term is being overused and misused. Industry salespeople use the term very freely when, attending a variety of meetings. However, in observing these people in an actual "networking" situation—very little of it is being done effectively.

Probably the best way to describe the term is to refer to it as an intermingling of people—or work formed in the same manner as a net. A net, of course, is an instrument formed of some material woven into meshes and used for touching things. If networking is done properly one can expect to catch a lot—of people, that is, who are or will become prospects for business. Networking is a very simple process. Salespeople just need to remember the purpose and objective of networking; to establish e few rules for themselves; and prepare themselves with some logical support tools.

The purpose of networking is to join in group gatherings—as many and as often as practical; to meet new people. These could be social as well as business—as long as the salesperson is meeting new people. It's almost like a cold calling, but more fun and exciting.

The objective is to develop a "network" of people that may be helpful to you in business as well as people to whom you may be helpful. Remember, a real key is getting to meet new people in a social and/or business environment. All too often we see salespeople attending a meeting or reception and head straight for the people they know—and spend most of their time only with those friends and associates. When it comes to selecting a company for meal functions—again it's with friends or people they already know.

In networking the hotel salesperson should establish n few rules to get the job done properly:

▶ Set a goal for the number of new people to be met at the event
▶ Set a goal for a percentage of the new people who may be business prospects (for every 10 new contacts 3 will be genuine prospects)
▶ Arrive on time or a little early and meet as many people as possible at the door (be an unofficial host)- get people to know who you are
▶ Stay until the end of the meeting or function—leaving early does not maximize the use of time.

▶ If alcoholic beverages are served—be careful, keep drinking to a minimum
▶ Keep moving! Getting stuck for too long in any one group is losing time available for new contacts.
▶ Be sure to ask for, and give business cards.
▶ Take notes (let's not walk around with a clipboard, but just taking notes on the back of someone's business card will help in follow-ups.
▶ Remember to follow-up with notes, letters, phone calls to the contacts that you want to pursue.

In preparation for networking, the hotel salesperson should:

▶ Have plenty of business cards—and remember to give them out freely.
▶ Dress properly for the event—be sure not to overdress.
▶ Be clean looking and have good posture.
▶ Eat something before going to a reception so you can spend more time moving around, meeting people rather than eating. Also, it's difficult networking with a mouthful of canapés.
▶ Get a good idea of who will be in attendance so you can target some of the people you want to be sure to meet.
▶ Prepare same opening lines to use if you are the shy type—It's OK to approach anyone at group functions—don't wait to be approached.
▶ Think about your handshake—firm—not too strong or too flimsy a grip
▶ Plan not to talk too much—listen more ... Remember, this is not the place for a

sales pitch—just make contact to see if a prospect exists.
▶ Carry a small notepad in a purse or jacket pocket with pen or pencil, for ready access in the event some long notes need to be taken (This is for info that won't fit on the back of a business card).

It is very important in networking not to be all business all the time. Don't get people to think of you like a vulture—just out to swoop down on anyone to get some business. People will avoid you if you do. Be casual, friendly, helpful, confident and interested in other people. Always be sure to follow-up with the information developed during the networking process.

Experiment at the next meeting you will attend. Afterward, count up the number of prospects developed. You may be surprised—IT DOES WORK!

IT IS TIME TO WORRY ABOUT COMPETITION

July 1990

With the segmentation that is going on in the lodging industry today, there is a bed, a meal and a meeting room for just about anyone, in any price category. Just about when we think we have segmented out, a new one surfaces. In the budget, medium price and high end we find multiple breakout subcategories. Represented in the budget group we'll find low/low, medium/low and high/low. In the mid-price group, we've got low/mid, mid/mid, high/mid, and high/high. And in the high end, we'll find low/high, mid-high. Of course, all categories are building suites.

Just about every sub-segment from low/low-to-high/high is in competition with each other for market share. It's probably unlikely that high/high customers would trade down to low/low, and vice-versa, but we have seen trading up and down and across in major segments, particularly when new properties open up in the same neighborhood. Haven't you noticed that whenever a new lodging operation opens in your area you lose some market share?

Naturally, the idea is to do all we can so that some of our customers won't trade up or down. But, let's face it, no matter what we do we're bound to lose a few. In order to combat losing market share we just have to do a better job in bringing in new business.

And that's what selling is all about bringing in new business.

There are many things that make a salesperson successful. One of these is KNOWLEDGE. Knowledge of your own property (what is it you have to sell), knowledge of your customer or your prospect (what they need and/or want), and knowledge of your competition (everything there is to know). Most salespeople who have already completed a marketing plan for the property will have included all this information. Perhaps it is time to take that plan off the shelf and brush it off and bring it up-to-date.

Since competition is all around us, creating problems by taking some market share, it would be well for salespeople to be sure they know their competition "inside—out." Here are some things salespeople need to know about the competition:

- How do their features compare with yours?
- What can your property do better than theirs?
- What do their guests like about their property?
- What do their guests dislike?
- Who/What is their target market?
- What accounts do they have that you want?
- What do they do right about sales?
- What don't they do well in sales?
- How do you compare in sales activity? How does your property compare on service?
- What is their rate structure?
- On what basis do they discount rates?
- Can you outsell them at a better rate?
- Is there a perception of a better price/value at your place?
- How does their market share of the compare to yours?

This is certainly not a complete listing of comparisons that need to be made, but you get the idea. Once you've got a good handle on your competition make a listing of the things you need to do right to beat them.

You can pretty well bet that your competition, if they are smart, will be doing the same to you.

LET'S MAKE SITE SELECTION EASIER FOR PROSPECTS

August 1990

Communications is still one of the biggest opportunities we have in our business. There are internal as well as external communication challenges that we must face. Common sense will resolve all the difficulties in this area. Take, for example, the simple inquiry coming into a lodging facility far, space availability and rates. The inquiry comes in the form of a walk-in, a phone call or a letter. The walk-in presents no problem, this is usually not difficult to handle on the spot. It is the telephone or mail inquiry that has a tendency to get lost, misplaced or gets answered "sometime later"—or worse yet, doesn't get lost, but doesn't get answered either.

We, in the business, have absolutely no idea of the amount of business lost as a result of this display of poor communications. Generally, we seem to do better responding to the inquiries for meetings or other group business. Although, motor coach tour operators keep complaining more and more about properties not responding promptly to their requests for information.

More and more we find meeting planners, travel agents, and incentive houses working through hotel representatives or subscribing to directories or electronic site selection services. The most popular directories used are the Hotel and Travel Index and Official Hotel and Resort Guide. Most electronic site selection services list only those hotel that are paid subscribers—they pay a fee to be listed.

A new free service for meeting planners and group travel organizers has recently surfaced. It is called FAX A MEETING/FAX A GROUP INC or FMG for short. U.S. and offshore Officials of this new service state that all lodging facilities are entered in a database at no charge to anyone. When a group planner needs to locate sites for a group movement, a fax, outlining its needs, is sent to FMG through a free 800 fax number. The database is searched to locate sites that best meet the needs of the inquirer. At that time appropriate properties are faxed a message from FMG requesting availability, rates, etc. Within 72 hours of the original inquiry, FMG will fax all required details to the inquirer, free of charge. Planners and group movers may then handle a booking directly with a property.

Additional services offered by FMG include a "hot sheet" detailing unusually good deals available for last minute meetings due to other groups' cancellations. "Gap period listings" give specific dates hotels need group business due to seasonal booking patterns.

According to Neil Strickland, Senior Vice-President of FMG: "Our goal is to simplify the meeting planner and group travel organizer booking process at no cost to them." There is no charge for lodging facilities to be listed in the database. However, a small service fee is charged if there is a booking as a result of this service, but it is not paid to FMG until the group departure. Not a bad deal.

HOW MEETING PLANNERS SELECT HOTEL PROPERTIES

November 1991

Recently the editors of THE MEETING MANAGER an in-house publication of Meeting Planners International, conducted a discussion session with some of their—members to find out what it is that meeting planners really want from the facilities they select for meetings. The planners selected were from a variety of work and geographic areas: association, corporate, consultant, travel, finance, and government. We have found that in reviewing some of the comments concerning the reason planners select and use hotels would be of interest to our readers from being able to use this information in planning their sales efforts to develop meetings business.

When asked the question: What influenced your decision to choose a particular property? Responses reflected, generally, it was the people—the sales and service staff the creativity and flexibility of the staff-staff personnel that listens. Developing a rapport with salespeople was very important. There did not appear to be much of a concern that salespeople move around. Location rated high, specifically a "safe" part of town. Another area explored involve was what could have been done to improve the quality of the meeting. Responses included: temperature control; too tight seating arrangements; what kind of a "receiving department" is available for shipping of supplies; cannot locate shipments of equipment and supplies sent in advance of the meeting; identifying key person in charge of a banquet; never enough wait staff to serve people within an expedient amount of time (some agreed to have more servers added and are willing to pay for the cost of it); cash bars sometimes do not have enough bartenders or enough bars set up to serve the group; catering staff trying to tell you what your people like for food.

When asked about, have expectations changed over the past five years and what is being expected these days, the meeting planners indicated that hotels are building better meeting facilities; meeting planners are asked to serve on advisory boards to hotels; hotels are getting more contract oriented . . . trust level has changed with the introduction of the paper contract; please that the design of sleeping rooms has changed to make them more pleasant, particularly for the female traveler (make-up mirrors, ironing boards, hair-dryers, etc.); the amenity situation has created some inconsistencies—not finding the same things in the rooms upon different visits; the cost of telephone calls from hotel rooms has become more expensive due to surcharges-very frustrating-it has become more of a profit center more time was spent on discussing this issue than all the others); more attention is now being paid to the disabled for barrier-free meetings; more business centers being introduced. One area that was surprising was the discussion about offering to pay hotel bills before departure. There was an indication that some meeting planners prefer to do this, however, they had a problem getting with a bill, the accounting department to be able to come up indicating that the bill was not ready. Some association groups have a lot of cash and want to get rid of it by paying off some of the master accounts and this works OK during the period of the meeting, but on the last day, or even the day after, there is no way to get a final accounting. The meeting planner involved in the discussion all agreed that they do want to meet with someone in accounting to review all the bills before they depart.

Overall the most important area that meeting planners wanted to get accomplished were: increased dialogue. Treat each other as partners and not as adversaries. Talk to each other on the same level.

HOW ABOUT THAT SALE THAT WAS LOST?

April 1992

None of us are 100% perfect in closing sales. We do not win them all everyone loses sales to competition at one time or another. We probably blame it on product or pricing, when in fact, we, really do not know why we lost a particular sale. An important part of a sales audit is to review the lost business reports to evaluate why we lose out to clients. There needs to be a very professional approach to finding out why. Then, do something to correct the situation so that next time we'll do a better job of closing. Because of the current economic, as well as the over-built situation it becomes too easy for salespeople to blame the loss of a sale on discounting by other properties (which may very well not be the case). Let's look at some reasons why some sales are lost to competitors:

▶ To begin with, did we qualify the prospect properly? Did we get enough information at first to determine that the prospect was, in fact, a probable user of our property? We do not want to sell champagne to someone who can only afford beer. Also, the property may not even fit the needs of the person or group.

- Did we do a good enough job in listening to needs and wants? Too often salespeople get very excited about the property and proceed into a long sales pitch without really finding pit what is needed and if there is a good "fit" between what we got and the prospect wants.
- Was the prospect able to get a word in "edgewise"?
- How about the follow-up. How long did it take from the time the lead was received and the salesperson contacted the prospect? If we do not jump on a lead right away, we're not doing our job. Even after we qualified the prospect did we follow through with delivering something we promised? Such as a proposal?
- Did we show enough enthusiasm for the prospect's needs and how the property can meet their expectations?
- Were we fully prepared to answer questions about the property, make comparisons with competitions (if the subject surfaced) and related features into benefits?
- Were we persistent? Never take too much for granted that the prospect will get back to us. They need to be prodded. Sometimes if we do not pursue the prospect they might think we are not interested in their business. It may even take 10 sales calls to close a deal.
- Did we close property? Many salespeople never get to trying to close the deal or asking for the business. Sometimes we assume that if they like us we'll get the contact. Did you expect the customer to ask you to buy?
- How about overcoming objections? Are you comfortable doing this? Are you capable of doing this? The more practice you have in this area the better you become in overcoming objections. It takes practice . . . and you may have lost a deal because you did not know how to do it.
- Did you take the prospect for granted because they have been using your property all along and you assumed you would get this piece of business again?
- Did we do anything to turn the prospect "off." You would be surprised how often salespeople say things they don't mean to say. You never want to complain about anything, about yourself, about your employees, about your boss, about your property. It doesn't look good in the eyes of the prospect.
- So what about price? This seems to be one of the major objections (whether real or not) of prospects. How was it handled? Were you able to make comparisons? Show value? Add benefits? If the proper job in selling was accomplished and the prospect was sold, even before the price was mentioned, this should not have been a barrier to a sale.

A good sale happens when a prospective buyer is willing to buy what you are willing to sell, at a price with which you are both happy.

PROPER NETWORKING WILL ADD TO YOUR PERSONAL DEVELOPMENT

October 1999

Unfortunately, many salespeople do not take advantage of networking opportunities while attending meetings and conventions. It seems that we attend these events with the idea of learning something from the speakers; and that is good. However, we may be missing some wonderful chances to network which could often add to our own personal development.

Most of the time we think of "networking" as getting a chance to meet some new prospects for business. True, but there's also a great chance to meet with peers, exchange thoughts, get new ideas, and return from the experience truly motivated to improve, not only sales, but your way of doing things. Here's a way to do it: Have a plan—make some notes on things you would like to discuss with people you will meet at the event who have the same type of job function as you; other salespeople at different types of properties. Subjects such as job function, office conflicts, time management, handling complaints, satisfying your boss, paperwork, etc. Select topics that may be common with others whose job responsibilities are similar to yours. Think about the challenges you face; you'll want to ask others if they face the same situations. You will probably find many others who do. If nothing else you will find you are not alone in these cases. Most others have the same challenges. And, you will probably leave the meeting with lots of new insights and approaches to solving the problems you have encountered at your home base.

Here's the best way to get yourself in this business of networking. You should find out from the event sponsor the names, addresses, and phone numbers of those who will be in attendance (at my regular Hotel & Motel Management two and a half day sales workshop I always send out a listing of all participants. This way, each can make contact in advance of our workshop should they chose to do so). Make contact with a half dozen or so in advance of the event so you could set up a meeting with them. You can arrange to meet for breakfast, or a drink one evening, just to schmooze with each other. There is nothing wrong with explaining why you want to meet. It may not be a bad idea either, to invite one of the speakers to join the group. The whole idea is to share information by asking each other questions about concerns and solutions to problems being faced in the job function.

It may very well be that what you get out of networking could be more valuable than what you'll get out of the meeting content. We do a lot of sharing at my workshops and attendees seem to like the idea (that's what gave me the idea for this column). Further, you'll find that your networking contacts will develop into long term relationships with continuous contact over a long period of time.

Try "working" this networking opportunity the next time you attend your local HSMAI (Hospitality Sales & Marketing Association International) meeting or any of the national HSMAI conferences around the country. Also, do the same at any events taking place where you be in contact with other hotel sales folks.

Of course, if you are attending conferences of MPI (Meeting Professionals International), STAG (Society of Travel Agents in Government), or any of the other events where prospective buyers are in attendance, you can do the same thing by inviting buyers to meet with you over coffee or a drink. You never know, you may even make a sale.

NETWORKING: OLD IDEA, NEW DRIVE

December 2003

Networking is a key to getting new referrals for business in our hospitality industry. It just has to be part of everyone's marketing effort, at every level of our business. It's easy, anyone can do it, and it does not have to be something that only salespeople are supposed to do. It is just taking schmoozing to a new level. There is nothing new about it, has nothing to do with technology, and most seasoned sales executives will tell us that some of their best pieces of business have come as a result of referrals from another person. That's what networking is all about, referrals. The more people you know, the more new people you can meet, through their connections. It is like a fish net. You can keep spreading it by tying additional pieces together with little knots to make it larger.

One of the most successful networking opportunities for sales folks is an organization called BNI (Business Networking International), with 2,900 chapters in 24 countries worldwide. In the United States, there are chapters in every state. BNI was organized in 1985 by Ivan Misner in San Dimas, California. He even wrote a book about networking, "The World's Best Known Marketing Secret." The whole idea behind BNI is to provide business and professional people a networking organization that allows its members an opportunity to share an idea, contacts, and most importantly referrals. There are a few other similar organizations around the country to which salespeople belong for networking purposes: Le Tip, Power-Care, and Leads Club. These organizations are referral groups to which sales representatives from a variety of local businesses belong to local chapters, and meet regularly to pass on referrals to other members. Membership is limited to only one company representative from an industry. For example, there could only be one member from a local hotel, auto dealership, real estate company, bank, etc.

Meetings are generally held over breakfast for one-hour, and at that time, each member passes on information to other members about possible prospects for business. Also, at the meeting members take turns in giving a "commercial' about their respective business. Members also specify an ideal client or customer. Sometimes members will ask others to introduce them to someone they have been wanting to meet for business. Some chapters provide incentives for the number of referrals passed on to other members.

The idea of the organization is to keep the chapters small in size. In my area (Roanoke Valley of Virginia) BNI has 15 chapters, each having about 17 members. This particular group has been averaging 54 referrals to each other a month. One member commented: "It's like having 17 other people on my payroll" Another member commented: "Thirty percent of my new business is generated from my leads here." Members refer to this referral system as "structured word of mouth." To some extent, this is something like the local "Business After Hours" socials organized by the Chamber of Commerce committees, but more structured, of course. You can contact BNI via its website: www.

bni.com, and you can even get a free subscription to their online newsletter.

THE SELLING PROCESS, LET'S MAKE IT BETTER

April 2004

We are really into the recovery right now and already things are getting better from a sales standpoint in our hospitality business. It looks like occupancy is up, as is the average room rate, and along with that, revenue per available room has improved. We are not yet back to where we were in 2000, but we are getting there. Some changes need to be changed to our approach to group sales. For the past couple of years, because of the drop in business, it seems that we were giving away the store. The competition was tough, granted, and we cut rates, gave in on cancellations clauses as well as attrition clauses in group contracts, did whatever we thought we had to do to get business. And, then came the third party, independent planners. They not only wanted 10% commission in what group bookings they gave to a property, but some wanted, and got, 5% up front, at the time of booking, with the balance after the group departed, and some even got some bonus added to that. Just recently, the some of those Internet travel organization who have been selling individual rooms to travelers at lower than group rates, have gotten into the group business, offering blocks of rooms at properties to meeting planners at reduced rates. Why hotel operators will want to go along with that is hard to believe. I guess they do not feel their own salespeople can do the job themselves.

So with still some challenges facing us, we need to figure how we can best work within our sales effort to not only bring in more business, but better business. What we really need to do is figure out what type of business, meaning market segments, should we put our time, effort and resources into, so as to take advantage of the improvement of the economy and the expected turn around in demand. Salespeople should work with their accounting department to take a look at the various segments of business that has historically brought in the best average daily rate. These segments need to be analyzed to determine what percentage of a property's total business those segments generated. Of course, what we should want to do is get more business from the best segments. As an example, if the corporate meeting market segment was giving us the best average daily rate, but only generated 5–10% of our business, then we should start coming up with a sales plan that would increase that share of the business to maybe 20'% of our business. The idea being that we would concentrate more with all kinds of sales activity to be sure that we do generate that much more business for our property. We just need to go down the list and improve sales, with a plan, to improve sales in those market segments that bring in the best dollar revenue. That is the best way to improve any bottom line profit. All too often, salespeople tend to go after the business that is most easily gotten but may not be the most profitable. The easiest business to get are the markets that pay the least; yes, it is harder to get the better-paying business, but that is the job we have to do.

MAKE YOUR HOSPITALITY SALES MEETINGS MORE EFFECTIVE

March 2006

To what extent are your usual sales department meetings effective? It seems that just about every hotel and resort property has a sales meeting at least weekly. Then there are also regional sales meetings as well as corporate office sales department meetings from time to time. There are ways to make these meetings more meaningful, interesting and most worthwhile and productive. To start with, every meeting should have a reason; there is no point in having any kind of meeting if we do not have a special reason for having the meeting. We begin with having a written purpose and objective with the idea that it should not be the same for every meeting. In the purpose, we indicate what is going to happen at the meeting (like what is going to be discussed) and the objective should be a description of what outcome is expected (what is the goal of that particular session). This, then, needs to be communicated, well in advance, and in writing, to those who are expected to participate. Along with it should be an agenda listing the issues to be covered at the meeting (listing who is to address each item). Only those persons who are directly concerned with the content of the meeting need be invited. The amount of time allotted for each issued should be indicated on the agenda; otherwise, the meeting could take longer than necessary, and sometimes does not get through to the end of the items to discuss.

For a property that usually has a weekly sales department meeting the following should be considered:

▶ The Director of Sales should prepare the agenda and run the meeting.
▶ The General Manager should be in attendance.
▶ Keep the meeting to not more than one hour.
▶ Specific details of weekly group functions set-up and catering handled at a separate meeting to include a catering manager, conference service manager, food & beverage manager, and sometimes the chef.
▶ Recorded minutes of the meeting should be distributed to participants.
▶ Create a method for measuring the objectives of the meeting.

Of course, regional sales team meetings, as well as corporate sales meetings, would take longer and would not be held as frequently as local property sales meetings. However, the format could very well be similar.

Sales personnel should look forward to attending these meetings as an opportunity to learn more about how to conduct business, get feedback on their performance, as well as be motivated to become more productive in their functions. However, generally, there is the opposite. It seems that so many sales department staff at local properties does not like the idea of attending meetings; feeling it is a waste of time.

In a survey taken of business people who attend meetings, conducted by MC (now Verizon Business) certain questions were asked regarding business meetings and here are the responses:

- Percentage of meeting time that professionals believe is wasted: 50%
- Percentage of professionals who admit to daydreaming in meetings: 91%
- Percentage who have brought other work to meetings: 73%
- Percentage who have fallen asleep: 40%

We need to wonder how we stack up in connection with our own sales meetings.

JUST KEEP SELLING IN 2009 TO MAINTAIN PROFITABILITY

January 2009

The news media and pundits are driving us crazy with bad news about travel and tourism being cut back all over the place. All we see and hear about in new broadcasts is how bad our hospitality industry business will be this coming year. OK, we can go along with the fact that projections reflect a drop in occupancy of a few points; however, has anyone indicated that the real cause in the drop probably has more to do with the increase of about 200,000 more rooms on market in 2008? Don't forget, every time a new property opens up in your hometown, right in your own territory, it may very well be helping themselves to some of your room night business. These new properties just do not generate many more room nights for your destination right away; they may, of course, depending on the type of property, brand, facilities, amenities, etc., develop new markets and new business. However, right off the cuff they will be able to do but will eat into your business and the other lodging operators in the area. There is nothing new about this. We've seen it done time and time again.

Some of the writers in the new media are telling the public that these are the worst times in history for the hotel business. Baloney! How wrong they are. Where were they in the early 70s and mid-80s? Now those were really pretty bad times for us. How about the period of 1986–1992. the worst years for hotels with 64% of the hotels could not make debt service, and the industry overall did not make a dime during that period. Today, even with an expected drop of occupancy our hotels will still be profitable, with RevPAR being down just a little. The whole idea, of course, would be to maintain the rate. Do not get scared into dropping rates to try to get more business. If rates get dropped by 10% you will need to get an additional 74Yo in occupancy to break even. So be careful about getting nervous because a property down the street is getting some of your business.

The real key to doing your best business is to provide customer service that ensures the return of a guest. Tony Prusak, Director of Sales at the Marriott in Cleveland says that is what they are doing: '"In a down economy, we believe that customer loyalty is our strongest asset! When you make customer satisfaction your competitive advantage you will perform better than the rest of your competitors. We are ensuring that our hotel is doing everything to make the customer experience we deliver significantly better than anyone else in the market. We know that our customers understand that quality has a price. If you start the fire sale mentality in your hotel, your product and service levels may suffer as a result too."

Let's be positive about the current economy, if, in fact, we do have somewhat of a "down" period in our industry, it will be short-lived, and pretty soon, as PKF has indicated in their projections, in 2010, and 2011, we'll be back with improved business. Maintain your sales effort, keep selling and provide that extraordinary customer service.

IMPROVE YOUR NETWORKING SKILLS TO BUILD BUSINESS

March 2009

A very well-known, international speaker and author on building customer relations and sales had the following to say on a recent blog:

"I'm doing what Mom said not to do . . . speaking to strangers. When I was little my parents advised me not to talk to strangers when my parents weren't around. That was good advice at the time, but the good advice has a lifespan and this one has run out. Yes, it is still good for children to be cautious but for adults, but we MUST talk to strangers! The only way that business will advance or most problems will get solved is if we intentionally and consistently talk to strangers. Our goal ought to be to reduce the number of people who are strangers to us by several people a day, just by getting to know them. Salespeople must be skilled at connecting quickly with people they don't know and building trust. New neighbors, coworkers, teachers, and service people arrive in our world all of the time. Let's become very good at talking to strangers and making new friends of them every day." (Jim Cathcart's Biogs have new posts and photos with his latest insights and stories. http://www.jimcathcartsblog.com and http://www.highvaluerelationships.net,

We talk a lot about the subject of Networking, but do most salespeople, in our hospitality industry, really know what it is all about? The whole idea, of course, is as Jim has reflected in his writings is strictly "to meet new people," with the thought of expanding your network. It is not that everyone you meet is expected to be in a position to give you business, although that would be great, but we need to know to whom they may be "connected» that could be in a position to be helpful to you in creating sales for your property.

Just prior to writing this column, one of our international graduate students approached me with a concern. He was getting ready to leave for Las Vegas to attend an educational conference where he did not know anyone else in attendance. His concern was being able to network to his advantage. This particular graduate student did not meet people easily. It was suggested that first, he understand why he is attending this conference; what were his goals to accomplish? He was seeking contacts for future employment in education at universities. Great, now we have an objective. Then was advised to find out how many people would be in attendance, to come up with a suggested goal of how many new people he needed with whom to make contact. From that he would want to come up with a goal for the number of those contacts who may be in a position to become a member of his "network;" that is, who may be helpful to him in seeking employment. Therefore, if 200 people were expected to be in attendance, and he can set a goal of meeting 100 within the three day period; perhaps 30% may be probable future contacts for his network. This means, that in the process of networking it is not enough just to greet, and exchange business cards. There must some conversation with each contact to get a "feel" for their value to becoming part of one's network.

To be good at networking there should be a plan, as indicated above, and one should try to exceed the goal of a number of contacts. You do this by arriving early at networking opportunities. This means at all events that will take place at a conference of any kind; whether it be for getting registered, attending a session, or a social event. Networking opportunities just go on all the time at conferences. Be the first to arrive and the last to leave. Do not be a wall-flower, waiting for someone to come over to meet. You must keep moving around seeking out people you do not know. It is also perfectly acceptable, in an affinity setting, to walk into a small group to introduce yourself. A group would automatically open to accept you that means they are waiting for you to introduce yourself. Make the rounds of introductions and offer your business card.

In networking, although individuals you meet may not be a position to personally be of any kind of help to you, they may be connected via their network via persons who could be helpful to you in building sales. It is important to stay in touch with your network from time to time, keep building relationships.

COOPETITION PREPARING FOR WHAT IS NEXT IN HOSPITALITY SALES

November 2010

The economic downturn, even as it is turning around, continues to negatively impact our hotel properties, so we have to start thinking of new ways to create opportunities for the development of more business. One potential way to improve business development is Coopetition. Perhaps, we are not ready for this strategy called Coopetition, but we need to think about it. Other industries have used Coopetition in response to the challenging economy. For example, Pharmaceutical companies use Coopetition to share research costs across the industry. A dozen large pharmaceutical companies share information on the results of clinical trials for Alzheimer's disease. This information sharing is coordinated by a unique third-party organization, the Coalition Against Major Diseases. Industry leaders decided to override competitive concerns in order to reduce research costs and accelerate the launch of new medicines. They then compete on sales of products derived from shared research.

Is it time for hotel companies to start sharing information with competitors for the development of new business? We see some of this now in some areas of the U.S., where hotels get together with their Convention and Visitors Bureaus to market their area at trade shows. Competing hotels have joined forces in some destinations to market themselves. Hotel sales forces of competing hotels gather to conduct a sales blitz in a certain location. So we are occasionally practicing Coopetition in small, informal ways, but not quite the most effective way.

Coopetition is effective when the competitors agree to formal rules and processes of cooperation. For example, most coopetition arrangements use a third party entity to manage the coopetition. According to David Green, President David Green Organization, a unique third-party organization is a key to fair, effective coordination of the information sharing between competing companies. Green says that "Information sharing builds better business intelligence to target emerging customers, and reduces the cost of building sophisticated intelligence on the customers." Sharing records allows for predictive analysis of the customers' probable behavior. For example, some meetings move around the country for various reasons. If they meet out West one time, what is the probability of the next meeting being held on the East Coast? What would happen if one property shared information with other properties in another regional area? This scenario exemplifies the main idea that Coopetition is a way for competitive hotels to share information for the development of new business for each.

Besides using third-party organizations, the other key Coopetition task is to manage data to improve understanding of customers' behavior patterns. Effective data management in the Coopetition model will produce for all properties access to new customers, improved customer profiles, and updated business intelligence. Now, the economy continues to challenge sales and bookings, so that at some point, perhaps sooner versus later, a variety of hoteliers have to look at the world a bit differently, and not through the typical lenses. We know that it is challenging for properties to release data, and there is concern about what is being done with the data, particularly those old "maybe" accounts. In most cases, these records are collecting dust. So when considering the Coopetition model, the key benefit is that it's a cost-effective approach to increasing the sales pie for all participating properties.

NETWORKING IS NOT SELLING

March 2012

All too often, in our business of bringing in more business, we have sales staffers attending a variety of gatherings where networking is critical. However, they go with the objective of making some bookings, and end up doing more selling than networking. Spending time selling at networking type events is the mistake that some sales folks make. If the whole idea of attending a Chamber of Commerce social event, or something similar, or even being an exhibitor at a trade show is to book business, right there, at the event we need to rethink what is our real objective.

If we indicate that we will be attending a particular event for the purpose of networking, then don't do any selling. The whole idea behind networking is to meet as many people as possible, and to determine how many of those we contact are in a position to provide business at some future date. What It is that we are doing is prospecting, not selling. In making contact with persons during a networking event, the idea is to spend as little time as possible with anyone. All that is necessary is to get enough information that would indicate whether that contact, or the organization represented, is in a position to do business with you, should they choose to do so. What is necessary then, is to be sure to do the follow-up necessary sometime following the event. That is the time when the selling is done. We must not forget that at an event the person to whom you may be in contact is probably also busy, making the rounds for their own interest in being there. Many times a salesperson gets "wound up" in trying to make a sales pitch that too much time is taken away from contacting other people. The best idea is to limit the time of a contact; using just enough time to get critical information. Using some sort of form would be helpful; that way one does not forget to ask critical questions. Spending five minutes with anyone contact is more than enough time to find out who, what, when and where, along with accepting a calling card (that helps make sure you have the correct spelling of the name, and Contact Information). This way you maximize your time in a networking operation.

Measuring performance is critical in the business of networking. In most cases, particularly if exhibiting in trade shows, you would want to show ROI (Return on Investment). In doing so the idea is to provide the amount of estimated business that would be generated for your property based on the number of qualified prospects identified. From that number, you can figure that about 10% would actually be booked at some later date. Considering the type of event being covered by the trade show, one would be able to estimate the dollar amount of business that particular market segment could produce. Your past figures of that segment of business should be able to reflect the dollars generated by one booking. That amount is then identified by the number represented by 10%. For example, at a trade show or corporate meeting planners; last year your property had 20 such meetings; the total revenue generated by all of those 20 meetings would be divided by 20 to determine the average amount produced by that market segment. If the average comes out to $10,000, as an example, and the estimated number of prospects that will turn into buyers is 5, then, of course, the estimated ROI for attending that convention and trade show would be $50,000. However, experience has reflected that in many occasions that would be a very low number.

Trade shows, in general, would have a much greater ROI, and are truly the best opportunity for networking in building a business. For more information on trade shows go to the Trade Show Exhibitors Association (www.TSEA.org).

REVENUE MANAGEMENT: THE GUIDE TO IMPROVING PROFITABLE SALES

November 2012

No question about it; we leave it up to the sales folks to bring in the business to build sales for a property. However, to what extent are the sales being brought in the most profitable? We see it more and more these days that owners and operators are starting to develop more formalized programs in revenue management to help guide the direction of sales. Of course, a key to getting this to work effectively is for the sales departments to concentrate on booking more future business. In revenue management, we forecast business based on what we already have on the books for future business as well as the history of business during future periods. With that in mind, it behooves sales departments to concentrate on booking business for the future, thereby blocking rooms for groups to reflect higher occupancies during those periods. In being able to project high occupancy during any period, revenue management then dictates that salespeople are in a better position in negotiating for better profitable sales.

There is a better chance for sales departments to improve negotiations in this coming year. With all the figures coming out reflecting an increase in demand, higher occupancies, as well increase is ADR, it certainly reflects that there should be an improvement in more profitable sales to be booked, especially in group meetings. A recent report on a study by Meeting Professionals International (www.mpiweb.org) reflects that there should be an increase in group meetings by companies as well as associations. Further, those budgets for these meetings will be increased. This tells us, right now, that sales staffers need to start working on putting the future business on the books.

Our best figures in occupancy and rates were in 2007. We started seeing a dip in 2008, not only due to the economy, but also with the growth of more rooms being produced in the marketplace. At that time we also had the highest ADR; however, it started dropping, and never got back up for the next few years, until this past year, when we saw an increase starting in occupancy in 2011 and 2012. So, things are looking pretty good for next year. We could very well get back to the occupancy we had in 2007, as well as reflect the best ADR ever. It is all up to the sales folks to deliver the product, and create a profitable business. Here is where revenue management comes in to play. Based on whatever occupancy is being projected, plus past history, plus the information from recent reports of increases expected in demand and occupancy, a guide should be produced for the sales staff.

Chapter 16

Understanding the Features of Your Product, Especially F&B

FACILITIES, NOT ROOM RATES, WILL NET MEETINGS BUSINESS

November 1984

The meetings business is on its way to a banner year, according to Mel Hosansky, with more than $30 billion to be spent this year on meetings, conferences, trade shows, training seminars, and conventions. This amount represents a sizable increase over the $9 billion spent 10 years ago.

Among the biggest recipients of those meeting dollars will be lodging operators who provide sleeping accommodations, meeting space, food functions, receptions, and equipment rental.

No type of location is out of the running for the meetings market, and with hotel construction adding more than 70,000 rooms in 1984, the competition for this market is fierce. Strong contenders for this business are still the downtown and airport locations, although suburban and roadside properties are beginning to get a good share. Resorts have always been and will continue to be very strong.

Location, location, location is still the cry of meeting planners in determining where conferences and conventions will take place, but in this writer's opinion, more than 50% of the meetings business gets to a property because of a buyer's inquiry rather than through a sales solicitation.

The next emphasis is on facilities. Room and meeting space availability are important, but product quality and guest satisfaction are also prime considerations.

MEETING THEIR NEEDS

For some hotel/motel sales personnel, room rates may be a major consideration, but sophisticated planners are concerned with meeting the needs of the group. Accessibility, facilities, guest services, friendliness of the staff, cleanliness of rooms and foodservice rate high on their checklist. Meeting attendees will remember good (or bad) foodservice longer than anything else about a meeting.

Trying to close on rate alone is not working anymore. Room rates are still important as they once were. Some rate negotiations will almost always take place, but salespeople should try to sell the product, staff, and value.

Maximizing sales means getting the best possible rate for a property at all times, and this means setting the average daily rate for group meeting business as the next highest rate after transient business.

Your salespeople should be convincing the meeting planner that, for a wide variety of reasons, the property is the best possible choice. And, when it is determined that the property meets the needs of the group, it's time to close the deal at the best possible rate.

A plan to go after the meetings market should be a must for 1985. This market will continue to grow at a rapid rate, and the largest growth will be in the small meetings market conferences for less than 50 persons. Cities convention bureaus, hotels/motels, conference centers, and even freestanding restaurants are scrambling for this business, but success will be determined by the best selling job.

SALESPEOPLE: THE MORE YOU KNOW, THE MORE YOU'LL SELL

March 1986

There is an amazing number of hotel/motel salespeople around who don't have complete knowledge of what they're selling. Ask a salesperson the capacity of a meeting room for a classroom-style setup and his answer will probably start with the word "about." The same holds true for banquet seating capacities, the number of trade show booths and even the number of rooms a property will put aside for a convention.

The successful salesperson has confidence in his knowledge of the various products being sold or presented in a property. From guest rooms to parking facilities, salespeople should know all they can know about the property they represent.

The following list is by no means complete, but it includes those areas of a hotel/motel operation essential to a salesperson's knowledge.

Rooms Division: Basic information about rooms should be known, such as the types of rooms available (kings, doubles, double-doubles, parlors, suites); their number and location; which rooms connect; number of rooms on each floor; size of rooms; in-room amenities; which rooms have hide-a-bed sofas; number of cots available; availability of cribs; which rooms have special attractions (view, poolside, balcony, etc.). It is important to know policies governing areas such as room blocks, deposits, cancellation, payment, check-cashing for members of a group, and check-in/check-out times.

Meeting and Banquet Space: Many formulas exist to determine capacities for meetings and banquets. But due to the availability of a variety of table sizes and shapes, and due to obstructions that may exist in some rooms, most formulas just don't work anymore. The only real way to know capacities is for salespeople to get involved with the catering and banquet staff. Have them set up each room in a variety of configurations. This will provide exact figures for capacities with different room setups.

Consideration must be given to certain requirements of the customer. A banquet requiring a head table and/or dance floor will reduce dining space considerably. A meeting room where rear view projection equipment is required will reduce the available seating space-as will a meeting that requires room for specialized demonstrations. Once all the "basic" capacities are known for each meeting and banquet room, the salespeople will still have to question the customer carefully on what will be going on in the room during the event.

Most every catering director can relate a story of setting up a beautiful room for a banquet when, without warning, in walks a 15-piece orchestra looking for a bandstand. Someone just forgot to tell someone else that there would be a band for dancing during the banquet, and the ballroom was set to capacity without leaving room for the band or dance floor. It really does happens.

Restaurants/Lounges: Salespeople need to know about capacities and hours of operation as well as prices of drinks and meals. It is also helpful to be aware of the sizes of drinks served in the lounges. When not in use for the public, are these places available to groups for meetings or special events?

Recreational Activities: If available at the property, then salespeople must know about the cost of recreation facilities, when available and starting times (for golf, for example). If such facilities are not available at the property, salespeople should know whether or not they exist near the property. If they do, how far away are they? How much will it cost a guest to use them? Will the property make arrangements?

The general idea is that to be successful in the sales effort, one must have complete knowledge of everything the property has to offer. That knowledge will be reflected in the salesperson's confidence as well as on the property's bottom line.

SALESPEOPLE SHOULD LEARN ABOUT LATEST TRENDS IN FOOD

March 1997

Today's meeting planner is having to pay much closer attention to satisfying the demands of conference and convention attendees. And that includes satisfying their increasingly selective appetites for certain foods and beverages. In hotel F&B departments, no longer does the rule say: "Give them bigger portions of heavy food." Today, light and healthy foods are in, heavy foods are out. The 5,000-calorie dinner no longer is necessary-nor do most people desire it.

At the same time, there are those who still prefer the shrimp cocktail appetizer, the steak or prime rib entree, and the fancy ice-cream parfait dessert. What it boils down to is this: Salespeople have to be tuned in to the menu preferences of the groups they are selling; they have to be aware of what they want to eat while at the meeting facility.

MORE KNOWLEDGE NEEDED

Salespeople need to improve their knowledge about what kinds of foods are available in the F&B department and the kinds of dishes the kitchen staff is capable of preparing. This knowledge can go a long way in closing a meetings deal.

Hotel General Managers should set up educational programs with the sales staff,

F&B director, catering manager and chef to discuss this. It would be a good idea to invite a corporate-meeting planner or trade association executive to join the meeting. One probable result of such a meeting could be the development of a supplemental banquet menu that includes lighter, healthier selections.

Meeting planners are more on top of the trend toward healthier food than many salespeople give them credit for. When a planner says he wants a light meal, he doesn't mean one that's relatively inexpensive (though salespeople sometimes interpret it that way). The planners are talking about meals with fewer calories filling but not heavy.

REFRESHMENT VARIETY

Refreshment breaks, too, are steering away from the traditional offering of coffee and Danish. We're learning that people want more variety: decaffeinated coffee and tea, soft drinks, diet drinks, diet drinks, juices, health-food snacks, and yogurt are becoming popular for breaks.

Also, we're beginning to notice fewer full breakfasts being to notice fewer full being served. Meeting planners are advising attendees who want full breakfasts to have them in the hotel coffee shop or dining room. We're seeing a trend toward coffee, and perhaps juices, being the only items available at the start of a meeting. Planners have discovered that this is a good budget-reducing technique, which in turn frees up additional funds for other F&B events.

Luncheon menus, too, are becoming lighter and healthier. Sale and F&B personnel need to become more knowledgeable to be able to meet the increasing demand for lighter lunches. Soups and salads, vegetable plates, sandwich bars, and poultry items are becoming popular items for lunch. Groups are not requesting red-meat items as much these days especially not for lunch and dinner.

MULTIPLE SPONSORS

Foodservice departments will need to be alerted to those types of meetings and conventions during which F&B events are sponsored by different groups. People involved with the sponsoring organization wouldn't necessarily know what another sponsor is providing for the menu. The food service staff should be aware of this and advise the sponsors accordingly, so that variety is achieved and balanced menu planning is effectively coordinated.

Remember this: when people attend meetings and conferences, they will recall the meals they had at their hotel long after they've forgotten the convention itself. You must make sure they remember those meals for the right reasons.

MORE CREATIVITY NEEDED IN SELLING HOTEL BAR FUNCTIONS

April 1987

It's really not all that surprising that most hotel/motel salespeople know very little about selling receptions and bar functions. After all, salespeople have the primary responsibility of selling group rooms. In connection with that, though, there is a certain amount of catering to be sold as well.

Perhaps all the pre-printed menus and prices lists for food and beverage functions have steered us away from being creative in this areas, or perhaps they do not provide the opportunity for creative negotiating to book a piece of business. And maybe General Managers have not taken the time to get their salespeople together with the catering department or F&B directors to discuss how to sell bar functions.

IMPROVING BEVERAGE-SALES RATIO

General Managers need to examine the ratio of banquet food sales to banquet beverage sales. In all probability, there is room for great improvement in the latter area. Pre-luncheon sand pre-dinner cash bars, for example, could create a very positive situation for profitability if more attention were paid to them. Many meeting planners may not have the budget for planned or "hosted" bar, so why not have salespeople try to get the planner to allow the property to put in a cash bar?

Here's another example: Are wines being sold along with luncheons or dinners? Examine this ratio as well-there are opportunities for improving it if proper selling techniques are used. If the meeting planner's budget can't handle the cost of serving wine with the meal, have the salesperson ask the planner if he or she would have any objection to waiters and waitresses selling bottles of wine to the tables.

ROOM FOR CREATIVITY

Our salespeople just need to know more about various methods of selling beverages for various F&B functions. There is plenty of room for creativity and negotiating in this area. Many good pieces of business have fallen through the cracks because the planner of a meeting loses confidence in the property, and very often that's due to the fact that the salesperson doesn't have sufficient knowledge to sell a bar function.

In many cases, a property has a catering person on staff who can handle the beverage side of a booking deal. That's the ideal way to ensure that the business doesn't get lost. But can we be sure why a piece of business is not booked? A seller's confidence has a direct bearing upon a buyer's confidence. People won't buy just anything without being confident that what they're buying is right for them.

So if in fact there is a problem, how do we correct it? In selling and negotiating, as in anything else, knowledge is power. And in sales, knowledge of the salesperson's own product is of utmost importance. If a hotel/motel salesperson is working on booking a convention and can quote reception packages, bottle prices, and drink prices, then he's ahead of the game. And if the prospect asks, "What size drink do I get for such-and-such a price?," the salesperson should be able to provide the answer. If bottle prices are quoted, the salesperson should have the knowledge and confidence to answer correctly if he's asked the size of the bottle.

DISCUSSION SESSION NEEDED

A good half-day session can solve the problem. The General Manager should take the initiative to institute such a session if it's needed (and in all probability, it is needed!). Get the catering department involved. Get the F&B people involved. Get creative about selling beverages for banquet and meeting functions.

You could even call the session, "Everything Salespeople Need To Know about Selling Booze." Then let the salespeople sell.

GROUP F&B FUNCTIONS HAVE POTENTIAL TO INCREASE PROFITS

May 1987

In today's lodging industry, food & beverage service is playing an increasingly important role as a significant contributor to a hotel's bottom line. This is especially true when it comes to meetings and conferences. In many cases, the master account billing covering the F&B service for a meeting adds up to more than 50% of the total room-sales billings for that meeting.

In addition to that, there is probably an additional 15% of total room sales being spent for a la carte food in the property's restaurants. And depending on the type of meeting, there is often a whopping amount of money being spent in the lounges for beverage services.

NOTICEABLE IMPACT

On the whole, most all types of meetings-seminars, training programs, conferences, whatever we choose to call them-can very well create a most noticeable impact on the F&B profit/loss statement for a property. Food and beverage sales in connection with a meeting will, in all probability, exceed the normal ratio of F&B sales to room sales.

The great impact this kind of business has on a hotel or motel relates to its profitability. Compared with sales at an a la carte restaurant or cocktail lounge, catering for meetings generally involves lower costs for food, beverage, and labor. The reason for this is that meals are planned in advance and require guaranteed counts at least two or three days ahead of times. With proper food purchasing and labor scheduling, there should be a significant improvement in savings.

SELL F&B TO GROUPS

Generally, hotel/motel people are concerned with group room-sales-and that's how it should be. The most important function for a salesperson is to generate new room-sales, preferably through group meeting or conference business. However, the salesperson should always remember this: Groups that have more planned food functions and receptions are groups that could be more profitable to the property. While those groups that only book rooms but eat meals elsewhere are certainly good for room revenue, they play havoc with the property's food & beverage labor costs. Our objective is to sell profitably, which means going after the kinds of groups that will eat "in-house," hopefully at affairs catered by the property's F&B department.

Two areas that are not generally being handled effectively are receptions prior to banquets and the selling of wines during meals. By reviewing the percentage of banquet beverage sales to banquet food sales, managers can tell how well they are doing in these areas. The secret to additional profitability is to improve that ratio. It's amazing how much reception business is discouraged because of pricing that is illogical to the meeting planner. It makes only a comparison of wine sales to food sales in catered functions to show how poorly we're doing with these types of sales.

The property's catering manager or banquet manager needs to work closely with the sales department to encourage more creative menu planning, more creative refreshment breaks, and more creative receptions and theme parties.

GET CUSTOMER ENTHUSED

Getting the meeting planner excited and enthusiastic about the property's capabilities will help to sell not only rooms, but all the food & beverage services that your hotel makes available.

Often, a meeting planner wants to see what your hotel can offer his group not only in rooms, but in food & beverage services.

So, let's show him what we have to offer, and how creative we can be in fulfilling his group's food and beverage requirements. In short, let's make it easy for prospects to buy.

PLANNERS NEED TO KNOW THE BENEFITS OF HOTEL FEATURES

July 1987

More and more frequently, we are seeing advertising and direct-response tools being used to sell the features of hotels and motels.

But the fact is that nobody is interested in the features themselves. We are all selfish when it comes to hotel features, we want to know, "What will they do for me?" Why should a prospective customer be interested in the 25,000-square-foot ballroom, the amphitheater or the gourmet restaurant unless there is some benefit to him?

Meeting planners, for examples, need to have successful meetings. In order for them to do so, the property they book must meet their specific needs and objectives-and they must know ahead of times that the property's features will play a significant role. As hotel salespeople, we need to do a better job of relating our properties' features to the benefits they can give our customers. The key, of course, is to develop information relative to their needs.

In selling, the idea is to come up with appropriate questions, the answers to which will reveal the planners' needs and how our hotels' features will help satisfy them. What this means is that more business can be developed by asking questions than by merely describing features.

To effectively sell to a meeting planner, we must determine his need before we tell him what the hotel has to offer. For instance, your hotel's 700 or 800 rooms won't mean anything to a planner if he only needs 30 or 40 for his meeting. The same applies to a meeting or banquet space and amenities.

Once you determine the planner's needs, the next step is to personalize your property's benefits. Here's an example: "I understand, Mr. Planner that your people drive in from all over the state. You realize, of course, that 200 cars in our 300 parking spaces, in addition to handling parking for our other guests."

As I said, the key to successful selling is meeting needs. Acting as a "consultant" helps determine what those needs are. Then, by personalizing, you show the prospect how his needs are to be met. Planners want their needs met and their problems solved-they don't need a sales pitch about features in which they aren't interested.

People Make It Happen

Here's something else: The property itself cannot relate a feature to a benefit, it doesn't conduct business with its customers, and it doesn't make an else-people do the relating, people conduct the business, people make the sales. (And remember: Everyone sells, so every staff member is a salesperson.)

A hotel's features serve only to support the benefits that meet the needs or resolve the problems of meeting planners. Those wonderful advertisements we've been seeing can announce what a property has to offer; they can what someone's appetite; they may even trigger some inquiries. But it takes people to follow-up on a lead and close the deal, and that can happen only if there are real or perceived benefits to help a planner arrange a successful meeting.

SUGGESTING RECEPTIONS CAN HELP IMPROVE GROUPS SALES

February 1988

At most lodging properties, salespeople concentrate on selling rooms-and rightfully so.

But when it comes to selling group meetings, the sales department should also concern itself with improving bar sales during meetings. This should go hand-in-hand with selling the group room-block.

Often, groups don't consider the possibility of holding a reception. Here's where the sales department can show how a reception can enhance the entire meeting. Along the way, of course, additional revenues can be generated. Salespeople should always suggest that the reception is held prior to the meeting's opening. Then, when the meeting starts the following day, attendees will already know each other and feel comfortable in each other's company.

Per-Person, Per-Hour

From a profit standpoint (as well as a service standpoint), it's usually best to sell a cocktail reception by the per-person, per-hour system. This involves the meeting planner paying a flat fee for each guest for a specific amount of time-usually an hour. The hotel provides an open area that offers a variety of cocktails, beer, wine, and soft drinks. A valuable selling point for this kind of reception is that the meeting planner knows exactly what the cost will be. In addition, the hotel benefits from this type of setup because bar costs are lower as compared with selling liquor by the drink or by the bottle.

With the per-guest, per-hour format, it's important that the cocktail reception start no earlier than the cocktail reception start no earlier than the designated hour and end no later than the designated hour. And it's always in everyone's best interest to serve brand-name beverages rather than the lesser-known varieties-even if it costs a bit more.

Foodservice during the cocktail reception can also be sold the same way-per-person, per-hour. An ideal way to handle this is to provide butler or tray service. This is where you have waiters or waitresses continuously walking around the room offering hot and cold "pick-up" food served from small trays. (In terms of the guest server ratio, the ideal situation is to provide one server for every 20 guests.)

This butler-style service gives the kitchen an opportunity to present a variety of foods that can be served in small quantities on individual trays. Also, it gives the kitchen staff a chance to be creative with the items they prepare for the reception. For pricing-out the food to be served at the reception, you can figure that each person will consume eight to 10 food items in an hour (given an evenly mixed group of men and women).

PRICING-OUT BAR SERVICE

Pricing-out a one-hour, open-bar reception, you can figure that the average person will consume about 2.5 drinks (again, given an evenly mixed group of men and women). A satisfactory and acceptable cocktail will contain approximately 1.25 ounces of liquor. From a bar-service standpoint, two bartenders should be sufficient to handle a group of 100 guests who arrive at the party at about the same time and stay for the entire hour.

It's important to note that when setting up such a reception, the sales department should always consult with the catering department of the food and beverage director.

Salespeople need to keep in mind that the hotel must help the meeting planner reach the objectives of the meeting. Sometimes, the suggestion of a reception/cocktail hour might not be in the group's plans in some cases, the idea might even turn them off. Therefore, salespeople need to be careful. They should be familiar with the background of the organization, the group, and the meeting planner before suggesting a cocktail reception.

Next to guest room sales, the most profitable part of the hotel business is the sales of hard liquor. By suggesting cocktail receptions for group meetings, salespeople can help maximize profits for their hotels.

BASIC KNOWLEDGE OF F&B IS BIG HELP IN BOOSTING SALES

May 1989

In an ideal world, hotel/motel salespeople would know everything there is to know about F&B

The fact is, however, that knowing all there is to know about F&B just isn't practical. After all, if salespeople were experts in food and beverage, they wouldn't be salespeople for long: Hoteliers would jump at the chance to give them food-and-beverage jobs in their hotels. Foodservice is an area in which many hotel operators could really use some assistance.

So perhaps we should be satisfied if our salespeople merely have some general F&B knowledge-enough, at least, to intelligently discuss such basic topics as catering functions and how to set up a bar. The fact is, however, that although salespeople are hungry for every bit of F&B training they can get, in most hotels operations it is just not available to them, in smaller lodging properties, it is usually the norm that a sales manager is also the catering manager. If there is any difficulty in handling food-and-beverage arrangements, the sales manager can go to a chef or food manager for assistance. In large hotels that have separate catering departments, salespeople generally get very little opportunity to discuss F&B "Someone in catering will be in touch with you" is usually what the hotel sales-person hears.

This is an area in which we may be able to improve our performance in closing sales. It seems apparent that people responsible for arranging meetings would feel better about working with salespeople who know something about meals and bar operations. It only stands to reason that a salesperson who can talk intelligently about F&B would have more credibility.

Having enough knowledge to be able to offer meal suggestions, discuss banquet setup or offer alternatives to the usual boring reception would help salespeople sell better-and sell more.

Many people base their purchases on personal feelings. If a salesperson can get excited about a group's coffee break, luncheon or reception, that enthusiasm becomes contagious. Getting a buyer to feel good to have confidence both in the salesperson and the product will help close the sale.

All things being equal, the more knowledge a salesperson has about food and beverage, the easier it will be to sell.

ACQUIRING KNOWLEDGE

What does it take to acquire a fair amount of F&B knowledge? You must start, of course, with dedication and commitment: Exposure to the product, a bit of on-the-job training and a great deal of interest on the part of the General Manager will get the job done in a relatively short period of time. Now, we're not talking about salespeople becoming chefs, and we're not talking about salespeople becoming bartenders, servers or hosts. We're just talking about getting some basic knowledge of catering and banquet functions.

With commitment on the parts of key people involved, some extra time put in by salespeople, some trade-magazine reading, some exposure to the kitchen and asking a lot of questions, our salespeople will be able to present themselves as knowledgeable to the prospects whom they are trying to sell.

BE HONEST, REALISTIC

A salesperson should never attempt to fool prospective buyers into believing that he or she is an expert in all areas of food-and-beverage operations. Remember: While it's true that our salespeople should have a basic understanding of F&B procedures, we should always, be honest and realistic about the extent of their knowledge regarding certain details.

After all, there's still the chef, the food-and-beverage director, the catering manager or the General Manager to call in for help.

The main point, of course, is this: We should never let a prospect for business get away.

MORE KNOWLEDGE OF BANQUET-BAR SERVICE CAN HIKE SALES

June 1989

In today's lodging industry, segmentation not only applies to the types of lodging properties available, but to the sales departments within these properties.

We have national-accounts salespeople, regional salespeople, and local salespeople. We have salespeople for corporate sales, incentive sales, travel-industry sales, association sales, international sales, and transient sales. Obviously, this must work well or we wouldn't continue to organize our sales departments in this way. Of course, there's always the small lodging property that has one or two salespeople who sell to all markets-and who do the conference servicing and catering as well.

Sound Advice

Maybe to be effective in sales it's not necessary for salespeople to be knowledgeable about food and beverage, but, as my Uncle Louie says, "It can't hoit." Managers of all types of properties need to look at the percentage of bar sales to food sales for all groups. I have the feeling we're missing the boat in not generating enough bar sales during our selling situations. Considering that most hotel operations lose money in the food-and-beverage department, they can pick up the bottom-line profit by increasing sales of liquor, beer, and wine.

Generally speaking, hotel salespeople are concerned mainly with filling guest rooms and I certainly have no argument with that. However, I'm convinced that with a better working knowledge of banquet-bar service, a salesperson comes off looking good in the eyes of the buyer and will close more deals and sell more alcoholic beverages.

Here is a list of bar-service questions to which all salespeople should know the answers:

▶ How many ways can we sell bar service?
▶ What are the prices for each?
▶ On the average, how many drinks will a person consume during a one-hour open-bar function? (The answer will vary according to the type of group and its mix.)
▶ When selling by the bottle, how many bottles of liquor, and of what amount and type, will be the average for a given number of people during a one-hour reception?
▶ What's the difference between fifths, quarts, and liters?
▶ What size drinks do your bartenders typically pour at a cash bar (or, when a purchase is made by the bottle at a hosted bar)?
▶ From a profit standpoint, what's the best way to sell a hosted bar?
▶ What are the advantages and disadvantages to the buyer of the different ways to buy bar services?
▶ At what point does the buyer pay for bartenders?
▶ How do you determine how many bartenders and bar setups to use for a reception?

DIVERGENT ANSWERS

Having said that it's a good idea for salespeople to know the answers to these questions, here's something to think about: It's almost guaranteed that if you talk with the General Manager, F&B director, and the managers of catering, bar service, and convention services, they will all have different answers to these questions. It might be instructive to have your sales staff present these questions to the various managers at your property. After comparing what probably will be widely divergent answers, everyone should come to an agreement on the responses they think are best. Meeting planners will do more business with salespeople who are knowledgeable about the products and services they're selling.

I've always been happy if total banquet-bar sales are about 25% of overall banquet sales. However, this is only one person's viewpoint. If you have another opinion-or can offer formulas or responses to the above list of questions-we'd like to hear from you. Drop us a line.

MORE CREATIVITY, AGGRESSIVENESS NEEDED IN F&B SALES

October 1989

We just don't seem to be doing enough revenue in banquet-bar sales as it relates to banquet food and group rooms. It's generally agreed that we do not make money in the hotel food department. Rooms are our best bet for profits-but after rooms come booze.

F&B AN AFTERTHOUGHT

It seems that when salespeople are out selling, the idea is to make the sale on the rooms and worry about selling food and booze later. There's nothing wrong with that. However, many times we accommodate the group for their needed meeting rooms, food functions, and planned receptions. That is, we sell them only what they need or ask for in food and booze.

CREATIVITY MUST FIGURE

We need to be more creative in the area of food and liquor, not only from the standpoint of doing a better job of closing a deal, but from the standpoint of necessary profit. All right, we agreed we do not make money on food, but there is no reason to shy away from selling it. Any General Manager or food-and-beverage director will tell you how important it is to include as many planned functions as possible in the group sale. in fact, there are thousands of stories out there that can be told by salespeople of how the closing of a large piece of group business was accomplished as a result of how the food-and-beverage functions were sold. The creative selling of food and liquor should always be part of a salesperson's pitch.

KNOW THAT BAR HISTORY

It is of the utmost importance for salespeople to be aware of a group's bar history. We need to find out the drinking habits of people attending. Do they drink beer, wine, and liquor? Are they heavy or light drinkers? Are they after-dinner cordial drinkers? Once that history is obtained, make sure to ask the planner and check with salespeople at hotels that the group has patronized in the past. The idea is to be well prepared during the sales pitch so that the person planning the meeting appreciates your knowledge and help. You certainly do not want to talk about a beer party when the group usually has champagne and vice versa.

Always try to get a group to book a welcome arrival party-whether it is with beer, wine, and cheese, open bar, whatever. Sometimes a planner will tell you that there is nothing in the budget for a party. if it could be profitable for the hotel, you may want to sell a cash bar with some charge for bartenders and cashiers, or even feature a sliding scale for that based on bar sales. in this case you must know the group history and habits: otherwise, it may not be a profitable thing to do. And, if there is a poor turnout, it may look bad for the planner.

SUGGEST A SPONSOR

Another opportunity exists if there is no budget. Help the group get a sponsor. It is not as hard as some people think. You can help them contact suppliers to their type of business who could be solicited to pick up the tab for a welcome reception. it works out well for the supplier, the meeting planner, and the hotel.

WINES, CORDIALS APPROPRIATE

The serving of wine with dinner is another area where we seem to be lacking in our sales efforts. Many times a group would love to have wine served with a meal, but the planner doesn't think of it and we fail to suggest it. The same goes for cordials after dinner. If there is no reception planned, a cash bar before lunch or dinner might work well if we suggest it.

Things are getting tougher all the time and salespeople have to provide profitable sales. Let's start picking up our ratio of bar sales to food sales to room sales and watch those dollars fall to the bottom line.

CATERING AN IMPORTANT PART OF THE SALES EFFORT

August 1991

Most lodging operators make very little, if any, profit on food and beverage. Only those with very heavy revenues in that area can make a decent bottom line, if expenses are properly controlled with high food and labor costs, the best shot for operators is to increase food/beverage sales. Most managers would agree that there agree that there is more profit to be made in catering sales than from revenues generated in dining rooms, coffee shops, and room service.

Salespeople have mainly concentrated on room sales and are accustomed to working against goals for room revenue and room nights. And, this is proper since without the occupied rooms there is not much chance for food/beverage sales-except, of course, for a local catering business. There's no question about it-the real profit is in the rooms. Along with that, however, the sales effort should always be concerned with the amount of food/beverage sales that could be generated from a group booking.

When lodging operators take a look at profit and loss statements they may accept a loss on the food statement as long there is a good profit on the room's statement. Today many operators are seeing heavy losses in F&B and are trying all kinds of promotions to increase revenue in restaurants and bars. Catering sales really needs their attention, whether it be from local catered functions or from groups that already in the house.

If, with increased efforts in catering sales, those showing losses can just break even it would be very meaningful to the whole operation. Increased food and beverage revenue are very critical these days to ensure the successful operations of lodging facilities. Regardless of the type of facility: whether a convention hotel, suburban property, resort, conference center whatever, it could very well be the food and beverage statement that dictates whether the facility can stay in business.

For the most part, sales department personnel could use additional training in "profitable F&B sales." When the rooms are sold out then the meeting and function rooms should be busy as well. Close working relationships have to be developed between sales departments and the catering department to make sure that catering sales are maximized. I am pretty much convinced that any property can increase its catering revenue by 15% strictly through upselling. Too often we just present banquet menus to clients, having them pick a menu. What has happened to the consultive selling idea of creating a menu for the client? Does anyone still do what we used to do in the days before we knew about printed menus? we started from scratch, with a blank sheet of paper.

Based on the purpose of the function, the needs of the group, the demographics of the attendees, history of former functions, and budget considerations, we let the creative juices flow and designed a "special" menu. We also thought about what we already had on the books for other functions in the house to minimize variety.

Now, there is nothing wrong with the pre-printed, pre-priced menu, but I wonder if we are not missing the boat in generating additional revenues by not providing the "extra touch" of making it a little better for the client with some consultive suggestions. Special appetizers and special desserts just add so much to a catered event (and adds to the price, as well).

Take a look at your own banquet menu. What is the highest price for a dinner item? Some might, but most will not have a $100 plus item listed. Probably most kitchens would be able to produce such a dinner, but we do not seem to offer it. Why limit yourself to a top dinner price of $29,95 or $49.95 or $79.50 if you can do something even better? Isn't it good to show that you can produce something like that? One way to upsell is to show higher prices than most people would want to spend. There is a better chance, this way, that people will select something other than the lower priced items. And, then there are people and groups that would be willing to pay the higher price, knowing you can produce that type of a menu.

Getting salespeople to understand how critical food/beverage sales are to a property and getting them more involved in catering sales will help get that bottom line in better shape.

ARE YOU SURE YOU WANT A "PODIUM"?

May 1992

Or is it a Rostrum, or a lectern. or a pulpit? We still seem to have a problem communicating. Not only with the public. But each other when it comes to terminology in our industry. For years we have been using terms which mean one thing to us and another to someone else. You would think by now we would have all the terms and phrases down pat so that we're all speaking from the same glossary.

Probably the most misused term, in-house (among ourselves), is "podium." How often have we seen banquet set-up sheets, as well as letters of confirmation to clients, referring to having a "podium" for a speaker? All too often, I m afraid what we really wanted was a "lectern" (either tabletop or floor model). According to the *Glossary of Hospitality Management Terms,* compiled by Professor Andrew Schwartz and David C. Dorf, a podium is "a small platform on which a speaker stands, usually for purposes of audience visibility. This is also properly called a rostrum." They define a lectern as "a reading desk either a table or floor model, used by convention and meeting speakers." Webster's New world Dictionary indicates substantially the same definition.

Many times there is confusion with the terms "connecting rooms" and "adjacent (or adjoining) rooms." And. by the way, what's a suite? The really big, confusing one, especially with the individual, leisure travelers, probably has to do with what we mean by "guaranteed" room. There is a big difference between that and a "confirmed" room. Just because we ask someone: "do you want us to guarantee the room" and they reply: "of course," this doesn't necessarily mean that the person understands there is a charge for a "no-show," whatever that means.'

I have to agree with Schwartz and Dorf when they state in their introduction to the book: "years ago there was so much less to learn (about hospitality industry terminology). But, especially during the past two decades, the application of modern management practices, the emphasis on marketing and servicing, and the widespread uses of technology introduced not only new processes and procedures, but a whole new lexicon of terms which are now part of our everyday working vocabulary."

Let's take a look at all the different publicist with whom we do business. Besides the leisure traveler, who, probably is the least likely to understand our terms and concepts, there is a business traveler, who is a little more knowledgeable. Then there is a whole bunch of people out there in different categories of market segment whom we solicit for business because they are in a position to direct individual travelers as well as groups to our properties. They include travel agents.

Corporate travel managers, motor coach tour operators, association executives, sports teams travel managers, relocation officers, corporate meeting planners, independent meeting planners, social club leaders, church group leaders, tour wholesalers, airlines crew housing managers, as well as a host of others with whom we want to do business.

We must consider that most of these people, with whom we communicate all the time, never worked in a lodging operation.

Whatever they know about our terminology they learn from us. Isn't it time we started using the same terms, with the same meanings, among ourselves? Then, let the rest of the world know what we mean when we use our own in-house jargon.

COCKTAILS ARE STILL IN AT MEETINGS—BUT SO IS DRINKING LIGHT

September 1994

The cocktail party isn't what it used to be. Two-hour open bars are giving way to wine receptions, and hard liquor is being replaced by mock cocktails. Wine and beer are becoming the beverages of choice and liquor are becoming more visible in hospitality suites following banquets. This doesn't mean we should stop selling the open bar for receptions or the cash bar prior to luncheon meetings or serving cocktails at banquets. we are still working toward maximizing profitable sales. Catering managers, as well as other sales personnel, should always consider offering these type of services to all meeting planners for all types of functions-it's still very profitable for properties.

Corporate and association decision-makers are also reevaluating the length of cocktail receptions. is the two-hour or even a one-hour reception before dinner still necessary? The trend seems to favor a 45-minute or, with some small groups, a 30-minute reception. In many cases, when a reception is shortened, the bar is kept open for a short time at the start of the banquet to take care of late arrivals or to accommodate guests who want a cocktail with dinner. When working with groups we must try to remember to sell something in the area of bar services-even if it is only short one as mentioned above. It is still profitable, if priced right.

The size of the pour, i.e., the amount of liquor used to make a drink has a lot to do with the cost of the reception. Whether the drink is a highball, straight shot, or on the rocks, the bar will usually use a standard measurement that can vary from seven-eighths of an ounce to two ounces. So if you are in a position to sell a bar set-up, be sure to be aware of what you will charge, depending, of course, on the size of the pour if you are selling it by the drink. This may have a lot to do with whether someone arranging a meeting or social function will be the bar service. The most profitable way to sell a bar function is to have a price per person, per hour. This will provide you with the best bar cost possible, if it is priced out right. The usual consumption will be about 2-1/2 drinks per hour per person. So based on the type of liquor, the size of the pour and length of the party, plus the profit you want, will determine what you need to charge.

Wine receptions are becoming more popular than ever. Per ounce, the alcohol content of most wines is considerably less than that of hard liquors. Generally, a wine reception also costs less than an open bar. For those who prefer not to drink at all, mock cocktails, liquor look-alike that contain no alcohol, help guests overcome the social pressure to drink. The drinks look just like Manhattans, martinis, or whiskey sours and even taste like the real thing. Salespeople need to check with the food and beverage director to be sure the property staff understands how to make these drinks.

LET'S START CHARGING FOR AUDIO VISUAL EQUIPMENT AND SUPPLIES

July 1995

There shouldn't be any good reason for supplying meeting groups with free microphones, overhead projectors, screens, slide projectors, easels, flip charts, marking pens or whatever. If you have included some sort of fee arrangement within a package deal, then you're probably covered. Or if, based on your negotiations, you have decided to provide something without cost as a trade-off, then that's OK too. All too often, properties have been providing a variety of audiovisual stuff, without cost, for no good reason at all. Professional meeting planners, for the most part, include a budgeted amount for a/v equipment and supplies and they do expect to pay for its use.

How often have you supplied free easels and flip charts for meetings and find that the flip charts disappear, along with your set of marking pens, after a meeting? Too often, I would guess. What about the wear and tear on equipment and the cost of maintenance and replacement? These factors should be considered in putting a price on everything you are asked to supply for a meeting. More and more hotels are starting to charge for these things—along with charging for meeting space. We're getting better at it. $25.00 here and $15.00 there adds up to a pretty hefty amount on your bottom line by the end of the year. We just need to get out of the business of giving things away for fear we may lose a meeting. You'll probably get more respect for charging, as well.

One of the most often requested pieces of audiovisual equipment is the overhead projector. It's probably the best tool for small meetings. And, it's also most available at properties than any other a/v device. We just don't do the proper job in storing or taking care of it and you're probably getting a lot of complaints on it not projecting a clean image (the stage is probably scratched and needs replacement) or the mirror is not clean or the arm is bent or the focus wheel doesn't work. If you start charging you'll have the funds to maintain it better or replace it when necessary. Also, the banquet set up person, or a/v technician, or the banquet manager or whoever works the room should understand how to operate the equipment—especially how to replace a burned out bulb. How often does someone who is in charge of a meeting ask for transparencies for an overhead (they didn't bring enough, or just needed an extra supply)? Do you have any on hand? You should! And, be sure to charge for them. Do you have a good supply of transparency marking pens—in different colors? Be sure to charge for them, and let them take them with them when they leave. Note that these types of pens are different than the marking pens used for flip charts. And, speaking of flip charts—these shouldn't come free either. Be sure to have masking tape available in case the meeting facilitator wants to hang the marked sheets up on a wall. Sell it by the roll. Next time you have a meeting and don't charge for most of these items—just add up the cost of everything you have supplied without cost. You'll be very surprised!

It would be important for you to have a pre-printed price listed for a piece of a/v equipment as well as supplies. At the time of booking a meeting be sure to inquire about what the needs may be and to explain the pricing structure. During negotiations you may decide not to charge for something, that's OK. At least the meeting planner would better appreciate the value of what is being received without charge.

As far as charging for meeting space—this is becoming more and more popular among hotel operators. And, it is being more understood by meeting planners. Sometimes you may want to provide space without cost due to a large number of sleeping rooms being sold. This is pretty typical. However, just be sure that a sliding scale is used based upon the number of room nights actually picked up at the time of the meeting. As an example: If 100 room nights are picked up—the cost of $500 for the ballroom could be waived; however, if only 50–100 room nights are used then the cost would be $250; and if less than 50 room nights are used the cost would be $500—or something like that. Just come up with some sort of formula and be sure it's in the contract.

HELPING WITH THE GROUP MENU PLANNING

March 2005

It is not just good anymore to present meeting or event planners with a printed menu and expect a prospect to make a decision only from what was presented to them. We just have to get into this business of doing more consulting when it comes to planning meals than just selling a pre-printed package of menus. Over the years, survey after survey, has always reflected that what is remembered most about meetings is the food and service rather than the speakers or content.

There is so much more to feed the troops than picking a menu based on price, or what the planner personally likes or dislikes. These days we really need to put some time on working with needs and wants of members of the group. The make-up of any group is so varied when it comes eating. In most groups we've got different age groups, genders, types of jobs, where they live, as well as differences in choice of what is eaten; and eating healthier is definitely "IN." This doesn't mean that food served at a meal function shouldn't be attractive or delicious or tasty or less than wholesome. It should be appealing to the eye, and taste you want more, but also healthy for the body.

We have to start thinking about meeting the feeding needs of specific groups we handle. Of course, finding out if the members of the group are into eating healthy, nutritious foods, or perhaps they do not care. There is every indication, these days most people want to eat healthier. Once we understand the needs and wants of those participating in the meeting, as it pertains to foods and special diets, we have to be able to convey this information to our food and beverage staff, and come up with some creative menus. Getting the chef involved in helping to create menus that meet the specific needs of group members is something that we will really need to do in order to provide outstanding service for our prospects. Granted it is a lot easier on our staff to just hand out pre-printed suggested menus and have someone pick one from column A and one from column B. But, does that really serve the needs of the group? Also, there is no question about it, the kitchen and service staff will perform better in these areas if you know just what your participants want by having them provide input through their meeting planner.

The event planner providing you with information such as group demographics (age, sex, children families, married or singles, type of work they do, retired, etc.) is always helpful. Don't forget—ditch diggers and oil riggers probably will require different menus than retirees. So, we need to be able to consult on these matters and help provide menus that make sense for participants in the group.

We also need to be prepared to advise on any special dietary problems that may come up; let's not wait for the last minute to find out that a group has some people who need a diabetic menu or low salt menu or kosher menu, etc. We need to ask all these kind of questions when booking meal functions.

The variety of food items that we can prepare is endless. We should not be limited only to the menu selections. Being more of a consultant on meal functions are really providing superior service to our clients; and isn't that why we are in business?

INCREASE CATERING SALES WITH BETTER PROSPECTING FUNCTIONS

October 2005

In the business of improving catering sales, it appears that the biggest challenge is to find enough time in a busy day to do some prospecting for new business. At most hotel locations the catering department stays pretty busy with a good deal of paperwork in carrying out the business that has already been booked by the property sales staff or the catering department staff. It would be most logical to try to work in some appropriate time to seek out additional profitable catering opportunities.

The easiest and most productive method of seeking new business is to go after some old business. Most properties have file systems, either in file folders or in computer software programs, with untold numbers of contacts who have provided catering business to the property in the past. Of course, we would like to think that files have been created for all previous catering customers and that a trace, or follow-up method would assure that contact would be made on a regular basis with persons who provided catering business previously. Because of a heavy work-load at most locations, we seem to fail in making these follow-up calls. It will be found that in searching the old files (and some files go back a number of years) no follow-up has been made to seek more business.

Time and time again, over a number of years, we have supplied our Hospitality and

Tourism Management students, as a laboratory opportunity, to hotel properties just for the purpose of reviewing old files, making appropriate follow-up calls and determining if there is any new business that might be available from those old file contacts. We have proven, without question, that there is a tremendous amount of new local catering business available that can be developed this way.

Seek out nearby universities that have hospitality type programs, make contacts and see if there is an opportunity to provide hands-on, prospecting experience, as described above. In all probability, contact could also be made with professors of marketing departments at colleges or junior colleges for the same purpose. Be aware that in using students, it would not be appropriate to think that students would also be in a position to make any sales. The opportunity is for the students to ask critical questions to determine in the contact is, a good prospect for a new catering business. The information that is developed should be put on some sort of Prospect Survey Sheet. The answers to questions asked by the students are reflected in this form. These forms are then turned over to catering staff personnel who should then make contact with these prospects to make the sale.

If a nearby university or college students are not available, then it may be very worthwhile to experiment with the employment of temporary, part-time help to do this function. The return of investment would be reflected by the amount of new catering business revenue against the cost of the help.

PRODUCT KNOWLEDGE IS A KEY IN IMPROVING GROUP SALES

May 2006

Knowledge is the key to making a sale, whether it is for an individual room or for a group. Of course, knowing everything about your own property is important, but we also need to understand how the property relates to our competitors. Another piece of knowledge is to understand the needs of prospect that would be in a position to rent a room or a group of rooms.

It would seem to be logical that every salesperson working at a property would want to make a personal product analysis at their location. This should require that visits be made to all areas of a property and notes taken on the variety of features available, and its impact on value to a customer. For example, what is the range of square footage in the sleeping rooms? It is amazing at the number of salespeople who are not aware of the size of rooms. It this important to all guests or groups? No, of course, not! However, it could be a value to some people to know that the rooms at your property are, in fact, larger than those at a competitor. The same goes for the number of connecting rooms. Because of a variety of reasons, there are some individuals, as well as groups, that will require such accommodations.

Just think about the variety of property features available that may need some explanation when it comes to selling:

▶ How many ADA rooms are available and do they have roll-in showers?

- ▶ Are the connecting rooms set up 'king bed' in one and 'double bed' in the other?
- ▶ What can you say about mattresses that ensure a good night's sleep?
- ▶ What in-room amenities do you have compared to the competition?
- ▶ What is the charge for in-room phone and internet use?
- ▶ What security devices are installed in sleeping rooms?
- ▶ How many smoking rooms are available/
- ▶ What is the availability of car parking number of spaces.
- ▶ If you have an exercise room, what equipment is available?

Of course, the listing can go on and on to include all the features in food and bar service, meeting space, recreation, business services, and other outlets. Not every guest or group buyer will want to know everything. However, understanding the specific needs of a prospect will help a salesperson understand what that person needs to know about some of the features.

How important is it to know the thread count on the bed linen? Well, hardly, anyone will ask about that. But, would it hurt to know, especially if the property was using some very expensive linen? It could be a big help in selling value to a particular guest or group.

When it comes to selling your meeting facilities it certainly would be critical to know specific seating capacities in all types of set-ups for meetings, conferences, socials, and banquets. The same would apply for booth or table space for exhibits. How to present the information about the property features is also critical. It is just not enough to just mention the feature, but it needs to be presented as a value to the prospective buyer.

IMPROVE CATERING SALES WITH MORE PROSPECTING

October 2006

Within our industry, we find that there is a very wide variety of organizational structure of sales offices at full-service hotels. Sales staffers are employed for the main purpose of developing at booking group business, the major concentration being in the area of room sales. Of course, it is expected that a certain amount of food and beverage business would be developed via the groups. Sales staffers are usually assigned to bring in business from either specific market segments or all segments within certain geographical territories. In most cases, once group business is booked, who it comes to the food and beverage portion of the booking, the catering office within a property then takes over and handles those arrangements. There is an opportunity at that time for catering staff help create additional sales for food and beverage functions for the group. With this in mind, we all understand that catering personnel is also in the business of selling. All too often, however, we lo upon the catering staff to handle group food and beverage business already booked by someone else From time to time there is an opportunity to create new catering business if a prospect would call I walks-in to a property to seek help in considering a catered event to be held.

Granted catering staff is always very busy handling the myriad of details in connection w group business already arranged via the sales office. However, the big question that needs to be answered is: to what extent are catering staffers involved in the development of new catering business via active solicitation locally? For the most part, a good portion of banquet facilities, at many hotels go empty, without being used for functions during some days of any week. In the business of operating a hotel, and the selling of rooms, we are very concerned with the occupancy percentage. We do not really hear much about being concerned with the occupancy percentage of banquet rooms. we all know it is tough getting great profits in the food and beverage business, and anytime we can develop additional banquet business it surely will help the bottom line. Therefore, management, sales staff and especially catering personnel should be interested in making sure that there is some kind of place to increase the occupancy of banquet space by selling more catered events in those areas.

We need to come up with strategies to ensure that the catering department staff gets involved in the business of prospecting for additional banquet business to fill up the function space that is no getting used. This means seeking short-term bookings that require meal and beverage service via local contacts. Hours need to be allotted, on a regular basis, for time to actively solicit this business is just not enough to wait for people to come by the property or for those who call on inquiries. The local paper, on a daily basis, will provide enough leads to keep a number of employees busy with follow-up phone calls to find new banquet prospects.

PRODUCT ANALYSIS: A KEY ACTIVITY OF SALES PERSONNEL

March 2011

When was the last time a salesperson, at a property, stayed overnight? It certainly is not unreasonable to expect people who sell things to have the experience of using the product. In the case of a hotel property, what is the product? Of course, the answer is the whole hotel. Experiencing the hotel product should take into account the exposure to all the features. So that means not only a night, as a guest, in a hotel room, but using all the amenities that are available. Taking meals in the hotel food and beverage outlets; experience in the recreational activities available, and all other features, adds to product knowledge.

There is no question about it; sales staffers, in any business, need to know everything there is to know about the product they are selling. What salespeople need to do during an exercise is to complete a Fact Sheet covering the hotel, for all areas such as: Rooms, Food/Beverage, Meeting Facilities, and other hotel features available, i.e.: Spa, parking, security, etc. Performing such an exercise could be so very valuable to a salesperson. Not only knowing all the features, but getting the "feel "of being exposed to them is what is most valuable.

So many different things about a hotel property appeal to so many different people. Spending the night in a bed reflects the real comfort of the mattress, pillows, sheets, etc. Expressing the comfort of a king size bed could be critical in the sales process, for some prospects. Or even something about the bathroom, the space, the shower, the soap, even maybe the quality of the toilet paper, could be of valuable in the case of some prospects. How many salespeople know the size of their sleeping rooms; or the real quality of the bedding, or the number of connecting rooms (not adjoining rooms), or the two, three or four, security devices on a sleeping room door? How about the view from the room window? How about space for someone to work on a laptop, and the details of the chair that would be used to do that work? What is known about the lighting in the room; and how does it affect reading from the bedside? Sampling the quality of the individual coffee\packets used for the coffee maker could be helpful in some cases.

When it comes to dining service, what do the salespeople really know about the quality of the food and service? What is it about the menu that differentiates the property from others? Experiencing breakfast, lunch, and dinner is important so that the salespeople get a real feeling for the quality and environment of the property's dining service. What do the salespeople really know about the meeting facilities? This is most critical in the business of selling to groups. It is not only how many people space can accommodate, but capacities for different types of needs, is what is important. Do salespeople really know the quality of the conference table and chairs being used in the meeting or banquet facilities "The comfort of attendees at meetings is critical when it comes to all-day workshops or seminars? What is it about the chairs that add to this comfort? Then, there is knowledge of the audiovisual equipment available. What about parking for conference attendees? How many parking spaces are available, and what security is available?

The product analyst list can go on and on, covering all the outside, recreational activities, as well all other features and amenities of the property. The key is getting salespeople exposed to all this so they really have product knowledge.

PRODUCT KNOWLEDGE = KNOWING ALL THE FEATURES OF A HOTEL

November 2013

Many operators and owners of hotel properties would really be surprised to learn that so many hotel sales executive, even those with a good deal of experience, lack a bunch of knowledge about their own product. Yes, there are even some General Managers in the same boat. The whole idea about being a great salesperson is to know everything there is to know about all the features within your product. There are really a very wide variety of facts of features that may not be known, for example: number of parking spaces in the lot; type of room door security; number of ADA rooms; ADA rooms with roll-in showers (no, not a portable shower that gets rolled in to the bathroom); size of rooms; number of connecting rooms (no, not the number of rooms that are adjoining (side by side); and the list goes on. Of course, not everyone one of these mentioned or other features need to be used to make a sale. Not everyone even needs to know any of these. However, there will be a time when prospective guest or group or group buyer will need to know something that would be important information for them to know for the sale to be made.

The way to accomplish the product knowledge is to have each sales associate, or maybe even do it in teams, complete a product Analysis. To do this there is no form, just a blank piece of paper (or use an iPad or iPhone to record information). Start with the rooms division and cover everything that needs to be known about the room and all its features. Then, follow with a complete review of the food and beverage division, including all the F&B outlets, room service, capacities of restaurants and lounges; search out all the features. Next could be the Meeting and Banquet spaces. What also needs to be recorded would be all the other property features such as parking facilities, security, recreational facilities, WIFI availability, include nearby facilities and activities about which guests may want to know. Also, not to forget emergency plans for medical aid, fire, or even threat.

This exercise is not easy; everyone doing this will be surprised at how much they have learned by accomplishing the product knowledge. It is important to know that when discussions are held with prospects for the business, they do not need to know everything about the hotel, just because the salesperson knows everything. In the business of selling, we must remember that what is needed is to meet the needs of the prospect. This means it is more important to get information than give it. And, only give the information that is necessary to meet the needs.

Chapter 17

HR in Connection with Sales Staff Employment

Outline Job Description and Compensation Before Recruiting

February 1980

The decision has been made. That extra person will be hired to develop additional sales for the smaller property. The lodging operator is convinced the investment made will bring back at least 10 times the dollars invested. Now, the next step is to recruit prospects for the job. The object in recruiting (for any job) is to find the right person who can fulfill a specific need. Therefore, there is a need to outline the job itself as well as the basic requirements to fill the job.

If the person is to go out and call on people then, obviously, a car is needed as well as the ability to be friendly and communicative. If the person will be typing letters to clients, then there is a need for typing skills, as well as a need for a good command of the language, and so on. In a small property, the salesperson should have excellent typing ability, and have a car. In larger properties where the sales staff consists of more than one person, then perhaps a secretary will handle typing.

Of course, appearance and personality will enter into the picture because the salesperson will be representing the property and presenting the best image on sales calls is important. Now, there is a need to outline the job and duties:

Job Description

Job Title: Sales Representative
Reports To: Manager (or lodging operator)
Function: Responsible for the development of maximum profitable sales of room, food, beverage, and space.
Primary Duties:

▶ Develops profitable sales through various methods such as outside sales calls, telephone solicitation and letter writing as well as developing and following up on leads.
▶ Follows through with management to ensure that the sales commitment is being fulfilled.
▶ Submits daily call sheets to the manager as a record of sales calls.

- Maintains up-to-date tracer files for follow-up on leads.
- Communicates, with all departments of the motel as it pertains to the sales commitment.
- Develops sales promotion ideas to submit to management.
- Responsible for proper communication between groups and motel management.
- Researches competition for management.
- Maintains files and records.

Once the basic requirements are known and the job description is written, there is one more decision to be made before recruiting begins. What is the compensation package available to the person who is to be hired? Compensation will include base pay (40 hours per week); incentive program; and employee benefits (insurance programs, sick pay, holiday pay, meals, car allowance, expense spending, etc.).

It is very important that all this be prepared in advance so that there is no misunderstanding during the interviews for employment. Experienced salespeople are expensive and if the money is there, it is probably best to hire someone with one-to-two years' experience.

Depending on the geographic area, this could cost $15,000–$18,000 a year in basic salary exclusive of incentives or benefits. Salaries for inexperienced people that can be trained vary by area. The starting wage should be budgeted so that there are additional funds available for three small raises the first year. After all, employees become more valuable each month as knowledge and experience are gained. A starting salary in most areas (other than large metropolitan areas) would range $9,000–$10,000 a year.

To start a new sales operation, a good guide to use would be up to two percent of total property sales (rooms, food, and beverage) for the direct sales budget. This would include wages, employee benefits incentive compensation, travel and entertainment.

Here's an example: A 100-room motel with a bar and restaurant has a 70% occupancy with an average rate of $27 per room per night. Annual room sales are $690,000 and annual restaurant sales are $285,000—total property sales are $970,000. The total budget for a salesperson, plus expenses, comes under 1.5% of total sales. With proper training and supervision that salesperson should bring in after the first year at least an additional $150,000 in sales for the property.

Once the budget and compensation package has been determined and a job description prepared and the basic qualifications outlined, the lodging operator is now ready to recruit.

RECRUIT SALESPEOPLE FROM EXISTING STAFF, FRIENDS, AND SUPPLIERS

March 1980

Recruiting for a salesperson means looking into every opportunity available to find people who may be qualified for the position. Recruiting opportunities include:

- **Existing Staff**: There might be a night auditor, desk clerk, bellman, restaurant

hostess, waitress or someone on the staff who could be well qualified for the job. This is by far the best source for recruiting the new salesperson.
- **Friends of Employees**: If the vacancy cannot be filled from existing staff, the employees should be made aware of the opening and encouraged to recommend people they know.
- **Competition**: This is always a good source of personnel. Although taking someone else's employee doesn't always make for the best relationships, it is still not a bad idea to recruit from this source.
- **Friends and Neighbors**: The lodging operator can get a good deal of recruiting help through this source.
- **Local Newspaper Advertising**: After all other opportunities for recruiting have been exhausted, the newspaper classified section may be used. By including the basic requirements and starting salary in the advertisement, a good deal of unnecessary interview time can be eliminated.

Always ask respondents to call for appointments. The telephone conversation itself can be an important part of the interview and can help eliminate persons who may not be prospects at all. Prior to being interviewed, prospects for the sales position should be requested to complete an application form. Any basic employment application will suffice; however, if none is available, the applicant should write a brief biographical sketch. It should include previous employment experience, education and some references that may be contacted. Many of the essential qualities of a good salesperson do not necessarily reveal themselves on an application. An interview is always necessary to evaluate any prospective employee.

It is just as important for the applicant to interview, or look over, the prospective employer as it is for the lodging operator to look over and interview the applicant. An interview location should be selected in advance that is clear of clutter and people. The applicant should not be kept waiting beyond the appointment time, should be greeted personally by the lodging operator and escorted to the interview location. A manager's office is fine and there should be no interruptions by staff or phone calls during the interview.

It is important for the property to make a good impression on the applicant. Interviews in the lobby area or restaurant do not work well and do not present a «confidential" enough atmosphere. Informal interviews are always best. If possible, a table or desk in the office should not present itself as a barrier between the two people. An employer sitting behind a desk creates an authoritarian atmosphere in the room, which can add to the stiffness of the interview relationship. Two easy chairs side-by-side do make the best set up for an interview.

The interview should start with the lodging operator clearly describing the sales position: its function, what qualifications are necessary, and what the compensation will be. Once this is understood and the applicant is still interested, he or she should be asked to explain why there is an interest in this job. It's always important for the applicant to know the growth potential of the job and of the company.

Based upon the interviews and application information, the applicants can be screened down to three or four potentials. Once this takes place, some checking of references and previous employers should be done.

HIRING: LOOK FOR CONFIDENCE AND AGGRESSIVENESS

May 1980

The lodging operator now has to choose among the many applicants for the sales position. The idea is to hire someone that will: (1) become a permanent employee; (2) learn to do the job as directed; (3) be productive for the company; (4) become promotable within the organization; (5) have a good return on investment (ten to one for the company); (6) get along well with the rest of the staff; and (7) someone who really wants the job.

It is important in the selection process not to let outstanding strengths outweigh weaknesses. For example, just because an applicant graduated with honors from a well-known university doesn't mean that person would be a good prospect. Having a degree alone doesn't make someone a capable salesperson. There are other important qualities to observe.

The salesperson is a key staff member who increases business; therefore, the one selected must represent the property in the best way possible. People judge appearances before personality even enter into the picture. Naturally, the person selected should have a good physical appearance, be neat, look clean and walk lively. During the interview process, the operator must look for the following:

▶ **Aggressiveness**: That certain quality that combines energy and really wanting to do the job successfully.
▶ **Sincerity and honesty**: People prefer doing business with someone that can be trusted and believed.
▶ **Confidence**: Does the applicant have faith in being able to do the job?
▶ **Communication Skills**: Is the applicant a good listener as well as a good talker?

The selection process includes: the review of applicants to screen down a number of applicants to interview, the interviews themselves, and reference and background check.

Some companies use psychological testing devices to determine intelligence, work preference, etc.; however, for the smaller lodging operator, this is not necessary. They are expensive and haven't conclusively been proven to reduce turn over or to increase sales. A good interview will suffice.

GO OVER DETAILS

When a decision is made, the job is offered and the operator must describe the compensation and the "rules of work" in detail. Although this information should have been covered during the interview, it is important to repeat during the hiring process.

Hours of work are from 8:30 a.m. to 5:00 p.m. daily, and 8:30 a.m. to 12:00 noon on Saturday plus some Sunday or night work when necessary. (It is important to conform to state and federal labor regulations; therefore, hours worked may have to be adjusted by the operator). The salesperson should type his or her own correspondence and reports. Written reports are to be made to the manager or operator. (Daily activity reports should be made verbally each day during training). Outside sales calls are to be made daily from 9:30 a.m. to 4:00 p.m. Monday through Friday, and Saturday mornings are inside workdays. Prospects should be entertained on property often for breakfast, lunch, and dinner.

The prices quoted on food and beverage and rooms by the salesperson should be approved by management. The salesperson may drink alcoholic beverages on the property only when entertaining prospects, and then it should be limited. No promise should be made on a salary increase for any specific future dates. Employment should be probationary for the first three months to determine how well the 'Program is going. After three months, the salesperson should be reviewed, and if kept on the payroll, a salary increase should be given. An incentive program should also be worked out at that time for improved sales.

The use of a personally owned automobile needs to be reimbursed at the going rate. (Check what other local industries and government agencies are paying. The local IRS office may also have are commendation.)

TRAIN NEW SALESPERSON WITH IMMEDIATE FIELD EXPERIENCE

June 1980

Once the new salesperson has been employed, the operator has the responsibility of providing direction and training. Many large companies have formal sales training programs; however, the small operator has no choice but to accept this task personally. This could cause a problem since many small operators had very little, if any, sales training or experience. A new, inexperienced person cannot be exposed to a sales position without some direction, instruction, and supervision. Here are some basic training principles:

In any sales activity, regardless of the type of business, one must know what it is that has to be sold. The lodging operator should not fall into the trap of thinking that in order to accomplish product knowledge the "trainee" must work at the front desk, night audit, housekeeping, restaurant, etc.

This is not necessary. Spending more than one day to become familiar with the house and what there is to sell is a waste of time and develops bad habits for the property. (Once the "trainee" knows how to do various operational jobs, it is too easy to peg him or her a "fill-in" for someone off duty. The "trainee" was hired for sales work, not to be a "jack-of-all-trades.")

The idea is to get the person "out on the road" as quickly as possible so he or she can be productive. The salesperson should have a basic knowledge of the number of rooms, types of rooms, rack rates, group rates and policy on rates; meeting room sizes and capacities, as well as available banquet menus, should also be known. The salesperson should be fully aware of the food and beverage service available on the property since that person also has the opportunity and responsibility for selling all the services of the property. The type of foodservice is important to know-coffee shop and/or full-service restaurant, meal hours, type of food served, restaurant prices and specialty items.

The type of lounge and beverage service is also important to know-hours of operation, alcoholic beverages available, entertainment, any "happy hour" or "special" nights. As salespeople become more experienced in making calls on local businesses, many opportunities will exist for developing additional food and beverage business for the property.

The beginner in sales has to learn that it is a waste of time to try to sell someone that is not a prospect. Therefore, the "trainee" needs to learn how to prospect. There are many ways to do this, but the best method for long-range production is for the person to get out on the road and ask. Most new people are nervous about making "sales" calls; taking "surveys" could create a more comfortable feeling. The operator should take the "trainee" out on' 'survey calls, "starting with easy calls which are personal visits made to people the operator knows (a banker, insurance man, supplier).

The first couple of days of "survey calls" should be set up by the lodging operator who calls these people and explains to them that a training program for a new salesperson is being conducted. He or she should say that the operator and salesperson would both like to drop in within the next couple of days to conduct a survey to see if they can develop business.

This type of approach will almost always receive positive results. Don't try to make an appointment. The object here is to train the salesperson to make "cold calls," without appointments.

On the cold calls, the lodging operator introduces the salesperson, but actually does the first couple of surveys personally. Each call is discussed thoroughly after each visit so that "kinks" may be worked out.

On the third call, the salesperson takes over the interview even to the point of making contact with the prospect and introducing the lodging operator. A thorough review session is held following the interview, and the salesperson continues to make the other interviews.

If the proper selection was made for filling this sales position, the salesperson will feel comfortable enough to continue after two or three interviews. When the "drop-in" leads provided by the lodging operator have been exhausted, the salesperson may start on establishing a procedure for cold call prospecting.

Use a simple form for asking key questions and making notations on the form as responses are given. (There is nothing wrong with taking notes during an interview.)

POOR HIRING PRACTICES AGGRAVATE EMPLOYEE TURNOVER RATE

August 1989

Turnover in sales personnel is a very real problem. Looking at it realistically, however, the salespeople themselves are not to blame. It's the system that is lacking.

Many studies have been conducted on the subject of turnover. I'm sure there are studies about why hotel salespeople change jobs. In all probability, the list of reasons reads something like this (though not in any specific rank or order):

- No opportunity for advancement;
- Poor direction-or none at all-from superiors;
- No on-the-job training;
- Not knowing what's expected;
- No incentive to produce;
- Personality clashes;
- No recognition;
- Too much paperwork;
- Required to do things other than sales;
- Burnout;
- Too much control;
- No review system;
- Not enough pay;
- Too many hours;

There surely are other reasons, and it would be interesting to see what a professional study might turn up.

PROPER SELECTION

Perhaps we should first explore the recruiting, selection and hiring process. General Managers or sales directors often hire a salesperson without knowing what is needed. It seems logical that hiring the right person for a specific slot would be a step in reducing turnover. But all too of ten, people who are hired for sales are just not right for the job.

Before a decision is made to employ someone in sales, there should be a written job description. Then sales goals for the particular job must be established, followed by selection criteria. A plan should be written so there's a guide for how to get people to apply for the job.

PROFESSIONAL SEARCH

The easiest, most professional route would be to use an executive-search company. Of course, there is a fee for such a service; however, there are many advantages mostly, the saving of many hours involved in screening applicants.

According to Phil Alfus, senior principal with Innkeeper's Management Corp., a New York-based executive-search firm, "Too often, hotels will place an advertisement in their local paper and hope it will attract the right candidate. This is fine if you're looking for trainees, but hoteliers should first look in-house. If the desire is to find an experienced candidate who can open new markets, then there is a need for a good search firm. It would canvass potential applicants, some of whom could be competitors, and present candidates who fit the specific profile."

PROPER HIRING PROCEDURES

Turnover could be reduced further with proper hiring practices. The job offer needs to be written and clearly defined. Another way to reduce turnover is to maintain communication between salesperson, supervisor and General Manager. It's been observed that the hotel properties that do not have sales turnover are the ones in which the General Manager is personally involved.

On-the-job training also is vital to reducing turnover. Salespeople generally will do a better job selling when they know more about a hotel's workings. From time to time, salespeople will have to be given more responsibility. This helps improve their self-esteem and goes a long way toward preventing burnout.

Probably one of the key concerns of less-experienced salespeople is expressed, in the "How am I doing?" syndrome. These people need feedback constantly, good or bad. What we do wrong is to provide annual reviews related to salary increases. Why don't we do periodic reviews, perhaps quarterly, that aren't tied to salary? They should be strictly performance reviews.

We'll always have to deal with hotels "stealing" salespeople from each other. With properties opening all over, there is a need to recruit experienced salespeople. Naturally, they have to come from another hotel. Just don't let it be yours.

PERFORMANCE APPRAISALS IN THE SALES OFFICE

February 1990

It is generally agreed that employee appraisals are very important in all types of businesses. In the hotel industry, we find a wide variety of methods used to evaluate our sales personnel and different reasons for performing these evaluations. For the most part, the reasons, with different priorities, include: salary increases, continued employment, training needed, promotion, company policy, and "we've always done it."

Companies and properties should really start out with defining the purpose and objective of having sales personnel appraisals. The logical purpose could very well be to find out how the employee is working out; if the employee improved performance over a period of time; is the job description being followed; are goals being achieved and other purposes that may be important to the operation. The objectives could be to make certain decisions. These decisions may include: salary increases; incentive awards; to give more responsibility; to promote or transfer; to provide more skill terminate employment. The appraisal is very personal and certainly the most important tool for the sales employee. It has a great deal to do with career path within the industry. But, more important, it affected an individual's mental state—either favorably or unfavorably, depending on the outcome of the appraisal. All the more reason for appraisals to be very fair, objective, and unbiased—since we are dealing here with people's lives—careers, and mental outlooks. Biases play a big role upon the part of the rater in completing Appraisal forms. We can't seem to get rid of Biases completely. Some of the biases in rating people are non-job related elements such as: sex, Appearances (thin, tall, heavy, short, etc.), How friendly the rate is with employees, similar work habits as Rater, Outside interests being similar as a rater, Family relation (of the boss), etc. We can help in ridding the rates of biases by improving the appraisal forms being used. Appropriate questions on the form are one way and another is proper training for the raters.

In conducting appraisals, there are some misconceptions:

▶ The supervisor is the best source for review. This is not always the case. There are other persons who probably are as good or better to rate salespeople such as: peers, other supervisors,

others who work with the employee, and customers.
- The appraisal form must be kept simple. Because of this rating being so important to the individual and the company, the form should not be simple, but more in detail so that better judgment may be made. Quick and easy ratings are not always fair.
- We tend to think that we can recall employee performance over an extended period of time. Actually, only recent information is ever recalled easily. This is a good reason for having more periodic reviews rather than just an annual review.
- Appraisals need to be done annually or semiannually. The best way to achieve the purpose of the review is to contact them on an unscheduled basis—quarterly, at least, and should not be done to determine pay raises. Salary reviews may be determined as a result of all the appraisals during a period of time, but not tied to a specific performance rating. The periodic appraisals should be used for development of the individual. This then deals with performance, as it happens, rather than waiting a full year before dealing with it.
- All rates are motivated to rate accurately. As indicated above, there are Biases that take place in performance appraisals and all raters definitely do not rate accurately. Some are more lenient or more severe.

In producing an appraisal form the rater should be given the opportunity to rate the employee on a scale form a low to a high number on each item on the form, as well as add a narrative section for comments. Here are some items that may be included in an appraisal form for sales personnel:

- Job Knowledge
- Leadership Qualities
- Dependability
- Working Relationship—Inside
- Working Relationship—Outside
- Initiative
- Personal Appearance
- Cost Control
- Planning and Organizing

Another area on the form could very well convert the measuring of performance such as goals for bookings, sales calls, etc.

THE SELECTION PROCESS

April 1991

The lodging operator now has to make a selection among the many applicants for the sales position. The idea, of course, is to hire someone that will: become a permanent employee; learn to do the job as directed; be productive for the company; became promotable within the organization; have a good return on investment (ten to one for the company); get along well with the rest of the staff; and someone who really wants the job.

It is important in the selection process not to let outstanding strengths outweigh weaknesses. For example, just because an applicant graduated with honors from a well-known university doesn't mean that the person would be a good prospect even though perhaps the person may be shy or show some other weakness to qualify for the sales position. Lodging operators should be cautious of college graduates . . . having a degree alone doesn't make someone a capable salesperson. There are other important qualities to observe.

The salesperson is a key staff person who will increase business; therefore, the one selected must represent the property in the best way possible. People judge on appearance before personality enters into the picture . . . Naturally, the person selected should have a good physical appearance, be neat, look clean and walk lively.

During the interview process, the operator must look for the following;

- *Aggressiveness*—That certain quality that combines energy and really wanting to do the job successfully.
- *Sincerity and honesty*—People prefer doing business with someone that can be trusted and believed.
- *Confidence*—Does the applicant have faith in being able to do the job?
- *Communication skills*—Is the applicant a good listener as well as a good talker?

The selection process includes:

- Review of application to screen down a number of applicants for interview.
- Interview.
- Reference and background check.
- Some companies use psychological testing devices to determine intelligence, work preference, etc.; however, for the smaller lodging operator, this is not necessary. It is expensive and hasn't conclusively proven to reduce turnover or proved to increase sales. A good interview will do just as well.

Hiring

When a decision is made, the job is offered and the operator must describe the compensation and the "rules of work" in detail. Although this information should be covered during the interview, it is important to repeat during the hiring process.

Rules of Work

- Hours of work are from 8:30 a.m. to 5:00 p.m. daily and 8:30 a.m. to 12:00 noon on Saturday plus some Sunday or night work when necessary. (It is important to conform to State and Federal labor Regulations; therefore, hours worked may have to be adjusted by the operator.)
- The salesperson types own correspondence and reports.
- Written reports are to be made to the manager or operator. (Daily activity reports should be made verbally each day during training.)
- Outside sales calls are to be made daily from 9:30 a.m. to 4:00 p.m. Monday through Friday and Saturday mornings are inside work days.
- Prospects should be entertained on property often (breakfast; lunch or dinner).
- The prices quoted on food and beverage and rooms by the salesperson should be approved by management.

▶ The salesperson may drink alcoholic beverages on the property only when entertaining prospects, and then it should be limited.

COMPENSATION

Salary—Amount (when and how paid).

Salary increases—No promise should be made on salary increases for any specific future dates. Employment should be probationary for the first three months (a feeling out time) to determine how well the program is going. After three months, the salesperson should be reviewed, and if kept on the payroll, a salary increase should be given. An incentive program should also be worked out at that time for improved sales.

Expenses—the use of a personally owned automobile needs to be reimbursed at the going rate. (Check what other local industries and government agencies are paying. The local IRS office may also have a recommendation.)

Benefits—Company benefits should be explained completely (insurance, holidays, vacation, sick pay, etc.).Dress code should be explained as well as what meals are allowed on the property.

A letter of employment should be written to the newly hired sales representative outlining the details of employment (rules of work, compensation). Prevent problems down the road by having a complete understanding at the time of hiring.

Calling cards and letterhead should be ordered and space made for a sales office with appropriate furnishings.

Next, the sales representative gets trained.

RECRUITING, SELECTION, AND JOB DESCRIPTION

May 1991

The decision has been made! That additional sales for the property extra person needed to develop will be hired. The lodging operator is convinced the investment made will bring back at least ten times the dollars invested.

Now, the next step is to recruit prospects for the job. The object in recruiting (for any job) is to find the right person who can fill a specific need. Therefore, there is a need to outline the job itself as well as the basic requirements to fill the job. If the person is to go out and call on people then obviously, a car is needed; as well as the ability to be friendly and communicative; if the person will be typing letters to clients, then there is a need for typing skills as well as a need for a good command of the language, and so on.

In a small property, the salesperson should have excellent typing ability, and have a car. In larger properties where the sales staff consists of more than one person then perhaps a secretary will handle the typing. Of course, appearance and personality will enter into the picture because the salesperson will be representing the property and presenting the best image on sales calls is important.

Now, there is a need to outline the job and duties:

JOB DESCRIPTION

▶ Job title: Sales representative
▶ Reports to: Manager (or lodging operator)

▶ Function: Responsible for the development of maximum profitable sales of room, food, beverage, and space.
▶ Primary duties: Develops profitable sales through various methods such as outside sales calls, telephone solicitation, and latter writing as well as developing and following up leads. Follows through with management to ensure that the sales commitment is being fulfilled. Submits daily call sheets to the manager as a record of sales calls.
  - Maintains up-to-date tracer fillies for follow-up on leads.
  - Communicates with all departments of the motel as it pertains to the sales commitment.
  - Develops sales promotion ideas to submit to management.
  - Responsible for proper communication between groups and motel management.
  - Researches competition for management.
  - Maintains files and records.

Once the decision is made with regard to the basic requirements needed in a person to fill this job and the job description is known, there is one more decision to be made before recruiting begins. What is the compensation package available to the person who is to be hired?

Compensation includes: Base pay (40 hours per week); any incentive program; employee benefits (insurance programs, sick pay, holiday pay, meals, car allowance, expense spending, etc.).

It is very important that all this be prepared in advance so that there is no misunderstanding during the interviews for employment.

Experience salespeople are expensive and if the money is there, it is probably best to hire someone with one to two years' experience. Depending on the geographic area, this could cost between $115,000 to $118,000 a year in basic salary exclusive of incentives or benefits. Salaries for inexperienced people that can be trained varies with areas. The starting wage should be budgeted so that there are additional funds available for three small raises the first year. After all, employees become more valuable each month as knowledge and experience are gained. A starting salary in most areas (other than large metropolitan areas) would range from $19,000 to $110,000 a year. To start a new sales operation, a good guide to use would be up to 2% or total property sales (rooms, food, and beverage) for the direct sales budget. This would include wages, employee benefits, incentive compensation, travel, and entertainment.

Here's an example: A 100 room motel with a bar and restaurant has a 70% occupancy with an average rate of $127.00 per room per nights. Annual room sales are $1,690,000 and annual restaurant sales are $1,285,000—total property sales are $1,970,000. The total budget for a salesperson, plus expenses would easily come under 1.5% of total sales. With proper training and supervision, that salesperson should bring in after the first year at least an additional $150,000 in sales for the property.

Once the budget and compensation package has been determined and a job description prepared and the basic qualifications outlined, the lodging operator is now ready to recruit.

Recruiting for a salesperson means looking at every opportunity available to find a number of people who may be qualified to apply for the position. Recruiting opportunities include:

- Existing Staff—there might be a night auditor, desk clerk, bellman, restaurant hostess, waitress or someone on the staff that could be well qualified for the job. This is by far the best source for recruiting the new salesperson.
- Friends of Property Employees—If the vacancy cannot be filled from existing staff then the employees should be aware of the opening and encouraged to recommend people they know.
- Employees of Suppliers—The lodging operator has salespeople constantly calling, trying to sell products to the motel. An excellent opportunity exists to look these people over to see how well they conduct themselves selling their products.
- Competition—is always a good source for personnel. Although taking someone else's employee doesn't always make for the best relationships it is still not a bad idea to recruit from this source.
- Friends and Neighbors—The lodging operator could get a good deal of recruiting help through this source once it is known that the property is looking for a salesperson.
- Local Newspaper Advertising—After all other opportunities for recruiting have been exhausted, the newspaper classified section may be used. By including the basic requirements and starting salary in the advertisement, a good deal of unnecessary interview time can be eliminated. Always ask respondents to call tor appointment. The telephone conversation itself would be an important part of the interview and it can help eliminate persons who may not be prospects at all.

Prior to any interviews, prospects for the sales position should be requested to complete an application form. Any basic employment application for should suffice; however, if none are available, the applicant should write a brief biographical sketch would include previous employment, experiences, and education as well as some references that may be contacted.

Many of the essential qualities of a good salesperson do not necessarily reveal themselves on an application. An interview by the lodging operator is always necessary to evaluate any prospective employee.

SALES REP COMPANIES CAN WORK WELL FOR YOU

September 1991

A good addition to any lodging operation may very well be a sales representative company. These types of organizations have been around for a number of years, but there are many hotels, motels and conference centers do not know they exist. Or, if they have heard of such they are not too familiar with what it is they do. Hotel "reps," as they calling usually are people who have been in the sales/marketing end of the lodging business for a period of time and have gained extensive experience in selling lodging products to various market segments. Many rep companies have started as a one-person show and with success have built the business to large organizations. Reps work in various fashions to represent individual lodging properties, or groups of properties in selling their facilities to different markets. Most do their work to solicit groups for their members. This means they will call on meeting planners, association executives, tour operators, incentive houses and other buyers of lodging space. In some way or other these representative have to get paid. Most work on a fee basis per month, or a fee plus commission and there are some that will work on commission only. Some reps work only on specific markets and specifically targeted groups; as an example—high-end insurance companies for meetings and incentives. Others work all across the board, wherever built a long-lasting relationship with buyers and use these connections to book business for their clients. There are rep companies that are rather large and have offices in major cities in the U.S. and are able to make national contact personally just about all over the country for their clients. Generally, they attend key trade shows and conferences to make contacts and develop prospects or leads for their member clients. For the most part, reps would make the contact with the prospective buyer, provide details of a particular property that meets the prospect's needs then turn the lead over to the hotel client for follow-up and make the deal. There are some however, that will go as far prospecting, selling and closing the deal for the hotel clients. Just about all the reps will work on a contract for a minimum of one year since there is a good deal of group work that needs to be accomplished by them before they can even get into soliciting and selling and as we all know it does take some time and a lot of work to close some deals.

What appears to be a new type of sales representation company has just appeared on the scene. Here is a brand new organization with 15 sales reps, all experienced salespeople with specific specialties. This is a collection of hotel salespeople who specialize in selling to various market segments for a select group of high-quality hotels. What they do is go out and look for high volume accounts that are in a position to book business in a variety of types of hotels in various geographic locations and make deals for their hotel clients over a long period of time. These people are not after one shot meetings or conventions.

To give you an idea of their specialties they include: long stay housing, travel industry group sales, athletic sales, corporate sales, religious markets, entertainment market and even a pretty new specialized are—hazardous wastes. Most of the room blocks are rather small, such as 15–20 nights, but they are either long-term or keep repeating at multiple properties. One recent booking is only 8 room nights at properties, but volume runs over $1 million in room revenue for the year. In their first five months of operations, they booked $5.2 million and are now averaging $1 million a month in bookings. They now represent 200 properties . . . most are in the mid-price category. The company is called Market Segments, Inc., and has as its President a long-term hotel marketing executive, with 20 years' experience; he is Jeff

Halpern, CHSE. Right now, the company does not require any monthly fee, but does work on a commission basis.

WE COULD BE DOING A BETTER JOB WORKING WITH OUR YOUNGER EMPLOYEES

November 1993

There is nothing new about younger people joining the ranks in the hospitality industry. We see them all over the place; the high scholars and college students working part-time in guest service at the front desk or at the bell stand. We see them as dining room and coffee shop servers, cashiers and hosts. They are in the kitchen washing dishes or helping to prep. And we even see them in housekeeping, cleaning and making beds. Here they are, happy to be working, enthusiastic, ready, willing, and able to learn our business and continue in this field upon graduation. But how many ends up getting back into this work after completing high school or college? Probably a small percentage. How about the graduates from Universities and Community Colleges that have come out of hotel and restaurant programs? After putting in some time in a management-training program with a hospitality organization or taking a job with a local hotel or restaurant how long, do they last before moving on to another company or even to another career?

One of the biggest problems we face, in not only hospitality sales, but throughout the industry, is the situation of turnover. For the most part, we are really way behind time, in fact probably very old fashioned in the way we relate to and treat the younger employees. We need to have a better understanding of what these young people really want . . . out of the job, out of life, out of relationships with the employer, and what turns them on and off. When we understand this we will find that we can produce better employees, improve productivity and reduce the turnover.

Generally, we don't do a good enough job in taking the new employee through the hiring, employment, and organization process. So many of the newer, younger employees do not understand the function of their jobs and how it relates to the rest of the operation or how it impacts the profitability of an organization. Granted, many hotel properties, management companies as well as chain organizations do a fair job in this area. But, overall we fail as an industry.

Some new employees rarely ever get to see or talk to the General Manager at a property. One of the biggest turn-ones for a young employee is to be able to spend time with upper management, particularly if the job is in the area of sales. There's nothing wrong with a director of sales taking on the responsibility of hiring, training and supervising; but there has to be some time for the younger recruit to talk with the "big" boss.

We fail in the area of providing recognition and praise, when deserved, or even providing some unexpected, small reward for doing an extra special job. Or how about a nice letter the employee when a compliment is due and sending a copy of the letter to the corporate office or personnel file? The younger employee also need a "report card" a "how'm I doin'"—not only once a year when it is time to review for salary increases, but perhaps quarterly or even monthly, with the brand new ones.

Is working fun for the younger employee? Is there a happy environment around the place of work? Does the boss have a good sense of humor and cracks a joke now and then? Are there a lot of smiles around the place? If the answer is not yes to these questions then look out. The younger employee will not stay long at the job. This is of particular importance in the sales department.

Younger employees continually want to learn. Give them the opportunity to keep learning about the job and about the business. Give them a little more responsibility from time to time, they'll eat it up. Show some respect for them and their feelings and their thoughts. Ask them for opinions and be sincere about doing so.

Overall the real key to keeping the younger employee, especially in sales, is to pay attention to what they are doing show an interest; provide direction, supervision, and motivation. Do all this and you will have developed an enthusiastic, hardworking and loyal, long-term employee?

TIPS ON HIRING THE RIGHT SALESPEOPLE

October 1994

Harvey Mackay has done it again. he already has two best-sellers, under his belt and now he has come out with another. "Shark proof: Get the Job You Want, Keep the Job You Love in Today's Frenzied Job Market. Mackay is the one that came up with the idea for the 66 point questionnaire he makes his sales personnel complete on their clients. He is the big believer in relationship selling and has built a tremendous business as a result of his style of leadership in his company.

Mackay offers some tips for hiring the right people, and together, with a few of my own, if followed you'll be successful in getting more productive salespeople on your team.

▶ I've always told our students that the best way to get a job or to find another job is through networking. I saw a survey some time back that indicated over 65% of the best sales jobs were found through networking. Nothing works as well. The same would be true if you are looking to hire someone. The best salespeople you want to hire may be sought through networking. Your next best bet is probably a search firm; once you have laid out your requirements, let them find the bodies for you.

▶ Don't always check only on the references provided in a resume. During the interview squeeze out a couple of names referred to in the conversation and contact those people as well. Some applicants love to name drop. You may even ask for a few other names to check as references.

▶ Look for characteristics that successful salespeople usually display. How about energy, enthusiasm, persistence, good listener, creative, friendly, good communication skills (verbal and written), credibility, and of course, a sense of humor?

- Has the applicant prepared for the interview by having some knowledge of your organization?
- How about their ability make sense over the phone? I used to make phone calls to all applicants in which I may have been interested. A telephone interview was always first with me (sometimes even very early in the morning).
- I've witnessed so many interviews by managers where they end up doing all the talking, asking very few questions of the applicant and end up thinking they had a terrific interview. A smart interviewee can easily turn the tables and get you to do all the talking and make you feel they are great because they are good listeners.
- Then, of course, we want people who are good listeners since that is a characteristic of successful salespeople. Do they show interest in what you are saying, can they relate by signals, do they maintain eye contact?
- Get out from behind the desk during the interview. Even get out of your office, take a walk, go get an ice cream cone, go to a coffee shop, put the prospect at ease. I'm not a believer in the formal settings for interviews that so many people think is necessary for a good interview. Let them relax and be casual. You'll get more out of the interview.
- I used to tell our managers ... if you ever run into someone who will make a great salesperson, and you don't have room for another body, hire them anyway. We can always find room for a good salesperson. In fact, I used to have a program for salespeople I would subsidize at a property until there was an opening someplace. And, considering the turnover in our industry it was a good idea and worked well for me.
- Don't worry about hiring anyone that appears to be stronger than you, or more knowledgeable, or has more experience. If they are great then that will make you look good.
- Before anyone is hired to be sure there is an understanding regarding the details of the job (have a job description ready), what they are expected to do, to whom they will report and generally what is expected of them from a performance standpoint. Are there measurable goals?

Mackay would want you to ask yourself: "How would I feel if this person were working for my competitor?"

GET SALES INCENTIVES TO WORK BETTER FOR THE PROPERTY

May 1998

Incentives for sales personnel today are all across the board; with probably hundreds of different programs and plans based on many different things. And, that's good—incentives do work, but do they work best for the property? We get lots of inquiries throughout the year from owners, operators, managers and sales directors about what works best. Probably the best way to look at any incentive program for sales folks would be to first understand ROI (return on investment). What is expected in sales production from someone to offset the cost of sales salary? Here, one way is to determine the dollar value of delivered sales as a ratio of a person's salary. It could be 5, 10 or even 20 times someone's salary, depending upon the type of property and by room rates. A resort charging $250 and up for a room, night may very well want to think about figuring 20 times an experienced salesperson's salary. While for a new employee, with limited experience, at a property charging $60-$75 per night, it may only be figured at 5% or even less. The formula is a judgment call and needs to be worked out at the local level. The formula could be based on revenue from rooms delivered or a combination of revenue generated from rooms, space, food/beverage or any combination, based on a person's job description. A similar formula could be worked out for catering sales personnel based on space and food/beverage sales generated. Just be sure the incentive is paid once payment for the business is received. It is continually amazing to me how many incentive programs are based on business booked rather than received.

Another thing to consider in coming up with an amount of revenue generated as a ratio of salary would be to determine if the goal is based on repeat business or new business, it may also be a combination. You see, it can get complicated! Market segments may also create a situation whereby the ratio of sales to salary may not appear to be fair among sales personnel. For example, someone selling to the SMERF market (Social, Military, Educational, Religious, Fraternal) may be selling rooms at discounted rates during offseason, and another person may be selling to a higher rated market such as insurance incentive groups. In such a case the ratio of room sales goal to salary probably should be different, with a lower goal for the SMERF market salesperson.

Once the goal is determined it is time to start figuring what type of an incentive program to introduce. After all, what we have only done to this point is to figure how much delivered sales do we need to be able to pay the salary of the salesperson. Incentives are then determined based on increases in sales above and beyond the goal. Anywhere from 1% to 5% of the additional revenue may be reasonable based on how profitable the additional revenue could be to the property.

Special short-term campaigns work very well, but they need to be reachable and payable immediately. For example, a three-month program to improve sales by a certain amount over the previous year during June, July and August could be very motivating for the sales staff. The incentive could be a percentage of the additional revenue received for rooms and/or catering sales. This needs to be planned well enough in advance to allow the staff to do its work in booking for those months. Incentives should be paid as soon as the goals are met for each month, only after the business is delivered and paid. Putting rewards in winners' hands as soon as possible is important and motivating for the recipients.

For all types of incentives: cash, trips, gifts, etc., remember to always make goals, procedures, and the pay-out very specific so there is no misunderstanding. If the rules of the game are not clear it may not work to your advantage.

HOW DO WE MOTIVATE OUR STAFF TO DO THE THINGS WE WANT TO BE DONE?

December 1999

How to motivate people to do the things we want them to do is a most often discussed topic among hotel owners, operators, and managers. Just about every management meeting contains some topic related to motivation: "How can we get our staff to provide better service?" "Why aren't our people motivated to do a better job?" "Let's do something to motivate our salespeople to bring in more business!," and the list goes on and on. We keep coming up with "gimmicks" to motivate the staff. We have staff member "of the month;" incentive compensation for salespeople; free something or other for employees who don't miss a day of work during a month.

When you think about it, isn't it true that we really can't motivate anyone to do anything? Ever tried to motivate a cat to do something? Cats are known to be stubborn and no matter what you try to get them to do, it just won't work unless, you create the environment for them to do a specific thing (but then it is their idea, not yours). The same is probably true with us. People don't motivate people we just set up the environment which motivates people to do a certain thing.

The "gimmicks" suggested above is the setting up of the environment. But, they don't work with everyone. Set up a bonus structure for salespeople to sell more and, some will do and others will not (or cannot). Offer awards for employee of the month, and not all employees will try for it.

This business of "motivation" (getting someone excited about doing something, then getting them to do it) is nothing more than setting up psychological factors which influence reactions or even attitudes. So we need to dig a little deeper and think about what it is that we can do to get people "motivated" to do something. There is no point in trying to set up psychological factors toward motivation without examining the process from recruiting employees through to their supervision. Here we need to look at how we recruit, select, employ, orient, direct and supervise. The whole idea is to start off in hiring the right people for the right job. Making sure, of course, that there is no confusion regarding the job description, salary/benefits, and what is expected from the new employee. A proper orientation to the job function and the property is always necessary. The key to getting people motivated is in the direction and supervision provided by management. People need to be told exactly what is expected, and how to do it. Good supervision means good follow-up as well good two-way communications. Jim Cathcart, well-known author, and speaker, says: "The easiest way to manage a person today is to give them as much information as possible. The more they understand the less supervision they need."

Most everyone would agree that people work better and are motivated to go above and beyond the call of duty if: they like their employer, are treated fairly, are given respect, are provided good direction, have two-way communication with management, are involved in decisions, and, to some extent, are empowered.

In Jim's newsletter, "The Acorn Letter" he writes: "When you teach people why thing are done in a certain way, you vastly broaden their understanding. When one understands why then they can see more than one way to get there. They can make more decisions and adjust to changes that arise, while still reaching their goal. They are then more likely to give more to each task. *A purpose provides a motive which manifests as motivation.*"

It all starts with getting the "right" person employed. Josh Greenberg, General Manager of the Red Lobster in Christiansburg, Virginia, told me one day: "We want our employees to be nice to our customers, so we hire nice people. We can teach them the techniques of the job, we just can't teach them to be nice."

Now, that has a lot to do with motivation.

RESOLVING CHALLENGES WITH MEETING AND EVENT PLANNERS

April 2000

No matter how good you are in working out details with prospects who plan meetings and events with you, regardless of the size of your property, you are bound to end up, during the process, with some problems that were unexpected. Most get solved, somehow, but not without some stress and anxiety on your part. Here are some very basic areas where problems seem to arise. We have also offered some suggestions on how to avoid these situations.

Sales personnel turnover: You certainly can't avoid this one area. We do have a turnover. Well, you can't do anything about salespeople leaving, if they decide to go elsewhere, but you can do something to protect the interests of the property, as well as a prospectus. Keeping good records of discussions with group customers is a good start. Even if customers do not contact the property from time to time, be sure that someone is in contact with them. If an event is booked for a year or two out, be sure to stay in touch every three months just to know "what's going on." There is also turnover from the other end, with people who plan meetings and events for all types of organizations. Everything needs to be put in writing, such as any changes in the program, guest counts, arrival times, meal functions, meeting room set-ups, etc. There should be no surprises when the group arrives at your property. Most sizable properties these days will have a conference service manager who will be the main contact for the organizer or meeting planner, so if the salesperson leaves, no big deal. But, the conference service manager, could also be the person leaving. You will always have someone at the property who will look after the details of a group function; however, they can't do it without accurate record keeping. Not only is it important to stay up to date on the needs of a group, but confirmation of details to the person making the booking is critical as well. Even though there may be an ironclad contract on the meeting or event there is a myriad of details and changes that take place between the booking time and when the actual event takes place.

What "they" are saying about hotels: There been somewhat of an uproar from meeting and event planners concerning hotels' servicing of events; at least according to the postings on the listserv to which I belong (mim.com). Many feel that hotel representatives are not getting information back to the organizers soon enough in advance of the date of a meeting. It would seem logical that hotels would make sure that all details would be sent at least two weeks before the arrival of a group. This way there is some time allowed for the planner to get back to the hotel just in case there are details which were not covered in the communication, or there is a misunderstanding of what's what. Whether it be a Conference Resume or BEO (Banquet Event Order), the document reflects everything which has been agreed upon by the two parties. Some planners have indicated that the first time they get to see the BEO is when they arrive at the property for the event.

Hotel bill disagreements: More times than not there are disagreements with regard to the final bill submitted by a property (another hot item discussed on the listserv). One way to avoid this very irritating thing is to make sure that you include in the contract a stipulation about who is authorized to sign to the master account. Also, arrangements should be made for the organizer or meeting planner to meet with the property controller daily during the meeting to "sign off" on all charges of the previous day. This way both parties can go over the previous day's charges and resolve any questions which may come up about the billing. The last visit with the controller should prepare the property to submit a bill in a timely manner that would not create any problems with the bill being paid without hesitation.

GOOD HELP IS HARD TO FIND, ESPECIALLY IN OUR BUSINESS

May 2000

Finding employees, at all levels in hospitality, is getting tougher and tougher, especially when it comes to filling the sales vacancy. Filling the sales ranks seems to be the most difficult and very challenging. Just by taking a look at the job postings on some websites will tell the story; there is a lot of hunting going around trying to find sales help. Low unemployment throughout has created greater competition for us with other industries. The relatively lower salaries within our industry for sales personnel has turned off some good talent who have chosen to seek out other types of businesses for employment. Even recent graduates from university hotel management programs are going into other higher paying fields. It's a problem alright, but we make mistakes ourselves in recruiting, selection, employment, orientation, providing direction/supervision.

One of the biggest mistakes we make in hotel sales hiring practices is panicking and filling a vacancy or new position too quickly. Probably, the first thing we need to do is look at the job function that needs to be filled. Is there a Job Description? Can you separate the sales function from the service function? Perhaps one support person can back-up two or three sales folks who dedicate themselves to only selling. All too often we hire people for sales and in reality, they end up being a support person, which would probably cost less than a salesperson. Instead of replacing a salesperson would a less expensive administrative hire allow your existing sales staff to spend more time selling?

Maybe, at certain types of properties, you don't need to restrict yourself to experienced salespeople. Trainees are less expensive to hire, and you won't have to train them out of bad habits. Of course, someone would have to get these types trained; and that may have to be you.

What about part-timers. Perhaps a couple of those could very well replace one full-timer. It cost less; benefits are less; and together both probably would be able to put in more time in actually selling than one person.

Use your networks, your local connections, existing employees, local businesses, your customers, your vendors, and of course, the internet. I've seen a good number of postings for a sales job on: *www.hotel-on-line.com, www.madsearch.com, www.mim.com, www.hospitality-lst.com.* Just do a good search yourself and you can probably come up with a pretty good list.

You need to remember that the current incoming workers are a lot different than those of us who have been around a while. We had no problem putting in 60 hour weeks. It just ain't so today, so be aware when you try to drive someone to put in overtime. Workers today are more family-oriented than job focused. Not that there is anything wrong with that.

It is probably a good idea to request applicants to send, fax or email resumes along with a cover letter. This gives you a chance to review their writing skills in reviewing the letter. Do you see any creativity? The next good step would be to place a telephone call to the applicant and spend some time just chatting about their qualifications. How they handle themselves on the phone will give you a good idea of their oral and telephone skills. If everything looks positive, so far, then it may be time set up an in-person interview. The face-to-face meeting will you the chance to judge overall presentation skills, personality, dress, as well as demeanor. And, don't forget to check references.

It's a tough job getting the "right" person, but it will be worth the time. Ask those managers and owners of properties who have had constant turnover in the area of sales. Just don't hire in haste.

WE NEED TO HONOR OUR SALES FOLKS WHO BRING IN THE BUSINESS

February 2005

The ultimate objective of our business is to increase profitable sales. We've said it so many times: "You gotta bring it in at the top for it to fall to the bottom line." Without our sales departments continually bringing in new business, we are sure to fail. If all business came to our properties because of our advertising, direct mail, Web site, Internet vendors, and all other independent sources; and if every guest always returned to the property because of the "wonderful" service they received, then why would we have sales personnel on the payroll? The fact is that sales folks will always be needed at locations, large and small, for the purpose of bringing in new business. How valuable a salesperson is to a property has more to do with the amount of revenue generated by new business than anything else. Sometimes sales department staff is not appreciated enough for the work they have to do to bring in the business. It is not always an eight hour day or a five day week. Selling has a great deal to do with research, phone calls, meetings with prospects, a great deal of time "on the road," relationship building and wearing a continual smile. To do a good job it takes perseverance, knowledge, a consultative attitude, time away from home and family, making presentations, overcoming objections, negotiating deals and hopefully, writing contracts. A lot more than that also goes into it when it comes to attending meetings, speaking in public, dressing appropriately, having a great personality, and overall, being likable.

It really is a great pleasure to attend some national associations or company meetings and observe the professional hospitality salesperson do his/her job, particularly during a trade show. You will find that most salespeople do understand that the reason they attend and work a trade show is to build awareness and visibility for the property, and at the same time develop qualified prospects with they can do business. What we find at trade shows these days are sales folks who dress and look exceptionally well, and represent their properties well. They meet as many people as possible and ask appropriate questions to determine who might be a prospect. They know that they are better off finding prospects than trying to make a sales pitch to everyone they meet. Of course, they know that not everyone who attends a trade show or visits a booth will be someone who is in a position to bring business to the property. So prospecting is their game. That is, find as many contacts that would be in a position to do business with the property sometime down the road. Most persons who attend hospitality trade shows are not in a position to make deals for rooms and meeting space and banquets while they are at the show. Of course, there are exceptions with some motor coach tour operator shows or some business travel shows or even some bridal fairs.

Owners, managers, operators in this business of hospitality should really take time to honor their sales personnel for the work they are doing to bring in business that helps the bottom line and ensure jobs for the rest of the staff.

BITS AND PIECES OF STUDIES AND REPORTS THAT AFFECT US

April 2005

A number of very favorable reports and studies have come out from a variety of sources reflecting a very good year for the meetings and convention industry, and of course, for the lodging industry as well.

A study on responses from a sampling of meeting professionals, conducted by Yesawich, Pepperdine, Brown and Russell indicated that 23% of corporate meeting planners, and 19% of association planners, anticipate more off-site meetings from 2005 This is very good news for all types and sizes of properties; and I would not be surprised if most of these meetings fall into the small meeting category.

According to the Future Watch Report produced by Meeting Professionals International and American Express corporate meeting planners, associations and independent meeting planning companies forecasted a 5% budget increase in 2005, which is a 2% increase of what was spent on meetings and conferences in 2004.

How about incentives for our lodging salespeople? Do incentives work? Well, according to Martiz Research, Inc., incentive programs play a big role in employee productivity and retention. It looks like if you want to take advantage of the better business coming in this year and next, maybe you need to start thinking about incentive compensation for the sales department staff. The Maritz study reflected that the majority of the respondents indicated that incentives affect continued employment. And, in our industry, we surely do have a pretty high turnover rate with employees, particularly in sales. In the survey it looked like younger employees felt stronger about incentives, 74% of 18–34-year-olds compared to 60% of the 45–54-year-olds. More (68%) agreed that additional award opportunities would motivate them to be more productive.

This question of incentives for salespeople always seems to come up during our Hotel & Motel Management magazine sponsored sales workshops. It is surprising to learn that at most of the properties represented by the attendees only a few indicate that they have an incentive program. It certainly is a great way to motivate sales folks to be more productive in bringing in a more profitable business.

The latest on APEX (Accepted Practices Exchange) from the Convention Industry Council reflects that their first product, an online searchable industry glossary and a post-event guide are already available. Apex recently released an event specification guide which provides a standard format to follow to make sure that a planner provides all of the information that their supplier partners need. All of the event specifications will come in the same format and in the same order. This should save a good deal of time in the conference service department and catering departments. Sometime later this year APEX will release two more projects; a guide for proposals and one on meeting and site profiles. A guide to meeting contracts may be ready for release by the end of this year. Readers should check out APEX to see how these initiatives can be helpful to anyone's operation. For more information go to: www.conventionindustry.org).

THE THREE R's OF MOTIVATION: RECOGNITION, REWARDS, RETENTION

June 2005

It all really has to do with motivation. The three R's have more to do with motivation, than anything else. And, this applies to all employees, but especially for those involved with the sales effort. All too often, when we talk about employing someone for the sales department we refer to trying to find someone who is self-motivated. What does that mean? We might be referring to finding someone who has great energy, a "get up and go" person. Motivation is another thing. It has all to do with the environment and the behavior of other people with whom an employee is connected. We motivate people to do things through our own behavior; how we treat people; our company policies and programs. Most of the time it is the very small things of recognition that we do that motivates salespeople; like a "pat on the back," a "thank you," a letter of recognition to a spouse, a small gift, recognition of a good job done being announced at a meeting with other staff people. All this is a kind of favorable behavior on the part of a General Manager or director of sales.

That is what motivation is all about, general recognition.

Recognition also comes along with rewards. Here we are referring to some sort of incentive program for sales personnel to reflect improvement in production. This could be in form of monetary rewards on a monthly or quarterly basis. It could also be in form of gifts, or extra time off from the work schedule, or invitations to complimentary dinner or weekends at another property. Some form of incentive compensation is important for sales personnel; however, it needs to be accomplished by providing a simple method of computation. All too often we see some very complicated methods of computing the amount of the incentive related to a variety of goals or combination of figures which may include increased in occupancy percentage, room sales revenue, number of rooms booked, average daily rate, profitability, etc. It seems the method making the most sense, in providing an incentive, should be based solely on increase in sales revenue produced by a salesperson, which is most cases would be reflected in group sales. Further, the payment should be made only after the group has departed the property and made payment on the account. This way, incentives could be paid to salespeople on a monthly basis on monies collected for group business during the previous month.

A by-product for the organization is the third R, which is Retention. The successful salesperson who books a good deal of advanced group sales will want to keep booking and will stay around to be able to collect on the incentive program. Those sales personnel who may not be as effective, with lower sales records, and receiving smaller incentives may not have much incentive to stay around and they may depart the company.

There is a good book out, titled *The 24 Carrot Manager*, written by Adrian Gostick and Chester Elton. Here is some very good, easy reading of how a leader can unleash human potential through recognition and rewards.

KEEP YOUR SALESPEOPLE FROM SAYING "GOODBYE"

March 2006

There is no question about it; employee turnover in our industry is very high, and very expensive for the hospitality industry. Of course, most of it is in the hourly paid employees in food/beverage and hotel operations. However, right now we are seeing heavy turnover in the area of sales and catering management. Overall, employees will be seeking better opportunities elsewhere with higher pay, better benefits, job security, additional training, or for whatever reason. It looks like the demand is greater than the supply for sales personnel. Within the next two years, because of the expected increase in hotel development, there will even be a heavier increase in demand for sales personnel. With more competition for business coming online, hotels and resorts will need to do more in sales activities in going after the various markets that generate rooms and catering sales.

With more opportunities coming up for salespeople to move on to other challenges what is it that property owners, managers, and management companies can do to keep their quality sales personnel to stay on the job? People leave their current employment for a variety of reasons, and salaries alone are not on top of that list. Salespeople who feel comfortable in their environment, with favorable working conditions, with their production being recognized are the one who will most likely think twice before leaving their position for other employment. So, how do we make our sales staff personnel feel comfortable, and have favorable working conditions, and are able to be more productive? Here are some tips:

- Salespeople need to feel that they are making a difference. Help them use their best talents and skills in doing their job. This means providing training, direction, supervision, and motivation. They need to be told that they are making a difference with the work that they are doing. A pat on the back works.
- Allow opportunities for feedback on a regular basis. Salespeople need to be treated as partners of the business. They are in the field always talking with prospects and should be able to provide helpful information to property management.
- Get to know the sales staffs' family situation so you can better understand when they can be more productive or when they need to attend to something personal.
- Provide opportunities for sales staff to interact, whether it is social or business, in community activities. Networking is critical to our business.
- Encourage participation with industry associations such as Hospitality Sales and Marketing Association International where they can have the opportunity of attending conferences and workshops.
- Encourage continuous learning opportunities for them to be able to develop themselves which would be an asset to the property; and paying for their training.
- Create incentive programs to reward sales personnel for achieving certain sales goals. A good, reachable, dollar incentive opportunity could well replace any expected salary increases.
- Provide frequent (perhaps quarterly) reviews on job performance. Salespeople love report cards.
- Be sure to praise and say "thank you" for good performance.
- Remember sales personnel do have employment choices; just be sure they are happy in their work.

WILL THE INDEPENDENT SALESPERSON REPLACE PROPERTY SALES STAFF?

September 2007

Estimates have it that today there are probably over 25,000 third party, independent meeting planners at work, bringing loads of group meeting business to properties around the country. These people work independently, by themselves or are independent contractors for meeting planner companies. For the most part, they work on 10% commission on room revenue generated to hotels by the groups they bring in. So, when you think of this setup, aren't these people really "Independent Salespeople"? Meeting planning companies like Experient (formerly Conferon) with 750–800 of these independent contractors, are working around the world, generating group business for hotel properties. Another company, Helms-Briscoe has 900–1000 contractors, and there are a few more companies like Conference Direct as well as others like these with probably similar numbers.

No question about it, this kind of a set-up works very well for all parties. Any new group business generated to a hotel is certainly worth the 10% commission paid. We see more and more property salespeople spending time, working directly with many different independent meeting planners to get them to bring business to their facilities. Some properties even arrange familiarization trips for independents to show off how well the product can meet the needs of groups. From the look of things, it appears that a heavy percentage of group business for many properties are brought in via these independents. It seems that more time is spent on property salespeople in working on getting this type of group business, via the third party, as well as prospecting for and booking business directly with prospects.

These days sales personnel, whether working directly for a single property, or representing a group of hotels have a major purpose in their jobs; protecting the owners' assets by making sure that quality business, with maximum revenue, is generated through a variety of channels. More and more effort will need to be directed to work with companies like Experient (www.experient-inc.com), HelmsBriscoe (www.helms-briscoe.com), and Conference Direct (www.conferencedirect.com) as well as others, because of their solid connection with hundreds of companies, associations and a variety of other organizations that have group meetings business.

Of course, there still needs to be work done by salespeople to work on their own to prospect for, and book group business without going through the third party, independent meeting planners. It may not be a bad idea for property salespeople to start measuring their own effectiveness for their managers and owners by reflecting on how much of the group business in brought in by them, personally, compared to what is being brought in by the independents. How long will it be before we start thinking of these independent meeting planners as independent salespeople? Since hotels are paying them for the business maybe they are more of a salesperson rather a meeting planner. If there are enough of them out there doing the "selling" for a property, maybe they will be able to replace some sales staffers? Something to think about!

YOU MAY NOT NEED A SALES DEPARTMENT

March 2010

Not unless you want to improve results. We all know that it is not just the effort that counts in a sales operation, but the results of the effort. Certainly, there are too many hotel salespeople, and sales departments that are not productive A number of factors create, and contribute to the situation, for example:

▶ Poor recruiting and hiring techniques
▶ Lack of direction and supervision
▶ No motivation offered via management
▶ No sales plan or procedures to follow for staff
▶ Poor time management practices
▶ Other functions delegated to sales staff
▶ Poor staff communications
▶ Untrained sales staff
▶ Too comfortable offices
▶ Non-responsible management
▶ Low sales budgets
▶ No job descriptions for staff

The solution to the problem is not for lodging owners or operators to accept the responsibility of being chiefs-in-charge of sales. Instead, they must establish objectives for the sales department; objectives that are results oriented. For example: measurable goals for improved room occupancy, room revenue, and for food/beverage functions need to be established. Pie in the sky doesn't work; all goals should be realistic. The objective must be an improvement. The measure of the quality of a sales team is its ability to improve sales, regardless of the obstacles that are in the way. The usual arguments from salespeople for not improving sales come from a long list of "reason why we can't get business":

▶ Our country is in a recession; high unemployment
▶ Additional competition in the area
▶ Other properties cutting rates
▶ Many complaints on food service
▶ Highways have changed traffic patterns
▶ Product is dated; need refurbishing
▶ Too much change-over in management
▶ Staff not courteous
▶ Other jobs delegated to sales staff
▶ Meeting rooms not sound-proofed; bad lighting
▶ Poor parking for guests, etc., and the list goes on and on.

In reality, hotel owners and operators do not believe everything they hear . . . or do they? The items on this listing sound like plausible reasons for not getting business, but many of the problems are correctable by management. However, if both management and the sales department are convinced that business cannot be developed, then there is no choice but to save some money by getting rid of the sales department and get business catch-as-catch-can.

WE ARE CUSTOMERS TO EACH OTHER IN THIS BUSINESS OF HOTEL SALES

February 2013

For her New Year message, Nancy Friedman, The Telephone Doctor, in her monthly newsletter, came up with the acronym WACTO (We Are Customers To Each Other), and it applied so much to our business of hospitality sales that we are adapting some it for this column, with her permission.

Here are some of Nancy Friedman's ground rules for WACTEO (We Are Customers to Each Other).

1. *Understand Your Role:* Each employee should know the mission of their hotel organization and the role they play. Those who are in a small sales department of a hotel can often time miss the big picture. You may start to understand the 'why' of the things you're asked to do sometimes, and 'why' internal customer service is everyone's responsibility from the General Manager to maintenance. If management isn't doing their part, often times the entire customer service program will go out the window. You do not want double standards. Remember it starts at the top!

2. *Respect Employee Differences:* Cub fan? Cardinal fan? Republican? Democrat? Rock music, classical, whatever. Just because we don't agree with someone doesn't make us right. Differences are crucial for an organization. Differences are the key to understanding people. If everyone thought the same way, most of us wouldn't be needed. It's not healthy to argue just because a co-worker isn't doing it the way you would or thinking the way you do. Learn to respect the differences. That's why we have chocolate and vanilla ice cream.

3. *Recognize the Personal Space of Others:* Simply put, this boils down to the golden rule. Those who can work with a radio playing music may disturb others around them who aren't able to concentrate. Loud voices around someone who's on the phone with an external customer can be annoying also. If you're working in a cubical or sharing an office or area, we need to recognize there are others around you. Be sensitive to their wishes, as you would hope they would be to yours.

4. *Work to Resolve Conflicts:* Who hasn't had unkind words with another employee? Or perhaps you and a co-worker strongly disagree on a project or idea. Not trying to make it work can only lead to more stress and frustration. Normally someone from HR or another trusted employee can usually be of help on conflict resolutions.

5. *Show Appreciation:* We saved this for last because being appreciated, showing you care with a genuine 'thank you,' is critical to WACTEO. It can be a note, a phone call or just stopping by an office and letting someone know they did a great job. This makes a huge difference in our internal relationships. There are surveys upon surveys that show how much a genuine pat on the back of appreciation is thought of as a way of special compensation.